INSTRUCTOR'S SOLUTIONS MAN
VOLUME 2

JAMES B. WHITENTON
Southern Polytechnic State University

to accompany

FUNDAMENTALS OF
PHYSICS

SIXTH EDITION

DAVID HALLIDAY
University of Pittsburgh

ROBERT RESNICK
Rensselaer Polytechnic Institute

JEARL WALKER
Cleveland State University

JOHN WILEY & SONS, INC.
New York • Chichester • Weinheim • Brisbane • Singapore • Toronto

To order books or for customer service call 1-800-CALL-WILEY (225-5945).

ISBN 0-471-37366-4

Printed in the United States of America

10 9 8 7 6 5 4 3 2 1

Printed and bound by Bradford & Bigelow, Inc.

Preface

This booklet includes the solutions relevant to the EXERCISES & PROBLEMS sections of the 6th edition of **Fundamentals of Physics**, by Halliday, Resnick, and Walker. We also include solutions to problems in the Problems Supplement. We have not included solutions or discussions which pertain to the Questions sections. These solutions have been typed using the LaTeX format, with some figures being included as postscript files and others being generated during the compiling steps by PiCTeX. The solutions files are available upon request. Additional information regarding TeX-based software is available at http://www.ctan.org/.

It is recognized that no solution manual is without errors, and it is encouraged that any errors be reported to the author at the email address jwhitent@spsu.edu.

I am very grateful for helpful input from Bodo Barth, J. Richard Christman, Meighan Dillon, Barbara Moore, Jearl Walker and Linda Whitenton regarding the development of this document.

Note to adopters regarding The Instructor's Solutions Manual for Fundamentals of Physics, 6e by Halliday/ Resnick/Walker

Thank you very much for adopting FUNDAMENTALS OF PHYSICS, 6e. We are pleased to be able to provide you with a variety of support material to help you in the teaching of your course. Please note that all this material—including the Instructor's Solutions Manual—is copyrighted by John Wiley & Sons, Inc. and explicitly intended for use only at your institution.

Please note that our providing these solutions does not carry with it permission to distribute them beyond your institution. Before putting any of the solutions on a web site, we ask that you request formal permission from us to do so. Please write to: Permissions Department, John Wiley & Sons, Inc., 605 Third Avenue, New York, NY 10158-0012. In most cases, we will grant such permission PROVIDED THAT THE WEB SITE IS PASSWORD PROTECTED.

Our goal is to prevent students from other campuses from being able to access your solutions. We trust that you can understand how that might undermine the efforts of your colleagues at other institutions.

We appreciate your support and understanding in this matter.

TABLE OF CONTENTS

Chapter 22

1. Eq. 22-1 gives Coulomb's Law, $F = k\,\frac{|q_1||q_2|}{r^2}$, which we solve for the distance:

$$\begin{aligned} r &= \sqrt{\frac{k|q_1||q_2|}{F}} \\ &= \sqrt{\frac{(8.99 \times 10^9\,\mathrm{N\cdot m^2/C^2})\,(26.0 \times 10^{-6}\,\mathrm{C})\,(47.0 \times 10^{-6}\,\mathrm{C})}{5.70\,\mathrm{N}}} = 1.39 \text{ m} . \end{aligned}$$

2. The magnitude of the mutual force of attraction at $r = 0.120$ m is

$$F = k\frac{|q_1||q_2|}{r^2} = \left(8.99 \times 10^9\right)\frac{\left(3.00 \times 10^{-6}\right)\left(1.50 \times 10^{-6}\right)}{0.120^2} = 2.81 \text{ N} .$$

3. (a) With a understood to mean the magnitude of acceleration, Newton's second and third laws lead to

$$m_2 a_2 = m_1 a_1 \implies m_2 = \frac{(6.3 \times 10^{-7}\,\mathrm{kg})\,(7.0\,\mathrm{m/s^2})}{9.0\,\mathrm{m/s^2}} = 4.9 \times 10^{-7} \text{ kg} .$$

(b) The magnitude of the (only) force on particle 1 is

$$F = m_1 a_1 = k\frac{|q_1||q_2|}{r^2} = \left(8.99 \times 10^9\right)\frac{|q|^2}{0.0032^2} .$$

Inserting the values for m_1 and a_1 (see part (a)) we obtain $|q| = 7.1 \times 10^{-11}$ C.

4. The fact that the spheres are identical allows us to conclude that when two spheres are in contact, they share equal charge. Therefore, when a charged sphere (q) touches an uncharged one, they will (fairly quickly) each attain half that charge ($q/2$). We start with spheres 1 and 2 each having charge q and experiencing a mutual repulsive force $F = kq^2/r^2$. When the neutral sphere 3 touches sphere 1, sphere 1's charge decreases to $q/2$. Then sphere 3 (now carrying charge $q/2$) is brought into contact with sphere 2, a total amount of $q/2 + q$ becomes shared equally between them. Therefore, the charge of sphere 3 is $3q/4$ in the final situation. The repulsive force between spheres 1 and 2 is finally

$$F' = k\frac{\left(\frac{q}{2}\right)\left(\frac{3q}{4}\right)}{r^2} = \frac{3}{8}k\frac{q^2}{r^2} = \frac{3}{8}F .$$

5. We put the origin of a coordinate system at the lower left corner of the square and take $+x$ rightward and $+y$ upward. The force exerted by the charge $+q$ on the charge $+2q$ is

$$\vec{F}_1 = k\frac{q(2q)}{a^2}\,(-\hat{\mathrm{j}}) .$$

The force exerted by the charge $-q$ on the $+2q$ charge is directed along the diagonal of the square and has magnitude

$$F_2 = k\frac{q(2q)}{(a\sqrt{2})^2}$$

which becomes, upon finding its components (and using the fact that $\cos 45° = 1/\sqrt{2}$),

$$\vec{F}_2 = k\frac{q(2q)}{2\sqrt{2}\,a^2}\,\hat{\imath} + k\frac{q(2q)}{2\sqrt{2}\,a^2}\,\hat{\jmath}\ .$$

Finally, the force exerted by the charge $-2q$ on $+2q$ is

$$\vec{F}_3 = k\frac{(2q)(2q)}{a^2}\,\hat{\imath}\ .$$

(a) Therefore, the horizontal component of the resultant force on $+2q$ is

$$\begin{aligned} F_x &= F_{1x} + F_{2x} + F_{3x} = k\frac{q^2}{a^2}\left(\frac{1}{\sqrt{2}} + 4\right) \\ &= \left(8.99 \times 10^9\right)\frac{\left(1.0 \times 10^{-7}\right)^2}{0.050^2}\left(\frac{1}{\sqrt{2}} + 4\right) = 0.17\text{ N}\ . \end{aligned}$$

(b) The vertical component of the net force is

$$F_y = F_{1y} + F_{2y} + F_{3y} = k\frac{q^2}{a^2}\left(-2 + \frac{1}{\sqrt{2}}\right) = -0.046\text{ N}\ .$$

6. (a) The individual force magnitudes (acting on Q) are, by Eq. 22-1,

$$k\,\frac{|q_1|\,Q}{\left(-a - \frac{a}{2}\right)^2} = k\,\frac{|q_2|\,Q}{\left(a - \frac{a}{2}\right)^2}$$

which leads to $|q_1| = 9\,|q_2|$. Since Q is located between q_1 and q_2, we conclude q_1 and q_2 are like-sign. Consequently, $q_1 = 9q_2$.

(b) Now we have

$$k\,\frac{|q_1|\,Q}{\left(-a - \frac{3a}{2}\right)^2} = k\,\frac{|q_2|\,Q}{\left(a - \frac{3a}{2}\right)^2}$$

which yields $|q_1| = 25\,|q_2|$. Now, Q is not located between q_1 and q_2, one of them must push and the other must pull. Thus, they are unlike-sign, so $q_1 = -25q_2$.

7. We assume the spheres are far apart. Then the charge distribution on each of them is spherically symmetric and Coulomb's law can be used. Let q_1 and q_2 be the original charges. We choose the coordinate system so the force on q_2 is positive if it is repelled by q_1. Then, the force on q_2 is

$$F_a = -\frac{1}{4\pi\varepsilon_0}\frac{q_1 q_2}{r^2} = -k\,\frac{q_1 q_2}{r^2}$$

where $r = 0.500$ m. The negative sign indicates that the spheres attract each other. After the wire is connected, the spheres, being identical, acquire the same charge. Since charge is conserved, the total charge is the same as it was originally. This means the charge on each sphere is $(q_1 + q_2)/2$. The force is now one of repulsion and is given by

$$F_b = \frac{1}{4\pi\varepsilon_0}\,\frac{\left(\frac{q_1+q_2}{2}\right)\left(\frac{q_1+q_2}{2}\right)}{r^2} = k\,\frac{(q_1 + q_2)^2}{4r^2}\ .$$

We solve the two force equations simultaneously for q_1 and q_2. The first gives the product

$$q_1 q_2 = -\frac{r^2 F_a}{k} = -\frac{(0.500\,\text{m})^2(0.108\,\text{N})}{8.99 \times 10^9\,\text{N}\cdot\text{m}^2/\text{C}^2} = -3.00 \times 10^{-12}\ \text{C}^2\ ,$$

and the second gives the sum

$$q_1 + q_2 = 2r\sqrt{\frac{F_b}{k}} = 2(0.500\,\text{m})\sqrt{\frac{0.0360\,\text{N}}{8.99 \times 10^9\,\text{N}\cdot\text{m}^2/\text{C}^2}} = 2.00 \times 10^{-6}\ \text{C}$$

where we have taken the positive root (which amounts to assuming $q_1 + q_2 \geq 0$). Thus, the product result provides the relation

$$q_2 = \frac{-(3.00 \times 10^{-12}\,\text{C}^2)}{q_1}$$

which we substitute into the sum result, producing

$$q_1 - \frac{3.00 \times 10^{-12}\,\text{C}^2}{q_1} = 2.00 \times 10^{-6}\ \text{C}\ .$$

Multiplying by q_1 and rearranging, we obtain a quadratic equation

$$q_1^2 - (2.00 \times 10^{-6}\ \text{C})q_1 - 3.00 \times 10^{-12}\,\text{C}^2 = 0\ .$$

The solutions are

$$q_1 = \frac{2.00 \times 10^{-6}\,\text{C} \pm \sqrt{(-2.00 \times 10^{-6}\,\text{C})^2 - 4(-3.00 \times 10^{-12}\,\text{C}^2)}}{2}\ .$$

If the positive sign is used, $q_1 = 3.00 \times 10^{-6}$ C, and if the negative sign is used, $q_1 = -1.00 \times 10^{-6}$ C. Using $q_2 = (-3.00 \times 10^{-12})/q_1$ with $q_1 = 3.00 \times 10^{-6}$ C, we get $q_2 = -1.00 \times 10^{-6}$ C. If we instead work with the $q_1 = -1.00 \times 10^{-6}$ C root, then we find $q_2 = 3.00 \times 10^{-6}$ C. Since the spheres are identical, the solutions are essentially the same: one sphere originally had charge -1.00×10^{-6} C and the other had charge $+3.00 \times 10^{-6}$ C. What if we had not made the assumption, above, that $q_1 + q_2 \geq 0$? If the signs of the charges were reversed (so $q_1 + q_2 < 0$), then the forces remain the same, so a charge of $+1.00 \times 10^{-6}$ C on one sphere and a charge of -3.00×10^{-6} C on the other also satisfies the conditions of the problem.

8. With rightwards positive, the net force on q_3 is

$$k\frac{q_1 q_3}{(2d)^2} + k\frac{q_2 q_3}{d^2}\ .$$

We note that each term exhibits the proper sign (positive for rightward, negative for leftward) for all possible signs of the charges. For example, the first term (the force exerted on q_3 by q_1) is negative if they are unlike charges, indicating that q_3 is being pulled toward q_1, and it is positive if they are like charges (so q_3 would be repelled from q_1). Setting the net force equal to zero and canceling k, q_3 and d^2 leads to

$$\frac{q_1}{4} + q_2 = 0 \implies q_1 = -4q_2\ .$$

9. (a) If the system of three charges is to be in equilibrium, the force on each charge must be zero. Let the third charge be q_0. It must lie between the other two or else the forces acting on it due to the other charges would be in the same direction and q_0 could not be in equilibrium. Suppose q_0 is a distance x from q, as shown on the diagram below. The force acting on q_0 is then given by

$$F_0 = \frac{1}{4\pi\varepsilon_0}\left(\frac{qq_0}{x^2} - \frac{4qq_0}{(L-x)^2}\right)$$

where the positive direction is rightward. We require $F_0 = 0$ and solve for x. Canceling common factors yields $1/x^2 = 4/(L-x)^2$ and taking the square root yields $1/x = 2/(L-x)$. The solution is $x = L/3$.

$$\longleftarrow x \longrightarrow\longleftarrow L-x \longrightarrow$$

$q \qquad q_0 \qquad 4q$

The force on q is

$$F_q = \frac{-1}{4\pi\varepsilon_0}\left(\frac{qq_0}{x^2} + \frac{4q^2}{L^2}\right) .$$

The signs are chosen so that a negative force value would cause q to move leftward. We require $F_q = 0$ and solve for q_0:

$$q_0 = -\frac{4qx^2}{L^2} = -\frac{4}{9}q$$

where $x = L/3$ is used. We now examine the force on $4q$:

$$\begin{aligned} F_{4q} &= \frac{1}{4\pi\varepsilon_0}\left(\frac{4q^2}{L^2} + \frac{4qq_0}{(L-x)^2}\right) = \frac{1}{4\pi\varepsilon_0}\left(\frac{4q^2}{L^2} + \frac{4(-4/9)q^2}{(4/9)L^2}\right) \\ &= \frac{1}{4\pi\varepsilon_0}\left(\frac{4q^2}{L^2} - \frac{4q^2}{L^2}\right) \end{aligned}$$

which we see is zero. Thus, with $q_0 = -(4/9)q$ and $x = L/3$, all three charges are in equilibrium.

(b) If q_0 moves toward q the force of attraction exerted by q is greater in magnitude than the force of attraction exerted by $4q$. This causes q_0 to continue to move toward q and away from its initial position. The equilibrium is unstable.

10. There is no equilibrium position for q_3 *between* the two fixed charges, because it is being pulled by one and pushed by the other (since q_1 and q_2 have different signs); in this region this means the two force arrows on q_3 are in the same direction and cannot cancel. It should also be clear that off-axis (with the axis defined as that which passes through the two fixed charges) there are no equilibrium positions. On the semi-infinite region of the axis which is nearest q_2 and furthest from q_1 an equilibrium position for q_3 cannot be found because $|q_1| < |q_2|$ and the magnitude of force exerted by q_2 is everywhere (in that region) stronger than that exerted by q_1 on q_3. Thus, we must look in the semi-infinite region of the axis which is nearest q_1 and furthest from q_2, where the net force on q_3 has magnitude

$$\left| k\frac{|q_1q_3|}{x^2} - k\frac{|q_2q_3|}{(d+x)^2}\right|$$

with $d = 10$ cm and x assumed positive. We set this equal to zero, as required by the problem, and cancel k and q_3. Thus, we obtain

$$\frac{|q_1|}{x^2} - \frac{|q_2|}{(d+x)^2} = 0 \implies \left(\frac{d+x}{x}\right)^2 = \left|\frac{q_2}{q_1}\right| = 3$$

which yields (after taking the square root)

$$\frac{d+x}{x} = \sqrt{3} \implies x = \frac{d}{\sqrt{3}-1} \approx 14 \text{ cm}$$

for the distance between q_3 and q_1, so $x + d$ (the distance between q_2 and q_3) is approximately 24 cm.

11. (a) The magnitudes of the gravitational and electrical forces must be the same:

$$\frac{1}{4\pi\varepsilon_0}\frac{q^2}{r^2} = G\frac{mM}{r^2}$$

where q is the charge on either body, r is the center-to-center separation of Earth and Moon, G is the universal gravitational constant, M is the mass of Earth, and m is the mass of the Moon. We solve for q:

$$q = \sqrt{4\pi\varepsilon_0 GmM} .$$

According to Appendix C of the text, $M = 5.98 \times 10^{24}$ kg, and $m = 7.36 \times 10^{22}$ kg, so (using $4\pi\varepsilon_0 = 1/k$) the charge is

$$q = \sqrt{\frac{(6.67 \times 10^{-11}\,\text{N} \cdot \text{m}^2/\text{kg}^2)(7.36 \times 10^{22}\,\text{kg})(5.98 \times 10^{24}\,\text{kg})}{8.99 \times 10^9\,\text{N} \cdot \text{m}^2/\text{C}^2}} = 5.7 \times 10^{13}\ \text{C} .$$

We note that the distance r cancels because both the electric and gravitational forces are proportional to $1/r^2$.

(b) The charge on a hydrogen ion is $e = 1.60 \times 10^{-19}$ C, so there must be

$$\frac{q}{e} = \frac{5.7 \times 10^{13}\,\text{C}}{1.6 \times 10^{-19}\,\text{C}} = 3.6 \times 10^{32}\,\text{ions} .$$

Each ion has a mass of 1.67×10^{-27} kg, so the total mass needed is

$$(3.6 \times 10^{32})(1.67 \times 10^{-27}\,\text{kg}) = 6.0 \times 10^5\ \text{kg} .$$

12. (a) The distance between q_1 and q_2 is

$$r_{12} = \sqrt{(x_2 - x_1)^2 + (y_2 - y_1)^2} = \sqrt{(-0.020 - 0.035)^2 + (0.015 - 0.005)^2} = 0.0559\ \text{m} .$$

The magnitude of the force exerted by q_1 on q_2 is

$$F_{21} = k\frac{|q_1 q_2|}{r_{12}^2} = \frac{(8.99 \times 10^9)\left(3.0 \times 10^{-6}\right)\left(4.0 \times 10^{-6}\right)}{0.0559^2} = 34.5\ \text{N} .$$

The vector $\vec{F}_{21}$ is directed towards q_1 and makes an angle θ with the $+x$ axis, where

$$\theta = \tan^{-1}\left(\frac{y_2 - y_1}{x_2 - x_1}\right) = \tan^{-1}\left(\frac{1.5 - 0.5}{-2.0 - 3.5}\right) = -10.3^\circ .$$

(b) Let the third charge be located at (x_3, y_3), a distance r from q_2. We note that q_1, q_2 and q_3 must be colinear; otherwise, an equilibrium position for any one of them would be impossible to find. Furthermore, we cannot place q_3 on the same side of q_2 where we also find q_1, since in that region both forces (exerted on q_2 by q_3 and q_1) would be in the same direction (since q_2 is attracted to both of them). Thus, in terms of the angle found in part (a), we have $x_3 = x_2 - r\cos\theta$ and $y_3 = y_2 - r\sin\theta$ (which means $y_3 > y_2$ since θ is negative). The magnitude of force exerted on q_2 by q_3 is $F_{23} = k|q_2 q_3|/r^2$, which must equal that of the force exerted on it by q_1 (found in part (a)). Therefore,

$$k\frac{|q_2 q_3|}{r^2} = k\frac{|q_1 q_2|}{r_{12}^2} \implies r = r_{12}\sqrt{\frac{q_3}{q_1}} = 0.0645\ \text{cm} .$$

Consequently, $x_3 = x_2 - r\cos\theta = -2.0\,\text{cm} - (6.45\,\text{cm})\cos(-10.3^\circ) = -8.4\,\text{cm}$ and $y_3 = y_2 - r\sin\theta = 1.5\,\text{cm} - (6.45\,\text{cm})\sin(-10.3^\circ) = 2.7\,\text{cm}$.

13. The magnitude of the force of either of the charges on the other is given by

$$F = \frac{1}{4\pi\varepsilon_0}\frac{q(Q - q)}{r^2}$$

where r is the distance between the charges. We want the value of q that maximizes the function $f(q) = q(Q - q)$. Setting the derivative df/dq equal to zero leads to $Q - 2q = 0$, or $q = Q/2$.

14. (a) We choose the coordinate axes as shown in the diagram below. For ease of presentation (of the computations below) we assume $Q > 0$ and $q < 0$ (although the final result does not depend on this particular choice). The repulsive force between the diagonally opposite Q's is along our (tilted)

x axis. The attractive force between each pair of Q and q is along the sides (of length a). In our drawing, the distance between the center to the corner is d, where $d = a/\sqrt{2}$, and the diagonal itself is therefore of length $2d = a\sqrt{2}$.

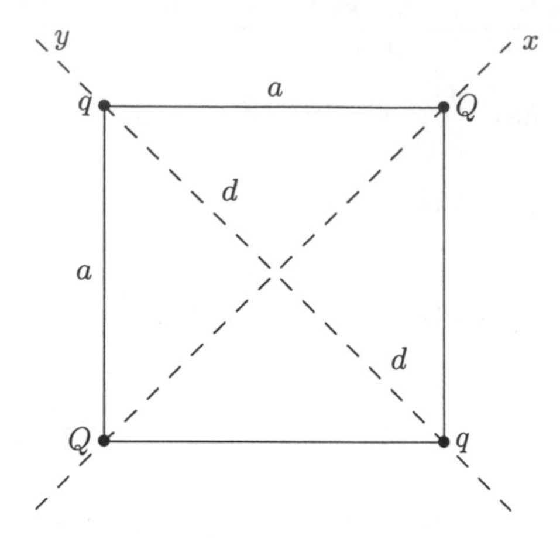

Since the angle between each attractive force and the x axis is 45° (note: $\cos 45° = 1/\sqrt{2}$), then the net force on Q is

$$\begin{aligned} F_x &= \frac{1}{4\pi\varepsilon_0}\left(\frac{(Q)(Q)}{(2d)^2} - 2\frac{(|q|)\,(Q)}{a^2}\cos 45°\right) \\ &= \frac{1}{4\pi\varepsilon_0}\left(\frac{Q^2}{2a^2} - 2\frac{|q|\cdot Q}{a^2}\frac{1}{\sqrt{2}}\right) \end{aligned}$$

which (upon requiring $F_x = 0$) leads to $|q| = Q/2\sqrt{2}$ or $q = -\frac{Q}{2\sqrt{2}}$.

(b) The net force on q, examined along the y axis is

$$\begin{aligned} F_y &= \frac{1}{4\pi\varepsilon_0}\left(\frac{q^2}{(2d)^2} - 2\frac{(|q|)\,(Q)}{a^2}\sin 45°\right) \\ &= \frac{1}{4\pi\varepsilon_0}\left(\frac{q^2}{2a^2} - 2\frac{|q|\cdot Q}{a^2}\frac{1}{\sqrt{2}}\right) \end{aligned}$$

which (if we demand $F_y = 0$) leads to $q = -2Q\sqrt{2}$ which is inconsistent with the result of part (a). Thus, we are unable to construct an equilibrium configuration with this geometry, where the only forces acting are given by Eq. 22-1.

15. (a) A force diagram for one of the balls is shown below. The force of gravity $m\vec{g}$ acts downward, the electrical force $\vec{F}_e$ of the other ball acts to the left, and the tension in the thread acts along the thread, at the angle θ to the vertical. The ball is in equilibrium, so its acceleration is zero. The y component of Newton's second law yields $T\cos\theta - mg = 0$ and the x component yields $T\sin\theta - F_e = 0$. We solve the first equation for T and obtain $T = mg/\cos\theta$. We substitute the

result into the second to obtain $mg\tan\theta - F_e = 0$.

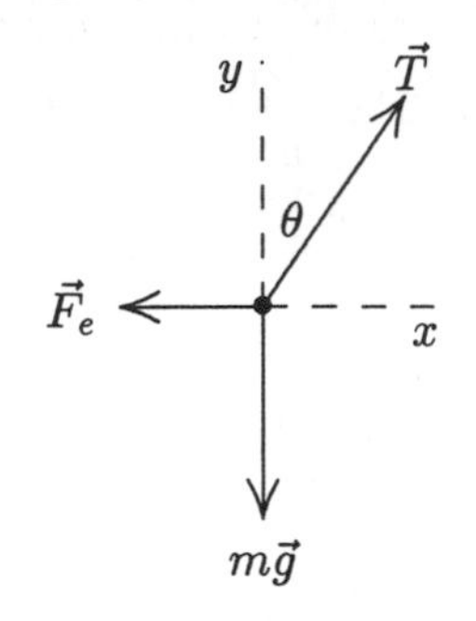

Examination of the geometry of Figure 22-19 leads to

$$\tan\theta = \frac{x/2}{\sqrt{L^2 - (x/2)^2}} \, .$$

If L is much larger than x (which is the case if θ is very small), we may neglect $x/2$ in the denominator and write $\tan\theta \approx x/2L$. This is equivalent to approximating $\tan\theta$ by $\sin\theta$. The magnitude of the electrical force of one ball on the other is

$$F_e = \frac{q^2}{4\pi\varepsilon_0 x^2}$$

by Eq. 22-4. When these two expressions are used in the equation $mg\tan\theta = F_e$, we obtain

$$\frac{mgx}{2L} \approx \frac{1}{4\pi\varepsilon_0}\frac{q^2}{x^2} \implies x \approx \left(\frac{q^2 L}{2\pi\varepsilon_0 mg}\right)^{1/3} .$$

(b) We solve $x^3 = 2kq^2L/mg)$ for the charge (using Eq. 22-5):

$$q = \sqrt{\frac{mgx^3}{2kL}} = \sqrt{\frac{(0.010\,\text{kg})(9.8\,\text{m/s}^2)(0.050\,\text{m})^3}{2(8.99\times10^9\,\text{N}\cdot\text{m}^2/\text{C}^2)(1.20\,\text{m})}} = \pm 2.4\times10^{-8}\ \text{C} .$$

16. If one of them is discharged, there would no electrostatic repulsion between the two balls and they would both come to the position $\theta = 0$, making contact with each other. A redistribution of the remaining charge would then occur, with each of the balls getting $q/2$. Then they would again be separated due to electrostatic repulsion, which results in the new equilibrium separation

$$x' = \left[\frac{(q/2)^2 L}{2\pi\varepsilon_0 mg}\right]^{1/3} = \left(\frac{1}{4}\right)^{1/3} x = \left(\frac{1}{4}\right)^{1/3} (5.0\,\text{cm}) = 3.1\ \text{cm} .$$

17. (a) Since the rod is in equilibrium, the net force acting on it is zero, and the net torque about any point is also zero. We write an expression for the net torque about the bearing, equate it to zero, and solve for x. The charge Q on the left exerts an upward force of magnitude $(1/4\pi\varepsilon_0)(qQ/h^2)$, at a distance $L/2$ from the bearing. We take the torque to be negative. The attached weight exerts a downward force of magnitude W, at a distance $x - L/2$ from the bearing. This torque is also negative. The charge Q on the right exerts an upward force of magnitude $(1/4\pi\varepsilon_0)(2qQ/h^2)$, at a distance $L/2$ from the bearing. This torque is positive. The equation for rotational equilibrium is

$$\frac{-1}{4\pi\varepsilon_0}\frac{qQ}{h^2}\frac{L}{2} - W\left(x - \frac{L}{2}\right) + \frac{1}{4\pi\varepsilon_0}\frac{2qQ}{h^2}\frac{L}{2} = 0 .$$

The solution for x is

$$x = \frac{L}{2}\left(1 + \frac{1}{4\pi\varepsilon_0}\frac{qQ}{h^2 W}\right) .$$

(b) If N is the magnitude of the upward force exerted by the bearing, then Newton's second law (with zero acceleration) gives

$$W - \frac{1}{4\pi\varepsilon_0}\frac{qQ}{h^2} - \frac{1}{4\pi\varepsilon_0}\frac{2qQ}{h^2} - N = 0 .$$

We solve for h so that $N = 0$. The result is

$$h = \sqrt{\frac{1}{4\pi\varepsilon_0}\frac{3qQ}{W}} .$$

18. The magnitude of the force is

$$F = k\frac{e^2}{r^2} = \left(8.99 \times 10^9 \, \frac{\text{N}\cdot\text{m}^2}{\text{C}^2}\right)\frac{(1.60 \times 10^{-19}\,\text{C})^2}{(2.82 \times 10^{-10}\,\text{m})^2} = 2.89 \times 10^{-9}\ \text{N} .$$

19. The mass of an electron is $m = 9.11 \times 10^{-31}$ kg, so the number of electrons in a collection with total mass $M = 75.0$ kg is

$$N = \frac{M}{m} = \frac{75.0\,\text{kg}}{9.11 \times 10^{-31}\,\text{kg}} = 8.23 \times 10^{31}\ \text{electrons} .$$

The total charge of the collection is

$$q = -Ne = -(8.23 \times 10^{31})(1.60 \times 10^{-19}\,\text{C}) = -1.32 \times 10^{13}\ \text{C} .$$

20. There are two protons (each with charge $q = +e$) in each molecule, so

$$Q = N_A q = (6.02 \times 10^{23})(2)(1.60 \times 10^{-19}\,\text{C}) = 1.9 \times 10^5\,\text{C} = 0.19\ \text{MC} .$$

21. (a) The magnitude of the force between the (positive) ions is given by

$$F = \frac{(q)(q)}{4\pi\varepsilon_0 r^2} = k\,\frac{q^2}{r^2}$$

where q is the charge on either of them and r is the distance between them. We solve for the charge:

$$q = r\sqrt{\frac{F}{k}} = (5.0 \times 10^{-10}\,\text{m})\sqrt{\frac{3.7 \times 10^{-9}\,\text{N}}{8.99 \times 10^9\,\text{N}\cdot\text{m}^2/\text{C}^2}} = 3.2 \times 10^{-19}\ \text{C} .$$

(b) Let N be the number of electrons missing from each ion. Then, $Ne = q$, or

$$N = \frac{q}{e} = \frac{3.2 \times 10^{-19}\,\text{C}}{1.6 \times 10^{-19}\,\text{C}} = 2 .$$

22. (a) Eq. 22-1 gives

$$F = \frac{\left(8.99 \times 10^9\,\text{N}\cdot\text{m}^2/\text{C}^2\right)\left(1.00 \times 10^{-16}\,\text{C}\right)^2}{\left(1.00 \times 10^{-2}\,\text{m}\right)^2} = 8.99 \times 10^{-19}\ \text{N} .$$

(b) If n is the number of excess electrons (of charge $-e$ each) on each drop then

$$n = -\frac{q}{e} = -\frac{-1.00 \times 10^{-16}\,\text{C}}{1.60 \times 10^{-19}\,\text{C}} = 625 .$$

23. Eq. 22-11 (in absolute value) gives

$$n = \frac{|q|}{e} = \frac{1.0 \times 10^{-7}\,\text{C}}{1.6 \times 10^{-19}\,\text{C}} = 6.3 \times 10^{11} .$$

24. With $F = m_e g$, Eq. 22-1 leads to

$$r^2 = \frac{ke^2}{m_e g} = \frac{\left(8.99 \times 10^9\,\text{N}\cdot\text{m}^2/\text{C}^2\right)\left(1.60 \times 10^{-19}\,\text{C}\right)^2}{(9.11 \times 10^{-31}\,\text{kg})\,(9.8\,\text{m/s}^2)}$$

which leads to $r = 5.1$ m. The second electron should be below the first one, so that the repulsive force (acting on the first) is in the direction opposite to the pull of Earth's gravity.

25. The unit Ampere is discussed in §22-4. The proton flux is given as 1500 protons per square meter per second, where each proton provides a charge of $q = +e$. The current through the spherical area $4\pi R^2 = 4\pi(6.37 \times 10^6\,\text{m})^2 = 5.1 \times 10^{14}\,\text{m}^2$ would be

$$i = (5.1 \times 10^{14}\,\text{m}^2)\left(1500\,\frac{\text{protons}}{\text{s}\cdot\text{m}^2}\right)(1.6 \times 10^{-19}\,\text{C/proton}) = 0.122\ \text{A}\ .$$

26. The volume of $250\,\text{cm}^3$ corresponds to a mass of $250\,\text{g}$ since the density of water is $1.0\,\text{g/cm}^3$. This mass corresponds to $250/18 = 14$ moles since the molar mass of water is 18. There are ten protons (each with charge $q = +e$) in each molecule of H_2O, so

$$Q = 14N_A q = 14(6.02 \times 10^{23})(10)(1.60 \times 10^{-19}\,\text{C}) = 1.3 \times 10^7\,\text{C} = 13\ \text{MC}\ .$$

27. (a) Every cesium ion at a corner of the cube exerts a force of the same magnitude on the chlorine ion at the cube center. Each force is a force of attraction and is directed toward the cesium ion that exerts it, along the body diagonal of the cube. We can pair every cesium ion with another, diametrically positioned at the opposite corner of the cube. Since the two ions in such a pair exert forces that have the same magnitude but are oppositely directed, the two forces sum to zero and, since every cesium ion can be paired in this way, the total force on the chlorine ion is zero.

(b) Rather than remove a cesium ion, we superpose charge $-e$ at the position of one cesium ion. This neutralizes the ion, and as far as the electrical force on the chlorine ion is concerned, it is equivalent to removing the ion. The forces of the eight cesium ions at the cube corners sum to zero, so the only force on the chlorine ion is the force of the added charge.
The length of a body diagonal of a cube is $\sqrt{3}a$, where a is the length of a cube edge. Thus, the distance from the center of the cube to a corner is $d = (\sqrt{3}/2)a$. The force has magnitude

$$F = k\,\frac{e^2}{d^2} = \frac{ke^2}{(3/4)a^2} = \frac{(8.99 \times 10^9\,\text{N}\cdot\text{m}^2/\text{C}^2)(1.60 \times 10^{-19}\,\text{C})^2}{(3/4)(0.40 \times 10^{-9}\,\text{m})^2} = 1.9 \times 10^{-9}\ \text{N}\ .$$

Since both the added charge and the chlorine ion are negative, the force is one of repulsion. The chlorine ion is pushed away from the site of the missing cesium ion.

28. If the relative difference between the proton and electron charges (in absolute value) were

$$\frac{q_p - |q_e|}{e} = 0.0000010$$

then the actual difference would be

$$q_p - |q_e| = 1.6 \times 10^{-25}\ \text{C}\ .$$

Amplified by a factor of $29 \times 3 \times 10^{22}$ as indicated in the problem, this amounts to a deviation from perfect neutrality of

$$\Delta q = \left(29 \times 3 \times 10^{22}\right)\left(1.6 \times 10^{-25}\ \text{C}\right) = 0.14\ \text{C}$$

in a copper penny. Two such pennies, at $r = 1.0$ m, would therefore experience a very large force. Eq. 22-1 gives

$$F = k\frac{(\Delta q)^2}{r^2} = 1.7 \times 10^8\ \text{N}\ .$$

29. None of the reactions given include a beta decay, so the number of protons, the number of neutrons, and the number of electrons are each conserved. Atomic numbers (numbers of protons and numbers of electrons) and molar masses (combined numbers of protons and neutrons) can be found in Appendix F of the text.

(a) ^{1}H has 1 proton, 1 electron, and 0 neutrons and ^{9}Be has 4 protons, 4 electrons, and $9 - 4 = 5$ neutrons, so X has $1 + 4 = 5$ protons, $1 + 4 = 5$ electrons, and $0 + 5 - 1 = 4$ neutrons. One of the neutrons is freed in the reaction. X must be boron with a molar mass of $5 + 4 = 9\,\text{g/mol}$: ^{9}B.

(b) ^{12}C has 6 protons, 6 electrons, and $12 - 6 = 6$ neutrons and ^{1}H has 1 proton, 1 electron, and 0 neutrons, so X has $6 + 1 = 7$ protons, $6 + 1 = 7$ electrons, and $6 + 0 = 6$ neutrons. It must be nitrogen with a molar mass of $7 + 6 = 13\,\text{g/mol}$: ^{13}N.

(c) ^{15}N has 7 protons, 7 electrons, and $15 - 7 = 8$ neutrons; ^{1}H has 1 proton, 1 electron, and 0 neutrons; and ^{4}He has 2 protons, 2 electrons, and $4 - 2 = 2$ neutrons; so X has $7 + 1 - 2 = 6$ protons, 6 electrons, and $8 + 0 - 2 = 6$ neutrons. It must be carbon with a molar mass of $6 + 6 = 12$: ^{12}C.

30. (a) The two charges are $q = \alpha Q$ (where α is a pure number presumably less than 1 and greater than zero) and $Q - q = (1 - \alpha)Q$. Thus, Eq. 22-4 gives

$$F = \frac{1}{4\pi\varepsilon_0}\frac{(\alpha Q)((1-\alpha)Q)}{d^2} = \frac{Q^2\alpha(1-\alpha)}{4\pi\varepsilon_0\,d^2} .$$

(b) The graph below, of F versus α, has been scaled so that the maximum is 1. In actuality, the maximum value of the force is $F_{\text{max}} = Q^2/16\pi\varepsilon_0\,d^2$.

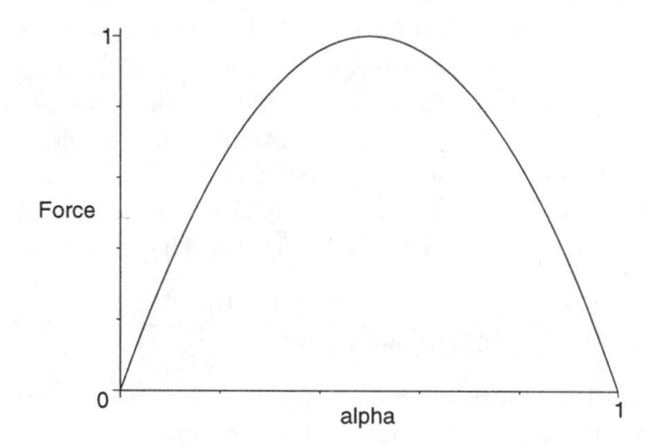

(c) It is clear that $\alpha = \frac{1}{2}$ gives the maximum value of F.

(d) Seeking the half-height points on the graph is difficult without grid lines or some of the special tracing features found in a variety of modern calculators. It is not difficult to algebraically solve for the half-height points (this involves the use of the quadratic formula). The results are

$$\alpha_1 = \frac{1}{2}\left(1 - \frac{1}{\sqrt{2}}\right) \approx 0.15 \quad \text{and}$$

$$\alpha_2 = \frac{1}{2}\left(1 + \frac{1}{\sqrt{2}}\right) \approx 0.85 .$$

31. (a) Eq. 22-11 (in absolute value) gives

$$n = \frac{|q|}{e} = \frac{2.00 \times 10^{-6}\,\text{C}}{1.60 \times 10^{-19}\,\text{C}} = 1.25 \times 10^{13} \text{ electrons} .$$

(b) Since you have the excess electrons (and electrons are lighter and more mobile than protons) then the electrons "leap" from you to the faucet instead of protons moving from the faucet to you (in the process of neutralizing your body).

(c) Unlike charges attract, and the faucet (which is grounded and is able to gain or lose any number of electrons due to its contact with Earth's large reservoir of mobile charges) becomes positively charged, especially in the region closest to your (negatively charged) hand, just before the spark.

(d) The cat is positively charged (before the spark), and by the reasoning given in part (b) the flow of charge (electrons) is from the faucet to the cat.

(e) If we think of the nose as a conducting sphere, then the side of the sphere closest to the fur is of one sign (of charge) and the side furthest from the fur is of the opposite sign (which, additionally, is oppositely charged from your bare hand which had stroked the cat's fur). The charges in your hand and those of the furthest side of the "sphere" therefore attract each other, and when close enough, manage to neutralize (due to the "jump" made by the electrons) in a painful spark.

32. (a) Using Coulomb's law, we obtain

$$F = \frac{q_1 q_2}{4\pi\varepsilon_0 r^2} = \frac{kq^2}{r^2} = \frac{\left(8.99 \times 10^9 \, \frac{\text{N}\cdot\text{m}^2}{\text{C}^2}\right)(1.00\,\text{C})^2}{(1.00\,\text{m})^2} = 8.99 \times 10^9 \text{ N} .$$

(b) If $r = 1000$ m, then

$$F = \frac{q_1 q_2}{4\pi\varepsilon_0 r^2} = \frac{kq^2}{r^2} = \frac{\left(8.99 \times 10^9 \, \frac{\text{N}\cdot\text{m}^2}{\text{C}^2}\right)(1.00\,\text{C})^2}{(1.00 \times 10^3\,\text{m})^2} = 8.99 \times 10^3 \text{ N} .$$

33. The unit Ampere is discussed in §22-4. Using i for current, the charge transferred is

$$q = it = \left(2.5 \times 10^4 \text{ A}\right)\left(20 \times 10^{-6} \text{ s}\right) = 0.50 \text{ C} .$$

34. Let the two charges be q_1 and q_2. Then $q_1 + q_2 = Q = 5.0 \times 10^{-5}$ C. We use Eq. 22-1:

$$1.0\,\text{N} = \frac{\left(8.99 \times 10^9 \, \frac{\text{N}\cdot\text{m}^2}{\text{C}^2}\right) q_1 q_2}{(2.0\,\text{m})^2} .$$

We substitute $q_2 = Q - q_1$ and solve for q_1 using the quadratic formula. The two roots obtained are the values of q_1 and q_2, since it does not matter which is which. We get 1.2×10^{-5} C and 3.8×10^{-5} C.

35. (a) Eq. 22-1 gives

$$F_{12} = k\frac{q_1 q_2}{d^2} = \left(8.99 \times 10^9 \, \frac{\text{N}\cdot\text{m}^2}{\text{C}^2}\right) \frac{\left(20.0 \times 10^{-6} \text{ C}\right)^2}{(1.50\,\text{m})^2} = 1.60 \text{ N} .$$

(b) A force diagram is shown as well as our choice of y axis (the dashed line).

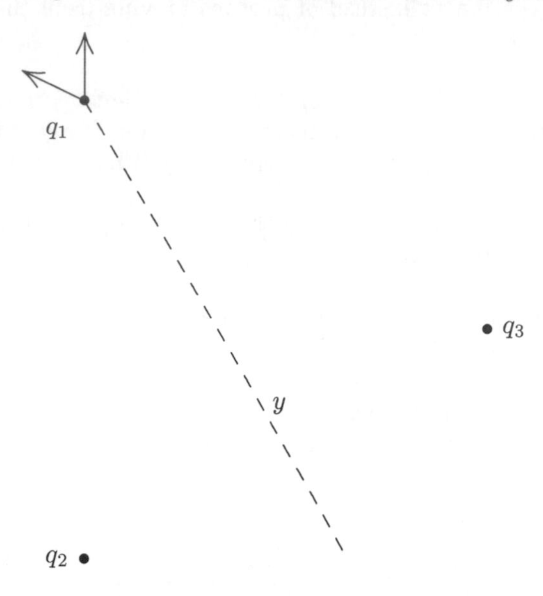

The y axis is meant to bisect the line between q_2 and q_3 in order to make use of the symmetry in the problem (equilateral triangle of side length d, equal-magnitude charges $q_1 = q_2 = q_3 = q$). We see that the resultant force is along this symmetry axis, and we obtain

$$|F_y| = 2\left(k\frac{q^2}{d^2}\right)\cos 30^\circ = 2.77 \text{ N} .$$

36. (a) Since $q_A = -2Q$ and $q_C = +8Q$, Eq. 22-4 leads to

$$\left|\vec{F}_{AC}\right| = \frac{|(-2Q)(+8Q)|}{4\pi\epsilon_0 d^2} = \frac{4Q^2}{\pi\epsilon_0 d^2} .$$

(b) After making contact with each other, both A and B have a charge of

$$\left(\frac{-2Q + (-4Q)}{2}\right) = -3Q .$$

When B is grounded its charge is zero. After making contact with C, which has a charge of $+8Q$, B acquires a charge of $[0 + (-8Q)]/2 = -4Q$, which charge C has as well. Finally, we have $Q_A = -3Q$ and $Q_B = Q_C = -4Q$. Therefore,

$$\left|\vec{F}_{AC}\right| = \frac{|(-3Q)(-4Q)|}{4\pi\epsilon_0 d^2} = \frac{3Q^2}{\pi\epsilon_0 d^2} .$$

(c) We also obtain

$$\left|\vec{F}_{BC}\right| = \frac{|(-4Q)(-4Q)|}{4\pi\epsilon_0 d^2} = \frac{4Q^2}{\pi\epsilon_0 d^2} .$$

37. The net charge carried by John whose mass is m is roughly

$$\begin{aligned} q &= (0.0001)\frac{mN_AZe}{M} \\ &= (0.0001)\frac{(90\,\text{kg})(6.02\times10^{23}\,\text{molecules/mol})(18\,\text{electron proton pairs/molecule})\,(1.6\times10^{-19}\,\text{C})}{0.018\,\text{kg/mol}} \\ &= 8.7\times10^5 \text{ C} , \end{aligned}$$

and the net charge carried by Mary is half of that. So the electrostatic force between them is estimated to be

$$F \approx k\frac{q(q/2)}{d^2} = \left(8.99 \times 10^9 \, \frac{\text{N} \cdot \text{m}^2}{\text{C}^2}\right) \frac{(8.7 \times 10^5 \, \text{C})^2}{2(30 \, \text{m})^2} \approx 4 \times 10^{18} \text{ N} .$$

38. Letting $kq^2/r^2 = mg$, we get

$$r = q\sqrt{\frac{k}{mg}} = (1.60 \times 10^{-19} \, \text{C}) \sqrt{\frac{8.99 \times 10^9 \, \frac{\text{Nm}^2}{\text{C}^2}}{(1.67 \times 10^{-27} \, \text{kg}) \, (9.8 \, \text{m/s}^2)}} = 0.119 \text{ m} .$$

39. Coulomb's law gives

$$F = \frac{|q| \cdot |q|}{4\pi\varepsilon_0 r^2} = \frac{k(e/3)^2}{r^2} = \frac{\left(8.99 \times 10^9 \, \frac{\text{N} \cdot \text{m}^2}{\text{C}^2}\right) (1.60 \times 10^{-19} \, \text{C})^2}{9(2.6 \times 10^{-15} \, \text{m})^2} = 3.8 \text{ N} .$$

40. We are concerned with the charges in the nucleus (not the "orbiting" electrons, if there are any). The nucleus of Helium has 2 protons and that of Thorium has 90.

   (a) Eq. 22-1 gives

$$F = k\frac{q^2}{r^2} = \frac{\left(8.99 \times 10^9 \, \frac{\text{N} \cdot \text{m}^2}{\text{C}^2}\right) (2(1.60 \times 10^{-19} \, \text{C})) \, (90(1.60 \times 10^{-19} \, \text{C}))}{(9.0 \times 10^{-15} \, \text{m})^2} = 5.1 \times 10^2 \text{ N} .$$

   (b) Estimating the helium nucleus mass as that of 4 protons (actually, that of 2 protons and 2 neutrons, but the neutrons have approximately the same mass), Newton's second law leads to

$$a = \frac{F}{m} = \frac{5.1 \times 10^2 \, \text{N}}{4(1.67 \times 10^{-27} \, \text{kg})} = 7.7 \times 10^{28} \text{ m/s}^2 .$$

41. Charge $q_1 = -80 \times 10^{-6}$ C is at the origin, and charge $q_2 = +40 \times 10^{-6}$ C is at $x = 0.20$ m. The force on $q_3 = +20 \times 10^{-6}$ C is due to the attractive and repulsive forces from q_1 and q_2, respectively. In symbols, $\vec{F}_{3\,\text{net}} = \vec{F}_{3\,1} + \vec{F}_{3\,2}$, where

$$|\vec{F}_{3\,1}| = k\,\frac{q_3\,|q_1|}{r_{3\,1}^2} \qquad \text{and} \qquad |\vec{F}_{3\,2}| = k\,\frac{q_3\,q_2}{r_{3\,2}^2} .$$

   (a) In this case $r_{3\,1} = 0.40$ m and $r_{3\,2} = 0.20$ m, with $\vec{F}_{3\,1}$ directed towards $-x$ and $\vec{F}_{3\,2}$ directed in the $+x$ direction. Using the value of k in Eq. 22-5, we obtain $\vec{F}_{3\,\text{net}} = 89.9 \approx 90$ N in the $+x$ direction.

   (b) In this case $r_{3\,1} = 0.80$ m and $r_{3\,2} = 0.60$ m, with $\vec{F}_{3\,1}$ directed towards $-x$ and $\vec{F}_{3\,2}$ towards $+x$. Now we obtain $\vec{F}_{3\,\text{net}} = 2.5$ N in the $-x$ direction.

   (c) Between the locations treated in parts (a) and (b), there must be one where $\vec{F}_{3\,\text{net}} = 0$. Writing $r_{3\,1} = x$ and $r_{3\,2} = x - 0.20$ m, we equate $|\vec{F}_{3\,1}|$ and $|\vec{F}_{3\,2}|$, and after canceling common factors, arrive at

$$\frac{|q_1|}{x^2} = \frac{q_2}{(x - 0.2)^2} .$$

   This can be further simplified to

$$\frac{(x - 0.2)^2}{x^2} = \frac{q_2}{|q_1|} = \frac{1}{2} .$$

   Taking the (positive) square root and solving, we obtain $x = 0.68$ m. If one takes the negative root and 'solves', one finds the location where the net force *would* be zero *if* q_1 and q_2 were of like sign (which is not the case here).

42. (a) Charge $Q_1 = +80 \times 10^{-9}$ C is on the y axis at $y = 0.003$ m, and charge $Q_2 = +80 \times 10^{-9}$ C is on the y axis at $y = -0.003$ m. The force on particle 3 (which has a charge of $q = +18 \times 10^{-9}$ C) is due to the vector sum of the repulsive forces from Q_1 and Q_2. In symbols, $\vec{F}_{3\,1} + \vec{F}_{3\,2} = \vec{F}_{3\text{ net}}$, where

$$|\vec{F}_{3\,1}| = k\,\frac{q_3\,|q_1|}{r_{3\,1}^2} \qquad \text{and} \quad |\vec{F}_{3\,2}| = k\,\frac{q_3\,q_2}{r_{3\,2}^2} \ .$$

Using the Pythagorean theorem, we have $r_{3\,1} = r_{3\,2} = 0.005$ m. In magnitude-angle notation (particularly convenient if one uses a vector capable calculator in polar mode), the indicated vector addition becomes

$$(0.518 \angle -37^\circ) + (0.518 \angle 37^\circ) = (0.829 \angle 0^\circ) \ .$$

Therefore, the net force is 0.829 N in the $+x$ direction.

(b) Switching the sign of Q_2 amounts to reversing the direction of its force on q. Consequently, we have

$$(0.518 \angle -37^\circ) + (0.518 \angle -143^\circ) = (0.621 \angle -90^\circ) \ .$$

Therefore, the net force is 0.621 N in the $-y$ direction.

43. (a) For the net force to be in the $+x$ direction, the y components of the individual forces must cancel. The angle of the force exerted by the $q_1 = 40\ \mu$C charge on $q = 20\ \mu$C is 45°, and the angle of force exerted on q by Q is at $-\theta$ where

$$\theta = \tan^{-1}\left(\frac{2.0}{3.0}\right) = 33.7^\circ \ .$$

Therefore, cancellation of y components requires

$$k\,\frac{q_1\,q}{\left(0.02\sqrt{2}\right)^2}\,\sin 45^\circ = k\,\frac{|Q|\,q}{\left(\sqrt{0.03^2 + 0.02^2}\right)^2}\,\sin\theta$$

from which we obtain $|Q| = 82.9\ \mu$C. Charge Q is "pulling" on q, so (since $q > 0$) we conclude $Q = -82.9\ \mu$C.

(b) Now, we require that the x components cancel, and we note that in this case, the angle of force on q exerted by Q is $+\theta$ (it is repulsive, and Q is positive-valued). Therefore,

$$k\,\frac{q_1\,q}{\left(0.02\sqrt{2}\right)^2}\,\cos 45^\circ = k\,\frac{Q\,q}{\left(\sqrt{0.03^2 + 0.02^2}\right)^2}\,\cos\theta$$

from which we obtain $Q = 55.2\ \mu$C.

44. We are looking for a charge q which, when placed at the origin, experiences $\vec{F}_{\text{net}} = 0$, where

$$\vec{F}_{\text{net}} = \vec{F}_1 + \vec{F}_2 + \vec{F}_3 \ .$$

The magnitude of these individual forces are given by Coulomb's law, Eq. 22-1, and without loss of generality we assume $q > 0$. The charges q_1 ($+6\ \mu$C), q_2 ($-4\ \mu$C), and q_3 (unknown), are located on the $+x$ axis, so that we know $\vec{F}_1$ points towards $-x$, $\vec{F}_2$ points towards $+x$, and $\vec{F}_3$ points towards $-x$ if $q_3 > 0$ and points towards $+x$ if $q_3 < 0$. Therefore, with $r_1 = 8$ m, $r_2 = 16$ m and $r_3 = 24$ m, we have

$$0 = -k\,\frac{q_1\,q}{r_1^2} + k\,\frac{|q_2|\,q}{r_2^2} - k\,\frac{q_3\,q}{r_3^2} \ .$$

Simplifying, this becomes

$$0 = -\frac{6}{8^2} + \frac{4}{16^2} - \frac{q_3}{24^2}$$

where q_3 is now understood to be in μC. Thus, we obtain $q_3 = -45\ \mu$C.

45. The magnitude of the net force on the $q = 42 \times 10^{-6}$ C charge is

$$k\,\frac{q_1\,q}{0.28^2} + k\,\frac{|q_2|\,q}{0.44^2}$$

where $q_1 = 30 \times 10^{-9}$ C and $|q_2| = 40 \times 10^{-9}$ C. This yields 0.22 N. Using Newton's second law, we obtain

$$m = \frac{F}{a} = \frac{0.22\,\text{N}}{100 \times 10^3\,\text{m/s}^2} = 2.2 \times 10^{-6}\ \text{kg} .$$

46. The charge dq within a thin shell of thickness dr is $\rho\,A\,dr$ where $A = 4\pi r^2$. Thus, with $\rho = b/r$, we have

$$q = \int dq = 4\pi b \int_{r_1}^{r_2} r\,dr = 2\pi b\left(r_2^2 - r_1^2\right) .$$

With $b = 3.0\,\mu\text{C/m}^2$, $r_2 = 0.06$ m and $r_1 = 0.04$ m, we obtain $q = 0.038\ \mu$C.

47. The charge dq within a thin section of the rod (of thickness dx) is $\rho\,A\,dx$ where $A = 4.00 \times 10^{-4}\ \text{m}^2$ and ρ is the charge per unit volume. The number of (excess) electrons in the rod (of length $L = 2.00$ m) is $N = q/(-e)$ where e is given in Eq. 22-14.

(a) In the case where $\rho = -4.00 \times 10^{-6}\ \text{C/m}^3$, we have

$$N = \frac{q}{-e} = \frac{\rho\,A}{-e}\int_0^L dx = \frac{|\rho|\,A\,L}{e}$$

which yields $N = 2.00 \times 10^{10}$.

(b) With $\rho = bx^2$ ($b = -2.00 \times 10^{-6}\ \text{C/m}^5$) we obtain

$$N = \frac{b\,A}{-e}\int_0^L x^2\,dx = \frac{|b|\,A\,L^3}{3\,e} = 1.33 \times 10^{10} .$$

48. When sphere C touches sphere A, they divide up their total charge ($Q/2$ plus Q) equally between them. Thus, sphere A now has charge $3Q/4$, and the magnitude of the force of attraction between A and B becomes

$$F = k\,\frac{\left(\frac{3Q}{4}\right)\left(\frac{Q}{4}\right)}{d^2} = 4.68 \times 10^{-19}\ \text{N} .$$

49. In experiment 1, sphere C first touches sphere A, and they divided up their total charge ($Q/2$ plus Q) equally between them. Thus, sphere A and sphere C each acquired charge $3Q/4$. Then, sphere C touches B and those spheres split up their total charge ($3Q/4$ plus $-Q/4$) so that B ends up with charge equal to $Q/4$. The force of repulsion between A and B is therefore

$$F_1 = k\,\frac{\left(\frac{3Q}{4}\right)\left(\frac{Q}{4}\right)}{d^2}$$

at the end of experiment 1. Now, in experiment 2, sphere C first touches B which leaves each of them with charge $Q/8$. When C next touches A, sphere A is left with charge $9Q/16$. Consequently, the force of repulsion between A and B is

$$F_2 = k\,\frac{\left(\frac{9Q}{16}\right)\left(\frac{Q}{8}\right)}{d^2}$$

at the end of experiment 2. The ratio is

$$\frac{F_2}{F_1} = \frac{\left(\frac{9}{16}\right)\left(\frac{1}{8}\right)}{\left(\frac{3}{4}\right)\left(\frac{1}{4}\right)} = 0.375 .$$

50. Regarding the forces on q_3 exerted by q_1 and q_2, one must "push" and the other must "pull" in order that the net force is zero; hence, q_1 and q_2 have opposite signs. For individual forces to cancel, their magnitudes must be equal:

$$k\frac{|q_1||q_3|}{(3d)^2} = k\frac{|q_2||q_3|}{(2d)^2}$$

which simplifies to

$$\frac{|q_1|}{9} = \frac{|q_2|}{4} .$$

Therefore, $q_1 = -\frac{9}{4}q_2$.

51. The individual force magnitudes are found using Eq. 22-1, with SI units (so $a = 0.02$ m) and k as in Eq. 22-5. We use magnitude-angle notation (convenient if ones uses a vector capable calculator in polar mode), listing the forces due to $+4.00q$, $+2.00q$, and $-2.00q$ charges:

$$(4.60 \times 10^{-24} \angle\ 180°) + (2.30 \times 10^{-24} \angle - 90°) + (1.02 \times 10^{-24} \angle - 145°) = (6.16 \times 10^{-24} \angle - 152°)$$

Therefore, the net force has magnitude 6.16×10^{-24} N and is at an angle of $-152°$ (or $208°$ measured counterclockwise from the $+x$ axis).

Chapter 23

1. (a) We note that the electric field points leftward at both points. Using $\vec{F} = q_0\vec{E}$, and orienting our x axis rightward (so $\hat{\text{i}}$ points right in the figure), we find

$$\vec{F} = (+1.6 \times 10^{-19}\,\text{C})\left(-40\,\frac{\text{N}}{\text{C}}\,\hat{\text{i}}\right) = -6.4 \times 10^{-18}\ \text{N}\ \hat{\text{i}}$$

which means the magnitude of the force on the proton is 6.4×10^{-18} N and its direction $(-\hat{\text{i}})$ is leftward.

(b) As the discussion in §23-2 makes clear, the field strength is proportional to the "crowdedness" of the field lines. It is seen that the lines are twice as crowded at A than at B, so we conclude that $E_A = 2E_B$. Thus, $E_B = 20$ N/C.

2. We note that the symbol q_2 is used in the problem statement to mean the absolute value of the negative charge which resides on the larger shell. The following sketch is for $q_1 = q_2$.

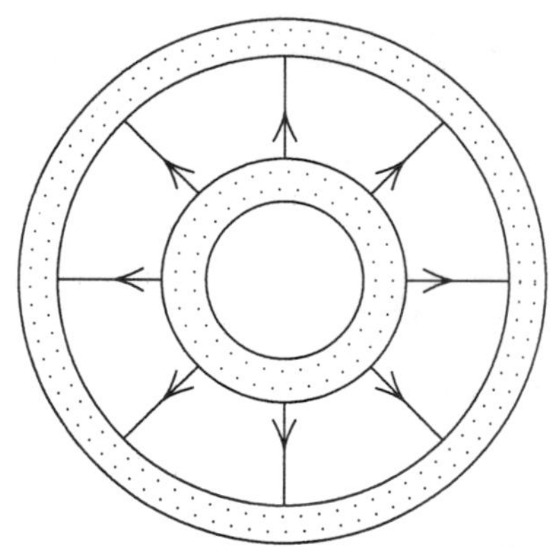

The following two sketches are for the cases $q_1 > q_2$ (left figure) and $q_1 < q_2$ (right figure).

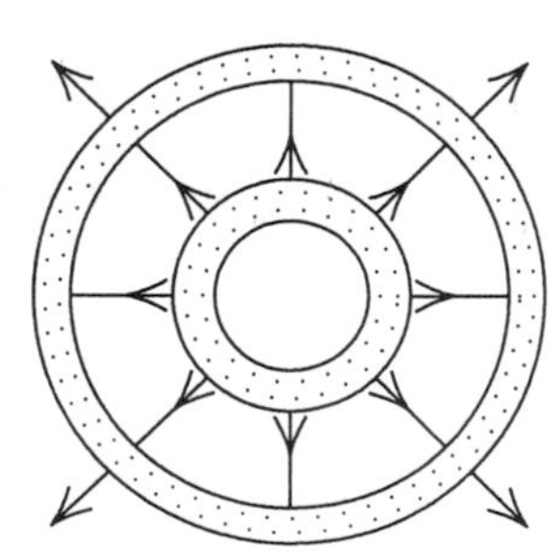

3. The diagram below is an edge view of the disk and shows the field lines above it. Near the disk, the lines are perpendicular to the surface and since the disk is uniformly charged, the lines are uniformly distributed over the surface. Far away from the disk, the lines are like those of a single point charge (the

charge on the disk). Extended back to the disk (along the dotted lines of the diagram) they intersect at the center of the disk.

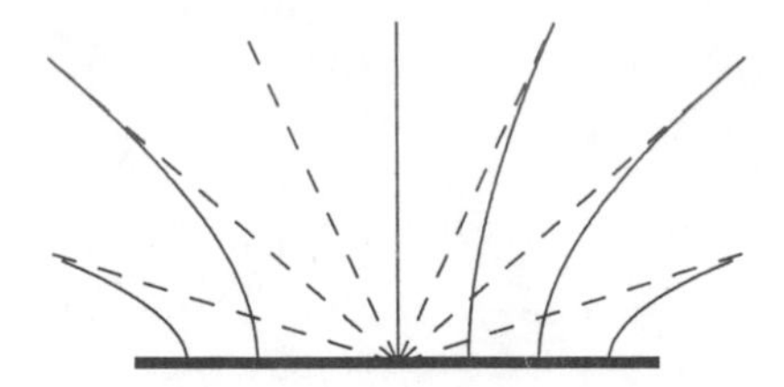

If the disk is positively charged, the lines are directed outward from the disk. If the disk is negatively charged, they are directed inward toward the disk. A similar set of lines is associated with the region below the disk.

4. We find the charge magnitude $|q|$ from $E = |q|/4\pi\varepsilon_0 r^2$:

$$q = 4\pi\varepsilon_0 E r^2 = \frac{(1.00\,\text{N/C})(1.00\,\text{m})^2}{8.99\times 10^9\,\frac{\text{N}\cdot\text{m}^2}{\text{C}^2}} = 1.11\times 10^{-10}\ \text{C}\ .$$

5. Since the magnitude of the electric field produced by a point charge q is given by $E = |q|/4\pi\varepsilon_0 r^2$, where r is the distance from the charge to the point where the field has magnitude E, the magnitude of the charge is

$$|q| = 4\pi\varepsilon_0 r^2 E = \frac{(0.50\,\text{m})^2(2.0\,\text{N/C})}{8.99\times 10^9\,\text{N}\cdot\text{m}^2/\text{C}^2} = 5.6\times 10^{-11}\ \text{C}\ .$$

6. For concreteness, consider that charge 2 lies 0.15 m east of charge 1, and the point at which we are asked to evaluate their net field is $r = 0.075$ m east of charge 1 and $r = 0.075$ m west of charge 2. The values of charge are $q_1 = -q_2 = 2.0\times 10^{-7}$ C. The magnitudes and directions of the individual fields are specified:

$$\left|\vec{E}_1\right| = \frac{q_1}{4\pi\varepsilon_0 r^2} = 3.2\times 10^5\ \text{N/C} \qquad \text{and} \quad \vec{E}_1 \text{ points east}$$

$$\left|\vec{E}_2\right| = \frac{|q_2|}{4\pi\varepsilon_0 r^2} = 3.2\times 10^5\ \text{N/C} \qquad \text{and} \quad \vec{E}_2 \text{ points east}$$

Since they point the same direction, the magnitude of the net field is the sum of their amplitudes, $\left|\vec{E}_{\text{net}}\right| = 6.4\times 10^5$ N/C, and it points east (that is, towards the negative charge).

7. Since the charge is uniformly distributed throughout a sphere, the electric field at the surface is exactly the same as it would be if the charge were all at the center. That is, the magnitude of the field is

$$E = \frac{q}{4\pi\varepsilon_0 R^2}$$

where q is the magnitude of the total charge and R is the sphere radius. The magnitude of the total charge is Ze, so

$$E = \frac{Ze}{4\pi\varepsilon_0 R^2} = \frac{(8.99\times 10^9\,\text{N}\cdot\text{m}^2/\text{C}^2)(94)(1.60\times 10^{-19}\,\text{C})}{(6.64\times 10^{-15}\,\text{m})^2} = 3.07\times 10^{21}\ \text{N/C}\ .$$

The field is normal to the surface and since the charge is positive, it points outward from the surface.

8. The individual magnitudes $\left|\vec{E}_1\right|$ and $\left|\vec{E}_2\right|$ are figured from Eq. 23-3, where the absolute value signs for q are unnecessary since these charges are both positive. Whether we add the magnitudes or subtract them depends on if $\vec{E}_1$ is in the same, or opposite, direction as $\vec{E}_2$. At points to the left of q_1 (along the

$-x$ axis) both fields point leftward, and at points right of q_2 (at $x > d$) both fields point rightward; in these regions the magnitude of the net field is the sum $\left|\vec{E}_1\right| + \left|\vec{E}_2\right|$. In the region between the charges $(0 < x < d)$ $\vec{E}_1$ points rightward and $\vec{E}_2$ points leftward, so the net field in this range is $\vec{E}_{\text{net}} = \left|\vec{E}_1\right| - \left|\vec{E}_2\right|$ in the $\hat{\text{i}}$ direction. Summarizing, we have

$$\vec{E}_{\text{net}} = \hat{\text{i}}\,\frac{1}{4\pi\varepsilon_0}\begin{cases} -\frac{q_1}{x^2} - \frac{q_2}{(d+|x|)^2} & \text{for } x < 0 \\ \frac{q_1}{x^2} - \frac{q_2}{(d-x)^2} & \text{for } 0 < x < d \\ \frac{q_1}{x^2} + \frac{q_2}{(x-d)^2} & \text{for } d < x \end{cases} .$$

We note that these can be written as a single expression applying to all three regions:

$$\vec{E}_{\text{net}} = \frac{1}{4\pi\varepsilon_0}\left(\frac{q_1 x}{|x|^3} + \frac{q_2(x-d)}{|x-d|^3}\right)\hat{\text{i}} .$$

For $-0.09 \le x \le 0.20$ m with $d = 0.10$ m and charge values as specified in the problem, we find

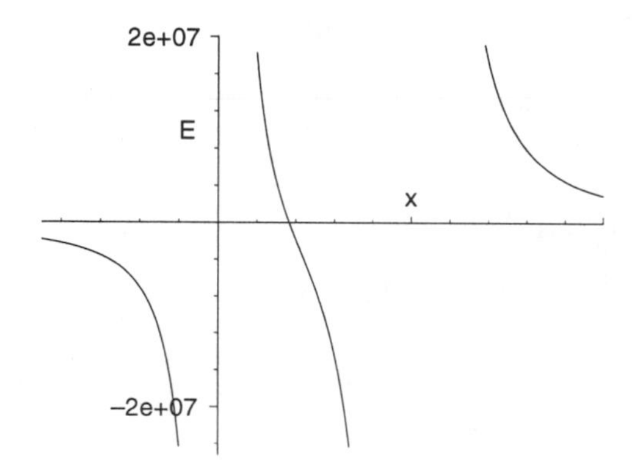

9. At points between the charges, the individual electric fields are in the same direction and do not cancel. Charge q_2 has a greater magnitude than charge q_1, so a point of zero field must be closer to q_1 than to q_2. It must be to the right of q_1 on the diagram.

d P

q_2 q_1

We put the origin at q_2 and let x be the coordinate of P, the point where the field vanishes. Then, the total electric field at P is given by

$$E = \frac{1}{4\pi\varepsilon_0}\left(\frac{q_2}{x^2} - \frac{q_1}{(x-d)^2}\right)$$

where q_1 and q_2 are the magnitudes of the charges. If the field is to vanish,

$$\frac{q_2}{x^2} = \frac{q_1}{(x-d)^2} .$$

We take the square root of both sides to obtain $\sqrt{q_2}/x = \sqrt{q_1}/(x-d)$. The solution for x is

$$x = \left(\frac{\sqrt{q_2}}{\sqrt{q_2} - \sqrt{q_1}}\right) d$$

$$= \left(\frac{\sqrt{4.0q_1}}{\sqrt{4.0q_1} - \sqrt{q_1}}\right) d$$
$$= \left(\frac{2.0}{2.0 - 1.0}\right) d = 2.0d$$
$$= (2.0)(50\,\text{cm}) = 100 \text{ cm} .$$

The point is 50 cm to the right of q_1.

10. The individual magnitudes $|\vec{E}_1|$ and $|\vec{E}_2|$ are figured from Eq. 23-3, where the absolute value signs for q_2 are unnecessary since this charge is positive. Whether we add the magnitudes or subtract them depends on if $\vec{E}_1$ is in the same, or opposite, direction as $\vec{E}_2$. At points left of q_1 (on the $-x$ axis) the fields point in opposite directions, but there is no possibility of cancellation (zero net field) since $|\vec{E}_1|$ is everywhere bigger than $|\vec{E}_2|$ in this region. In the region between the charges ($0 < x < d$) both fields point leftward and there is no possibility of cancellation. At points to the right of q_2 (where $x > d$), $\vec{E}_1$ points leftward and $\vec{E}_2$ points rightward so the net field in this range is

$$\vec{E}_{\text{net}} = |\vec{E}_2| - |\vec{E}_1| \quad \text{in the } \hat{\imath} \text{ direction.}$$

Although $|q_1| > q_2$ there is the possibility of $\vec{E}_{\text{net}} = 0$ since these points are closer to q_2 than to q_1. Thus, we look for the zero net field point in the $x > d$ region:

$$|\vec{E}_1| = |\vec{E}_2|$$
$$\frac{1}{4\pi\varepsilon_0}\frac{|q_1|}{x^2} = \frac{1}{4\pi\varepsilon_0}\frac{q_2}{(x-d)^2}$$

which leads to

$$\frac{x-d}{x} = \sqrt{\frac{q_2}{|q_1|}} = \sqrt{\frac{2}{5}} .$$

Thus, we obtain $x = \frac{d}{1-\sqrt{2/5}} \approx 2.7d$. A sketch of the field lines is shown below.

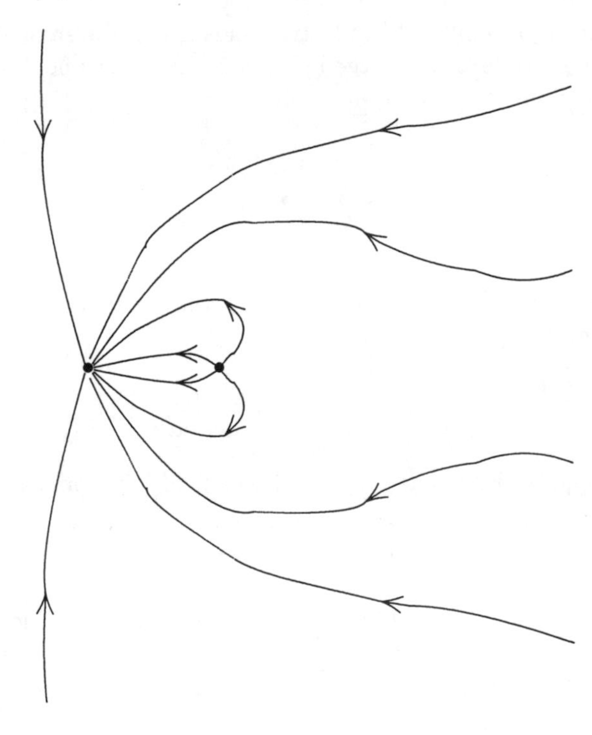

11. We place the origin of our coordinate system at point P and orient our y axis in the direction of the $q_4 = -12q$ charge (passing through the $q_3 = +3q$ charge). The x axis is perpendicular to the y axis, and thus passes through the identical $q_1 = q_2 = +5q$ charges. The individual magnitudes $|\vec{E}_1|$, $|\vec{E}_2|$, $|\vec{E}_3|$, and $|\vec{E}_4|$ are figured from Eq. 23-3, where the absolute value signs for q_1, q_2, and q_3 are unnecessary since those charges are positive (assuming $q > 0$). We note that the contribution from q_1 cancels that of q_2 (that is, $|\vec{E}_1| = |\vec{E}_2|$), and the net field (if there is any) should be along the y axis, with magnitude equal to

$$\vec{E}_{\rm net} = \frac{1}{4\pi\varepsilon_0}\left(\frac{|q_4|}{(2d)^2} - \frac{q_3}{d^2}\right)\hat{\jmath} = \frac{1}{4\pi\varepsilon_0}\left(\frac{12q}{4d^2} - \frac{3q}{d^2}\right)\hat{\jmath}$$

which is seen to be zero. A rough sketch of the field lines is shown below:

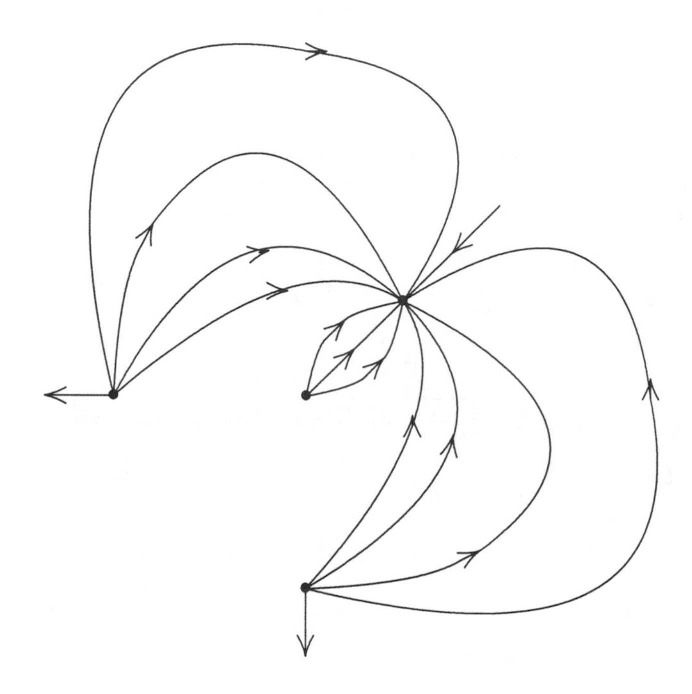

12. By symmetry we see the contributions from the $+q$ charges cancel each other, and we simply use Eq. 23-3 to compute magnitude of the field due to the $+2q$ charge (this field points at 45°, which is clear from the figure in the textbook).

$$\left|\vec{E}_{\rm net}\right| = \frac{1}{4\pi\varepsilon_0}\frac{2q}{r^2}$$

where $r = a/\sqrt{2}$. Thus, we obtain $\left|\vec{E}_{\rm net}\right| = q/\pi\varepsilon_0 a^2$.

13. We choose the coordinate axes as shown on the diagram below. At the center of the square, the electric fields produced by the charges at the lower left and upper right corners are both along the x axis and each points away from the center and toward the charge that produces it. Since each charge is a distance $d = \sqrt{2}a/2 = a/\sqrt{2}$ away from the center, the net field due to these two charges is

$$\begin{aligned} E_x &= \frac{1}{4\pi\varepsilon_0}\left(\frac{2q}{a^2/2} - \frac{q}{a^2/2}\right) = \frac{1}{4\pi\varepsilon_0}\frac{q}{a^2/2} \\ &= \frac{\left(8.99\times10^9\,\mathrm{N\cdot m^2/C^2}\right)\left(1.0\times10^{-8}\,\mathrm{C}\right)}{(0.050\,\mathrm{m})^2/2} = 7.19\times10^4\ \mathrm{N/C}\ . \end{aligned}$$

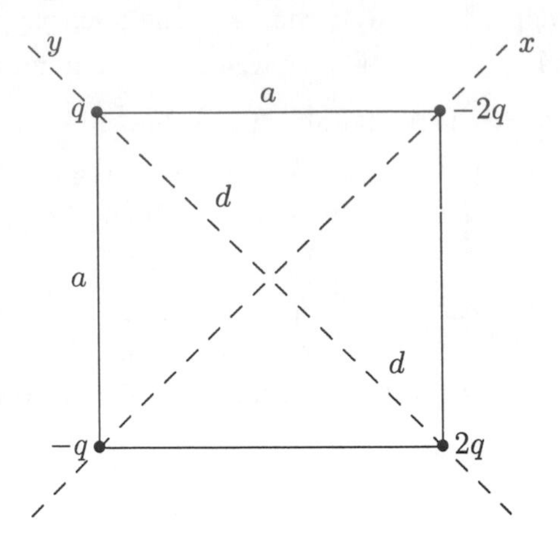

At the center of the square, the field produced by the charges at the upper left and lower right corners are both along the y axis and each points away from the charge that produces it. The net field produced at the center by these charges is

$$E_y = \frac{1}{4\pi\varepsilon_0}\left[\frac{2q}{a^2/2} - \frac{q}{a^2/2}\right] = \frac{1}{4\pi\varepsilon_0}\frac{q}{a^2/2} = 7.19 \times 10^4 \text{ N/C} .$$

The magnitude of the field is

$$E = \sqrt{E_x^2 + E_y^2} = \sqrt{2(7.19 \times 10^4 \text{ N/C})^2} = 1.02 \times 10^5 \text{ N/C}$$

and the angle it makes with the x axis is

$$\theta = \tan^{-1}\frac{E_y}{E_x} = \tan^{-1}(1) = 45^\circ .$$

It is upward in the diagram, from the center of the square toward the center of the upper side.

14. Since both charges are positive (and aligned along the z axis) we have

$$\left|\vec{E}_{\text{net}}\right| = \frac{1}{4\pi\varepsilon_0}\left[\frac{q}{(z-d/2)^2} + \frac{q}{(z+d/2)^2}\right] .$$

For $z \gg d$ we have $(z \pm d/2)^{-2} \approx z^{-2}$, so

$$\left|\vec{E}_{\text{net}}\right| \approx \frac{1}{4\pi\varepsilon_0}\left(\frac{q}{z^2} + \frac{q}{z^2}\right) = \frac{2q}{4\pi\varepsilon_0 z^2} .$$

15. The magnitude of the dipole moment is given by $p = qd$, where q is the positive charge in the dipole and d is the separation of the charges. For the dipole described in the problem, $p = (1.60 \times 10^{-19}\text{ C})(4.30 \times 10^{-9}\text{ m}) = 6.88 \times 10^{-28}\text{ C}\cdot\text{m}$. The dipole moment is a vector that points from the negative toward the positive charge.

16. From the figure below it is clear that the net electric field at point P points in the $-\hat{\text{j}}$ direction. Its magnitude is

$$\begin{aligned}\left|\vec{E}_{\text{net}}\right| &= 2E_1 \sin\theta = 2\left[k\frac{q}{(d/2)^2 + r^2}\right]\frac{d/2}{\sqrt{(d/2)^2 + r^2}} \\ &= k\frac{qd}{[(d/2)^2 + r^2]^{3/2}}\end{aligned}$$

where we use k for $1/4\pi\varepsilon_0$ for brevity. For $r \gg d$, we write $[(d/2)^2 + r^2]^{3/2} \approx r^3$ so the expression above reduces to

$$\left|\vec{E}_{\rm net}\right| \approx k\frac{qd}{r^3} .$$

Since $\vec{p} = (qd)\hat{\rm j}$,

$$\vec{E}_{\rm net} \approx -k\frac{\vec{p}}{r^3} .$$

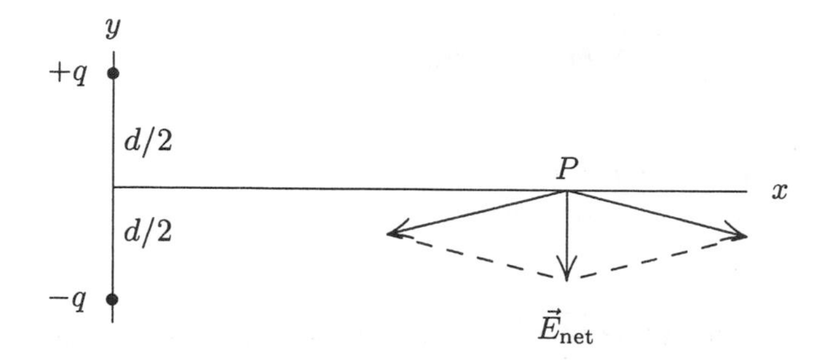

17. Think of the quadrupole as composed of two dipoles, each with dipole moment of magnitude $p = qd$. The moments point in opposite directions and produce fields in opposite directions at points on the quadrupole axis. Consider the point P on the axis, a distance z to the right of the quadrupole center and take a rightward pointing field to be positive. Then, the field produced by the right dipole of the pair is $qd/2\pi\varepsilon_0(z - d/2)^3$ and the field produced by the left dipole is $-qd/2\pi\varepsilon_0(z + d/2)^3$. Use the binomial expansions $(z - d/2)^{-3} \approx z^{-3} - 3z^{-4}(-d/2)$ and $(z + d/2)^{-3} \approx z^{-3} - 3z^{-4}(d/2)$ to obtain

$$E = \frac{qd}{2\pi\varepsilon_0}\left[\frac{1}{z^3} + \frac{3d}{2z^4} - \frac{1}{z^3} + \frac{3d}{2z^4}\right] = \frac{6qd^2}{4\pi\varepsilon_0 z^4} .$$

Let $Q = 2qd^2$. Then,

$$E = \frac{3Q}{4\pi\varepsilon_0 z^4} .$$

18. We use Eq. 23-3, assuming both charges are positive.

$$\begin{aligned} E_{\text{left ring}} &= E_{\text{right ring}} \qquad \text{evaluated at } P \\ \frac{q_1 R}{4\pi\varepsilon_0\left(R^2 + R^2\right)^{3/2}} &= \frac{q_2(2R)}{4\pi\varepsilon_0\left((2R)^2 + R^2\right)^{3/2}} \end{aligned}$$

Simplifying, we obtain

$$\frac{q_1}{q_2} = 2\left(\frac{2}{5}\right)^{3/2} \approx 0.51 .$$

19. The electric field at a point on the axis of a uniformly charged ring, a distance z from the ring center, is given by

$$E = \frac{qz}{4\pi\varepsilon_0(z^2 + R^2)^{3/2}}$$

where q is the charge on the ring and R is the radius of the ring (see Eq. 23–16). For q positive, the field points upward at points above the ring and downward at points below the ring. We take the positive direction to be upward. Then, the force acting on an electron on the axis is

$$F = -\frac{eqz}{4\pi\varepsilon_0(z^2 + R^2)^{3/2}} .$$

For small amplitude oscillations $z \ll R$ and z can be neglected in the denominator. Thus,

$$F = -\frac{eqz}{4\pi\varepsilon_0 R^3} .$$

The force is a restoring force: it pulls the electron toward the equilibrium point $z = 0$. Furthermore, the magnitude of the force is proportional to z, just as if the electron were attached to a spring with spring constant $k = eq/4\pi\varepsilon_0 R^3$. The electron moves in simple harmonic motion with an angular frequency given by

$$\omega = \sqrt{\frac{k}{m}} = \sqrt{\frac{eq}{4\pi\varepsilon_0 m R^3}}$$

where m is the mass of the electron.

20. From symmetry, we see that the net field at P is twice the field caused by the upper semicircular charge $+q = \lambda \cdot \pi R$ (and that it points downward). Adapting the steps leading to Eq. 23-21, we find

$$\vec{E}_{\text{net}} = 2\left(-\hat{\text{j}}\right)\frac{\lambda}{4\pi\varepsilon_0 R}\Big[\sin\theta\Big]_{-90^\circ}^{90^\circ} = -\frac{q}{\varepsilon_0 \pi^2 R^2}\,\hat{\text{j}} .$$

21. Studying Sample Problem 23-3, we see that the field evaluated at the center of curvature due to a charged distribution on a circular arc is given by

$$\vec{E} = \frac{\lambda}{4\pi\varepsilon_0 r}\Big[\sin\theta\Big]_{-\theta/2}^{\theta/2} \qquad \text{along the symmetry axis}$$

where $\lambda = q/r\theta$ with θ in radians. In this problem, each charged quarter-circle produces a field of magnitude

$$\left|\vec{E}\right| = \frac{|q|}{r\pi/2}\frac{1}{4\pi\varepsilon_0 r}\Big[\sin\theta\Big]_{-\pi/4}^{\pi/4} = \frac{|q|}{\varepsilon_0 \pi^2 r^2 \sqrt{2}} .$$

That produced by the positive quarter-circle points at $-45°$, and that of the negative quarter-circle points at $+45°$. By symmetry, we conclude that their net field is horizontal (and rightward in the textbook figure) with magnitude

$$E_x = 2\left(\frac{|q|}{\varepsilon_0 \pi^2 r^2 \sqrt{2}}\right)\cos 45^\circ = \frac{|q|}{\varepsilon_0 \pi^2 r^2} .$$

22. We find the maximum by differentiating Eq. 23-16 and setting the result equal to zero.

$$\frac{d}{dz}\left(\frac{qz}{4\pi\varepsilon_0\left(z^2+R^2\right)^{3/2}}\right) = \frac{q}{4\pi\varepsilon_0}\frac{R^2 - 2z^2}{\left(z^2+R^2\right)^{5/2}} = 0$$

which leads to $z = R/\sqrt{2}$.

23. (a) The linear charge density is the charge per unit length of rod. Since the charge is uniformly distributed on the rod, $\lambda = -q/L$.

(b) We position the x axis along the rod with the origin at the left end of the rod, as shown in the diagram. Let dx be an infinitesimal length of rod at x. The charge in this segment is $dq = \lambda\, dx$. The charge dq may be considered to be a point charge. The electric field it produces at point P has only an x component and this component is given by

$$dE_x = \frac{1}{4\pi\varepsilon_0}\frac{\lambda\, dx}{(L+a-x)^2} .$$

The total electric field produced at P by the whole rod is the integral

$$\begin{aligned} E_x &= \frac{\lambda}{4\pi\varepsilon_0}\int_0^L \frac{dx}{(L+a-x)^2} \\ &= \frac{\lambda}{4\pi\varepsilon_0}\left.\frac{1}{L+a-x}\right|_0^L \\ &= \frac{\lambda}{4\pi\varepsilon_0}\left(\frac{1}{a}-\frac{1}{L+a}\right) \\ &= \frac{\lambda}{4\pi\varepsilon_0}\frac{L}{a(L+a)}\,. \end{aligned}$$

When $-q/L$ is substituted for λ the result is

$$E_x = -\frac{1}{4\pi\varepsilon_0}\frac{q}{a(L+a)}\,.$$

The negative sign indicates that the field is toward the rod.

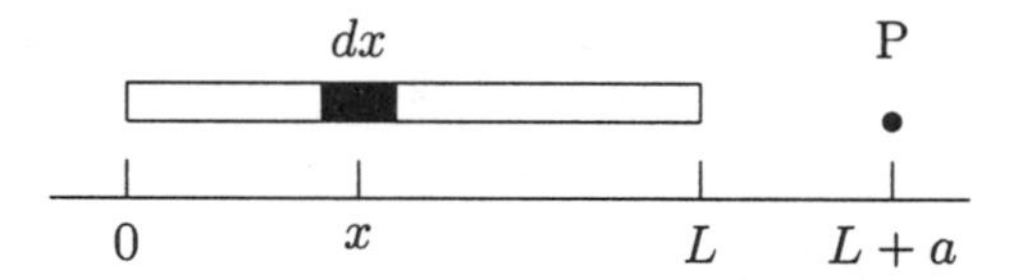

(c) If a is much larger than L, the quantity $L + a$ in the denominator can be approximated by a and the expression for the electric field becomes

$$E_x = -\frac{q}{4\pi\varepsilon_0 a^2}\,.$$

This is the expression for the electric field of a point charge at the origin.

24. We assume $q > 0$. Using the notation $\lambda = q/L$ we note that the (infinitesimal) charge on an element dx of the rod contains charge $dq = \lambda\,dx$. By symmetry, we conclude that all horizontal field components (due to the dq's) cancel and we need only "sum" (integrate) the vertical components. Symmetry also allows us to integrate these contributions over only half the rod ($0 \le x \le L/2$) and then simply double the result. In that regard we note that $\sin\theta = y/r$ where $r = \sqrt{x^2+y^2}$. Using Eq. 23-3 (with the 2 and $\sin\theta$ factors just discussed) we obtain

$$\begin{aligned} \left|\vec{E}\right| &= 2\int_0^{L/2}\left(\frac{dq}{4\pi\varepsilon_0 r^2}\right)\sin\theta \\ &= \frac{2}{4\pi\varepsilon_0}\int_0^{L/2}\left(\frac{\lambda dx}{x^2+y^2}\right)\left(\frac{y}{\sqrt{x^2+y^2}}\right) \\ &= \frac{\lambda y}{2\pi\varepsilon_0}\int_0^{L/2}\frac{dx}{(x^2+y^2)^{3/2}} \\ &= \frac{(q/L)y}{2\pi\varepsilon_0}\left[\frac{x}{y^2\sqrt{x^2+y^2}}\right]_0^{L/2} \\ &= \frac{q}{2\pi\varepsilon_0 Ly}\frac{L/2}{\sqrt{(L/2)^2+y^2}} \\ &= \frac{q}{2\pi\varepsilon_0\, y}\frac{1}{\sqrt{L^2+4y^2}} \end{aligned}$$

where the integral may be evaluated by elementary means or looked up in Appendix E (item #19 in the list of integrals).

25. Consider an infinitesimal section of the rod of length dx, a distance x from the left end, as shown in the diagram below. It contains charge $dq = \lambda\, dx$ and is a distance r from P. The magnitude of the field it produces at P is given by

$$dE = \frac{1}{4\pi\varepsilon_0}\frac{\lambda\, dx}{r^2}\ .$$

The x component is $dE_x = -\frac{1}{4\pi\varepsilon_0}\frac{\lambda\, dx}{r^2}\sin\theta$

and the y component is $dE_y = -\frac{1}{4\pi\varepsilon_0}\frac{\lambda\, dx}{r^2}\cos\theta$.

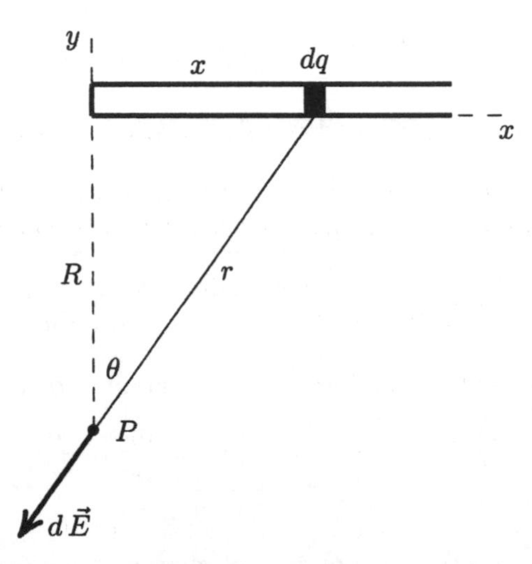

We use θ as the variable of integration and substitute $r = R/\cos\theta$, $x = R\tan\theta$ and $dx = (R/\cos^2\theta)\, d\theta$. The limits of integration are 0 and $\pi/2$ rad. Thus,

$$E_x = -\frac{\lambda}{4\pi\varepsilon_0 R}\int_0^{\pi/2}\sin\theta\, d\theta = \frac{\lambda}{4\pi\varepsilon_0 R}\cos\theta\Big|_0^{\pi/2} = -\frac{\lambda}{4\pi\varepsilon_0 R}$$

and

$$E_y = -\frac{\lambda}{4\pi\varepsilon_0 R}\int_0^{\pi/2}\cos\theta\, d\theta = -\frac{\lambda}{4\pi\varepsilon_0 R}\sin\theta\Big|_0^{\pi/2} = -\frac{\lambda}{4\pi\varepsilon_0 R}\ .$$

We notice that $E_x = E_y$ no matter what the value of R. Thus, $\vec{E}$ makes an angle of 45° with the rod for all values of R.

26. From Eq. 23-26

$$\begin{aligned} E &= \frac{\sigma}{2\varepsilon_0}\left(1 - \frac{z}{\sqrt{z^2+R^2}}\right) \\ &= \frac{5.3\times10^{-6}\,\mathrm{C/m^2}}{2\left(8.85\times10^{-12}\,\frac{\mathrm{C^2}}{\mathrm{N\cdot m^2}}\right)}\left[1 - \frac{12\,\mathrm{cm}}{\sqrt{(12\,\mathrm{cm})^2 + (2.5\,\mathrm{cm})^2}}\right] = 6.3\times10^3\ \mathrm{N/C}\ . \end{aligned}$$

27. At a point on the axis of a uniformly charged disk a distance z above the center of the disk, the magnitude of the electric field is

$$E = \frac{\sigma}{2\varepsilon_0}\left[1 - \frac{z}{\sqrt{z^2+R^2}}\right]$$

where R is the radius of the disk and σ is the surface charge density on the disk. See Eq. 23-26. The magnitude of the field at the center of the disk ($z = 0$) is $E_c = \sigma/2\varepsilon_0$. We want to solve for the value of z such that $E/E_c = 1/2$. This means

$$1 - \frac{z}{\sqrt{z^2 + R^2}} = \frac{1}{2} \implies \frac{z}{\sqrt{z^2 + R^2}} = \frac{1}{2} .$$

Squaring both sides, then multiplying them by z^2+R^2, we obtain $z^2 = (z^2/4)+(R^2/4)$. Thus, $z^2 = R^2/3$ and $z = R/\sqrt{3}$.

28. Eq. 23-28 gives

$$\vec{E} = \frac{\vec{F}}{q} = \frac{m\vec{a}}{(-e)} = -\left(\frac{m}{e}\right)\vec{a}$$

using Newton's second law. Therefore, with *east* being the $\hat{\imath}$ direction,

$$\vec{E} = -\left(\frac{9.11 \times 10^{-31}\,\text{kg}}{1.60 \times 10^{-19}\,\text{C}}\right)\left(1.80 \times 10^{9}\,\text{m/s}^2\,\hat{\imath}\right) = -0.0102\ \text{N/C}\ \hat{\imath}$$

which means the field has a magnitude of 0.0102 N/C and is directed westward.

29. The magnitude of the force acting on the electron is $F = eE$, where E is the magnitude of the electric field at its location. The acceleration of the electron is given by Newton's second law:

$$a = \frac{F}{m} = \frac{eE}{m} = \frac{(1.60 \times 10^{-19}\,\text{C})(2.00 \times 10^{4}\,\text{N/C})}{9.11 \times 10^{-31}\,\text{kg}} = 3.51 \times 10^{15}\ \text{m/s}^2 .$$

30. Vertical equilibrium of forces leads to the equality

$$q\left|\vec{E}\right| = mg \implies \left|\vec{E}\right| = \frac{mg}{2e} .$$

Using the mass given in the problem, we obtain $\left|\vec{E}\right| = 2.03 \times 10^{-7}$ N/C. Since the force of gravity is downward, then $q\vec{E}$ must point upward. Since $q > 0$ in this situation, this implies $\vec{E}$ must itself point upward.

31. We combine Eq. 23-9 and Eq. 23-28 (in absolute values).

$$F = |q|E = |q|\left(\frac{p}{2\pi\varepsilon_0 z^3}\right) = \frac{2kep}{z^3}$$

where we use Eq. 22-5 in the last step. Thus, we obtain

$$F = \frac{2\left(8.99 \times 10^{9}\,\text{N}\cdot\text{m}^2/\text{C}^2\right)\left(1.60 \times 10^{-19}\,\text{C}\right)\left(3.6 \times 10^{-29}\,\text{C}\cdot\text{m}\right)}{\left(25 \times 10^{-9}\,\text{m}\right)^3}$$

which yields a force of magnitude 6.6×10^{-15} N. If the dipole is oriented such that $\vec{p}$ is in the $+z$ direction, then $\vec{F}$ points in the $-z$ direction.

32. (a) $F_e = Ee = (3.0 \times 10^{6}\,\text{N/C})(1.6 \times 10^{-19}\,\text{C}) = 4.8 \times 10^{-13}\,\text{N}$.

    (b) $F_i = Eq_{\text{ion}} = Ee = 4.8 \times 10^{-13}\,\text{N}$.

33. (a) The magnitude of the force on the particle is given by $F = qE$, where q is the magnitude of the charge carried by the particle and E is the magnitude of the electric field at the location of the particle. Thus,

$$E = \frac{F}{q} = \frac{3.0 \times 10^{-6}\,\text{N}}{2.0 \times 10^{-9}\,\text{C}} = 1.5 \times 10^{3}\ \text{N/C} .$$

The force points downward and the charge is negative, so the field points upward.

(b) The magnitude of the electrostatic force on a proton is

$$F_e = eE = (1.60 \times 10^{-19}\,\text{C})(1.5 \times 10^3\,\text{N/C}) = 2.4 \times 10^{-16}\ \text{N}\ .$$

A proton is positively charged, so the force is in the same direction as the field, upward.

(c) The magnitude of the gravitational force on the proton is

$$F_g = mg = (1.67 \times 10^{-27}\,\text{kg})(9.8\,\text{m/s}^2) = 1.64 \times 10^{-26}\,\text{N}\ .$$

The force is downward.

(d) The ratio of the forces is

$$\frac{F_e}{F_g} = \frac{2.4 \times 10^{-16}\,\text{N}}{1.64 \times 10^{-26}\,\text{N}} = 1.5 \times 10^{10}\ .$$

34. (a) Since $\vec{E}$ points down and we need an upward electric force (to cancel the downward pull of gravity), then we require the charge of the sphere to be negative. The magnitude of the charge is found by working with the absolute value of Eq. 23-28:

$$|q| = \frac{F}{E} = \frac{mg}{E} = \frac{4.4\,\text{N}}{150\,\text{N/C}} = 0.029\ \text{C}\ .$$

(b) The feasibility of this experiment may be studied by using Eq. 23-3 (using k for $1/4\pi\varepsilon_0$).

$$E = k\frac{|q|}{r^2} \qquad \text{where} \quad \rho_{\text{sulfur}}\left(\frac{4}{3}\pi r^3\right) = m_{\text{sphere}}$$

Since the mass of the sphere is $4.4/9.8 \approx 0.45$ kg and the density of sulfur is about 2.1×10^3 kg/m^3 (see Appendix F), then we obtain

$$r = \left(\frac{3m_{\text{sphere}}}{4\pi\rho_{\text{sulfur}}}\right)^{1/3} = 0.037\,\text{m} \implies E = k\frac{|q|}{r^2} \approx 2 \times 10^{11}\ \text{N/C}$$

which is much too large a field to maintain in air (see problem #32).

35. (a) The magnitude of the force acting on the proton is $F = eE$, where E is the magnitude of the electric field. According to Newton's second law, the acceleration of the proton is $a = F/m = eE/m$, where m is the mass of the proton. Thus,

$$a = \frac{(1.60 \times 10^{-19}\,\text{C})(2.00 \times 10^4\,\text{N/C})}{1.67 \times 10^{-27}\,\text{kg}} = 1.92 \times 10^{12}\ \text{m/s}^2\ .$$

(b) We assume the proton starts from rest and use the kinematic equation $v^2 = v_0^2 + 2ax$ (or else $x = \frac{1}{2}at^2$ and $v = at$) to show that

$$v = \sqrt{2ax} = \sqrt{2(1.92 \times 10^{12}\,\text{m/s}^2)(0.0100\,\text{m})} = 1.96 \times 10^5\ \text{m/s}\ .$$

36. (a) The initial direction of motion is taken to be the $+x$ direction (this is also the direction of $\vec{E}$). We use $v_f^2 - v_i^2 = 2a\Delta x$ with $v_f = 0$ and $\vec{a} = \vec{F}/m = -e\vec{E}/m_e$ to solve for distance Δx:

$$\Delta x = \frac{-v_i^2}{2a} = \frac{-m_e v_i^2}{-2eE} = \frac{-(9.11 \times 10^{-31}\,\text{kg})(5.00 \times 10^6\,\text{m/s})^2}{-2(1.60 \times 10^{-19}\,\text{C})(1.00 \times 10^3\,\text{N/C})} = 7.12 \times 10^{-2}\ \text{m}\ .$$

(b) Eq. 2-17 leads to

$$t = \frac{\Delta x}{v_{\text{avg}}} = \frac{2\Delta x}{v_i} = \frac{2(7.12 \times 10^{-2}\,\text{m})}{5.00 \times 10^6\,\text{m/s}} = 2.85 \times 10^{-8}\ \text{s}\ .$$

(c) Using $\Delta v^2 = 2a\Delta x$ with the new value of Δx, we find

$$\begin{aligned} \frac{\Delta K}{K_i} &= \frac{\Delta(\frac{1}{2}m_e v^2)}{\frac{1}{2}m_e v_i^2} = \frac{\Delta v^2}{v_i^2} = \frac{2a\Delta x}{v_i^2} = \frac{-2eE\Delta x}{m_e v_i^2} \\ &= \frac{-2(1.60 \times 10^{-19}\,\mathrm{C})(1.00 \times 10^3\,\mathrm{N/C})(8.00 \times 10^{-3}\,\mathrm{m})}{(9.11 \times 10^{-31}\,\mathrm{kg})(5.00 \times 10^6\,\mathrm{m/s})^2} = -11.2\% \,. \end{aligned}$$

37. When the drop is in equilibrium, the force of gravity is balanced by the force of the electric field: $mg = qE$, where m is the mass of the drop, q is the charge on the drop, and E is the magnitude of the electric field. The mass of the drop is given by $m = (4\pi/3)r^3\rho$, where r is its radius and ρ is its mass density. Thus,

$$\begin{aligned} q &= \frac{mg}{E} = \frac{4\pi r^3 \rho g}{3E} \\ &= \frac{4\pi(1.64 \times 10^{-6}\,\mathrm{m})^3(851\,\mathrm{kg/m^3})(9.8\,\mathrm{m/s^2})}{3(1.92 \times 10^5\,\mathrm{N/C})} = 8.0 \times 10^{-19}\ \mathrm{C} \end{aligned}$$

and $q/e = (8.0 \times 10^{-19}\,\mathrm{C})/(1.60 \times 10^{-19}\,\mathrm{C}) = 5$.

38. Our approach (based on Eq. 23-29) consists of several steps. The first is to find an *approximate* value of e by taking differences between all the given data. The smallest difference is between the fifth and sixth values: $18.08 \times 10^{-19}\,\mathrm{C} - 16.48 \times 10^{-19}\,\mathrm{C} = 1.60 \times 10^{-19}\,\mathrm{C}$ which we denote e_{approx}. The goal at this point is to assign integers n using this approximate value of e:

$$\begin{array}{lcl} \text{datum 1} & \dfrac{6.563 \times 10^{-19}\,\mathrm{C}}{e_{\mathrm{approx}}} = 4.10 & \Longrightarrow \quad n_1 = 4 \\ \text{datum 2} & \dfrac{8.204 \times 10^{-19}\,\mathrm{C}}{e_{\mathrm{approx}}} = 5.13 & \Longrightarrow \quad n_2 = 5 \\ \text{datum 3} & \dfrac{11.50 \times 10^{-19}\,\mathrm{C}}{e_{\mathrm{approx}}} = 7.19 & \Longrightarrow \quad n_3 = 7 \\ \text{datum 4} & \dfrac{13.13 \times 10^{-19}\,\mathrm{C}}{e_{\mathrm{approx}}} = 8.21 & \Longrightarrow \quad n_4 = 8 \\ \text{datum 5} & \dfrac{16.48 \times 10^{-19}\,\mathrm{C}}{e_{\mathrm{approx}}} = 10.30 & \Longrightarrow \quad n_5 = 10 \\ \text{datum 6} & \dfrac{18.08 \times 10^{-19}\,\mathrm{C}}{e_{\mathrm{approx}}} = 11.30 & \Longrightarrow \quad n_6 = 11 \\ \text{datum 7} & \dfrac{19.71 \times 10^{-19}\,\mathrm{C}}{e_{\mathrm{approx}}} = 12.32 & \Longrightarrow \quad n_7 = 12 \\ \text{datum 8} & \dfrac{22.89 \times 10^{-19}\,\mathrm{C}}{e_{\mathrm{approx}}} = 14.31 & \Longrightarrow \quad n_8 = 14 \\ \text{datum 9} & \dfrac{26.13 \times 10^{-19}\,\mathrm{C}}{e_{\mathrm{approx}}} = 16.33 & \Longrightarrow \quad n_9 = 16 \end{array}$$

Next, we construct a new data set $(e_1, e_2, e_3 \ldots)$ by dividing the given data by the respective exact integers n_i (for $i = 1, 2, 3 \ldots$):

$$(e_1, e_2, e_3 \ldots) = \left(\frac{6.563 \times 10^{-19}\,\mathrm{C}}{n_1}, \frac{8.204 \times 10^{-19}\,\mathrm{C}}{n_2}, \frac{11.50 \times 10^{-19}\,\mathrm{C}}{n_3} \ldots \right)$$

which gives (carrying a few more figures than are significant)

$$(1.64075 \times 10^{-19}\,\mathrm{C},\ 1.6408 \times 10^{-19}\,\mathrm{C},\ 1.64286 \times 10^{-19}\,\mathrm{C} \ldots)$$

as the new data set (our experimental values for e). We compute the average and standard deviation of this set, obtaining

$$e_{\text{exptal}} = e_{\text{avg}} \pm \Delta e = (1.641 \pm 0.004) \times 10^{-19}\ \text{C}$$

which does not agree (to within one standard deviation) with the modern accepted value for e. The lower bound on this spread is $e_{\text{avg}} - \Delta e = 1.637 \times 10^{-19}$ C which is still about 2% too high.

39. (a) We use $\Delta x = v_{\text{avg}} t = vt/2$:

$$v = \frac{2\Delta x}{t} = \frac{2(2.0 \times 10^{-2}\,\text{m})}{1.5 \times 10^{-8}\,\text{s}} = 2.7 \times 10^6\ \text{m/s}\ .$$

(b) We use $\Delta x = \frac{1}{2}at^2$ and $E = F/e = ma/e$:

$$E = \frac{ma}{e} = \frac{2\Delta x m}{et^2} = \frac{2(2.0 \times 10^{-2}\,\text{m})(9.11 \times 10^{-31}\,\text{kg})}{(1.60 \times 10^{-19}\,\text{C})(1.5 \times 10^{-8}\,\text{s})^2} = 1.0 \times 10^3\ \text{N/C}\ .$$

40. We assume there are no forces or force-components along the x direction. We combine Eq. 23-28 with Newton's second law, then use Eq. 4-21 to determine time t followed by Eq. 4-23 to determine the final velocity (with $-g$ replaced by the a_y of this problem); for these purposes, the velocity components *given* in the problem statement are re-labeled as v_{0x} and v_{0y} respectively.

(a) We have $\vec{a} = \frac{q\vec{E}}{m} = -\left(\frac{e}{m}\right)\vec{E}$ which leads to

$$\vec{a} = -\left(\frac{1.60 \times 10^{-19}\,\text{C}}{9.11 \times 10^{-31}\,\text{kg}}\right)\left(120\ \frac{\text{N}}{\text{C}}\right)\hat{\text{j}} = -2.1 \times 10^{13}\,\text{m/s}^2\ \hat{\text{j}}\ .$$

(b) Since $v_x = v_{0x}$ in this problem (that is, $a_x = 0$), we obtain

$$\begin{aligned} t &= \frac{\Delta x}{v_{0x}} = \frac{0.020\,\text{m}}{1.5 \times 10^5\,\text{m/s}} = 1.3 \times 10^{-7}\ \text{s} \\ v_y &= v_{0y} + a_y t = 3.0 \times 10^3\,\text{m/s} + \left(-2.1 \times 10^{13}\,\text{m/s}^2\right)\left(1.3 \times 10^{-7}\,\text{s}\right) \end{aligned}$$

which leads to $v_y = -2.8 \times 10^6$ m/s. Therefore, in unit vector notation (with SI units understood) the final velocity is

$$\vec{v} = 1.5 \times 10^5\,\hat{\text{i}} - 2.8 \times 10^6\,\hat{\text{j}}\ .$$

41. We take the positive direction to be to the right in the figure. The acceleration of the proton is $a_p = eE/m_p$ and the acceleration of the electron is $a_e = -eE/m_e$, where E is the magnitude of the electric field, m_p is the mass of the proton, and m_e is the mass of the electron. We take the origin to be at the initial position of the proton. Then, the coordinate of the proton at time t is $x = \frac{1}{2}a_p t^2$ and the coordinate of the electron is $x = L + \frac{1}{2}a_e t^2$. They pass each other when their coordinates are the same, or $\frac{1}{2}a_p t^2 = L + \frac{1}{2}a_e t^2$. This means $t^2 = 2L/(a_p - a_e)$ and

$$\begin{aligned} x &= \frac{a_p}{a_p - a_e}L = \frac{eE/m_p}{(eE/m_p) + (eE/m_e)}L = \frac{m_e}{m_e + m_p}L \\ &= \frac{9.11 \times 10^{-31}\,\text{kg}}{9.11 \times 10^{-31}\,\text{kg} + 1.67 \times 10^{-27}\,\text{kg}}(0.050\,\text{m}) \\ &= 2.7 \times 10^{-5}\ \text{m}\ . \end{aligned}$$

42. (a) Using Eq. 23-28, we find

$$\begin{aligned} \vec{F} &= (8.00 \times 10^{-5}\,\text{C})(3.00 \times 10^3\,\text{N/C})\hat{\text{i}} + (8.00 \times 10^{-5}\,\text{C})(-600\,\text{N/C})\hat{\text{j}} \\ &= (0.240\,\text{N})\hat{\text{i}} - (0.0480\,\text{N})\hat{\text{j}}\ . \end{aligned}$$

Therefore, the force has magnitude equal to

$$F = \sqrt{(0.240\,\text{N})^2 + (0.0480\,\text{N})^2} = 0.245\ \text{N}\ ,$$

and makes an angle θ (which, if negative, means clockwise) measured from the $+x$ axis, where

$$\theta = \tan^{-1}\left(\frac{F_y}{F_x}\right) = \tan^{-1}\left(\frac{-0.0480\,\text{N}}{0.240\,\text{N}}\right) = -11.3^\circ\ .$$

(b) With $m = 0.0100$ kg, the coordinates (x, y) at $t = 3.00\,\text{s}$ are found by combining Newton's second law with the kinematics equations of Chapters 2-4:

$$\begin{aligned} x &= \frac{1}{2}a_x t^2 = \frac{F_x t^2}{2m} = \frac{(0.240)(3.00)^2}{2(0.0100)} = 108\ \text{m}\ , \\ y &= \frac{1}{2}a_y t^2 = \frac{F_y t^2}{2m} = \frac{(-0.0480)(3.00)^2}{2(0.0100)} = -21.6\ \text{m}\ . \end{aligned}$$

43. (a) The electric field is upward in the diagram and the charge is negative, so the force of the field on it is downward. The magnitude of the acceleration is $a = eE/m$, where E is the magnitude of the field and m is the mass of the electron. Its numerical value is

$$a = \frac{(1.60 \times 10^{-19}\,\text{C})(2.00 \times 10^{3}\,\text{N/C})}{9.11 \times 10^{-31}\,\text{kg}} = 3.51 \times 10^{14}\ \text{m/s}^2\ .$$

We put the origin of a coordinate system at the initial position of the electron. We take the x axis to be horizontal and positive to the right; take the y axis to be vertical and positive toward the top of the page. The kinematic equations are

$$x = v_0 t\cos\theta\ ,\quad y = v_0 t\sin\theta - \frac{1}{2}at^2\ ,\quad \text{and}\ \ v_y = v_0\sin\theta - at\ .$$

First, we find the greatest y coordinate attained by the electron. If it is less than d, the electron does not hit the upper plate. If it is greater than d, it will hit the upper plate if the corresponding x coordinate is less than L. The greatest y coordinate occurs when $v_y = 0$. This means $v_0\sin\theta - at = 0$ or $t = (v_0/a)\sin\theta$ and

$$\begin{aligned} y_{\text{max}} &= \frac{v_0^2\sin^2\theta}{a} - \frac{1}{2}a\frac{v_0^2\sin^2\theta}{a^2} = \frac{1}{2}\frac{v_0^2\sin^2\theta}{a} \\ &= \frac{(6.00 \times 10^6\,\text{m/s})^2\sin^2 45^\circ}{2(3.51 \times 10^{14}\,\text{m/s}^2)} = 2.56 \times 10^{-2}\,\text{m}\ . \end{aligned}$$

Since this is greater than $d = 2.00\,\text{cm}$, the electron might hit the upper plate.

(b) Now, we find the x coordinate of the position of the electron when $y = d$. Since

$$v_0\sin\theta = (6.00 \times 10^6\,\text{m/s})\sin 45^\circ = 4.24 \times 10^6\,\text{m/s}$$

and

$$2ad = 2(3.51 \times 10^{14}\,\text{m/s}^2)(0.0200\,\text{m}) = 1.40 \times 10^{13}\,\text{m}^2/\text{s}^2$$

the solution to $d = v_0 t\sin\theta - \frac{1}{2}at^2$ is

$$\begin{aligned} t &= \frac{v_0\sin\theta - \sqrt{v_0^2\sin^2\theta - 2ad}}{a} \\ &= \frac{4.24 \times 10^6\,\text{m/s} - \sqrt{(4.24 \times 10^6\,\text{m/s})^2 - 1.40 \times 10^{13}\,\text{m}^2/\text{s}^2}}{3.51 \times 10^{14}\,\text{m/s}^2} \\ &= 6.43 \times 10^{-9}\ \text{s}\ . \end{aligned}$$

The negative root was used because we want the *earliest* time for which $y = d$. The x coordinate is

$$\begin{aligned} x &= v_0 t \cos\theta \\ &= (6.00 \times 10^6 \text{ m/s})(6.43 \times 10^{-9}\text{ s}) \cos 45° = 2.72 \times 10^{-2} \text{ m} . \end{aligned}$$

This is less than L so the electron hits the upper plate at $x = 2.72\,\text{cm}$.

44. (a) The magnitude of the dipole moment is

$$p = qd = \left(1.50 \times 10^{-9}\text{ C}\right)\left(6.20 \times 10^{-6}\text{ m}\right) = 9.30 \times 10^{-15} \text{ C·m} .$$

(b) Following the solution to part (c) of Sample Problem 23-5, we find

$$U(180°) - U(0) = 2pE = 2\left(9.30 \times 10^{-15}\right)(1100) = 2.05 \times 10^{-11} \text{ J} .$$

45. (a) Eq. 23-33 leads to $\tau = pE \sin 0° = 0$.

(b) With $\theta = 90°$, the equation gives

$$\tau = pE = \left(2(1.6 \times 10^{-19}\text{ C})(0.78 \times 10^{-9}\text{ m})\right)\left(3.4 \times 10^6\text{ N/C}\right) = 8.5 \times 10^{-22} \text{ N·m} .$$

(c) Now the equation gives $\tau = pE \sin 180° = 0$.

46. Following the solution to part (c) of Sample Problem 23-5, we find

$$W = U(\theta_0 + \pi) - U(\theta_0) = -pE\left(\cos(\theta_0 + \pi) - \cos(\theta_0)\right) = 2pE\cos\theta_0 .$$

47. Eq. 23-35 ($\tau = -pE\sin\theta$) captures the sense as well as the magnitude of the effect. That is, this is a restoring torque, trying to bring the tilted dipole back to its aligned equilibrium position. If the amplitude of the motion is small, we may replace $\sin\theta$ with θ in radians. Thus, $\tau \approx -pE\theta$. Since this exhibits a simple negative proportionality to the angle of rotation, the dipole oscillates in simple harmonic motion, like a torsional pendulum with torsion constant $\kappa = pE$. The angular frequency ω is given by

$$\omega^2 = \frac{\kappa}{I} = \frac{pE}{I}$$

where I is the rotational inertia of the dipole. The frequency of oscillation is

$$f = \frac{\omega}{2\pi} = \frac{1}{2\pi}\sqrt{\frac{pE}{I}} .$$

48. (a) Using $k = 1/4\pi\varepsilon_0$, we estimate the field at $r = 0.02$ m using Eq. 23-3:

$$E = k\frac{q}{r^2} = \left(8.99 \times 10^9 \frac{\text{N} \cdot \text{m}^2}{\text{C}^2}\right)\frac{45 \times 10^{-12}\text{ C}}{(0.02\text{ m})^2} \approx 1 \times 10^3 \text{ N/C} .$$

(b) The field described by Eq. 23-3 is nonuniform.

(c) As the positively charged bee approaches the grain, a concentration of negative charge is induced on the closest side of the grain, leading to a force of attraction which makes the grain jump to the bee. Although in physical contact, it is not in electrical contact with the bee, or else it would acquire a net positive charge causing it to be repelled from the bee. As the bee (with grain) approaches the stigma, a concentration of negative charge is induced on the closest side of the stigma which is presumably highly nonuniform. In some configurations, the field from the stigma (acting on the positive side of the grain) will overcome the field from the bee acting on the negative side, and the grain will jump to the stigma.

49. We consider pairs of diametrically opposed charges. The net field due to just the charges in the one o'clock ($-q$) and seven o'clock ($-7q$) positions is clearly equivalent to that of a single $-6q$ charge sitting at the seven o'clock position. Similarly, the net field due to just the charges in the six o'clock ($-6q$) and twelve o'clock ($-12q$) positions is the same as that due to a single $-6q$ charge sitting at the twelve o'clock position. Continuing with this line of reasoning, we see that there are six equal-magnitude electric field vectors pointing at the seven o'clock, eight o'clock ... twelve o'clock positions. Thus, the resultant field of all of these points, by symmetry, is directed toward the position midway between seven and twelve o'clock. Therefore, $\vec{E}_{\text{resultant}}$ points towards the nine-thirty position.

50. (a) From Eq. 23-38 (and the facts that $\hat{\text{i}} \cdot \hat{\text{i}} = 1$ and $\hat{\text{j}} \cdot \hat{\text{i}} = 0$), the potential energy is

$$\begin{aligned} U &= -\vec{p} \cdot \vec{E} = -[(3.00\hat{\text{i}} + 4.00\hat{\text{j}})(1.24 \times 10^{-30}\,\text{C}\cdot\text{m})] \cdot [(4000\,\text{N/C})\hat{\text{i}}] \\ &= -1.49 \times 10^{-26}\ \text{J}\,. \end{aligned}$$

(b) From Eq. 23-34 (and the facts that $\hat{\text{i}} \times \hat{\text{i}} = 0$ and $\hat{\text{j}} \times \hat{\text{i}} = -\hat{\text{k}}$), the torque is

$$\begin{aligned} \vec{\tau} &= \vec{p} \times \vec{E} = [(3.00\hat{\text{i}} + 4.00\hat{\text{j}})(1.24 \times 10^{-30}\,\text{C}\cdot\text{m})] \times [(4000\,\text{N/C})\hat{\text{i}}] \\ &= (-1.98 \times 10^{-26}\,\text{N}\cdot\text{m})\hat{\text{k}}\,. \end{aligned}$$

(c) The work done is

$$\begin{aligned} W &= \Delta U = \Delta(-\vec{p} \cdot \vec{E}) = (\vec{p}_i - \vec{p}_f) \cdot \vec{E} \\ &= [(3.00\hat{\text{i}} + 4.00\hat{\text{j}}) - (-4.00\hat{\text{i}} + 3.00\hat{\text{j}})](1.24 \times 10^{-30}\,\text{C}\cdot\text{m})] \cdot [(4000\,\text{N/C})\hat{\text{i}}] \\ &= 3.47 \times 10^{-26}\ \text{J}\,. \end{aligned}$$

51. The point at which we are evaluating the net field is denoted by P. The contributions to the net field caused by the two electrons nearest P (the two electrons on the side of the triangle shared by P) are seen to cancel, so that we only need to compute the field (using Eq. 23-3) caused by the electron at the far corner, at a distance $r = 0.17$ m from P. Using $1/4\pi\varepsilon_0 = k$, we obtain

$$\left|\vec{E}_{\text{net}}\right| = k\frac{e}{r^2} = 4.8 \times 10^{-8}\ \text{N/C}\,.$$

52. Let q_1 denote the charge at $y = d$ and q_2 denote the charge at $y = -d$. The individual magnitudes $\left|\vec{E}_1\right|$ and $\left|\vec{E}_2\right|$ are figured from Eq. 23-3, where the absolute value signs for q are unnecessary since these charges are both positive. The distance from q_1 to a point on the x axis is the same as the distance from q_2 to a point on the x axis: $r = \sqrt{x^2 + d^2}$. By symmetry, the y component of the net field along the x axis is zero. The x component of the net field, evaluated at points on the positive x axis, is

$$E_x = 2\left(\frac{1}{4\pi\varepsilon_0}\right)\left(\frac{q}{x^2 + d^2}\right)\left(\frac{x}{\sqrt{x^2 + d^2}}\right)$$

where the last factor is $\cos\theta = x/r$ with θ being the angle for each individual field as measured from the x axis.

(a) If we simplify the above expression, and plug in $x = \alpha d$, we obtain

$$E_x = \frac{q}{2\pi\varepsilon_0\, d^2}\left(\frac{\alpha}{(\alpha^2 + 1)^{3/2}}\right)\,.$$

(b) The graph of $E = E_x$ versus α is shown below. For the purposes of graphing, we set $d = 1$ m and $q = 5.56 \times 10^{-11}$ C.

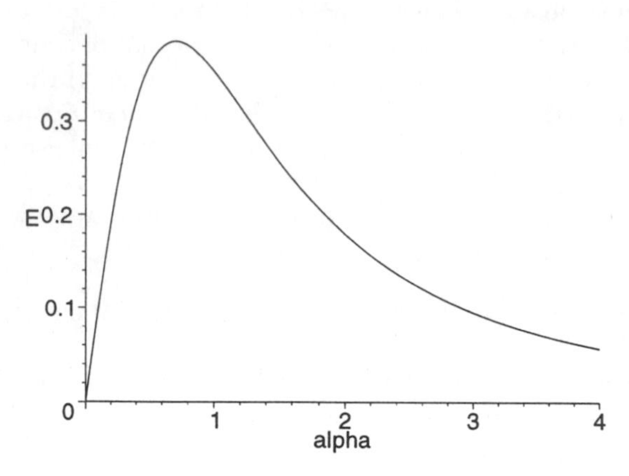

(c) From the graph, we estimate $E_{\max}$ occurs at about $\alpha = 0.7$. More accurate computation shows that the maximum occurs at $\alpha = 1/\sqrt{2}$.

(d) The graph suggests that "half-height" points occur at $\alpha \approx 0.2$ and $\alpha \approx 1.9$. Further numerical exploration leads to the values: $\alpha = 0.2047$ and $\alpha = 1.9864$.

53. (a) We combine Eq. 23-28 (in absolute value) with Newton's second law:

$$a = \frac{|q|E}{m} = \left(\frac{1.60 \times 10^{-19}\,\text{C}}{9.11 \times 10^{-31}\,\text{kg}}\right)\left(1.40 \times 10^{6}\,\frac{\text{N}}{\text{C}}\right) = 2.46 \times 10^{17}\ \text{m/s}^2 .$$

(b) With $v = \frac{c}{10} = 3.00 \times 10^7$ m/s, we use Eq. 2-11 to find

$$t = \frac{v - v_o}{a} = \frac{3.00 \times 10^7}{2.46 \times 10^{17}} = 1.22 \times 10^{-10}\ \text{s} .$$

(c) Eq. 2-16 gives

$$\Delta x = \frac{v^2 - v_o^2}{2a} = \frac{(3.00 \times 10^7)^2}{2\,(2.46 \times 10^{17})} = 1.83 \times 10^{-3}\ \text{m} .$$

54. Studying Sample Problem 23-3, we see that the field evaluated at the center of curvature due to a charged distribution on a circular arc is given by

$$\vec{E} = \frac{\lambda}{4\pi\varepsilon_0 r}\Big[\sin\theta\Big]_{-\theta/2}^{\theta/2} \qquad \text{along the symmetry axis}$$

where $\lambda = q/\ell = q/r\theta$ with θ in radians. Here ℓ is the length of the arc, given as $\ell = 4.0$ m. Therefore, $\theta = \ell/r = 4.0/2.0 = 2.0$ rad. Thus, with $q = 20 \times 10^{-9}$ C, we obtain

$$\left|\vec{E}\right| = \frac{q}{\ell}\,\frac{1}{4\pi\varepsilon_0 r}\Big[\sin\theta\Big]_{-1.0\,\text{rad}}^{1.0\,\text{rad}} = 38\ \text{N/C} .$$

55. A small section of the distribution has charge dq is $\lambda\,dx$, where $\lambda = 9.0 \times 10^{-9}$ C/m. Its contribution to the field at $x_P = 4.0$ m is

$$d\vec{E} = \frac{dq}{4\pi\,\varepsilon_0\,(x - x_P)^2}$$

pointing in the $+x$ direction. Thus, we have

$$\vec{E} = \int_0^{3.0\,\text{m}} \frac{\lambda\,dx}{4\pi\,\varepsilon_0\,(x - x_P)^2}\ \hat{\text{i}}$$

which becomes, using the substitution $u = x - x_P$,

$$\vec{E} = \frac{\lambda}{4\pi\,\varepsilon_0}\int_{-4.0\,\mathrm{m}}^{-1.0\,\mathrm{m}} \frac{du}{u^2}\,\hat{\imath} \;=\; \frac{\lambda}{4\pi\,\varepsilon_0}\left(\frac{-1}{-1.0\,\mathrm{m}} - \frac{-1}{-4.0\,\mathrm{m}}\right)\hat{\imath}$$

which yields 61 N/C in the $+x$ direction.

56. Let $q_1 = -4Q < 0$ and $q_2 = +2Q > 0$ (where we make the assumption that $Q > 0$). Also, let $d = 2.00$ m, the distance that separates the charges. The individual magnitudes $\left|\vec{E}_1\right|$ and $\left|\vec{E}_2\right|$ are figured from Eq. 23-3, where the absolute value signs for q_2 are unnecessary since this charge is positive. Whether we add the magnitudes or subtract them depends on if $\vec{E}_1$ is in the same, or opposite, direction as $\vec{E}_2$. At points left of q_1 (on the $-x$ axis) the fields point in opposite directions, but there is no possibility of cancellation (zero net field) since $\left|\vec{E}_1\right|$ is everywhere bigger than $\left|\vec{E}_2\right|$ in this region. In the region between the charges ($0 < x < d$) both fields point leftward and there is no possibility of cancellation. At points to the right of q_2 (where $x > d$), $\vec{E}_1$ points leftward and $\vec{E}_2$ points rightward so the net field in this range is

$$\vec{E}_{\text{net}} = \left|\vec{E}_2\right| - \left|\vec{E}_1\right| \quad \text{in the } \hat{\imath} \text{ direction.}$$

Although $|q_1| > q_2$ there is the possibility of $\vec{E}_{\text{net}} = 0$ since these points are closer to q_2 than to q_1. Thus, we look for the zero net field point in the $x > d$ region:

$$\begin{aligned}\left|\vec{E}_1\right| &= \left|\vec{E}_2\right| \\ \frac{1}{4\pi\varepsilon_0}\frac{|q_1|}{x^2} &= \frac{1}{4\pi\varepsilon_0}\frac{q_2}{(x-d)^2}\end{aligned}$$

which leads to

$$\frac{x-d}{x} = \sqrt{\frac{q_2}{|q_1|}} = \sqrt{\frac{1}{2}}\ .$$

Therefore, $x = \frac{d\sqrt{2}}{\sqrt{2}-1} = 6.8$ m specifies the position where $\vec{E}_{\text{net}} = 0$.

57. We note that the contributions to the field from the pair of $-2q$ charges exactly cancel, and we are left with the (opposing) contributions from the $4q$ (at $r = 2d$) and $-q$ (at $r = d$) charges. Therefore, using $k = 1/4\pi\varepsilon_0$

$$|\vec{E}_{\text{net}}| = k\,\frac{4q}{(2d)^2} - k\,\frac{q}{d^2} = 0\ .$$

The net field at P vanishes completely.

58. The field of each charge has magnitude

$$E = k\,\frac{e}{(0.020\,\mathrm{m})^2} = 3.6 \times 10^{-6}\ \mathrm{N/C}\ .$$

The directions are indicated in standard format below. We use the magnitude-angle notation (convenient if one is using a vector capable calculator in polar mode) and write (starting with the proton on the left and moving around clockwise) the contributions to $\vec{E}_{\text{net}}$ as follows:

$$(E\,\angle -20^\circ) + (E\,\angle\,130^\circ) + (E\,\angle -100^\circ) + (E\,\angle -150^\circ) + (E\,\angle\,0^\circ)\ .$$

This yields $(3.93 \times 10^{-6}\,\angle -76.4^\circ)$, with the N/C unit understood.

59. Eq. 23-38 gives $U = -\vec{p}\cdot\vec{E} = -pE\cos\theta$. We note that $\theta_i = 110^\circ$ and $\theta_f = 70^\circ$. Therefore,

$$\Delta U = -pE\,(\cos 70^\circ - \cos 110^\circ) = -3.3 \times 10^{-21}\ \mathrm{J}\ .$$

60. (a) Suppose the pendulum is at the angle θ with the vertical. The force diagram is shown to the right. $\vec{T}$ is the tension in the thread, mg is the magnitude of the force of gravity, and qE is the magnitude of the electric force. The field points upward and the charge is positive, so the force is upward. Taking the angle shown to be positive, then the torque on the sphere about the point where the thread is attached to the upper plate is $\tau = -(mg - qE)\ell \sin\theta$. If $mg > qE$ then the torque is a restoring torque; it tends to pull the pendulum back to its equilibrium position.

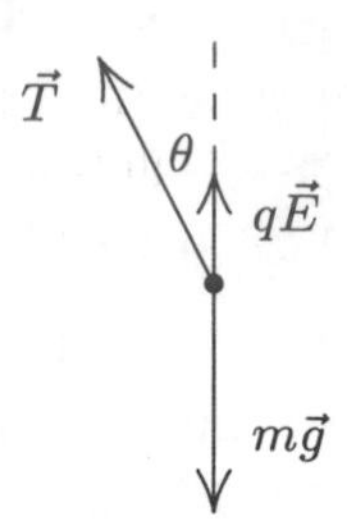

If the amplitude of the oscillation is small, $\sin\theta$ can be replaced by θ in radians and the torque is $\tau = -(mg - qE)\ell\theta$. The torque is proportional to the angular displacement and the pendulum moves in simple harmonic motion. Its angular frequency is $\omega = \sqrt{(mg - qE)\ell/I}$, where I is the rotational inertia of the pendulum. Since $I = m\ell^2$ for a simple pendulum,

$$\omega = \sqrt{\frac{(mg - qE)\ell}{m\ell^2}} = \sqrt{\frac{g - qE/m}{\ell}}$$

and the period is

$$T = \frac{2\pi}{\omega} = 2\pi\sqrt{\frac{\ell}{g - qE/m}} .$$

If $qE > mg$ the torque is not a restoring torque and the pendulum does not oscillate.

(b) The force of the electric field is now downward and the torque on the pendulum is $\tau = -(mg+qE)\ell\theta$ if the angular displacement is small. The period of oscillation is

$$T = 2\pi\sqrt{\frac{\ell}{g + qE/m}} .$$

61. (a) Using the density of water ($\rho = 1000\,\text{kg/m}^3$), the weight mg of the spherical drop (of radius $r = 6.0 \times 10^{-7}$ m) is

$$W = \rho V g = (1000\,\text{kg/m}^3)\left(\frac{4\pi}{3}(6.0 \times 10^{-7}\,\text{m})^3\right)(9.8\,\text{m/s}^2) = 8.87 \times 10^{-15}\,\text{N} .$$

(b) Vertical equilibrium of forces leads to $mg = qE = neE$, which we solve for n, the number of excess electrons:

$$n = \frac{mg}{eE} = \frac{8.87 \times 10^{-15}\,\text{N}}{(1.60 \times 10^{-19}\,\text{C})(462\,\text{N/C})} = 120 .$$

62. (a) Let $E = \sigma/2\varepsilon_0 = 3 \times 10^6$ N/C. With $\sigma = |q|/A$, this leads to

$$|q| = \pi R^2\sigma = 2\pi\varepsilon_0 R^2 E = \frac{R^2 E}{2k} = \frac{(2.5 \times 10^{-2}\,\text{m})^2 (3.0 \times 10^6\,\text{N/C})}{2\left(8.99 \times 10^9\,\frac{\text{N}\cdot\text{m}^2}{\text{C}^2}\right)} = 1.0 \times 10^{-7}\,\text{C} .$$

(b) Setting up a simple proportionality (with the areas), the number of atoms is estimated to be

$$N = \frac{\pi(2.5 \times 10^{-2}\,\text{m})^2}{0.015 \times 10^{-18}\,\text{m}^2} = 1.3 \times 10^{17} .$$

(c) Therefore, the fraction is

$$\frac{q}{Ne} = \frac{1.0 \times 10^{-7}\,\text{C}}{(1.3 \times 10^{17})(1.6 \times 10^{-19}\,\text{C})} \approx 5 \times 10^{-6} .$$

63. On the one hand, the conclusion (that $Q = +1.0\,\mu\text{C}$) is clear from symmetry. If a more in-depth justification is desired, one should use Eq. 23-3 for the electric field magnitudes of the three charges (each at the same distance $r = a/\sqrt{3}$ from C) and then find field components along suitably chosen axes, requiring each component-sum to be zero. If the y axis is vertical, then (assuming $Q > 0$) the component-sum along that axis leads to $2kq\sin 30°/r^2 = kQ/r^2$ where q refers to either of the charges at the bottom corners. This yields $Q = 2q\sin 30° = q$ and thus to the conclusion mentioned above.

64. From symmetry, the only two pairs of charges which produce a non-vanishing field $\vec{E}_{\text{net}}$ are: pair 1, which is in the middle of the two vertical sides of the square (the $+q$, $-2q$ pair); and pair 2, the $+5q$, $-5q$ pair. We denote the electric fields produced by each pair as $\vec{E}_1$ and $\vec{E}_2$, respectively. We set up a coordinate system as shown to the right, with the origin at the center of the square. Now,

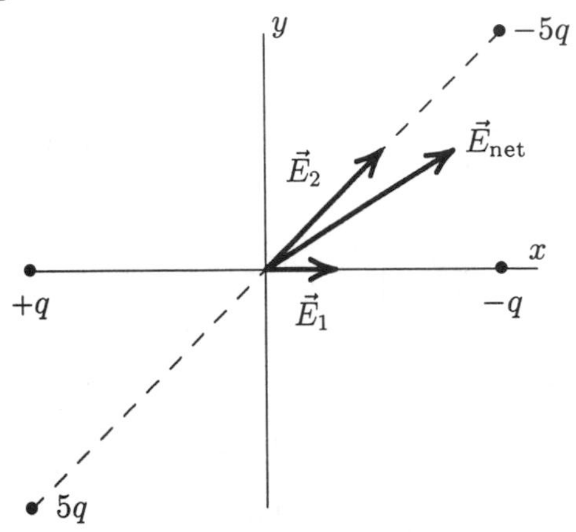

$$E_1 = \frac{1}{4\pi\varepsilon_0}\left(\frac{q}{d^2} + \frac{2q}{d^2}\right) = \frac{3q}{4\pi\varepsilon_0 d^2} \qquad \text{and} \qquad E_2 = k\left[\frac{5q}{(\sqrt{2}d)^2} + \frac{5q}{(\sqrt{2}d)^2}\right] = \frac{5q}{4\pi\varepsilon_0 d^2} .$$

Therefore, the components of $\vec{E}_{\text{net}}$ are given by

$$\begin{aligned} E_x &= E_{1x} + E_{2x} = E_1 + E_2\cos 45° \\ &= \frac{3q}{4\pi\varepsilon_0 d^2} + \left(\frac{5q}{4\pi\varepsilon_0 d^2}\right)\cos 45° = 6.536\left(\frac{q}{4\pi\varepsilon_0 d^2}\right) , \end{aligned}$$

and

$$E_y = E_{1y} + E_{2y} = E_2\sin 45° = \left(\frac{5q}{4\pi\varepsilon_0 d^2}\right)\sin 45° = 3.536\left(\frac{q}{4\pi\varepsilon_0 d^2}\right) .$$

Thus, the magnitude of $\vec{E}_{\text{net}}$ is

$$E = \sqrt{E_x^2 + E_y^2} = \sqrt{(6.536)^2 + (3.536)^2}\left(\frac{q}{4\pi\varepsilon_0 d^2}\right) = \frac{7.43q}{4\pi\varepsilon_0 d^2} ,$$

and $\vec{E}_{\text{net}}$ makes an angle θ with the positive x axis, where

$$\theta = \tan^{-1}\left(\frac{E_y}{E_x}\right) = \tan^{-1}\left(\frac{3.536}{6.536}\right) = 28.4° .$$

65. We denote the electron with subscript e

and the proton with p. From the figure to the right we see that

$$\left|\vec{E}_e\right| = \left|\vec{E}_p\right| = \frac{e}{4\pi\varepsilon_0 d^2}$$

where $d = 2.0 \times 10^{-6}$ m. We note that the components along the y axis cancel during the vector summation. With $k = 1/4\pi\varepsilon_0$ and $\theta = 60°$, the magnitude of the net electric field is obtained as follows:

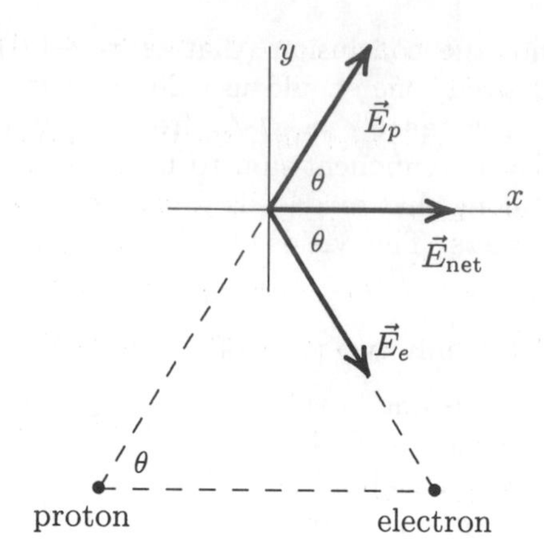

$$\begin{aligned}\left|\vec{E}_{\rm net}\right| &= E_x = 2E_e\cos\theta \\ &= 2\left(\frac{e}{4\pi\varepsilon_0 d^2}\right)\cos\theta = 2k\left[\frac{e}{d^2}\right]\cos\theta \\ &= 2\left(8.99\times10^9\,\frac{\rm N\cdot m^2}{\rm C^2}\right)\left[\frac{(1.6\times10^{-19}\,{\rm C})}{(2.0\times10^{-6}\,{\rm m})^2}\right]\cos 60° \\ &= 3.6\times10^2\ {\rm N/C}\,.\end{aligned}$$

66. (a) Since the two charges in question are of the same sign, the point $x = 2.0$ mm should be located in between them (so that the field vectors point in the opposite direction). Let the coordinate of the second particle be x' ($x' > 0$). Then, the magnitude of the field due to the charge $-q_1$ evaluated at x is given by $E = q_1/4\pi\varepsilon_0 x^2$, while that due to the second charge $-4q_1$ is $E' = 4q_1/4\pi\varepsilon_0(x'-x)^2$. We set the net field equal to zero:

$$\vec{E}_{\rm net} = 0 \implies E = E'$$

so that

$$\frac{q_1}{4\pi\varepsilon_0 x^2} = \frac{4q_1}{4\pi\varepsilon_0(x'-x)^2}\,.$$

Thus, we obtain $x' = 3x = 3(2.0\,{\rm mm}) = 6.0$ mm.

(b) In this case, with the second charge now positive, the electric field vectors produced by both charges are in the negative x direction, when evaluated at $x = 2.0$ mm. Therefore, the net field points in the negative x direction.

67. The distance from Q to P is $5a$, and the distance from q to P is $3a$. Therefore, the magnitudes of the individual electric fields are, using Eq. 23-3 (writing $1/4\pi\varepsilon_0 = k$),

$$|\vec{E}_Q| = \frac{k\,|Q|}{25\,a^2}\,, \qquad |\vec{E}_q| = \frac{k\,|q|}{9\,a^2}\,.$$

We note that $\vec{E}_q$ is along the y axis (directed towards $\pm y$ in accordance with the sign of q), and $\vec{E}_Q$ has x and y components, with $\vec{E}_{Q\,x} = \pm\frac{4}{5}|\vec{E}_Q|$ and $\vec{E}_{Q\,y} = \pm\frac{3}{5}|\vec{E}_Q|$ (signs corresponding to the sign of Q). Consequently, we can write the addition of components in a simple way (basically, by dropping the absolute values):

$$\begin{aligned}\vec{E}_{{\rm net}\,x} &= \frac{4\,k\,Q}{125\,a^2} \\ \vec{E}_{{\rm net}\,y} &= \frac{3\,k\,Q}{125\,a^2} + \frac{k\,q}{9\,a^2}\end{aligned}$$

(a) Equating $\vec{E}_{\text{net}\,x}$ and $\vec{E}_{\text{net}\,y}$, it is straightforward to solve for the relation between Q and q. We obtain $Q = \frac{125}{9}q \approx 14q$.

(b) We set $\vec{E}_{\text{net}\,y} = 0$ and find the necessary relation between Q and q. We obtain $Q = -\frac{125}{27}q \approx -4.6q$.

68. (a) From the second measurement (at (2.0, 0)) we see that the charge must be somewhere on the x axis. A line passing through (3.0, 3.0) with slope $\tan^{-1} 3/4$ will intersect the x axis at $x = -1.0$. Thus, the location of the particle is specified by the coordinates (in cm): $(-1.0,\ 0)$.

(b) Using $k = 1/4\pi\varepsilon_0$, the field magnitude measured at (2.0, 0) (which is $r = 0.030$ m from the charge) is

$$|\vec{E}| = k\,\frac{q}{r^2} = 100\ \text{N/C}\ .$$

Therefore, $q = 1.0 \times 10^{-11}$ C.

Chapter 24

1. (a) The mass flux is $wd\rho v = (3.22\,\text{m})(1.04\,\text{m})\left(1000\,\text{kg/m}^3\right)(0.207\,\text{m/s}) = 693\,\text{kg/s}$.

   (b) Since water flows only through area wd, the flux through the larger area is still $693\,\text{kg/s}$.

   (c) Now the mass flux is $(wd/2)\rho v = (693\,\text{kg/s})/2 = 347\,\text{kg/s}$.

   (d) Since the water flows through an area $(wd/2)$, the flux is $347\,\text{kg/s}$.

   (e) Now the flux is $(wd\cos\theta)\rho v = (693\,\text{kg/s})(\cos 34°) = 575\,\text{kg/s}$.

2. The vector area $\vec{A}$ and the electric field $\vec{E}$ are shown on the diagram below. The angle θ between them is $180° - 35° = 145°$, so the electric flux through the area is

$$\Phi = \vec{E}\cdot\vec{A} = EA\cos\theta = (1800\,\text{N/C})(3.2\times10^{-3}\,\text{m})^2\cos 145° = -1.5\times10^{-2}\ \text{N}\cdot\text{m}^2/\text{C}\ .$$

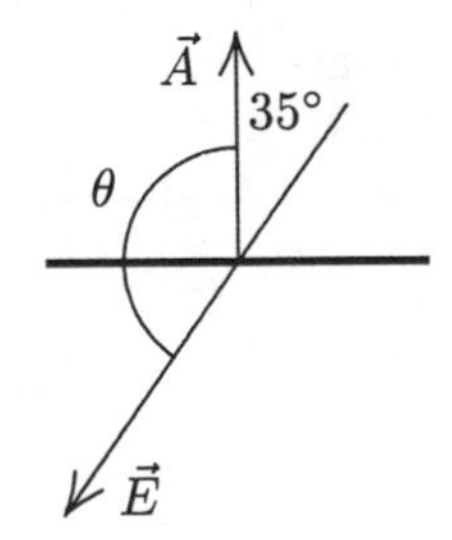

3. We use $\Phi = \vec{E}\cdot\vec{A}$, where $\vec{A} = A\hat{\jmath} = (1.40\,\text{m})^2\,\hat{\jmath}$.

   (a) $\Phi = (6.00\,\text{N/C})\hat{\imath}\cdot(1.40\,\text{m})^2\hat{\jmath} = 0$.

   (b) $\Phi = (-2.00\,\text{N/C})\hat{\jmath}\cdot(1.40\,\text{m})^2\hat{\jmath} = -3.92\,\text{N}\cdot\text{m}^2/\text{C}$.

   (c) $\Phi = [(-3.00\,\text{N/C})\hat{\imath} + (4.00\,\text{N/C})\hat{k}]\cdot(1.40\,\text{m})^2\hat{\jmath} = 0$.

   (d) The total flux of a uniform field through a closed surface is always zero.

4. We use the fact that electric flux relates to the enclosed charge: $\Phi = q_{\text{enclosed}}/\varepsilon_0$.

   (a) A surface which encloses the charges $2q$ and $-2q$, or all four charges.

   (b) A surface which encloses the charges $2q$ and q.

   (c) The maximum amount of negative charge we can enclose by any surface which encloses the charge $2q$ is $-q$, so it is impossible to get a flux of $-2q/\varepsilon_0$.

5. We use Gauss' law: $\varepsilon_0\Phi = q$, where Φ is the total flux through the cube surface and q is the net charge inside the cube. Thus,

$$\Phi = \frac{q}{\varepsilon_0} = \frac{1.8\times10^{-6}\,\text{C}}{8.85\times10^{-12}\,\text{C}^2/\text{N}\cdot\text{m}^2} = 2.0\times10^5\ \text{N}\cdot\text{m}^2/\text{C}\ .$$

6. The flux through the flat surface encircled by the rim is given by $\Phi = \pi a^2 E$. Thus, the flux through the netting is $\Phi' = -\Phi = -\pi a^2 E$.

7. (a) Let $A = (1.40\,\mathrm{m})^2$. Then

$$\begin{aligned} \Phi &= (3.00y\,\hat{\mathrm{j}}) \cdot (-A\,\hat{\mathrm{j}})|_{y=0} + (3.00y\,\hat{\mathrm{j}}) \cdot (A\,\hat{\mathrm{j}})|_{y=1.40} \\ &= (3.00)(1.40)(1.40)^2 = 8.23\ \mathrm{N{\cdot}m^2/C}\ . \end{aligned}$$

(b) The electric field can be re-written as $\vec{E} = 3.00y\hat{\mathrm{j}} + \vec{E}_0$, where $\vec{E}_0 = -4.00\hat{\mathrm{i}} + 6.00\hat{\mathrm{j}}$ is a constant field which does not contribute to the net flux through the cube. Thus Φ is still $8.23\,\mathrm{N{\cdot}m^2/C}$.

(c) The charge is given by

$$q = \varepsilon_0 \Phi = \left(8.85 \times 10^{-12}\,\frac{\mathrm{C}^2}{\mathrm{N{\cdot}m^2}}\right)(8.23\,\mathrm{N{\cdot}m^2/C}) = 7.29 \times 10^{-11}\ \mathrm{C}$$

in each case.

8. (a) The total surface area bounding the bathroom is

$$A = 2\,(2.5 \times 3.0) + 2\,(3.0 \times 2.0) + 2\,(2.0 \times 2.5) = 37\ \mathrm{m}^2\ .$$

The absolute value of the total electric flux, with the assumptions stated in the problem, is $|\Phi| = |\sum \vec{E} \cdot \vec{A}| = |\vec{E}|\,A = (600)(37) = 22 \times 10^3\ \mathrm{N{\cdot}m^2/C}$. By Gauss' law, we conclude that the enclosed charge (in absolute value) is $|q_{\mathrm{enc}}| = \varepsilon_0\,|\Phi| = 2.0 \times 10^{-7}$ C. Therefore, with volume $V = 15\ \mathrm{m}^3$, and recognizing that we are dealing with negative charges (see problem), we find the charge density is $q_{\mathrm{enc}}/V = -1.3 \times 10^{-8}\ \mathrm{C/m^3}$.

(b) We find $(|q_{\mathrm{enc}}|/e)/V = (2.0 \times 10^{-7}/1.6 \times 10^{-19})/15 = 8.2 \times 10^{10}$ excess electrons per cubic meter.

9. Let A be the area of one face of the cube, E_u be the magnitude of the electric field at the upper face, and E_ℓ be the magnitude of the field at the lower face. Since the field is downward, the flux through the upper face is negative and the flux through the lower face is positive. The flux through the other faces is zero, so the total flux through the cube surface is $\Phi = A(E_\ell - E_u)$. The net charge inside the cube is given by Gauss' law:

$$\begin{aligned} q &= \varepsilon_0 \Phi = \varepsilon_0 A(E_\ell - E_u) = (8.85 \times 10^{-12}\,\mathrm{C^2/N{\cdot}m^2})(100\,\mathrm{m})^2(100\,\mathrm{N/C} - 60.0\,\mathrm{N/C}) \\ &= 3.54 \times 10^{-6}\ \mathrm{C} = 3.54\ \mu\mathrm{C}\ . \end{aligned}$$

10. There is no flux through the sides, so we have two "inward" contributions to the flux, one from the top (of magnitude $(34)(3.0)^2$) and one from the bottom (of magnitude $(20)(3.0)^2$). With "inward" flux conventionally negative, the result is $\Phi = -486\ \mathrm{N{\cdot}m^2/C}$. Gauss' law, then, leads to $q_{\mathrm{enc}} = \varepsilon_0\,\Phi = -4.3 \times 10^{-9}$ C.

11. The total flux through any surface that completely surrounds the point charge is q/ε_0. If we stack identical cubes side by side and directly on top of each other, we will find that eight cubes meet at any corner. Thus, one-eighth of the field lines emanating from the point charge pass through a cube with a corner at the charge, and the total flux through the surface of such a cube is $q/8\varepsilon_0$. Now the field lines are radial, so at each of the three cube faces that meet at the charge, the lines are parallel to the face and the flux through the face is zero. The fluxes through each of the other three faces are the same, so the flux through each of them is one-third of the total. That is, the flux through each of these faces is $(1/3)(q/8\varepsilon_0) = q/24\varepsilon_0$.

12. Using Eq. 24-11, the surface charge density is

$$\sigma = E\varepsilon_0 = (2.3 \times 10^5\ \mathrm{N/C})\left(8.85 \times 10^{-12}\,\frac{\mathrm{C}^2}{\mathrm{N{\cdot}m^2}}\right) = 2.0 \times 10^{-6}\ \mathrm{C/m^2}\ .$$

13. (a) The charge on the surface of the sphere is the product of the surface charge density σ and the surface area of the sphere (which is $4\pi r^2$, where r is the radius). Thus,

$$q = 4\pi r^2 \sigma = 4\pi \left(\frac{1.2\,\text{m}}{2}\right)^2 (8.1 \times 10^{-6}\,\text{C/m}^2) = 3.66 \times 10^{-5}\ \text{C}\ .$$

(b) We choose a Gaussian surface in the form of a sphere, concentric with the conducting sphere and with a slightly larger radius. The flux is given by Gauss' law:

$$\Phi = \frac{q}{\varepsilon_0} = \frac{3.66 \times 10^{-5}\,\text{C}}{8.85 \times 10^{-12}\,\text{C}^2/\text{N}\cdot\text{m}^2} = 4.1 \times 10^6\ \text{N}\cdot\text{m}^2/\text{C}\ .$$

14. (a) The area of a sphere may be written $4\pi R^2 = \pi D^2$. Thus,

$$\sigma = \frac{q}{\pi D^2} = \frac{2.4 \times 10^{-6}\,\text{C}}{\pi (1.3\,\text{m})^2} = 4.5 \times 10^{-7}\ \text{C/m}^2\ .$$

(b) Eq. 24-11 gives

$$E = \frac{\sigma}{\varepsilon_o} = \frac{4.5 \times 10^{-7}\,\text{C/m}^2}{8.85 \times 10^{-12}\,\text{C}^2/\text{N}\cdot\text{m}^2} = 5.1 \times 10^4\ \text{N/C}\ .$$

15. (a) Consider a Gaussian surface that is completely within the conductor and surrounds the cavity. Since the electric field is zero everywhere on the surface, the net charge it encloses is zero. The net charge is the sum of the charge q in the cavity and the charge q_w on the cavity wall, so $q + q_w = 0$ and $q_w = -q = -3.0 \times 10^{-6}$ C.

(b) The net charge Q of the conductor is the sum of the charge on the cavity wall and the charge q_s on the outer surface of the conductor, so $Q = q_w + q_s$ and

$$q_s = Q - q_w = (10 \times 10^{-6}\,\text{C}) - (-3.0 \times 10^{-6}\,\text{C}) = +1.3 \times 10^{-5}\ \text{C}\ .$$

16. (a) The side surface area A for the drum of diameter D and length h is given by $A = \pi D h$. Thus

$$\begin{aligned} q &= \sigma A = \sigma \pi D h = \pi \varepsilon_0 E D h \\ &= \pi \left(8.85 \times 10^{-12}\,\frac{\text{C}^2}{\text{N}\cdot\text{m}^2}\right) (2.3 \times 10^5\,\text{N/C})\,(0.12\,\text{m})(0.42\,\text{m}) \\ &= 3.2 \times 10^{-7}\ \text{C}\ . \end{aligned}$$

(b) The new charge is

$$\begin{aligned} q' &= q\left(\frac{A'}{A}\right) = q\left(\frac{\pi D' h'}{\pi D h}\right) \\ &= (3.2 \times 10^{-7}\text{C})\left[\frac{(8.0\,\text{cm})(28\,\text{cm})}{(12\,\text{cm})(42\,\text{cm})}\right] = 1.4 \times 10^{-7}\ \text{C}\ . \end{aligned}$$

17. The magnitude of the electric field produced by a uniformly charged infinite line is $E = \lambda/2\pi\varepsilon_0 r$, where λ is the linear charge density and r is the distance from the line to the point where the field is measured. See Eq. 24-12. Thus,

$$\lambda = 2\pi\varepsilon_0 E r = 2\pi(8.85 \times 10^{-12}\,\text{C}^2/\text{N}\cdot\text{m}^2)(4.5 \times 10^4\,\text{N/C})(2.0\,\text{m}) = 5.0 \times 10^{-6}\ \text{C/m}\ .$$

18. We imagine a cylindrical Gaussian surface A of radius r and unit length concentric with the metal tube. Then by symmetry

$$\oint_A \vec{E} \cdot d\vec{A} = 2\pi r E = \frac{q_{\text{enclosed}}}{\varepsilon_0}\ .$$

(a) For $r > R$, $q_{\text{enclosed}} = \lambda$, so $E(r) = \lambda/2\pi r\varepsilon_0$.

(b) For $r < R$, $q_{\text{enclosed}} = 0$, so $E = 0$. The plot of E vs r is shown below. Here, the maximum value is

$$E_{\text{max}} = \frac{\lambda}{2\pi r\varepsilon_0} = \frac{(2.0 \times 10^{-8}\,\text{C/m})}{2\pi(0.030\,\text{m})\,(8.85 \times 10^{-12}\,\text{C}^2/\text{N}\cdot\text{m}^2)} = 1.2 \times 10^4\ \text{N/C}\ .$$

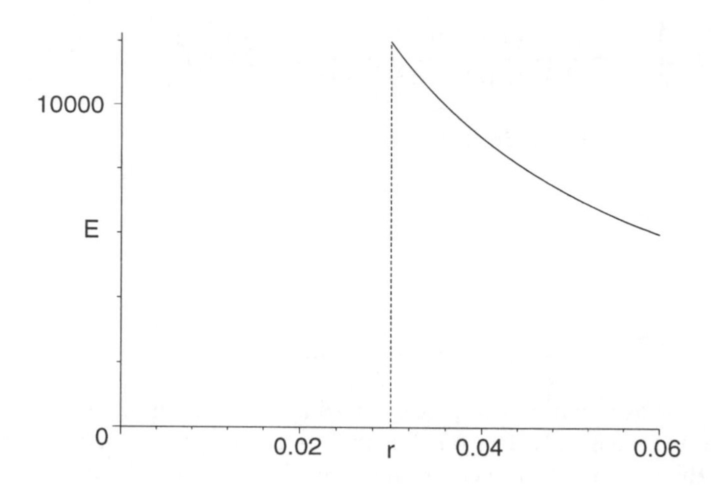

19. We assume the charge density of both the conducting cylinder and the shell are uniform, and we neglect fringing. Symmetry can be used to show that the electric field is radial, both between the cylinder and the shell and outside the shell. It is zero, of course, inside the cylinder and inside the shell.

(a) We take the Gaussian surface to be a cylinder of length L, coaxial with the given cylinders and of larger radius r than either of them. The flux through this surface is $\Phi = 2\pi rLE$, where E is the magnitude of the field at the Gaussian surface. We may ignore any flux through the ends. Now, the charge enclosed by the Gaussian surface is $q - 2q = -q$. Consequently, Gauss' law yields $2\pi r\varepsilon_0 LE = -q$, so

$$E = -\frac{q}{2\pi\varepsilon_0 Lr}\ .$$

The negative sign indicates that the field points inward.

(b) Next, we consider a cylindrical Gaussian surface whose radius places it within the shell itself. The electric field is zero at all points on the surface since any field within a conducting material would lead to current flow (and thus to a situation other than the electrostatic ones being considered here), so the total electric flux through the Gaussian surface is zero and the net charge within it is zero (by Gauss' law). Since the central rod is known to have charge q, then the inner surface of the shell must have charge $-q$. And since the shell is known to have total charge $-2q$, it must therefore have charge $-q$ on its outer surface.

(c) Finally, we consider a cylindrical Gaussian surface whose radius places it between the outside of conducting rod and inside of the shell. Similarly to part (a), the flux through the Gaussian surface is $\Phi = 2\pi rLE$, where E is the field at this Gaussian surface, in the region between the rod and the shell. The charge enclosed by the Gaussian surface is only the charge q on the rod. Therefore, Gauss' law yields

$$2\pi\varepsilon_0 rLE = q \implies E = \frac{q}{2\pi\varepsilon_0 Lr}\ .$$

The positive sign indicates that the field points outward.

20. We denote the radius of the thin cylinder as $R = 0.015$ m. Using Eq. 24-12, the net electric field for $r > R$ is given by

$$E_{\text{net}} = E_{\text{wire}} + E_{\text{cylinder}} = \frac{-\lambda}{2\pi\varepsilon_0 r} + \frac{\lambda'}{2\pi\varepsilon_0 r}$$

where $-\lambda = -3.6\,\text{nC/m}$ is the linear charge density of the wire and λ' is the linear charge density of the thin cylinder. We note that the surface and linear charge densities of the thin cylinder are related by

$$q_{\text{cylinder}} = \lambda' L = \sigma(2\pi R L) \implies \lambda' = \sigma(2\pi R) .$$

Now, E_{net} outside the cylinder will equal zero, provided that $2\pi R\sigma = \lambda$, or

$$\sigma = \frac{\lambda}{2\pi R} = \frac{3.6\times 10^{-9}\,\text{C/m}}{(2\pi)(0.015\,\text{m})} = 3.8\times 10^{-8}\ \text{C/m}^2 .$$

21. We denote the inner and outer cylinders with subscripts i and o, respectively.

(a) Since $r_i < r = 4.0\,\text{cm} < r_o$,

$$E(r) = \frac{\lambda_i}{2\pi\varepsilon_0 r} = \frac{5.0\times 10^{-6}\,\text{C/m}}{2\pi\,(8.85\times 10^{-12}\,\text{C}^2/\text{N}\cdot\text{m}^2)\,(4.0\times 10^{-2}\,\text{m})} = 2.3\times 10^{6}\ \text{N/C} .$$

$\vec{E}(r)$ points radially outward.

(b) Since $r > r_o$,

$$E(r) = \frac{\lambda_i + \lambda_o}{2\pi\varepsilon_0 r} = \frac{5.0\times 10^{-6}\,\text{C/m} - 7.0\times 10^{-6}\,\text{C/m}}{2\pi\,(8.85\times 10^{-12}\,\text{C}^2/\text{N}\cdot\text{m}^2)\,(8.0\times 10^{-2}\,\text{m})} = -4.5\times 10^{5}\ \text{N/C} ,$$

where the minus sign indicates that $\vec{E}(r)$ points radially inward.

22. To evaluate the field using Gauss' law, we employ a cylindrical surface of area $2\pi\, r\, L$ where L is very large (large enough that contributions from the ends of the cylinder become irrelevant to the calculation). The volume within this surface is $V = \pi\, r^2 L$, or expressed more appropriate to our needs: $dV = 2\pi\, r\, L\, dr$. The charge enclosed is, with $A = 2.5\times 10^{-6}\ \text{C/m}^5$,

$$q_{\text{enc}} = \int_0^r A\, r^2\, 2\pi\, r\, L\, dr = \frac{\pi}{2}\, A\, L\, r^4 .$$

By Gauss' law, we find $\Phi = |\vec{E}|(2\pi r L) = q_{\text{enc}}/\varepsilon_0$; we thus obtain

$$\left|\vec{E}\right| = \frac{A\, r^3}{4\,\varepsilon_0} .$$

(a) With $r = 0.030$ m, we find $|\vec{E}| = 1.9$ N/C.

(b) Once outside the cylinder, Eq. 24-12 is obeyed. To find $\lambda = q/L$ we must find the total charge q. Therefore,

$$\frac{q}{L} = \frac{1}{L}\int_0^{0.04} A\, r^2\, 2\pi\, r\, L\, dr = 1.0\times 10^{-11}\ \text{C/m} .$$

And the result, for $r = 0.050$ m, is $|\vec{E}| = \lambda/2\pi\varepsilon_0\, r = 3.6$ N/C.

23. The electric field is radially outward from the central wire. We want to find its magnitude in the region between the wire and the cylinder as a function of the distance r from the wire. Since the magnitude of the field at the cylinder wall is known, we take the Gaussian surface to coincide with the wall. Thus, the Gaussian surface is a cylinder with radius R and length L, coaxial with the wire. Only the charge on the wire is actually enclosed by the Gaussian surface; we denote it by q. The area of the Gaussian

surface is $2\pi RL$, and the flux through it is $\Phi = 2\pi RLE$. We assume there is no flux through the ends of the cylinder, so this Φ is the total flux. Gauss' law yields $q = 2\pi\varepsilon_0 RLE$. Thus,

$$q = 2\pi\left(8.85 \times 10^{-12}\,\frac{\mathrm{C}^2}{\mathrm{N\cdot m^2}}\right)(0.014\,\mathrm{m})(0.16\,\mathrm{m})\left(2.9 \times 10^4\,\mathrm{N/C}\right) = 3.6 \times 10^{-9}\ \mathrm{C}\ .$$

24. (a) In Eq. 24-12, $\lambda = q/L$ where q is the net charge enclosed by a cylindrical Gaussian surface of radius r. The field is being measured outside the system (the charged rod coaxial with the neutral cylinder) so that the net enclosed charge is only that which is on the rod. Consequently,

$$|\vec{E}| = \frac{\lambda}{2\pi\varepsilon_0\, r} = \frac{2.0 \times 10^{-9}}{2\pi\varepsilon_0\,(0.15)} = 240\ \mathrm{N/C}\ .$$

(b) and (c) Since the field is zero inside the conductor (in an electrostatic configuration), then there resides on the inner surface charge $-q$, and on the outer surface, charge $+q$ (where q is the charge on the rod at the center). Therefore, with $r_i = 0.05$ m, the surface density of charge is

$$\sigma_{\mathrm{inner}} = \frac{-q}{2\pi r_i L} = -\frac{\lambda}{2\pi r_i} = -6.4 \times 10^{-9}\ \mathrm{C/m^2}$$

for the inner surface. And, with $r_o = 0.10$ m, the surface charge density of the outer surface is

$$\sigma_{\mathrm{outer}} = \frac{+q}{2\pi r_o L} = \frac{\lambda}{2\pi r_o} = +3.2 \times 10^{-9}\ \mathrm{C/m^2}\ .$$

25. (a) The diagram below shows a cross section (or, perhaps more appropriately, "end view") of the charged cylinder (solid circle). Consider a Gaussian surface in the form of a cylinder with radius r and length ℓ, coaxial with the charged cylinder. An "end view" of the Gaussian surface is shown as a dotted circle. The charge enclosed by it is $q = \rho V = \pi r^2 \ell\rho$, where $V = \pi r^2\ell$ is the volume of the cylinder.

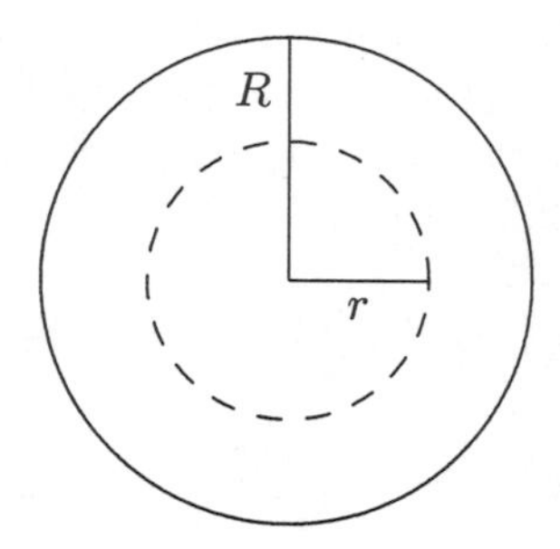

If ρ is positive, the electric field lines are radially outward, normal to the Gaussian surface and distributed uniformly along it. Thus, the total flux through the Gaussian cylinder is $\Phi = EA_{\mathrm{cylinder}} = E(2\pi r\ell)$. Now, Gauss' law leads to

$$2\pi\varepsilon_0 r\ell E = \pi r^2\ell\rho \implies E = \frac{\rho r}{2\varepsilon_0}\ .$$

(b) Next, we consider a cylindrical Gaussian surface of radius $r > R$. If the external field E_{ext} then the flux is $\Phi = 2\pi r\ell E_{\mathrm{ext}}$. The charge enclosed is the total charge in a section of the charged cylinder with length ℓ. That is, $q = \pi R^2\ell\rho$. In this case, Gauss' law yields

$$2\pi\varepsilon_0 r\ell E_{\mathrm{ext}} = \pi R^2\ell\rho \implies E_{\mathrm{ext}} = \frac{R^2\rho}{2\varepsilon_0 r}\ .$$

26. According to Eq. 24-13 the electric field due to either sheet of charge with surface charge density σ is perpendicular to the plane of the sheet (pointing *away* from the sheet if the charge is positive) and has magnitude $E = \sigma/2\varepsilon_0$. Using the superposition principle, we conclude:

(a) $E = \sigma/\varepsilon_0$, pointing up;
(b) $E = 0$;
(c) and, $E = \sigma/\varepsilon_0$, pointing down.

27. (a) To calculate the electric field at a point very close to the center of a large, uniformly charged conducting plate, we may replace the finite plate with an infinite plate with the same area charge density and take the magnitude of the field to be $E = \sigma/\varepsilon_0$, where σ is the area charge density for the surface just under the point. The charge is distributed uniformly over both sides of the original plate, with half being on the side near the field point. Thus,

$$\sigma = \frac{q}{2A} = \frac{6.0 \times 10^{-6}\,\text{C}}{2(0.080\,\text{m})^2} = 4.69 \times 10^{-4}\ \text{C/m}^2\ .$$

The magnitude of the field is

$$E = \frac{4.69 \times 10^{-4}\,\text{C/m}^2}{8.85 \times 10^{-12}\,\text{C}^2/\text{N}\cdot\text{m}^2} = 5.3 \times 10^7\ \text{N/C}\ .$$

The field is normal to the plate and since the charge on the plate is positive, it points away from the plate.

(b) At a point far away from the plate, the electric field is nearly that of a point particle with charge equal to the total charge on the plate. The magnitude of the field is $E = q/4\pi\varepsilon_0 r^2 = kq/r^2$, where r is the distance from the plate. Thus,

$$E = \frac{(8.99 \times 10^9\,\text{N}\cdot\text{m}^2/\text{C}^2)\,(6.0 \times 10^{-6}\,\text{C})}{(30\,\text{m})^2} = 60\ \text{N/C}\ .$$

28. The charge distribution in this problem is equivalent to that of an infinite sheet of charge with surface charge density σ plus a small circular pad of radius R located at the middle of the sheet with charge density $-\sigma$. We denote the electric fields produced by the sheet and the pad with subscripts 1 and 2, respectively. The net electric field $\vec{E}$ is then

$$\begin{aligned} \vec{E} &= \vec{E}_1 + \vec{E}_2 = \left(\frac{\sigma}{2\varepsilon_0}\right)\hat{\text{k}} + \frac{(-\sigma)}{2\varepsilon_0}\left(1 - \frac{z}{\sqrt{z^2+R^2}}\right)\hat{\text{k}} \\ &= \frac{\sigma z}{2\varepsilon_0\sqrt{z^2+R^2}}\,\hat{\text{k}} \end{aligned}$$

where Eq. 23-26 is used for $\vec{E}_2$.

29. The forces acting on the ball are shown in the diagram below. The gravitational force has magnitude mg, where m is the mass of the ball; the electrical force has magnitude qE, where q is the charge on the ball and E is the magnitude of the electric field at the position of the ball; and, the tension in the thread is denoted by T. The electric field produced by the plate is normal to the plate and points to the right. Since the ball is positively charged, the electric force on it also points to the right. The tension in the thread makes the angle θ $(= 30°)$ with the vertical.

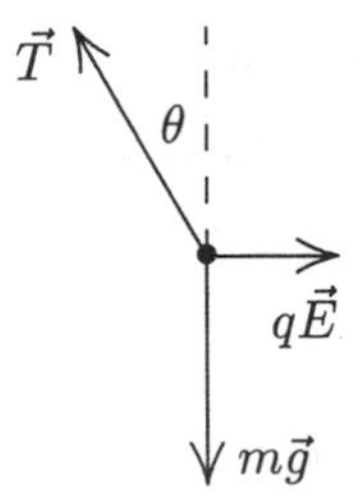

Since the ball is in equilibrium the net force on it vanishes. The sum of the horizontal components yields $qE - T\sin\theta = 0$ and the sum of the vertical components yields $T\cos\theta - mg = 0$. The expression

$T = qE/\sin\theta$, from the first equation, is substituted into the second to obtain $qE = mg\tan\theta$. The electric field produced by a large uniform plane of charge is given by $E = \sigma/2\varepsilon_0$, where σ is the surface charge density. Thus,

$$\frac{q\sigma}{2\varepsilon_0} = mg\tan\theta$$

and

$$\begin{aligned}\sigma &= \frac{2\varepsilon_0 mg\tan\theta}{q} \\ &= \frac{2(8.85\times10^{-12}\,\mathrm{C^2/N\cdot m^2})(1.0\times10^{-6}\,\mathrm{kg})(9.8\,\mathrm{m/s^2})\tan30^\circ}{2.0\times10^{-8}\,\mathrm{C}} \\ &= 5.0\times10^{-9}\ \mathrm{C/m^2}\ .\end{aligned}$$

30. Let $\hat{\mathrm{i}}$ be a unit vector pointing to the left. We use Eq. 24-13.

(a) To the left of the plates: $\vec{E} = (\sigma/2\varepsilon_0)\hat{\mathrm{i}}$ (from the right plate)$+(-\sigma/2\varepsilon_0)\hat{\mathrm{i}}$ (from the left one)$= 0$.

(b) To the right of the plates: $\vec{E} = (\sigma/2\varepsilon_0)(-\hat{\mathrm{i}})$ (from the right plate)$+(-\sigma/2\varepsilon_0)(-\hat{\mathrm{i}})$ (from the left one)$= 0$.

(c) Between the plates:

$$\begin{aligned}\vec{E} &= \left(\frac{\sigma}{2\varepsilon_0}\right)\hat{\mathrm{i}} + \left(\frac{-\sigma}{2\varepsilon_0}\right)(-\hat{\mathrm{i}}) = \left(\frac{\sigma}{\varepsilon_0}\right)\hat{\mathrm{i}} \\ &= \left(\frac{7.0\times10^{-22}\,\mathrm{C/m^2}}{8.85\times10^{-12}\,\frac{\mathrm{N\cdot m^2}}{\mathrm{C^2}}}\right)\hat{\mathrm{i}} = (7.9\times10^{-11}\ \mathrm{N/C})\ \hat{\mathrm{i}}\ .\end{aligned}$$

31. The charge on the metal plate, which is negative, exerts a force of repulsion on the electron and stops it. First find an expression for the acceleration of the electron, then use kinematics to find the stopping distance. We take the initial direction of motion of the electron to be positive. Then, the electric field is given by $E = \sigma/\varepsilon_0$, where σ is the surface charge density on the plate. The force on the electron is $F = -eE = -e\sigma/\varepsilon_0$ and the acceleration is

$$a = \frac{F}{m} = -\frac{e\sigma}{\varepsilon_0 m}$$

where m is the mass of the electron. The force is constant, so we use constant acceleration kinematics. If v_0 is the initial velocity of the electron, v is the final velocity, and x is the distance traveled between the initial and final positions, then $v^2 - v_0^2 = 2ax$. Set $v = 0$ and replace a with $-e\sigma/\varepsilon_0 m$, then solve for x. We find

$$x = -\frac{v_0^2}{2a} = \frac{\varepsilon_0 m v_0^2}{2e\sigma}\ .$$

Now $\frac{1}{2}mv_0^2$ is the initial kinetic energy K_0, so

$$x = \frac{\varepsilon_0 K_0}{e\sigma}\ .$$

We convert the given value of K_0 to Joules. Since $1.00\,\mathrm{eV} = 1.60\times10^{-19}\,\mathrm{J}$, $100\,\mathrm{eV} = 1.60\times10^{-17}\,\mathrm{J}$. Thus,

$$x = \frac{(8.85\times10^{-12}\,\mathrm{C^2/N\cdot m^2})(1.60\times10^{-17}\,\mathrm{J})}{(1.60\times10^{-19}\,\mathrm{C})(2.0\times10^{-6}\,\mathrm{C/m^2})} = 4.4\times10^{-4}\ \mathrm{m}\ .$$

32. We use the result of part (c) of problem 30 to obtain the surface charge density.

$$E = \sigma/\varepsilon_0 \implies \sigma = \varepsilon_0 E = \left(8.85\times10^{-12}\,\frac{\mathrm{C^2}}{\mathrm{N\cdot m^2}}\right)(55\,\mathrm{N/C}) = 4.9\times10^{-10}\ \mathrm{C/m^2}\ .$$

33. (a) We use a Gaussian surface in the form of a box with rectangular sides. The cross section is shown with dashed lines in the diagram below. It is centered at the central plane of the slab, so the left and right faces are each a distance x from the central plane. We take the thickness of the rectangular solid to be a, the same as its length, so the left and right faces are squares. The electric field is normal to the left and right faces and is uniform over them. If ρ is positive, it points outward at both faces: toward the left at the left face and toward the right at the right face. Furthermore, the magnitude is the same at both faces. The electric flux through each of these faces is Ea^2. The field is parallel to the other faces of the Gaussian surface and the flux through them is zero. The total flux through the Gaussian surface is $\Phi = 2Ea^2$.

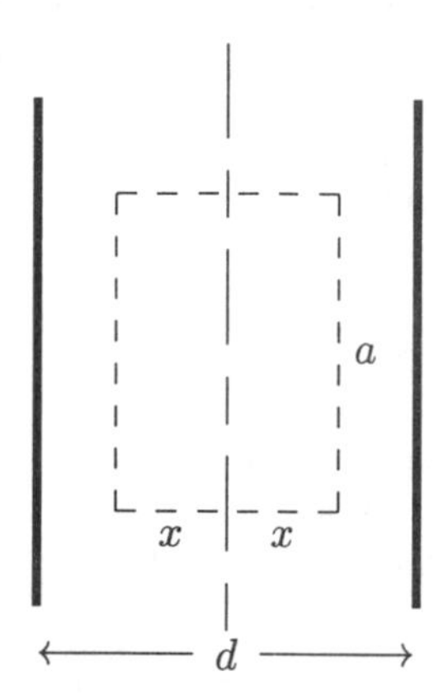

The volume enclosed by the Gaussian surface is $2a^2x$ and the charge contained within it is $q = 2a^2x\rho$. Gauss' law yields $2\varepsilon_0 Ea^2 = 2a^2x\rho$. We solve for the magnitude of the electric field:

$$E = \frac{\rho x}{\varepsilon_0} .$$

(b) We take a Gaussian surface of the same shape and orientation, but with $x > d/2$, so the left and right faces are outside the slab. The total flux through the surface is again $\Phi = 2Ea^2$ but the charge enclosed is now $q = a^2 d\rho$. Gauss' law yields $2\varepsilon_0 Ea^2 = a^2 d\rho$, so

$$E = \frac{\rho d}{2\varepsilon_0} .$$

34. (a) The flux is still $-750\,\text{N}\cdot\text{m}^2/\text{C}$, since it depends only on the amount of charge enclosed.

(b) We use $\Phi = q/\varepsilon_0$ to obtain the charge q:

$$q = \varepsilon_0 \Phi = \left(8.85 \times 10^{-12}\,\frac{\text{C}^2}{\text{N}\cdot\text{m}^2}\right)(-750\,\text{N}\cdot\text{m}^2/\text{C}) = -6.64 \times 10^{-10}\ \text{C} .$$

35. Charge is distributed uniformly over the surface of the sphere and the electric field it produces at points outside the sphere is like the field of a point particle with charge equal to the net charge on the sphere. That is, the magnitude of the field is given by $E = q/4\pi\varepsilon_0 r^2$, where q is the magnitude of the charge on the sphere and r is the distance from the center of the sphere to the point where the field is measured. Thus,

$$q = 4\pi\varepsilon_0 r^2 E = \frac{(0.15\,\text{m})^2(3.0 \times 10^3\,\text{N/C})}{8.99 \times 10^9\,\text{N}\cdot\text{m}^2/\text{C}^2} = 7.5 \times 10^{-9}\ \text{C} .$$

The field points inward, toward the sphere center, so the charge is negative: -7.5×10^{-9} C.

36. (a) Since $r_1 = 10.0\,\text{cm} < r = 12.0\,\text{cm} < r_2 = 15.0\,\text{cm}$,

$$E(r) = \frac{1}{4\pi\varepsilon_0}\frac{q_1}{r^2} = \frac{(8.99 \times 10^9\,\text{N}\cdot\text{m}^2/\text{C}^2)(4.00 \times 10^{-8}\,\text{C})}{(0.120\,\text{m})^2} = 2.50 \times 10^4\ \text{N/C} .$$

(b) Since $r_1 < r_2 < r = 20.0\,\text{cm}$,

$$\begin{aligned} E(r) &= \frac{1}{4\pi\varepsilon_0}\frac{q_1+q_2}{r^2} = \frac{(8.99\times10^9\,\text{N}\cdot\text{m}^2/\text{C}^2)(4.00+2.00)(1\times10^{-8}\,\text{C})}{(0.200\,\text{m})^2} \\ &= 1.35\times10^4\ \text{N/C}\ . \end{aligned}$$

37. The field is radially outward and takes on equal magnitude-values over the surface of any sphere centered at the atom's center. We take the Gaussian surface to be such a sphere (of radius r). If E is the magnitude of the field, then the total flux through the Gaussian sphere is $\Phi = 4\pi r^2 E$. The charge enclosed by the Gaussian surface is the positive charge at the center of the atom plus that portion of the negative charge within the surface. Since the negative charge is uniformly distributed throughout the large sphere of radius R, we can compute the charge inside the Gaussian sphere using a ratio of volumes. That is, the negative charge inside is $-Zer^3/R^3$. Thus, the total charge enclosed is $Ze - Zer^3/R^3$ for $r \le R$. Gauss' law now leads to

$$4\pi\varepsilon_0 r^2 E = Ze\left(1-\frac{r^3}{R^3}\right) \implies E = \frac{Ze}{4\pi\varepsilon_0}\left(\frac{1}{r^2}-\frac{r}{R^3}\right)\ .$$

38. We interpret the question as referring to the field *just* outside the sphere (that is, at locations roughly equal to the radius r of the sphere). Since the area of a sphere is $A = 4\pi r^2$ and the surface charge density is $\sigma = q/A$ (where we assume q is positive for brevity), then

$$E = \frac{\sigma}{\varepsilon_0} = \frac{1}{\varepsilon_0}\left(\frac{q}{4\pi r^2}\right) = \frac{1}{4\pi\varepsilon_0}\frac{q}{r^2}$$

which we recognize as the field of a point charge (see Eq. 23-3).

39. The proton is in uniform circular motion, with the electrical force of the sphere on the proton providing the centripetal force. According to Newton's second law, $F = mv^2/r$, where F is the magnitude of the force, v is the speed of the proton, and r is the radius of its orbit, essentially the same as the radius of the sphere. The magnitude of the force on the proton is $F = eq/4\pi\varepsilon_0 r^2$, where q is the magnitude of the charge on the sphere. Thus,

$$\frac{1}{4\pi\varepsilon_0}\frac{eq}{r^2} = \frac{mv^2}{r}$$

so

$$\begin{aligned} q &= \frac{4\pi\varepsilon_0 mv^2 r}{e} = \frac{(1.67\times10^{-27}\,\text{kg})(3.00\times10^5\,\text{m/s})^2(0.0100\,\text{m})}{(8.99\times10^9\,\text{N}\cdot\text{m}^2/\text{C}^2)(1.60\times10^{-19}\,\text{C})} \\ &= 1.04\times10^{-9}\ \text{C}\ . \end{aligned}$$

The force must be inward, toward the center of the sphere, and since the proton is positively charged, the electric field must also be inward. The charge on the sphere is negative: $q = -1.04\times10^{-9}\,\text{C}$.

40. We imagine a spherical Gaussian surface of radius r centered at the point charge $+q$. From symmetry consideration E is the same throughout the surface, so

$$\oint \vec{E}\cdot d\vec{A} = 4\pi r^2 E = \frac{q_{\text{encl}}}{\varepsilon_0}\ ,$$

which gives

$$E(r) = \frac{q_{\text{encl}}}{4\pi\varepsilon_0 r^2}\ ,$$

where q_{encl} is the net charge enclosed by the Gaussian surface.

(a) Now $a < r < b$, where $E = 0$. Thus $q_{\text{encl}} = 0$, so the charge on the inner surface of the shell is $q_i = -q$.

(b) The shell as a whole is electrically neutral, so the outer shell must carry a charge of $q_o = +q$.

(c) For $r < a$ $\quad q_{\text{encl}} = +q$, so

$$E\bigg|_{r<a} = \frac{q}{4\pi\varepsilon_0 r^2}\ .$$

(d) For $b > r > a$ $\quad E = 0$, since this region is inside the metallic part of the shell.

(e) For $r > b$ $\quad q_{\text{encl}} = +q$, so

$$E\bigg|_{r<a} = \frac{q}{4\pi\varepsilon_0 r^2}\ .$$

The field lines are sketched to the right.

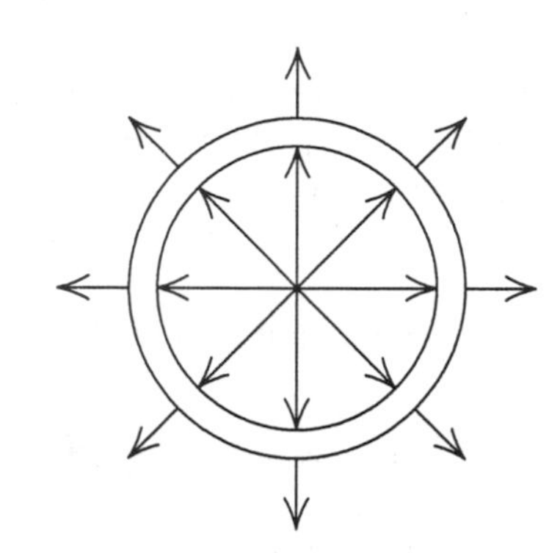

(f) The net charge of the central point charge-inner surface combination is zero. Thus the electric field it produces is also zero.

(g) The outer shell has a spherically symmetric charge distribution with a net charge $+q$. Thus the field it produces for $r > b$ is $E = q/(4\pi\varepsilon_0 r^2)$.

(h) Yes. In fact there will be a distribution of induced charges on the outer shell, as a result of a flow of positive charges toward the side of the surface that is closer to the negative point charge outside the shell.

(i) No. The change in the charge distribution on the outer shell cancels the effect of the negative point charge. The field lines are sketched below.

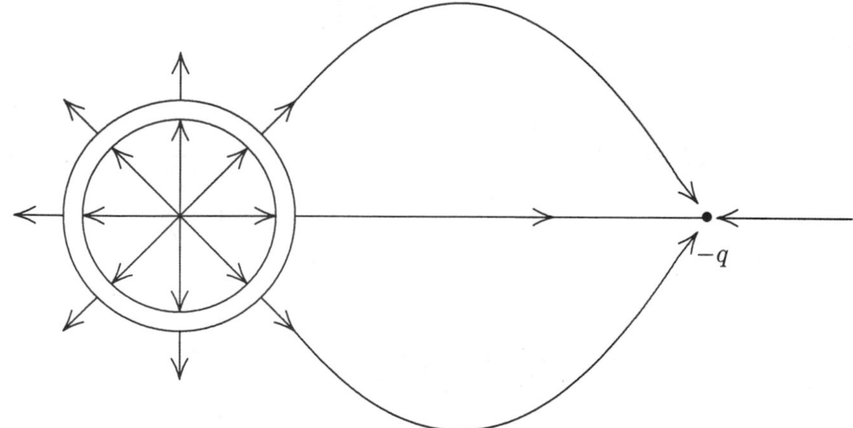

(j) Yes, there is a force on the $-q$ point charge, as expected from Eq. 22-4.

(k) The field lines around the first charge at the center of the spherical shell is unchanged. The implication, then, is that there is still no net force on that charge.

(l) We assume there is some non-electrical force holding the spherical shell in place, which compensates for the force of the $-q$ point charge exerted on the outside surface charges on the shell. Newton's third law applies to this situation, as far as the $-q$ point charge and the surface charges on the sphere are concerned. There is no direct force between the central $+q$ charge and the external $-q$ point charge, so we would not apply Newton's third law to their interaction.

41. (a) We integrate the volume charge density over the volume and require the result be equal to the total charge:

$$\int dx \int dy \int dz\, \rho = 4\pi \int_0^R dr\, r^2\, \rho = Q\ .$$

Substituting the expression $\rho = \rho_s r/R$ and performing the integration leads to

$$4\pi \left(\frac{\rho_s}{R}\right)\left(\frac{R^4}{4}\right) = Q \implies Q = \pi\rho_s R^3\ .$$

(b) At a certain point within the sphere, at some distance r_o from the center, the field (see Eq. 24-8 through Eq. 24-10) is given by Gauss' law:

$$E = \frac{1}{4\pi\varepsilon_0}\frac{q_{\rm enc}}{r_o^2}$$

where $q_{\rm enc}$ is given by an integral similar to that worked in part (a):

$$q_{\rm enc} = 4\pi \int_0^{r_o} dr\, r^2\, \rho = 4\pi \left(\frac{\rho_s}{R}\right)\left(\frac{r_o^4}{4}\right)\ .$$

Therefore,

$$E = \frac{1}{4\pi\varepsilon_0}\frac{\pi\rho_s r_o^4}{R r_o^2}$$

which (using the relation between ρ_s and Q derived in part (a)) becomes

$$E = \frac{1}{4\pi\varepsilon_0}\frac{\pi\left(\frac{Q}{\pi R^3}\right) r_o^2}{R}$$

and simplifies to the desired result (shown in the problem statement) if we change notation $r_o \to r$.

42. (a) We note that the symbol "e" stands for the elementary charge in the manipulations below. From

$$-e = \int_0^\infty \rho(r) 4\pi r^2\, dr = \int_0^\infty A \exp\left(-2r/a_0\right) 4\pi r^2\, dr = \pi a_0^3 A$$

we get $A = -e/\pi a_0^3$.

(b) The magnitude of the field is

$$\begin{aligned} E &= \frac{q_{\rm encl}}{4\pi\varepsilon_0 a_0^2} = \frac{1}{4\pi\varepsilon_0 a_0^2}\left(e + \int_0^{a_0} \rho(r) 4\pi r^2\, dr\right) \\ &= \frac{e}{4\pi\varepsilon_0 a_0^2}\left(1 - \frac{4}{a_0^3}\int_0^{a_0} \exp(-2r/a_0)\, r^2\, dr\right) \\ &= \frac{5\, e \exp(-2)}{4\pi\varepsilon_0\, a_0^2}\ . \end{aligned}$$

We note that $\vec{E}$ points radially outward.

43. At all points where there is an electric field, it is radially outward. For each part of the problem, use a Gaussian surface in the form of a sphere that is concentric with the sphere of charge and passes through the point where the electric field is to be found. The field is uniform on the surface, so

$$\oint \vec{E}\cdot d\vec{A} = 4\pi r^2 E$$

where r is the radius of the Gaussian surface.

(a) Here r is less than a and the charge enclosed by the Gaussian surface is $q(r/a)^3$. Gauss' law yields

$$4\pi r^2 E = \left(\frac{q}{\varepsilon_0}\right)\left(\frac{r}{a}\right)^3 \implies E = \frac{qr}{4\pi\varepsilon_0 a^3} \, .$$

(b) In this case, r is greater than a but less than b. The charge enclosed by the Gaussian surface is q, so Gauss' law leads to

$$4\pi r^2 E = \frac{q}{\varepsilon_0} \implies E = \frac{q}{4\pi\varepsilon_0 r^2} \, .$$

(c) The shell is conducting, so the electric field inside it is zero.

(d) For $r > c$, the charge enclosed by the Gaussian surface is zero (charge q is inside the shell cavity and charge $-q$ is on the shell). Gauss' law yields

$$4\pi r^2 E = 0 \implies E = 0 \, .$$

(e) Consider a Gaussian surface that lies completely within the conducting shell. Since the electric field is everywhere zero on the surface, $\oint \vec{E} \cdot d\vec{A} = 0$ and, according to Gauss' law, the net charge enclosed by the surface is zero. If Q_i is the charge on the inner surface of the shell, then $q + Q_i = 0$ and $Q_i = -q$. Let Q_o be the charge on the outer surface of the shell. Since the net charge on the shell is $-q$, $Q_i + Q_o = -q$. This means $Q_o = -q - Q_i = -q - (-q) = 0$.

44. The field is zero for $0 \le r \le a$ as a result of Eq. 24-16. Since q_{enc} (for $a \le r \le b$) is related to the volume by

$$q_{\text{enc}} = \rho\left(\frac{4\pi r^3}{3} - \frac{4\pi a^3}{3}\right)$$

then

$$E = \frac{1}{4\pi\varepsilon_0}\frac{q_{\text{enc}}}{r^2} = \frac{\rho}{4\pi\varepsilon_0 r^2}\left(\frac{4\pi r^3}{3} - \frac{4\pi a^3}{3}\right) = \frac{\rho}{3\varepsilon_0}\frac{r^3 - a^3}{r^2}$$

for $a \le r \le b$. And for $r \ge b$ we have $E = q_{\text{total}}/4\pi\varepsilon_0 r^2$ or

$$E = \frac{\rho}{3\varepsilon_0}\frac{b^3 - a^3}{r^2} \qquad r \ge b \, .$$

This is plotted below for r in meters from 0 to 0.30 m. The peak value of the electric field, reached at $r = b = 0.20$ m, is 6.6×10^3 N/C.

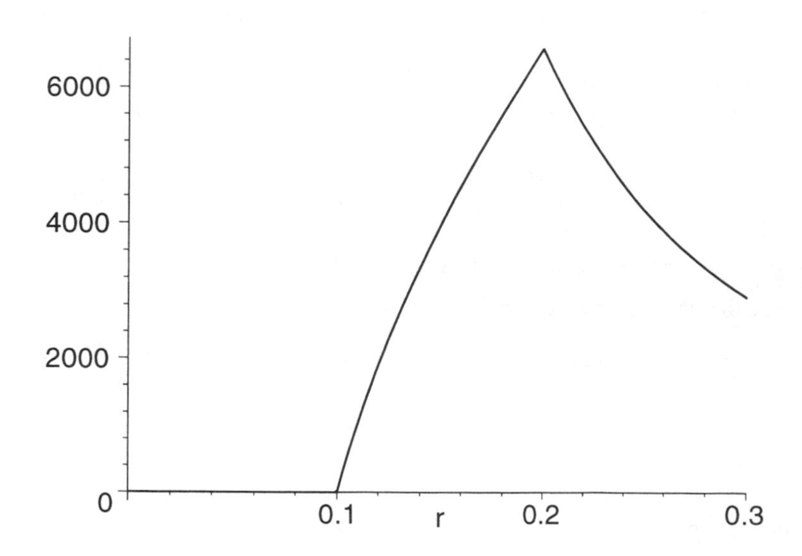

45. To find an expression for the electric field inside the shell in terms of A and the distance from the center of the shell, select A so the field does not depend on the distance. We use a Gaussian surface in the form of a sphere with radius r_g, concentric with the spherical shell and within it ($a < r_g < b$). Gauss' law will be used to find the magnitude of the electric field a distance r_g from the shell center. The charge that is both in the shell and within the Gaussian sphere is given by the integral $q_s = \int \rho \, dV$ over the portion of the shell within the Gaussian surface. Since the charge distribution has spherical symmetry, we may take dV to be the volume of a spherical shell with radius r and infinitesimal thickness dr: $dV = 4\pi r^2 \, dr$. Thus,

$$q_s = 4\pi \int_a^{r_g} \rho r^2 \, dr = 4\pi \int_a^{r_g} \frac{A}{r} r^2 \, dr = 4\pi A \int_a^{r_g} r \, dr = 2\pi A(r_g^2 - a^2) \, .$$

The total charge inside the Gaussian surface is $q + q_s = q + 2\pi A(r_g^2 - a^2)$. The electric field is radial, so the flux through the Gaussian surface is $\Phi = 4\pi r_g^2 E$, where E is the magnitude of the field. Gauss' law yields

$$4\pi\varepsilon_0 E r_g^2 = q + 2\pi A(r_g^2 - a^2) \, .$$

We solve for E:

$$E = \frac{1}{4\pi\varepsilon_0}\left[\frac{q}{r_g^2} + 2\pi A - \frac{2\pi A a^2}{r_g^2}\right] \, .$$

For the field to be uniform, the first and last terms in the brackets must cancel. They do if $q - 2\pi A a^2 = 0$ or $A = q/2\pi a^2$.

46. (a) From Gauss' law,

$$\vec{E}(\vec{r}) = \frac{1}{4\pi\varepsilon_0}\frac{q_{\text{encl}}}{r^3}\vec{r} = \frac{1}{4\pi\varepsilon_0}\frac{(4\pi\rho r^3/3)\vec{r}}{r^3} = \frac{\rho\vec{r}}{3\varepsilon_0} \, .$$

(b) The charge distribution in this case is equivalent to that of a whole sphere of charge density ρ plus a smaller sphere of charge density $-\rho$ which fills the void. By superposition

$$\vec{E}(\vec{r}) = \frac{\rho\vec{r}}{3\varepsilon_0} + \frac{(-\rho)(\vec{r} - \vec{a})}{3\varepsilon_0} = \frac{\rho\vec{a}}{3\varepsilon_0} \, .$$

47. We use

$$E(r) = \frac{q_{\text{encl}}}{4\pi\varepsilon_0 r^2} = \frac{1}{4\pi\varepsilon_0 r^2}\int_0^r \rho(r) 4\pi r^2 \, dr$$

to solve for $\rho(r)$:

$$\rho(r) = \frac{\varepsilon_0}{r^2}\frac{d}{dr}[r^2 E(r)] = \frac{\varepsilon_0}{r^2}\frac{d}{dr}(K r^6) = 6K\varepsilon_0 r^3 \, .$$

48. (a) We consider the radial field produced at points within a uniform cylindrical distribution of charge. The volume enclosed by a Gaussian surface in this case is $L\pi r^2$. Thus, Gauss' law leads to

$$E = \frac{|q_{\text{enc}}|}{\varepsilon_0 A_{\text{cylinder}}} = \frac{|\rho| \, (L\pi r^2)}{\varepsilon_0 \, (2\pi r L)} = \frac{|\rho| r}{2\varepsilon_0} \, .$$

(b) We note from the above expression that the magnitude of the radial field grows with r.

(c) Since the charged powder is negative, the field points radially inward.

(d) The largest value of r which encloses charged material is $r_{\max} = R$. Therefore, with $|\rho| = 0.0011 \text{ C/m}^3$ and $R = 0.050$ m, we obtain

$$E_{\max} = \frac{|\rho| R}{2\varepsilon_0} = 3.1 \times 10^6 \text{ N/C} \, .$$

(e) According to condition 1 mentioned in the problem, the field is high enough to produce an electrical discharge (at $r = R$).

49. (a) At A, the only field contribution is from the $+5.00Q$ particle in the hollow (this follows from Gauss' law – it is the only charge enclosed by a Gaussian spherical surface passing through point A, concentric with the shell). Thus, using k for $1/4\pi\varepsilon_0$, we have $\vec{E} = k(5Q)/(0.5)^2 = 20kQ$ directed radially outward.

(b) Point B is in the conducting material, where the field must be zero in any electrostatic situation.

(c) Point C is outside the sphere where the net charge at smaller values of radius is $-3.00Q + 5.00Q = 2.00Q$. Therefore, we have $\vec{E} = k(2Q)/(2)^2 = \frac{1}{2}kQ$ directed radially outward.

50. Since the fields involved are uniform, the precise location of P are not relevant. Since the sheets are oppositely charged (though not equally so), the field contributions are additive (since P is between them). Using Eq. 24-13, we obtain

$$\vec{E} = \frac{\sigma_1}{2\varepsilon_0} + \frac{3\sigma_1}{2\varepsilon_0} = \frac{2\sigma_1}{\varepsilon_0}$$

directed towards the negatively charged sheet.

51. (a) We imagine a Gaussian surface A which is just outside the inner surface of the spherical shell. Then $\vec{E}$ is zero everywhere on surface A. Thus

$$\oint_A \vec{E} \cdot d\vec{A} = \frac{(Q' + Q)}{\varepsilon_0} = 0 ,$$

where Q' is the charge on the inner surface of the shell. This gives $Q' = -Q$.

(b) Since $\vec{E}$ remains zero on surface A the result is unchanged.

(c) Now,

$$\oint_A \vec{E} \cdot d\vec{A} = \frac{(Q' + q + Q)}{\varepsilon_0} = 0 ,$$

so $Q' = -(Q + q)$.

(d) Yes, since $\vec{E}$ remains zero on surface A regardless of where you place the sphere inside the shell.

52. We choose a coordinate system whose origin is at the center of the flat base, such that the base is in the xy plane and the rest of the hemisphere is in the $z > 0$ half space.

(a) $\Phi = \pi R^2(-\hat{k}) \cdot E\hat{k} = -\pi R^2 E$.

(b) Since the flux through the entire hemisphere is zero, the flux through the curved surface is $\vec{\Phi}_c = -\Phi_{\text{base}} = \pi R^2 E$.

53. Let $\Phi_0 = 10^3\,\text{N}\cdot\text{m}^2/\text{C}$. The net flux through the entire surface of the dice is given by

$$\Phi = \sum_{n=1}^{6} \Phi_n = \sum_{n=1}^{6} (-1)^n n\, \Phi_0 = \Phi_0(-1 + 2 - 3 + 4 - 5 + 6) = 3\Phi_0 .$$

Thus, the net charge enclosed is

$$q = \varepsilon_0 \Phi = 3\varepsilon_0 \Phi_0 = 3\left(8.85 \times 10^{-12}\,\frac{\text{C}^2}{\text{N}\cdot\text{m}^2}\right)(10^3\,\text{N}\cdot\text{m}^2/\text{C}) = 2.66 \times 10^{-8}\,\text{C} .$$

54. We use $\Phi = \int \vec{E} \cdot d\vec{A}$. We note that the side length of the cube is $3.0\,\text{m} - 1.0\,\text{m} = 2.0\,\text{m}$.

(a) On the top face of the cube $y = 2.0\,\text{m}$ and $d\vec{A} = (dA)\hat{j}$. So $\vec{E} = 4\hat{i} - 3((2.0)^2 + 2)\hat{j} = 4\hat{i} - 18\hat{j}$. Thus the flux is

$$\begin{aligned}\Phi &= \int_{\text{top}} \vec{E} \cdot d\vec{A} = \int_{\text{top}} (4\hat{i} - 18\hat{j}) \cdot (dA)\hat{j} \\ &= -18 \int_{\text{top}} dA = (-18)(2.0)^2\,\text{N}\cdot\text{m}^2/\text{C} = -72\,\text{N}\cdot\text{m}^2/\text{C} .\end{aligned}$$

(b) On the bottom face of the cube $y = 0$ and $d\vec{A} = (dA)(-\hat{\jmath})$. So $\vec{E} = 4\hat{\imath} - 3(0^2+2)\hat{\jmath} = 4\hat{\imath} - 6\hat{\jmath}$. Thus, the flux is

$$\begin{aligned}\Phi &= \int_{\text{bottom}} \vec{E}\cdot d\vec{A} = \int_{\text{bottom}} (4\hat{\imath} - 6\hat{\jmath})\cdot(dA)(-\hat{\jmath}) \\ &= 6\int_{\text{bottom}} dA = 6(2.0)^2\,\text{N}\cdot\text{m}^2/\text{C} = +24\,\text{N}\cdot\text{m}^2/\text{C}\,.\end{aligned}$$

(c) On the left face of the cube $d\vec{A} = (dA)(-\hat{\imath})$. So

$$\begin{aligned}\Phi &= \int_{\text{left}} \vec{E}\cdot d\vec{A} = \int_{\text{left}} (4\hat{\imath} + E_y\hat{\jmath})\cdot(dA)(-\hat{\imath}) \\ &= -4\int_{\text{bottom}} dA = -4(2.0)^2\,\text{N}\cdot\text{m}^2/\text{C} = -16\,\text{N}\cdot\text{m}^2/\text{C}\,.\end{aligned}$$

(d) On the back face of the cube $d\vec{A} = (dA)(-\hat{k})$. But since $\vec{E}$ has no z component $\vec{E}\cdot d\vec{A} = 0$. Thus, $\Phi = 0$.

(e) We now have to add the flux through all six faces. One can easily verify that the flux through the front face is zero, while that through the right face is the opposite of that through the left one, or $+16\,\text{N}\cdot\text{m}^2/\text{C}$. Thus the net flux through the cube is $\Phi = (-72 + 24 - 16 + 0 + 0 + 16)\,\text{N}\cdot\text{m}^2/\text{C} = -48\,\text{N}\cdot\text{m}^2/\text{C}$.

55. The net enclosed charge q is given by

$$q = \varepsilon_0\Phi = \left(8.85\times10^{-12}\,\frac{\text{C}^2}{\text{N}\cdot\text{m}^2}\right)(-48\,\text{N}\cdot\text{m}^2/\text{C}) = -4.2\times10^{-10}\ \text{C}\,.$$

56. Since the fields involved are uniform, the precise location of P is not relevant; what is important is it is above the three sheets, with the positively charged sheets contributing upward fields and the negatively charged sheet contributing a downward field, which conveniently conforms to usual conventions (of upward as positive and downward as negative). The net field is directed upward, and (from Eq. 24-13) is magnitude is

$$\left|\vec{E}\right| = \frac{\sigma_1}{2\varepsilon_0} + \frac{\sigma_2}{2\varepsilon_0} + \frac{\sigma_3}{2\varepsilon_0} = \frac{1.0\times10^{-6}}{2\times8.85\times10^{-12}} = 5.6\times10^4\ \text{N/C}\,.$$

57. (a) Outside the sphere, we use Eq. 24-15 and obtain

$$\vec{E} = \frac{1}{4\pi\varepsilon_0}\frac{q}{r^2} = 1.5\times10^4\,\text{N/C outward}\,.$$

(b) With $q = +6.00\times10^{-12}$ C, Eq. 24-20 leads to $\vec{E} = 2.5\times10^4$ N/C directed outward.

58. (a) and (b) There is no flux through the sides, so we have two contributions to the flux, one from the $x = 2$ end (with $\Phi_2 = +(2+2)(\pi(0.20)^2) = 0.50\ \text{N}\cdot\text{m}^2/\text{C}$) and one from the $x = 0$ end (with $\Phi_0 = -(2)(\pi(0.20)^2)$). By Gauss' law we have $q_{\text{enc}} = \varepsilon_0(\Phi_2 + \Phi_0) = 2.2\times10^{-12}$ C.

59. (a) The cube is totally within the spherical volume, so the charge enclosed is $\rho V_{\text{cube}} = (500\times10^{-9})(0.040)^3 = 3.2\times10^{-11}$ C. By Gauss' law, we find $\Phi = q_{\text{enc}}/\varepsilon_0 = 3.6\ \text{N}\cdot\text{m}^2/\text{C}$.

(b) Now the sphere is totally contained within the cube (note that the radius of the sphere is less than half the side-length of the cube). Thus, the total charge is $q_{\text{enc}}\rho V_{\text{sphere}} = 4.5\times10^{-10}$ C. By Gauss' law, we find $\Phi = q_{\text{enc}}/\varepsilon_0 = 51\ \text{N}\cdot\text{m}^2/\text{C}$.

60. We use $\Phi = q_{\text{enclosed}}/\varepsilon_0$ and the fact that the amount of positive (negative) charges on the left (right) side of the conductor is $q\,(-q)$. Thus, $\Phi_1 = q/\varepsilon_0$, $\Phi_2 = -q/\varepsilon_0$, $\Phi_3 = q/\varepsilon_0$, $\Phi_4 = (q-q)/\varepsilon_0 = 0$, and $\Phi_5 = (q+q-q)/\varepsilon_0 = q/\varepsilon_0$.

61. (a) For $r < R$, $E = 0$ (see Eq. 24-16).

(b) For r slightly greater than R,

$$\begin{aligned} E_R &= \frac{1}{4\pi\varepsilon_0}\frac{q}{r^2} \approx \frac{q}{4\pi\varepsilon_0 R^2} = \frac{(8.99\times10^9\,\text{N}\cdot\text{m}^2/\text{C}^2)(2.0\times10^{-7}\text{C})}{(0.25\,\text{m})^2} \\ &= 2.9\times10^4\ \text{N/C}\ . \end{aligned}$$

(c) For $r > R$,

$$E = \frac{1}{4\pi\varepsilon_0}\frac{q}{r^2} = E_R\left(\frac{R}{r}\right)^2 = (2.9\times10^4\,\text{N/C})\left(\frac{0.25\,\text{m}}{3.0\,\text{m}}\right)^2 = 200\ \text{N/C}\ .$$

62. The field due to a sheet of charge is given by Eq. 24-13. Both sheets are horizontal (parallel to the xy plane), producing vertical fields (parallel to the z axis). At points above the $z = 0$ sheet (sheet A), its field points upward (towards $+z$); at points above the $z = 2.0$ sheet (sheet B), its field does likewise. However, below the $z = 2.0$ sheet, its field is oriented downward.

(a) The magnitude of the net field in the region between the sheets is

$$\left|\vec{E}\right| = \frac{\sigma_A}{2\varepsilon_0} - \frac{\sigma_B}{2\varepsilon_0} = 2.8\times10^2\ \text{N/C}\ .$$

(b) The magnitude of the net field at points above both sheets is

$$\left|\vec{E}\right| = \frac{\sigma_A}{2\varepsilon_0} + \frac{\sigma_B}{2\varepsilon_0} = 6.2\times10^2\ \text{N/C}\ .$$

63. To exploit the symmetry of the situation, we imagine a closed Gaussian surface in the shape of a cube, of edge length d, with the charge q situated at the inside center of the cube. The cube has six faces, and we expect an equal amount of flux through each face. The total amount of flux is $\Phi_{\text{net}} = q/\varepsilon_0$, and we conclude that the flux through the square is one-sixth of that. Thus, $\Phi = q/6\varepsilon_0$.

64. (a) At $x = 0.040$ m, the net field has a rightward ($+x$) contribution (computed using Eq. 24-13) from the charge lying between $x = -0.050$ m and $x = 0.040$ m, and a leftward ($-x$) contribution (again computed using Eq. 24-13) from the charge in the region from $x = 0.040$ m to $x = 0.050$ m. Thus, since $\sigma = q/A = \rho V/A = \rho\Delta x$ in this situation, we have

$$\left|\vec{E}\right| = \frac{\rho(0.090\ \text{m})}{2\varepsilon_0} - \frac{\rho(0.010\ \text{m})}{2\varepsilon_0} = 5.4\ \text{N/C}\ .$$

(b) In this case, the field contributions from all layers of charge point rightward, and we obtain

$$\left|\vec{E}\right| = \frac{\rho(0.100\ \text{m})}{2\varepsilon_0} = 6.8\ \text{N/C}\ .$$

65. (a) The direction of the electric field at P_1 is away from q_1 and its magnitude is

$$\left|\vec{E}\right| = \frac{q}{4\pi\varepsilon_0 r_1^2} = \frac{(8.99\times10^9\,\text{N}\cdot\text{m}^2/\text{C}^2)(1.0\times10^{-7}\text{C})}{(0.015\,\text{m})^2} = 4.0\times10^6\,\text{N/C}\ .$$

(b) $\vec{E} = 0$, since P_2 is inside the metal.

66. We use Eqs. 24-15, 24-16 and the superposition principle.

(a) $E = 0$ in the region inside the shell.

(b) $E = (1/4\pi\varepsilon_0)(q_a/r^2)$.

(c) $E = (1/4\pi\varepsilon_0)(q_a + q_b)/r^2$.

(d) Since $E = 0$ for $r < a$ the charge on the inner surface of the inner shell is always zero. The charge on the outer surface of the inner shell is therefore q_a. Since $E = 0$ inside the metallic outer shell the net charge enclosed in a Gaussian surface that lies in between the inner and outer surfaces of the outer shell is zero. Thus the inner surface of the outer shell must carry a charge $-q_a$, leaving the charge on the outer surface of the outer shell to be $q_b + q_a$.

67. (a) We use $m_e g = eE = e\sigma/\varepsilon_0$ to obtain the surface charge density.

$$\sigma = \frac{m_e g \varepsilon_0}{e} = \frac{(9.11 \times 10^{-31}\,\text{kg})\,(9.8\,\text{m/s}^2)\left(8.85 \times 10^{-12}\,\frac{\text{C}^2}{\text{N}\cdot\text{m}^2}\right)}{1.60 \times 10^{-19}\,\text{C}} = 4.9 \times 10^{-22}\ \text{C/m}^2 \ .$$

(b) Downward (since the electric force exerted on the electron must be upward).

68. (a) In order to have net charge $-10\ \mu$C when $-14\ \mu$C is known to be on the outer surface, then there must be $+4\ \mu$C on the inner surface (since charges reside on the surfaces of a conductor in electrostatic situations).

(b) In order to cancel the electric field inside the conducting material, the contribution from the $+4\ \mu$C on the inner surface must be canceled by that of the charged particle in the hollow. Thus, the particle's charge is $-4\ \mu$C.

Chapter 25

1. (a) An Ampere is a Coulomb per second, so

$$84\,\text{A}\cdot\text{h} = \left(84\,\frac{\text{C}\cdot\text{h}}{\text{s}}\right)\left(3600\,\frac{\text{s}}{\text{h}}\right) = 3.0\times10^{5}\ \text{C}\ .$$

(b) The change in potential energy is $\Delta U = q\,\Delta V = (3.0\times10^5\,\text{C})(12\,\text{V}) = 3.6\times10^6\,\text{J}$.

2. The magnitude is $\Delta U = e\Delta V = 1.2\times10^9\,\text{eV} = 1.2\,\text{GeV}$.

3. (a) When charge q moves through a potential difference ΔV, its potential energy changes by $\Delta U = q\,\Delta V$. In this case, $\Delta U = (30\,\text{C})(1.0\times10^9\,\text{V}) = 3.0\times10^{10}\,\text{J}$.

(b) We equate the final kinetic energy $\frac{1}{2}mv^2$ of the automobile to the energy released by the lightning, denoted by $U_{\text{lightning}}$.

$$v = \sqrt{\frac{2\,U_{\text{lightning}}}{m}} = \sqrt{\frac{2(3.0\times10^{10}\,\text{J})}{1000\,\text{kg}}} = 7.7\times10^{3}\ \text{m/s}\ .$$

(c) We equate the energy required to melt mass m of ice to the energy released by the lightning: $\Delta U = mL_F$, where L_F is the heat of fusion for ice. Thus,

$$m = \frac{\Delta U}{L_F} = \frac{3.0\times10^{10}\,\text{J}}{3.33\times10^{5}\,\text{J/kg}} = 9.0\times10^{4}\ \text{kg}\ .$$

4. (a) $V_B - V_A = \Delta U/(-e) = (3.94\times10^{-19}\,\text{J})/(-1.60\times10^{-19}\,\text{C}) = -2.46\,\text{V}$.

(b) $V_C - V_A = V_B - V_A = -2.46\,\text{V}$.

(c) $V_C - V_B = 0$ (Since C and B are on the same equipotential line).

5. The electric field produced by an infinite sheet of charge has magnitude $E = \sigma/2\varepsilon_0$, where σ is the surface charge density. The field is normal to the sheet and is uniform. Place the origin of a coordinate system at the sheet and take the x axis to be parallel to the field and positive in the direction of the field. Then the electric potential is

$$V = V_s - \int_0^x E\,dx = V_s - Ex\ ,$$

where V_s is the potential at the sheet. The equipotential surfaces are surfaces of constant x; that is, they are planes that are parallel to the plane of charge. If two surfaces are separated by Δx then their potentials differ in magnitude by $\Delta V = E\Delta x = (\sigma/2\varepsilon_0)\Delta x$. Thus,

$$\Delta x = \frac{2\varepsilon_0\,\Delta V}{\sigma} = \frac{2\left(8.85\times10^{-12}\,\text{C}^2/\text{N}\cdot\text{m}^2\right)(50\,\text{V})}{0.10\times10^{-6}\,\text{C/m}^2} = 8.8\times10^{-3}\ \text{m}\ .$$

6. (a) $E = F/e = (3.9 \times 10^{-15}\,\text{N})/(1.60 \times 10^{-19}\,\text{C}) = 2.4 \times 10^4\,\text{N/C}$.

   (b) $\Delta V = E\Delta s = (2.4 \times 10^4\,\text{N/C})(0.12\,\text{m}) = 2.9 \times 10^3\,\text{V}$.

7. The potential difference between the wire and cylinder is given, not the linear charge density on the wire. We use Gauss' law to find an expression for the electric field a distance r from the center of the wire, between the wire and the cylinder, in terms of the linear charge density. Then integrate with respect to r to find an expression for the potential difference between the wire and cylinder in terms of the linear charge density. We use this result to obtain an expression for the linear charge density in terms of the potential difference and substitute the result into the equation for the electric field. This will give the electric field in terms of the potential difference and will allow you to compute numerical values for the field at the wire and at the cylinder. For the Gaussian surface use a cylinder of radius r and length ℓ, concentric with the wire and cylinder. The electric field is normal to the rounded portion of the cylinder's surface and its magnitude is uniform over that surface. This means the electric flux through the Gaussian surface is given by $2\pi r\ell E$, where E is the magnitude of the electric field. The charge enclosed by the Gaussian surface is $q = \lambda\ell$, where λ is the linear charge density on the wire. Gauss' law yields $2\pi\varepsilon_0 r\ell E = \lambda\ell$. Thus,

$$E = \frac{\lambda}{2\pi\varepsilon_0 r} .$$

Since the field is radial, the difference in the potential V_c of the cylinder and the potential V_w of the wire is

$$\Delta V = V_w - V_c = -\int_{r_c}^{r_w} E\,dr = \int_{r_w}^{r_c} \frac{\lambda}{2\pi\varepsilon_0 r}\,dr = \frac{\lambda}{2\pi\varepsilon_0}\ln\frac{r_c}{r_w} ,$$

where r_w is the radius of the wire and r_c is the radius of the cylinder. This means that

$$\lambda = \frac{2\pi\varepsilon_0\,\Delta V}{\ln(r_c/r_w)}$$

and

$$E = \frac{\lambda}{2\pi\varepsilon_0 r} = \frac{\Delta V}{r\ln(r_c/r_w)} .$$

(a) We substitute r_c for r to obtain the field at the surface of the wire:

$$\begin{aligned} E &= \frac{\Delta V}{r_w\ln(r_c/r_w)} = \frac{850\,\text{V}}{(0.65 \times 10^{-6}\,\text{m})\ln[(1.0 \times 10^{-2}\,\text{m})/(0.65 \times 10^{-6}\,\text{m})]} \\ &= 1.36 \times 10^8\ \text{V/m} . \end{aligned}$$

(b) We substitute r_c for r to find the field at the surface of the cylinder:

$$\begin{aligned} E &= \frac{\Delta V}{r_c\ln(r_c/r_w)} = \frac{850\,\text{V}}{(1.0 \times 10^{-2}\,\text{m})\ln[(1.0 \times 10^{-2}\,\text{m})/(0.65 \times 10^{-6}\,\text{m})]} \\ &= 8.82 \times 10^3\ \text{V/m} . \end{aligned}$$

8. (a) The potential as a function of r is

$$V(r) = V(0) - \int_0^r E(r)\,dr = 0 - \int_0^r \frac{qr}{4\pi\varepsilon_0 R^3}\,dr = -\frac{qr^2}{8\pi\varepsilon_0 R^3} .$$

   (b) $\Delta V = V(0) - V(R) = q/8\pi\varepsilon_0 R$.

   (c) Since $\Delta V = V(0) - V(R) > 0$, the potential at the center of the sphere is higher.

9. (a) We use Gauss' law to find expressions for the electric field inside and outside the spherical charge distribution. Since the field is radial the electric potential can be written as an integral of the field along a sphere radius, extended to infinity. Since different expressions for the field apply in different regions the integral must be split into two parts, one from infinity to the surface of the distribution and one from the surface to a point inside. Outside the charge distribution the magnitude of the field is $E = q/4\pi\varepsilon_0 r^2$ and the potential is $V = q/4\pi\varepsilon_0 r$, where r is the distance from the center of the distribution. This is the same as the field and potential of a point charge at the center of the spherical distribution. To find an expression for the magnitude of the field inside the charge distribution, we use a Gaussian surface in the form of a sphere with radius r, concentric with the distribution. The field is normal to the Gaussian surface and its magnitude is uniform over it, so the electric flux through the surface is $4\pi r^2 E$. The charge enclosed is qr^3/R^3. Gauss' law becomes

$$4\pi\varepsilon_0 r^2 E = \frac{qr^3}{R^3} ,$$

so

$$E = \frac{qr}{4\pi\varepsilon_0 R^3} .$$

If V_s is the potential at the surface of the distribution ($r = R$) then the potential at a point inside, a distance r from the center, is

$$V = V_s - \int_R^r E\,dr = V_s - \frac{q}{4\pi\varepsilon_0 R^3}\int_R^r r\,dr = V_s - \frac{qr^2}{8\pi\varepsilon_0 R^3} + \frac{q}{8\pi\varepsilon_0 R} .$$

The potential at the surface can be found by replacing r with R in the expression for the potential at points outside the distribution. It is $V_s = q/4\pi\varepsilon_0 R$. Thus,

$$V = \frac{q}{4\pi\varepsilon_0}\left[\frac{1}{R} - \frac{r^2}{2R^3} + \frac{1}{2R}\right] = \frac{q}{8\pi\varepsilon_0 R^3}(3R^2 - r^2) .$$

(b) In problem 8 the electric potential was taken to be zero at the center of the sphere. In this problem it is zero at infinity. According to the expression derived in part (a) the potential at the center of the sphere is $V_c = 3q/8\pi\varepsilon_0 R$. Thus $V - V_c = -qr^2/8\pi\varepsilon_0 R^3$. This is the result of problem 8.

(c) The potential difference is

$$\Delta V = V_s - V_c = \frac{2q}{8\pi\varepsilon_0 R} - \frac{3q}{8\pi\varepsilon_0 R} = -\frac{q}{8\pi\varepsilon_0 R} .$$

The expression obtained in problem 8 would give this same value.

(d) Only potential differences have physical significance, not the value of the potential at any particular point. The same value can be added to the potential at every point without changing the electric field, for example. Changing the reference point from the center of the distribution to infinity changes the value of the potential at every point but it does not change any potential differences.

10. (a)

$$W = \int_i^f q_0\vec{E}\cdot d\vec{s} = \frac{q_0\sigma}{2\varepsilon_0}\int_0^z dz = \frac{q_0\sigma z}{2\varepsilon_0} .$$

(b) Since $V - V_0 = -W/q_0 = -\sigma z/2\varepsilon_0$,

$$V = V_0 - \frac{\sigma z}{2\varepsilon_0} .$$

11. (a) For $r > r_2$ the field is like that of a point charge and

$$V = \frac{1}{4\pi\varepsilon_0}\frac{Q}{r} ,$$

where the zero of potential was taken to be at infinity.

(b) To find the potential in the region $r_1 < r < r_2$, first use Gauss's law to find an expression for the electric field, then integrate along a radial path from r_2 to r. The Gaussian surface is a sphere of radius r, concentric with the shell. The field is radial and therefore normal to the surface. Its magnitude is uniform over the surface, so the flux through the surface is $\Phi = 4\pi r^2 E$. The volume of the shell is $(4\pi/3)(r_2^3 - r_1^3)$, so the charge density is

$$\rho = \frac{3Q}{4\pi(r_2^3 - r_1^3)} ,$$

and the charge enclosed by the Gaussian surface is

$$q = \left(\frac{4\pi}{3}\right)(r^3 - r_1^3)\rho = Q\left(\frac{r^3 - r_1^3}{r_2^3 - r_1^3}\right) .$$

Gauss' law yields

$$4\pi\varepsilon_0 r^2 E = Q\left(\frac{r^3 - r_1^3}{r_2^3 - r_1^3}\right) \implies E = \frac{Q}{4\pi\varepsilon_0}\frac{r^3 - r_1^3}{r^2(r_2^3 - r_1^3)} .$$

If V_s is the electric potential at the outer surface of the shell ($r = r_2$) then the potential a distance r from the center is given by

$$\begin{aligned} V &= V_s - \int_{r_2}^{r} E\,dr = V_s - \frac{Q}{4\pi\varepsilon_0}\frac{1}{r_2^3 - r_1^3}\int_{r_2}^{r}\left(r - \frac{r_1^3}{r^2}\right)dr \\ &= V_s - \frac{Q}{4\pi\varepsilon_0}\frac{1}{r_2^3 - r_1^3}\left(\frac{r^2}{2} - \frac{r_2^2}{2} + \frac{r_1^3}{r} - \frac{r_1^3}{r_2}\right) . \end{aligned}$$

The potential at the outer surface is found by placing $r = r_2$ in the expression found in part (a). It is $V_s = Q/4\pi\varepsilon_0 r_2$. We make this substitution and collect terms to find

$$V = \frac{Q}{4\pi\varepsilon_0}\frac{1}{r_2^3 - r_1^3}\left(\frac{3r_2^2}{2} - \frac{r^2}{2} - \frac{r_1^3}{r}\right) .$$

Since $\rho = 3Q/4\pi(r_2^3 - r_1^3)$ this can also be written

$$V = \frac{\rho}{3\varepsilon_0}\left(\frac{3r_2^2}{2} - \frac{r^2}{2} - \frac{r_1^3}{r}\right) .$$

(c) The electric field vanishes in the cavity, so the potential is everywhere the same inside and has the same value as at a point on the inside surface of the shell. We put $r = r_1$ in the result of part (b). After collecting terms the result is

$$V = \frac{Q}{4\pi\varepsilon_0}\frac{3(r_2^2 - r_1^2)}{2(r_2^3 - r_1^3)} ,$$

or in terms of the charge density

$$V = \frac{\rho}{2\varepsilon_0}(r_2^2 - r_1^2) .$$

(d) The solutions agree at $r = r_1$ and at $r = r_2$.

12. The charge is

$$q = 4\pi\varepsilon_0 RV = \frac{(10\,\text{m})(-1.0\,\text{V})}{8.99 \times 10^9\,\frac{\text{N}\cdot\text{m}^2}{\text{C}^2}} = -1.1 \times 10^{-9}\ \text{C} .$$

13. (a) The potential difference is

$$\begin{aligned} V_A - V_B &= \frac{q}{4\pi\varepsilon_0 r_A} - \frac{q}{4\pi\varepsilon_0 r_B} \\ &= (1.0 \times 10^{-6}\,\mathrm{C})\left(8.99 \times 10^9\,\frac{\mathrm{N\cdot m^2}}{\mathrm{C^2}}\right)\left(\frac{1}{2.0\,\mathrm{m}} - \frac{1}{1.0\,\mathrm{m}}\right) = -4500\ \mathrm{V}\ . \end{aligned}$$

(b) Since $V(r)$ depends only on the magnitude of $\vec{r}$, the result is unchanged.

14. In the sketches shown below, the lines with the arrows are field lines and those without are the equipotentials (which become more circular the closer one gets to the individual charges) . In all pictures, q_2 is on the left and q_1 is on the right (which is reversed from the way it is shown in the textbook).

(a)

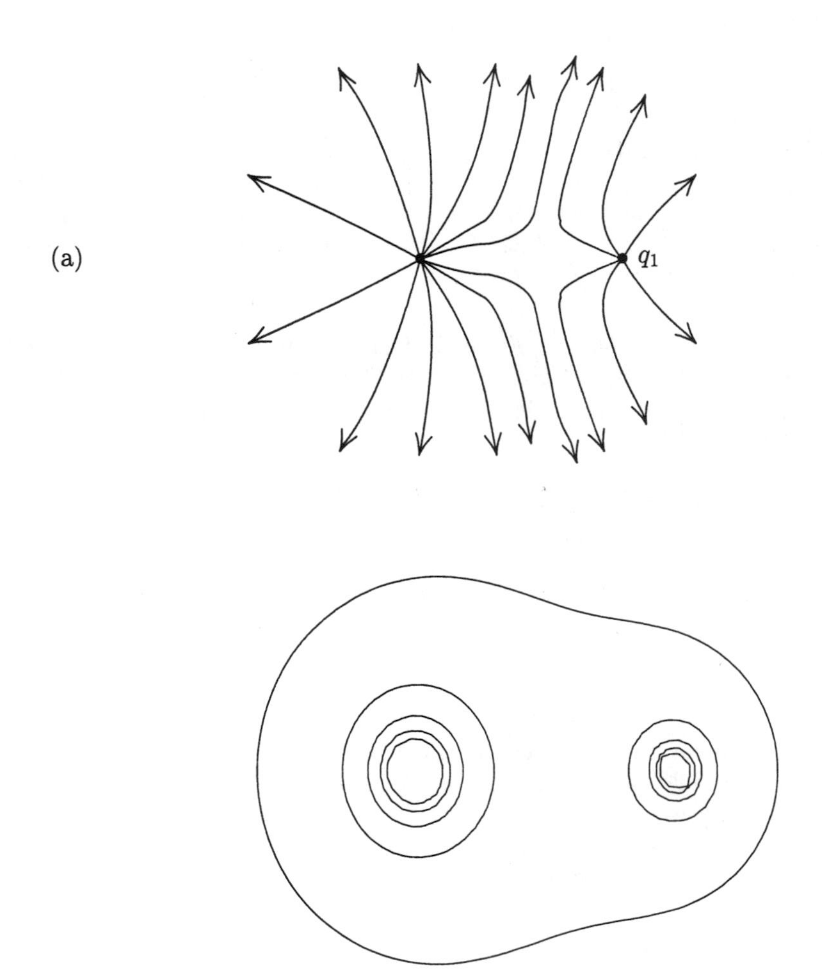

(b)

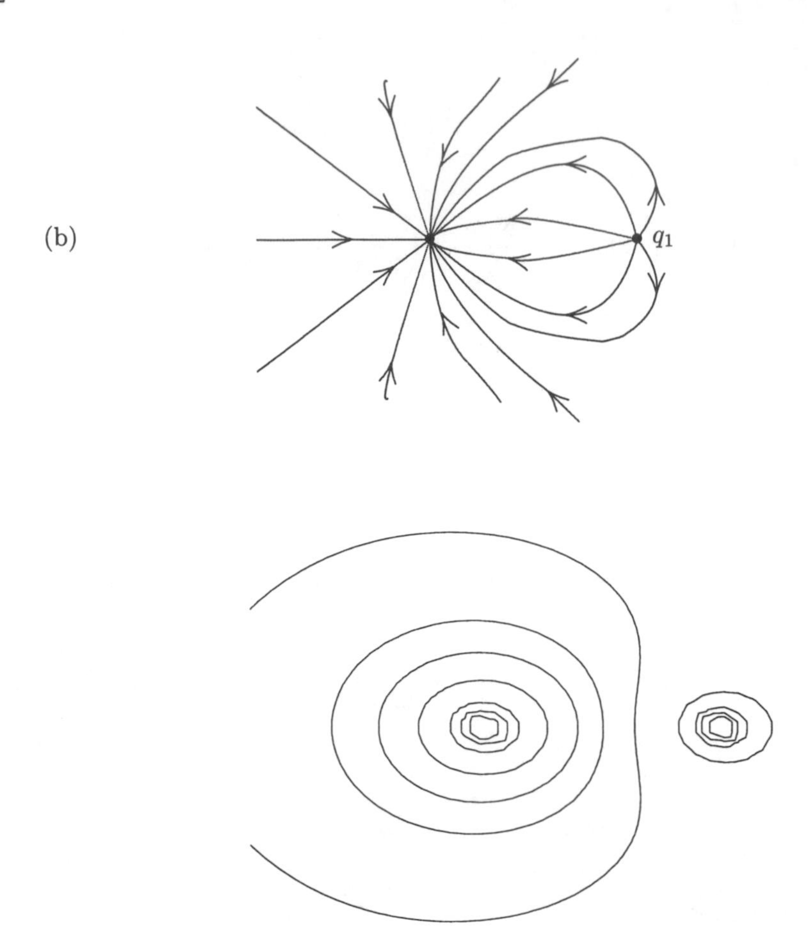

15. First, we observe that $V(x)$ cannot be equal to zero for $x > d$. In fact $V(x)$ is always negative for $x > d$. Now we consider the two remaining regions on the x axis: $x < 0$ and $0 < x < d$. For $x < 0$ the separation between q_1 and a point on the x axis whose coordinate is x is given by $d_1 = -x$; while the corresponding separation for q_2 is $d_2 = d - x$. We set

$$V(x) = k\left(\frac{q_1}{d_1} + \frac{q_2}{d_2}\right) = \frac{q}{4\pi\varepsilon_0}\left(\frac{1}{-x} + \frac{-3}{d-x}\right) = 0$$

to obtain $x = -d/2$. Similarly, for $0 < x < d$ we have $d_1 = x$ and $d_2 = d - x$. Let

$$V(x) = k\left(\frac{q_1}{d_1} + \frac{q_2}{d_2}\right) = \frac{q}{4\pi\varepsilon_0}\left(\frac{1}{x} + \frac{-3}{d-x}\right) = 0$$

and solve: $x = d/4$.

16. Since according to the problem statement there is a point in between the two charges on the x axis where the net electric field is zero, the fields at that point due to q_1 and q_2 must be directed opposite to each other. This means that q_1 and q_2 must have the same sign (i.e., either both are positive or both negative). Thus, the potentials due to either of them must be of the same sign. Therefore, the net electric potential cannot possibly be zero anywhere except at infinity.

17. (a) The electric potential V at the surface of the drop, the charge q on the drop, and the radius R of the drop are related by $V = q/4\pi\varepsilon_0 R$. Thus

$$R = \frac{q}{4\pi\varepsilon_0 V} = \frac{(8.99 \times 10^9\ \mathrm{N\cdot m^2/C^2})(30 \times 10^{-12}\ \mathrm{C})}{500\ \mathrm{V}} = 5.4 \times 10^{-4}\ \mathrm{m}\ .$$

(b) After the drops combine the total volume is twice the volume of an original drop, so the radius R' of the combined drop is given by $(R')^3 = 2R^3$ and $R' = 2^{1/3}R$. The charge is twice the charge of original drop: $q' = 2q$. Thus,

$$V' = \frac{1}{4\pi\varepsilon_0}\frac{q'}{R'} = \frac{1}{4\pi\varepsilon_0}\frac{2q}{2^{1/3}R} = 2^{2/3}V = 2^{2/3}(500\,\text{V}) \approx 790\ \text{V}\ .$$

18. (a) The charge on the sphere is

$$q = 4\pi\varepsilon_0 VR = \frac{(200\,\text{V})(0.15\,\text{m})}{8.99\times 10^9\ \frac{\text{N}\cdot\text{m}^2}{\text{C}^2}} = 3.3\times 10^{-9}\ \text{C}\ .$$

(b) The (uniform) surface charge density (charge divided by the area of the sphere) is

$$\sigma = \frac{q}{4\pi R^2} = \frac{3.3\times 10^{-9}\,\text{C}}{4\pi(0.15\,\text{m})^2} = 1.2\times 10^{-8}\ \text{C/m}^2\ .$$

19. Assume the charge on Earth is distributed with spherical symmetry. If the electric potential is zero at infinity then at the surface of Earth it is $V = q/4\pi\varepsilon_0 R$, where q is the charge on Earth and $R = 6.37\times 10^6$ m is the radius of Earth. The magnitude of the electric field at the surface is $E = q/4\pi\varepsilon_0 R^2$, so $V = ER = (100\,\text{V/m})(6.37\times 10^6\,\text{m}) = 6.4\times 10^8\,\text{V}$.

20. The net electric potential at point P is the sum of those due to the six charges:

$$\begin{aligned} V_P &= \sum_{i=1}^{6} V_{Pi} = \sum_{i=1}^{6}\frac{q_i}{4\pi\varepsilon_0 r_i} \\ &= \frac{1}{4\pi\varepsilon_0}\left[\frac{5.0q}{\sqrt{d^2+(d/2)^2}} + \frac{-2.0q}{d/2} + \frac{-3.0q}{\sqrt{d^2+(d/2)^2}}\right. \\ &\qquad \left. + \frac{3.0q}{\sqrt{d^2+(d/2)^2}} + \frac{-2.0q}{d/2} + \frac{-5.0q}{\sqrt{d^2+(d/2)^2}}\right] \\ &= \frac{-0.94q}{4\pi\varepsilon_0 d}\ . \end{aligned}$$

21. A charge $-5q$ is a distance $2d$ from P, a charge $-5q$ is a distance d from P, and two charges $+5q$ are each a distance d from P, so the electric potential at P is

$$V = \frac{q}{4\pi\varepsilon_0}\left[-\frac{5}{2d} - \frac{5}{d} + \frac{5}{d} + \frac{5}{d}\right] = \frac{5q}{8\pi\varepsilon_0}\ .$$

The zero of the electric potential was taken to be at infinity.

22. We use Eq. 25-20:

$$V = \frac{1}{4\pi\varepsilon_0}\frac{p}{r^2} = \frac{\left(8.99\times 10^9\ \frac{\text{N}\cdot\text{m}^2}{\text{C}^2}\right)(1.47\times 3.34\times 10^{-30}\,\text{C}\cdot\text{m})}{(52.0\times 10^{-9}\,\text{m})^2} = 1.63\times 10^{-5}\ \text{V}\ .$$

23. A positive charge q is a distance $r-d$ from P, another positive charge q is a distance r from P, and a negative charge $-q$ is a distance $r+d$ from P. Sum the individual electric potentials created at P to find the total:

$$V = \frac{q}{4\pi\varepsilon_0}\left[\frac{1}{r-d} + \frac{1}{r} - \frac{1}{r+d}\right]\ .$$

We use the binomial theorem to approximate $1/(r-d)$ for r much larger than d:

$$\frac{1}{r-d} = (r-d)^{-1} \approx (r)^{-1} - (r)^{-2}(-d) = \frac{1}{r} + \frac{d}{r^2}\ .$$

Similarly,

$$\frac{1}{r+d} \approx \frac{1}{r} - \frac{d}{r^2}.$$

Only the first two terms of each expansion were retained. Thus,

$$V \approx \frac{q}{4\pi\varepsilon_0}\left[\frac{1}{r} + \frac{d}{r^2} + \frac{1}{r} - \frac{1}{r} + \frac{d}{r^2}\right] = \frac{q}{4\pi\varepsilon_0}\left[\frac{1}{r} + \frac{2d}{r^2}\right] = \frac{q}{4\pi\varepsilon_0 r}\left[1 + \frac{2d}{r}\right].$$

24. (a) From Eq. 25-35

$$V = 2\frac{\lambda}{4\pi\varepsilon_0}\ln\left[\frac{L/2 + \sqrt{(L^2/4) + d^2}}{d}\right].$$

(b) The potential at P is $V = 0$ due to superposition.

25. (a) All the charge is the same distance R from C, so the electric potential at C is

$$V = \frac{1}{4\pi\varepsilon_0}\left[\frac{Q}{R} - \frac{6Q}{R}\right] = -\frac{5Q}{4\pi\varepsilon_0 R},$$

where the zero was taken to be at infinity.

(b) All the charge is the same distance from P. That distance is $\sqrt{R^2 + z^2}$, so the electric potential at P is

$$V = \frac{1}{4\pi\varepsilon_0}\left[\frac{Q}{\sqrt{R^2+z^2}} - \frac{6Q}{\sqrt{R^2+z^2}}\right] = -\frac{5Q}{4\pi\varepsilon_0\sqrt{R^2+z^2}}.$$

26. The potential is

$$V_P = \frac{1}{4\pi\varepsilon_0}\int_{\text{rod}}\frac{dq}{R} = \frac{1}{4\pi\varepsilon_0 R}\int_{\text{rod}} dq = \frac{-Q}{4\pi\varepsilon_0 R}.$$

We note that the result is exactly what one would expect for a point-charge $-Q$ at a distance R. This "coincidence" is due, in part, to the fact that V is a scalar quantity.

27. The disk is uniformly charged. This means that when the full disk is present each quadrant contributes equally to the electric potential at P, so the potential at P due to a single quadrant is one-fourth the potential due to the entire disk. First find an expression for the potential at P due to the entire disk. We consider a ring of charge with radius r and (infinitesimal) width dr. Its area is $2\pi r\,dr$ and it contains charge $dq = 2\pi\sigma r\,dr$. All the charge in it is a distance $\sqrt{r^2+z^2}$ from P, so the potential it produces at P is

$$dV = \frac{1}{4\pi\varepsilon_0}\frac{2\pi\sigma r\,dr}{\sqrt{r^2+z^2}} = \frac{\sigma r\,dr}{2\varepsilon_0\sqrt{r^2+z^2}}.$$

The total potential at P is

$$V = \frac{\sigma}{2\varepsilon_0}\int_0^R \frac{r\,dr}{\sqrt{r^2+z^2}} = \frac{\sigma}{2\varepsilon_0}\sqrt{r^2+z^2}\bigg|_0^R = \frac{\sigma}{2\varepsilon_0}\left[\sqrt{R^2+z^2} - z\right].$$

The potential V_{sq} at P due to a single quadrant is

$$V_{sq} = \frac{V}{4} = \frac{\sigma}{8\varepsilon_0}\left[\sqrt{R^2+z^2} - z\right].$$

28. Consider an infinitesimal segment of the rod, located between x and $x+dx$. It has length dx and contains charge $dq = \lambda\,dx$, where $\lambda = Q/L$ is the linear charge density of the rod. Its distance from P_1 is $d+x$ and the potential it creates at P_1 is

$$dV = \frac{1}{4\pi\varepsilon_0}\frac{dq}{d+x} = \frac{1}{4\pi\varepsilon_0}\frac{\lambda\,dx}{d+x}.$$

To find the total potential at P_1, integrate over the rod:

$$V = \frac{\lambda}{4\pi\varepsilon_0}\int_0^L \frac{dx}{d+x} = \frac{\lambda}{4\pi\varepsilon_0}\ln(d+x)\Big|_0^L = \frac{Q}{4\pi\varepsilon_0 L}\ln\left(1+\frac{L}{d}\right) .$$

29. Consider an infinitesimal segment of the rod, located between x and $x+dx$. It has length dx and contains charge $dq = \lambda\,dx = cx\,dx$. Its distance from P_1 is $d+x$ and the potential it creates at P_1 is

$$dV = \frac{1}{4\pi\varepsilon_0}\frac{dq}{d+x} = \frac{1}{4\pi\varepsilon_0}\frac{cx\,dx}{d+x} .$$

To find the total potential at P_1, integrate over the rod:

$$V = \frac{c}{4\pi\varepsilon_0}\int_0^L \frac{x\,dx}{d+x} = \frac{c}{4\pi\varepsilon_0}[x - d\ln(x+d)]\Big|_0^L = \frac{c}{4\pi\varepsilon_0}\left[L - d\ln\left(1+\frac{L}{d}\right)\right] .$$

30. The magnitude of the electric field is given by

$$|E| = \left|-\frac{\Delta V}{\Delta x}\right| = \frac{2(5.0\,\text{V})}{0.015\,\text{m}} = 6.7\times 10^2\ \text{V/m} .$$

At any point in the region between the plates, $\vec{E}$ points away from the positively charged plate, directly towards the negatively charged one.

31. We use Eq. 25-41:

$$\begin{aligned} E_x(x,y) &= -\frac{\partial V}{\partial x} = -\frac{\partial}{\partial x}\left((2.0\,\text{V/m}^2)x^2 - (3.0\,\text{V/m}^2)y^2\right) = -2(2.0\,\text{V/m}^2)x ; \\ E_y(x,y) &= -\frac{\partial V}{\partial y} = -\frac{\partial}{\partial y}\left((2.0\,\text{V/m}^2)x^2 - (3.0\,\text{V/m}^2)y^2\right) = 2(3.0\,\text{V/m}^2)y . \end{aligned}$$

We evaluate at $x = 3.0\,\text{m}$ and $y = 2.0\,\text{m}$ to obtain the magnitude of $\vec{E}$:

$$E = \sqrt{E_x^2 + E_y^2} = 17\ \text{V/m} .$$

$\vec{E}$ makes an angle θ with the positive x axis, where

$$\theta = \tan^{-1}\left(\frac{E_y}{E_x}\right) = 135^\circ .$$

32. We use Eq. 25-41. This is an ordinary derivative since the potential is a function of only one variable.

$$\begin{aligned} \vec{E} &= -\left(\frac{dV}{dx}\right)\hat{\imath} = -\frac{d}{dx}(1500x^2)\hat{\imath} = (-3000x)\hat{\imath} \\ &= (-3000\,\text{V/m}^2)(0.0130\,\text{m})\hat{\imath} = (-39\,\text{V/m})\hat{\imath} . \end{aligned}$$

33. (a) The charge on every part of the ring is the same distance from any point P on the axis. This distance is $r = \sqrt{z^2+R^2}$, where R is the radius of the ring and z is the distance from the center of the ring to P. The electric potential at P is

$$V = \frac{1}{4\pi\varepsilon_0}\int\frac{dq}{r} = \frac{1}{4\pi\varepsilon_0}\int\frac{dq}{\sqrt{z^2+R^2}} = \frac{1}{4\pi\varepsilon_0}\frac{1}{\sqrt{z^2+R^2}}\int dq = \frac{1}{4\pi\varepsilon_0}\frac{q}{\sqrt{z^2+R^2}} .$$

(b) The electric field is along the axis and its component is given by

$$\begin{aligned} E &= -\frac{\partial V}{\partial z} = -\frac{q}{4\pi\varepsilon_0}\frac{\partial}{\partial z}(z^2+R^2)^{-1/2} \\ &= \frac{q}{4\pi\varepsilon_0}\left(\frac{1}{2}\right)(z^2+R^2)^{-3/2}(2z) = \frac{q}{4\pi\varepsilon_0}\frac{z}{(z^2+R^2)^{3/2}} . \end{aligned}$$

This agrees with Eq. 23-16.

34. (a) Consider an infinitesimal segment of the rod from x to $x + dx$. Its contribution to the potential at point P_2 is

$$dV = \frac{1}{4\pi\varepsilon_0}\frac{\lambda(x)dx}{\sqrt{x^2+y^2}} = \frac{1}{4\pi\varepsilon_0}\frac{cx}{\sqrt{x^2+y^2}}\,dx .$$

Thus,

$$V = \int_{\text{rod}} dV_P = \frac{c}{4\pi\varepsilon_0}\int_0^L \frac{x}{\sqrt{x^2+y^2}}\,dx = \frac{c}{4\pi\varepsilon_0}(\sqrt{L^2+y^2}-y) .$$

(b) The y component of the field there is

$$E_y = -\frac{\partial V_P}{\partial y} = -\frac{c}{4\pi\varepsilon_0}\frac{d}{dy}(\sqrt{L^2+y^2}-y) = \frac{c}{4\pi\varepsilon_0}\left(1-\frac{y}{\sqrt{L^2+y^2}}\right) .$$

(c) We obtained above the value of the potential at any point P strictly on the y-axis. In order to obtain $E_x(x, y)$ we need to first calculate $V(x, y)$. That is, we must find the potential for an arbitrary point located at (x, y). Then $E_x(x, y)$ can be obtained from $E_x(x, y) = -\partial V(x, y)/\partial x$.

35. (a) According to the result of problem 28, the electric potential at a point with coordinate x is given by

$$V = \frac{Q}{4\pi\varepsilon_0 L}\ln\left(\frac{x-L}{x}\right) .$$

We differentiate the potential with respect to x to find the x component of the electric field:

$$\begin{aligned} E_x &= -\frac{\partial V}{\partial x} = -\frac{Q}{4\pi\varepsilon_0 L}\frac{\partial}{\partial x}\ln\left(\frac{x-L}{x}\right) = -\frac{Q}{4\pi\varepsilon_0 L}\frac{x}{x-L}\left(\frac{1}{x}-\frac{x-L}{x^2}\right) \\ &= -\frac{Q}{4\pi\varepsilon_0 x(x-L)} . \end{aligned}$$

At $x = -d$ we obtain

$$E_x = -\frac{Q}{4\pi\varepsilon_0 d(d+L)} .$$

(b) Consider two points an equal infinitesimal distance on either side of P_1, along a line that is perpendicular to the x axis. The difference in the electric potential divided by their separation gives the transverse component of the electric field. Since the two points are situated symmetrically with respect to the rod, their potentials are the same and the potential difference is zero. Thus the transverse component of the electric field is zero.

36. (a) We use Eq. 25-43 with $q_1 = q_2 = -e$ and $r = 2.00\,\text{nm}$:

$$U = k\frac{q_1 q_2}{r} = k\frac{e^2}{r} = \frac{\left(8.99\times10^9\,\frac{\text{N}\cdot\text{m}^2}{\text{C}^2}\right)(1.60\times10^{-19}\,\text{C})^2}{2.00\times10^{-9}\,\text{m}} = 1.15\times10^{-19}\ \text{J} .$$

(b) Since $U > 0$ and $U \propto r^{-1}$ the potential energy U decreases as r increases.

37. We choose the zero of electric potential to be at infinity. The initial electric potential energy U_i of the system before the particles are brought together is therefore zero. After the system is set up the final potential energy is

$$\begin{aligned} U_f &= \frac{q^2}{4\pi\varepsilon_0}\left(-\frac{1}{a}-\frac{1}{a}+\frac{1}{\sqrt{2}a}-\frac{1}{a}-\frac{1}{a}+\frac{1}{\sqrt{2}a}\right) \\ &= \frac{2q^2}{4\pi\varepsilon_0 a}\left(\frac{1}{\sqrt{2}}-2\right) = -\frac{0.21q^2}{\varepsilon_0 a} . \end{aligned}$$

Thus the amount of work required to set up the system is given by $W = \Delta U = U_f - U_i = -0.21q^2/(\varepsilon_0 a)$.

38. The electric potential energy is

$$\begin{aligned} U &= k\sum_{i\neq j}\frac{q_i q_j}{r_{ij}} = \frac{1}{4\pi\varepsilon_0 d}\left(q_1q_2 + q_1q_3 + q_2q_4 + q_3q_4 + \frac{q_1q_4}{\sqrt{2}} + \frac{q_2q_3}{\sqrt{2}}\right) \\ &= \frac{\left(8.99\times10^9\,\frac{\text{N}\cdot\text{m}^2}{\text{C}^2}\right)}{1.3\,\text{m}}\Bigg[(12)(-24) + (12)(31) + (-24)(17) + (31)(17) \\ &\quad + \frac{(12)(17)}{\sqrt{2}} + \frac{(-24)(31)}{\sqrt{2}}\Bigg](10^{-19}\,\text{C})^2 \\ &= -1.2\times10^{-6}\ \text{J} . \end{aligned}$$

39. (a) Let $\ell = 0.15$ m be the length of the rectangle and $w = 0.050$ m be its width. Charge q_1 is a distance ℓ from point A and charge q_2 is a distance w, so the electric potential at A is

$$\begin{aligned} V_A &= \frac{1}{4\pi\varepsilon_0}\left[\frac{q_1}{\ell}+\frac{q_2}{w}\right] \\ &= (8.99\times10^9\,\text{N}\cdot\text{m}^2/\text{C}^2)\left[\frac{-5.0\times10^{-6}\,\text{C}}{0.15\,\text{m}}+\frac{2.0\times10^{-6}\,\text{C}}{0.050\,\text{m}}\right] \\ &= 6.0\times10^4\ \text{V} . \end{aligned}$$

(b) Charge q_1 is a distance w from point b and charge q_2 is a distance ℓ, so the electric potential at B is

$$\begin{aligned} V_B &= \frac{1}{4\pi\varepsilon_0}\left[\frac{q_1}{w}+\frac{q_2}{\ell}\right] \\ &= (8.99\times10^9\,\text{N}\cdot\text{m}^2/\text{C}^2)\left[\frac{-5.0\times10^{-6}\,\text{C}}{0.050\,\text{m}}+\frac{2.0\times10^{-6}\,\text{C}}{0.15\,\text{m}}\right] \\ &= -7.8\times10^5\ \text{V} . \end{aligned}$$

(c) Since the kinetic energy is zero at the beginning and end of the trip, the work done by an external agent equals the change in the potential energy of the system. The potential energy is the product of the charge q_3 and the electric potential. If U_A is the potential energy when q_3 is at A and U_B is the potential energy when q_3 is at B, then the work done in moving the charge from B to A is $W = U_A - U_B = q_3(V_A - V_B) = (3.0\times10^{-6}\,\text{C})(6.0\times10^4\,\text{V} + 7.8\times10^5\,\text{V}) = 2.5\,\text{J}$.

(d) The work done by the external agent is positive, so the energy of the three-charge system increases.

(e) and (f) The electrostatic force is conservative, so the work is the same no matter which path is used.

40. The work required is

$$W = \Delta U = \frac{1}{4\pi\varepsilon_0}\left[\frac{(4q)(5q)}{2d}+\frac{(5q)(-2q)}{d}\right] = 0 .$$

41. The particle with charge $-q$ has both potential and kinetic energy, and both of these change when the radius of the orbit is changed. We first find an expression for the total energy in terms of the orbit radius r. Q provides the centripetal force required for $-q$ to move in uniform circular motion. The magnitude of the force is $F = Qq/4\pi\varepsilon_0 r^2$. The acceleration of $-q$ is v^2/r, where v is its speed. Newton's second law yields

$$\frac{Qq}{4\pi\varepsilon_0 r^2} = \frac{mv^2}{r} \implies mv^2 = \frac{Qq}{4\pi\varepsilon_0 r} ,$$

and the kinetic energy is $K = \frac{1}{2}mv^2 = Qq/8\pi\varepsilon_0 r$. The potential energy is $U = -Qq/4\pi\varepsilon_0 r$, and the total energy is

$$E = K + U = \frac{Qq}{8\pi\varepsilon_0 r} - \frac{Qq}{4\pi\varepsilon_0 r} = -\frac{Qq}{8\pi\varepsilon_0 r} .$$

When the orbit radius is r_1 the energy is $E_1 = -Qq/8\pi\varepsilon_0 r_1$ and when it is r_2 the energy is $E_2 = -Qq/8\pi\varepsilon_0 r_2$. The difference $E_2 - E_1$ is the work W done by an external agent to change the radius:

$$W = E_2 - E_1 = -\frac{Qq}{8\pi\varepsilon_0}\left(\frac{1}{r_2} - \frac{1}{r_1}\right) = \frac{Qq}{8\pi\varepsilon_0}\left(\frac{1}{r_1} - \frac{1}{r_2}\right) .$$

42. (a) The potential is

$$\begin{aligned} V(r) &= \frac{1}{4\pi\varepsilon_0}\frac{e}{r} \\ &= \frac{\left(8.99 \times 10^9 \frac{\text{N}\cdot\text{m}^2}{\text{C}^2}\right)(1.60 \times 10^{-19}\text{ C})}{5.29 \times 10^{-11}\text{ m}} = 27.2\text{ V} . \end{aligned}$$

(b) The potential energy is $U = -eV(r) = -27.2\,\text{eV}$.

(c) Since $m_e v^2/r = -e^2/4\pi\varepsilon_0 r^2$,

$$K = \frac{1}{2}mv^2 = -\frac{1}{2}\left(\frac{e^2}{4\pi\varepsilon_0 r}\right) = -\frac{1}{2}V(r) = \frac{27.2\,\text{eV}}{2} = 13.6\text{ eV} .$$

(d) The energy required is

$$\Delta E = 0 - [V(r) + K] = 0 - (-27.2\,\text{eV} + 13.6\,\text{eV}) = 13.6\text{ eV} .$$

43. We use the conservation of energy principle. The initial potential energy is $U_i = q^2/4\pi\varepsilon_0 r_1$, the initial kinetic energy is $K_i = 0$, the final potential energy is $U_f = q^2/4\pi\varepsilon_0 r_2$, and the final kinetic energy is $K_f = \frac{1}{2}mv^2$, where v is the final speed of the particle. Conservation of energy yields

$$\frac{q^2}{4\pi\varepsilon_0 r_1} = \frac{q^2}{4\pi\varepsilon_0 r_2} + \frac{1}{2}mv^2 .$$

The solution for v is

$$\begin{aligned} v &= \sqrt{\frac{2q^2}{4\pi\varepsilon_0 m}\left(\frac{1}{r_1} - \frac{1}{r_2}\right)} \\ &= \sqrt{\frac{(8.99 \times 10^9\text{N}\cdot\text{m}^2/\text{C}^2)(2)(3.1 \times 10^{-6}\text{ C})^2}{20 \times 10^{-6}\text{ kg}}\left(\frac{1}{0.90 \times 10^{-3}\text{ m}} - \frac{1}{2.5 \times 10^{-3}\text{ m}}\right)} \\ &= 2.5 \times 10^3\text{ m/s} . \end{aligned}$$

44. Let $r = 1.5\,\text{m}$, $x = 3.0\,\text{m}$, $q_1 = -9.0\,\text{nC}$, and $q_2 = -6.0\,\text{pC}$. The work done by an external agent is given by

$$\begin{aligned} W &= \Delta U = \frac{q_1 q_2}{4\pi\varepsilon_0}\left(\frac{1}{r} - \frac{1}{\sqrt{r^2+x^2}}\right) \\ &= (-9.0\times10^{-9}\,\text{C})(-6.0\times10^{-12}\,\text{C})\left(8.99\times10^9\,\frac{\text{N}\cdot\text{m}^2}{\text{C}^2}\right)\cdot\left[\frac{1}{1.5\,\text{m}} - \frac{1}{\sqrt{(1.5\,\text{m})^2+(3.0\,\text{m})^2}}\right] \\ &= 1.8\times10^{-10}\,\text{J}\,. \end{aligned}$$

45. (a) The potential energy is

$$U = \frac{q^2}{4\pi\varepsilon_0 d} = \frac{(8.99\times10^9\,\text{N}\cdot\text{m}^2/\text{C}^2)(5.0\times10^{-6}\,\text{C})^2}{1.00\,\text{m}} = 0.225\,\text{J}$$

relative to the potential energy at infinite separation.

(b) Each sphere repels the other with a force that has magnitude

$$F = \frac{q^2}{4\pi\varepsilon_0 d^2} = \frac{(8.99\times10^9\,\text{N}\cdot\text{m}^2/\text{C}^2)(5.0\times10^{-6}\,\text{C})^2}{(1.00\,\text{m})^2} = 0.225\,\text{N}\,.$$

According to Newton's second law the acceleration of each sphere is the force divided by the mass of the sphere. Let m_A and m_B be the masses of the spheres. The acceleration of sphere A is

$$a_A = \frac{F}{m_A} = \frac{0.225\,\text{N}}{5.0\times10^{-3}\,\text{kg}} = 45.0\,\text{m/s}^2$$

and the acceleration of sphere B is

$$a_B = \frac{F}{m_B} = \frac{0.225\,\text{N}}{10\times10^{-3}\,\text{kg}} = 22.5\,\text{m/s}^2\,.$$

(c) Energy is conserved. The initial potential energy is $U = 0.225\,\text{J}$, as calculated in part (a). The initial kinetic energy is zero since the spheres start from rest. The final potential energy is zero since the spheres are then far apart. The final kinetic energy is $\frac{1}{2}m_A v_A^2 + \frac{1}{2}m_B v_B^2$, where v_A and v_B are the final velocities. Thus,

$$U = \frac{1}{2}m_A v_A^2 + \frac{1}{2}m_B v_B^2\,.$$

Momentum is also conserved, so

$$0 = m_A v_A + m_B v_B\,.$$

These equations may be solved simultaneously for v_A and v_B. Substituting $v_B = -(m_A/m_B)v_A$, from the momentum equation into the energy equation, and collecting terms, we obtain $U = \frac{1}{2}(m_A/m_B)(m_A+m_B)v_A^2$. Thus,

$$\begin{aligned} v_A &= \sqrt{\frac{2Um_B}{m_A(m_A+m_B)}} \\ &= \sqrt{\frac{2(0.225\,\text{J})(10\times10^{-3}\,\text{kg})}{(5.0\times10^{-3}\,\text{kg})(5.0\times10^{-3}\,\text{kg}+10\times10^{-3}\,\text{kg})}} = 7.75\,\text{m/s}\,. \end{aligned}$$

We thus obtain

$$v_B = -\frac{m_A}{m_B}v_A = -\left(\frac{5.0\times10^{-3}\,\text{kg}}{10\times10^{-3}\,\text{kg}}\right)(7.75\,\text{m/s}) = -3.87\,\text{m/s}\,.$$

46. The change in electric potential energy of the electron-shell system as the electron starts from its initial position and just reaches the shell is $\Delta U = (-e)(-V) = eV$. Thus from $\Delta U = K = \frac{1}{2}m_e v_i^2$ we find the initial electron speed to be

$$v_i = \sqrt{\frac{2\Delta U}{m_e}} = \sqrt{\frac{2eV}{m_e}} \ .$$

47. We use conservation of energy, taking the potential energy to be zero when the moving electron is far away from the fixed electrons. The final potential energy is then $U_f = 2e^2/4\pi\varepsilon_0 d$, where d is half the distance between the fixed electrons. The initial kinetic energy is $K_i = \frac{1}{2}mv^2$, where m is the mass of an electron and v is the initial speed of the moving electron. The final kinetic energy is zero. Thus $K_i = U_f$ or $\frac{1}{2}mv^2 = 2e^2/4\pi\varepsilon_0 d$. Hence

$$v = \sqrt{\frac{4e^2}{4\pi\varepsilon_0\, dm}} = \sqrt{\frac{(8.99\times10^9\,\mathrm{N\cdot m^2/C^2})(4)(1.60\times10^{-19}\,\mathrm{C})^2}{(0.010\,\mathrm{m})(9.11\times10^{-31}\,\mathrm{kg})}} = 3.2\times10^2\ \mathrm{m/s}\ .$$

48. The initial speed v_i of the electron satisfies $K_i = \frac{1}{2}m_e v_i^2 = e\Delta V$, which gives

$$v_i = \sqrt{\frac{2e\Delta V}{m_e}} = \sqrt{\frac{2(1.60\times10^{-19}\,\mathrm{J})(625\,\mathrm{V})}{9.11\times10^{-31}\,\mathrm{kg}}} = 1.48\times10^7\ \mathrm{m/s}\ .$$

49. Let the distance in question be r. The initial kinetic energy of the electron is $K_i = \frac{1}{2}m_e v_i^2$, where $v_i = 3.2\times10^5\,\mathrm{m/s}$. As the speed doubles, K becomes $4K_i$. Thus

$$\Delta U = \frac{-e^2}{4\pi\varepsilon_0 r} = -\Delta K = -(4K_i - K_i) = -3K_i = -\frac{3}{2}m_e v_i^2\ ,$$

or

$$\begin{aligned} r &= \frac{2e^2}{3(4\pi\varepsilon_0)m_e v_i^2} = \frac{2(1.6\times10^{-19}\,\mathrm{C})^2\left(8.99\times10^9\,\frac{\mathrm{N\cdot m^2}}{\mathrm{C^2}}\right)}{3(9.11\times10^{-19}\,\mathrm{kg})(3.2\times10^5\,\mathrm{m/s})^2} \\ &= 1.6\times10^{-9}\ \mathrm{m}\ . \end{aligned}$$

50. Since the electric potential throughout the entire conductor is a constant, the electric potential at its center is also $+400\,\mathrm{V}$.

51. If the electric potential is zero at infinity, then the potential at the surface of the sphere is given by $V = q/4\pi\varepsilon_0 r$, where q is the charge on the sphere and r is its radius. Thus

$$q = 4\pi\varepsilon_0 rV = \frac{(0.15\,\mathrm{m})(1500\,\mathrm{V})}{8.99\times10^9\,\mathrm{N\cdot m^2/C^2}} = 2.5\times10^{-8}\ \mathrm{C}\ .$$

52. (a) Since the two conductors are connected V_1 and V_2 must be the same.
   (b) Let $V_1 = q_1/4\pi\varepsilon_0 R_1 = V_2 = q_2/4\pi\varepsilon_0 R_2$ and note that $q_1 + q_2 = q$ and $R_2 = 2R_1$. We solve for q_1 and q_2: $q_1 = q/3$, $q_2 = 2q/3$.
   (c) The ratio of surface charge densities is

$$\frac{\sigma_1}{\sigma_2} = \frac{q_1/4\pi R_1^2}{q_2/4\pi R_2^2} = \left(\frac{q_1}{q_2}\right)\left(\frac{R_2}{R_1}\right)^2 = 2\ .$$

53. (a) The electric potential is the sum of the contributions of the individual spheres. Let q_1 be the charge on one, q_2 be the charge on the other, and d be their separation. The point halfway between them is the same distance $d/2$ ($= 1.0\,\mathrm{m}$) from the center of each sphere, so the potential at the halfway point is

$$V = \frac{q_1 + q_2}{4\pi\varepsilon_0 d/2} = \frac{(8.99\times10^9\,\mathrm{N\cdot m^2/C^2})(1.0\times10^{-8}\,\mathrm{C} - 3.0\times10^{-8}\,\mathrm{C})}{1.0\,\mathrm{m}} = -1.80\times10^2\ \mathrm{V}\ .$$

(b) The distance from the center of one sphere to the surface of the other is $d-R$, where R is the radius of either sphere. The potential of either one of the spheres is due to the charge on that sphere and the charge on the other sphere. The potential at the surface of sphere 1 is

$$\begin{aligned} V_1 &= \frac{1}{4\pi\varepsilon_0}\left[\frac{q_1}{R}+\frac{q_2}{d-R}\right] \\ &= (8.99\times10^9\,\mathrm{N\cdot m^2/C^2})\left[\frac{1.0\times10^{-8}\,\mathrm{C}}{0.030\,\mathrm{m}}-\frac{3.0\times10^{-8}\,\mathrm{C}}{2.0\,\mathrm{m}-0.030\,\mathrm{m}}\right] \\ &= 2.9\times10^3\ \mathrm{V}\ . \end{aligned}$$

The potential at the surface of sphere 2 is

$$\begin{aligned} V_2 &= \frac{1}{4\pi\varepsilon_0}\left[\frac{q_1}{d-R}+\frac{q_2}{R}\right] \\ &= (8.99\times10^9\,\mathrm{N\cdot m^2/C^2})\left[\frac{1.0\times10^{-8}\,\mathrm{C}}{2.0\,\mathrm{m}-0.030\,\mathrm{m}}-\frac{3.0\times10^{-8}\,\mathrm{C}}{0.030\,\mathrm{m}}\right] \\ &= -8.9\times10^3\ \mathrm{V}\ . \end{aligned}$$

54. (a) The magnitude of the electric field is

$$E=\frac{\sigma}{\varepsilon_0}=\frac{q}{4\pi\varepsilon_0R^2}=\frac{(3.0\times10^{-8}\,\mathrm{C})\left(8.99\times10^9\,\frac{\mathrm{N\cdot m^2}}{\mathrm{C^2}}\right)}{(0.15\,\mathrm{m})^2}=1.2\times10^4\,\mathrm{N/C}\ .$$

(b) $V=RE=(0.15\,\mathrm{m})(1.2\times10^4\,\mathrm{N/C})=1.8\times10^3\,\mathrm{V}$.

(c) Let the distance be x. Then

$$\Delta V=V(x)-V=\frac{q}{4\pi\varepsilon_0}\left(\frac{1}{R+x}-\frac{1}{R}\right)=-500\,\mathrm{V}\ ,$$

which gives

$$x=\frac{R\Delta V}{-V-\Delta V}=\frac{(0.15\,\mathrm{m})(-500\,\mathrm{V})}{-1800\,\mathrm{V}+500\,\mathrm{V}}=5.8\times10^{-2}\ \mathrm{m}\ .$$

55. (a) The potential would be

$$\begin{aligned} V_e &= \frac{Q_e}{4\pi\varepsilon_0R_e}=\frac{4\pi R_e^2\sigma_e}{4\pi\varepsilon_0R_e}=4\pi R_e\sigma_ek \\ &= 4\pi(6.37\times10^6\,\mathrm{m})(1.0\,\mathrm{electron/\,m^2})(-1.6\times10^{-19}\,\mathrm{C/\,electron})\left(8.99\times10^9\,\frac{\mathrm{N\cdot m^2}}{\mathrm{C^2}}\right) \\ &= -0.12\ \mathrm{V}\ . \end{aligned}$$

(b) The electric field is

$$E=\frac{\sigma_e}{\varepsilon_0}=\frac{V_e}{R_e}=-\frac{0.12\,\mathrm{V}}{6.37\times10^6\,\mathrm{m}}=-1.8\times10^{-8}\ \mathrm{N/C}\ ,$$

where the minus sign indicates that $\vec{E}$ is radially inward.

56. Since the charge distribution is spherically symmetric we may write

$$E(r)=\frac{1}{4\pi\varepsilon_0}\frac{q_{\mathrm{encl}}}{r}\ ,$$

where q_{encl} is the charge enclosed in a sphere of radius r centered at the origin. Also, Eq. 25-18 is implemented in the form: $V(r) - V(r') = \int_r^{r'} E(r)\,dr$. The results are as follows: For $r > R_2 > R_1$

$$V(r) = \frac{q_1 + q_2}{4\pi\varepsilon_0 r} \quad \text{and} \quad E(r) = \frac{q_1 + q_2}{4\pi\varepsilon_0 r^2} \; .$$

For $R_2 > r > R_1$

$$V(r) = \frac{1}{4\pi\varepsilon_0}\left(\frac{q_1}{r} + \frac{q_2}{R_2}\right) \quad \text{and} \quad E(r) = \frac{q_1}{4\pi\varepsilon_0 r^2} \; .$$

Finally, for $R_2 > R_1 > r$

$$V = \frac{1}{4\pi\varepsilon_0}\left(\frac{q_1}{R_1} + \frac{q_2}{R_2}\right) \quad \text{and} \quad E = 0 \; .$$

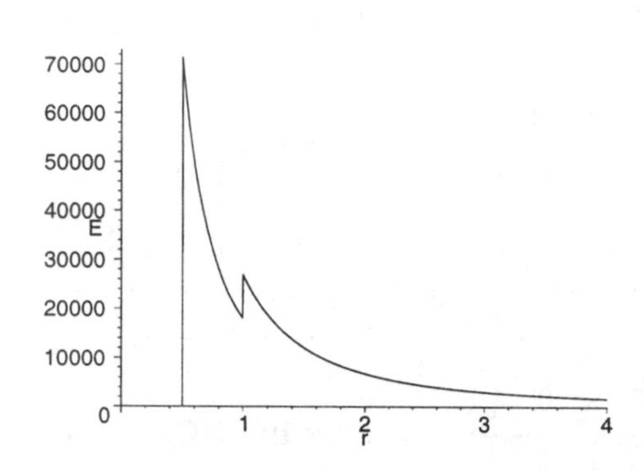

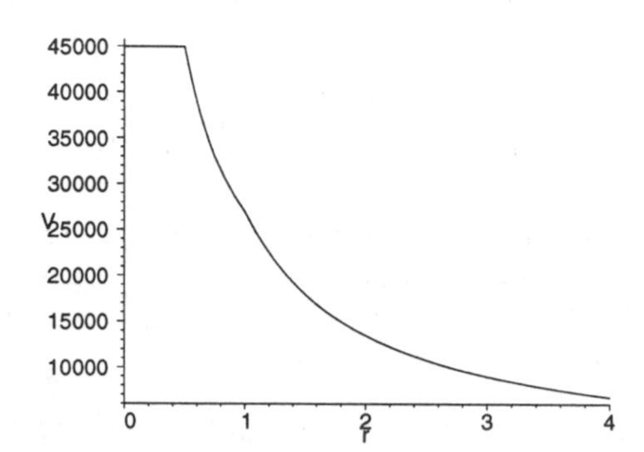

57. (a) We use Eq. 25-18 to find the potential:

$$\begin{aligned} V_{\text{wall}} - V &= -\int_r^R E\,dr \\ 0 - V &= -\int_r^R \left(\frac{\rho r}{2\varepsilon_0}\right) \\ -V &= -\frac{\rho}{4\varepsilon_0}\left(R^2 - r^2\right) . \end{aligned}$$

Consequently, $V = \frac{\rho}{4\varepsilon_0}\left(R^2 - r^2\right)$.

(b) The value at $r = 0$ is

$$V_{\text{center}} = \frac{-1.1 \times 10^{-3}\,\text{C/m}^3}{4\,(8.85 \times 10^{-12}\,\text{C/V·m})}\left((0.05\,\text{m})^2 - 0\right) = -7.8 \times 10^4\ \text{V} \; .$$

58. We treat the system as a superposition of a disk of surface charge density σ and radius R and a smaller, oppositely charged, disk of surface charge density $-\sigma$ and radius r. For each of these, Eq 25-37 applies (for $z > 0$)

$$V = \frac{\sigma}{2\varepsilon_0}\left(\sqrt{z^2 + R^2} - z\right) + \frac{-\sigma}{2\varepsilon_0}\left(\sqrt{z^2 + r^2} - z\right) \; .$$

This expression does vanish as $r \to \infty$, as the problem requires. Substituting $r = R/5$ and $z = 2R$ and simplifying, we obtain

$$V = \frac{\sigma R}{\varepsilon_0}\left(\frac{5\sqrt{5} - \sqrt{101}}{10}\right) \approx \frac{\sigma R}{\varepsilon_0}(0.113) \; .$$

59. We use $q = 1.37 \times 10^5$ C from Sample Problem 22-7 and $k = 1/4\pi\varepsilon_0$ to find the potential:

$$V = \frac{q}{4\pi\varepsilon_0 R_e} = \frac{(1.37 \times 10^5\ \text{C})\left(8.99 \times 10^9\ \frac{\text{N·m}^2}{\text{C}^2}\right)}{6.37 \times 10^6\ \text{m}} = 1.93 \times 10^8\ \text{V} \; .$$

60. (a) The potential on the surface is

$$V = \frac{q}{4\pi\varepsilon_0 R} = \frac{(4.0 \times 10^{-6}\,\text{C})\left(8.99 \times 10^9\,\frac{\text{N}\cdot\text{m}^2}{\text{C}^2}\right)}{0.10\,\text{m}} = 3.6 \times 10^5\ \text{V} .$$

(b) The field just outside the sphere would be

$$E = \frac{q}{4\pi\varepsilon_0 R^2} = \frac{V}{R} = \frac{3.6 \times 10^5\,\text{V}}{0.10\,\text{m}} = 3.6 \times 10^6\ \text{V/m} ,$$

which would have exceeded 3.0 MV/m. So this situation cannot occur.

61. If the electric potential is zero at infinity then at the surface of a uniformly charged sphere it is $V = q/4\pi\varepsilon_0 R$, where q is the charge on the sphere and R is the sphere radius. Thus $q = 4\pi\varepsilon_0 RV$ and the number of electrons is

$$N = \frac{|q|}{e} = \frac{4\pi\varepsilon_0 R|V|}{e} = \frac{(1.0 \times 10^{-6}\,\text{m})(400\text{V})}{(8.99 \times 10^9\,\text{N}\cdot\text{m}^2/\text{C}^2)(1.60 \times 10^{-19}\,\text{C})} = 2.8 \times 10^5 .$$

62. This can be approached more than one way, but the simplest is to observe that the net potential (using Eq. 25-27) due to the $+2q$ and $-2q$ charges is zero at both the initial and final positions of the movable charge ($+5q$). This implies that no work is necessary to effect its change of position, which, in turn, implies there is no resulting change in potential energy of the configuration. Hence, the ratio is unity.

63. We imagine moving all the charges on the surface of the sphere to the center of the the sphere. Using Gauss' law, we see that this would not change the electric field *outside* the sphere. The magnitude of the electric field E of the uniformly charged sphere as a function of r, the distance from the center of the sphere, is thus given by $E(r) = q/(4\pi\varepsilon_0 r^2)$ for $r > R$. Here R is the radius of the sphere. Thus, the potential V at the surface of the sphere (where $r = R$) is given by

$$\begin{aligned} V(R) &= V\Big|_{r=\infty} + \int_R^\infty E(r)\,dr = \int_\infty^R \frac{q}{4\pi\varepsilon_0 r^2}\,dr = \frac{q}{4\pi\varepsilon_0 R} \\ &= \frac{\left(8.99 \times 10^9\,\frac{\text{N}\cdot\text{m}^2}{\text{C}^2}\right)(1.50 \times 10^8\,\text{C})}{0.160\,\text{m}} = 8.43 \times 10^2\ \text{V} . \end{aligned}$$

64. We use $E_x = -dV/dx$, where dV/dx is the local slope of the V vs. x curve depicted in Fig. 25-54. The results are: $E_x(ab) = -6.0\,\text{V/m}$, $E_x(bc) = 0$, $E_x(cd) = E_x(de) = 3.0\,\text{V/m}$, $E_x(ef) = 15\,\text{V/m}$, $E_x(fg) = 0$, $E_x(gh) = -3.0\,\text{V/m}$. Since these values are constant during their respective time-intervals, their graph consists of several disconnected line-segments (horizontal) and is not shown here in the interest of saving space.

65. On the dipole axis $\theta = 0$ or π, so $|\cos\theta| = 1$. Therefore, magnitude of the electric field is

$$|E(r)| = \left|-\frac{\partial V}{\partial r}\right| = \frac{p}{4\pi\varepsilon_0}\left|\frac{d}{dr}\left(\frac{1}{r^2}\right)\right| = \frac{p}{2\pi\varepsilon_0 r^3} .$$

66. (a) We denote the surface charge density of the disk as σ_1 for $0 < r < R/2$, and as σ_2 for $R/2 < r < R$. Thus the total charge on the disk is given by

$$\begin{aligned} q &= \int_{\text{disk}} dq = \int_0^{R/2} 2\pi\sigma_1 r\,dr + \int_{R/2}^R 2\pi\sigma_2 r\,dr = \frac{\pi}{4}R^2(\sigma_1 + 3\sigma_2) \\ &= \frac{\pi}{4}(2.20 \times 10^{-2}\,\text{m})^2[1.50 \times 10^{-6}\,\text{C/m}^2 + 3(8.00 \times 10^{-7}\,\text{C/m}^2)] \\ &= 1.48 \times 10^{-9}\,\text{C} . \end{aligned}$$

(b) We use Eq. 25-36:

$$\begin{aligned} V(z) &= \int_{\text{disk}} dV = k\left[\int_0^{R/2} \frac{\sigma_1(2\pi R')dR'}{\sqrt{z^2+R'^2}} + \int_{R/2}^{R} \frac{\sigma_2(2\pi R')dR'}{\sqrt{z^2+R'^2}}\right] \\ &= \frac{\sigma_1}{2\varepsilon_0}\left(\sqrt{z^2+\frac{R^2}{4}}-z\right) + \frac{\sigma_2}{2\varepsilon_0}\left(\sqrt{z^2+R^2}-\sqrt{z^2+\frac{R^2}{4}}\right) . \end{aligned}$$

Substituting the numerical values of σ_1, σ_2, R and z, we obtain $V(z) = 7.95 \times 10^2$ V.

67. From the previous chapter, we know that the radial field due to an infinite line-source is

$$E = \frac{\lambda}{2\pi\varepsilon_0 r}$$

which integrates, using Eq. 25-18, to obtain

$$V_i = V_f + \frac{\lambda}{2\pi\varepsilon_0}\int_{r_i}^{r_f} \frac{dr}{r} = V_f + \frac{\lambda}{2\pi\varepsilon_0}\ln\left(\frac{r_f}{r_i}\right) .$$

The subscripts i and f are somewhat arbitrary designations, and we let $V_i = V$ be the potential of some point P at a distance $r_i = r$ from the wire and $V_f = V_o$ be the potential along some reference axis (which will be the z axis described in this problem) at a distance $r_f = a$ from the wire. In the "end-view" presented below, the wires and the z axis appear as points as they intersect the xy plane. The potential due to the wire on the left (intersecting the plane at $x = -a$) is

$$V_{\text{negative wire}} = V_o + \frac{(-\lambda)}{2\pi\varepsilon_0}\ln\left(\frac{a}{\sqrt{(x+a)^2+y^2}}\right) ,$$

and the potential due to the wire on the right (intersecting the plane at $x = +a$) is

$$V_{\text{positive wire}} = V_o + \frac{(+\lambda)}{2\pi\varepsilon_0}\ln\left(\frac{a}{\sqrt{(x-a)^2+y^2}}\right) .$$

Since potential is a scalar quantity, the net potential at point P is the addition of $V_{-\lambda}$ and $V_{+\lambda}$ which simplifies to

$$V_{\text{net}} = 2V_o + \frac{\lambda}{2\pi\varepsilon_0}\left(\ln\left(\frac{a}{\sqrt{(x-a)^2+y^2}}\right) - \ln\left(\frac{a}{\sqrt{(x+a)^2+y^2}}\right)\right) = \frac{\lambda}{4\pi\varepsilon_0}\ln\left(\frac{(x+a)^2+y^2}{(x-a)^2+y^2}\right)$$

where we have set the potential along the z axis equal to zero ($V_o = 0$) in the last step (which we are free to do). This is the expression used to obtain the equipotentials shown below. The center dot in the figure is the intersection of the z axis with the xy plane, and the dots on either side are the intersections of the wires with the plane.

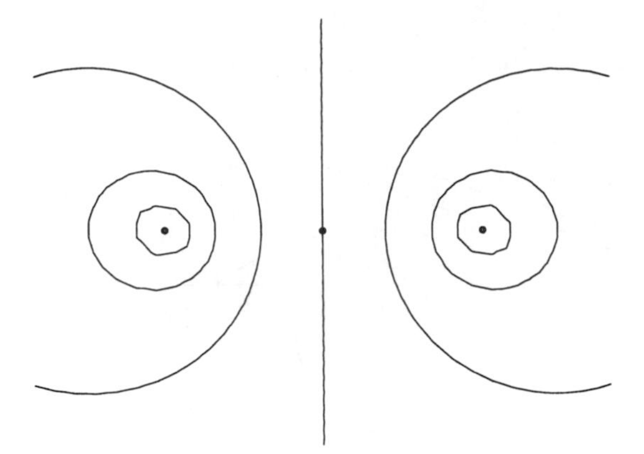

68. The potential difference is $\Delta V = E\Delta s = (1.92 \times 10^5\,\text{N/C})(0.0150\,\text{m}) = 2.90 \times 10^3\,\text{V}$.

69. Since the charge distribution on the arc is equidistant from the point where V is evaluated, its contribution is identical to that of a point charge at that distance. We assume $V \to 0$ as $r \to \infty$ and apply Eq. 25-27:

$$V = \frac{1}{4\pi\varepsilon_0}\frac{+Q}{R} + \frac{1}{4\pi\varepsilon_0}\frac{+4Q}{2R} + \frac{1}{4\pi\varepsilon_0}\frac{-2Q}{R}$$

which simplifies to $Q/4\pi\varepsilon_0 R$.

70. From the previous chapter, we know that the radial field due to an infinite line-source is

$$E = \frac{\lambda}{2\pi\varepsilon_0 r}$$

which integrates, using Eq. 25-18, to obtain

$$V_i = V_f + \frac{\lambda}{2\pi\varepsilon_0}\int_{r_i}^{r_f}\frac{dr}{r} = V_f + \frac{\lambda}{2\pi\varepsilon_0}\ln\left(\frac{r_f}{r_i}\right) .$$

The subscripts i and f are somewhat arbitrary designations, and we let $V_i = V$ be the potential of some point P at a distance $r_i = r$ from the wire and $V_f = V_o$ be the potential along some reference axis (which intersects the plane of our figure, shown below, at the xy coordinate origin, placed midway between the bottom two line charges – that is, the midpoint of the bottom side of the equilateral triangle) at a distance $r_f = a$ from each of the bottom wires (and a distance $a\sqrt{3}$ from the topmost wire). Thus, each side of the triangle is of length $2a$. Skipping some steps, we arrive at an expression for the net potential created by the three wires (where we have set $V_o = 0$):

$$V_{\text{net}} = \frac{\lambda}{4\pi\varepsilon_0}\ln\left(\frac{(x^2 + (y - a\sqrt{3})^2)^2}{((x+a)^2 + y^2)((x-a)^2 + y^2)}\right)$$

which forms the basis of our contour plot shown below. On the same plot we have shown four electric field lines, which have been sketched (as opposed to rigorously calculated) and are not meant to be as accurate as the equipotentials. The $\pm 2\lambda$ by the top wire in our figure should be -2λ (the $\pm$ typo is an artifact of our plotting routine).

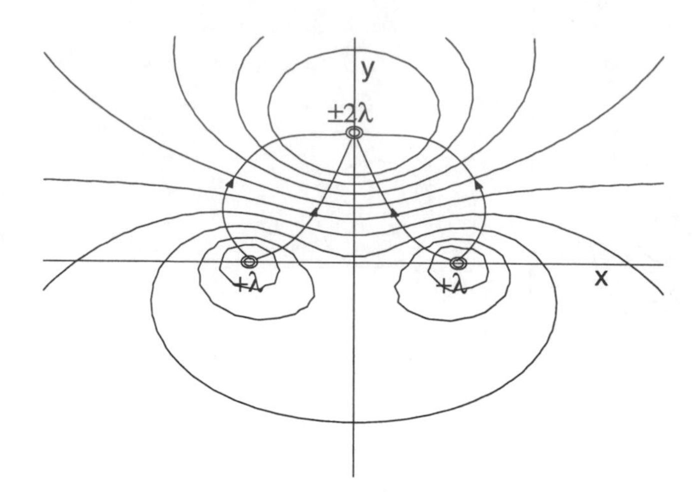

71. The charges are equidistant from the point where we are evaluating the potential – which is computed using Eq. 25-27 (or its integral equivalent). Eq. 25-27 implicitly assumes $V \to 0$ as $r \to \infty$. Thus, we have

$$V = \frac{1}{4\pi\varepsilon_0}\frac{+Q}{R} + \frac{1}{4\pi\varepsilon_0}\frac{-2Q}{R} + \frac{1}{4\pi\varepsilon_0}\frac{+3Q}{R}$$

which simplifies to $Q/2\pi\varepsilon_0 R$.

72. The radius of the cylinder (0.020 m, the same as r_B) is denoted R, and the field magnitude there (160 N/C) is denoted E_B. The electric field beyond the surface of the sphere follows Eq. 24-12, which expresses inverse proportionality with r:

$$\frac{|\vec{E}|}{E_B} = \frac{R}{r} \qquad \text{for } r \geq R\ .$$

(a) Thus, if $r = r_C = 0.050$ m, we obtain $|\vec{E}| = (160)(0.020)/(0.050) = 64$ N/C.

(b) Integrating the above expression (where the variable to be integrated, r, is now denoted ϱ) gives the potential difference between V_B and V_C.

$$V_B - V_C = \int_R^r \frac{E_B\,R}{\varrho}\,d\varrho = E_B\,R\,\ln\!\left(\frac{r}{R}\right) = 2.9\text{ V}\ .$$

(c) The electric field throughout the conducting volume is zero, which implies that the potential there is constant and equal to the value it has on the surface of the charged cylinder: $V_A - V_B = 0$.

73. The net potential (at point A or B) is computed using Eq. 25-27. Thus, using k for $1/4\pi\varepsilon_0$, the difference is

$$\begin{aligned} V_B - V_A &= \left(\frac{kq}{2d} + \frac{k(-5q)}{2d}\right) - \left(\frac{kq}{d} + \frac{k(-5q)}{5d}\right) \\ &= -\frac{4kq}{2d} \end{aligned}$$

which simplifies to $-q/2\pi\varepsilon_0$ in SI units (with $d = 1$ m).

74. Eq. 25-32 applies with $dq = \lambda\,dx = bx\,dx$ (along $0 \leq x \leq 0.20$ m).

(a) Here $r = x > 0$, so that

$$V = \frac{1}{4\pi\varepsilon_0}\int_0^{0.20}\frac{bx\,dx}{x} = \frac{b(0.20)}{4\pi\varepsilon_0}$$

which yields $V = 36$ V.

(b) Now $r = \sqrt{x^2 + d^2}$ where $d = 0.15$ m, so that

$$V = \frac{1}{4\pi\varepsilon_0}\int_0^{0.20}\frac{bx\,dx}{\sqrt{x^2+d^2}} = \frac{b}{4\pi\varepsilon_0}\left(\sqrt{x^2+d^2}\right)\Big|_0^{0.20}$$

which yields $V = 18$ V.

75. (a) Using Eq. 25-26, we calculate the radius r of the sphere representing the 30 V equipotential surface:

$$r = \frac{q}{4\pi\varepsilon_0 V} = 4.5 \text{ m}.$$

(b) If the potential were a linear function of r then it would have equally spaced equipotentials, but since $V \propto 1/r$ they are spaced more and more widely apart as r increases.

76. We denote $q = 25 \times 10^{-9}$ C, $y = 0.6$ m, $x = 0.8$ m, with $V =$ the net potential (assuming $V \to 0$ as $r \to \infty$). Then,

$$V_A = \frac{1}{4\pi\varepsilon_0}\frac{q}{y} + \frac{1}{4\pi\varepsilon_0}\frac{(-q)}{x}$$

$$V_B = \frac{1}{4\pi\varepsilon_0}\frac{q}{x} + \frac{1}{4\pi\varepsilon_0}\frac{(-q)}{y}$$

leads to

$$V_B - V_A = \frac{2}{4\pi\varepsilon_0}\frac{q}{x} - \frac{2}{4\pi\varepsilon_0}\frac{q}{y} = \frac{q}{2\pi\varepsilon_0}\left(\frac{1}{x} - \frac{1}{y}\right)$$

which yields $\Delta V = -187 \approx -190$ V.

77. (a) By Eq. 25-18, the change in potential is the negative of the "area" under the curve. Thus, using the area-of-a-triangle formula, we have

$$V - 10 = -\int_0^{x=2}\vec{E}\cdot d\vec{s} = \frac{1}{2}(2)(20)$$

which yields $V = 30$ V.

(b) For any region within $0 < x < 3$ m, $-\int \vec{E}\cdot d\vec{s}$ is positive, but for any region for which $x > 3$ m it is negative. Therefore, $V = V_{\text{max}}$ occurs at $x = 3$ m.

$$V - 10 = -\int_0^{x=3}\vec{E}\cdot d\vec{s} = \frac{1}{2}(3)(20)$$

which yields $V_{\text{max}} = 40$ V.

(c) In view of our result in part (b), we see that now (to find $V = 0$) we are looking for some $X > 3$ m such that the "area" from $x = 3$ m to $x = X$ is 40 V. Using the formula for a triangle ($3 < x < 4$) and a rectangle ($4 < x < X$), we require

$$\frac{1}{2}(1)(20) + (X - 4)(20) = 40 .$$

Therefore, $X = 5.5$ m.

78. In the "inside" region between the plates, the individual fields (given by Eq. 24.13) are in the same direction ($-\hat{\imath}$):

$$\vec{E}_{\rm in} = -\left(\frac{50 \times 10^{-9}}{2\varepsilon_0} + \frac{25 \times 10^{-9}}{2\varepsilon_0}\right)\hat{\imath} = -4.2 \times 10^3\,\hat{\imath}$$

in SI units (N/C or V/m). And in the "outside" region where $x > 0.5$ m, the individual fields point in opposite directions:

$$\vec{E}_{\rm out} = -\frac{50 \times 10^{-9}}{2\varepsilon_0}\hat{\imath} + \frac{25 \times 10^{-9}}{2\varepsilon_0}\hat{\imath} = -1.4 \times 10^3\,\hat{\imath}\ .$$

Therefore, by Eq. 25-18, we have

$$\begin{aligned}\Delta V = -\int_0^{0.8} \vec{E}\cdot d\vec{s} &= -\int_0^{0.5} |\vec{E}|_{\rm in}\,dx - \int_{0.5}^{0.8} |\vec{E}|_{\rm out}\,dx \\ &= -\left(4.2 \times 10^3\right)(0.5) - \left(1.4 \times 10^3\right)(0.3) \\ &= 2.5 \times 10^3\ \text{V}\ .\end{aligned}$$

79. We connect A to the origin with a line along the y axis, along which there is no change of potential (Eq. 25-18: $\int \vec{E}\cdot d\vec{s} = 0$). Then, we connect the origin to B with a line along the x axis, along which the change in potential is

$$\Delta V = -\int_0^{x=4} \vec{E}\cdot d\vec{s} = -4.00\int_0^4 x\,dx = -4.00\left(\frac{4^2}{2}\right)$$

which yields $V_B - V_A = -32$ V.

80. (a) The charges are equal and are the same distance from C. We use the Pythagorean theorem to find the distance $r = \sqrt{(d/2)^2 + (d/2)^2} = d/\sqrt{2}$. The electric potential at C is the sum of the potential due to the individual charges but since they produce the same potential, it is twice that of either one:

$$\begin{aligned}V &= \frac{2q}{4\pi\varepsilon_0}\frac{\sqrt{2}}{d} = \frac{2\sqrt{2}q}{4\pi\varepsilon_0 d} \\ &= \frac{(8.99 \times 10^9\,\text{N}\cdot\text{m}^2/\text{C}^2)(2)\sqrt{2}(2.0 \times 10^{-6}\,\text{C})}{0.020\,\text{m}} = 2.54 \times 10^6\ \text{V}\ .\end{aligned}$$

(b) As you move the charge into position from far away the potential energy changes from zero to qV, where V is the electric potential at the final location of the charge. The change in the potential energy equals the work you must do to bring the charge in:

$$W = qV = \left(2.0 \times 10^{-6}\,\text{C}\right)\left(2.54 \times 10^6\,\text{V}\right) = 5.1\ \text{J}\ .$$

(c) The work calculated in part (b) represents the potential energy of the interactions between the charge brought in from infinity and the other two charges. To find the total potential energy of the three-charge system you must add the potential energy of the interaction between the fixed charges. Their separation is d so this potential energy is $q^2/4\pi\varepsilon_0 d$. The total potential energy is

$$\begin{aligned}U &= W + \frac{q^2}{4\pi\varepsilon_0 d} \\ &= 5.1\,\text{J} + \frac{(8.99 \times 10^9\,\text{N}\cdot\text{m}^2/\text{C}^2)(2.0 \times 10^{-6}\,\text{C})^2}{0.020\,\text{m}} = 6.9\ \text{J}\ .\end{aligned}$$

81. (a) Let the quark-quark separation be r. To "naturally" obtain the eV unit, we only plug in for one of the e values involved in the computation:

$$\begin{aligned} U_{\text{up-up}} &= \frac{1}{4\pi\varepsilon_0}\frac{\left(\frac{2e}{3}\right)\left(\frac{2e}{3}\right)}{r} = \frac{4ke}{9r}e \\ &= \frac{4\left(8.99\times10^9\,\frac{\text{N}\cdot\text{m}^2}{\text{C}^2}\right)\left(1.60\times10^{-19}\,\text{C}\right)}{9\left(1.32\times10^{-15}\,\text{m}\right)}e \\ &= 4.84\times10^5\,\text{eV} = 0.484\ \text{MeV}\ . \end{aligned}$$

(b) The total consists of all pair-wise terms:

$$U = \frac{1}{4\pi\varepsilon_0}\left[\frac{\left(\frac{2e}{3}\right)\left(\frac{2e}{3}\right)}{r} + \frac{\left(\frac{-e}{3}\right)\left(\frac{2e}{3}\right)}{r} + \frac{\left(\frac{-e}{3}\right)\left(\frac{2e}{3}\right)}{r}\right] = 0\ .$$

82. (a) At the smallest center-to-center separation r_{min} the initial kinetic energy K_i of the proton is entirely converted to the electric potential energy between the proton and the nucleus. Thus,

$$K_i = \frac{1}{4\pi\varepsilon_0}\frac{eq_{\text{lead}}}{r_{\text{min}}} = \frac{82e^2}{4\pi\varepsilon_0 r_{\text{min}}}\ .$$

In solving for r_{min} using the eV unit, we note that a factor of e cancels in the middle line:

$$\begin{aligned} r_{\text{min}} &= \frac{82e^2}{4\pi\varepsilon_0 K_i} = k\frac{82e^2}{4.80\times10^6\,\text{eV}} \\ &= \left(8.99\times10^9\,\frac{\text{N}\cdot\text{m}^2}{\text{C}^2}\right)\frac{82(1.6\times10^{-19}\,\text{C})}{4.80\times10^6\,\text{V}} \\ &= 2.5\times10^{-14}\,\text{m} = 25\ \text{fm}\ . \end{aligned}$$

It is worth recalling that a volt is a Newton·meter/Coulomb, in making sense of the above manipulations.

(b) An alpha particle has 2 protons (as well as 2 neutrons). Therefore, using r'_{min} for the new separation, we find

$$K_i = \frac{1}{4\pi\varepsilon_0}\frac{q_\alpha q_{\text{lead}}}{r'_{\text{min}}} = 2\left(\frac{82e^2}{4\pi\varepsilon_0 r'_{\text{min}}}\right) = \frac{82e^2}{4\pi\varepsilon_0 r_{\text{min}}}$$

which leads to $r'_{\text{min}} = 2r_{\text{min}} = 50\,\text{fm}$.

83. The potential energy of the two-charge system is

$$\begin{aligned} U &= \frac{1}{4\pi\varepsilon_0}\left[\frac{q_1 q_2}{\sqrt{(x_1-x_2)^2+(y_1-y_2)^2}}\right] \\ &= \frac{\left(8.99\times10^9\,\frac{\text{N}\cdot\text{m}^2}{\text{C}^2}\right)(3.0\times10^{-6}\,\text{C})(-4.0\times10^{-6}\,\text{C})}{\sqrt{(3.5+2.0)^2+(0.50-1.5)^2}\ \text{cm}} \end{aligned}$$

$$= -1.9\ \text{J}\ .$$

Thus, -1.9 J of work is needed.

84. For a point on the axis of the ring the potential (assuming $V \to 0$ as $r \to \infty$) is

$$V = \frac{q}{4\pi\varepsilon_0\sqrt{z^2+R^2}}$$

where $q = 16\times10^{-6}$ C and $R = 0.030$ m. Therefore,

$$V_B - V_A = \frac{q}{4\pi\varepsilon_0}\left(\frac{1}{\sqrt{z_B^2+R^2}} - \frac{1}{R}\right)$$

where $z_B = 0.040$ m. The result is -1.92×10^6 V.

85. We apply Eq. 25-41:

$$\begin{aligned} E_x &= -\frac{\partial V}{\partial x} = -2yz^2 \\ E_y &= -\frac{\partial V}{\partial y} = -2xz^2 \\ E_z &= -\frac{\partial V}{\partial z} = -4xyz \end{aligned}$$

which, at $(x, y, z) = (3, -2, 4)$, gives $(E_x, E_y, E_z) = (64, -96, 96)$ in SI units. The magnitude of the field is therefore

$$\left|\vec{E}\right| = \sqrt{E_x^2 + E_y^2 + E_z^2} = 150 \text{ V/m} = 150 \text{ N/C} .$$

86. We note that for two points on a circle, separated by angle θ (in radians), the direct-line distance between them is $r = 2R\sin(\theta/2)$. Using this fact, distinguishing between the cases where $N =$ odd and $N =$ even, and counting the pair-wise interactions very carefully, we arrive at the following results for the total potential energies. We use $k = 1/4\pi\varepsilon_0$. For configuration 1 (where all N electrons are on the circle), we have

$$\begin{aligned} U_{1,N=\text{even}} &= \frac{Nke^2}{2R}\left(\sum_{j=1}^{\frac{N}{2}-1} \frac{1}{\sin(j\theta/2)} + \frac{1}{2}\right) \\ U_{1,N=\text{odd}} &= \frac{Nke^2}{2R}\left(\sum_{j=1}^{\frac{N-1}{2}} \frac{1}{\sin(j\theta/2)}\right) \end{aligned}$$

where $\theta = \frac{2\pi}{N}$. For configuration 2, we find

$$\begin{aligned} U_{2,N=\text{even}} &= \frac{(N-1)ke^2}{2R}\left(\sum_{j=1}^{\frac{N}{2}-1} \frac{1}{\sin(j\theta'/2)} + 2\right) \\ U_{2,N=\text{odd}} &= \frac{(N-1)ke^2}{2R}\left(\sum_{j=1}^{\frac{N-3}{2}} \frac{1}{\sin(j\theta'/2)} + \frac{5}{2}\right) \end{aligned}$$

where $\theta' = \frac{2\pi}{N-1}$. The results are all of the form

$$U_{1 \text{ or } 2} = \frac{ke^2}{2R} \times \text{a pure number} .$$

In our table, below, we have the results for those "pure numbers" as they depend on N and on which configuration we are considering. The values listed in the U rows are the potential energies divided by $ke^2/2R$.

N	4	5	6	7	8	9	10	11	12	13	14	15
U_1	3.83	6.88	10.96	16.13	22.44	29.92	38.62	48.58	59.81	72.35	86.22	101.5
U_2	4.73	7.83	11.88	16.96	23.13	30.44	39.92	48.62	59.58	71.81	85.35	100.2

We see that the potential energy for configuration 2 is greater than that for configuration 1 for $N < 12$, but for $N \geq 12$ it is configuration 1 that has the greatest potential energy.

(a) Configuration 1 has the smallest U for $2 \leq N \leq 11$, and configuration 2 has the smallest U for $12 \leq N \leq 15$.

(b) $N = 12$ is the smallest value such that $U_2 < U_1$.

(c) For $N = 12$, configuration 2 consists of 11 electrons distributed at equal distances around the circle, and one electron at the center. A specific electron e_0 on the circle is R distance from the one in the center, and is

$$r = 2R\sin\left(\frac{\pi}{11}\right) \approx 0.56R$$

distance away from its nearest neighbors on the circle (of which there are two – one on each side). Beyond the nearest neighbors, the next nearest electron on the circle is

$$r = 2R\sin\left(\frac{2\pi}{11}\right) \approx 1.1R$$

distance away from e_0. Thus, we see that there are only two electrons closer to e_0 than the one in the center.

87. (First problem of **Cluster**)

(a) The field between the plates is uniform; we apply Eq. 25-42 to find the magnitude of the (horizontal) field: $|\vec{E}| = \Delta V/D$ (assuming $\Delta V > 0$). This produces a horizontal acceleration from Eq. 23-1 and Newton's second law (applied along the x axis):

$$a_x = \frac{|\vec{F}_x|}{m} = \frac{q|\vec{E}|}{m} = \frac{q\Delta V}{m\,D}$$

where $q > 0$ has been assumed; the problem indicates that the acceleration is rightward, which constitutes our choice for the $+x$ direction. If we choose upward as the $+y$ direction then $a_y = -g$, and we apply the free-fall equations of Chapter 2 to the y motion while applying the constant (a_x) acceleration equations of Table 2-1 to the x motion. The displacement is defined by $\Delta x = +D/2$ and $\Delta y = -d$, and the initial velocity is zero. Simultaneous solution of

$$\begin{aligned} \Delta x &= v_{0x}\,t + \frac{1}{2}\,a_x\,t^2 \quad \text{and} \\ \Delta y &= v_{0y}\,t + \frac{1}{2}\,a_y\,t^2 \ , \end{aligned}$$

leads to

$$d = \frac{gD}{2a_x} = \frac{gmD^2}{2q\Delta V} \ .$$

(b) We can continue along the same lines as in part (a) (using Table 2-1) to find v, or we can use energy conservation – which we feel is more instructive. The gain in kinetic energy derives from two potential energy changes: from gravity comes mgd and from electric potential energy comes $q|\vec{E}|\Delta x = q\Delta V/2$. Consequently,

$$\frac{1}{2}mv^2 = mgd + \frac{1}{2}q\Delta V$$

which (upon using the expression for d above) yields

$$v = \sqrt{\frac{mg^2D^2}{q\Delta V} + \frac{q\Delta V}{m}} \ .$$

(c) and (d) Using SI units (so $q = 1.0 \times 10^{-10}$ C, $m = 1.0 \times 10^{-9}$ kg) we plug into our results to obtain $d = 0.049$ m and $v = 1.4$ m/s.

88. (Second problem of **Cluster**)

(a) We argue by symmetry that of the total potential energy in the initial configuration, a third converts into the kinetic energy of each of the particles. And, because the total potential energy consists of three equal contributions

$$U = \frac{1}{4\pi\varepsilon_0}\frac{q^2}{d}$$

then any of the particle's final kinetic energy is equal to this U. Therefore, using k for$1/4\pi\varepsilon_0$, we obtain

$$v = \sqrt{\frac{2U}{m}} = |q|\sqrt{\frac{2k}{m\,d}} \ .$$

(b) In this case, two of the U contributions to the total potential energy are converted into a single kinetic term:

$$v = \sqrt{\frac{2(2U)}{m}} = 2|q|\sqrt{\frac{k}{m\,d}} \ .$$

(c) Now it is clear that the one remaining U contribution is converted into a particle's kinetic energy:

$$v = \sqrt{\frac{2U}{m}} = |q|\sqrt{\frac{2k}{m\,d}} \ .$$

(d) This leaves no potential energy to convert into kinetic for the last particle that is released. It maintains zero speed.

89. (Third problem of **Cluster**)

(a) By momentum conservation we see that their final speeds are the same. We use energy conservation (where the "final" subscript refers to when they are infinitely far away from each other):

$$\begin{aligned} U_i &= K_f \\ \frac{1}{4\pi\varepsilon_0}\frac{2Q^2}{D} &= 2\left(\frac{1}{2}mv^2\right) \end{aligned}$$

which (using $k = 1/4\pi\varepsilon_0$) yields

$$v = |Q|\sqrt{\frac{2k}{m\,D}} \ .$$

(b) As noted above, this result is the same as that of part (a).

(c) We use energy conservation (where the "final" subscript refers to when their surfaces have made contact):

$$\begin{aligned} U_i &= K_f + U_f \\ \frac{1}{4\pi\varepsilon_0}\frac{-2Q^2}{D} &= 2\left(\frac{1}{2}mv^2\right) + \frac{1}{4\pi\varepsilon_0}\frac{-2Q^2}{2r} \end{aligned}$$

which (using $k = 1/4\pi\varepsilon_0$) yields

$$v = |Q|\sqrt{\frac{k}{m\,r} - \frac{2k}{m\,D}} \approx |Q|\sqrt{\frac{k}{m\,r}}$$

since $r \ll D$.

(d) As before, the speeds of the particles are equal (by momentum conservation).

(e) and (f) The collision being elastic means no kinetic energy is lost (or gained), so they are able to return to their original positions (climbing back up that potential "hill") whereupon their potential energy is again U_i and their kinetic energies (hence, speeds) are zero.

90. (Fourth problem of **Cluster**)

(a) At its displaced position, its potential energy (using $k = 1/4\pi\varepsilon_0$) is

$$U_i = k\frac{qQ}{d - x_0} + k\frac{qQ}{d + x_0} = \frac{2kqQd}{d^2 - x_0^2} .$$

And at A, the potential energy is

$$U_A = 2\left(k\frac{qQ}{d}\right) .$$

Setting this difference equal to the kinetic energy of the particle ($\frac{1}{2}mv^2$) and solving for the speed yields

$$v = \sqrt{\frac{2\,(U_i - U_A)}{m}} = 2\,x_0\sqrt{\frac{k\,q\,Q}{m\,d\,(d^2 - x_0^2)}} .$$

(b) It is straightforward to consider small x_0 (more precisely, $x_0/d \ll 1$) in the above expression (so that $d^2 - x_0^2 \approx d^2$). The result is

$$v \approx 2\,\frac{x_0}{d}\sqrt{\frac{k\,q\,Q}{m\,d}} .$$

(c) Plugging in the given values (converted to SI units) yields $v \approx 19$ m/s.

(d) Using the Pythagorean theorem, we now have

$$U_i = 2k\frac{-qQ}{\sqrt{d^2 + x_0^2}} .$$

Therefore, (with U_A in this part equal to the negative of U_A in the previous part)

$$v = \sqrt{\frac{2\,(U_i - U_A)}{m}} = 2\sqrt{\frac{k\,q\,Q}{m}\left(\frac{1}{d} - \frac{1}{\sqrt{d^2 + x_0^2}}\right)} .$$

To simplify, the binomial theorem (Appendix E) is employed:

$$\frac{1}{\sqrt{d^2 + x_0^2}} \approx \frac{1}{d}\left(1 - \frac{1}{2}\frac{x_0^2}{d^2}\right)$$

which leads to

$$v \approx \frac{x_0}{d}\sqrt{\frac{2k\,q\,Q}{m\,d}} .$$

Chapter 26

1. The minimum charge measurable is

$$q_{\text{min}} = CV_{\text{min}} = (50\,\text{pF})(0.15\,\text{V}) = 7.5\ \text{pC}\ .$$

2. (a) The capacitance of the system is

$$C = \frac{q}{\Delta V} = \frac{70\,\text{pC}}{20\,\text{V}} = 3.5\ \text{pF}\ .$$

(b) The capacitance is independent of q; it is still 3.5 pF.

(c) The potential difference becomes

$$\Delta V = \frac{q}{C} = \frac{200\,\text{pC}}{3.5\,\text{pF}} = 57\ \text{V}\ .$$

3. Charge flows until the potential difference across the capacitor is the same as the potential difference across the battery. The charge on the capacitor is then $q = CV$, and this is the same as the total charge that has passed through the battery. Thus, $q = (25 \times 10^{-6}\,\text{F})(120\,\text{V}) = 3.0 \times 10^{-3}\,\text{C}$.

4. We verify the units relationship as follows:

$$[\varepsilon_0] = \frac{\text{F}}{\text{m}} = \frac{\text{C}}{\text{V}\cdot\text{m}} = \frac{\text{C}}{(\text{N}\cdot\text{m}/\text{C})\,\text{m}} = \frac{\text{C}^2}{\text{N}\cdot\text{m}^2}\ .$$

5. (a) The capacitance of a parallel-plate capacitor is given by $C = \varepsilon_0 A/d$, where A is the area of each plate and d is the plate separation. Since the plates are circular, the plate area is $A = \pi R^2$, where R is the radius of a plate. Thus,

$$C = \frac{\varepsilon_0 \pi R^2}{d} = \frac{(8.85 \times 10^{-12}\,\text{F/m})\pi(8.2 \times 10^{-2}\,\text{m})^2}{1.3 \times 10^{-3}\,\text{m}} = 1.4 \times 10^{-10}\,\text{F} = 140\ \text{pF}\ .$$

(b) The charge on the positive plate is given by $q = CV$, where V is the potential difference across the plates. Thus, $q = (1.4 \times 10^{-10}\,\text{F})(120\,\text{V}) = 1.7 \times 10^{-8}\,\text{C} = 17\,\text{nC}$.

6. We use $C = A\varepsilon_0/d$. Thus

$$d = \frac{A\varepsilon_0}{C} = \frac{(1.00\,\text{m}^2)\left(8.85 \times 10^{-12}\,\frac{\text{C}^2}{\text{N}\cdot\text{m}^2}\right)}{1.00\,\text{F}} = 8.85 \times 10^{-12}\ \text{m}\ .$$

Since d is much less than the size of an atom ($\sim 10^{-10}\,\text{m}$), this capacitor cannot be constructed.

7. Assuming conservation of volume, we find the radius of the combined spheres, then use $C = 4\pi\varepsilon_0 R$ to find the capacitance. When the drops combine, the volume is doubled. It is then $V = 2(4\pi/3)R^3$. The new radius R' is given by

$$\frac{4\pi}{3}(R')^3 = 2\frac{4\pi}{3}R^3 ,$$

so

$$R' = 2^{1/3}R .$$

The new capacitance is

$$C' = 4\pi\varepsilon_0 R' = 4\pi\varepsilon_0 2^{1/3}R = 5.04\pi\varepsilon_0 .$$

8. (a) We use Eq. 26-17:

$$C = 4\pi\varepsilon_0\frac{ab}{b-a} = \frac{(40.0\,\text{mm})(38.0\,\text{mm})}{(8.99\times10^9\,\frac{\text{N}\cdot\text{m}^2}{\text{C}^2})\,(40.0\,\text{mm}-38.0\,\text{mm})} = 84.5\text{ pF} .$$

(b) Let the area required be A. Then $C = \varepsilon_0 A/(b-a)$, or

$$A = \frac{C(b-a)}{\varepsilon_0} = \frac{(84.5\,\text{pF})(40.0\,\text{mm}-38.0\,\text{mm})}{(8.85\times10^{-12}\,\frac{\text{C}^2}{\text{N}\cdot\text{m}^2})} = 191\text{ cm}^2 .$$

9. According to Eq. 26-17 the capacitance of a spherical capacitor is given by

$$C = 4\pi\varepsilon_0\frac{ab}{b-a} ,$$

where a and b are the radii of the spheres. If a and b are nearly the same then $4\pi ab$ is nearly the surface area of either sphere. Replace $4\pi ab$ with A and $b-a$ with d to obtain

$$C \approx \frac{\varepsilon_0 A}{d} .$$

10. The equivalent capacitance is

$$C_{\text{eq}} = C_3 + \frac{C_1C_2}{C_1+C_2} = 4.00\,\mu\text{F} + \frac{(10.0\,\mu\text{F})(5.00\,\mu\text{F})}{10.0\,\mu\text{F}+5.00\,\mu\text{F}} = 7.33\,\mu\text{F} .$$

11. The equivalent capacitance is given by $C_{\text{eq}} = q/V$, where q is the total charge on all the capacitors and V is the potential difference across any one of them. For N identical capacitors in parallel, $C_{\text{eq}} = NC$, where C is the capacitance of one of them. Thus, $NC = q/V$ and

$$N = \frac{q}{VC} = \frac{1.00\,\text{C}}{(110\,\text{V})(1.00\times10^{-6}\,\text{F})} = 9090 .$$

12. The charge that passes through meter A is

$$q = C_{\text{eq}}V = 3CV = 3(25.0\,\mu\text{F})(4200\,\text{V}) = 0.315\text{ C} .$$

13. The equivalent capacitance is

$$C_{\text{eq}} = \frac{(C_1+C_2)C_3}{C_1+C_2+C_3} = \frac{(10.0\,\mu\text{F}+5.00\,\mu\text{F})(4.00\,\mu\text{F})}{10.0\,\mu\text{F}+5.00\,\mu\text{F}+4.00\,\mu\text{F}} = 3.16\,\mu\text{F} .$$

14. (a) and (b) The original potential difference V_1 across C_1 is

$$V_1 = \frac{C_{\text{eq}}V}{C_1+C_2} = \frac{(3.16\,\mu\text{F})(100\,\text{V})}{10.0\,\mu\text{F}+5.00\,\mu\text{F}} = 21.1\text{ V} .$$

Thus $\Delta V_1 = 100\,\text{V} - 21.1\,\text{V} = 79\,\text{V}$ and $\Delta q_1 = C_1\Delta V_1 = (10.0\,\mu\text{F})(79\,\text{V}) = 7.9\times10^{-4}\,\text{C}$.

15. Let x be the separation of the plates in the lower capacitor. Then the plate separation in the upper capacitor is $a - b - x$. The capacitance of the lower capacitor is $C_\ell = \varepsilon_0 A/x$ and the capacitance of the upper capacitor is $C_u = \varepsilon_0 A/(a - b - x)$, where A is the plate area. Since the two capacitors are in series, the equivalent capacitance is determined from

$$\frac{1}{C_{\text{eq}}} = \frac{1}{C_\ell} + \frac{1}{C_u} = \frac{x}{\varepsilon_0 A} + \frac{a-b-x}{\varepsilon_0 A} = \frac{a-b}{\varepsilon_0 A} .$$

Thus, the equivalent capacitance is given by $C_{\text{eq}} = \varepsilon_0 A/(a - b)$ and is independent of x.

16. (a) The potential difference across C_1 is $V_1 = 10\,\text{V}$. Thus, $q_1 = C_1 V_1 = (10\,\mu\text{F})(10\,\text{V}) = 1.0 \times 10^{-4}\,\text{C}$.

(b) Let $C = 10\,\mu\text{F}$. We first consider the three-capacitor combination consisting of C_2 and its two closest neighbors, each of capacitance C. The equivalent capacitance of this combination is

$$C_{\text{eq}} = C + \frac{C_2 C}{C + C_2} = 1.5C .$$

Also, the voltage drop across this combination is

$$V = \frac{CV_1}{C + C_{eq}} = \frac{CV_1}{C + 1.5C} = \frac{2}{5}V_1 .$$

Since this voltage difference is divided equally between C_2 and the one connected in series with it, the voltage difference across C_2 satisfies $V_2 = V/2 = V_1/5$. Thus

$$q_2 = C_2 V_2 = (10\,\mu\text{F})\left(\frac{10\,\text{V}}{5}\right) = 2.0 \times 10^{-5}\ \text{V} .$$

17. The charge initially on the charged capacitor is given by $q = C_1 V_0$, where $C_1 = 100\,\text{pF}$ is the capacitance and $V_0 = 50\,\text{V}$ is the initial potential difference. After the battery is disconnected and the second capacitor wired in parallel to the first, the charge on the first capacitor is $q_1 = C_1 V$, where $v = 35\,\text{V}$ is the new potential difference. Since charge is conserved in the process, the charge on the second capacitor is $q_2 = q - q_1$, where C_2 is the capacitance of the second capacitor. Substituting $C_1 V_0$ for q and $C_1 V$ for q_1, we obtain $q_2 = C_1(V_0 - V)$. The potential difference across the second capacitor is also V, so the capacitance is

$$C_2 = \frac{q_2}{V} = \frac{V_0 - V}{V} C_1 = \frac{50\,\text{V} - 35\,\text{V}}{35\,\text{V}}(100\,\text{pF}) = 3\ \text{pF} .$$

18. (a) First, the equivalent capacitance of the two $4.0\,\mu\text{F}$ capacitors connected in series is given by $4.0\,\mu\text{F}/2 = 2.0\,\mu\text{F}$. This combination is then connected in parallel with two other 2.0-μF capacitors (one on each side), resulting in an equivalent capacitance $C = 3(2.0\,\mu\text{F}) = 6.0\,\mu\text{F}$. This is now seen to be in series with another combination, which consists of the two 3.0-μF capacitors connected in parallel (which are themselves equivalent to $C' = 2(3.0\,\mu\text{F}) = 6.0\,\mu\text{F}$). Thus, the equivalent capacitance of the circuit is

$$C_{\text{eq}} = \frac{CC'}{C + C'} = \frac{(6.0\,\mu\text{F})(6.0\,\mu\text{F})}{6.0\,\mu\text{F} + 6.0\,\mu\text{F}} = 3.0\,\mu\text{F} .$$

(b) Let $V = 20\,\text{V}$ be the potential difference supplied by the battery. Then $q = C_{\text{eq}} V = (3.0\,\mu\text{F})(20\,\text{V}) = 6.0 \times 10^{-5}\,\text{C}$.

(c) The potential difference across C_1 is given by

$$V_1 = \frac{CV}{C + C'} = \frac{(6.0\,\mu\text{F})(20\,\text{V})}{6.0\,\mu\text{F} + 6.0\,\mu\text{F}} = 10\,\text{V} ,$$

and the charge carried by C_1 is $q_1 = C_1 V_1 = (3.0\,\mu\text{F})(10\,\text{V}) = 3.0 \times 10^{-5}\,\text{C}$.

(d) The potential difference across C_2 is given by $V_2 = V - V_1 = 20\,\text{V} - 10\,\text{V} = 10\,\text{V}$. Consequently, the charge carried by C_2 is $q_2 = C_2V_2 = (2.0\,\mu\text{F})(10\,\text{V}) = 2.0 \times 10^{-5}$ C.

(e) Since this voltage difference V_2 is divided equally between C_3 and the other 4.0-μF capacitors connected in series with it, the voltage difference across C_3 is given by $V_3 = V_2/2 = 10\,\text{V}/2 = 5.0\,\text{V}$. Thus, $q_3 = C_3V_3 = (4.0\,\mu\text{F})(5.0\,\text{V}) = 2.0 \times 10^{-5}$ C.

19. (a) After the switches are closed, the potential differences across the capacitors are the same and the two capacitors are in parallel. The potential difference from a to b is given by $V_{ab} = Q/C_{\text{eq}}$, where Q is the net charge on the combination and C_{eq} is the equivalent capacitance. The equivalent capacitance is $C_{\text{eq}} = C_1 + C_2 = 4.0 \times 10^{-6}$ F. The total charge on the combination is the net charge on either pair of connected plates. The charge on capacitor 1 is

$$q_1 = C_1V = (1.0 \times 10^{-6}\,\text{F})(100\,\text{V}) = 1.0 \times 10^{-4}\,\text{C}$$

and the charge on capacitor 2 is

$$q_2 = C_2V = (3.0 \times 10^{-6}\,\text{F})(100\,\text{V}) = 3.0 \times 10^{-4}\,\text{C}\ ,$$

so the net charge on the combination is $3.0 \times 10^{-4}\,\text{C} - 1.0 \times 10^{-4}\,\text{C} = 2.0 \times 10^{-4}$ C. The potential difference is

$$V_{ab} = \frac{2.0 \times 10^{-4}\,\text{C}}{4.0 \times 10^{-6}\,\text{F}} = 50\,\text{V}\ .$$

(b) The charge on capacitor 1 is now $q_1 = C_1V_{ab} = (1.0 \times 10^{-6}\,\text{F})(50\,\text{V}) = 5.0 \times 10^{-5}$ C.

(c) The charge on capacitor 2 is now $q_2 = C_2V_{ab} = (3.0 \times 10^{-6}\,\text{F})(50\,\text{V}) = 1.5 \times 10^{-4}$ C.

20. (a) In this situation, capacitors 1 and 3 are in series, which means their charges are necessarily the same:

$$q_1 = q_3 = \frac{C_1C_3V}{C_1 + C_3} = \frac{(1.0\,\mu\text{F})(3.0\,\mu\text{F})(12\,\text{V})}{1.0\,\mu\text{F} + 3.0\,\mu\text{F}} = 9.0\,\mu\text{C}\ .$$

Also, capacitors 2 and 4 are in series:

$$q_2 = q_4 = \frac{C_2C_4V}{C_2 + C_4} = \frac{(2.0\,\mu\text{F})(4.0\,\mu\text{F})(12\,\text{V})}{2.0\,\mu\text{F} + 4.0\,\mu\text{F}} = 16\,\mu\text{C}\ .$$

(b) With switch 2 also closed, the potential difference V_1 across C_1 must equal the potential difference across C_2 and is

$$V_1 = \frac{C_3 + C_4}{C_1 + C_2 + C_3 + C_4}V = \frac{(3.0\,\mu\text{F} + 4.0\,\mu\text{F})(12\,\text{V})}{1.0\,\mu\text{F} + 2.0\,\mu\text{F} + 3.0\,\mu\text{F} + 4.0\,\mu\text{F}} = 8.4\,\text{V}\ .$$

Thus, $q_1 = C_1V_1 = (1.0\,\mu\text{F})(8.4\,\text{V}) = 8.4\,\mu\text{C}$, $q_2 = C_2V_1 = (2.0\,\mu\text{F})(8.4\,\text{V}) = 17\,\mu\text{C}$, $q_3 = C_3(V - V_1) = (3.0\,\mu\text{F})(12\,\text{V} - 8.4\,\text{V}) = 11\,\mu\text{C}$, and $q_4 = C_4(V - V_1) = (4.0\,\mu\text{F})(12\,\text{V} - 8.4\,\text{V}) = 14\,\mu\text{C}$.

21. The charges on capacitors 2 and 3 are the same, so these capacitors may be replaced by an equivalent capacitance determined from

$$\frac{1}{C_{\text{eq}}} = \frac{1}{C_2} + \frac{1}{C_3} = \frac{C_2 + C_3}{C_2C_3}\ .$$

Thus, $C_{\text{eq}} = C_2C_3/(C_2 + C_3)$. The charge on the equivalent capacitor is the same as the charge on either of the two capacitors in the combination and the potential difference across the equivalent capacitor is given by q_2/C_{eq}. The potential difference across capacitor 1 is q_1/C_1, where q_1 is the charge on this capacitor. The potential difference across the combination of capacitors 2 and 3 must be the same as the potential difference across capacitor 1, so $q_1/C_1 = q_2/C_{\text{eq}}$. Now some of the charge originally on capacitor 1 flows to the combination of 2 and 3. If q_0 is the original charge, conservation of charge yields

$q_1 + q_2 = q_0 = C_1 V_0$, where V_0 is the original potential difference across capacitor 1. Solving the two equations

$$\frac{q_1}{C_1} = \frac{q_2}{C_{\text{eq}}} \quad \text{and} \quad q_1 + q_2 = C_1 V_0$$

for q_1 and q_2, we find

$$q_2 = C_1 V_0 - q_1 \quad \text{and} \quad q_1 = \frac{C_1^2 V_0}{C_{\text{eq}} + C_1} = \frac{C_1^2 V_0}{\frac{C_2 C_3}{C_2 + C_3} + C_1} = \frac{C_1^2 (C_2 + C_3) V_0}{C_1 C_2 + C_1 C_3 + C_2 C_3} .$$

The charges on capacitors 2 and 3 are

$$q_2 = q_3 = C_1 V_0 - q_1 = C_1 V_0 - \frac{C_1^2 (C_2 + C_3) V_0}{C_1 C_2 + C_1 C_3 + C_2 C_3} = \frac{C_1 C_2 C_3 V_0}{C_1 C_2 + C_1 C_3 + C_2 C_3} .$$

22. Let $\mathcal{V} = 1.00\,\text{m}^3$. Using Eq. 26-23, the energy stored is

$$\begin{aligned} U &= u\mathcal{V} = \frac{1}{2}\varepsilon_0 E^2 \mathcal{V} \\ &= \frac{1}{2}\left(8.85 \times 10^{-12}\,\frac{\text{C}^2}{\text{N}\cdot\text{m}^2}\right)(150\,\text{V/m})^2(1.00\,\text{m}^3) \\ &= 9.96 \times 10^{-8}\ \text{J} . \end{aligned}$$

23. The energy stored by a capacitor is given by $U = \frac{1}{2}CV^2$, where V is the potential difference across its plates. We convert the given value of the energy to Joules. Since a Joule is a watt·second, we multiply by $(10^3\,\text{W/kW})(3600\,\text{s/h})$ to obtain $10\,\text{kW}\cdot\text{h} = 3.6 \times 10^7\,\text{J}$. Thus,

$$C = \frac{2U}{V^2} = \frac{2(3.6 \times 10^7\,\text{J})}{(1000\,\text{V})^2} = 72\ \text{F} .$$

24. (a) The capacitance is

$$C = \frac{\varepsilon_0 A}{d} = \frac{\left(8.85 \times 10^{-12}\,\frac{\text{C}^2}{\text{N}\cdot\text{m}^2}\right)(40 \times 10^{-4}\,\text{m}^2)}{1.0 \times 10^{-3}\,\text{m}} = 3.5 \times 10^{-11}\,\text{F} = 35\ \text{pF} .$$

(b) $q = CV = (35\,\text{pF})(600\,\text{V}) = 2.1 \times 10^{-8}\,\text{C} = 21\,\text{nC}$.

(c) $U = \frac{1}{2}CV^2 = \frac{1}{2}(35\,\text{pF})(21\,\text{nC})^2 = 6.3 \times 10^{-6}\,\text{J} = 6.3\,\mu\text{J}$.

(d) $E = V/d = 600\,\text{V}/1.0 \times 10^{-3}\,\text{m} = 6.0 \times 10^5\,\text{V/m}$.

(e) The energy density (energy per unit volume) is

$$u = \frac{U}{Ad} = \frac{6.3 \times 10^{-6}\,\text{J}}{(40 \times 10^{-4}\,\text{m}^2)(1.0 \times 10^{-3}\,\text{m})} = 1.6\ \text{J/m}^3 .$$

25. The total energy is the sum of the energies stored in the individual capacitors. Since they are connected in parallel, the potential difference V across the capacitors is the same and the total energy is

$$U = \frac{1}{2}(C_1 + C_2)V^2 = \frac{1}{2}\left(2.0 \times 10^{-6}\,\text{F} + 4.0 \times 10^{-6}\,\text{F}\right)(300\,\text{V})^2 = 0.27\ \text{J} .$$

26. The total energy stored in the capacitor bank is

$$U = \frac{1}{2}C_{\text{total}}V^2 = \frac{1}{2}(2000)(5.00 \times 10^{-6}\,\text{F})(50000\,\text{V})^2 = 1.3 \times 10^7\ \text{J} .$$

Thus, the cost is

$$\frac{(1.3 \times 10^7\,\text{J})(3.0\,\text{cent/kW}\cdot\text{h})}{3.6 \times 10^6\,\text{J/kW}\cdot\text{h}} = 10\,\text{cents} .$$

27. (a) In the first case $U = q^2/2C$, and in the second case $U = 2(q/2)^2/2C = q^2/4C$. So the energy is now $4.0\,\mathrm{J}/2 = 2.0\,\mathrm{J}$.

(b) It becomes the thermal energy generated in the wire connecting the capacitors during the discharging process (although a small fraction of it is probably radiated away in the form of radio waves).

28. (a) The potential difference across C_1 (the same as across C_2) is given by

$$V_1 = V_2 = \frac{C_3 V}{C_1 + C_2 + C_3} = \frac{(4.00\,\mu\mathrm{F})(100\,\mathrm{V})}{10.0\,\mu\mathrm{F} + 5.00\,\mu\mathrm{F} + 4.00\,\mu\mathrm{F}} = 21.1\ \mathrm{V}\ .$$

Also, $V_3 = V - V_1 = V - V_2 = 100\,\mathrm{V} - 21.1\,\mathrm{V} = 78.9\,\mathrm{V}$. Thus,

$$\begin{aligned} q_1 &= C_1 V_1 = (10.0\,\mu\mathrm{F})(21.1\,\mathrm{V}) = 2.11 \times 10^{-4}\ \mathrm{C} \\ q_2 &= C_2 V_2 = (5.00\,\mu\mathrm{F})(21.1\,\mathrm{V}) = 1.05 \times 10^{-4}\ \mathrm{C} \\ q_3 &= q_1 + q_2 = 2.11 \times 10^{-4}\,\mathrm{C} + 1.05 \times 10^{-4}\,\mathrm{C} = 3.16 \times 10^{-4}\ \mathrm{C}\ . \end{aligned}$$

(b) The potential differences were found in the course of solving for the charges in part (a).

(c) The stored energies are as follows:

$$\begin{aligned} U_1 &= \frac{1}{2} C_1 V_1^2 = \frac{1}{2}(10.0\,\mu\mathrm{F})(21.1\,\mathrm{V})^2 = 2.22 \times 10^{-3}\ \mathrm{J}\ , \\ U_2 &= \frac{1}{2} C_2 V_2^2 = \frac{1}{2}(5.00\,\mu\mathrm{F})(21.1\,\mathrm{V})^2 = 1.11 \times 10^{-3}\ \mathrm{J}\ , \\ U_3 &= \frac{1}{2} C_3 V_3^2 = \frac{1}{2}(4.00\,\mu\mathrm{F})(78.9\,\mathrm{V})^2 = 1.25 \times 10^{-2}\ \mathrm{J}\ . \end{aligned}$$

29. (a) Let q be the charge on the positive plate. Since the capacitance of a parallel-plate capacitor is given by $\varepsilon_0 A/d$, the charge is $q = CV = \varepsilon_0 AV/d$. After the plates are pulled apart, their separation is $2d$ and the potential difference is V'. Then $q = \varepsilon_0 AV'/2d$ and

$$V' = \frac{2d}{\varepsilon_0 A} q = \frac{2d}{\varepsilon_0 A} \frac{\varepsilon_0 A}{d} V = 2V\ .$$

(b) The initial energy stored in the capacitor is

$$U_i = \frac{1}{2} CV^2 = \frac{\varepsilon_0 AV^2}{2d}$$

and the final energy stored is

$$U_f = \frac{1}{2} \frac{\varepsilon_0 A}{2d} (V')^2 = \frac{1}{2} \frac{\varepsilon_0 A}{2d} 4V^2 = \frac{\varepsilon_0 AV^2}{d}\ .$$

This is twice the initial energy.

(c) The work done to pull the plates apart is the difference in the energy: $W = U_f - U_i = \varepsilon_0 AV^2/2d$.

30. (a) The charge in the Figure is

$$q_3 = C_3 V = (4.00\,\mu\mathrm{F})(100\,\mathrm{V}) = 4.00 \times 10^{-4}\,\mathrm{mC}\ ,$$

$$q_1 = q_2 = \frac{C_1 C_2 V}{C_1 + C_2} = \frac{(10.0\,\mu\mathrm{F})(5.00\,\mu\mathrm{F})(100\,\mathrm{V})}{10.0\,\mu\mathrm{F} + 5.00\,\mu\mathrm{F}} = 3.33 \times 10^{-4}\ \mathrm{C}\ .$$

(b) $V_1 = q_1/C_1 = 3.33 \times 10^{-4}\,\mathrm{C}/10.0\,\mu\mathrm{F} = 33.3\,\mathrm{V}$, $V_2 = V - V_1 = 100\,\mathrm{V} - 33.3\,\mathrm{V} = 66.7\,\mathrm{V}$, and $V_3 = V = 100\,\mathrm{V}$.

(c) We use $U_i = \frac{1}{2}C_iV_i^2$, where $i = 1, 2, 3$. The answers are $U_1 = 5.6\,\text{mJ}$, $U_1 = 11\,\text{mJ}$, and $U_1 = 20\,\text{mJ}$.

31. We first need to find an expression for the energy stored in a cylinder of radius R and length L, whose surface lies between the inner and outer cylinders of the capacitor ($a < R < b$). The energy density at any point is given by $u = \frac{1}{2}\varepsilon_0 E^2$, where E is the magnitude of the electric field at that point. If q is the charge on the surface of the inner cylinder, then the magnitude of the electric field at a point a distance r from the cylinder axis is given by

$$E = \frac{q}{2\pi\varepsilon_0 Lr}$$

(see Eq. 26-12), and the energy density at that point is given by

$$u = \frac{1}{2}\varepsilon_0 E^2 = \frac{q^2}{8\pi^2\varepsilon_0 L^2 r^2} .$$

The energy in the cylinder is the volume integral

$$U_R = \int u \, d\mathcal{V} .$$

Now, $d\mathcal{V} = 2\pi r L\, dr$, so

$$U_R = \int_a^R \frac{q^2}{8\pi^2\varepsilon_0 L^2 r^2}\, 2\pi r L\, dr = \frac{q^2}{4\pi\varepsilon_0 L}\int_a^R \frac{dr}{r} = \frac{q^2}{4\pi\varepsilon_0 L}\ln\frac{R}{a} .$$

To find an expression for the total energy stored in the capacitor, we replace R with b:

$$U_b = \frac{q^2}{4\pi\varepsilon_0 L}\ln\frac{b}{a} .$$

We want the ratio U_R/U_b to be 1/2, so

$$\ln\frac{R}{a} = \frac{1}{2}\ln\frac{b}{a}$$

or, since $\frac{1}{2}\ln(b/a) = \ln(\sqrt{b/a})$, $\ln(R/a) = \ln(\sqrt{b/a})$. This means $R/a = \sqrt{b/a}$ or $R = \sqrt{ab}$.

32. We use $E = q/4\pi\varepsilon_0 R^2 = V/R$. Thus

$$u = \frac{1}{2}\varepsilon_0 E^2 = \frac{1}{2}\varepsilon_0\left(\frac{V}{R}\right)^2 = \frac{1}{2}\left(8.85\times10^{-12}\,\frac{\text{C}^2}{\text{N}\cdot\text{m}^2}\right)\left(\frac{8000\,\text{V}}{0.050\,\text{m}}\right)^2 = 0.11\ \text{J/m}^3 .$$

33. The charge is held constant while the plates are being separated, so we write the expression for the stored energy as $U = q^2/2C$, where q is the charge and C is the capacitance. The capacitance of a parallel-plate capacitor is given by $C = \varepsilon_0 A/x$, where A is the plate area and x is the plate separation, so

$$U = \frac{q^2 x}{2\varepsilon_0 A} .$$

If the plate separation increases by dx, the energy increases by $dU = (q^2/2\varepsilon_0 A)\,dx$. Suppose the agent pulling the plate apart exerts force F. Then the agent does work $F\,dx$ and if the plates begin and end at rest, this must equal the increase in stored energy. Thus,

$$F\,dx = \left(\frac{q^2}{2\varepsilon_0 A}\right) dx$$

and

$$F = \frac{q^2}{2\varepsilon_0 A} .$$

The net force on a plate is zero, so this must also be the magnitude of the force one plate exerts on the other. The force can also be computed as the product of the charge q on one plate and the electric field E_1 due to the charge on the other plate. Recall that the field produced by a uniform plane surface of charge is $E_1 = q/2\varepsilon_0 A$. Thus, $F = q^2/2\varepsilon_0 A$.

34. If the original capacitance is given by $C = \varepsilon_0 A/d$, then the new capacitance is $C' = \varepsilon_0 \kappa A/2d$. Thus $C'/C = \kappa/2$ or $\kappa = 2C'/C = 2(2.6\,\text{pF}/1.3\,\text{pF}) = 4.0$.

35. The capacitance with the dielectric in place is given by $C = \kappa C_0$, where C_0 is the capacitance before the dielectric is inserted. The energy stored is given by $U = \frac{1}{2}CV^2 = \frac{1}{2}\kappa C_0 V^2$, so

$$\kappa = \frac{2U}{C_0 V^2} = \frac{2(7.4 \times 10^{-6}\,\text{J})}{(7.4 \times 10^{-12}\,\text{F})(652\,\text{V})^2} = 4.7 \; .$$

According to Table 26-1, you should use Pyrex.

36. (a) We use $C = \varepsilon_0 A/d$ to solve for d:

$$d = \frac{\varepsilon_0 A}{C} = \frac{\left(8.85 \times 10^{-12}\,\frac{\text{C}^2}{\text{N}\cdot\text{m}^2}\right)(0.35\,\text{m}^2)}{50 \times 10^{-12}\,\text{F}} = 6.2 \times 10^{-2}\,\text{m} \; .$$

(b) We use $C \propto \kappa$. The new capacitance is $C' = C(\kappa/\kappa_{\text{air}}) = (50\,\text{pf})(5.6/1.0) = 280\,\text{pF}$.

37. The capacitance of a cylindrical capacitor is given by

$$C = \kappa C_0 = \frac{2\pi\kappa\varepsilon_0 L}{\ln(b/a)} \, ,$$

where C_0 is the capacitance without the dielectric, κ is the dielectric constant, L is the length, a is the inner radius, and b is the outer radius. The capacitance per unit length of the cable is

$$\frac{C}{L} = \frac{2\pi\kappa\varepsilon_0}{\ln(b/a)} = \frac{2\pi(2.6)(8.85 \times 10^{-12}\,\text{F/m})}{\ln\left[(0.60\,\text{mm})/(0.10\,\text{mm})\right]} = 8.1 \times 10^{-11}\,\text{F/m} = 81\ \text{pF/m} \; .$$

38. (a) We use Eq. 26-14:

$$C = 2\pi\varepsilon_0\kappa\frac{L}{\ln(b/a)} = \frac{(4.7)(0.15\,\text{m})}{2\left(8.99 \times 10^9\,\frac{\text{N}\cdot\text{m}^2}{\text{C}^2}\right)\ln(3.8\,\text{cm}/3.6\,\text{cm})} = 0.73\ \text{nF} \; .$$

(b) The breakdown potential is $(14\,\text{kV/mm})(3.8\,\text{cm} - 3.6\,\text{cm}) = 28\,\text{kV}$.

39. The capacitance is given by $C = \kappa C_0 = \kappa\varepsilon_0 A/d$, where C_0 is the capacitance without the dielectric, κ is the dielectric constant, A is the plate area, and d is the plate separation. The electric field between the plates is given by $E = V/d$, where V is the potential difference between the plates. Thus, $d = V/E$ and $C = \kappa\varepsilon_0 AE/V$. Thus,

$$A = \frac{CV}{\kappa\varepsilon_0 E} \, .$$

For the area to be a minimum, the electric field must be the greatest it can be without breakdown occurring. That is,

$$A = \frac{(7.0 \times 10^{-8}\,\text{F})(4.0 \times 10^3\,\text{V})}{2.8(8.85 \times 10^{-12}\,\text{F/m})(18 \times 10^6\,\text{V/m})} = 0.63\ \text{m}^2 \; .$$

40. The capacitor can be viewed as two capacitors C_1 and C_2 in parallel, each with surface area $A/2$ and plate separation d, filled with dielectric materials with dielectric constants κ_1 and κ_2, respectively. Thus

$$C = C_1 + C_2 = \frac{\varepsilon_0(A/2)\kappa_1}{d} + \frac{\varepsilon_0(A/2)\kappa_2}{d} = \frac{\varepsilon_0 A}{d}\left(\frac{\kappa_1 + \kappa_2}{2}\right) .$$

41. We assume there is charge q on one plate and charge $-q$ on the other. The electric field in the lower half of the region between the plates is

$$E_1 = \frac{q}{\kappa_1 \varepsilon_0 A},$$

where A is the plate area. The electric field in the upper half is

$$E_2 = \frac{q}{\kappa_2 \varepsilon_0 A}.$$

Let $d/2$ be the thickness of each dielectric. Since the field is uniform in each region, the potential difference between the plates is

$$V = \frac{E_1 d}{2} + \frac{E_2 d}{2} = \frac{qd}{2\varepsilon_0 A}\left[\frac{1}{\kappa_1} + \frac{1}{\kappa_2}\right] = \frac{qd}{2\varepsilon_0 A}\frac{\kappa_1 + \kappa_2}{\kappa_1 \kappa_2},$$

so

$$C = \frac{q}{V} = \frac{2\varepsilon_0 A}{d}\frac{\kappa_1 \kappa_2}{\kappa_1 + \kappa_2}.$$

This expression is exactly the same as the that for C_{eq} of two capacitors in series, one with dielectric constant κ_1 and the other with dielectric constant κ_2. Each has plate area A and plate separation $d/2$. Also we note that if $\kappa_1 = \kappa_2$, the expression reduces to $C = \kappa_1 \varepsilon_0 A/d$, the correct result for a parallel-plate capacitor with plate area A, plate separation d, and dielectric constant κ_1.

42. Let $C_1 = \varepsilon_0(A/2)\kappa_1/2d = \varepsilon_0 A\kappa_1/4d$, $C_2 = \varepsilon_0(A/2)\kappa_2/d = \varepsilon_0 A\kappa_2/2d$, and $C_3 = \varepsilon_0 A\kappa_3/2d$. Note that C_2 and C_3 are effectively connected in series, while C_1 is effectively connected in parallel with the C_2-C_3 combination. Thus,

$$\begin{aligned} C &= C_1 + \frac{C_2 C_3}{C_2 + C_3} = \frac{\varepsilon_0 A\kappa_1}{4d} + \frac{(\varepsilon_0 A/d)(\kappa_2/2)(\kappa_3/2)}{\kappa_2/2 + \kappa_3/2} \\ &= \frac{\varepsilon_0 A}{4d}\left(\kappa_1 + \frac{2\kappa_2\kappa_3}{\kappa_2 + \kappa_3}\right). \end{aligned}$$

43. (a) The electric field in the region between the plates is given by $E = V/d$, where V is the potential difference between the plates and d is the plate separation. The capacitance is given by $C = \kappa\varepsilon_0 A/d$, where A is the plate area and κ is the dielectric constant, so $d = \kappa\varepsilon_0 A/C$ and

$$E = \frac{VC}{\kappa\varepsilon_0 A} = \frac{(50\,\text{V})(100\times10^{-12}\,\text{F})}{5.4(8.85\times10^{-12}\,\text{F/m})(100\times10^{-4}\,\text{m}^2)} = 1.0\times10^4\,\text{V/m}.$$

(b) The free charge on the plates is $q_f = CV = (100\times10^{-12}\,\text{F})(50\,\text{V}) = 5.0\times10^{-9}\,\text{C}$.

(c) The electric field is produced by both the free and induced charge. Since the field of a large uniform layer of charge is $q/2\varepsilon_0 A$, the field between the plates is

$$E = \frac{q_f}{2\varepsilon_0 A} + \frac{q_f}{2\varepsilon_0 A} - \frac{q_i}{2\varepsilon_0 A} - \frac{q_i}{2\varepsilon_0 A},$$

where the first term is due to the positive free charge on one plate, the second is due to the negative free charge on the other plate, the third is due to the positive induced charge on one dielectric surface, and the fourth is due to the negative induced charge on the other dielectric surface. Note that the field due to the induced charge is opposite the field due to the free charge, so they tend to cancel. The induced charge is therefore

$$\begin{aligned} q_i &= q_f - \varepsilon_0 AE \\ &= 5.0\times10^{-9}\,\text{C} - (8.85\times10^{-12}\,\text{F/m})(100\times10^{-4}\,\text{m}^2)(1.0\times10^4\,\text{V/m}) \\ &= 4.1\times10^{-9}\,\text{C} = 4.1\,\text{nC}. \end{aligned}$$

44. (a) The electric field E_1 in the free space between the two plates is $E_1 = q/\varepsilon_0 A$ while that inside the slab is $E_2 = E_1/\kappa = q/\kappa\varepsilon_0 A$. Thus,

$$V_0 = E_1(d-b) + E_2 b = \left(\frac{q}{\varepsilon_0 A}\right)\left(d - b + \frac{b}{\kappa}\right) ,$$

and the capacitance is

$$\begin{aligned} C &= \frac{q}{V_0} = \frac{\varepsilon_0 A\kappa}{\kappa(d-b)+b} \\ &= \frac{\left(8.85\times 10^{-12}\,\frac{\mathrm{C}^2}{\mathrm{N\cdot m^2}}\right)(115\times 10^{-4}\,\mathrm{m}^2)(2.61)}{(2.61)(0.0124\,\mathrm{m} - 0.00780\,\mathrm{m}) + (0.00780\,\mathrm{m})} \\ &= 13.4\ \mathrm{pF}\ . \end{aligned}$$

(b) $q = CV = (13.4\times 10^{-12}\,\mathrm{F})(85.5\,\mathrm{V}) = 1.15\,\mathrm{nC}$.

(c) The magnitude of the electric field in the gap is

$$E_1 = \frac{q}{\varepsilon_0 A} = \frac{1.15\times 10^{-9}\,\mathrm{C}}{\left(8.85\times 10^{-12}\,\frac{\mathrm{C}^2}{\mathrm{N\cdot m^2}}\right)(115\times 10^{-4}\,\mathrm{m}^2)} = 1.13\times 10^4\ \mathrm{N/C}\ .$$

(d) Using Eq. 26-32, we obtain

$$E_2 = \frac{E_1}{\kappa} = \frac{1.13\times 10^4\,\mathrm{N/C}}{2.61} = 4.33\times 10^3\ \mathrm{N/C}\ .$$

45. (a) According to Eq. 26-17 the capacitance of an air-filled spherical capacitor is given by

$$C_0 = 4\pi\varepsilon_0 \frac{ab}{b-a}\ .$$

When the dielectric is inserted between the plates the capacitance is greater by a factor of the dielectric constant κ. Consequently, the new capacitance is

$$C = 4\pi\kappa\varepsilon_0 \frac{ab}{b-a}\ .$$

(b) The charge on the positive plate is

$$q = CV = 4\pi\kappa\varepsilon_0 \frac{ab}{b-a} V\ .$$

(c) Let the charge on the inner conductor to be $-q$. Immediately adjacent to it is the induced charge q'. Since the electric field is less by a factor $1/\kappa$ than the field when no dielectric is present, then $-q + q' = -q/\kappa$. Thus,

$$q' = \frac{\kappa - 1}{\kappa} q = 4\pi(\kappa - 1)\varepsilon_0 \frac{ab}{b-a} V\ .$$

46. (a) We apply Gauss's law with dielectric: $q/\varepsilon_0 = \kappa E A$, and solve for κ:

$$\kappa = \frac{q}{\varepsilon_0 E A} = \frac{8.9\times 10^{-7}\,\mathrm{C}}{\left(8.85\times 10^{-12}\,\frac{\mathrm{C}^2}{\mathrm{N\cdot m^2}}\right)(1.4\times 10^{-6}\,\mathrm{V/m})(100\times 10^{-4}\,\mathrm{m}^2)} = 7.2\ .$$

(b) The charge induced is

$$q' = q\left(1 - \frac{1}{\kappa}\right) = (8.9\times 10^{-7}\,\mathrm{C})\left(1 - \frac{1}{7.2}\right) = 7.7\times 10^{-7}\ \mathrm{C}\ .$$

47. Assuming the charge on one plate is $+q$ and the charge on the other plate is $-q$, we find an expression for the electric field in each region, in terms of q, then use the result to find an expression for the potential difference V between the plates. The capacitance is

$$C = \frac{q}{V} \; .$$

The electric field in the dielectric is $E_d = q/\kappa\varepsilon_0 A$, where κ is the dielectric constant and A is the plate area. Outside the dielectric (but still between the capacitor plates) the field is $E = q/\varepsilon_0 A$. The field is uniform in each region so the potential difference across the plates is

$$V = E_d b + E(d-b) = \frac{qb}{\kappa\varepsilon_0 A} + \frac{q(d-b)}{\varepsilon_0 A} = \frac{q}{\varepsilon_0 A}\frac{b+\kappa(d-b)}{\kappa} \; .$$

The capacitance is

$$C = \frac{q}{V} = \frac{\kappa\varepsilon_0 A}{\kappa(d-b)+b} = \frac{\kappa\varepsilon_0 A}{\kappa d - b(\kappa-1)} \; .$$

The result does not depend on where the dielectric is located between the plates; it might be touching one plate or it might have a vacuum gap on each side.

For the capacitor of Sample Problem 26-8, $\kappa = 2.61$, $A = 115\,\text{cm}^2 = 115 \times 10^{-4}\,\text{m}^2$, $d = 1.24\,\text{cm} = 1.24 \times 10^{-2}\,\text{m}$, and $b = 0.78\,\text{cm} = 0.78 \times 10^{-2}\,\text{m}$, so

$$\begin{aligned} C &= \frac{2.61(8.85 \times 10^{-12}\,\text{F/m})(115 \times 10^{-4}\,\text{m}^2)}{2.61(1.24 \times 10^{-2}\,\text{m}) - (0.780 \times 10^{-2}\,\text{m})(2.61-1)} \\ &= 1.34 \times 10^{-11}\,\text{F} = 13.4\,\text{pF} \end{aligned}$$

in agreement with the result found in the sample problem. If $b = 0$ and $\kappa = 1$, then the expression derived above yields $C = \varepsilon_0 A/d$, the correct expression for a parallel-plate capacitor with no dielectric. If $b = d$, then the derived expression yields $C = \kappa\varepsilon_0 A/d$, the correct expression for a parallel-plate capacitor completely filled with a dielectric.

48. (a) Eq. 26-22 yields

$$U = \frac{1}{2}CV^2 = \frac{1}{2}\left(200 \times 10^{-12}\,\text{F}\right)\left(7.0 \times 10^3\,\text{V}\right)^2 = 4.9 \times 10^{-3}\,\text{J} \; .$$

(b) Our result from part (a) is much less than the required 150 mJ, so such a spark should not have set off an explosion.

49. (a) With the potential difference equal to 600 V, a capacitance of $2.5 \times 10^{-10}\,\text{F}$ can only store energy equal to $U = \frac{1}{2}CV^2 = 4.5 \times 10^{-5}\,\text{J}$.

(b) No, our result from part (a) is only about 20% of that needed to produce a spark.

(c) Considering the charge as a constant, then voltage should be inversely proportional to the capacitance. Therefore, if the capacitance drops by a factor of ten, then we expect the voltage to increase by that same factor: $V_f = 6000$ V.

(d) Now the energy stored is $U' = \frac{1}{2}C_f V_f^2 = 4.5 \times 10^{-4}\,\text{J}$, a factor of ten greater than the value we obtained in part (a).

(e) Yes, this new value of energy is nearly double that needed for a spark.

50. (a) We calculate the charged surface area of the cylindrical volume as follows:

$$A = 2\pi rh + \pi r^2 = 2\pi(0.20\,\text{m})(0.10\,\text{m}) + \pi(0.20\,\text{m})^2 = 0.25\ \text{m}^2$$

where we note from the figure that although the bottom is charged, the top is not. Therefore, the charge is $q = \sigma A = -0.50\,\mu\text{C}$ on the exterior surface, and consequently (according to the assumptions in the problem) that same charge q is induced in the interior of the fluid.

(b) By Eq. 26-21, the energy stored is

$$U = \frac{q^2}{2C} = \frac{\left(5.0 \times 10^{-7}\,\text{C}\right)^2}{2\,(35 \times 10^{-12}\,\text{F})} = 3.6 \times 10^{-3}\ \text{J} .$$

(c) Our result is within a factor of three of that needed to cause a spark. Our conclusion is that it will probably not cause a spark; however, there is not enough of a safety factor to be sure.

51. (a) We know from Eq. 26-7 that the magnitude of the electric field is directly proportional to the surface charge density:

$$E = \frac{\sigma}{\varepsilon_0} = \frac{15 \times 10^{-6}\,\text{C/m}^2}{8.85 \times 10^{-12}\,\text{C}^2/\text{N}\cdot\text{m}^2} = 1.7 \times 10^6\ \text{V/m} .$$

Regarding the units, it is worth noting that a Volt is equivalent to a N·m/C.

(b) Eq. 26-23 yields

$$u = \frac{1}{2}\varepsilon_0 E^2 = 13\ \text{J/m}^3 .$$

(c) The energy U is the energy-per-unit-volume multiplied by the (variable) volume of the region between the layers of plastic food wrap. Since the distance between the layers is x, and we use A for the area over which the (say, positive) charge is spread, then that volume is Ax. Thus,

$$U = uAx \qquad \text{where} \qquad u = 13\ \text{J/m}^3 .$$

(d) The magnitude of force is

$$\left|\vec{F}\right| = \frac{dU}{dx} = uA .$$

(e) The force per unit area is

$$\frac{\left|\vec{F}\right|}{A} = u = 13\ \text{N/m}^2 .$$

Regarding units, it is worth noting that a Joule is equivalent to a N·m, which explains how J/m^3 may be set equal to N/m^2 in the above manipulation. We note, too, that the pressure unit N/m^2 is generally known as a Pascal (Pa).

(f) Combining our steps in parts (a) through (e), we have

$$\begin{aligned} \frac{\left|\vec{F}\right|}{A} &= u = \frac{1}{2}\varepsilon_0 E^2 \\ 6.0\ \text{N/m}^2 &= \frac{1}{2}\varepsilon_0 \left(\frac{\sigma}{\varepsilon_0}\right)^2 = \frac{\sigma^2}{2\varepsilon_0} \end{aligned}$$

which leads to $\sigma = \sqrt{2(8.85 \times 10^{-12})(6.0)} = 1.0 \times 10^{-5}\ \text{C/m}^2$.

52. (a) We do not employ energy conservation since, in reaching equilibrium, some energy is dissipated either as heat or radio waves. Charge is conserved; therefore, if $Q = C_1 V_{\text{bat}} = 40\ \mu\text{C}$, and q_1 and q_2 are the charges on C_1 and C_2 after the switch is thrown to the right and equilibrium is reached, then

$$Q = q_1 + q_2 .$$

Reducing the right portion of the circuit (the C_3, C_4 parallel pair which are in series with C_2) we have an equivalent capacitance of $C' = 8.0\ \mu\text{F}$ which has charge $q' = q_2$ and potential difference equal to that of C_1. Thus,

$$\begin{aligned} V_1 &= V' \\ \frac{q_1}{C_1} &= \frac{q_2}{C'} \end{aligned}$$

which yields $4q_1 = q_2$. Therefore,

$$Q = q_1 + 4q_1$$

leads to $q_1 = 8.0\,\mu C$ and consequently to $q_2 = 32\,\mu C$.

(b) From Eq. 26-1, we have $V_2 = (32\,\mu C)(16\,\mu F) = 2.0$ V.

53. Using Eq. 26-27, with $\sigma = q/A$, we have

$$\left|\vec{E}\right| = \frac{q}{\kappa\varepsilon_0 A} = 200 \times 10^3 \text{ N/C}$$

which yields $q = 3.3 \times 10^{-7}$ C. Eq. 26-21 and Eq. 26-25 therefore lead to

$$U = \frac{q^2}{2C} = \frac{q^2 d}{2\kappa\varepsilon_0 A} = 6.6 \times 10^{-5} \text{ J} .$$

54. (a) The potential across capacitor 1 is 10 V, so the charge on it is

$$q_1 = C_1 V_1 = (10\,\mu\text{F})(10\text{ V}) \;=\; 100\,\mu\text{C} .$$

(b) Reducing the right portion of the circuit produces an equivalence equal to $6.0\,\mu$F, with 10 V across it. Thus, a charge of $60\,\mu$C is on it – and consequently also on the bottom right capacitor. The bottom right capacitor has, as a result, a potential across it equal to

$$V = \frac{q}{C} = \frac{60\,\mu\text{C}}{10\,\mu\text{F}} \;=\; 6.0\text{ V} ,$$

which leaves $10 - 6 = 4.0$ V across the group of capacitors in the upper right portion of the circuit. Inspection of the arrangement (and capacitance values) of that group reveals that this 4.0 V must be equally divided by C_2 and the capacitor directly below it (in series with it). Therefore, with 2.0 V across capacitor 2, we find

$$q_2 = C_2 V_2 = (10\,\mu\text{F})(2.0\text{ V}) = 20\,\mu\text{C} .$$

55. (a) We use $q = CV = \varepsilon_0 AV/d$ to solve for A:

$$A = \frac{Cd}{\varepsilon_0} = \frac{(10 \times 10^{-12}\,\text{F})(1.0 \times 10^{-3}\,\text{m})}{\left(8.85 \times 10^{-12}\,\frac{\text{C}^2}{\text{N}\cdot\text{m}^2}\right)} = 1.1 \times 10^{-3}\,\text{m}^2 .$$

(b) Now,

$$C' = C\left(\frac{d}{d'}\right) = (10\,\text{pF})\left(\frac{1.0\,\text{mm}}{0.9\,\text{mm}}\right) = 11\,\text{pF} .$$

(c) The new potential difference is $V' = q/C' = CV/C'$. Thus,

$$\Delta V = V' - V = \frac{(10\,\text{pF})(12\,\text{V})}{11\,\text{pF}} - 12\,\text{V} = 1.2\text{ V} .$$

In a microphone, mechanical pressure applied to the aluminum foil as a result of sound can cause the capacitance of the foil to change, thereby inducing a variable ΔV in response to the sound signal.

56. (a) Here D is not attached to anything, so that the $6C$ and $4C$ capacitors are in series (equivalent to $2.4C$). This is then in parallel with the $2C$ capacitor, which produces an equivalence of $4.4C$. Finally the $4.4C$ is in series with C and we obtain

$$C_{\text{eq}} = \frac{(C)\,(4.4C)}{C + 4.4C} = 0.82C = 41\,\mu\text{F}$$

where we have used the fact that $C = 50\,\mu$F.

(b) Now, B is the point which is not attached to anything, so that the $6C$ and $2C$ capacitors are now in series (equivalent to $1.5C$), which is then in parallel with the $4C$ capacitor (and thus equivalent to $5.5C$). The $5.5C$ is then in series with the C capacitor; consequently,

$$C_{\text{eq}} = \frac{(C)(5.5C)}{C + 5.5C} = 0.85C = 42\,\mu\text{F} .$$

57. In the first case the two capacitors are effectively connected in series, so the output potential difference is $V_{\text{out}} = CV_{\text{in}}/2C = V_{\text{in}}/2 = 50.0\,\text{V}$. In the second case the lower diode acts as a wire so $V_{\text{out}} = 0$.

58. For maximum capacitance the two groups of plates must face each other with maximum area. In this case the whole capacitor consists of $(n-1)$ identical single capacitors connected in parallel. Each capacitor has surface area A and plate separation d so its capacitance is given by $C_0 = \varepsilon_0 A/d$. Thus, the total capacitance of the combination is

$$C = (n-1)C_0 = \frac{(n-1)\varepsilon_0 A}{d} .$$

59. The voltage across capacitor 1 is

$$V_1 = \frac{q_1}{C_1} = \frac{30\,\mu\text{C}}{10\,\mu\text{F}} = 3.0\text{ V} .$$

Since $V_1 = V_2$, the total charge on capacitor 2 is

$$q_2 = C_2V_2 = (20\,\mu\text{F})\,(2\text{ V}) = 60\ \mu\text{C} ,$$

which means a total of $90\,\mu\text{C}$ of charge is on the pair of capacitors C_1 and C_2. This implies there is a total of $90\,\mu\text{C}$ of charge also on the C_3 and C_4 pair. Since $C_3 = C_4$, the charge divides equally between them, so $q_3 = q_4 = 45\,\mu\text{C}$. Thus, the voltage across capacitor 3 is

$$V_3 = \frac{q_3}{C_3} = \frac{45\,\mu\text{C}}{20\,\mu\text{F}} = 2.3\text{ V} .$$

Therefore, $|V_A - V_B| = V_1 + V_3 \; = \; 5.3\text{ V}$.

60. (a) The equivalent capacitance is

$$C_{\text{eq}} = \frac{C_1C_2}{C_1 + C_2} = \frac{(6.00\,\mu\text{F})(4.00\,\mu\text{F})}{6.00\,\mu\text{F} + 4.00\,\mu\text{F}} = 2.40\,\mu\text{F} .$$

(b) $q = C_{\text{eq}}V = (2.40\,\mu\text{F})(200\,\text{V}) = 4.80 \times 10^4\,\text{C}$.

(c) $V_1 = q/C_1 = 4.80 \times 10^4\,\text{C}/2.40\,\mu\text{F} = 120\,\text{V}$, and $V_2 = V - V_1 = 200\,\text{V} - 120\,\text{V} = 80\,\text{V}$.

61. (a) Now $C_{\text{eq}} = C_1 + C_2 = 6.00\,\mu\text{F} + 4.00\,\mu\text{F} = 10.0\,\mu\text{F}$.

(b) $q_1 = C_1V = (6.00\,\mu\text{F})(200\,\text{V}) = 1.20 \times 10^{-3}\,\text{C}$, $q_2 = C_2V = (4.00\,\mu\text{F})(200\,\text{V}) = 8.00 \times 10^{-4}\,\text{C}$.

(c) $V_1 = V_2 = 200\,\text{V}$.

62. We cannot expect simple energy conservation to hold since energy is presumably dissipated either as heat in the hookup wires or as radio waves while the charge oscillates in the course of the system "settling down" to its final state (of having 40 V across the parallel pair of capacitors C and $60\,\mu\text{F}$). We do expect charge to be conserved. Thus, if Q is the charge originally stored on C and q_1, q_2 are the charges on the parallel pair after "setting down," then

$$\begin{aligned} Q &= q_1 + q_2 \\ C(100\text{ V}) &= C(40\text{ V}) + (60\,\mu\text{F})\,(40\text{ V}) \end{aligned}$$

which leads to the solution $C = 40\,\mu\text{F}$.

63. (a) Put five such capacitors in series. Then, the equivalent capacitance is $2.0\,\mu\text{F}/5 = 0.40\,\mu\text{F}$. With each capacitor taking a 200-V potential difference, the equivalent capacitor can withstand 1000 V.

(b) As one possibility, you can take three identical arrays of capacitors, each array being a five-capacitor combination described in part (a) above, and hook up the arrays in parallel. The equivalent capacitance is now $C_{\text{eq}} = 3(0.40\,\mu\text{F}) = 1.2\,\mu\text{F}$. With each capacitor taking a 200-V potential difference the equivalent capacitor can withstand 1000 V.

64. (a) The energy per unit volume is

$$u = \frac{1}{2}\varepsilon_0 E^2 = \frac{1}{2}\varepsilon_0 \left(\frac{e}{4\pi\varepsilon_0 r^2}\right)^2 = \frac{e^2}{32\pi^2\varepsilon_0 r^4} \ .$$

(b) From the expression above $u \propto r^{-4}$. So for $r \to 0 \quad u \to \infty$.

65. (a) They each store the same charge, so the maximum voltage is across the smallest capacitor. With 100 V across 10 μF, then the voltage across the 20 μF capacitor is 50 V and the voltage across the 25 μF capacitor is 40 V. Therefore, the voltage across the arrangement is 190 V.

(b) Using Eq. 26-21 or Eq. 26-22, we sum the energies on the capacitors and obtain $U_{\text{total}} = 0.095$ J.

66. (a) Since the field is constant and the capacitors are in parallel (each with 600 V across them) with identical distances ($d = 0.00300$ m) between the plates, then the field in A is equal to the field in B:

$$\left|\vec{E}\right| = \frac{V}{d} = 2.00 \times 10^5 \text{ V/m} \ .$$

(b) See the note in part (a).

(c) For the air-filled capacitor, Eq. 26-4 leads to

$$\sigma = \frac{q}{A} = \varepsilon_0 \left|\vec{E}\right| = 1.77 \times 10^{-6} \text{ C/m}^2 \ .$$

(d) For the dielectric-filled capacitor, we use Eq. 26-27:

$$\sigma = \kappa\varepsilon_0 \left|\vec{E}\right| = 4.60 \times 10^{-6} \text{ C/m}^2 \ .$$

(e) Although the discussion in the textbook (§26-8) is in terms of the charge being held fixed (while a dielectric is inserted), it is readily adapted to this situation (where comparison is made of two capacitors which have the same *voltage* and are identical except for the fact that one has a dielectric). The fact that capacitor B has a relatively large charge but only produces the field that A produces (with its smaller charge) is in line with the point being made (in the text) with Eq. 26-32 and in the material that follows. Adapting Eq. 26-33 to this problem, we see that the difference in charge densities between parts (c) and (d) is due, in part, to the (negative) layer of charge at the top surface of the dielectric; consequently,

$$\sigma' = \left(1.77 \times 10^{-6}\right) - \left(4.60 \times 10^{-6}\right) = -2.83 \times 10^{-6} \text{ C/m}^2 \ .$$

67. (a) The equivalent capacitance is $C_{\text{eq}} = C_1C_2/(C_1 + C_2)$. Thus the charge q on each capacitor is

$$q = C_{\text{eq}}V = \frac{C_1C_2V}{C_1 + C_2} = \frac{(2.0\,\mu\text{F})(8.0\,\mu\text{F})(300\text{ V})}{2.0\,\mu\text{F} + 8.0\,\mu\text{F}} = 4.8 \times 10^{-4} \text{ C} \ .$$

The potential differences are: $V_1 = q/C_1 = 4.8 \times 10^{-4}\text{ C}/2.0\,\mu\text{F} = 240\text{ V}$, $V_2 = V - V_1 = 300\text{ V} - 240\text{ V} = 60\text{ V}$.

(b) Now we have $q_1'/C_1 = q_2'/C_2 = V'$ (V' being the new potential difference across each capacitor) and $q_1' + q_2' = 2q$. We solve for q_1', q_2' and V:

$$\begin{aligned} q_1' &= \frac{2C_1 q}{C_1 + C_2} = \frac{2(2.0\,\mu\text{F})(4.8\times 10^{-4}\text{ C})}{2.0\,\mu\text{F} + 8.0\,\mu\text{F}} = 1.9\times 10^{-4}\text{ C}\ , \\ q_2' &= 2q - q_1 = 7.7\times 10^{-4}\text{ C}\ , \\ V' &= \frac{q_1'}{C_1} = \frac{1.92\times 10^{-4}\text{ C}}{2.0\,\mu\text{F}} = 96\text{ V}\ . \end{aligned}$$

(c) In this circumstance, the capacitors will simply discharge themselves, leaving $q_1 = q_2 = 0$ and $V_1 = V_2 = 0$.

68. We use $U = \frac{1}{2}CV^2$. As V is increased by ΔV, the energy stored in the capacitor increases correspondingly from U to $U + \Delta U$: $U + \Delta U = \frac{1}{2}C(V + \Delta V)^2$. Thus, $(1 + \Delta V/V)^2 = 1 + \Delta U/U$, or

$$\frac{\Delta V}{V} = \sqrt{1 + \frac{\Delta U}{U}} - 1 = \sqrt{1 + 10\%} - 1 = 4.9\%\ .$$

69. (a) The voltage across C_1 is 12 V, so the charge is

$$q_1 = C_1 V_1 = 24\,\mu\text{C}\ .$$

(b) We reduce the circuit, starting with C_4 and C_3 (in parallel) which are equivalent to $4\,\mu$F. This is then in series with C_2, resulting in an equivalence equal to $\frac{4}{3}\,\mu$F which would have 12 V across it. The charge on this $\frac{4}{3}\,\mu$F capacitor (and therefore on C_2) is $(\frac{4}{3}\,\mu\text{F})(12\text{ V}) = 16\,\mu$C. Consequently, the voltage across C_2 is

$$V_2 = \frac{q_2}{C_2} = \frac{16\,\mu\text{C}}{2\,\mu\text{F}} = 8\text{ V}\ .$$

This leaves $12 - 8 = 4$ V across C_4 (similarly for C_3).

70. (a) The energy stored is

$$U = \frac{1}{2}CV^2 = \frac{1}{2}(130\times 10^{-12}\text{ F})(56.0\text{ V})^2 = 2.04\times 10^{-7}\text{ J}\ .$$

(b) No, because we don't know the volume of the space inside the capacitor where the electric field is present.

71. We reduce the circuit, starting with C_1 and C_2 (in series) which are equivalent to $4\,\mu$F. This is then parallel to C_3 and results in a total of $8\,\mu$F, which is now in series with C_4 and can be further reduced. However, the final step in the reduction is not necessary, as we observe that the $8\,\mu$F equivalence from the top 3 capacitors has the same capacitance as C_4 and therefore the same voltage; since they are in series, that voltage is then $12/2 = 6$ V.

72. We use $C = \varepsilon_0 \kappa A/d \propto \kappa/d$. To maximize C we need to choose the material with the greatest value of κ/d. It follows that the mica sheet should be chosen.

73. (a) After reducing the pair of $4\,\mu$F capacitors to a series equivalence of $2\,\mu$F, we have three 2 μF capacitors in the upper right portion of the circuit all in parallel – and thus equivalent to $6\,\mu$F. In the lower right portion of the circuit are two $3\,\mu$F capacitors in parallel, equivalent also to 6 μF. These two 6 μF equivalent-capacitors are then in series, so that the full reduction leads to an equivalence of $3.0\,\mu$F.

(b) With 20 V across the result of part (a), we have a charge equal to $q = CV = (3.0\,\mu\text{F})(20\text{ V}) = 60\,\mu$C.

74. (a) The length d is effectively shortened by b so $C' = \varepsilon_0 A/(d - b)$.

(b) The energy before, divided by the energy after inserting the slab is

$$\frac{U}{U'} = \frac{q^2/2C}{q^2/2C'} = \frac{C'}{C} = \frac{\varepsilon_0 A/(d-b)}{\varepsilon_0 A/d} = \frac{d}{d-b} \ .$$

(c) The work done is

$$W = \Delta U = U' - U = \frac{q^2}{2}\left(\frac{1}{C'} - \frac{1}{C}\right) = \frac{q^2}{2\varepsilon_0 A}(d-b-d) = -\frac{q^2 b}{2\varepsilon_0 A} \ .$$

Since $W < 0$ the slab is sucked in.

75. (a) $C' = \varepsilon_0 A/(d-b)$, the same as part (a) in problem 74.

(b) Now,

$$\frac{U}{U'} = \frac{\frac{1}{2}CV^2}{\frac{1}{2}C'V^2} = \frac{C}{C'} = \frac{\varepsilon_0 A/d}{\varepsilon_0 A/(d-b)} = \frac{d-b}{d} \ .$$

(c) The work done is

$$W = \Delta U = U' - U = \frac{1}{2}(C' - C)V^2 = \frac{\varepsilon_0 A}{2}\left(\frac{1}{d-b} - \frac{1}{d}\right)V^2 = \frac{\varepsilon_0 AbV^2}{2d(d-b)} \ .$$

Since $W > 0$ the slab must be pushed in.

76. We do not employ energy conservation since, in reaching equilibrium, some energy is dissipated either as heat or radio waves. Charge is conserved; therefore, if $Q = 48\,\mu\text{C}$, and q_1 and q_3 are the charges on C_1 and C_3 after the switch is thrown to the right (and equilibrium is reached), then

$$Q = q_1 + q_3 \ .$$

We note that $V_{1\text{ and }2} = V_3$ because of the parallel arrangement, and $V_1 = \frac{1}{2}V_{1\text{ and }2}$ since they are identical capacitors. This leads to

$$\begin{aligned} 2V_1 &= V_3 \\ 2\frac{q_1}{C_1} &= \frac{q_3}{C_3} \\ 2q_1 &= q_3 \end{aligned}$$

where the last step follows from multiplying both sides by $2.00\,\mu\text{F}$. Therefore,

$$Q = q_1 + (2q_1)$$

which yields $q_1 = 16\,\mu\text{C}$ and $q_3 = 32\,\mu\text{C}$.

77. (a) Since $u = \frac{1}{2}\kappa\varepsilon_0 E^2$, we select the material with the greatest value of κE_{max}^2, where E_{max} is its dielectric strength. We therefore choose strontium titanate, with the corresponding minimum volume

$$\mathcal{V}_{\text{min}} = \frac{U}{U_{\text{max}}} = \frac{2U}{\kappa\varepsilon_0 E_{\text{max}}^2} = \frac{2(250\,\text{kJ})}{(310)\left(8.85\times10^{-12}\,\frac{\text{C}^2}{\text{N}\cdot\text{m}^2}\right)(8\,\text{kV/mm})^2} = 2.85\ \text{m}^3 \ .$$

(b) We solve for κ' from $U = \frac{1}{2}\kappa'\varepsilon_0 E_{\text{max}}^2 \mathcal{V}'_{\text{min}}$:

$$\kappa' = \frac{2U}{\varepsilon_0 \mathcal{V}' E_{\text{max}}^2} = \frac{2(250\,\text{kJ})}{\left(8.85\times10^{-12}\,\frac{\text{C}^2}{\text{N}\cdot\text{m}^2}\right)(0.0870\,\text{m}^3)(8\,\text{kV/mm})^2} = 1.01\times10^4 \ .$$

78. (a) Initially, the capacitance is

$$C_0 = \frac{\varepsilon_0 A}{d} = \frac{\left(8.85 \times 10^{-12} \frac{\mathrm{C}^2}{\mathrm{N \cdot m^2}}\right)(0.12\,\mathrm{m}^2)}{1.2 \times 10^{-2}\,\mathrm{m}} = 89\ \mathrm{pF}\ .$$

(b) Working through Sample Problem 26-6 algebraically, we find:

$$C = \frac{\varepsilon_0 A \kappa}{\kappa(d-b)+b} = \frac{\left(8.85 \times 10^{-12} \frac{\mathrm{C}^2}{\mathrm{N \cdot m^2}}\right)(0.12\,\mathrm{m}^2)(4.8)}{(4.8)(1.2-0.40)(10^{-2}\,\mathrm{m}) + (4.0 \times 10^{-3}\,\mathrm{m})} = 120\ \mathrm{pF}\ .$$

(c) Before the insertion, $q = C_0 V (89\,\mathrm{pF})(120\,\mathrm{V}) = 11\,\mathrm{nC}$. Since the battery is disconnected, q will remain the same after the insertion of the slab.

(d) $E = q/\varepsilon_0 A = 11 \times 10^{-9}\,\mathrm{C}/\left(8.85 \times 10^{-12} \frac{\mathrm{C}^2}{\mathrm{N \cdot m^2}}\right)(0.12\,\mathrm{m}^2) = 10\,\mathrm{kV/m}$.

(e) $E' = E/\kappa = (10\,\mathrm{kV/m})/4.8 = 2.1\,\mathrm{kV/m}$.

(f) $V = E(d-b) + E'b = (10\,\mathrm{kV/m})(0.012\,\mathrm{m} - 0.0040\,\mathrm{m}) + (2.1\,\mathrm{kV/m})(0.40 \times 10^{-3}\,\mathrm{m}) = 88\,\mathrm{V}$.

(g) The work done is

$$\begin{aligned} W_{\text{ext}} &= \Delta U = \frac{q^2}{2}\left(\frac{1}{C} - \frac{1}{C_0}\right) \\ &= \frac{(11 \times 10^{-9}\,\mathrm{C})^2}{2}\left(\frac{1}{89 \times 10^{-12}\,\mathrm{F}} - \frac{1}{120 \times 10^{-12}\,\mathrm{F}}\right) \\ &= -1.7 \times 10^{-7}\ \mathrm{J}\ . \end{aligned}$$

79. (a) Since $u = \frac{1}{2}\kappa\varepsilon_0 E^2$, $E_{\text{slab}} = E_{\text{air}}/\kappa_{\text{slab}}$, and $U = u\mathcal{V}$ (where $\mathcal{V}$ = volume), then the fraction of energy stored in the air gaps is

$$\begin{aligned} \frac{U_{\text{air}}}{U_{\text{total}}} &= \frac{E_{\text{air}}^2 A(d-b)}{E_{\text{air}}^2 A(d-b) + \kappa_{\text{slab}} E_{\text{slab}}^2 Ab} = \frac{1}{1 + \kappa_{\text{slab}}(E_{\text{slab}}/E_{\text{air}})^2[b/(d-b)]} \\ &= \frac{1}{1 + (2.61)(1/2.61)^2[0.780/(1.24-0.780)]} = 0.606\ . \end{aligned}$$

(b) The fraction of energy stored in the slab is $1 - 0.606 = 0.394$.

80. (a) The equivalent capacitance of the three capacitors connected in parallel is $C_{\text{eq}} = 3C = 3\varepsilon_0 A/d = \varepsilon_0 A/(d/3)$. Thus, the required spacing is $d/3$.

(b) Now, $C_{\text{eq}} = C/3 = \varepsilon_0 A/3d$, so the spacing should be $3d$.

81. We do not employ energy conservation since, in reaching equilibrium, some energy is dissipated either as heat or radio waves. Charge is conserved; therefore, if $Q = C_1 V_{\text{bat}} = 24\,\mu\mathrm{C}$, and q_1 and q_3 are the charges on C_1 and C_3 after the switch is thrown to the right (and equilibrium is reached), then

$$Q = q_1 + q_3\ .$$

We reduce the series pair C_2 and C_3 to $C' = 4/3\ \mu\mathrm{F}$ which has charge $q' = q_3$ and the same voltage that we find across C_1. Therefore,

$$\begin{aligned} V_1 &= V' \\ \frac{q_1}{C_1} &= \frac{q_3}{C'} \end{aligned}$$

which leads to $q_1 = 1.5q_3$. Hence,

$$Q = (1.5q_3) + q_3$$

leads to $q_3 = 9.6\,\mu\mathrm{C}$.

82. (First problem of **Cluster**)

(a) We do not employ energy conservation since, in reaching equilibrium, some energy is dissipated either as heat or radio waves. Charge is conserved; therefore, if $Q = C_1 V_{\text{bat}} = 400\,\mu\text{C}$, and q_1 and q_2 are the charges on C_1 and C_2 after the switch S is closed (and equilibrium is reached), then

$$Q = q_1 + q_2 \quad .$$

After switch S is closed, the capacitor voltages are equal, so that

$$\begin{aligned} V_1 &= V_2 \\ \frac{q_1}{C_1} &= \frac{q_2}{C_2} \end{aligned}$$

which yields $\frac{3}{4}q_1 = q_2$. Therefore,

$$Q = q_1 + \left(\frac{3}{4}q_1\right)$$

which gives the result $q_1 = 229\,\mu\text{C}$.

(b) The relation $\frac{3}{4}q_1 = q_2$ gives the result $q_2 = 171\,\mu\text{C}$.

(c) We apply Eq. 27-1: $V_1 = q_1/C_1 = 5.71$ V.

(d) Similarly, $V_2 = q_2/C_2 = 5.71$ V (which is equal to V_1, of course – since that fact was used in the solution to part (a)).

(e) When C_1 had charge Q and was connected to the battery, the energy stored was $\frac{1}{2}C_1 V_{\text{bat}}^2 = 2.00 \times 10^{-3}$ J. The energy stored after S is closed is $\frac{1}{2}C_1 V_1^2 + \frac{1}{2}C_2 V_2^2 = 1.14 \times 10^{-3}$ J. The *decrease* is therefore 8.6×10^{-4} J.

83. (Second problem of **Cluster**)

(a) The change (from the previous problem) is that the initial charge (before switch S is closed) is $Q + Q'$ where Q is as before but $Q' = C_2(10\text{ V}) = 600\,\mu\text{C}$. We assume the polarities of these capacitor charges are the same. With this modification, we follow steps similar to those in the previous solution:

$$\begin{aligned} Q + Q' &= q_1 + q_2 \\ &= q_1 + \left(\frac{3}{4}q_1\right) \end{aligned}$$

which yields $q_1 = 571\,\mu\text{C}$.

(b) The relation $\frac{3}{4}q_1 = q_2$ gives the result $q_2 = 429\,\mu\text{C}$.

(c) We apply Eq. 27-1: $V_1 = q_1/C_1 = 14.3$ V.

(d) Similarly, $V_2 = q_2/C_2 = 14.3$ V.

(e) The initial energy now includes $\frac{1}{2}C_2(20\text{ V})^2$ in addition to the $\frac{1}{2}C_1 V_{\text{bat}}^2$ computed in the previous case. Thus, the total initial energy is 8.00×10^{-3} J. And the final stored energy is $\frac{1}{2}C_1 V_1^2 + \frac{1}{2}C_2 V_2^2 = 7.14 \times 10^{-3}$ J. The *decrease* is therefore 8.6×10^{-4} J, as it was in the previous problem.

84. (Third problem of **Cluster**)

(a) With the series pair C_2 and C_3 reduced to a single $C' = 10\,\mu\text{F}$ capacitor, this becomes very similar to problem 82. Noting for later use that $q' = q_2 = q_3$, and using notation similar to that used in the solution to problem 82, we have

$$Q = q_1 + q'$$

where $Q = C_1 V_{\rm bat} = 400\,\mu$C. Also, after switch S is closed,

$$\begin{aligned} V_1 &= V' \\ \frac{q_1}{C_1} &= \frac{q'}{C'} \end{aligned}$$

which yields $\frac{1}{4}q_1 = q'$. Therefore,

$$Q = q_1 + \left(\frac{1}{4}q_1\right)$$

which gives the result $q_1 = 320\,\mu$C.

(b) We use $q_2 = q_3 = \frac{1}{4}q_1$ to obtain the result $80\,\mu$C.

(c) See part (b).

(d) (e) and (f) Eq. 26-1 yields

$$V = \frac{q}{C} = \begin{cases} 8.0 \text{ V} & \text{for } C_1 \\ 5.3 \text{ V} & \text{for } C_2 \\ 2.7 \text{ V} & \text{for } C_3 \end{cases}$$

85. (Fourth problem of **Cluster**)

(a) With the parallel pair C_2 and C_3 reduced to a single $C' = 45\,\mu$F capacitor, this becomes very similar to problem 82. Using notation similar to that used in the solution to 82, we have

$$Q = q_1 + q'$$

where $Q = C_1 V_{\rm bat} = 400\,\mu$C. Also, after switch S is closed,

$$\begin{aligned} V_1 &= V' \\ \frac{q_1}{C_1} &= \frac{q'}{C'} \end{aligned}$$

which yields $\frac{9}{8}q_1 = q'$. Therefore,

$$Q = q_1 + \left(\frac{9}{8}q_1\right)$$

which gives the result $q_1 = 188\,\mu$C.

(b) We find the voltage across capacitor 1 from q_1/C_1 (see below) and (since the capacitors are in parallel) use the fact that $V_1 = V_2 = V_3$ with $q = CV$ to obtain the charges: $q_2 = 71\,\mu$C and $q_3 = 141\,\mu$C.

(c) See part (b).

(d) (e) and (f) The capacitors all have the same voltage. $V = q_1/C_1 = 4.7$ V.

86. (Fifth problem of **Cluster**)

(a) To begin with, the charge on capacitor 1 is $Q_1 = C_1 V_{\rm bat} = 400\,\mu$C, and the charge on capacitor 2 is $Q_2 = C_2 V_{\rm bat} = 150\,\mu$C. After the rearrangement and closing of the switch, the total charge in the upper portion of the circuit is $Q_1 - Q_2 = Q = 250\,\mu$C. With notation similar to that in the previous problems,

$$\begin{aligned} Q &= q_1 + q_2 \\ &= C_1 V + C_2 V \end{aligned}$$

which yields $V = 4.55$ V, which, in turn implies $q_1 = C_1 V = 182\,\mu$C and $q_2 = C_2 V = 68\,\mu$C. To achieve this distribution (with $+182\,\mu$C on one upper plate and $+68\,\mu$C on the other upper plate) from the arrangement right before closing the switch (with $+400\,\mu$C on one upper plate and $-150\,\mu$C on the other upper plate), it is necessary for $218\,\mu$C to flow through the switch.

(b) As shown above, $V = 4.55 \text{ V} = V_1 = V_2$.

Chapter 27

1. (a) The charge that passes through any cross section is the product of the current and time. Since $4.0\,\text{min} = (4.0\,\text{min})(60\,\text{s/min}) = 240\,\text{s}$, $q = it = (5.0\,\text{A})(240\,\text{s}) = 1200\,\text{C}$.

   (b) The number of electrons N is given by $q = Ne$, where e is the magnitude of the charge on an electron. Thus $N = q/e = (1200\,\text{C})/(1.60 \times 10^{-19}\,\text{C}) = 7.5 \times 10^{21}$.

2. We adapt the discussion in the text to a moving two-dimensional collection of charges. Using σ for the charge per unit area and w for the belt width, we can see that the transport of charge is expressed in the relationship $i = \sigma v w$, which leads to

$$\sigma = \frac{i}{vw} = \frac{100 \times 10^{-6}\,\text{A}}{(30\,\text{m/s})(50 \times 10^{-2}\,\text{m})} = 6.7 \times 10^{-6}\ \text{C/m}^2\ .$$

3. Suppose the charge on the sphere increases by Δq in time Δt. Then, in that time its potential increases by

$$\Delta V = \frac{\Delta q}{4\pi\varepsilon_0 r}\ ,$$

where r is the radius of the sphere. This means

$$\Delta q = 4\pi\varepsilon_0 r\,\Delta V\ .$$

Now, $\Delta q = (i_{\text{in}} - i_{\text{out}})\,\Delta t$, where i_{in} is the current entering the sphere and i_{out} is the current leaving. Thus,

$$\begin{aligned}\Delta t &= \frac{\Delta q}{i_{\text{in}} - i_{\text{out}}} = \frac{4\pi\varepsilon_0 r\,\Delta V}{i_{\text{in}} - i_{\text{out}}} \\ &= \frac{(0.10\,\text{m})(1000\,\text{V})}{(8.99 \times 10^9\,\text{F/m})(1.0000020\,\text{A} - 1.0000000\,\text{A})} = 5.6 \times 10^{-3}\ \text{s}\ .\end{aligned}$$

4. (a) The magnitude of the current density vector is

$$J = \frac{i}{A} = \frac{i}{\pi d^2/4} = \frac{4(1.2 \times 10^{-10}\,\text{A})}{\pi(2.5 \times 10^{-3}\,\text{m})^2} = 2.4 \times 10^{-5}\ \text{A/m}^2\ .$$

   (b) The drift speed of the current-carrying electrons is

$$v_d = \frac{J}{ne} = \frac{2.4 \times 10^{-5}\,\text{A/m}^2}{(8.47 \times 10^{28}/\text{m}^3)\,(1.60 \times 10^{-19}\,\text{C})} = 1.8 \times 10^{-15}\ \text{m/s}\ .$$

5. (a) The magnitude of the current density is given by $J = nqv_d$, where n is the number of particles per unit volume, q is the charge on each particle, and v_d is the drift speed of the particles. The particle

concentration is $n = 2.0 \times 10^8/\text{cm}^3 = 2.0 \times 10^{14}\,\text{m}^{-3}$, the charge is $q = 2e = 2(1.60 \times 10^{-19}\,\text{C}) = 3.20 \times 10^{-19}\,\text{C}$, and the drift speed is $1.0 \times 10^5\,\text{m/s}$. Thus,

$$J = (2 \times 10^{14}/\text{m})(3.2 \times 10^{-19}\,\text{C})(1.0 \times 10^5\,\text{m/s}) = 6.4\ \text{A/m}^2 .$$

Since the particles are positively charged the current density is in the same direction as their motion, to the north.

(b) The current cannot be calculated unless the cross-sectional area of the beam is known. Then $i = JA$ can be used.

6. We express the magnitude of the current density vector in SI units by converting the diameter values in mils to inches (by dividing by 1000) and then converting to meters (by multiplying by 0.0254) and finally using

$$J = \frac{i}{A} = \frac{i}{\pi R^2} = \frac{4i}{\pi D^2} .$$

For example, the gauge 14 wire with $D = 64\,\text{mil} = 0.0016$ m is found to have a (maximum safe) current density of $J = 7.2 \times 10^6\ \text{A/m}^2$. In fact, this is the wire with the largest value of J allowed by the given data. The values of J in SI units are plotted below as a function of their diameters in mils.

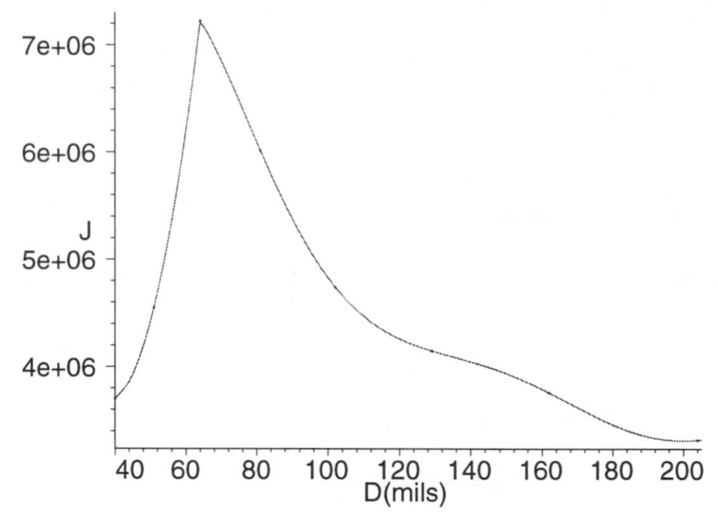

7. The cross-sectional area of wire is given by $A = \pi r^2$, where r is its radius (half its thickness). The magnitude of the current density vector is $J = i/A = i/\pi r^2$, so

$$r = \sqrt{\frac{i}{\pi J}} = \sqrt{\frac{0.50\,\text{A}}{\pi(440 \times 10^4\,\text{A/m}^2)}} = 1.9 \times 10^{-4}\,\text{m} .$$

The diameter of the wire is therefore $d = 2r = 2(1.9 \times 10^{-4}\,\text{m}) = 3.8 \times 10^{-4}\,\text{m}$.

8. (a) Since $1\,\text{cm}^3 = 10^{-6}\,\text{m}^3$, the magnitude of the current density vector is

$$J = nev = \left(\frac{8.70}{10^{-6}\,\text{m}^3}\right)(1.60 \times 10^{-19}\,\text{C})\left(470 \times 10^3\,\text{m/s}\right) = 6.54 \times 10^{-7}\ \text{A/m}^2 .$$

(b) Although the total surface area of Earth is $4\pi R_E^2$ (that of a sphere), the area to be used in a computation of how many protons in an approximately unidirectional beam (the solar wind) will be captured by Earth is its projected area. In other words, for the beam, the encounter is with a "target" of circular area πR_E^2. The rate of charge transport implied by the influx of protons is

$$i = AJ = \pi R_E^2 J = \pi\left(6.37 \times 10^6\,\text{m}\right)^2\left(6.54 \times 10^{-7}\,\text{A/m}^2\right) = 8.34 \times 10^7\ \text{A} .$$

9. (a) The charge that strikes the surface in time Δt is given by $\Delta q = i\,\Delta t$, where i is the current. Since each particle carries charge $2e$, the number of particles that strike the surface is

$$N = \frac{\Delta q}{2e} = \frac{i\,\Delta t}{2e} = \frac{(0.25 \times 10^{-6}\,\text{A})(3.0\,\text{s})}{2(1.6 \times 10^{-19}\,\text{C})} = 2.3 \times 10^{12} .$$

(b) Now let N be the number of particles in a length L of the beam. They will all pass through the beam cross section at one end in time $t = L/v$, where v is the particle speed. The current is the charge that moves through the cross section per unit time. That is, $i = 2eN/t = 2eNv/L$. Thus $N = iL/2ev$. To find the particle speed, we note the kinetic energy of a particle is

$$K = 20\,\text{MeV} = (20 \times 10^6\,\text{eV})(1.60 \times 10^{-19}\,\text{J/eV}) = 3.2 \times 10^{-12}\,\text{J} .$$

Since $K = \frac{1}{2}mv^2$, then the speed is $v = \sqrt{2K/m}$. The mass of an alpha particle is (very nearly) 4 times the mass of a proton, or $m = 4(1.67 \times 10^{-27}\,\text{kg}) = 6.68 \times 10^{-27}\,\text{kg}$, so

$$v = \sqrt{\frac{2(3.2 \times 10^{-12}\,\text{J})}{6.68 \times 10^{-27}\,\text{kg}}} = 3.1 \times 10^7\,\text{m/s}$$

and

$$N = \frac{iL}{2ev} = \frac{(0.25 \times 10^{-6})(20 \times 10^{-2}\,\text{m})}{2(1.60 \times 10^{-19}\,\text{C})(3.1 \times 10^7\,\text{m/s})} = 5.0 \times 10^3 .$$

(c) We use conservation of energy, where the initial kinetic energy is zero and the final kinetic energy is $20\,\text{MeV} = 3.2 \times 10^{-12}\,\text{J}$. We note, too, that the initial potential energy is $U_i = qV = 2eV$, and the final potential energy is zero. Here V is the electric potential through which the particles are accelerated. Consequently,

$$K_f = U_i = 2eV \implies V = \frac{K_f}{2e} = \frac{3.2 \times 10^{-12}\,\text{J}}{2(1.60 \times 10^{-19}\text{C})} = 10 \times 10^6\,\text{V} .$$

10. (a) The current resulting from this non-uniform current density is

$$i = \int_{\text{cylinder}} J\,dA = \int_0^R J_0\left(1 - \frac{r}{R}\right) 2\pi r dr = \frac{1}{3}\pi R^2 J_0 = \frac{1}{3} A J_0 .$$

(b) In this case,

$$i = \int_{\text{cylinder}} J\,dA = \frac{J_0}{R}\int_0^R r \cdot 2\pi r dr = \frac{2}{3}\pi R^2 J_0 = \frac{2}{3} A J_0 .$$

The result is different from that in part (a) because the current density in part (b) is lower near the center of the cylinder (where the area is smaller for the same radial interval) and higher outward, resulting in a higher average current density over the cross section and consequently a greater current than that in part (a).

11. We use $v_d = J/ne = i/Ane$. Thus,

$$\begin{aligned} t &= \frac{L}{v_d} = \frac{L}{i/Ane} = \frac{LAne}{i} \\ &= \frac{(0.85\,\text{m})(0.21 \times 10^{-4}\,\text{m}^2)(8.47 \times 10^{28}/\,\text{m}^3)(1.60 \times 10^{-19}\,\text{C})}{300\,\text{A}} \\ &= 8.1 \times 10^2\,\text{s} = 13\,\text{min} . \end{aligned}$$

12. We find the conductivity of Nichrome (the reciprocal of its resistivity) as follows:

$$\sigma = \frac{1}{\rho} = \frac{L}{RA} = \frac{L}{(V/i)A} = \frac{Li}{VA} = \frac{(1.0\,\text{m})(4.0\,\text{A})}{(2.0\,\text{V})(1.0 \times 10^{-6}\,\text{m}^2)} = 2.0 \times 10^6/\Omega{\cdot}\text{m} .$$

13. The resistance of the wire is given by $R = \rho L/A$, where ρ is the resistivity of the material, L is the length of the wire, and A is its cross-sectional area. In this case,

$$A = \pi r^2 = \pi(0.50 \times 10^{-3}\,\text{m})^2 = 7.85 \times 10^{-7}\ \text{m}^2 \ .$$

Thus,

$$\rho = \frac{RA}{L} = \frac{(50 \times 10^{-3}\,\Omega)(7.85 \times 10^{-7}\,\text{m}^2)}{2.0\,\text{m}} = 2.0 \times 10^{-8}\ \Omega\cdot\text{m} \ .$$

14. Since $100\,\text{cm} = 1\,\text{m}$, then $10^4\,\text{cm}^2 = 1\,\text{m}^2$. Thus,

$$R = \frac{\rho L}{A} = \frac{(3.00 \times 10^{-7}\,\Omega\cdot\text{m})(10.0 \times 10^3\,\text{m})}{56.0 \times 10^{-4}\,\text{m}^2} = 0.536\ \Omega \ .$$

15. Since the potential difference V and current i are related by $V = iR$, where R is the resistance of the electrician, the fatal voltage is $V = (50 \times 10^{-3}\,\text{A})(2000\,\Omega) = 100\,\text{V}$.

16. (a) $i = V/R = 23.0\,\text{V}/15.0 \times 10^{-3}\,\Omega = 1.53 \times 10^3\,\text{A}$.

(b) The cross-sectional area is $A = \pi r^2 = \frac{1}{4}\pi D^2$. Thus, the magnitude of the current density vector is

$$J = \frac{i}{A} = \frac{4i}{\pi D^2} = \frac{4(1.53 \times 10^{-3}\,\text{A})}{\pi(6.00 \times 10^{-3}\,\text{m})^2} = 5.41 \times 10^7\ \text{A/m}^2 \ .$$

(c) The resistivity is $\rho = RA/L = (15.0 \times 10^{-3}\,\Omega)(\pi)(6.00 \times 10^{-3}\,\text{m})^2/[4(4.00\,\text{m})] = 10.6 \times 10^{-8}\,\Omega\cdot\text{m}$. The material is platinum.

17. The resistance of the coil is given by $R = \rho L/A$, where L is the length of the wire, ρ is the resistivity of copper, and A is the cross-sectional area of the wire. Since each turn of wire has length $2\pi r$, where r is the radius of the coil, then $L = (250)2\pi r = (250)(2\pi)(0.12\,\text{m}) = 188.5\,\text{m}$. If r_w is the radius of the wire itself, then its cross-sectional area is $A = \pi r_w^2 = \pi(0.65 \times 10^{-3}\,\text{m})^2 = 1.33 \times 10^{-6}\,\text{m}^2$. According to Table 27-1, the resistivity of copper is $1.69 \times 10^{-8}\,\Omega\cdot\text{m}$. Thus,

$$R = \frac{\rho L}{A} = \frac{(1.69 \times 10^{-8}\,\Omega\cdot\text{m})(188.5\,\text{m})}{1.33 \times 10^{-6}\,\text{m}^2} = 2.4\ \Omega \ .$$

18. In Eq. 27-17, we let $\rho = 2\rho_0$ where ρ_0 is the resistivity at $T_0 = 20°\text{C}$:

$$\rho - \rho_0 = 2\rho_0 - \rho_0 = \rho_0\alpha\,(T - T_0) \ ,$$

and solve for the temperature T:

$$T = T_0 + \frac{1}{\alpha} = 20°\text{C} + \frac{1}{4.3 \times 10^{-3}/\text{K}} \approx 250°\text{C} \ .$$

Since a change in Celsius is equivalent to a change on the Kelvin temperature scale, the value of α used in this calculation is not inconsistent with the other units involved. It is worth nothing that this agrees well with Fig. 27-10.

19. Since the mass and density of the material do not change, the volume remains the same. If L_0 is the original length, L is the new length, A_0 is the original cross-sectional area, and A is the new cross-sectional area, then $L_0A_0 = LA$ and $A = L_0A_0/L = L_0A_0/3L_0 = A_0/3$. The new resistance is

$$R = \frac{\rho L}{A} = \frac{\rho 3L_0}{A_0/3} = 9\frac{\rho L_0}{A_0} = 9R_0 \ ,$$

where R_0 is the original resistance. Thus, $R = 9(6.0\,\Omega) = 54\,\Omega$.

20. The thickness (diameter) of the wire is denoted by D. We use $R \propto L/A$ (Eq. 27-16) and note that $A = \frac{1}{4}\pi D^2 \propto D^2$. The resistance of the second wire is given by

$$R_2 = R\left(\frac{A_1}{A_2}\right)\left(\frac{L_2}{L_1}\right) = R\left(\frac{D_1}{D_2}\right)^2\left(\frac{L_2}{L_1}\right) = R(2)^2\left(\frac{1}{2}\right) = 2R \ .$$

21. The resistance of conductor A is given by

$$R_A = \frac{\rho L}{\pi r_A^2} \ ,$$

where r_A is the radius of the conductor. If r_o is the outside diameter of conductor B and r_i is its inside diameter, then its cross-sectional area is $\pi(r_o^2 - r_i^2)$, and its resistance is

$$R_B = \frac{\rho L}{\pi\left(r_o^2 - r_i^2\right)} \ .$$

The ratio is

$$\frac{R_A}{R_B} = \frac{r_o^2 - r_i^2}{r_A^2} = \frac{(1.0\,\text{mm})^2 - (0.50\,\text{mm})^2}{(0.50\,\text{mm})^2} = 3.0 \ .$$

22. (a) The current in each strand is $i = 0.750\,\text{A}/125 = 6.00 \times 10^{-3}\,\text{A}$.
(b) The potential difference is $V = iR = (6.00 \times 10^{-3}\,\text{A})(2.65 \times 10^{-6}\,\Omega) = 1.59 \times 10^{-8}\,\text{V}$.
(c) The resistance is $R_{\text{total}} = 2.65 \times 10^{-6}\,\Omega/125 = 2.12 \times 10^{-8}\,\Omega$.

23. We use $J = E/\rho$, where E is the magnitude of the (uniform) electric field in the wire, J is the magnitude of the current density, and ρ is the resistivity of the material. The electric field is given by $E = V/L$, where V is the potential difference along the wire and L is the length of the wire. Thus $J = V/L\rho$ and

$$\rho = \frac{V}{LJ} = \frac{115\,\text{V}}{(10\,\text{m})\left(1.4 \times 10^4\,\text{A/m}^2\right)} = 8.2 \times 10^{-4}\,\Omega\cdot\text{m} \ .$$

24. (a) $i = V/R = 35.8\,\text{V}/935\,\Omega = 3.83 \times 10^{-2}\,\text{A}$.
(b) $J = i/A = (3.83 \times 10^{-2}\,\text{A})/(3.50 \times 10^{-4}\,\text{m}^2) = 109\,\text{A/m}^2$.
(c) $v_d = J/ne = (109\,\text{A/m}^2)/[(5.33 \times 10^{22}/\text{m}^3)(1.60 \times 10^{-19}\,\text{C})] = 1.28 \times 10^{-2}\,\text{m/s}$.
(d) $E = V/L = 35.8\,\text{V}/0.158\,\text{m} = 227\,\text{V/m}$.

25. The resistance at operating temperature T is $R = V/i = 2.9\,\text{V}/0.30\,\text{A} = 9.67\,\Omega$. Thus, from $R - R_0 = R_0\alpha(T - T_0)$, we find

$$\begin{aligned} T &= T_0 + \frac{1}{\alpha}\left(\frac{R}{R_0} - 1\right) \\ &= 20°\text{C} + \left(\frac{1}{4.5 \times 10^{-3}/\text{K}}\right)\left(\frac{9.67\,\Omega}{1.1\,\Omega} - 1\right) \end{aligned}$$

which yields approximately 1900°C. Since a change in Celsius is equivalent to a change on the Kelvin temperature scale, the value of α used in this calculation is not inconsistent with the other units involved. Table 27-1 has been used.

26. We use $J = \sigma E = (n_+ + n_-)ev_d$, which combines Eq. 27-13 and Eq. 27-7.

(a) The drift velocity is

$$v_d = \frac{\sigma E}{(n_+ + n_-)e} = \frac{(2.70 \times 10^{-14}/\Omega\cdot\text{m})(120\,\text{V/m})}{[(620 + 550)/\text{cm}^3](1.60 \times 10^{-19}\,\text{C})} = 1.73\ \text{cm/s} \ .$$

(b) $J = \sigma E = (2.70 \times 10^{-14}/\,\Omega\cdot\text{m})(120\,\text{V/m}) = 3.24 \times 10^{-12}\,\text{A/m}^2$.

27. (a) Let ΔT be the change in temperature and κ be the coefficient of linear expansion for copper. Then $\Delta L = \kappa L\,\Delta T$ and

$$\frac{\Delta L}{L} = \kappa\,\Delta T = (1.7 \times 10^{-5}/\text{K})(1.0°\text{C}) = 1.7 \times 10^{-5} \ .$$

This is equivalent to 0.0017%. Since a change in Celsius is equivalent to a change on the Kelvin temperature scale, the value of κ used in this calculation is not inconsistent with the other units involved. Incorporating a factor of 2 for the two-dimensional nature of A, the fractional change in area is

$$\frac{\Delta A}{A} = 2\kappa\,\Delta T = 2(1.7 \times 10^{-5}/\text{K})(1.0°\text{C}) = 3.4 \times 10^{-5}$$

which is 0.0034%. For small changes in the resistivity ρ, length L, and area A of a wire, the change in the resistance is given by

$$\Delta R = \frac{\partial R}{\partial \rho}\,\Delta\rho + \frac{\partial R}{\partial L}\,\Delta L + \frac{\partial R}{\partial A}\,\Delta A \ .$$

Since $R = \rho L/A$, $\partial R/\partial\rho = L/A = R/\rho$, $\partial R/\partial L = \rho/A = R/L$, and $\partial R/\partial A = -\rho L/A^2 = -R/A$. Furthermore, $\Delta\rho/\rho = \alpha\,\Delta T$, where α is the temperature coefficient of resistivity for copper ($4.3 \times 10^{-3}/\text{K} = 4.3 \times 10^{-3}/\text{C}°$, according to Table 27-1). Thus,

$$\begin{aligned}\frac{\Delta R}{R} &= \frac{\Delta\rho}{\rho} + \frac{\Delta L}{L} - \frac{\Delta A}{A} \\ &= (\alpha + \kappa - 2\kappa)\,\Delta T = (\alpha - \kappa)\,\Delta T \\ &= (4.3 \times 10^{-3}/\text{C}° - 1.7 \times 10^{-5}/\text{C}°)\,(1.0\,\text{C}°) = 4.3 \times 10^{-3} \ .\end{aligned}$$

This is 0.43%, which we note (for the purposes of the next part) is primarily determined by the $\Delta\rho/\rho$ term in the above calculation.

(b) The fractional change in resistivity is much larger than the fractional change in length and area. Changes in length and area affect the resistance much less than changes in resistivity.

28. We use $R \propto L/A$. The diameter of a 22-gauge wire is 1/4 that of a 10-gauge wire. Thus from $R = \rho L/A$ we find the resistance of 25 ft of 22-gauge copper wire to be $R = (1.00\,\Omega)(25\,\text{ft}/1000\,\text{ft})(4)^2 = 0.40\,\Omega$.

29. (a) The current i is shown in Fig. 27-22 entering the truncated cone at the left end and leaving at the right. This is our choice of positive x direction. We make the assumption that the current density J at each value of x may be found by taking the ratio i/A where $A = \pi r^2$ is the cone's cross-section area at that particular value of x. The direction of $\vec{J}$ is identical to that shown in the figure for i (our $+x$ direction). Using Eq. 27-11, we then find an expression for the electric field at each value of x, and next find the potential difference V by integrating the field along the x axis, in accordance with the ideas of Chapter 25. Finally, the resistance of the cone is given by $R = V/i$. Thus,

$$J = \frac{i}{\pi r^2} = \frac{E}{\rho}$$

where we must deduce how r depends on x in order to proceed. We note that the radius increases linearly with x, so (with c_1 and c_2 to be determined later) we may write

$$r = c_1 + c_2\,x \ .$$

Choosing the origin at the left end of the truncated cone, the coefficient c_1 is chosen so that $r = a$ (when $x = 0$); therefore, $c_1 = a$. Also, the coefficient c_2 must be chosen so that (at the right end of the truncated cone) we have $r = b$ (when $x = L$); therefore, $c_2 = (b - a)/L$. Our expression, then, becomes

$$r = a + \left(\frac{b-a}{L}\right)\,x \ .$$

Substituting this into our previous statement and solving for the field, we find

$$E = \frac{i\rho}{\pi}\left(a + \frac{b-a}{L}x\right)^{-2} .$$

Consequently, the potential difference between the faces of the cone is

$$\begin{aligned} V &= -\int_0^L E\,dx = -\frac{i\rho}{\pi}\int_0^L \left(a + \frac{b-a}{L}x\right)^{-2} dx \\ &= \frac{i\rho}{\pi}\frac{L}{b-a}\left(a + \frac{b-a}{L}x\right)^{-1}\Bigg|_0^L = \frac{i\rho}{\pi}\frac{L}{b-a}\left(\frac{1}{a} - \frac{1}{b}\right) \\ &= \frac{i\rho}{\pi}\frac{L}{b-a}\frac{b-a}{ab} = \frac{i\rho L}{\pi ab} . \end{aligned}$$

The resistance is therefore

$$R = \frac{V}{i} = \frac{\rho L}{\pi ab} .$$

(b) If $b = a$, then $R = \rho L/\pi a^2 = \rho L/A$, where $A = \pi a^2$ is the cross-sectional area of the cylinder.

30. From Eq. 27-20, $\rho \propto \tau^{-1} \propto v_{\text{eff}}$. The connection with v_{eff} is indicated in part (b) of Sample Problem 27-5, which contains useful insight regarding the problem we are working now. According to Chapter 20, $v_{\text{eff}} \propto \sqrt{T}$. Thus, we may conclude that $\rho \propto \sqrt{T}$.

31. The power dissipated is given by the product of the current and the potential difference:

$$P = iV = (7.0 \times 10^{-3}\,\text{A})(80 \times 10^3\,\text{V}) = 560\text{W} .$$

32. Since $P = iV$, $q = it = Pt/V = (7.0\,\text{W})(5.0\,\text{h})(3600\,\text{s/h})/9.0\,\text{V} = 1.4 \times 10^4\,\text{C}$.

33. (a) Electrical energy is converted to heat at a rate given by

$$P = \frac{V^2}{R} ,$$

where V is the potential difference across the heater and R is the resistance of the heater. Thus,

$$P = \frac{(120\,\text{V})^2}{14\,\Omega} = 1.0 \times 10^3\,\text{W} = 1.0\,\text{kW} .$$

(b) The cost is given by

$$(1.0\,\text{kW})(5.0\,\text{h})(5.0\,\text{cents/kW}\cdot\text{h}) = 25\ \text{cents} .$$

34. The resistance is $R = P/i^2 = (100\,\text{W})/(3.00\,\text{A})^2 = 11.1\,\Omega$.

35. The relation $P = V^2/R$ implies $P \propto V^2$. Consequently, the power dissipated in the second case is

$$P = \left(\frac{1.50\,\text{V}}{3.00\,\text{V}}\right)^2 (0.540\,\text{W}) = 0.135\ \text{W} .$$

36. (a) From $P = V^2/R$ we find $R = V^2/P = (120\,\text{V})^2/500\,\text{W} = 28.8\,\Omega$.

(b) Since $i = P/V$, the rate of electron transport is

$$\frac{i}{e} = \frac{P}{eV} = \frac{500\,\text{W}}{(1.60 \times 10^{-19}\,\text{C})(120\,\text{V})} = 2.60 \times 10^{19}/\text{s} .$$

37. (a) The power dissipated, the current in the heater, and the potential difference across the heater are related by $P = iV$. Therefore,

$$i = \frac{P}{V} = \frac{1250\,\text{W}}{115\,\text{V}} = 10.9\ \text{A} .$$

(b) Ohm's law states $V = iR$, where R is the resistance of the heater. Thus,

$$R = \frac{V}{i} = \frac{115\,\text{V}}{10.9\,\text{A}} = 10.6\ \Omega .$$

(c) The thermal energy E generated by the heater in time $t = 1.0\,\text{h} = 3600\,\text{s}$ is

$$E = Pt = (1250\,\text{W})(3600\,\text{s}) = 4.5 \times 10^6\ \text{J} .$$

38. (a) From $P = V^2/R = AV^2/\rho L$, we solve for the length:

$$L = \frac{AV^2}{\rho P} = \frac{(2.60 \times 10^{-6}\,\text{m}^2)(75.0\,\text{V})^2}{(5.00 \times 10^{-7}\,\Omega{\cdot}\text{m})(5000\,\text{W})} = 5.85\ \text{m} .$$

(b) Since $L \propto V^2$ the new length should be

$$L' = L\left(\frac{V'}{V}\right)^2 = (5.85\,\text{m})\left(\frac{100\,\text{V}}{75.0\,\text{V}}\right)^2 = 10.4\ \text{m} .$$

39. Let R_H be the resistance at the higher temperature (800°C) and let R_L be the resistance at the lower temperature (200°C). Since the potential difference is the same for the two temperatures, the power dissipated at the lower temperature is $P_L = V^2/R_L$, and the power dissipated at the higher temperature is $P_H = V^2/R_H$, so $P_L = (R_H/R_L)P_H$. Now $R_L = R_H + \alpha R_H\,\Delta T$, where ΔT is the temperature difference $T_L - T_H = -600\,\text{C}° = -600\,\text{K}$. Thus,

$$P_L = \frac{R_H}{R_H + \alpha R_H\,\Delta T}P_H = \frac{P_H}{1 + \alpha\,\Delta T} = \frac{500\,\text{W}}{1 + (4.0 \times 10^{-4}/\text{K})(-600\,\text{K})} = 660\ \text{W} .$$

40. (a) The monthly cost is $(100\,\text{W})(24\,\text{h/day})(31\,\text{day/month})(6\,\text{cents/kW}{\cdot}\text{h}) = 446\,\text{cents} = \$4.46$, assuming a 31-day month.

(b) $R = V^2/P = (120\,\text{V})^2/100\,\text{W} = 144\,\Omega$.

(c) $i = P/V = 100\,\text{W}/120\,\text{V} = 0.833\,\text{A}$.

41. (a) The charge q that flows past any cross section of the beam in time Δt is given by $q = i\,\Delta t$, and the number of electrons is $N = q/e = (i/e)\,\Delta t$. This is the number of electrons that are accelerated. Thus

$$N = \frac{(0.50\,\text{A})(0.10 \times 10^{-6}\,\text{s})}{1.60 \times 10^{-19}\,\text{C}} = 3.1 \times 10^{11} .$$

(b) Over a long time t the total charge is $Q = nqt$, where n is the number of pulses per unit time and q is the charge in one pulse. The average current is given by $i_{\text{avg}} = Q/t = nq$. Now $q = i\,\Delta t = (0.50\,\text{A})(0.10 \times 10^{-6}\,\text{s}) = 5.0 \times 10^{-8}\,\text{C}$, so

$$i_{\text{avg}} = (500/\text{s})(5.0 \times 10^{-8}\,\text{C}) = 2.5 \times 10^{-5}\ \text{A} .$$

(c) The accelerating potential difference is $V = K/e$, where K is the final kinetic energy of an electron. Since $K = 50\,\text{MeV}$, the accelerating potential is $V = 50\,\text{kV} = 5.0 \times 10^7\,\text{V}$. During a pulse the power output is

$$P = iV = (0.50\,\text{A})(5.0 \times 10^7\,\text{V}) = 2.5 \times 10^7\ \text{W} .$$

This is the peak power. The average power is

$$P_{\text{avg}} = i_{\text{avg}}V = (2.5 \times 10^{-5}\,\text{A})(5.0 \times 10^7\,\text{V}) = 1.3 \times 10^3\ \text{W} .$$

42. (a) Since $P = i^2R = J^2A^2R$, the current density is

$$J = \frac{1}{A}\sqrt{\frac{P}{R}} = \frac{1}{A}\sqrt{\frac{P}{\rho L/A}} = \sqrt{\frac{P}{\rho LA}}$$
$$= \sqrt{\frac{1.0\,\mathrm{W}}{\pi(3.5\times10^{-5}\,\Omega\cdot\mathrm{m})(2.0\times10^{-2}\,\mathrm{m})(5.0\times10^{-3}\,\mathrm{m})^2}} = 1.3\times10^5\ \mathrm{A/m^2}\ .$$

(b) From $P = iV = JAV$ we get

$$V = \frac{P}{AJ} = \frac{P}{\pi r^2 J} = \frac{1.0\,\mathrm{W}}{\pi(5.0\times10^{-3}\,\mathrm{m})^2(1.3\times10^5\,\mathrm{A/m^2})} = 9.4\times10^{-2}\ \mathrm{V}\ .$$

43. (a) Using Table 27-1 and Eq. 27-10 (or Eq. 27-11), we have

$$\left|\vec{E}\right| = \rho\left|\vec{J}\right| = (1.69\times10^{-8}\,\Omega\cdot\mathrm{m})\left(\frac{2.0\,\mathrm{A}}{2.0\times10^{-6}\,\mathrm{m^2}}\right) = 1.7\times10^{-2}\ \mathrm{V/m}\ .$$

(b) Using $L = 4.0$ m, the resistance is found from Eq. 27-16: $R = \rho L/A = 0.034\,\Omega$. The rate of thermal energy generation is found from Eq. 27-22: $P = i^2R = 0.14$ W. Assuming a steady rate, the thermal energy generated in 30 minutes is $(0.14\,\mathrm{J/s})(30\times60\,\mathrm{s}) = 2.4\times10^2$ J.

44. (a) Current is the transport of charge; here it is being transported "in bulk" due to the volume rate of flow of the powder. From Chapter 15, we recall that the volume rate of flow is the product of the cross-sectional area (of the stream) and the (average) stream velocity. Thus, $i = \rho Av$ where ρ is the charge per unit volume. If the cross-section is that of a circle, then $i = \rho\pi R^2 v$.

(b) Recalling that a Coulomb per second is an Ampere, we obtain

$$i = \left(1.1\times10^{-3}\,\mathrm{C/m^3}\right)\pi(0.050\,\mathrm{m})^2(2.0\,\mathrm{m/s}) = 1.7\times10^{-5}\ \mathrm{A}\ .$$

(c) The motion of charge is not in the same direction as the potential difference computed in problem 57 of Chapter 25. It might be useful to think of (by analogy) Eq. 7-48; there, the scalar (dot) product in $P = \vec{F}\cdot\vec{v}$ makes it clear that $P = 0$ if $\vec{F}\perp\vec{v}$. This suggests that a radial potential difference and an axial flow of charge will not together produce the needed transfer of energy (into the form of a spark).

(d) With the assumption that there is (at least) a voltage equal to that computed in problem 57 of Chapter 25, in the proper direction to enable the transference of energy (into a spark), then we use our result from that problem in Eq. 27-21:

$$P = iV = \left(1.7\times10^{-5}\,\mathrm{A}\right)\left(7.8\times10^4\,\mathrm{V}\right) = 1.3\ \mathrm{W}\ .$$

(e) Recalling that a Joule per second is a Watt, we obtain $(1.3\,\mathrm{W})(0.20\,\mathrm{s}) = 0.27\,\mathrm{J}$ for the energy that can be transferred at the exit of the pipe.

(f) This result is greater than the 0.15 J needed for a spark, so we conclude that the spark was likely to have occurred at the exit of the pipe, going into the silo.

45. (a) Since the area of a hemisphere is $2\pi r^2$ then the magnitude of the current density vector is

$$\left|\vec{J}\right| = \frac{i}{A} = \frac{I}{2\pi r^2}\ .$$

(b) Eq. 27-11 yields $\left|\vec{E}\right| = \rho\left|\vec{J}\right| = \rho I/2\pi r^2$.

(c) Eq. 25-18 leads to

$$\Delta V = V_r - V_b = -\int_b^r \vec{E}\cdot d\vec{r} = -\int_b^r \left(\frac{\rho I}{2\pi r^2}\right) dr = \frac{\rho I}{2\pi}\left(\frac{1}{r} - \frac{1}{b}\right) .$$

(d) Using the given values, we obtain $\left|\vec{J}\right| = \frac{100}{2\pi(10)^2} = 0.16\ \text{A/m}^2$.

(e) Also, $\left|\vec{E}\right| = 16$ V/m (or 16 N/C).

(f) With $b = 0.010$ m, the voltage is $\Delta V = -1.6 \times 10^5$ V.

46\. (a) Using Eq. 27-11 and Eq. 25-42, we obtain

$$\left|\vec{J}_A\right| = \frac{\left|\vec{E}_A\right|}{\rho} = \frac{|\Delta V_A|}{\rho L} = \frac{40 \times 10^{-6}\ \text{V}}{(100\ \Omega\cdot\text{m})(20\ \text{m})} = 2.0 \times 10^{-8}\ \text{A/m}^2 .$$

(b) Similarly, in region B we find

$$\left|\vec{J}_B\right| = \frac{|\Delta V_B|}{\rho L} = \frac{60 \times 10^{-6}\ \text{V}}{(100\ \Omega\cdot\text{m})(20\ \text{m})} = 3.0 \times 10^{-8}\ \text{A/m}^2 .$$

(c) With $w = 1.0$ m and $d_A = 3.8$ m (so that the cross-section area is $d_A w$) we have (using Eq. 27-5)

$$i_A = \left|\vec{J}_A\right| d_A w = \left(2.0 \times 10^{-8}\ \text{A/m}^2\right)(1.0\ \text{m})(3.8\ \text{m}) = 7.6 \times 10^{-8}\ \text{A} .$$

(d) Assuming $i_A = i_B$ we obtain

$$d_B = \frac{i_B}{\left|\vec{J}_B\right| w} = \frac{7.6 \times 10^{-8}\ \text{A}}{(3.0 \times 10^{-8}\ \text{A/m}^2)(1.0\ \text{m})} = 2.5\ \text{m} .$$

(e) We do not show the graph-and-figure here, but describe it briefly. To be meaningful (as a function of x) we would plot $V(x)$ measured relative to $V(0)$ (the voltage at, say, the left edge of the figure, which we are effectively setting equal to 0). From the problem statement, we note that $V(x)$ would grow linearly in region A, increasing by 40 μV for each 20 m distance. Once we reach the transition region (between A and B) we might assume a parabolic shape for $V(x)$ as it changes from the 40 μV-per-20 m slope to a 60 μV-per-20 m slope (which becomes its constant slope once we are into region B, where the function is again linear). The figure goes further than region B, so as we leave region B, we might assume again a parabolic shape for the function as it tends back down toward some lower slope value.

47\. (a) It is useful to read the whole problem before considering the sketch here in part (a) (which we do not show, but briefly describe). We find in part (d) and part (f), below, that $J_A > J_B$ which suggests that the streamlines should be closer together in region A than in B (at least for portions of those regions which lie close to the pipe). Associated with this (see part (g)) the sketch of the streamlines should reflect that fact that some of the conduction charge-carriers are entering the pipe walls during the transition from region A to region B.

(b) Eq. 27-16 yields

$$\rho_{\text{pipe}} = R\frac{A}{L} = (6.0\ \Omega)\left(\frac{0.010\ \text{m}^2}{1.0 \times 10^6\ \text{m}}\right) = 6.0 \times 10^{-8}\ \Omega\cdot\text{m} .$$

(c) If the resistance of 1000 km of pipe is 6.0 Ω then the resistance of $L = 1.0$ km of pipe is $R = 6.0$ mΩ. Thus in region A, Ohm's law leads to

$$i_{\text{pipe}} = \frac{V_{ab}}{R} = \frac{8.0\ \text{mV}}{6.0\ \text{m}\Omega} = 1.3\ \text{A} .$$

(d) Using Eq. 27-11 and Eq. 25-42 (in absolute value), we find the magnitude of the current density vector in region A:

$$\left|\vec{J}_{\text{ground}}\right| = \frac{V_{ab}}{\rho_{\text{ground}} L} = \frac{0.0080\ \text{V}}{(500\ \Omega\cdot\text{m})(1000\ \text{m})} = 1.6 \times 10^{-8}\ \text{A/m}^2 .$$

(e) Similarly, in region B we obtain

$$i_{\text{pipe}} = \frac{V_{cd}}{R} = \frac{9.5\ \text{mV}}{6.0\ \text{m}\Omega} = 1.6\ \text{A} ,$$

(f) and

$$\left|\vec{J}_{\text{ground}}\right| = \frac{V_{cd}}{\rho_{\text{ground}} L} = \frac{0.0095\ \text{V}}{(1000\ \Omega\cdot\text{m})(1000\ \text{m})} = 9.5 \times 10^{-9}\ \text{A/m}^2 .$$

(g) These results suggest that the pipe walls, in leaving region A and entering region B, have "absorbed" some of the current, leaving the current density in the nearby ground somewhat "depleted" of the telluric flows.

(h) We assume the transition $B \to A$ is the reverse of that discussed in part (g). Here, some current leaves the pipe walls and joins in the ground-supported telluric flows.

(i) There is no current here, because there is no potential difference along this section of pipe. The reason $V_{gh} = 0$ is best seen using Eq. 27-11 and Eq. 25-18 (and remembering that the scalar dot product gives zero for perpendicular vectors). The arrows shown in the figure for current actually refer, in the technical sense, to the direction of $\vec{J}$. We refer to this as the x direction. The pipe section gh is oriented in what we will refer to as the y direction. Eq. 27-11 implies that $\vec{J}$ and $\vec{E}$ must be in the same direction (x). But a nonzero voltage difference here would require (by Eq. 25-18) $\int \vec{E} \cdot d\vec{s} \neq 0$. But since $d\vec{s} = dy$ for this section of pipe, then $\vec{E} \cdot d\vec{s}$ vanishes identically.

(j) Our discussion in part (j) serves also to motivate the fact that the current in section fg is less than that in section ef by a factor of $\cos 45° = 1/\sqrt{2}$. To see this, one may consider the component of the electric field which would "drive" the current (in the sense of Eq. 27-11) along section fg; it is less than the field responsible for the current in section ef by exactly the factor just mentioned. Thus,

$$i_{fg} = i_{ef} \cos 45° = \frac{1.0\ \text{A}}{\sqrt{2}} = 0.71\ \text{A} .$$

(k) The answers to the previous parts indicate that current leaves the pipe at point f and

(l) at point g.

48. (a) We use Eq. 27-16. The new area is $A' = AL/L' = A/2$.

(b) The new resistance is $R' = R(A/A')(L'/L) = 4R$.

49. We use $P = i^2R = i^2\rho L/A$, or $L/A = P/i^2\rho$. So the new values of L and A satisfy

$$\left(\frac{L}{A}\right)_{\text{new}} = \left(\frac{P}{i^2\rho}\right)_{\text{new}} = \frac{30}{4^2}\left(\frac{P}{i^2\rho}\right)_{\text{old}} = \frac{30}{16}\left(\frac{L}{A}\right)_{\text{old}} .$$

Consequently, $(L/A)_{\text{new}} = 1.875(L/A)_{\text{old}}$. Note, too, that $(LA)_{\text{new}} = (LA)_{\text{old}}$. We solve the above two equations for L_{new} and A_{new}:

$$\begin{aligned} L_{\text{new}} &= \sqrt{1.875}L_{\text{old}} = 1.369L_{\text{old}} \\ A_{\text{new}} &= \sqrt{1/1.875}A_{\text{old}} = 0.730A_{\text{old}} . \end{aligned}$$

50. (a) We denote the copper wire with subscript c and the aluminum wire with subscript a.

$$R = \rho_a \frac{L}{A} = \frac{(2.75 \times 10^{-8}\ \Omega\cdot\text{m})(1.3\ \text{m})}{(5.2 \times 10^{-3}\ \text{m})^2} = 1.3 \times 10^{-3}\ \Omega .$$

(b) Let $R = \rho_c L/(\pi d^2/4)$ and solve for the diameter d of the copper wire:

$$d = \sqrt{\frac{4\rho_c L}{\pi R}} = \sqrt{\frac{4(1.69 \times 10^{-8}\,\Omega\cdot\text{m})(1.3\,\text{m})}{\pi(1.3 \times 10^{-3}\,\Omega)}} = 4.6 \times 10^{-3}\ \text{m} .$$

51. We use Eq. 27-17: $\rho - \rho_0 = \rho\alpha(T - T_0)$, and solve for T:

$$T = T_0 + \frac{1}{\alpha}\left(\frac{\rho}{\rho_0} - 1\right) = 20°\text{C} + \frac{1}{4.3 \times 10^{-3}/\text{K}}\left(\frac{58\,\Omega}{50\,\Omega} - 1\right) = 57°\text{C} .$$

We are assuming that $\rho/\rho_0 = R/R_0$.

52. Since values from the referred-to graph can only be crudely estimated, we do not present a graph here, but rather indicate a few values. Since $R = V/i$ then we see $R = \infty$ when $i = 0$ (which the graph seems to show throughout the range $-\infty < V < 2$ V) and $V \neq 0$. For voltages values larger than 2 V, the resistance changes rapidly according to the ratio V/i. For instance, $R \approx 3.1/0.002 = 1550\,\Omega$ when $V = 3.1$ V, and $R \approx 3.8/0.006 = 633\,\Omega$ when $V = 3.8$ V

53. (a)

$$V = iR = i\rho\frac{L}{A} = \frac{(12\,\text{A})(1.69 \times 10^{-8}\,\Omega\cdot\text{m})(4.0 \times 10^{-2}\,\text{m})}{\pi(5.2 \times 10^{-3}\,\text{m}/2)^2} = 3.8 \times 10^{-4}\,\text{V} .$$

(b) Since it moves in the direction of the electron drift which is against the direction of the current, its tail is negative compared to its head.

(c) The time of travel relates to the drift speed:

$$\begin{aligned} t &= \frac{L}{v_d} = \frac{lAne}{i} = \frac{\pi L d^2 ne}{4i} \\ &= \frac{\pi(1.0 \times 10^{-2}\,\text{m})(5.2 \times 10^{-3}\,\text{m})^2(8.47 \times 10^{28}/\text{m}^3)(1.60 \times 10^{-19}\,\text{C})}{4(12\,\text{A})} \\ &= 238\,\text{s} = 3\,\text{min}\,58\,\text{s} . \end{aligned}$$

54. Using Eq. 7-48 and Eq. 27-22, the rate of change of mechanical energy of the piston-Earth system, mgv, must be equal to the rate at which heat is generated from the coil: $mgv = i^2R$. Thus

$$v = \frac{i^2R}{mg} = \frac{(0.240\,\text{A})^2(550\,\Omega)}{(12\,\text{kg})\,(9.8\,\text{m/s}^2)} = 0.27\ \text{m/s} .$$

55. Eq. 27-21 gives the rate of thermal energy production:

$$P = iV = (10\ \text{A})(120\ \text{V}) = 1.2\ \text{kW} .$$

Dividing this into the 180 kJ necessary to cook the three hot-dogs leads to the result $t = 150$ s.

56. We find the drift speed from Eq. 27-7:

$$v_d = \frac{|\vec{J}|}{ne} = 1.5 \times 10^{-4}\,\text{m/s} .$$

At this (average) rate, the time required to travel $L = 5.0$ m is

$$t = \frac{L}{v_d} = 3.4 \times 10^4\,\text{s} .$$

57. (a) $i = (n_h + n_e)e = (2.25 \times 10^{15}/\text{s} + 3.50 \times 10^{15}/\text{s})(1.60 \times 10^{-19}\,\text{C}) = 9.20 \times 10^{-4}$ A.

(b) The magnitude of the current density vector is

$$\left|\vec{J}\right| = \frac{i}{A} = \frac{9.20 \times 10^{-4}\,\text{A}}{\pi(0.165 \times 10^{-3}\,\text{m})^2} = 1.08 \times 10^4\ \text{A/m}^2\ .$$

58. (a) Since $\rho = RA/L = \pi Rd^2/4L = \pi(1.09 \times 10^{-3}\,\Omega)(5.50 \times 10^{-3}\,\text{m})^2/[4(1.60\,\text{m})] = 1.62 \times 10^{-8}\,\Omega\cdot\text{m}$, the material is silver.

(b) The resistance of the round disk is

$$R = \rho\frac{L}{A} = \frac{4\rho L}{\pi d^2} = \frac{4(1.62 \times 10^{-8}\,\Omega\cdot\text{m})(1.00 \times 10^{-3}\,\text{m})}{\pi(2.00 \times 10^{-2}\,\text{m})^2} = 5.16 \times 10^{-8}\ \Omega\ .$$

59. The horsepower required is

$$P = \frac{iV}{0.80} = \frac{(10\,\text{A})(12\,\text{V})}{(0.80)(746\,\text{W/hp})} = 0.20\ \text{hp}\ .$$

60. (a) The current is

$$i = \frac{V}{R} = \frac{V}{\rho L/A} = \frac{\pi V d^2}{4\rho L} = \frac{\pi(1.20\,\text{V})[(0.0400\,\text{in.})(2.54 \times 10^{-2}\,\text{m/in.})]^2}{4(1.69 \times 10^{-8}\,\Omega\cdot\text{m})(33.0\,\text{m})} = 1.74\,\text{A}\ .$$

(b) The magnitude of the current density vector is

$$\left|\vec{J}\right| = \frac{i}{A} = \frac{4i}{\pi d^2} = \frac{4(1.74\,\text{A})}{\pi[(0.0400\,\text{in.})(2.54 \times 10^{-2}\,\text{m/in.})]^2} = 2.15 \times 10^6\ \text{A/m}^2.$$

(c) $E = V/L = 1.20\,\text{V}/33.0\,\text{m} = 3.63 \times 10^{-2}\,\text{V/m}$.

(d) $P = Vi = (1.20\,\text{V})(1.74\,\text{A}) = 2.09\,\text{W}$.

61. We use $R/L = \rho/A = 0.150\,\Omega/\text{km}$.

(a) For copper $J = i/A = (60.0\,\text{A})(0.150\,\Omega/\text{km})/(1.69 \times 10^{-8}\,\Omega\cdot\text{m}) = 5.32 \times 10^5\,\text{A/m}^2$; and for aluminum $J = (60.0\,\text{A})(0.150\,\Omega/\text{km})/(2.75 \times 10^{-8}\,\Omega\cdot\text{m}) = 3.27 \times 10^5\,\text{A/m}^2$.

(b) We denote the mass densities as ρ_m. For copper $(m/L)_c = (\rho_m A)_c = (8960\,\text{kg/m}^3)\,(1.69 \times 10^{-8}\,\Omega\cdot\text{m})/(0.150\,\Omega/\text{km}) = 1.01\,\text{kg/m}$; and for aluminum $(m/L)_a = (\rho_m A)_a = (2700\,\text{kg/m}^3)(2.75 \times 10^{-8}\,\Omega\cdot\text{m})/(0.150\,\Omega/\text{km}) = 0.495\,\text{kg/m}$.

62. (a) We use $P = V^2/R \propto V^2$, which gives $\Delta P \propto \Delta V^2 \approx 2V\Delta V$. The percentage change is roughly $\Delta P/P = 2\Delta V/V = 2(110 - 115)/115 = -8.6\%$.

(b) A drop in V causes a drop in P, which in turn lowers the temperature of the resistor in the coil. At a lower temperature R is also decreased. Since $P \propto R^{-1}$ a decrease in R will result in an increase in P, which partially offsets the decrease in P due to the drop in V. Thus, the actual drop in P will be smaller when the temperature dependency of the resistance is taken into consideration.

63. Using $A = \pi r^2$ with $r = 5 \times 10^{-4}$ m with Eq. 27-5 yields

$$|\vec{J}| = \frac{i}{A} = 2.5 \times 10^6\ \text{A/m}^2\ .$$

Then, with $|\vec{E}| = 5.3$ V/m, Eq. 27-10 leads to

$$\rho = \frac{5.3\,\text{V/m}}{2.5 \times 10^6\ \text{A/m}^2} = 2.1 \times 10^{-6}\ \Omega\cdot\text{m}\ .$$

64. A least squares fit of the data gives $R = \frac{537}{5} + \frac{1111}{1750}T$ with T in degrees Celsius.

(a) At $T = 20°\text{C}$, our expression gives $R = \frac{21017}{175} \approx 120\,\Omega$.

(b) At $T = 0°\text{C}$, our expression gives $R = \frac{537}{5} \approx 107\,\Omega$.

(c) Defining α_R by

$$\alpha_R = \frac{R - R_{20}}{R_{20}\,(T - 20°\text{C})}$$

then we are effectively requiring $\alpha_R R_{20}$ to equal the $\frac{1111}{1750}$ factor in our least squares fit. This implies that $\alpha_R = 1111/210170 = 0.00529/\text{C}°$ if $R_{20} = \frac{21017}{175} \approx 120\,\Omega$ is used as the reference.

(d) Now we define α_R by

$$\alpha_R = \frac{R - R_0}{R_0\,(T - 0°\text{C})} ,$$

which means we require $\alpha_R R_0$ to equal the $\frac{1111}{1750}$ factor in our least squares fit. In this case, $\alpha_R = 1111/187950 = 0.00591/\text{C}°$ if $R_0 = \frac{537}{5} \approx 107\,\Omega$ is used as the reference.

(e) Our least squares fit expression predicts $R = 96473/350 \approx 276\,\Omega$ at $T = 265°\text{C}$.

65. The electric field points towards lower values of potential (see Eq. 25-40) so $\vec{E}$ is directed towards point B (which we take to be the $\hat{\imath}$ direction in our calculation). Since the field is considered to be uniform inside the wire, then its magnitude is, by Eq. 25-42,

$$|\vec{E}| = \frac{|\Delta V|}{L} = \frac{50}{200} = 0.25\ \text{V/m} .$$

Using Eq. 27-11, with $\rho = 1.7 \times 10^{-8}\,\Omega\cdot\text{m}$, we obtain

$$\vec{E} = \rho\vec{J} \implies \vec{J} = 1.5 \times 10^7\ \hat{\text{i}}$$

in SI units (A/m^2).

66. Assuming $\vec{J}$ is directed along the wire (with no radial flow) we integrate, starting with Eq. 27-4,

$$i = \int \left|\vec{J}\right| dA = \int_{R/2}^{R} kr\, 2\pi r\, dr = \frac{2}{3}k\pi\left(R^3 - \frac{R^3}{8}\right)$$

where $k = 3.0 \times 10^8$ and SI units understood. Therefore, if $R = 0.00200$ m, we obtain $i = 4.40$ A.

67. (First problem of **Cluster**)

(a) We are told that $r_B = \frac{1}{2}r_A$ and $L_B = 2L_A$. Thus, using Eq. 27-16,

$$R_B = \rho\,\frac{L_B}{\pi r_B^2} = \rho\,\frac{2L_A}{\frac{1}{4}\pi r_A^2} = 8R_A = 64\ \Omega .$$

(b) The current-densities are assumed uniform.

$$\frac{J_A}{J_B} = \frac{\frac{i}{\pi r_A^2}}{\frac{i}{\pi r_B^2}} = \frac{\frac{i}{\pi r_A^2}}{\frac{i}{\frac{1}{4}\pi r_A^2}} = \frac{1}{4} .$$

68. (Second problem of **Cluster**)

(a) We use Eq. 27-16 to compute the resistances in SI units:

$$\begin{aligned} R_C &= \rho_C \frac{L_C}{\pi r_C^2} = \left(2 \times 10^{-6}\right) \frac{1}{\pi (0.0005)^2} = 2.5\ \Omega \\ R_D &= \rho_D \frac{L_D}{\pi r_D^2} = \left(1 \times 10^{-6}\right) \frac{1}{\pi (0.00025)^2} = 5.1\ \Omega\ . \end{aligned}$$

The voltages follow from Ohm's law:

$$\begin{aligned} |V_1 - V_2| = V_C &= iR_C = 5.1\ \text{V} \\ |V_2 - V_3| = V_D &= iR_D = 10\ \text{V}\ . \end{aligned}$$

(b) See solution for part (a).

(c) and (d) The power is calculated from Eq. 27-22:

$$P = i^2 R = \begin{cases} 10\ \text{W} & \text{for } R = R_C \\ 20\ \text{W} & \text{for } R = R_D \end{cases}$$

69. (Third problem of **Cluster**)

(a) We use Eq. 27-17 with $\rho = \frac{10}{8}\rho_0$ (we are neglecting any thermal expansion of the material) and $T - T_0 = 100$ K in order to obtain $\alpha = 2.5 \times 10^{-3}$/K. Now with this value of α but $T = 600$ K (so $T - T_0 = 300$ K) we find $\rho = 1.75\rho_0 \rightarrow R = 1.75(8.0\,\Omega) = 14\,\Omega$.

(b) We are assuming the wires have unknown but equal length (not the lengths shown in Figure 27-33). With $\alpha_D = 5.0 \times 10^{-3}$/K, we find $\rho = 2.5\rho_0$ for $T - T_0 = 300$ K. With the same assumptions as in part (a), this implies $R = 2.5R_0$ where $R_0 = 16\,\Omega$ (that the resistance of D is twice that of C at 300 K is evident in part (a) of the *previous* solution. Therefore, $R = 2.5(16\,\Omega) = 40\,\Omega$ for wire D at $T = 600$ K.

70. (Fourth problem of **Cluster**)
From Eq. 27-23, we obtain the resistance at temperature T:

$$R = \frac{V^2}{P} = \frac{12^2}{10} = 14.4\ \Omega\ .$$

Thus, the ratio R/R_0 with R_0 representing the resistance at 300 K is 7.2, which we take to equal the ratio of resistivities (ignoring any thermal expansion of the filament). Eq. 27-17, then, leads to

$$\frac{\rho}{\rho_0} = 7.2 = 1 + \alpha\,(T - 300)\quad .$$

Using Table 27-1 ($\alpha = 4.5 \times 10^{-3}$/K) we find $T = 1.7 \times 10^3$ K.

Chapter 28

1. (a) The cost is $(100\,\text{W}\cdot 8.0\,\text{h}/2.0\,\text{W}\cdot\text{h})(\$0.80) = \$320$.
   (b) The cost is $(100\,\text{W}\cdot 8.0\,\text{h}/10^3\,\text{W}\cdot\text{h})(\$0.06) = \$0.048 = 4.8\,\text{cents}$.

2. The chemical energy of the battery is reduced by $\Delta E = q\mathcal{E}$, where q is the charge that passes through in time $\Delta t = 6.0\,\text{min}$, and $\mathcal{E}$ is the emf of the battery. If i is the current, then $q = i\,\Delta t$ and $\Delta E = i\mathcal{E}\,\Delta t = (5.0\,\text{A})(6.0\,\text{V})(6.0\,\text{min})(60\,\text{s/min}) = 1.1\times 10^4\,\text{J}$. We note the conversion of time from minutes to seconds.

3. If P is the rate at which the battery delivers energy and Δt is the time, then $\Delta E = P\,\Delta t$ is the energy delivered in time Δt. If q is the charge that passes through the battery in time Δt and $\mathcal{E}$ is the emf of the battery, then $\Delta E = q\mathcal{E}$. Equating the two expressions for ΔE and solving for Δt, we obtain

$$\Delta t = \frac{q\mathcal{E}}{P} = \frac{(120\,\text{A}\cdot\text{h})(12\,\text{V})}{100\,\text{W}} = 14.4\,\text{h} = 14\,\text{h}\ 24\,\text{min}\ .$$

4. (a) Since $\mathcal{E}_1 > \mathcal{E}_2$ the current flows counterclockwise.
   (b) Battery 1, since the current flows through it from its negative terminal to the positive one.
   (c) Point B, since the current flows from B to A.

5. (a) Let i be the current in the circuit and take it to be positive if it is to the left in R_1. We use Kirchhoff's loop rule: $\mathcal{E}_1 - iR_2 - iR_1 - \mathcal{E}_2 = 0$. We solve for i:

$$i = \frac{\mathcal{E}_1 - \mathcal{E}_2}{R_1 + R_2} = \frac{12\,\text{V} - 6.0\,\text{V}}{4.0\,\Omega + 8.0\,\Omega} = 0.50\ \text{A}\ .$$

   A positive value is obtained, so the current is counterclockwise around the circuit.
   (b) If i is the current in a resistor R, then the power dissipated by that resistor is given by $P = i^2R$. For R_1, $P_1 = (0.50\,\text{A})^2(4.0\,\Omega) = 1.0\,\text{W}$ and for R_2, $P_2 = (0.50\,\text{A})^2(8.0\,\Omega) = 2.0\,\text{W}$.
   (c) If i is the current in a battery with emf $\mathcal{E}$, then the battery supplies energy at the rate $P = i\mathcal{E}$ provided the current and emf are in the same direction. The battery absorbs energy at the rate $P = i\mathcal{E}$ if the current and emf are in opposite directions. For $\mathcal{E}_1$, $P_1 = (0.50\,\text{A})(12\,\text{V}) = 6.0\,\text{W}$ and for $\mathcal{E}_2$, $P_2 = (0.50\,\text{A})(6.0\,\text{V}) = 3.0\,\text{W}$. In battery 1 the current is in the same direction as the emf. Therefore, this battery supplies energy to the circuit; the battery is discharging. The current in battery 2 is opposite the direction of the emf, so this battery absorbs energy from the circuit. It is charging.

6. (a) The energy transferred is

$$U = Pt = \frac{\mathcal{E}^2 t}{r+R} = \frac{(2.0\,\text{V})^2(2.0\,\text{min})(60\,\text{s/min})}{1.0\,\Omega + 5.0\,\Omega} = 80\ \text{J}\ .$$

   (b) The amount of thermal energy generated is

$$U' = i^2Rt = \left(\frac{\mathcal{E}}{r+R}\right)^2 Rt = \left(\frac{2.0\,\text{V}}{1.0\,\Omega + 5.0\,\Omega}\right)^2(5.0\,\Omega)(2.0\,\text{min})(60\,\text{s/min}) = 67\ \text{J}\ .$$

(c) The difference between U and U', which is equal to 13 J, is the thermal energy that is generated in the battery due to its internal resistance.

7. (a) The potential difference is $V = \mathcal{E} + ir = 12\,\text{V} + (0.040\,\Omega)(50\,\text{A}) = 14\,\text{V}$.

   (b) $P = i^2 r = (50\,\text{A})^2(0.040\,\Omega) = 100\,\text{W}$.

   (c) $P' = iV = (50\,\text{A})(12\,\text{V}) = 600\,\text{W}$.

   (d) In this case $V = \mathcal{E} - ir = 12\,\text{V} - (0.040\,\Omega)(50\,\text{A}) = 10\,\text{V}$ and $P = i^2 r = 100\,\text{W}$.

8. (a) Below, we graph Eq. 28-4 (scaled by a factor of 100) for $\mathcal{E} = 2.0\,\text{V}$ and $r = 100\,\Omega$ over the range $0 \le R \le 500\,\Omega$. We multiplied the SI output of Eq. 28-4 by 100 so that this graph would not be vanishingly small with the other graph (see part (b)) when they are plotted together.

   (b) In the same graph, we show $V_R = iR$ over the same range. The graph of current i is the one that starts at 2 (which corresponds to 0.02 A in SI units) and the graph of voltage V_R is the one that starts at 0 (when $R = 0$). The value of V_R are in SI units (not scaled by any factor).

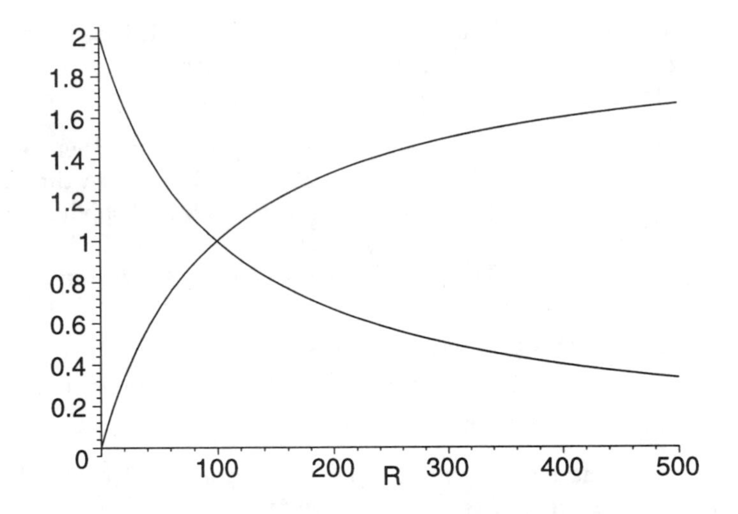

   (c) In our final graph, we show the dependence of power $P = iV_R$ (dissipated in resistor R) as a function of R. The units of the vertical axis are Watts. We note that it is maximum when $R = r$.

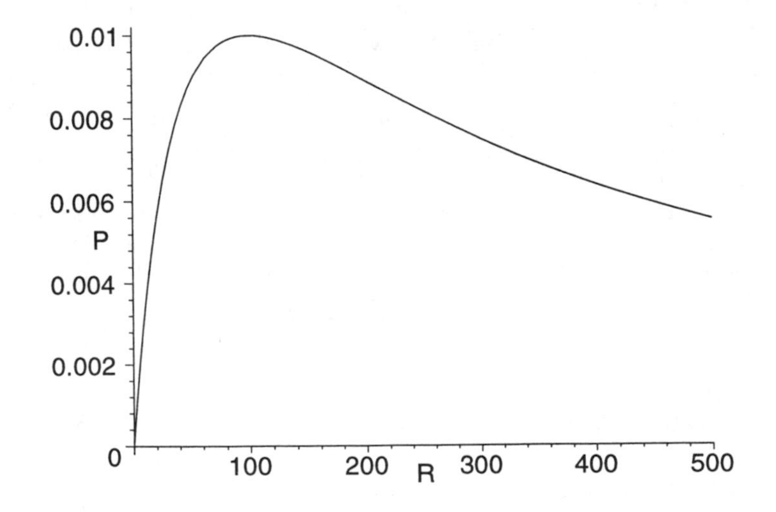

9. (a) If i is the current and ΔV is the potential difference, then the power absorbed is given by $P = i\,\Delta V$. Thus,

$$\Delta V = \frac{P}{i} = \frac{50\,\text{W}}{1.0\,\text{A}} = 50\ \text{V}\ .$$

Since the energy of the charge decreases, point A is at a higher potential than point B; that is, $V_A - V_B = 50\,\text{V}$.

(b) The end-to-end potential difference is given by $V_A - V_B = +iR + \mathcal{E}$, where $\mathcal{E}$ is the emf of element C and is taken to be positive if it is to the left in the diagram. Thus, $\mathcal{E} = V_A - V_B - iR = 50\,\text{V} - (1.0\,\text{A})(2.0\,\Omega) = 48\,\text{V}$.

(c) A positive value was obtained for $\mathcal{E}$, so it is toward the left. The negative terminal is at B.

10. The current in the circuit is $i = (150\,\text{V} - 50\,\text{V})/(3.0\,\Omega + 2.0\,\Omega) = 20\,\text{A}$. So from $V_Q + 150\,\text{V} - (2.0\,\Omega)i = V_P$, we get $V_Q = 100\,\text{V} + (2.0\,\Omega)(20\,\text{A}) - 150\,\text{V} = -10\,\text{V}$.

11. From $V_a - \mathcal{E}_1 = V_c - ir_1 - iR$ and $i = (\mathcal{E}_1 - \mathcal{E}_2)/(R + r_1 + r_2)$, we get

$$\begin{aligned} V_a - V_c &= \mathcal{E}_1 - i\,(r_1 + R) \\ &= \mathcal{E}_1 - \left(\frac{\mathcal{E}_1 - \mathcal{E}_2}{R + r_1 + r_2}\right)(r_1 + R) \\ &= 4.4\,\text{V} - \left(\frac{4.4\,\text{V} - 2.1\,\text{V}}{5.5\,\Omega + 1.8\,\Omega + 2.3\,\Omega}\right)(2.3\,\Omega + 5.5\,\Omega) \\ &= 2.5\ \text{V}\ . \end{aligned}$$

12. (a) We solve $i = (\mathcal{E}_2 - \mathcal{E}_1)/(r_1 + r_2 + R)$ for R:

$$R = \frac{\mathcal{E}_2 - \mathcal{E}_1}{i} - r_1 - r_2 = \frac{3.0\,\text{V} - 2.0\,\text{V}}{1.0 \times 10^{-3}\,\text{A}} - 3.0\,\Omega - 3.0\,\Omega = 9.9 \times 10^2\,\Omega\ .$$

(b) $P = i^2R = (1.0 \times 10^{-3}\,\text{A})^2(9.9 \times 10^2\,\Omega) = 9.9 \times 10^{-4}\,\text{W}$.

13. Let the emf be V. Then $V = iR = i'(R + R')$, where $i = 5.0\,\text{A}$, $i' = 4.0\,\text{A}$ and $R' = 2.0\,\Omega$. We solve for R:

$$R = \frac{i'R'}{i - i'} = \frac{(4)(2)}{5 - 4} = 8.0\ \Omega\ .$$

14. The internal resistance of the battery is $r = (12\,\text{V} - 11.4\,\text{V})/50\,\text{A} = 0.012\,\Omega < 0.020\,\Omega$, so the battery is OK. The resistance of the cable is $R = 3.0\,\text{V}/50\,\text{A} = 0.060\,\Omega > 0.040\,\Omega$, so the cable is defective.

15. To be as general as possible, we refer to the individual emf's as $\mathcal{E}_1$ and $\mathcal{E}_2$ and wait until the latter steps to equate them ($\mathcal{E}_1 = \mathcal{E}_2 = \mathcal{E}$). The batteries are placed in series in such a way that their voltages add; that is, they do not "oppose" each other. The total resistance in the circuit is therefore $R_{\text{total}} = R + r_1 + r_2$ (where the problem tells us $r_1 > r_2$), and the "net emf" in the circuit is $\mathcal{E}_1 + \mathcal{E}_2$. Since battery 1 has the higher internal resistance, it is the one capable of having a zero terminal voltage, as the computation in part (a) shows.

(a) The current in the circuit is

$$i = \frac{\mathcal{E}_1 + \mathcal{E}_2}{r_1 + r_2 + R}\ ,$$

and the requirement of zero terminal voltage leads to

$$\mathcal{E}_1 = ir_1 \implies R = \frac{\mathcal{E}_2 r_1 - \mathcal{E}_1 r_2}{\mathcal{E}_1}$$

which reduces to $R = r_1 - r_2$ when we set $\mathcal{E}_1 = \mathcal{E}_2$.

(b) As mentioned above, this occurs in battery 1.

16. (a) Let the emf of the solar cell be $\mathcal{E}$ and the output voltage be V. Thus,

$$V = \mathcal{E} - ir = \mathcal{E} - \left(\frac{V}{R}\right) r$$

for both cases. Numerically, we get $0.10\,\text{V} = \mathcal{E} - (0.10\,\text{V}/500\,\Omega)r$ and $0.15\,\text{V} = \mathcal{E} - (0.15\,\text{V}/1000\,\Omega)r$. We solve for $\mathcal{E}$ and r: $\mathcal{E} = 0.30\,\text{V}$, $r = 1000\,\Omega$.

(b) The efficiency is

$$\frac{V^2/R}{P_{\text{received}}} = \frac{0.15\,\text{V}}{(1000\,\Omega)(5.0\,\text{cm}^2)(2.0\times10^{-3}\,\text{W/cm}^2)} = 2.3\times10^{-3}\ .$$

17. (a) Using Eq. 28-4, we take the derivative of the power $P = i^2R$ with respect to R and set the result equal to zero:

$$\frac{dP}{dR} = \frac{d}{dR}\left(\frac{\mathcal{E}^2 R}{(R+r)^2}\right) = \frac{\mathcal{E}^2(r-R)}{(R+r)^3} = 0$$

which clearly has the solution $R = r$.

(b) When $R = r$, the power dissipated in the external resistor equals

$$P_{\max} = \left.\frac{\mathcal{E}^2 R}{(R+r)^2}\right|_{R=r} = \frac{\mathcal{E}^2}{4r}\ .$$

18. Let the resistances of the two resistors be R_1 and R_2. Note that the smallest value of the possible R_{eq} must be the result of connecting R_1 and R_2 in parallel, while the largest one must be that of connecting them in series. Thus, $R_1R_2/(R_1+R_2) = 3.0\,\Omega$ and $R_1 + R_2 = 16\,\Omega$. So R_1 and R_2 must be $4.0\,\Omega$ and $12\,\Omega$, respectively.

19. The potential difference across each resistor is $V = 25.0\,\text{V}$. Since the resistors are identical, the current in each one is $i = V/R = (25.0\,\text{V})/(18.0\,\Omega) = 1.39\,\text{A}$. The total current through the battery is then $i_{\text{total}} = 4(1.39\,\text{A}) = 5.56\,\text{A}$. One might alternatively use the idea of equivalent resistance; for four identical resistors in parallel the equivalent resistance is given by

$$\frac{1}{R_{\text{eq}}} = \sum \frac{1}{R} = \frac{4}{R}\ .$$

When a potential difference of 25.0 V is applied to the equivalent resistor, the current through it is the same as the total current through the four resistors in parallel. Thus $i_{\text{total}} = V/R_{\text{eq}} = 4V/R = 4(25.0\,\text{V})/(18.0\,\Omega) = 5.56\,\text{A}$.

20. We note that two resistors in parallel, say R_1 and R_2, are equivalent to

$$R_{\text{parallel pair}} = \frac{1}{\frac{1}{R_1} + \frac{1}{R_2}} = \frac{R_1R_2}{R_1+R_2}\ .$$

This situation (Figure 28-27) consists of a parallel pair which are then in series with a single $2.50\,\Omega$ resistor. Thus, the situation has an equivalent resistance of

$$R_{\text{eq}} = 2.50\,\Omega + \frac{(4.00\,\Omega)(4.00\,\Omega)}{4.00\,\Omega + 4.00\,\Omega} = 4.50\,\Omega\ .$$

21. Let i_1 be the current in R_1 and take it to be positive if it is to the right. Let i_2 be the current in R_2 and take it to be positive if it is upward. When the loop rule is applied to the lower loop, the result is

$$\mathcal{E}_2 - i_1 R_1 = 0 \ .$$

and when it is applied to the upper loop, the result is

$$\mathcal{E}_1 - \mathcal{E}_2 - \mathcal{E}_3 - i_2 R_2 = 0 \ .$$

The first equation yields

$$i_1 = \frac{\mathcal{E}_2}{R_1} = \frac{5.0\,\text{V}}{100\,\Omega} = 0.050\ \text{A} \ .$$

The second yields

$$i_2 = \frac{\mathcal{E}_1 - \mathcal{E}_2 - \mathcal{E}_3}{R_2} = \frac{6.0\,\text{V} - 5.0\,\text{V} - 4.0\,\text{V}}{50\,\Omega} = -0.060\ \text{A} \ .$$

The negative sign indicates that the current in R_2 is actually downward. If V_b is the potential at point b, then the potential at point a is $V_a = V_b + \mathcal{E}_3 + \mathcal{E}_2$, so $V_a - V_b = \mathcal{E}_3 + \mathcal{E}_2 = 4.0\,\text{V} + 5.0\,\text{V} = 9.0\,\text{V}$.

22.
- S_1, S_2 and S_3 all open: $i_a = 0.00\,\text{A}$.
- S_1 closed, S_2 and S_3 open: $i_a = \mathcal{E}/2R_1 = 120\,\text{V}/40.0\,\Omega = 3.00\,\text{A}$.
- S_2 closed, S_1 and S_3 open: $i_a = \mathcal{E}/(2R_1 + R_2) = 120\,\text{V}/50.0\,\Omega = 2.40\,\text{A}$.
- S_3 closed, S_1 and S_2 open: $i_a = \mathcal{E}/(2R_1 + R_2) = 120\,\text{V}/60.0\,\Omega = 2.00\,\text{A}$.
- S_1 open, S_2 and S_3 closed: $R_{\text{eq}} = R_1 + R_2 + R_1(R_1 + R_2)/(2R_1 + R_2) = 20.0\,\Omega + 10.0\,\Omega + (20.0\,\Omega)(30.0\,\Omega)/(50.0\,\Omega) = 42.0\,\Omega$, so $i_a = \mathcal{E}/R_{\text{eq}} = 120\,\text{V}/42.0\,\Omega = 2.86\,\text{A}$.
- S_2 open, S_1 and S_3 closed: $R_{\text{eq}} = R_1 + R_1(R_1 + 2R_2)/(2R_1 + 2R_2) = 20.0\,\Omega + (20.0\,\Omega)\times(40.0\,\Omega)/(60.0\,\Omega) = 33.3\,\Omega$, so $i_a = \mathcal{E}/R_{\text{eq}} = 120\,\text{V}/33.3\,\Omega = 3.60\,\text{A}$.
- S_3 open, S_1 and S_2 closed: $R_{\text{eq}} = R_1 + R_1(R_1 + R_2)/(2R_1 + R_2) = 20.0\,\Omega + (20.0\,\Omega)\times(30.0\,\Omega)/(50.0\,\Omega) = 32.0\,\Omega$, so $i_a = \mathcal{E}/R_{\text{eq}} = 120\,\text{V}/32.0\,\Omega = 3.75\,\text{A}$.
- S_1, S_2 and S_3 all closed: $R_{\text{eq}} = R_1 + R_1R'/(R_1 + R')$ where $R' = R_2 + R_1(R_1 + R_2)/(2R_1 + R_2) = 22.0\,\Omega$, i.e., $R_{\text{eq}} = 20.0\,\Omega + (20.0\,\Omega)(22.0\,\Omega)/(20.0\,\Omega + 22.0\,\Omega) = 30.5\,\Omega$, so $i_a = \mathcal{E}/R_{\text{eq}} = 120\,\text{V}/30.5\,\Omega = 3.94\,\text{A}$.

23. (a) Let $\mathcal{E}$ be the emf of the battery. When the bulbs are connected in parallel, the potential difference across them is the same and is also the same as the emf of the battery. The power dissipated by bulb 1 is $P_1 = \mathcal{E}^2/R_1$, and the power dissipated by bulb 2 is $P_2 = \mathcal{E}^2/R_2$. Since R_1 is greater than R_2, bulb 2 dissipates energy at a greater rate than bulb 1 and is the brighter of the two.

(b) When the bulbs are connected in series the current in them is the same. The power dissipated by bulb 1 is now $P_1 = i^2 R_1$ and the power dissipated by bulb 2 is $P_2 = i^2 R_2$. Since R_1 is greater than R_2 greater power is dissipated by bulb 1 than by bulb 2 and bulb 1 is the brighter of the two.

24. The currents i_1, i_2 and i_3 are obtained from Eqs. 28-15 through 28-17:

$$\begin{aligned}
i_1 &= \frac{\mathcal{E}_1(R_2 + R_3) - \mathcal{E}_2 R_3}{R_1R_2 + R_2R_3 + R_1R_3} = \frac{(4.0\,\text{V})(10\,\Omega + 5.0\,\Omega) - (1.0\,\text{V})(5.0\,\Omega)}{(10\,\Omega)(10\,\Omega) + (10\,\Omega)(5.0\,\Omega) + (10\,\Omega)(5.0\,\Omega)} \\
&= 0.275\ \text{A} \ , \\
i_2 &= \frac{\mathcal{E}_1 R_3 - \mathcal{E}_2(R_1 + R_2)}{R_1R_2 + R_2R_3 + R_1R_3} = \frac{(4.0\,\text{V})(5.0\,\Omega) - (1.0\,\text{V})(10\,\Omega + 5.0\,\Omega)}{(10\,\Omega)(10\,\Omega) + (10\,\Omega)(5.0\,\Omega) + (10\,\Omega)(5.0\,\Omega)} \\
&= 0.025\ \text{A} \ , \\
i_3 &= i_2 - i_1 = 0.025\,\text{A} - 0.275\,\text{A} = -0.250\ \text{A} \ .
\end{aligned}$$

$V_d - V_c$ can now be calculated by taking various paths. Two examples: from $V_d - i_2R_2 = V_c$ we get $V_d - V_c = i_2R_2 = (0.0250\,\text{A})(10\,\Omega) = +0.25\,\text{V}$; from $V_d + i_3R_3 + \mathcal{E}_2 = V_c$ we get $V_d - V_c = -i_3R_3 - \mathcal{E}_2 = -(-0.250\,\text{A})(5.0\,\Omega) - 1.0\,\text{V} = +0.25\,\text{V}$.

25. Let r be the resistance of each of the narrow wires. Since they are in parallel the resistance R of the composite is given by

$$\frac{1}{R} = \frac{9}{r} ,$$

or $R = r/9$. Now $r = 4\rho\ell/\pi d^2$ and $R = 4\rho\ell/\pi D^2$, where ρ is the resistivity of copper. $A = \pi d^2/4$ was used for the cross-sectional area of a single wire, and a similar expression was used for the cross-sectional area of the thick wire. Since the single thick wire is to have the same resistance as the composite,

$$\frac{4\rho\ell}{\pi D^2} = \frac{4\rho\ell}{9\pi d^2} \implies D = 3d .$$

26. (a) $R_{eq}(FH) = (10.0\,\Omega)(10.0\,\Omega)(5.00\,\Omega)/[(10.0\,\Omega)(10.0\,\Omega) + 2(10.0\,\Omega)(5.00\,\Omega)] = 2.50\,\Omega$.
    (b) $R_{eq}(FG) = (5.00\,\Omega)R/(R+5.00\,\Omega)$, where $R = 5.00\,\Omega+(5.00\,\Omega)(10.0\,\Omega)/(5.00\,\Omega+10.0\,\Omega) = 8.33\,\Omega$. So $R_{eq}(FG) = (5.00\,\Omega)(8.33\,\Omega)/(5.00\,\Omega + 8.33\,\Omega) = 3.13\,\Omega$.

27. Let the resistors be divided into groups of n resistors each, with all the resistors in the same group connected in series. Suppose there are m such groups that are connected in parallel with each other. Let R be the resistance of any one of the resistors. Then the equivalent resistance of any group is nR, and R_{eq}, the equivalent resistance of the whole array, satisfies

$$\frac{1}{R_{eq}} = \sum_1^m \frac{1}{nR} = \frac{m}{nR} .$$

Since the problem requires $R_{eq} = 10\,\Omega = R$, we must select $n = m$. Next we make use of Eq. 28-13. We note that the current is the same in every resistor and there are $n \cdot m = n^2$ resistors, so the maximum total power that can be dissipated is $P_{total} = n^2 P$, where $P = 1.0\,\text{W}$ is the maximum power that can be dissipated by any one of the resistors. The problem demands $P_{total} \geq 5.0P$, so n^2 must be at least as large as 5.0. Since n must be an integer, the smallest it can be is 3. The least number of resistors is $n^2 = 9$.

28. (a) R_2, R_3 and R_4 are in parallel. By finding a common denominator and simplifying, the equation $1/R = 1/R_2 + 1/R_3 + 1/R_4$ gives an equivalent resistance of

$$R = \frac{R_2R_3R_4}{R_2R_3 + R_2R_4 + R_3R_4} = \frac{(50\,\Omega)(50\,\Omega)(75\,\Omega)}{(50\,\Omega)(50\,\Omega) + (50\,\Omega)(75\,\Omega) + (50\,\Omega)(75\,\Omega)} = 19\,\Omega .$$

Thus, considering the series contribution of resistor R_1, the equivalent resistance for the network is $R_{eq} = R_1 + R = 100\,\Omega + 19\,\Omega = 1.2 \times 10^2\,\Omega$.
    (b) $i_1 = \mathcal{E}/R_{eq} = 6.0\,\text{V}/(1.1875 \times 10^2\,\Omega) = 5.1 \times 10^{-2}\,\text{A}$; $i_2 = (\mathcal{E} - V_1)/R_2 = (\mathcal{E} - i_1R_1)/R_2 = [6.0\,\text{V} - (5.05 \times 10^{-2}\,\text{A})(100\,\Omega)]/50\,\Omega = 1.9 \times 10^{-2}\,\text{A}$; $i_3 = (\mathcal{E} - V_1)/R_3 = i_2R_2/R_3 = (1.9 \times 10^{-2}\,\text{A})(50\,\Omega/50\,\Omega) = 1.9\times10^{-2}\,\text{A}$; $i_4 = i_1 - i_2 - i_3 = 5.0\times10^{-2}\,\text{A} - 2(1.895\times10^{-2}\,\text{A}) = 1.2\times10^{-2}\,\text{A}$.

29. (a) The batteries are identical and, because they are connected in parallel, the potential differences across them are the same. This means the currents in them are the same. Let i be the current in either battery and take it to be positive to the left. According to the junction rule the current in R is $2i$ and it is positive to the right. The loop rule applied to either loop containing a battery and R yields

$$\mathcal{E} - ir - 2iR = 0 \implies i = \frac{\mathcal{E}}{r + 2R} .$$

The power dissipated in R is

$$P = (2i)^2 R = \frac{4\mathcal{E}^2 R}{(r + 2R)^2} .$$

We find the maximum by setting the derivative with respect to R equal to zero. The derivative is

$$\frac{dP}{dR} = \frac{4\mathcal{E}^2}{(r + 2R)^2} - \frac{16\mathcal{E}^2 R}{(r + 2R)^3} = \frac{4\mathcal{E}^2(r - 2R)}{(r + 2R)^3} .$$

The derivative vanishes (and P is a maximum) if $R = r/2$.

(b) We substitute $R = r/2$ into $P = 4\mathcal{E}^2R/(r+2R)^2$ to obtain

$$P_{\max} = \frac{4\mathcal{E}^2(r/2)}{[r+2(r/2)]^2} = \frac{\mathcal{E}^2}{2r} .$$

30. (a) By symmetry, when the two batteries are connected in parallel the current i going through either one is the same. So from $\mathcal{E} = ir + (2i)R$ we get $i_R = 2i = 2\mathcal{E}/(r+2R)$. When connected in series $2\mathcal{E} - i_R r - i_R r - i_R R = 0$, or $i_R = 2\mathcal{E}/(2r+R)$.

(b) In series, since $R > r$.

(c) In parallel, since $R < r$.

31. (a) We first find the currents. Let i_1 be the current in R_1 and take it to be positive if it is upward. Let i_2 be the current in R_2 and take it to be positive if it is to the left. Let i_3 be the current in R_3 and take it to be positive if it is to the right. The junction rule produces

$$i_1 + i_2 + i_3 = 0 .$$

The loop rule applied to the left-hand loop produces

$$\mathcal{E}_1 - i_3R_3 + i_1R_1 = 0$$

and applied to the right-hand loop produces

$$\mathcal{E}_2 - i_2R_2 + i_1R_1 = 0 .$$

We substitute $i_1 = -i_2 - i_3$, from the first equation, into the other two to obtain

$$\mathcal{E}_1 - i_3R_3 - i_2R_1 - i_3R_1 = 0$$

and

$$\mathcal{E}_2 - i_2R_2 - i_2R_1 - i_3R_1 = 0 .$$

The first of these yields

$$i_3 = \frac{\mathcal{E}_1 - i_2R_1}{R_1 + R_3} .$$

Substituting this into the second equation and solving for i_2, we obtain

$$\begin{aligned} i_2 &= \frac{\mathcal{E}_2(R_1+R_3) - \mathcal{E}_1R_1}{R_1R_2 + R_1R_3 + R_2R_3} \\ &= \frac{(1.00\,\text{V})(5.00\,\Omega + 4.00\,\Omega) - (3.00\,\text{V})(5.00\,\Omega)}{(5.00\,\Omega)(2.00\,\Omega) + (5.00\,\Omega)(4.00\,\Omega) + (2.00\,\Omega)(4.00\,\Omega)} = -0.158\,\text{A} . \end{aligned}$$

We substitute into the expression for i_3 to obtain

$$i_3 = \frac{\mathcal{E}_1 - i_2R_1}{R_1 + R_3} = \frac{3.00\,\text{V} - (-0.158\,\text{A})(5.00\,\Omega)}{5.00\,\Omega + 4.00\,\Omega} = 0.421\,\text{A} .$$

Finally,

$$i_1 = -i_2 - i_3 = -(-0.158\,\text{A}) - (0.421\,\text{A}) = -0.263\,\text{A} .$$

Note that the current in R_1 is actually downward and the current in R_2 is to the right. The current in R_3 is also to the right. The power dissipated in R_1 is $P_1 = i_1^2R_1 = (-0.263\,\text{A})^2(5.00\,\Omega) = 0.346\,\text{W}$.

(b) The power dissipated in R_2 is $P_2 = i_2^2R_2 = (-0.158\,\text{A})^2(2.00\,\Omega) = 0.0499\,\text{W}$.

(c) The power dissipated in R_3 is $P_3 = i_3^2R_3 = (0.421\,\text{A})^2(4.00\,\Omega) = 0.709\,\text{W}$.

(d) The power supplied by $\mathcal{E}_1$ is $i_3\mathcal{E}_1 = (0.421\,\text{A})(3.00\,\text{V}) = 1.26\,\text{W}$.

(e) The power "supplied" by $\mathcal{E}_2$ is $i_2\mathcal{E}_2 = (-0.158\,\text{A})(1.00\,\text{V}) = -0.158\,\text{W}$. The negative sign indicates that $\mathcal{E}_2$ is actually absorbing energy from the circuit.

32. (a) We use $P = \mathcal{E}^2/R_{\text{eq}}$, where

$$R_{\text{eq}} = 7.00\,\Omega + \frac{(12.0\,\Omega)(4.00\,\Omega)R}{(12.0\,\Omega)(4.0\,\Omega) + (12.0\,\Omega)R + (4.00\,\Omega)R} .$$

Put $P = 60.0\,\text{W}$ and $\mathcal{E} = 24.0\,\text{V}$ and solve for R: $R = 19.5\,\Omega$.

(b) Since $P \propto R_{\text{eq}}$, we must minimize R_{eq}, which means $R = 0$.

(c) Now we must maximize R_{eq}, or set $R = \infty$.

(d) Since $R_{\text{eq, max}} = 7.00\,\Omega + (12.0\,\Omega)(4.00\,\Omega)/(12.0\,\Omega + 4.00\,\Omega) = 10.0\,\Omega$, $P_{\text{min}} = \mathcal{E}^2/R_{\text{eq, max}} = (24.0\,\text{V})^2/10.0\,\Omega = 57.6\,\text{W}$. Since $R_{\text{eq, min}} = 7.00\,\Omega$, $P_{\text{max}} = \mathcal{E}^2/R_{\text{eq, min}} = (24.0\,\text{V})^2/7.00\,\Omega = 82.3\,\text{W}$.

33. (a) We note that the R_1 resistors occur in series pairs, contributing net resistance $2R_1$ in each branch where they appear. Since $\mathcal{E}_2 = \mathcal{E}_3$ and $R_2 = 2R_1$, from symmetry we know that the currents through $\mathcal{E}_2$ and $\mathcal{E}_3$ are the same: $i_2 = i_3 = i$. Therefore, the current through $\mathcal{E}_1$ is $i_1 = 2i$. Then from $V_b - V_a = \mathcal{E}_2 - iR_2 = \mathcal{E}_1 + (2R_1)(2i)$ we get

$$i = \frac{\mathcal{E}_2 - \mathcal{E}_1}{4R_1 + R_2} = \frac{4.0\,\text{V} - 2.0\,\text{V}}{4(1.0\,\Omega) + 2.0\,\Omega} = 0.33\,\text{A} .$$

Therefore, the current through $\mathcal{E}_1$ is $i_1 = 2i = 0.67\,\text{A}$, flowing downward. The current through $\mathcal{E}_2$ is 0.33 A, flowing upward; the same holds for $\mathcal{E}_3$.

(b) $V_a - V_b = -iR_2 + \mathcal{E}_2 = -(0.333\,\text{A})(2.0\,\Omega) + 4.0\,\text{V} = 3.3\,\text{V}$.

34. The voltage difference across R is $V_R = \mathcal{E}R'/(R' + 2.00\,\Omega)$, where $R' = (5.00\,\Omega R)/(5.00\,\Omega + R)$. Thus,

$$\begin{aligned} P_R &= \frac{V_R^2}{R} = \frac{1}{R}\left(\frac{\mathcal{E}R'}{R' + 2.00\,\Omega}\right)^2 = \frac{1}{R}\left(\frac{\mathcal{E}}{1 + 2.00\,\Omega/R'}\right)^2 \\ &= \frac{\mathcal{E}^2}{R}\left[1 + \frac{(2.00\,\Omega)(5.00\,\Omega + R)}{(5.00\,\Omega)R}\right]^{-2} \equiv \frac{\mathcal{E}^2}{f(R)} \end{aligned}$$

where we use the equivalence symbol $\equiv$ to define the expression $f(R)$. To maximize P_R we need to minimize the expression $f(R)$. We set

$$\frac{df(R)}{dR} = -\frac{4.00\,\Omega^2}{R^2} + \frac{49}{25} = 0$$

to obtain $R = \sqrt{(4.00\,\Omega^2)(25)/49} = 1.43\,\Omega$.

35. (a) The copper wire and the aluminum sheath are connected in parallel, so the potential difference is the same for them. Since the potential difference is the product of the current and the resistance, $i_C R_C = i_A R_A$, where i_C is the current in the copper, i_A is the current in the aluminum, R_C is the resistance of the copper, and R_A is the resistance of the aluminum. The resistance of either component is given by $R = \rho L/A$, where ρ is the resistivity, L is the length, and A is the cross-sectional area. The resistance of the copper wire is $R_C = \rho_C L/\pi a^2$, and the resistance of the aluminum sheath is $R_A = \rho_A L/\pi(b^2 - a^2)$. We substitute these expressions into $i_C R_C = i_A R_A$, and cancel the common factors L and π to obtain

$$\frac{i_C \rho_C}{a^2} = \frac{i_A \rho_A}{b^2 - a^2} .$$

We solve this equation simultaneously with $i = i_C + i_A$, where i is the total current. We find

$$i_C = \frac{r_C^2 \rho_C i}{(r_A^2 - r_C^2)\rho_C + r_C^2 \rho_A}$$

and

$$i_A = \frac{(r_A^2 - r_C^2)\rho_C i}{(r_A^2 - r_C^2)\rho_C + r_C^2\rho_A} \, .$$

The denominators are the same and each has the value

$$\begin{aligned}(b^2 - a^2)\rho_C + a^2\rho_A &= \left[(0.380 \times 10^{-3}\,\text{m})^2 - (0.250 \times 10^{-3}\,\text{m})^2\right](1.69 \times 10^{-8}\,\Omega\cdot\text{m}) \\ &\quad + (0.250 \times 10^{-3}\,\text{m})^2(2.75 \times 10^{-8}\,\Omega\cdot\text{m}) \\ &= 3.10 \times 10^{-15}\,\Omega\cdot\text{m}^3 \, .\end{aligned}$$

Thus,

$$i_C = \frac{(0.250 \times 10^{-3}\,\text{m})^2(2.75 \times 10^{-8}\Omega\cdot\text{m})(2.00\,\text{A})}{3.10 \times 10^{-15}\,\Omega\cdot\text{m}^3} = 1.11\ \text{A}$$

and

$$\begin{aligned}i_A &= \frac{\left[(0.380 \times 10^{-3}\,\text{m})^2 - (0.250 \times 10^{-3}\,\text{m})^2\right](1.69 \times 10^{-8}\,\Omega\cdot\text{m})(2.00\,\text{A})}{3.10 \times 10^{-15}\,\Omega\cdot\text{m}^3} \\ &= 0.893\,\text{A} \, .\end{aligned}$$

(b) Consider the copper wire. If V is the potential difference, then the current is given by $V = i_C R_C = i_C \rho_C L/\pi a^2$, so

$$L = \frac{\pi a^2 V}{i_C \rho_C} = \frac{(\pi)(0.250 \times 10^{-3}\,\text{m})^2(12.0\,\text{V})}{(1.11\,\text{A})(1.69 \times 10^{-8}\,\Omega\cdot\text{m})} = 126\,\text{m} \, .$$

36. (a) Since $i = \mathcal{E}/(r + R_{\text{ext}})$ and $i_{\text{max}} = \mathcal{E}/r$, we have $R_{\text{ext}} = R(i_{\text{max}}/i - 1)$ where $r = 1.50\,\text{V}/1.00\,\text{mA} = 1.50 \times 10^3\,\Omega$. Thus, $R_{\text{ext}} = (1.5 \times 10^3\,\Omega)(1/0.10 - 1) = 1.35 \times 10^4\,\Omega$;

(b) $R_{\text{ext}} = (1.5 \times 10^3\,\Omega)(1/0.50 - 1) = 1.50 \times 10^3\,\Omega$;

(c) $R_{\text{ext}} = (1.5 \times 10^3\,\Omega)(1/0.90 - 1) = 167\,\Omega$.

(d) Since $r = 20.0\,\Omega + R$, $R = 1.50 \times 10^3\,\Omega - 20.0\,\Omega = 1.48 \times 10^3\,\Omega$.

37. (a) The current in R_1 is given by

$$i_1 = \frac{\mathcal{E}}{R_1 + R_2 R_3/(R_2 + R_3)} = \frac{5.0\,\text{V}}{2.0\,\Omega + (4.0\,\Omega)(6.0\,\Omega)/(4.0\,\Omega + 6.0\,\Omega)} = 1.14\,\text{A} \, .$$

Thus

$$i_3 = \frac{\mathcal{E} - V_1}{R_3} = \frac{\mathcal{E} - i_1 R_1}{R_3} = \frac{5.0\,\text{V} - (1.14\,\text{A})(2.0\,\Omega)}{6.0\,\Omega} = 0.45\,\text{A} \, .$$

(b) We simply interchange subscripts 1 and 3 in the equation above. Now

$$\begin{aligned}i_3 &= \frac{\mathcal{E}}{R_3 + (R_2 R_1/(R_2 + R_1))} \\ &= \frac{5.0\,\text{V}}{6.0\,\Omega + ((2.0\,\Omega)(4.0\,\Omega)/(2.0\,\Omega + 4.0\,\Omega))} \\ &= 0.6818\ \text{A}\end{aligned}$$

and

$$i_1 = \frac{5.0\,\text{V} - (0.6818\,\text{A})(6.0\,\Omega)}{2.0\,\Omega} = 0.45\ \text{A} \, ,$$

the same as before.

38. (a) $\mathcal{E} = V + ir = 12\,\text{V} + (10\,\text{A})(0.050\,\Omega) = 12.5\,\text{V}$.

(b) Now $\mathcal{E} = V' + (i_{\rm motor} + 8.0\,{\rm A})r$, where $V' = i'_A R_{\rm light} = (8.0\,{\rm A})(12\,{\rm V}/10\,{\rm A}) = 9.6\,{\rm V}$. Therefore,

$$i_{\rm motor} = \frac{\mathcal{E} - V'}{r} - 8.0\,{\rm A} = \frac{12.5\,{\rm V} - 9.6\,{\rm V}}{0.050\,\Omega} - 8.0\,{\rm A} = 50\ {\rm A}\ .$$

39. The current in R_2 is i. Let i_1 be the current in R_1 and take it to be downward. According to the junction rule the current in the voltmeter is $i - i_1$ and it is downward. We apply the loop rule to the left-hand loop to obtain

$$\mathcal{E} - iR_2 - i_1R_1 - ir = 0\ .$$

We apply the loop rule to the right-hand loop to obtain

$$i_1R_1 - (i - i_1)R_V = 0\ .$$

The second equation yields

$$i = \frac{R_1 + R_V}{R_V} i_1\ .$$

We substitute this into the first equation to obtain

$$\mathcal{E} - \frac{(R_2 + r)(R_1 + R_V)}{R_V} i_1 + R_1 i_1 = 0\ .$$

This has the solution

$$i_1 = \frac{\mathcal{E}R_V}{(R_2 + r)(R_1 + R_V) + R_1R_V}\ .$$

The reading on the voltmeter is

$$\begin{aligned} i_1R_1 &= \frac{\mathcal{E}R_VR_1}{(R_2 + r)(R_1 + R_V) + R_1R_V} \\ &= \frac{(3.0\,{\rm V})(5.0 \times 10^3\,\Omega)(250\,\Omega)}{(300\,\Omega + 100\,\Omega)(250\,\Omega + 5.0 \times 10^3\,\Omega) + (250\,\Omega)(5.0 \times 10^3\,\Omega)} = 1.12\,{\rm V}\ . \end{aligned}$$

The current in the absence of the voltmeter can be obtained by taking the limit as R_V becomes infinitely large. Then

$$i_1R_1 = \frac{\mathcal{E}R_1}{R_1 + R_2 + r} = \frac{(3.0\,{\rm V})(250\,\Omega)}{250\,\Omega + 300\,\Omega + 100\,\Omega} = 1.15\,{\rm V}\ .$$

The fractional error is $(1.12 - 1.15)/(1.15) = -0.030$, or -3.0%.

40. The currents in R and R_V are i and $i' - i$, respectively. Since $V = iR = (i' - i)R_V$ we have, by dividing both sides by V, $1 = (i'/V - i/V)R_V = (1/R' - 1/R)R_V$. Thus,

$$\frac{1}{R} = \frac{1}{R'} - \frac{1}{R_V}\ .$$

41. Let the current in the ammeter be i'. We have $V = i'(R + R_A)$, or $R = V/i' - R_A = R' - R_A$, where $R' = V/i'$ is the apparent reading of the resistance.

42. (a) In the first case

$$\begin{aligned} i' &= \frac{\mathcal{E}}{R_{\rm eq}} = \frac{\mathcal{E}}{R_A + R_0 + R_VR/(R + R_V)} \\ &= \frac{12.0\,{\rm V}}{3.00\,\Omega + 100\,\Omega + (300\,\Omega)(85.0\,\Omega)/(300\,\Omega + 85.0\,\Omega)} \\ &= 7.09 \times 10^{-2}\ {\rm A}\ , \end{aligned}$$

and $V = \mathcal{E} - i'(R_A + R_0) = 12.0\,\text{V} - (0.0709\,\text{A})(103.00\,\Omega) = 4.70\,\text{V}$. In the second case $V = \mathcal{E}R'/(R' + R_0)$, where

$$R' = \frac{R_V\,(R + R_A)}{R_V + R + R_A} = \frac{(300\,\Omega)(300\,\Omega + 85.0\,\Omega)}{300\,\Omega + 85.0\,\Omega + 3.00\,\Omega} = 68.0\,\Omega\ .$$

So $V = (12.0\,\text{V})(68.0\,\Omega)/(68.0\,\Omega{+}100\,\Omega) = 4.86\,\text{V}$, and $i' = V(R{+}R_A) = 4.86\,\text{V}/(300\,\Omega{+}85.0\,\Omega) = 5.52 \times 10^{-2}\,\text{A}$.

(b) In the first case $R' = V/i' = 4.70\,\text{V}/(7.09 \times 10^{-2}\,\text{A}) = 66.3\,\Omega$. In the second case $R' = V/i' = 4.86\,\text{V}/(5.52 \times 10^{-2}\,\text{A}) = 88.0\,\Omega$.

43. Let i_1 be the current in R_1 and R_2, and take it to be positive if it is toward point a in R_1. Let i_2 be the current in R_s and R_x, and take it to be positive if it is toward b in R_s. The loop rule yields $(R_1 + R_2)i_1 - (R_x + R_s)i_2 = 0$. Since points a and b are at the same potential, $i_1 R_1 = i_2 R_s$. The second equation gives $i_2 = i_1 R_1/R_s$, which is substituted into the first equation to obtain

$$(R_1 + R_2)i_1 = (R_x + R_s)\frac{R_1}{R_s}i_1 \implies R_x = \frac{R_2 R_s}{R_1}\ .$$

44. (a) We use $q = q_0 e^{-t/\tau}$, or $t = \tau \ln(q_0/q)$, where $\tau = RC$ is the capacitive time constant. Thus, $t_{1/3} = \tau \ln[q_0/(2q_0/3)] = \tau \ln(3/2) = 0.41\tau$.

(b) $t_{2/3} = \tau \ln[q_0/(q_0/3)] = \tau \ln 3 = 1.1\tau$.

45. During charging, the charge on the positive plate of the capacitor is given by

$$q = C\mathcal{E}(1 - e^{-t/\tau})\ ,$$

where C is the capacitance, $\mathcal{E}$ is applied emf, and $\tau = RC$ is the capacitive time constant. The equilibrium charge is $q_{\text{eq}} = C\mathcal{E}$. We require $q = 0.99q_{\text{eq}} = 0.99C\mathcal{E}$, so

$$0.99 = 1 - e^{-t/\tau}\ .$$

Thus,

$$e^{-t/\tau} = 0.01\ .$$

Taking the natural logarithm of both sides, we obtain $t/\tau = -\ln 0.01 = 4.6$ and $t = 4.6\tau$.

46. (a) $\tau = RC = (1.40 \times 10^6\,\Omega)(1.80 \times 10^{-6}\,\text{F}) = 2.52\,\text{s}$.

(b) $q_o = \mathcal{E}C = (12.0\,\text{V})(1.80\,\mu\text{F}) = 21.6\,\mu\text{C}$.

(c) The time t satisfies $q = q_0(1 - e^{-t/RC})$, or

$$t = RC \ln\left(\frac{q_0}{q_0 - q}\right) = (2.52\,\text{s}) \ln\left(\frac{21.6\,\mu\text{C}}{21.6\,\mu\text{C} - 16.0\,\mu\text{C}}\right) = 3.40\ \text{s}\ .$$

47. (a) The voltage difference V across the capacitor varies with time as $V(t) = \mathcal{E}(1 - e^{-t/RC})$. At $t = 1.30\,\mu\text{s}$ we have $V(t) = 5.00\,\text{V}$, so $5.00\,\text{V} = (12.0\,\text{V})(1 - e^{-1.30\,\mu\text{s}/RC})$, which gives $\tau = (1.30\,\mu\text{s})/\ln(12/7) = 2.41\,\mu\text{s}$.

(b) $C = \tau/R = 2.41\,\mu\text{s}/15.0\,\text{k}\Omega = 161\,\text{pF}$.

48. The potential difference across the capacitor varies as a function of time t as $V(t) = V_0 e^{-t/RC}$. Using $V = V_0/4$ at $t = 2.0\,\text{s}$, we find

$$R = \frac{t}{C\ln(V_0/V)} = \frac{2.0\,\text{s}}{(2.0 \times 10^{-6}\,\text{F})\ln 4} = 7.2 \times 10^5\,\Omega\ .$$

49. (a) The charge on the positive plate of the capacitor is given by

$$q = C\mathcal{E}\left(1 - e^{-t/\tau}\right) ,$$

where $\mathcal{E}$ is the emf of the battery, C is the capacitance, and τ is the time constant. The value of τ is $\tau = RC = (3.00 \times 10^6\,\Omega)(1.00 \times 10^{-6}\,\text{F}) = 3.00\,\text{s}$. At $t = 1.00\,\text{s}$, $t/\tau = (1.00\,\text{s})/(3.00\,\text{s}) = 0.333$ and the rate at which the charge is increasing is

$$\frac{dq}{dt} = \frac{C\mathcal{E}}{\tau}e^{-t/\tau} = \frac{(1.00 \times 10^{-6})(4.00\,\text{V})}{3.00\,\text{s}}e^{-0.333} = 9.55 \times 10^{-7}\,\text{C/s} .$$

(b) The energy stored in the capacitor is given by

$$U_C = \frac{q^2}{2C} .$$

and its rate of change is

$$\frac{dU_C}{dt} = \frac{q}{C}\frac{dq}{dt} .$$

Now

$$q = C\mathcal{E}\left(1 - e^{-t/\tau}\right) = (1.00 \times 10^{-6})(4.00\,\text{V})(1 - e^{-0.333}) = 1.13 \times 10^{-6}\,\text{C} ,$$

so

$$\frac{dU_C}{dt} = \left(\frac{1.13 \times 10^{-6}\,\text{C}}{1.00 \times 10^{-6}\,\text{F}}\right)(9.55 \times 10^{-7}\,\text{C/s}) = 1.08 \times 10^{-6}\,\text{W} .$$

(c) The rate at which energy is being dissipated in the resistor is given by $P = i^2R$. The current is 9.55×10^{-7}A, so

$$P = (9.55 \times 10^{-7}\text{A})^2(3.00 \times 10^6\,\Omega) = 2.74 \times 10^{-6}\,\text{W} .$$

(d) The rate at which energy is delivered by the battery is

$$i\mathcal{E} = (9.55 \times 10^{-7}\text{A})(4.00\,\text{V}) = 3.82 \times 10^{-6}\,\text{W} .$$

The energy delivered by the battery is either stored in the capacitor or dissipated in the resistor. Conservation of energy requires that $i\mathcal{E} = (q/C)\,(dq/dt) + i^2R$. Except for some round-off error the numerical results support the conservation principle.

50. (a) The charge q on the capacitor as a function of time is $q(t) = (\mathcal{E}C)(1 - e^{-t/RC})$, so the charging current is $i(t) = dq/dt = (\mathcal{E}/R)e^{-t/RC}$. The energy supplied by the emf is then

$$U = \int_0^\infty \mathcal{E}i\,dt = \frac{\mathcal{E}^2}{R}\int_0^\infty e^{-t/RC}dt = C\mathcal{E}^2 = 2U_C$$

where $U_C = \frac{1}{2}C\mathcal{E}^2$ is the energy stored in the capacitor.

(b) By directly integrating i^2R we obtain

$$U_R = \int_0^\infty i^2Rdt = \frac{\mathcal{E}^2}{R}\int_0^\infty e^{-2t/RC}\,dt = \frac{1}{2}C\mathcal{E}^2 .$$

51. (a) The potential difference V across the plates of a capacitor is related to the charge q on the positive plate by $V = q/C$, where C is capacitance. Since the charge on a discharging capacitor is given by $q = q_0\,e^{-t/\tau}$, this means $V = V_0\,e^{-t/\tau}$ where V_0 is the initial potential difference. We solve for the time constant τ by dividing by V_0 and taking the natural logarithm:

$$\tau = -\frac{t}{\ln(V/V_0)} = -\frac{10.0\,\text{s}}{\ln[(1.00\,\text{V})/(100\,\text{V})]} = 2.17\,\text{s} .$$

(b) At $t = 17.0\,\text{s}$, $t/\tau = (17.0\,\text{s})/(2.17\,\text{s}) = 7.83$, so

$$V = V_0\, e^{-t/\tau} = (100\,\text{V})\, e^{-7.83} = 3.96 \times 10^{-2}\ \text{V}\ .$$

52. The time it takes for the voltage difference across the capacitor to reach V_L is given by $V_L = \mathcal{E}(1-e^{-t/RC})$. We solve for R:

$$R = \frac{t}{C\ln[\mathcal{E}/(\mathcal{E}-V_L)]} = \frac{0.500\,\text{s}}{(0.150\times10^{-6}\,\text{F})\ln[95.0\,\text{V}/(95.0\,\text{V}-72.0\,\text{V})]} = 2.35\times10^{6}\,\Omega$$

where we used $t = 0.500\,\text{s}$ given (implicitly) in the problem.

53. (a) The initial energy stored in a capacitor is given by

$$U_C = \frac{q_0^2}{2C}\ ,$$

where C is the capacitance and q_0 is the initial charge on one plate. Thus

$$q_0 = \sqrt{2CU_C} = \sqrt{2(1.0\times10^{-6}\,\text{F})(0.50\,\text{J})} = 1.0\times10^{-3}\ \text{C}\ .$$

(b) The charge as a function of time is given by $q = q_0\, e^{-t/\tau}$, where τ is the capacitive time constant. The current is the derivative of the charge

$$i = -\frac{dq}{dt} = \frac{q_0}{\tau}\, e^{-t/\tau}\ ,$$

and the initial current is $i_0 = q_0/\tau$. The time constant is $\tau = RC = (1.0\times10^{-6}\,\text{F})(1.0\times10^{6}\,\Omega) = 1.0\,\text{s}$. Thus $i_0 = (1.0\times10^{-3}\,\text{C})/(1.0\,\text{s}) = 1.0\times10^{-3}\,\text{A}$.

(c) We substitute $q = q_0\, e^{-t/\tau}$ into $V_C = q/C$ to obtain

$$V_C = \frac{q_0}{C}\, e^{-t/\tau} = \left(\frac{1.0\times10^{-3}\,\text{C}}{1.0\times10^{-6}\,\text{F}}\right) e^{-t/1.0\,\text{s}} = (1.0\times10^{3}\,\text{V})\, e^{-1.0t}\ ,$$

where t is measured in seconds. We substitute $i = (q_0/\tau)\, e^{-t/\tau}$ into $V_R = iR$ to obtain

$$V_R = \frac{q_0 R}{\tau}\, e^{-t/\tau} = \frac{(1.0\times10^{-3}\,\text{C})(1.0\times10^{6}\,\Omega)}{1.0\,\text{s}}\, e^{-t/1.0\,\text{s}} = (1.0\times10^{3}\,\text{V})\, e^{-1.0t}\ ,$$

where t is measured in seconds.

(d) We substitute $i = (q_0/\tau)\, e^{-t/\tau}$ into $P = i^2 R$ to obtain

$$P = \frac{q_0^2 R}{\tau^2}\, e^{-2t/\tau} = \frac{(1.0\times10^{-3}\,\text{C})^2(1.0\times10^{6}\,\Omega)}{(1.0\,\text{s})^2}\, e^{-2t/1.0\,\text{s}} = (1.0\,\text{W})\, e^{-2.0t}\ ,$$

where t is again measured in seconds.

54. We use the result of problem 48: $R = t/[C\ln(V_0/V)]$. Then, for $t_{\text{min}} = 10.0\,\mu\text{s}$

$$R_{\text{min}} = \frac{10.0\,\mu\text{s}}{(0.220\,\mu\text{F})\ln(5.00/0.800)} = 24.8\,\Omega.$$

For $t_{\text{max}} = 6.00\,\text{ms}$,

$$R_{\text{max}} = \left(\frac{6.00\,\text{ms}}{10.0\,\mu\text{s}}\right)(24.8\,\Omega) = 1.49\times10^{4}\,\Omega\ ,$$

where in the last equation we used $\tau = RC$.

55. (a) At $t = 0$ the capacitor is completely uncharged and the current in the capacitor branch is as it would be if the capacitor were replaced by a wire. Let i_1 be the current in R_1 and take it to be positive if it is to the right. Let i_2 be the current in R_2 and take it to be positive if it is downward. Let i_3 be the current in R_3 and take it to be positive if it is downward. The junction rule produces

$$i_1 = i_2 + i_3 \ ,$$

the loop rule applied to the left-hand loop produces

$$\mathcal{E} - i_1 R_1 - i_2 R_2 = 0 \ ,$$

and the loop rule applied to the right-hand loop produces

$$i_2 R_2 - i_3 R_3 = 0 \ .$$

Since the resistances are all the same we can simplify the mathematics by replacing R_1, R_2, and R_3 with R. The solution to the three simultaneous equations is

$$i_1 = \frac{2\mathcal{E}}{3R} = \frac{2(1.2 \times 10^3\,\mathrm{V})}{3(0.73 \times 10^6\,\Omega)} = 1.1 \times 10^{-3}\,\mathrm{A}$$

and

$$i_2 = i_3 = \frac{\mathcal{E}}{3R} = \frac{1.2 \times 10^3\,\mathrm{V}}{3(0.73 \times 10^6\,\Omega)} = 5.5 \times 10^{-4}\,\mathrm{A} \ .$$

At $t = \infty$ the capacitor is fully charged and the current in the capacitor branch is 0. Thus, $i_1 = i_2$, and the loop rule yields

$$\mathcal{E} - i_1 R_1 - i_1 R_2 = 0 \ .$$

The solution is

$$i_1 = i_2 = \frac{\mathcal{E}}{2R} = \frac{1.2 \times 10^3\,\mathrm{V}}{2(0.73 \times 10^6\,\Omega)} = 8.2 \times 10^{-4}\,\mathrm{A} \ .$$

(b) We take the upper plate of the capacitor to be positive. This is consistent with current flowing into that plate. The junction equation is $i_1 = i_2 + i_3$, and the loop equations are

$$\mathcal{E} - i_1 R - i_2 R = 0 \quad \text{and} \quad -\frac{q}{C} - i_3 R + i_2 R = 0 \ .$$

We use the first equation to substitute for i_1 in the second and obtain $\mathcal{E} - 2i_2 R - i_3 R = 0$. Thus $i_2 = (\mathcal{E} - i_3 R)/2R$. We substitute this expression into the third equation above to obtain $-(q/C) - (i_3 R) + (\mathcal{E}/2) - (i_3 R/2) = 0$. Now we replace i_3 with dq/dt to obtain

$$\frac{3R}{2}\frac{dq}{dt} + \frac{q}{C} = \frac{\mathcal{E}}{2} \ .$$

This is just like the equation for an RC series circuit, except that the time constant is $\tau = 3RC/2$ and the impressed potential difference is $\mathcal{E}/2$. The solution is

$$q = \frac{C\mathcal{E}}{2}\left(1 - e^{-2t/3RC}\right) \ .$$

The current in the capacitor branch is

$$i_3 = \frac{dq}{dt} = \frac{\mathcal{E}}{3R} e^{-2t/3RC} \ .$$

The current in the center branch is

$$\begin{aligned} i_2 &= \frac{\mathcal{E}}{2R} - \frac{i_3}{2} = \frac{\mathcal{E}}{2R} - \frac{\mathcal{E}}{6R} e^{-2t/3RC} \\ &= \frac{\mathcal{E}}{6R}\left(3 - e^{-2t/3RC}\right) \end{aligned}$$

and the potential difference across R_2 is

$$V_2 = i_2 R = \frac{\mathcal{E}}{6}\left(3 - e^{-2t/3RC}\right) .$$

This is shown in the following graph.

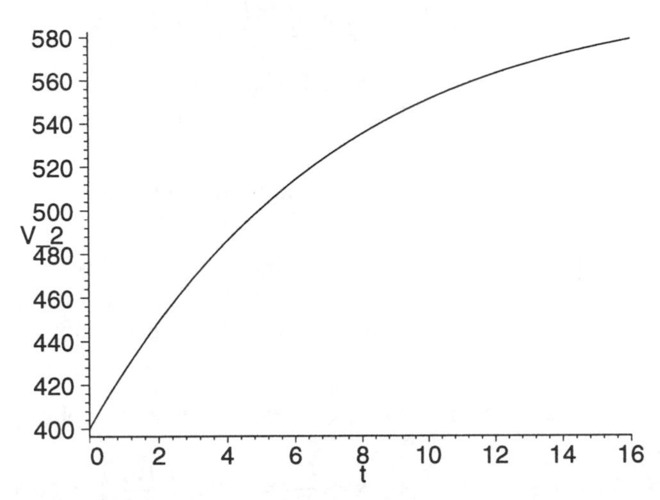

(c) For $t = 0$, $e^{-2t/3RC}$ is 1 and $V_R = \mathcal{E}/3 = (1.2 \times 10^3\,\text{V})/3 = 400\,\text{V}$. For $t = \infty$, $e^{-2t/3RC}$ is 0 and $V_R = \mathcal{E}/2 = (1.2 \times 20^3\,\text{V})/2 = 600\,\text{V}$.

(d) After "a long time" means after several time constants. Then, the current in the capacitor branch is very small and can be approximated by 0.

56. (a) We found in part (e) of problem 45 in Chapter 27 that the magnitude of the electric field is $E = 16$ V/m. Taking this to be roughly constant over the small distance ($\ell = 0.50$ m) involved here, then we approximate the potential difference between the man's feet as

$$\Delta V \approx E\ell = 8 \text{ V} .$$

(b) The voltage found in part (a) drives a current i through the two feet (each represented by $R_f = 300\,\Omega$) and the torso (represented by $R_t = 1000\,\Omega$). Thus,

$$i = \frac{\Delta V}{2R_f + R_t} = \frac{8\,\text{V}}{2(300\,\Omega) + 1000\,\Omega}$$

which yields $i \approx 5$ mA.

(c) Our value for i is far less than the stated 100 mA minimum required to put the heart into fibrillation.

57. (a) The four tires act as resistors in parallel, with an equivalent value given by

$$\frac{1}{R_{\text{eq}}} = \sum_{n=1}^{4} \frac{1}{R_{\text{tire}}} = \frac{4}{R_{\text{tire}}} \implies R_{\text{eq}} = \frac{R_{\text{tire}}}{4} .$$

Using the stated values ($C = 5.0 \times 10^{-10}\,\text{F}$ and $10^8\,\Omega < R_{\text{tire}} < 10^{11}\,\Omega$) we find the capacitive time constant $\tau = R_{\text{eq}}C$ in the range $0.012\,\text{s} < \tau < 13\,\text{s}$.

(b) Eq. 26-22 leads to

$$U_0 = \frac{1}{2}CV^2 = \frac{1}{2}\left(5.00 \times 10^{-10}\,\text{F}\right)\left(30.0 \times 10^3\,\text{V}\right)^2 = 0.225 \text{ J} .$$

(c) As demonstrated in Sample Problem 28-5, the energy "decays" exponentially according to

$$U = U_0 e^{-2t/\tau} .$$

Solving for the time which gives $U = 0.050$ J, we find

$$t = \frac{\tau}{2} \ln\left(\frac{U_0}{U}\right) = \frac{\tau}{2} \ln\left(\frac{0.225}{0.050}\right)$$

which yields, for the range of time constants found in part (a), values of t in the range $0.094\,\mathrm{s} < t < 9.4\,\mathrm{s}$. To obtain these particular values, we used 3-figure versions of the part (a) results ($0.0125\,\mathrm{s} < \tau < 12.5\,\mathrm{s}$).

(d) The lower range of resistance leads to the smaller times to discharge, which is the more desirable situation. Based on this criterion, low resistance tires are favored.

(e) There are a variety of ways to safely and quickly ground a large charged object. A large metal cable connected to, say, the (metal) building frame and held at the end of, say, a long lucite rod might be used (to touch a part of the car that does not have much paint or grease on it) to make the car safe to handle.

58. (a) In the process described in the problem, no charge is gained or lost. Thus, $q =$ constant. Hence,

$$q = C_1 V_1 = C_2 V_2 \implies V_2 = V_1 \frac{C_1}{C_2} = (200)\left(\frac{150}{10}\right) = 3000 \text{ V} .$$

(b) Eq. 28-36, with $\tau = RC$, describes not only the discharging of q but also of V. Thus,

$$V = V_0 e^{-t/\tau} \implies t = RC \ln\left(\frac{V_0}{V}\right) = (300 \times 10^9\,\Omega)\,(10 \times 10^{-12}\,\mathrm{F}) \ln\left(\frac{3000}{100}\right)$$

which yields $t = 10$ s. This is a longer time than most people are inclined to wait before going on to their next task (such as handling the sensitive electronic equipment).

(c) We solve $V = V_0 e^{-t/RC}$ for R with the new values $V_0 = 1400$ V and $t = 0.30$ s. Thus,

$$R = \frac{t}{C \ln(V_0/V)} = \frac{0.30\,\mathrm{s}}{(10 \times 10^{-12}\,\mathrm{F}) \ln(1400/100)} = 1.1 \times 10^{10}\ \Omega .$$

59. (a) Since $R_{\text{tank}} = 140\,\Omega$, $i = 12\,\mathrm{V}/(10\,\Omega + 140\,\Omega) = 8.0 \times 10^{-2}\,\mathrm{A}$.

(b) Now, $R_{\text{tank}} = (140\,\Omega + 20\,\Omega)/2 = 80\,\Omega$, so $i = 12\,\mathrm{V}/(10\,\Omega + 80\,\Omega) = 0.13\,\mathrm{A}$.

(c) When full, $R_{\text{tank}} = 20\,\Omega$ so $i = 12\,\mathrm{V}/(10\,\Omega + 20\,\Omega) = 0.40\,\mathrm{A}$.

60. (a) The magnitude of the current density vector is

$$\begin{aligned} J_A = J_B &= \frac{i}{A} = \frac{V}{(R_1 + R_2)A} = \frac{4\,\mathrm{V}}{(R_1 + R_2)\pi D^2} \\ &= \frac{4(60.0\,\mathrm{V})}{\pi(0.127\,\Omega + 0.729\,\Omega)(2.60 \times 10^{-3}\,\mathrm{m})^2} \\ &= 1.32 \times 10^7\ \mathrm{A/m^2} . \end{aligned}$$

(b) $V_A = VR_1/(R_1 + R_2) = (60.0\,\mathrm{V})(0.127\,\Omega)/(0.127\,\Omega + 0.729\,\Omega) = 8.90\,\mathrm{V}$, and $V_B = V - V_A = 60.0\,\mathrm{V} - 8.9\,\mathrm{V} = 51.1\,\mathrm{V}$.

(c) Calculate the resistivity from $R = \rho L/A$ for both materials: $\rho_A = R_A A/L_A = \pi R_A D^2/4L_A = \pi(0.127\,\Omega)(2.60 \times 10^{-3}\,\mathrm{m})^2/[4(40.0\,\mathrm{m})] = 1.69 \times 10^{-8}\,\Omega\cdot\mathrm{m}$. So A is made of copper. Similarly we find $\rho_B = 9.68 \times 10^{-8}\,\Omega\cdot\mathrm{m}$, so B is made of iron.

61. We denote silicon with subscript s and iron with i. Let $T_0 = 20°$. If

$$\begin{aligned} R(T) &= R_s(T) + R_i(T) = R_s(T_0)[1+\alpha(T-T_0)] + R_i(T_0)[1+\alpha_i(T-T_0)] \\ &= (R_s(T)_0\alpha_s + R_i(T_0)\alpha_i) + (\text{temperature independentterms}) \end{aligned}$$

is to be temperature-independent, we must require that $R_s(T_0)\alpha_s + R_i(T_0)\alpha_i = 0$. Also note that $R_s(T_0) + R_i(T_0) = R = 1000\,\Omega$. We solve for $R_s(T_0)$ and $R_i(T_0)$ to obtain

$$R_s(T_0) = \frac{R\alpha_i}{\alpha_i - \alpha_s} = \frac{(1000\,\Omega)(6.5\times 10^{-3})}{6.5\times 10^{-3} + 70\times 10^{-3}} = 85.0\ \Omega\ ,$$

and $R_i(T_0) = 1000\,\Omega - 85.0\,\Omega = 915\,\Omega$.

62. The potential difference across R_2 is

$$V_2 = iR_2 = \frac{\mathcal{E}\,R_2}{R_1 + R_2 + R_3} = \frac{(12\,\text{V})(4.0\,\Omega)}{3.0\,\Omega + 4.0\,\Omega + 5.0\,\Omega} = 4.0\ \text{V}\ .$$

63. Since $R_{\text{eq}} < R$, the two resistors ($R = 12.0\,\Omega$ and R_x) must be connected in parallel:

$$R_{\text{eq}} = 3.00\,\Omega = \frac{R_x R}{R + R_x} = \frac{R_x(12.0\,\Omega)}{12.0\,\Omega + R_x}\ .$$

We solve for R_x: $R_x = R_{\text{eq}}R/(R - R_{\text{eq}}) = (3.00\,\Omega)(12.0\,\Omega)/(12.0\,\Omega - 3.00\,\Omega) = 4.00\,\Omega$.

64. Consider the lowest branch with the two resistors $R_1 = 3.0\,\Omega$ and $R_2 = 5.0\,\Omega$. The voltage difference across the $5.0\,\Omega$ resistor is

$$V = i_2 R_2 = \frac{\mathcal{E}\,R_2}{R_1 + R_2} = \frac{(120\,\text{V})(5.0\,\Omega)}{3.0\,\Omega + 5.0\,\Omega} = 7.5\ \text{V}\ .$$

65. When all the batteries are connected in parallel, each supplies a current i; thus, $i_R = Ni$. Then from $\mathcal{E} = ir + i_R R = ir + Nir$, we get $i_R = N\mathcal{E}/[(N+1)r]$. When all the batteries are connected in series, $i_r = i_R$ and $\mathcal{E}_{\text{total}} = N\mathcal{E} = Ni_r r + i_R R = Ni_R r + i_R r$, so $i_R = N\mathcal{E}/[(N+1)r]$.

66. (a) They are in parallel and the portions of A and B between the load and their respective sliding contacts have the same potential difference. It is clearly important not to "short" the system (particularly if the load turns out to have very little resistance) by having the sliding contacts too close to the load-ends of A and B to start with. Thus, we suggest putting the contacts roughly in the middle of each. Since $R_1 > R_2$, larger currents generally go through B (depending on the position of the sliding contact) than through A. Therefore, B is analogous to a "coarse" control, as A is to a "fine control." Hence, we recommend adjusting the current roughly with B, and then making fine adjustments with A.

    (b) Relatively large percentage changes in A cause only small percentage charges in the resistance of the parallel combination, thus permitting fine adjustment; any change in A causes half as much change in this combination.

67. When connected in series, the rate at which electric energy dissipates is $P_s = \mathcal{E}^2/(R_1 + R_2)$. When connected in parallel, the corresponding rate is $P_p = \mathcal{E}^2(R_1 + R_2)/R_1R_2$. Letting $P_p/P_s = 5$, we get $(R_1 + R_2)^2/R_1R_2 = 5$, where $R_1 = 100\,\Omega$. We solve for R_2: $R_2 = 38\,\Omega$ or $260\,\Omega$.

68. (a) Placing a wire (of resistance r) with current i running directly from point a to point b in Fig. 28-41 divides the top of the picture into a left and a right triangle. If we label the currents through each resistor with the corresponding subscripts (for instance, i_s goes toward the lower right through R_s and i_x goes toward the upper right through R_x), then the currents must be related as follows:

$$\begin{aligned} i_0 &= i_1 + i_s \quad &\text{and}\quad && i_1 = i + i_2 \\ i_s + i &= i_x \quad &\text{and}\quad && i_2 + i_x = i_0 \end{aligned}$$

where the last relation is not independent of the previous three. The loop equations for the two triangles and also for the bottom loop (containing the battery and point b) lead to

$$\begin{aligned} i_s R_s - i_1 R_1 - ir &= 0 \\ i_2 R_2 - i_x R_x - ir &= 0 \\ \mathcal{E} - i_0 R_0 - i_s R_s - i_x R_x &= 0 \ . \end{aligned}$$

We incorporate the current relations from above into these loop equations in order to obtain three well-posed "simultaneous" equations, for three unknown currents (i_s, i_1 and i):

$$\begin{aligned} i_s R_s - i_1 R_1 - ir &= 0 \\ i_1 R_2 - i_s R_x - i\,(r + R_x + R_2) &= 0 \\ \mathcal{E} - i_s\,(R_0 + R_s + R_x) - i_1 R_0 - i R_x &= 0 \end{aligned}$$

The problem statement further specifies $R_1 = R_2 = R$ and $R_0 = 0$, which causes our solution for i to simplify significantly. It becomes

$$i = \frac{\mathcal{E}\,(R_s - R_x)}{2rR_s + 2R_xR_s + R_sR + 2rR_x + R_xR}$$

which is equivalent to the result shown in the problem statement.

(b) Examining the numerator of our final result in part (a), we see that the condition for $i = 0$ is $R_s = R_x$. Since $R_1 = R_2 = R$, this is equivalent to $R_x = R_sR_2/R_1$, consistent with the result of Problem 43.

69. The voltage across the rightmost resistors is $V_{12} = (1.4\text{ A})(8.0\ \Omega + 4.0\ \Omega) = 16.8$ V, which is equal to V_{16} (the voltage across the 16 Ω resistor, which has current equal to $V_{16}/(16\ \Omega) = 1.05$ A). By the junction rule, the current in the rightmost 2.0 Ω resistor is $1.05 + 1.4 = 2.45$ A, so its voltage is $V_2 = (2.0\ \Omega)(2.45\text{ A}) = 4.9$ V. The loop rule tells us the voltage across the 2.0 Ω resistor (the one going "downward" in the figure) is $V_2' = V_2 + V_{16} = 21.7$ V (implying that the current through it is $i_2' = V_2'/(2.0\ \Omega) = 10.85$ A). The junction rule now gives the current in the leftmost 2.0 Ω resistor as $10.85 + 2.45 = 13.3$ A, implying that the voltage across it is $V_2'' = (13.3\text{ A})(2.0\ \Omega) = 26.6$ V. Therefore, by the loop rule, $\mathcal{E} = V_2'' + V_2' = 48.3$ V.

70. In the steady state situation, the capacitor voltage will equal the voltage across the 15 kΩ resistor:

$$V_0 = (15\text{ k}\Omega)\left(\frac{20\text{ V}}{10\text{ k}\Omega + 15\text{ k}\Omega}\right) = 12\text{ V} \ .$$

Now, multiplying Eq. 28-36 by the capacitance leads to $V = V_0 e^{-t/RC}$ describing the voltage across the capacitor (and across the $R = 15$ kΩ resistor) after the switch is opened (at $t = 0$). Thus, with $t = 0.00400$ s, we obtain

$$V = (12)e^{-0.004/(15000)(0.4\times10^{-6})} = 6.16\text{ V} \ .$$

Therefore, using Ohm's law, the current through the 15 kΩ resistor is $6.16/15000 = 4.11 \times 10^{-4}$ A.

71. (a) By symmetry, we see that i_1 is half the current that goes through the battery. The battery current is found by dividing $\mathcal{E}$ by the equivalent resistance of the circuit, which is easily found to be 6.0 Ω. Thus,

$$i_1 = \frac{1}{2}\,i_{\text{bat}} = \frac{1}{2}\,\frac{12\text{ V}}{6.0\ \Omega} = 1.0\text{ A}$$

and is clearly downward (in the figure).

(b) We use Eq. 28-14: $P = i_{\text{bat}}\mathcal{E} = 24$ W.

72. The series pair of 2.0 Ω resistors on the right reduce to $R' = 4.0\ \Omega$, and the parallel pair of identical 4.0 Ω resistors on the left reduce to $R'' = 2.0\ \Omega$. The voltage across R' must equal that across R''; thus,

$$\begin{aligned} V' &= V'' \\ i'R' &= i''R'' \\ i' &= \frac{1}{2}i'' \end{aligned}$$

where in the last step we divide by R' and simplify. This relation, plus the junction rule condition $6.0\text{ A} = i' + i''$ leads to the solution $i'' = 4.0$ A. It is clear by symmetry that $i = \frac{1}{2}i''$, so we conclude $i = 2.0$ A.

73. (a) We reduce the parallel pair of identical 2.0 Ω resistors (on the right side) to $R' = 1.0\ \Omega$, and we reduce the series pair of identical 2.0 Ω resistors (on the upper left side) to $R'' = 4.0\ \Omega$. With R denoting the 2.0 Ω resistor at the bottom (between V_2 and V_1), we now have three resistors in series which are equivalent to

$$R + R' + R'' = 7.0\ \Omega$$

across which the voltage is 7.0 V (by the loop rule, this is 12 V − 5.0 V), implying that the current is 1.0 A (clockwise). Thus, the voltage across R' is (1.0 A)(1.0 Ω) = 1.0 V, which means that (examining the right side of the circuit) the voltage difference between *ground* and V_1 is $12 - 1 = 11$ V. Noting the orientation of the battery, we conclude $V_1 = -11$ V.

(b) The voltage across R'' is (1.0 A)(4.0 Ω) = 4.0 V, which means that (examining the left side of the circuit) the voltage difference between *ground* and V_2 is $5.0 + 4.0 = 9.0$ V. Noting the orientation of the battery, we conclude $V_2 = -9.0$ V. This can be verified by considering the voltage across R and the value we obtained for V_1.

74. (a) From symmetry we see that the current through the top set of batteries (i) is the same as the current through the second set. This implies that the current through the $R = 4.0\,\Omega$ resistor at the bottom is $i_R = 2i$. Thus, with r denoting the internal resistance of each battery (equal to $4.0\,\Omega$) and $\mathcal{E}$ denoting the 20 V emf, we consider one loop equation (the outer loop), proceeding counterclockwise:

$$3(\mathcal{E} - ir) - (2i)R = 0\ .$$

This yields $i = 3.0$ A. Consequently, $i_R = 6.0$ A.

(b) The terminal voltage of each battery is $\mathcal{E} - ir = 8.0$ V.

(c) Using Eq. 28-14, we obtain $P = i\mathcal{E} = (3)(20) = 60$ W.

(d) Using Eq. 27-22, we have $P = i^2r = 36$ W.

75. (a) The work done by the battery relates to the potential energy change:

$$q\Delta V = eV = e(12\,\text{V}) = 12\,\text{eV} = (12\,\text{eV})(1.6\times10^{-19}\,\text{J/eV}) = 1.9\times10^{-18}\text{ J}\ .$$

(b) $P = iV = neV = (3.4\times10^{18}/\text{s})(1.6\times10^{-19}\,\text{C})(12\,\text{V}) = 6.5\,\text{W}$.

76. (a) We denote $L = 10\,\text{km}$ and $\alpha = 13\,\Omega/\text{km}$. Measured from the east end we have $R_1 = 100\,\Omega = 2\alpha(L - x) + R$, and measured from the west end $R_2 = 200\,\Omega = 2\alpha x + R$. Thus,

$$x = \frac{R_2 - R_1}{4\alpha} + \frac{L}{2} = \frac{200\,\Omega - 100\,\Omega}{4(13\,\Omega/\text{km})} + \frac{10\,\text{km}}{2} = 6.9\text{ km}\ .$$

(b) Also, we obtain

$$R = \frac{R_1 + R_2}{2} - \alpha L = \frac{100\,\Omega + 200\,\Omega}{2} - (13\,\Omega/\text{km})(10\,\text{km}) = 20\ \Omega\ .$$

77. (a) From $P = V^2/R$ we find $V = \sqrt{PR} = \sqrt{(10\,\text{W})(0.10\,\Omega)} = 1.0\,\text{V}$.

(b) From $i = V/R = (\mathcal{E} - V)/r$ we find

$$r = R\left(\frac{\mathcal{E} - V}{V}\right) = (0.10\,\Omega)\left(\frac{1.5\,\text{V} - 1.0\,\text{V}}{1.0\,\text{V}}\right) = 0.050\,\Omega\ .$$

78. (a) The power delivered by the motor is $P = (2.00\,\text{V})(0.500\,\text{m/s}) = 1.00\,\text{W}$. From $P = i^2 R_{\text{motor}}$ and $\mathcal{E} = i(r + R_{\text{motor}})$ we then find $i^2 r - i\mathcal{E} + P = 0$ (which also follows directly from the conservation of energy principle). We solve for i:

$$i = \frac{\mathcal{E} \pm \sqrt{\mathcal{E}^2 - 4rP}}{2r} = \frac{2.00\,\text{V} \pm \sqrt{(2.00\,\text{V})^2 - 4(0.500\,\Omega)(1.00\,\text{W})}}{2(0.500\,\Omega)}\ .$$

The answer is either 3.41 A or 0.586 A.

(b) We use $V = \mathcal{E} - ir = 2.00\,\text{V} - i(0.500\,\Omega)$. We substitute the two values of i obtained in part (a) into the above formula to get $V = 0.293\,\text{V}$ or $1.71\,\text{V}$.

(c) The power P delivered by the motor is the same for either solution. Since $P = iV$ we may have a lower i and higher V or, alternatively, a lower V and higher i. One can check that the two sets of solutions for i and V above do yield the same power $P = iV$.

79. Let the power supplied be P_s and that dissipated be P_d. Since $P_d = i^2 R$ and $i = P_s/\mathcal{E}$, we have $P_d = P_s^2/\mathcal{E}^2 R \propto \mathcal{E}^{-2}$. The ratio is then

$$\frac{P_d(\mathcal{E} = 110,000\,\text{V})}{P_d(\mathcal{E} = 110\,\text{V})} = \left(\frac{110\,\text{V}}{110,000\,\text{V}}\right)^2 = 1.0 \times 10^{-6}\ .$$

80. (a) $R_{\text{eq}}(AB) = 20.0\,\Omega/3 = 6.67\,\Omega$ (three $20.0\,\Omega$ resistors in parallel).

(b) $R_{\text{eq}}(AC) = 20.0\,\Omega/3 = 6.67\,\Omega$ (three $20.0\,\Omega$ resistors in parallel).

(c) $R_{\text{eq}}(BC) = 0$ (as B and C are connected by a conducting wire).

81. The maximum power output is $(120\,\text{V})(15\,\text{A}) = 1800\,\text{W}$. Since $1800\,\text{W}/500\,\text{W} = 3.6$, the maximum number of 500 W lamps allowed is 3.

82. The part of R_0 connected in parallel with R is given by $R_1 = R_0 x/L$, where $L = 10\,\text{cm}$. The voltage difference across R is then $V_R = \mathcal{E}R'/R_{\text{eq}}$, where $R' = RR_1/(R + R_1)$ and $R_{\text{eq}} = R_0(1 - x/L) + R'$. Thus

$$P_R = \frac{V_R^2}{R} = \frac{1}{R}\left(\frac{\mathcal{E}RR_1/(R + R_1)}{R_0(1 - x/L) + RR_1/(R + R_1)}\right)^2\ .$$

Algebraic manipulation then leads to

$$P_R = \frac{100R(\mathcal{E}x/R_0)^2}{(100R/R_0 + 10x - x^2)^2}$$

where x is measured in cm.

83. (a) Since $P = \mathcal{E}^2/R_{\text{eq}}$, the higher the power rating the smaller the value of R_{eq}. To achieve this, we can let the low position connect to the larger resistance (R_1), middle position connect to the smaller resistance (R_2), and the high position connect to both of them in parallel.

(b) For $P = 100\,\text{W}$, $R_{\text{eq}} = R_1 = \mathcal{E}^2/P = (120\,\text{V})^2/100\,\text{W} = 144\,\Omega$; for $P = 300\,\text{W}$, $R_{\text{eq}} = R_1R_2/(R_1 + R_2) = (144\,\Omega)R_2/(144\,\Omega + R_2) = (120\,\text{V})^2/300\,\text{W}$. We obtain $R_2 = 72\,\Omega$.

84. Note that there is no voltage drop across the ammeter. Thus, the currents in the bottom resistors are the same, which we call i (so the current through the battery is $2i$ and the voltage drop across each of the bottom resistors is iR). The resistor network can be reduced to an equivalence of

$$R_{\text{eq}} = \frac{(2R)(R)}{2R + R} + \frac{(R)(R)}{R + R} = \frac{7}{6}R$$

which means that we can determine the current through the battery (and also through each of the bottom resistors):

$$2i = \frac{\mathcal{E}}{R_{\text{eq}}} \implies i = \frac{3\mathcal{E}}{7R} .$$

By the loop rule (going around the left loop, which includes the battery, resistor $2R$ and one of the bottom resistors), we have

$$\mathcal{E} - i_{2R}(2R) - iR = 0 \implies i_{2R} = \frac{\mathcal{E} - iR}{2R} .$$

Substituting $i = 3\mathcal{E}/7R$, this gives $i_{2R} = 2\mathcal{E}/7R$. The difference between i_{2R} and i is the current through the ammeter. Thus,

$$i_{\text{ammeter}} = i - i_{2R} = \frac{3\mathcal{E}}{7R} - \frac{2\mathcal{E}}{7R} = \frac{\mathcal{E}}{7R} .$$

85. The current in the ammeter is given by $i_A = \mathcal{E}/(r + R_1 + R_2 + R_A)$. The current in R_1 and R_2 without the ammeter is $i = \mathcal{E}/(r + R_1 + R_2)$. The percent error is then

$$\begin{aligned} \frac{\Delta i}{i} &= \frac{i - i_A}{i} = 1 - \frac{r + R_1 + R_2}{r + R_1 + R_2 + R_A} = \frac{R_A}{r + R_1 + R_2 + R_A} \\ &= \frac{0.10\,\Omega}{2.0\,\Omega + 5.0\,\Omega + 4.0\,\Omega + 0.10\,\Omega} = 0.90\% . \end{aligned}$$

86. When S is open for a long time, the charge on C is $q_i = \mathcal{E}_2 C$. When S is closed for a long time, the current i in R_1 and R_2 is $i = (\mathcal{E}_2 - \mathcal{E}_1)/(R_1 + R_2) = (3.0\,\text{V} - 1.0\,\text{V})/(0.20\,\Omega + 0.40\,\Omega) = 3.33\,\text{A}$. The voltage difference V across the capacitor is then $V = \mathcal{E}_2 - iR_2 = 3.0\,\text{V} - (3.33\,\text{A})(0.40\,\Omega) = 1.67\,\text{V}$. Thus the final charge on C is $q_f = VC$. So the change in the charge on the capacitor is $\Delta q = q_f - q_i = (V - \mathcal{E}_2)C = (1.67\,\text{V} - 3.0\,\text{V})(10\,\mu\text{F}) = -13\,\mu\text{C}$.

87. Requiring no current through the $10.0\,\Omega$ resistor means that 20.0 V will be across R (which has current i_R). The current through the $20.0\,\Omega$ resistor is also i_R, so the loop rule leads to

$$50.0\,\text{V} - 20.0\,\text{V} - i_R(20.0\,\Omega) = 0$$

which yields $i_R = 1.5$ A. Therefore,

$$R = \frac{20.0\,\text{V}}{i_R} = 13.3\ \Omega .$$

88. (a) The capacitor is *initially* uncharged, which implies (by the loop rule) that there is zero voltage (at $t = 0$) across the 10 kΩ resistor, and that 30 V is across the 20 kΩ resistor. Therefore, by Ohm's law, $i_{10} = 0$,

(b) and $i_{20} = (30\,\text{V})/(20\,\text{k}\Omega) = 1.5 \times 10^{-3}$ A.

(c) As $t \to \infty$ the current to the capacitor reduces to zero and the 20 kΩ and 10 kΩ resistors behave more like a series pair (having the same current), equivalent to 30 kΩ. The current through them, then, at long times, is $i = (30\,\text{V})/(30\,\text{k}\Omega) = 1.0 \times 10^{-3}$ A.

89. (a) The six resistors to the left of $\mathcal{E}_1 = 16$ V battery can be reduced to a single resistor $R = 8.0\ \Omega$, through which the current must be $i_R = \mathcal{E}_1/R = 2.0$ A. Now, by the loop rule, the current through the 3.0 Ω and 1.0 Ω resistors at the upper right corner is

$$i' = \frac{16.0\,\mathrm{V} - 8.0\,\mathrm{V}}{3.0\ \Omega + 1.0\ \Omega} = 2.0\ \mathrm{A}$$

in a direction that is "backward" relative to the $\mathcal{E}_2 = 8.0$ V battery. Thus, by the junction rule,

$$i_1 = i_R + i' = 4.0\ \mathrm{A}$$

and is upward (that is, in the "forward" direction relative to $\mathcal{E}_1$).

(b) The current i_2 derives from a succession of symmetric splittings of i_R (reversing the procedure of reducing those six resistors to find R in part (a)). We find

$$i_2 = \frac{1}{2}\left(\frac{1}{2} i_R\right) = 0.50\ \mathrm{A}$$

and is clearly downward.

(c) Using our conclusions from part (a) in Eq. 28-14, we obtain $P = i_1\mathcal{E}_1 = (4)(16) = 64$ W supplied.

(d) Using results calculated in part (a) in Eq. 28-14, we obtain $P = i'\mathcal{E}_2 = (2)(8) = 16$ W absorbed.

90. We reduce the parallel pair of identical 4.0 Ω resistors to $R' = 2.0\ \Omega$, which has current $i = 2i_1$ going through it. It is in series with a 2.0 Ω resistor, which leads to an equivalence of $R = 4.0\ \Omega$ with current i. We find a path (for use with the loop rule) that goes through this R, the 4.0 V battery, and the 20 V battery, and proceed counterclockwise (assuming i goes rightward through R):

$$20\,\mathrm{V} + 4.0\,\mathrm{V} - iR = 0$$

which leads to $i = 6.0$ A. Consequently, $i_1 = \frac{1}{2}i = 3.0$ A going rightward.

91. With the unit Ω understood, the equivalent resistance for this circuit is

$$R_{\mathrm{eq}} = \frac{20R + 100}{R + 10} .$$

Therefore, the power supplied by the battery (equal to the power dissipated in the resistors) is

$$P = \frac{V^2}{R} = V^2\,\frac{R + 10}{20R + 100}$$

where $V = 12$ V. We attempt to extremize the expression by working through the $dP/dR = 0$ condition and do not find a value of R that satisfies it. We note, then, that the function is a monotonically decreasing function of R, with $R = 0$ giving the maximum possible value (since $R < 0$ values are not being allowed). With the value $R = 0$, we obtain $P = 14.4$ W.

92. The resistor by the letter i is above three other resistors; together, these four resistors are equivalent to a resistor $R = 10\ \Omega$ (with current i). As if we were presented with a maze, we find a path through R that passes through any number of batteries (10, it turns out) but no other resistors, which – as in any good maze – winds "all over the place." Some of the ten batteries are opposing each other (particularly the ones along the outside), so that their net emf is only $\mathcal{E} = 40$ V. The current through R is then $i = \mathcal{E}/R = 4.0$ A, and is directed upward in the figure.

93. (First problem of **Cluster**)

(a) R_2 and R_3 are in parallel; their equivalence is in series with R_1. Therefore,

$$R_{\mathrm{eq}} = R_1 + \frac{R_2 R_3}{R_2 + R_3} = 300\ \Omega .$$

(b) The current through the battery is $\mathcal{E}/R_{\text{eq}} = 0.0200$ A, which is also the current through R_1. Hence, the voltage across R_1 is $V_1 = (0.0200 \text{ A})(100 \ \Omega) = 2.00$ V.

(c) From the loop rule,

$$\mathcal{E} - V_1 - i_3 R_3 = 0$$

which yields $i_3 = 6.67 \times 10^{-3}$ A.

94. (Second problem of **Cluster**)

(a) The loop rule (proceeding counterclockwise around the right loop) leads to $\mathcal{E}_2 - i_1 R_1 = 0$ (where i_1 was assumed downward). This yields $i_1 = 0.060$ A (downward).

(b) The loop rule (counterclockwise around the left loop) gives

$$(+\mathcal{E}_1) + (+i_1 R_1) + (-i_3 R_3) = 0$$

where i_3 has been assumed leftward. This yields $i_3 = 0.180$ A (leftward).

(c) The junction rule tells us that the current through the 12 V battery is $0.180 + 0.060 = 0.240$ A upward.

95. (Third problem of **Cluster**)

(a) Using the junction rule ($i_1 = i_2 + i_3$) we write two loop rule equations:

$$\begin{aligned} \mathcal{E}_1 - i_2 R_2 - (i_2 + i_3)\, R_1 &= 0 \\ \mathcal{E}_2 - i_3 R_3 - (i_2 + i_3)\, R_1 &= 0 \ . \end{aligned}$$

Solving, we find $i_2 = 0.0109$ A (rightward, as was assumed in writing the equations as we did), $i_3 = 0.0273$ A (leftward), and $i_1 = i_2 + i_3 = 0.0382$ A (downward).

(b) See the results in part (a).

(c) See the results in part (a).

(d) The voltage across R_1 equals V_A: $(0.0382 \text{ A})(100 \ \Omega) = +3.82$ V.

96. (Fourth problem of **Cluster**)

(a) The symmetry of the problem allows us to use i_2 as the current in *both* of the R_2 resistors and i_1 for the R_1 resistors. We see from the junction rule that $i_3 = i_1 - i_2$. There are only two independent loop rule equations:

$$\begin{aligned} \mathcal{E} - i_2 R_2 - i_1 R_1 &= 0 \\ \mathcal{E} - 2 i_1 R_1 - (i_1 - i_2)\, R_3 &= 0 \ . \end{aligned}$$

where in the latter equation, a zigzag path through the bridge has been taken. Solving, we find $i_1 = 0.002625$ A , $i_2 = 0.00225$ A and $i_3 = i_1 - i_2 = 0.000375$ A. Therefore, $V_A - V_B = i_1 R_1 = 5.25$ V.

(b) It follows also that $V_B - V_C = i_3 R_3 = 1.50$ V.

(c) We find $V_C - V_D = i_1 R_1 = 5.25$ V.

(d) Finally, $V_A - V_C = i_2 R_2 = 6.75$ V.

Chapter 29

1. (a) We use Eq. 29-3: $F_B = |q|vB\sin\phi = (+3.2\times10^{-19}\,\text{C})(550\,\text{m/s})(0.045\,\text{T})(\sin 52°) = 6.2\times10^{-18}\,\text{N}$.
   (b) $a = F_B/m = (6.2\times10^{-18}\,\text{N})/(6.6\times10^{-27}\,\text{kg}) = 9.5\times10^{8}\,\text{m/s}^2$.
   (c) Since it is perpendicular to $\vec{v}$, $\vec{F}_B$ does not do any work on the particle. Thus from the work-energy theorem both the kinetic energy and the speed of the particle remain unchanged.

2. (a) The largest value of force occurs if the velocity vector is perpendicular to the field. Using Eq. 29-3,

$$F_{B,\,\text{max}} = |q|vB\sin(90°) = evB = (1.60\times10^{-19}\,\text{C})(7.20\times10^{6}\,\text{m/s})(83.0\times10^{-3}\,\text{T}) = 9.56\times10^{-14}\,\text{N}\ .$$

   The smallest value occurs if they are parallel: $F_{B,\,\text{min}} = |q|vB\sin(0) = 0$.
   (b) By Newton's second law, $a = F_B/m_e = |q|vB\sin\theta/m_e$, so the angle θ between $\vec{v}$ and $\vec{B}$ is

$$\theta = \sin^{-1}\left(\frac{m_e a}{|q|vB}\right) = \sin^{-1}\left[\frac{(9.11\times10^{-31}\,\text{kg})(4.90\times10^{14}\,\text{m/s}^2)}{(1.60\times10^{-16}\,\text{C})(7.20\times10^{6}\,\text{m/s})(83.0\times10^{-3}\,\text{T})}\right] = 0.267°\ .$$

3. (a) Eq. 29-3 leads to

$$v = \frac{F_B}{eB\sin\phi} = \frac{6.50\times10^{-17}\,\text{N}}{(1.60\times10^{-19}\,\text{C})(2.60\times10^{-3}\,\text{T})\sin 23.0°} = 4.00\times10^{5}\ \text{m/s}\ .$$

   (b) The kinetic energy of the proton is

$$K = \frac{1}{2}mv^2 = \frac{1}{2}(1.67\times10^{-27}\,\text{kg})(4.00\times10^{5}\,\text{m/s})^2 = 1.34\times10^{-16}\ \text{J}\ .$$

   This is $(1.34\times10^{-16}\,\text{J})/(1.60\times10^{-19}\,\text{J/eV}) = 835\ \text{eV}$.

4. (a) The force on the electron is

$$\begin{aligned}\vec{F}_B &= q\vec{v}\times\vec{B} = q(v_x\hat{\text{i}} + v_y\hat{\text{j}})\times(B_x\hat{\text{i}} + B_y\hat{j})\\ &= q(v_xB_y - v_yB_x)\hat{\text{k}}\\ &= (-1.6\times10^{-19}\,\text{C})[(2.0\times10^{6}\,\text{m/s})(-0.15\,\text{T}) - (3.0\times10^{6}\,\text{m/s})(0.030\,\text{T})]\\ &= (6.2\times10^{-14}\ \text{N})\ \hat{\text{k}}\ .\end{aligned}$$

   Thus, the magnitude of $\vec{F}_B$ is $6.2\times10^{14}\,\text{N}$, and $\vec{F}_B$ points in the positive z direction.
   (b) This amounts to repeating the above computation with a change in the sign in the charge. Thus, $\vec{F}_B$ has the same magnitude but points in the negative z direction.

5. (a) The textbook uses "geomagnetic north" to refer to Earth's magnetic pole lying in the northern hemisphere. Thus, the electrons are traveling northward. The vertical component of the magnetic field is downward. The right-hand rule indicates that $\vec{v}\times\vec{B}$ is to the west, but since the electron is negatively charged (and $\vec{F} = q\vec{v}\times\vec{B}$), the magnetic force on it is to the east.

(b) We combine $F = m_e a$ with $F = evB\sin\phi$. Here, $B\sin\phi$ represents the downward component of Earth's field (given in the problem). Thus, $a = evB/m_e$. Now, the electron speed can be found from its kinetic energy. Since $K = \frac{1}{2}mv^2$,

$$v = \sqrt{\frac{2K}{m_e}} = \sqrt{\frac{2(12.0 \times 10^3\,\text{eV})(1.60 \times 10^{-19}\ \text{J/eV})}{9.11 \times 10^{-31}\,\text{kg}}} = 6.49 \times 10^7\ \text{m/s}\ .$$

Therefore,

$$a = \frac{evB}{m_e} = \frac{(1.60 \times 10^{-19}\,\text{C})(6.49 \times 10^7\,\text{m/s})(55.0 \times 10^{-6}\,\text{T})}{9.11 \times 10^{-31}\,\text{kg}} = 6.27 \times 10^{14}\ \text{m/s}^2\ .$$

(c) We ignore any vertical deflection of the beam which might arise due to the horizontal component of Earth's field. Technically, then, the electron should follow a circular arc. However, the deflection is so small that many of the technicalities of circular geometry may be ignored, and a calculation along the lines of projectile motion analysis (see Chapter 4) provides an adequate approximation:

$$\Delta x = vt \implies t = \frac{\Delta x}{v} = \frac{0.200\,\text{m}}{6.49 \times 10^7\,\text{m/s}}$$

which yields a time of $t = 3.08 \times 10^{-9}$ s. Then, with our y axis oriented eastward,

$$\Delta y = \frac{1}{2}at^2 = \frac{1}{2}\left(6.27 \times 10^{14}\right)\left(3.08 \times 10^{-9}\right)^2 = 0.00298\ \text{m}\ .$$

6. (a) The net force on the proton is given by

$$\begin{aligned} \vec{F} &= \vec{F}_E + \vec{F}_B = q\vec{E} + q\vec{v} \times \vec{B} \\ &= (1.6 \times 10^{-19}\,\text{C})[(4.0\,\text{V/m})\hat{\text{k}} + (2000\,\text{m/s})\hat{\text{j}} \times (-2.5\,\text{mT})\hat{\text{i}}] \\ &= (1.4 \times 10^{-18}\ \text{N})\ \hat{\text{k}}\ . \end{aligned}$$

(b) In this case

$$\begin{aligned} \vec{F} &= \vec{F}_E + \vec{F}_B = q\vec{E} + q\vec{v} \times \vec{B} \\ &= (1.6 \times 10^{-19}\,\text{C})[(-4.0\,\text{V/m})\hat{\text{k}} + (2000\,\text{m/s})\hat{\text{j}} \times (-2.5\,\text{mT})\hat{\text{i}}] \\ &= (1.6 \times 10^{-19}\ \text{N})\ \hat{\text{k}}\ . \end{aligned}$$

(c) In the final case,

$$\begin{aligned} \vec{F} &= \vec{F}_E + \vec{F}_B = q\vec{E} + q\vec{v} \times \vec{B} \\ &= (1.6 \times 10^{-19}\,\text{C})[(4.0\,\text{V/m})\hat{\text{i}} + (2000\,\text{m/s})\hat{\text{j}} \times (-2.5\,\text{mT})\hat{\text{i}}] \\ &= (6.4 \times 10^{-19}\,\text{N})\,\hat{\text{i}} + (8.0 \times 10^{-19}\ \text{N})\,\hat{\text{k}}\ . \end{aligned}$$

The magnitude of the force is now

$$\sqrt{F_x^2 + F_y^2 + F_z^2} = \sqrt{(6.4 \times 10^{-19}\,\text{N})^2 + 0 + (8.0 \times 10^{-19}\,\text{N})^2} = 1.0 \times 10^{-18}\ \text{N}\ .$$

7. (a) Equating the magnitude of the electric force ($F = eE$) with that of the magnetic force (Eq. 29-3), we obtain $B = E/v\sin\phi$. The field is smallest when the $\sin\phi$ factor is at its largest value; that is, when $\phi = 90°$. Now, we use $K = \frac{1}{2}mv^2$ to find the speed:

$$v = \sqrt{\frac{2K}{m_e}} = \sqrt{\frac{2(2.5 \times 10^3\,\text{eV})(1.60 \times 10^{-19}\ \text{J/eV})}{9.11 \times 10^{-31}\,\text{kg}}} = 2.96 \times 10^7\ \text{m/s}\ .$$

Thus,

$$B = \frac{E}{v} = \frac{10 \times 10^3 \text{V/m}}{2.96 \times 10^7 \text{ m/s}} = 3.4 \times 10^{-4} \text{ T} .$$

The magnetic field must be perpendicular to both the electric field and the velocity of the electron.

(b) A proton will pass undeflected if its velocity is the same as that of the electron. Both the electric and magnetic forces reverse direction, but they still cancel.

8. (a) Letting $\vec{F} = q(\vec{E} + \vec{v} \times \vec{B}) = 0$, we get $vB \sin\phi = E$. We note that (for given values of the fields) this gives a minimum value for speed whenever the $\sin\phi$ factor is at its maximum value (which is 1, corresponding to $\phi = 90°$). So $v_{\text{min}} = E/B = (1.50 \times 10^3 \text{ V/m})/(0.400 \text{ T}) = 3.75 \times 10^3 \text{ m/s}$.

   (b) Having noted already that $\vec{v} \perp \vec{B}$, we now point out that $\vec{v} \times \vec{B}$ (which direction is given by the right-hand rule) must be in the direction opposite to $\vec{E}$. Thus, we can use the *left* hand to indicate the arrangement of vectors: if one points the thumb, index finger, and middle finger on the left hand so that all three are mutually perpendicular, then the thumb represents $\vec{v}$, the index finger indicates $\vec{B}$, and the middle finger represents $\vec{E}$.

9. Straight line motion will result from zero net force acting on the system; we ignore gravity. Thus, $\vec{F} = q(\vec{E} + \vec{v} \times \vec{B}) = 0$. Note that $\vec{v} \perp \vec{B}$ so $|\vec{v} \times \vec{B}| = vB$. Thus, obtaining the speed from the formula for kinetic energy, we obtain

$$\begin{aligned} B &= \frac{E}{v} = \frac{E}{\sqrt{2m_e K}} \\ &= \frac{100 \text{ V}/(20 \times 10^{-3} \text{ m})}{\sqrt{2(9.11 \times 10^{-31} \text{ kg})(1.0 \times 10^3 \text{ V})(1.60 \times 10^{-19} \text{ C})}} \\ &= 2.7 \times 10^{-4} \text{ T} . \end{aligned}$$

10. We apply $\vec{F} = q(\vec{E} + \vec{v} \times \vec{B}) = m_e\vec{a}$ to solve for $\vec{E}$:

$$\begin{aligned} \vec{E} &= \frac{m_e\vec{a}}{q} + \vec{B} \times \vec{v} \\ &= \frac{(9.11 \times 10^{-31} \text{ kg})(2.00 \times 10^{12} \text{ m/s}^2)\hat{\text{i}}}{-1.60 \times 10^{-19} \text{ C}} + (400 \,\mu\text{T})\hat{\text{i}} \times [(12.0 \text{ km/s})\hat{\text{j}} + (15.0 \text{ km/s})\hat{\text{k}}] \\ &= (-11.4\hat{\text{i}} - 6.00\hat{\text{j}} + 4.80\hat{\text{k}}) \text{ V/m} . \end{aligned}$$

11. Since the total force given by $\vec{F} = e(\vec{E} + \vec{v} \times \vec{B})$ vanishes, the electric field $\vec{E}$ must be perpendicular to both the particle velocity $\vec{v}$ and the magnetic field $\vec{B}$. The magnetic field is perpendicular to the velocity, so $\vec{v} \times \vec{B}$ has magnitude vB and the magnitude of the electric field is given by $E = vB$. Since the particle has charge e and is accelerated through a potential difference V, $\frac{1}{2}mv^2 = eV$ and $v = \sqrt{2eV/m}$. Thus,

$$E = B\sqrt{\frac{2eV}{m}} = (1.2\text{T})\sqrt{\frac{2(1.60 \times 10^{-19} \text{ C})(10 \times 10^3 \text{ V})}{(6.0 \text{ u})(1.661 \times 10^{-27} \text{ kg/u})}} = 6.8 \times 10^5 \text{ V/m} .$$

12. We use Eq. 29-12 to solve for V:

$$V = \frac{iB}{nle} = \frac{(23 \text{ A})(0.65 \text{ T})}{(8.47 \times 10^{28}/\text{m}^3)(150 \,\mu\text{m})(1.6 \times 10^{-19} \text{ C})} = 7.4 \times 10^{-6} \text{ V} .$$

13. (a) In Chapter 27, the electric field (called E_C in this problem) which "drives" the current through the resistive material is given by Eq. 27-11, which (in magnitude) reads $E_C = \rho J$. Combining this with Eq. 27-7, we obtain

$$E_C = \rho n e v_d .$$

Now, regarding the Hall effect, we use Eq. 29-10 to write $E = v_d B$. Dividing one equation by the other, we get $E/E_c = B/ne\rho$.

(b) Using the value of copper's resistivity given in Chapter 27, we obtain

$$\frac{E}{E_c} = \frac{B}{ne\rho} = \frac{0.65\,\mathrm{T}}{(8.47\times10^{28}/\,\mathrm{m}^3)(1.60\times10^{-19}\,\mathrm{C})(1.69\times10^{-8}\,\Omega\cdot\mathrm{m})} = 2.84\times10^{-3}\ .$$

14. For a free charge q inside the metal strip with velocity $\vec{v}$ we have $\vec{F} = q(\vec{E} + \vec{v}\times\vec{B})$. We set this force equal to zero and use the relation between (uniform) electric field and potential difference. Thus,

$$v = \frac{E}{B} = \frac{|V_x - V_y|/d_{xy}}{B} = \frac{(3.90\times10^{-9}\,\mathrm{V})}{(1.20\times10^{-3}\,\mathrm{T})(0.850\times10^{-2}\,\mathrm{m})} = 0.382\ \mathrm{m/s}\ .$$

15. From Eq. 29-16, we find

$$B = \frac{m_e v}{er} = \frac{(9.11\times10^{-31}\,\mathrm{kg})(1.3\times10^{6}\,\mathrm{m/s})}{(1.60\times10^{-19}\,\mathrm{C})(0.35\,\mathrm{m})} = 2.1\times10^{-5}\ \mathrm{T}\ .$$

16. (a) The accelerating process may be seen as a conversion of potential energy eV into kinetic energy. Since it starts from rest, $\frac{1}{2}m_e v^2 = eV$ and

$$v = \sqrt{\frac{2eV}{m_e}} = \sqrt{\frac{2(1.60\times10^{-19}\,\mathrm{C})(350\,\mathrm{V})}{9.11\times10^{-31}\ \mathrm{kg}}} = 1.11\times10^{7}\ \mathrm{m/s}\ .$$

(b) Eq. 29-16 gives

$$r = \frac{m_e v}{eB} = \frac{(9.11\times10^{-31}\ \mathrm{kg})(1.11\times10^{7}\ \mathrm{m/s})}{(1.60\times10^{-19}\,\mathrm{C})(200\times10^{-3}\,\mathrm{T})} = 3.16\times10^{-4}\ \mathrm{m}\ .$$

17. (a) From $K = \frac{1}{2}m_e v^2$ we get

$$v = \sqrt{\frac{2K}{m_e}} = \sqrt{\frac{2(1.20\times10^{3}\,\mathrm{eV})(1.60\times10^{-19}\,\mathrm{eV/J})}{9.11\times10^{-31}\,\mathrm{kg}}} = 2.05\times10^{7}\ \mathrm{m/s}\ .$$

(b) From $r = m_e v/qB$ we get

$$B = \frac{m_e v}{qr} = \frac{(9.11\times10^{-31}\,\mathrm{kg})(2.05\times10^{7}\ \mathrm{m/s})}{(1.60\times10^{-19}\,\mathrm{C})(25.0\times10^{-2}\,\mathrm{m})} = 4.67\times10^{-4}\ \mathrm{T}\ .$$

(c) The "orbital" frequency is

$$f = \frac{v}{2\pi r} = \frac{2.07\times10^{7}\,\mathrm{m/s}}{2\pi(25.0\times10^{-2}\,\mathrm{m})} = 1.31\times10^{7}\ \mathrm{Hz}\ .$$

(d) $T = 1/f = (1.31\times10^{7}\,\mathrm{Hz})^{-1} = 7.63\times10^{-8}$ s.

18. The period of revolution for the iodine ion is $T = 2\pi r/v = 2\pi m/Bq$, which gives

$$m = \frac{BqT}{2\pi} = \frac{(45.0\times10^{-3}\,\mathrm{T})(1.60\times10^{-19}\,\mathrm{C})(1.29\times10^{-3}\,\mathrm{s})}{(7)(2\pi)(1.66\times10^{-27}\,\mathrm{kg/u})} = 127\ \mathrm{u}\ .$$

19. (a) The frequency of revolution is

$$f = \frac{Bq}{2\pi m_e} = \frac{(35.0\times10^{-6}\,\mathrm{T})(1.60\times10^{-19}\,\mathrm{C})}{2\pi(9.11\times10^{-31}\,\mathrm{kg})} = 9.78\times10^{5}\,\mathrm{Hz}\ .$$

(b) Using Eq. 29-16, we obtain

$$r = \frac{m_e v}{qB} = \frac{\sqrt{2m_e K}}{qB} = \frac{\sqrt{2(9.11 \times 10^{-31}\,\text{kg})(100\,\text{eV})(1.60 \times 10^{-19}\,\text{J/eV})}}{(1.60 \times 10^{-19}\,\text{C})(35.0 \times 10^{-6}\,\text{T})} = 0.964\,\text{m} .$$

20. (a) Using Eq. 29-16, we obtain

$$v = \frac{rqB}{m_\alpha} = \frac{2eB}{4.00\,\text{u}} = \frac{2(4.50 \times 10^{-2}\,\text{m})(1.60 \times 10^{-19}\,\text{C})(1.20\,\text{T})}{(4.00\,\text{u})(1.66 \times 10^{-27}\,\text{kg/u})} = 2.60 \times 10^6\ \text{m/s} .$$

(b) $T = 2\pi r/v = 2\pi(4.50 \times 10^{-2}\,\text{m})/(2.60 \times 10^6\,\text{m/s}) = 1.09 \times 10^{-7}\,\text{s}$.

(c) The kinetic energy of the alpha particle is

$$K = \frac{1}{2}m_\alpha v^2 = \frac{(4.00\,\text{u})(1.66 \times 10^{-27}\,\text{kg/u})(2.60 \times 10^6\,\text{m/s})^2}{2(1.60 \times 10^{-19}\,\text{J/eV})} = 1.40 \times 10^5\ \text{eV} .$$

(d) $\Delta V = K/q = 1.40 \times 10^5\,\text{eV}/2e = 7.00 \times 10^4\,\text{V}$.

21. So that the magnetic field has an effect on the moving electrons, we need a non-negligible component of $\vec{B}$ to be perpendicular to $\vec{v}$ (the electron velocity). It is most efficient, therefore, to orient the magnetic field so it is perpendicular to the plane of the page. The magnetic force on an electron has magnitude $F_B = evB$, and the acceleration of the electron has magnitude $a = v^2/r$. Newton's second law yields $evB = m_e v^2/r$, so the radius of the circle is given by $r = m_e v/eB$ in agreement with Eq. 29-16. The kinetic energy of the electron is $K = \frac{1}{2}m_e v^2$, so $v = \sqrt{2K/m_e}$. Thus,

$$r = \frac{m_e}{eB}\sqrt{\frac{2K}{m_e}} = \sqrt{\frac{2m_e K}{e^2B^2}} .$$

This must be less than d, so

$$\sqrt{\frac{2m_e K}{e^2B^2}} \le d$$

or

$$B \ge \sqrt{\frac{2m_e K}{e^2d^2}} .$$

If the electrons are to travel as shown in Fig. 29-33, the magnetic field must be out of the page. Then the magnetic force is toward the center of the circular path, as it must be (in order to make the circular motion possible).

22. Let $v_\parallel = v\cos\theta$. The electron will proceed with a uniform speed $v_\parallel$ in the direction of $\vec{B}$ while undergoing uniform circular motion with frequency f in the direction perpendicular to B: $f = eB/2\pi m_e$. The distance d is then

$$\begin{aligned} d &= v_\parallel T = \frac{v_\parallel}{f} = \frac{(v\cos\theta)2\pi m_e}{eB} \\ &= \frac{2\pi(1.5 \times 10^7\,\text{m/s})(9.11 \times 10^{-31}\,\text{kg})(\cos 10^\circ)}{(1.60 \times 10^{-19}\,\text{C})(1.0 \times 10^{-3}\,\text{T})} = 0.53\,\text{m} . \end{aligned}$$

23. Referring to the solution of problem 19 part (b), we see that $r = \sqrt{2mK}/qB$ implies $K = (rqB)^2/2m \propto q^2m^{-1}$. Thus,

(a) $K_\alpha = (q_\alpha/q_p)^2(m_p/m_\alpha)K_p = (2)^2(1/4)K_p = K_p = 1.0\,\text{MeV}$;

(b) $K_d = (q_d/q_p)^2(m_p/m_d)K_p = (1)^2(1/2)K_p = 1.0\,\text{MeV}/2 = 0.50\,\text{MeV}$.

24. Referring to the solution of problem 19 part (b), we see that $r = \sqrt{2mK}/qB$ implies the proportionality: $r \propto \sqrt{mK}/qB$. Thus,

$$r_\alpha = \sqrt{\frac{m_\alpha K_\alpha}{m_p K_p}}\frac{q_p}{q_\alpha} r_p = \sqrt{\frac{4.0\,\text{u}}{1.0\,\text{u}}}\frac{e r_p}{2e} = r_p\,;$$

$$r_d = \sqrt{\frac{m_d K_d}{m_p K_p}}\frac{q_p}{q_d} r_d = \sqrt{\frac{2.0\,\text{u}}{1.0\,\text{u}}}\frac{e r_d}{e} = \sqrt{2} r_p\,.$$

25. (a) We solve for B from $m = B^2 q x^2/8V$ (see Sample Problem 29-3):

$$B = \sqrt{\frac{8Vm}{qx^2}}\,.$$

We evaluate this expression using $x = 2.00$ m:

$$B = \sqrt{\frac{8(100\times 10^3\,\text{V})(3.92\times 10^{-25}\,\text{kg})}{(3.20\times 10^{-19}\,\text{C})(2.00\,\text{m})^2}} = 0.495\ \text{T}\,.$$

(b) Let N be the number of ions that are separated by the machine per unit time. The current is $i = qN$ and the mass that is separated per unit time is $M = mN$, where m is the mass of a single ion. M has the value

$$M = \frac{100\times 10^{-6}\,\text{kg}}{3600\,\text{s}} = 2.78\times 10^{-8}\,\text{kg/s}\,.$$

Since $N = M/m$ we have

$$i = \frac{qM}{m} = \frac{(3.20\times 10^{-19}\,\text{C})(2.78\times 10^{-8}\,\text{kg/s})}{3.92\times 10^{-25}\,\text{kg}} = 2.27\times 10^{-2}\ \text{A}\,.$$

(c) Each ion deposits energy qV in the cup, so the energy deposited in time Δt is given by

$$E = NqV\,\Delta t = \frac{iqV}{q}\,\Delta t = iV\,\Delta t\,.$$

For $\Delta t = 1.0$h,

$$E = (2.27\times 10^{-2}\,\text{A})(100\times 10^3\,\text{V})(3600\,\text{s}) = 8.17\times 10^6\ \text{J}\,.$$

To obtain the second expression, i/q is substituted for N.

26. The equation of motion for the proton is

$$\begin{aligned}\vec{F} &= q\vec{v}\times\vec{B} = q(v_x\hat{\text{i}} + v_y\hat{\text{j}} + v_z\hat{\text{k}})\times B\hat{\text{i}} = qB(v_z\hat{\text{j}} - v_y\hat{\text{k}}) \\ &= m_p\vec{a} = m_p\left[\left(\frac{dv_x}{dt}\right)\hat{\text{i}} + \left(\frac{dv_y}{dt}\right)\hat{\text{j}} + \left(\frac{dv_z}{dt}\right)\hat{\text{k}}\right]\,.\end{aligned}$$

Thus,

$$\begin{aligned}\frac{dv_x}{dt} &= 0 \\ \frac{dv_y}{dt} &= \omega v_z \\ \frac{dv_z}{dt} &= -\omega v_y\,,\end{aligned}$$

where $\omega = eB/m_p$. The solution is $v_x = v_{0x}$, $v_y = v_{0y}\cos\omega t$ and $v_z = -v_{0y}\sin\omega t$. In summary, we have $\vec{v}(t) = v_{0x}\hat{\text{i}} + v_{0y}\cos(\omega t)\hat{\text{j}} - v_{0y}(\sin\omega t)\hat{\text{k}}$.

27. (a) If v is the speed of the positron then $v \sin\phi$ is the component of its velocity in the plane that is perpendicular to the magnetic field. Here ϕ is the angle between the velocity and the field (89°). Newton's second law yields $eBv\sin\phi = m_e(v\sin\phi)^2/r$, where r is the radius of the orbit. Thus $r = (m_e v/eB)\sin\phi$. The period is given by

$$T = \frac{2\pi r}{v\sin\phi} = \frac{2\pi m_e}{eB} = \frac{2\pi(9.11\times10^{-31}\,\text{kg})}{(1.60\times10^{-19}\,\text{C})(0.10\,\text{T})} = 3.6\times10^{-10}\ \text{s}\ .$$

The equation for r is substituted to obtain the second expression for T.

(b) The pitch is the distance traveled along the line of the magnetic field in a time interval of one period. Thus $p = vT\cos\phi$. We use the kinetic energy to find the speed: $K = \frac{1}{2}m_e v^2$ means

$$v = \sqrt{\frac{2K}{m_e}} = \sqrt{\frac{2(2.0\times10^3\,\text{eV})(1.60\times10^{-19}\ \text{J/eV})}{9.11\times10^{-31}\,\text{kg}}} = 2.651\times10^7\ \text{m/s}\ .$$

Thus

$$p = (2.651\times10^7\,\text{m/s})(3.58\times10^{-10}\,\text{s})\cos 89° = 1.7\times10^{-4}\ \text{m}\ .$$

(c) The orbit radius is

$$R = \frac{m_e v\sin\phi}{eB} = \frac{(9.11\times10^{-31}\,\text{kg})\,(2.651\times10^7\,\text{m/s})\sin 89°}{(1.60\times10^{-19}\,\text{C})\,(0.10\text{T})} = 1.5\times10^{-3}\ \text{m}\ .$$

28. We consider the point at which it enters the field-filled region, velocity vector pointing downward. The field points out of the page so that $\vec{v}\times\vec{B}$ points leftward, which indeed seems to be the direction it is "pushed"; therefore, $q > 0$ (it is a proton).

(a) Eq. 29-17 becomes

$$\begin{aligned} T &= \frac{2\pi m_p}{e\,|\vec{B}|} \\ 2\left(130\times10^{-9}\right) &= \frac{2\pi\left(1.67\times10^{-27}\right)}{(1.60\times10^{-19})\,|\vec{B}|} \end{aligned}$$

which yields $|\vec{B}| = 0.252$ T.

(b) Doubling the kinetic energy implies multiplying the speed by $\sqrt{2}$. Since the period T does not depend on speed, then it remains the same (even though the radius increases by a factor of $\sqrt{2}$). Thus, $t = T/2 = 130$ ns, again.

29. (a) $-q$, from conservation of charges.

(b) Each of the two particles will move in the same circular path, initially going in the opposite direction. After traveling half of the circular path they will collide. So the time is given by $t = T/2 = \pi m/Bq$ (where Eq. 29-17 has been used).

30. (a) Using Eq. 29-23 and Eq. 29-18, we find

$$f_{\text{osc}} = \frac{qB}{2\pi m_p} = \frac{(1.60\times10^{-19}\,\text{C})(1.2\,\text{T})}{2\pi(1.67\times10^{-27}\,\text{kg})} = 1.8\times10^7\ \text{Hz}\ .$$

(b) From $r = m_p v/qB = \sqrt{2m_p K}/qB$ we have

$$K = \frac{(rqB)^2}{2m_p} = \frac{[(0.50\,\text{m})(1.60\times10^{-19}\,\text{C})(1.2\,\text{T})]^2}{2(1.67\times10^{-27}\,\text{kg})(1.60\times10^{-19}\,\text{J/eV})} = 1.7\times10^7\ \text{eV}\ .$$

31. We approximate the total distance by the number of revolutions times the circumference of the orbit corresponding to the average energy. This should be a good approximation since the deuteron receives the same energy each revolution and its period does not depend on its energy. The deuteron accelerates twice in each cycle, and each time it receives an energy of $qV = 80 \times 10^3$ eV. Since its final energy is 16.6 MeV, the number of revolutions it makes is

$$n = \frac{16.6 \times 10^6 \text{ eV}}{2(80 \times 10^3 \text{ eV})} = 104 \ .$$

Its average energy during the accelerating process is 8.3 MeV. The radius of the orbit is given by $r = mv/qB$, where v is the deuteron's speed. Since this is given by $v = \sqrt{2K/m}$, the radius is

$$r = \frac{m}{qB}\sqrt{\frac{2K}{m}} = \frac{1}{qB}\sqrt{2Km} \ .$$

For the average energy

$$r = \frac{\sqrt{2(8.3 \times 10^6 \text{ eV})(1.60 \times 10^{-19} \text{ J/eV})(3.34 \times 10^{-27} \text{ kg})}}{(1.60 \times 10^{-19} \text{ C})(1.57 \text{ T})} = 0.375 \text{ m} \ .$$

The total distance traveled is about $n2\pi r = (104)(2\pi)(0.375) = 2.4 \times 10^2$ m.

32. (a) Since $K = \frac{1}{2}mv^2 = \frac{1}{2}m(2\pi R f_{osc})^2 \propto m$,

$$K_p = \left(\frac{m_p}{m_d}\right) K_d = \frac{1}{2}K_d = \frac{1}{2}(17 \text{ MeV}) = 8.5 \text{ MeV} \ .$$

(b) We require a magnetic field of strength

$$B_p = \frac{1}{2}B_d = \frac{1}{2}(1.6 \text{ T}) = 0.80 \text{ T} \ .$$

(c) Since $K \propto B^2/m$,

$$K_p' = \left(\frac{m_d}{m_p}\right) K_d = 2K_d = 2(17 \text{ MeV}) = 34 \text{ MeV} \ .$$

(d) Since $f_{osc} = Bq/(2\pi m) \propto m^{-1}$,

$$f_{osc,\,d} = \left(\frac{m_d}{m_P}\right) f_{osc,\,p} = 2(12 \times 10^6 \text{ s}^{-1}) = 2.4 \times 10^7 \text{ Hz} \ .$$

(e) Now,

$$\begin{aligned} K_\alpha &= \left(\frac{m_\alpha}{m_d}\right) K_d = 2K_d = 2(17 \text{ MeV}) = 34 \text{ MeV} \ , \\ B_\alpha &= \left(\frac{m_\alpha}{m_d}\right)\left(\frac{q_d}{q_\alpha}\right) B_d = 2\left(\frac{1}{2}\right)(1.6 \text{ T}) = 1.6 \text{ T} \ , \\ K_\alpha' &= K_\alpha = 34 \text{ MeV} \qquad \text{(Since } B_\alpha = B_d = 1.6 \text{ T)} \ , \end{aligned}$$

and

$$f_{osc,\,\alpha} = \left(\frac{q_\alpha}{a_d}\right)\left(\frac{m_d}{m_\alpha}\right) f_{osc,\,d} = 2\left(\frac{2}{4}\right)(12 \times 10^6 \text{ s}^{-1}) = 1.2 \times 10^7 \text{ Hz} \ .$$

33. The magnitude of the magnetic force on the wire is given by $F_B = iLB\sin\phi$, where i is the current in the wire, L is the length of the wire, B is the magnitude of the magnetic field, and ϕ is the angle between the current and the field. In this case $\phi = 70°$. Thus,

$$F_B = (5000 \text{ A})(100 \text{ m})(60.0 \times 10^{-6} \text{ T})\sin 70° = 28.2 \text{ N} \ .$$

We apply the right-hand rule to the vector product $\vec{F}_B = i\vec{L} \times \vec{B}$ to show that the force is to the west.

34. The magnetic force on the (straight) wire is

$$F_B = iBL\sin\theta = (13.0\,\text{A})(1.50\,\text{T})(1.80\,\text{m})(\sin 35.0°) = 20.1\,\text{N}\ .$$

35. The magnetic force on the wire must be upward and have a magnitude equal to the gravitational force mg on the wire. Applying the right-hand rule reveals that the current must be from left to right. Since the field and the current are perpendicular to each other the magnitude of the magnetic force is given by $F_B = iLB$, where L is the length of the wire. Thus,

$$iLB = mg \implies i = \frac{mg}{LB} = \frac{(0.0130\,\text{kg})(9.8\,\text{m/s}^2)}{(0.620\,\text{m})(0.440\,\text{T})} = 0.467\ \text{A}\ .$$

36. The magnetic force on the wire is

$$\begin{aligned}\vec{F}_B &= i\vec{L}\times\vec{B} = iL\hat{\text{i}}\times(B_y\hat{\text{j}} + B_z\hat{\text{k}}) = iL(-B_z\hat{\text{j}} + B_y\hat{\text{k}}) \\ &= (0.50\,\text{A})(0.50\,\text{m})[-(0.010\,\text{T})\hat{\text{j}} + (0.0030\,\text{T})\hat{\text{k}}] \\ &= (-2.5\times 10^{-3}\hat{\text{j}} + 0.75\times 10^{-3}\hat{\text{k}})\,\text{N}\ .\end{aligned}$$

37. The magnetic force must push horizontally on the rod to overcome the force of friction, but it can be oriented so that it also pulls up on the rod and thereby reduces both the normal force and the force of friction. The forces acting on the rod are: $\vec{F}$, the force of the magnetic field; mg, the magnitude of the (downward) force of gravity; $\vec{N}$, the normal force exerted by the stationary rails upward on the rod; and $\vec{f}$, the (horizontal) force of friction. For definiteness, we assume the rod is on the verge of moving eastward, which means that $\vec{f}$ points westward (and is equal to its maximum possible value $\mu_s N$). Thus, $\vec{F}$ has an eastward component F_x and an upward component F_y, which can be related to the components of the magnetic field once we assume a direction for the current in the rod. Thus, again for definiteness, we assume the current flows northward. Then, by the righthand rule, a downward component (B_d) of $\vec{B}$ will produce the eastward F_x, and a westward component (B_w) will produce the upward F_y. Specifically,

$$F_x = iLB_d \qquad \text{and} \qquad F_y = iLB_w\ .$$

Considering forces along a vertical axis, we find

$$N = mg - F_y = mg - iLB_w$$

so that

$$f = f_{s,\text{max}} = \mu_s\,(mg - iLB_w)\ .$$

It is on the verge of motion, so we set the horizontal acceleration to zero:

$$F_x - f = 0 \implies iLB_d = \mu_s\,(mg - iLB_w)\ .$$

The angle of the field components is adjustable, and we can minimize with respect to it. Defining the angle by $B_w = B\sin\theta$ and $B_d = B\cos\theta$ (which means θ is being measured from a vertical axis) and writing the above expression in these terms, we obtain

$$iLB\cos\theta = \mu_s\,(mg - iLB\sin\theta) \implies B = \frac{\mu_s mg}{iL(\cos\theta + \mu_s\sin\theta)}$$

which we differentiate (with respect to θ) and set the result equal to zero. This provides a determination of the angle:

$$\theta = \tan^{-1}(\mu_s) = \tan^{-1}(0.60) = 31°\ .$$

Consequently,

$$B_{\text{min}} = \frac{0.60(1.0\,\text{kg})(9.8\,\text{m/s}^2)}{(50\,\text{A})(1.0\,\text{m})(\cos 31° + 0.60\sin 31°)} = 0.10\ \text{T}\ .$$

38. (a) From $F_B = iLB$ we get

$$i = \frac{F_B}{LB} = \frac{10 \times 10^3\,\mathrm{N}}{(3.0\,\mathrm{m})(10 \times 10^{-6}\,\mathrm{T})} = 3.3 \times 10^8\ \mathrm{A}\ .$$

(b) $P = i^2R = (3.3 \times 10^8\,\mathrm{A})^2(1.0\,\Omega) = 1.0 \times 10^{17}\,\mathrm{W}$.

(c) It is totally unrealistic because of the huge current and the accompanying high power loss.

39. The applied field has two components: $B_x > 0$ and $B_z > 0$. Considering each straight-segment of the rectangular coil, we note that Eq. 29-26 produces a non-zero force only for the component of $\vec{B}$ which is perpendicular to that segment; we also note that the equation is effectively multiplied by $N = 20$ due to the fact that this is a 20-turn coil. Since we wish to compute the torque about the hinge line, we can ignore the force acting on the straight-segment of the coil which lies along the y axis (forces acting at the axis of rotation produce no torque about that axis). The top and bottom straight-segments experience forces due to Eq. 29-26 (caused by the B_z component), but these forces are (by the right-hand rule) in the $\pm y$ directions and are thus unable to produce a torque about the y axis. Consequently, the torque derives completely from the force exerted on the straight-segment located at $x = 0.050$ m, which has length $L = 0.10$ m and is shown in Figure 29-36 carrying current in the $-y$ direction. Now, the B_z component will produce a force on this straight-segment which points in the $-x$ direction (back towards the hinge) and thus will exert no torque about the hinge. However, the B_x component (which is equal to $B\cos\theta$ where $B = 0.50$ T and $\theta = 30°$) produces a force equal to $NiLB_x$ which points (by the right-hand rule) in the $+z$ direction. Since the action of this force is perpendicular to the plane of the coil, and is located a distance x away from the hinge, then the torque has magnitude

$$\tau = (NiLB_x)\,(x) = NiLxB\cos\theta = (20)(0.10)(0.10)(0.050)(0.50)\cos 30° = 0.0043$$

in SI units (N·m). Since $\vec{\tau} = \vec{r} \times \vec{F}$, the direction of the torque is $-y$. An alternative way to do this problem is through the use of Eq. 29-37. We do not show those details here, but note that the magnetic moment vector (a necessary part of Eq. 29-37) has magnitude

$$|\vec{\mu}| = NiA = (20)(0.10\,\mathrm{A})(0.0050\,\mathrm{m}^2)$$

and points in the $-z$ direction. At this point, Eq. 3-30 may be used to obtain the result for the torque vector.

40. We establish coordinates such that the two sides of the right triangle meet at the origin, and the $\ell_y = 50$ cm side runs along the $+y$ axis, while the $\ell_x = 120$ cm side runs along the $+x$ axis. The angle made by the hypotenuse (of length 130 cm) is $\theta = \tan^{-1}(50/120) = 22.6°$, relative to the 120 cm side. If one measures the angle counterclockwise from the $+x$ direction, then the angle for the hypotenuse is $180° - 22.6° = +157°$. Since we are only asked to find the magnitudes of the forces, we have the freedom to assume the current is flowing, say, counterclockwise in the triangular loop (as viewed by an observer on the $+z$ axis. We take $\vec{B}$ to be in the same direction as that of the current flow in the hypotenuse. Then, with $B = |\vec{B}| = 0.0750$ T,

$$B_x = -B\cos\theta = -0.0692\ \mathrm{T} \qquad \text{and} \qquad B_y = B\sin\theta = 0.0288\ \mathrm{T}\ .$$

(a) Eq. 29-26 produces zero force when $\vec{L} \parallel \vec{B}$ so there is no force exerted on the hypotenuse. On the 50 cm side, the B_x component produces a force $i\ell_yB_x\hat{k}$, and there is no contribution from the B_y component. Using SI units, the magnitude of the force on the ℓ_y side is therefore

$$(4.00\,\mathrm{A})(0.500\,\mathrm{m})(0.0692\,\mathrm{T}) = 0.138\ \mathrm{N}\ .$$

On the 120 cm side, the B_y component produces a force $i\ell_xB_y\hat{k}$, and there is no contribution from the B_x component. Using SI units, the magnitude of the force on the ℓ_x side is also

$$(4.00\,\mathrm{A})(1.20\,\mathrm{m})(0.0288\,\mathrm{T}) = 0.138\ \mathrm{N}\ .$$

(b) The net force is

$$i\ell_y B_x \hat{k} + i\ell_x B_y \hat{k} = 0 ,$$

keeping in mind that $B_x < 0$ due to our initial assumptions. If we had instead assumed $\vec{B}$ went the opposite direction of the current flow in the hypotenuse, then $B_x > 0$ but $B_y < 0$ and a zero net force would still be the result.

41. If N closed loops are formed from the wire of length L, the circumference of each loop is L/N, the radius of each loop is $R = L/2\pi N$, and the area of each loop is $A = \pi R^2 = \pi(L/2\pi N)^2 = L^2/4\pi N^2$. For maximum torque, we orient the plane of the loops parallel to the magnetic field, so the dipole moment is perpendicular to the field. The magnitude of the torque is then

$$\tau = NiAB = (Ni)\left(\frac{L^2}{4\pi N^2}\right)B = \frac{iL^2B}{4\pi N} .$$

To maximize the torque, we take N to have the smallest possible value, 1. Then $\tau = iL^2B/4\pi$.

42. We replace the current loop of arbitrary shape with an assembly of small adjacent rectangular loops filling the same area which was enclosed by the original loop (as nearly as possible). Each rectangular loop carries a current i flowing in the same sense as the original loop. As the sizes of these rectangles shrink to infinitesimally small values, the assembly gives a current distribution equivalent to that of the original loop. The magnitude of the torque $\Delta\vec{\tau}$ exerted by $\vec{B}$ on the nth rectangular loop of area ΔA_n is given by $\Delta\tau_n = NiB\sin\theta\Delta A_n$. Thus, for the whole assembly

$$\tau = \sum_n \Delta\tau_n = NiB\sum_n \Delta A_n = NiAB\sin\theta .$$

43. Consider an infinitesimal segment of the loop, of length ds. The magnetic field is perpendicular to the segment, so the magnetic force on it is has magnitude $dF = iB\,ds$. The horizontal component of the force has magnitude $dF_h = (iB\cos\theta)\,ds$ and points inward toward the center of the loop. The vertical component has magnitude $dF_v = (iB\sin\theta)\,ds$ and points upward. Now, we sum the forces on all the segments of the loop. The horizontal component of the total force vanishes, since each segment of wire can be paired with another, diametrically opposite, segment. The horizontal components of these forces are both toward the center of the loop and thus in opposite directions. The vertical component of the total force is

$$F_v = iB\sin\theta \int ds = (iB\sin\theta)2\pi a .$$

We note the i, B, and θ have the same value for every segment and so can be factored from the integral.

44. The total magnetic force on the loop L is

$$\vec{F}_B = i\oint_L (d\vec{L} \times \vec{B}) = i(\oint_L d\vec{L}) \times \vec{B} = 0 .$$

We note that $\oint_L d\vec{L} = 0$. If $\vec{B}$ is not a constant, however, then the equality

$$\oint_L (d\vec{L} \times \vec{B}) = (\oint_L d\vec{L}) \times \vec{B}$$

is not necessarily valid, so $\vec{F}_B$ is not always zero.

45. (a) The current in the galvanometer should be 1.62 mA when the potential difference across the resistor-galvanometer combination is 1.00 V. The potential difference across the galvanometer alone is $iR_g = (1.62 \times 10^{-3}\,\text{A})(75.3\,\Omega) = 0.122\,\text{V}$, so the resistor must be in series with the galvanometer and the potential difference across it must be $1.00\,\text{V} - 0.122\,\text{V} = 0.878\text{V}$. The resistance should be $R = (0.878\,\text{V})/(1.62 \times 10^{-3}\,\text{A}) = 542\,\Omega$.

(b) The current in the galvanometer should be 1.62 mA when the total current in the resistor and galvanometer combination is 50.0 mA. The resistor should be in parallel with the galvanometer, and the current through it should be 50.0 mA − 1.62 mA = 48.38 mA. The potential difference across the resistor is the same as that across the galvanometer, 0.122 V, so the resistance should be $R = (0.122\,\text{V})/(48.38\times 10^{-3}\,\text{A}) = 2.52\,\Omega$.

46. We use $\tau_{\text{max}} = |\vec{\mu}\times\vec{B}|_{\text{max}} = \mu B = i\pi a^2 B$, and note that $i = qf = qv/2\pi a$. So

$$\tau_{\text{max}} = \left(\frac{qv}{2\pi a}\right)\pi a^2 B = \frac{1}{2}qvaB\ .$$

47. We use Eq. 29-37 where $\vec{\mu}$ is the magnetic dipole moment of the wire loop and $\vec{B}$ is the magnetic field, as well as Newton's second law. Since the plane of the loop is parallel to the incline the dipole moment is normal to the incline. The forces acting on the cylinder are the force of gravity mg, acting downward from the center of mass, the normal force of the incline N, acting perpendicularly to the incline through the center of mass, and the force of friction f, acting up the incline at the point of contact. We take the x axis to be positive down the incline. Then the x component of Newton's second law for the center of mass yields

$$mg\sin\theta - f = ma\ .$$

For purposes of calculating the torque, we take the axis of the cylinder to be the axis of rotation. The magnetic field produces a torque with magnitude $\mu B\sin\theta$, and the force of friction produces a torque with magnitude fr, where r is the radius of the cylinder. The first tends to produce an angular acceleration in the counterclockwise direction, and the second tends to produce an angular acceleration in the clockwise direction. Newton's second law for rotation about the center of the cylinder, $\tau = I\alpha$, gives

$$fr - \mu B\sin\theta = I\alpha\ .$$

Since we want the current that holds the cylinder in place, we set $a = 0$ and $\alpha = 0$, and use one equation to eliminate f from the other. The result is $mgr = \mu B$. The loop is rectangular with two sides of length L and two of length $2r$, so its area is $A = 2rL$ and the dipole moment is $\mu = NiA = 2NirL$. Thus, $mgr = 2NirLB$ and

$$i = \frac{mg}{2NLB} = \frac{(0.250\,\text{kg})(9.8\,\text{m/s}^2)}{2(10.0)(0.100\,\text{m})(0.500\,\text{T})} = 2.45\ \text{A}\ .$$

48. From $\mu = NiA = i\pi r^2$ we get

$$i = \frac{\mu}{\pi r^2} = \frac{8.00\times 10^{22}\,\text{J/T}}{\pi(3500\times 10^3\,\text{m})^2} = 2.08\times 10^9\ \text{A}\ .$$

49. (a) The magnitude of the magnetic dipole moment is given by $\mu = NiA$, where N is the number of turns, i is the current in each turn, and A is the area of a loop. In this case the loops are circular, so $A = \pi r^2$, where r is the radius of a turn. Thus

$$i = \frac{\mu}{N\pi r^2} = \frac{2.30\,\text{A}\cdot\text{m}^2}{(160)(\pi)(0.0190\,\text{m})^2} = 12.7\ \text{A}\ .$$

(b) The maximum torque occurs when the dipole moment is perpendicular to the field (or the plane of the loop is parallel to the field). It is given by

$$\tau_{\text{max}} = \mu B = \left(2.30\,\text{A}\cdot\text{m}^2\right)\left(35.0\times 10^{-3}\,\text{T}\right) = 8.05\times 10^{-2}\ \text{N}\cdot\text{m}\ .$$

50. (a) $\mu = NAi = \pi r^2 i = \pi(0.150\,\text{m})^2(2.60\,\text{A}) = 0.184\,\text{A}\cdot\text{m}^2$.

(b) The torque is

$$\tau = |\vec{\mu}\times\vec{B}| = \mu B\sin\theta = \left(0.184\,\text{A}\cdot\text{m}^2\right)(12.0\,\text{T})\sin 41.0^\circ = 1.45\ \text{N}\cdot\text{m}\ .$$

51. (a) The area of the loop is $A = \frac{1}{2}(30\,\text{cm})(40\,\text{cm}) = 6.0 \times 10^2\,\text{cm}^2$, so

$$\mu = iA = (5.0\,\text{A})\left(6.0 \times 10^{-2}\,\text{m}^2\right) = 0.30\ \text{A}\cdot\text{m}^2 \ .$$

(b) The torque on the loop is

$$\tau = \mu B \sin\theta = \left(0.30\,\text{A}\cdot\text{m}^2\right)\left(80 \times 10^3\,\text{T}\right)\sin 90° = 2.4 \times 10^{-2}\ \text{N}\cdot\text{m} \ .$$

52. (a) We use $\vec{\tau} = \vec{\mu} \times \vec{B}$, where $\vec{\mu}$ points into the wall (since the current goes clockwise around the clock). Since $\vec{B}$ points towards the one-hour (or "5-minute") mark, and (by the properties of vector cross products) $\vec{\tau}$ must be perpendicular to it, then (using the right-hand rule) we find $\vec{\tau}$ points at the 20-minute mark. So the time interval is 20 min.

(b) The torque is given by

$$\begin{aligned}\tau &= \left|\vec{\mu} \times \vec{B}\right| = \mu B \sin 90° \\ &= NiAB = \pi N i r^2 B \\ &= 6\pi(2.0\,\text{A})(0.15\,\text{m})^2(70 \times 10^{-3}\,\text{T}) \\ &= 5.9 \times 10^{-2}\ \text{N}\cdot\text{m} \ .\end{aligned}$$

53. (a) The magnitude of the magnetic moment vector is

$$\mu = \sum_n i_n A_n = \pi r_1^2 i_1 + \pi r_2^2 i_2 = \pi(7.00\,\text{A})[(0.300\,\text{m})^2 + (0.200\,\text{m})^2] = 2.86\ \text{A}\cdot\text{m}^2 \ .$$

(b) Now,

$$\mu = \pi r_1^2 i_1 - \pi r_2^2 i_2 = \pi(7.00\,\text{A})[(0.300\,\text{m})^2 - (0.200\,\text{m})^2] = 1.10\ \text{A}\cdot\text{m}^2 \ .$$

54. Let $a = 30.0\,\text{cm}$, $b = 20.0\,\text{cm}$, and $c = 10.0\,\text{cm}$. From the given hint, we write

$$\begin{aligned}\vec{\mu} &= \vec{\mu}_1 + \vec{\mu}_2 = iab(-\hat{\text{k}}) + iac(\hat{\text{j}}) \\ &= ia(c\hat{\text{j}} - b\hat{\text{k}}) \\ &= (5.00\,\text{A})(0.300\,\text{m})[(0.100\,\text{m})\hat{\text{j}} - (0.200\,\text{m})\hat{\text{k}}] \\ &= (0.150\hat{\text{j}} - 0.300\hat{\text{k}})\ \text{A}\cdot\text{m}^2.\end{aligned}$$

Thus, using the Pythagorean theorem,

$$\mu = \sqrt{(0.150)^2 + (0.300)^2} = 0.335\ \text{A}\cdot\text{m}^2 \ ,$$

and $\vec{\mu}$ is in the yz plane at angle θ to the $+y$ direction, where

$$\theta = \tan^{-1}\left(\frac{\mu_y}{\mu_x}\right) = \tan^{-1}\left(\frac{-0.300}{0.150}\right) = -63.4° \ .$$

55. The magnetic dipole moment is $\vec{\mu} = \mu(0.60\,\hat{\text{i}} - 0.80\,\hat{\text{j}})$, where $\mu = NiA = Ni\pi r^2 = 1(0.20\text{A})\pi(0.080\text{m})^2 = 4.02 \times 10^{-4}\,\text{A}\cdot\text{m}^2$. Here i is the current in the loop, N is the number of turns, A is the area of the loop, and r is its radius.

(a) The torque is

$$\begin{aligned}\vec{\tau} &= \vec{\mu} \times \vec{B} = \mu(0.60\,\hat{\text{i}} - 0.80\,\hat{\text{j}}) \times (0.25\,\hat{\text{i}} + 0.30\,\hat{\text{k}}) \\ &= \mu\left[(0.60)(0.30)(\hat{\text{i}} \times \hat{\text{k}}) - (0.80)(0.25)(\hat{\text{j}} \times \hat{\text{i}}) - (0.80)(0.30)(\hat{\text{j}} \times \hat{\text{k}})\right] \\ &= \mu[-0.18\,\hat{\text{j}} + 0.20\,\hat{\text{k}} - 0.24\,\hat{\text{i}}] \ .\end{aligned}$$

Here $\hat{\mathrm{i}} \times \hat{\mathrm{k}} = -\hat{\mathrm{j}}$, $\hat{\mathrm{j}} \times \hat{\mathrm{i}} = -\hat{\mathrm{k}}$, and $\hat{\mathrm{j}} \times \hat{\mathrm{k}} = \hat{\mathrm{i}}$ are used. We also use $\hat{\mathrm{i}} \times \hat{\mathrm{i}} = 0$. Now, we substitute the value for μ to obtain

$$\vec{\tau} = \left(-0.97 \times 10^{-4}\,\hat{\mathrm{i}} - 7.2 \times 10^{-4}\,\hat{\mathrm{j}} + 8.0 \times 10^{-4}\,\hat{\mathrm{k}}\right) \mathrm{N{\cdot}m} .$$

(b) The potential energy of the dipole is given by

$$\begin{aligned} U &= -\vec{\mu}\cdot\vec{B} = -\mu(0.60\,\hat{\mathrm{i}} - 0.80\,\hat{\mathrm{j}})\cdot(0.25\,\hat{\mathrm{i}} + 0.30\,\hat{\mathrm{k}}) \\ &= -\mu(0.60)(0.25) = -0.15\mu = -6.0 \times 10^{-4}\ \mathrm{J} . \end{aligned}$$

Here $\hat{\mathrm{i}}\cdot\hat{\mathrm{i}} = 1$, $\hat{\mathrm{i}}\cdot\hat{\mathrm{k}} = 0$, $\hat{\mathrm{j}}\cdot\hat{\mathrm{i}} = 0$, and $\hat{\mathrm{j}}\cdot\hat{\mathrm{k}} = 0$ are used.

56. The unit vector associated with the current element (of magnitude $d\ell$) is $-\hat{\mathrm{j}}$. The (infinitesimal) force on this element is

$$d\vec{F} = i\,d\ell(-\hat{\mathrm{j}}) \times (0.3y\hat{\mathrm{i}} + 0.4y\hat{\mathrm{j}})$$

with SI units (and 3 significant figures) understood.

(a) Since $\hat{\mathrm{j}} \times \hat{\mathrm{i}} = -\hat{\mathrm{k}}$ and $\hat{\mathrm{j}} \times \hat{\mathrm{j}} = 0$, we obtain

$$d\vec{F} = 0.3iy\,d\ell\,\hat{\mathrm{k}} = \left(6.00 \times 10^{-4}\,\mathrm{N/m^2}\right) y\,d\ell\,\hat{\mathrm{k}} .$$

(b) We integrate the force element found in part (a), using the symbol ξ to stand for the coefficient $6.00 \times 10^{-4}\,\mathrm{N/m^2}$, and obtain

$$\vec{F} = \int d\vec{F} = \xi\hat{\mathrm{k}} \int_0^{0.25} y\,dy = \xi\hat{\mathrm{k}}\left(\frac{0.25^2}{2}\right) = 1.88 \times 10^{-5}\,\mathrm{N}\,\hat{\mathrm{k}} .$$

57. Since the velocity is constant, the net force on the proton vanishes. Using Eq. 29-2 and Eq. 23-28, we obtain the requirement (Eq. 29-7) for the proton's speed in terms of the crossed fields:

$$v = \frac{E}{B} \implies E = (50\,\mathrm{m/s})(0.0020\,\mathrm{T}) = 0.10\ \mathrm{V/m} .$$

By the right-hand rule, the magnetic force points in the $\hat{\mathrm{k}}$ direction. To cancel this, the electric force must be in the $-\hat{\mathrm{k}}$ direction. Since $q > 0$ for the proton, we conclude $\vec{E} = -0.10\,\mathrm{V/m}\,\hat{\mathrm{k}}$.

58. (a) The kinetic energy gained is due to the potential energy decrease as the dipole swings from a position specified by angle θ to that of being aligned (zero angle) with the field. Thus,

$$K = U_i - U_f = -\mu B\cos\theta - (-\mu B\cos 0^\circ) .$$

Therefore, using SI units, the angle is

$$\theta = \cos^{-1}\left(1 - \frac{K}{\mu B}\right) = \cos^{-1}\left(1 - \frac{0.00080}{(0.020)(0.052)}\right) = 77^\circ .$$

(b) Since we are making the assumption that no energy is dissipated in this process, then the dipole will continue its rotation (similar to a pendulum) until it reaches an angle $\theta = 77^\circ$ on the other side of the alignment axis.

59. Using Eq. 29-2 and Eq. 3-30, we obtain

$$\vec{F} = q\,(v_x B_y - v_y B_x)\,\hat{\mathrm{k}} = q\,(v_x(3B_x) - v_y B_x)\,\hat{\mathrm{k}}$$

where we use the fact that $B_y = 3B_x$. Since the force (at the instant considered) is $F_z\,\hat{\mathrm{k}}$ where $F_z = 6.4 \times 10^{-19}\,\mathrm{N}$, then we are led to the condition

$$q\,(3v_x - v_y)\,B_x = F_z \implies B_x = \frac{F_z}{q\,(3v_x - v_y)} .$$

Substituting $V_x = 2.0\,\mathrm{m/s}$, $v_y = 4.0\,\mathrm{m/s}$ and $q = -1.6 \times 10^{-19}\,\mathrm{C}$, we obtain $B_x = -2.0$ T.

60. The current is in the $+\hat{\imath}$ direction. Thus, the $\hat{\imath}$ component of $\vec{B}$ has no effect, and (with x in meters) we evaluate

$$\begin{aligned}\vec{F} &= (3.00\ \text{A})\int_0^1 \left(-0.600\ \text{T/m}^2\right) x^2\,dx\,\left(\hat{\imath}\times\hat{\jmath}\right)\\ &= -1.80\hat{k}\left(\frac{1^3}{3}\right)\ \text{A}\cdot\text{T}\cdot\text{m}\\ &= -0.600\,\text{N}\ \hat{k}\ .\end{aligned}$$

61. (a) We seek the electrostatic field established by the separation of charges (brought on by the magnetic force). We use the ideas discussed in §29-4; especially, see SAMPLE PROBLEM 29-2. With Eq. 29-10, we define the magnitude of the electric field as $|\vec{E}| = v|\vec{B}| = (20)(0.03) = 0.6$ V/m. Its direction may be inferred from Figure 29-8; its direction is opposite to that defined by $\vec{v}\times\vec{B}$. In summary,

$$\vec{E} = -0.600\,\text{V/m}\ \hat{k}$$

which insures that $\vec{F} = q(\vec{E} + \vec{v}\times\vec{B})$ vanishes.

(b) Eq. 29-9 yields $V = (0.6\ \text{V/m})(2\ \text{m}) = 1.20$ V.

62. With the $\vec{B}$ pointing "out of the page," we evaluate the force (using the right-hand rule) at, say, the dot shown on the left edge of the particle's path, where its velocity is down. If the particle were positively charged, then the force at the dot would be toward the left, which is at odds with the figure (showing it being bent towards the right). Therefore, the particle is negatively charged; it is an electron.

(a) Using Eq. 29-3 (with angle ϕ equal to 90°), we obtain

$$v = \frac{|\vec{F}|}{e\,|\vec{B}|} = 4.99\times 10^6\ \text{m/s}\ .$$

(b) Using either Eq. 29-14 or Eq. 29-16, we find $r = 0.00710$ m.

(c) Using Eq. 29-17 (in either its first or last form) readily yields $T = 8.93\times 10^{-9}$ s.

63. (a) We are given $\vec{B} = B_x\hat{\imath} = 6\times 10^{-5}\,\hat{\imath}$ T, so that $\vec{v}\times\vec{B} = -v_yB_x\hat{k}$ where $v_y = 4\times 10^4$ m/s. We note that the magnetic force on the electron is $(-e)(-v_yB_x\hat{k})$ and therefore points in the $+\hat{k}$ direction, at the instant the electron enters the field-filled region. In these terms, Eq. 29-16 becomes

$$r = \frac{m_e\,v_y}{e\,B_x} = 0.0038\ \text{m}\ .$$

(b) One revolution takes $T = 2\pi r/v_y = 0.60\ \mu$s, and during that time the "drift" of the electron in the x direction (which is the *pitch* of the helix) is $\Delta x = v_xT = 0.019$ m where $v_x = 32\times 10^3$ m/s.

(c) Returning to our observation of force direction made in part (a), we consider how this is perceived by an observer at some point on the $-x$ axis. As the electron moves away from him, he sees it enter the region with positive v_y (which he might call "upward") but "pushed" in the $+z$ direction (to his right). Hence, he describes the electron's spiral as clockwise.

64. The force associated with the magnetic field must point in the $\hat{\jmath}$ direction in order to cancel the force of gravity in the $-\hat{\jmath}$ direction. By the right-hand rule, $\vec{B}$ points in the $-\hat{k}$ direction (since $\hat{\imath}\times(-\hat{k}) = \hat{\jmath}$). Note that the charge is positive; also note that we need to assume $B_y = 0$. The magnitude $|B_z|$ is given by Eq. 29-3 (with $\phi = 90°$). Therefore, with $m = 10\times 10^{-3}$ kg, $v = 2.0\times 10^4$ m/s and $q = 80\times 10^{-6}$ C, we find

$$\vec{B} = B_z\hat{k} = -\left(\frac{mg}{qv}\right)\hat{k} = -0.061\hat{k}$$

in SI units (Tesla).

65. By the right-hand rule, we see that $\vec{v} \times \vec{B}$ points along $-\hat{k}$. From Eq. 29-2 ($\vec{F} = q\vec{v} \times \vec{B}$), we find that for the force to point along $+\hat{k}$, we must have $q < 0$. Now, examining the magnitudes (in SI units) in Eq. 29-3, we find

$$\begin{aligned} |\vec{F}| &= |q|\, v\, |\vec{B}| \sin\phi \\ 0.48 &= |q|(4000)(0.0050)\sin 35° \end{aligned}$$

which yields $|q| = 0.040$ C. In summary, then, $q = -40\,\text{mC}$.

66. (a) Since $K = qV$ we have $K_p = K_d = \frac{1}{2}K_\alpha$ (as $q_\alpha = 2K_d = 2K_p$).

(b) and (c) Since $r = \sqrt{2mK}/qB \propto \sqrt{mK}/q$, we have

$$\begin{aligned} r_d &= \sqrt{\frac{m_d K_d}{m_p K_p}}\frac{q_p r_p}{q_d} = \sqrt{\frac{(2.00\,\text{u})K_p}{(1.00\,\text{u})K_p}}\, r_p = 10\sqrt{2}\,\text{cm} = 14\text{ cm} , \\ r_\alpha &= \sqrt{\frac{m_\alpha K_\alpha}{m_p K_p}}\frac{q_p r_p}{q_\alpha} = \sqrt{\frac{(4.00\,\text{u})K_\alpha}{(1.00\,\text{u})(K_\alpha/2)}}\frac{e r_p}{2e} = 10\sqrt{2}\,\text{cm} = 14\text{ cm} . \end{aligned}$$

67. (a) The radius of the cyclotron dees should be

$$r = \frac{m_p v}{qB} = \frac{(1.67 \times 10^{-27}\,\text{kg})(3.00 \times 10^8\,\text{m/s})/10}{(1.60 \times 10^{-19}\,\text{C})(1.4\,\text{T})} = 0.22\text{ m} .$$

(b) The frequency should be

$$f_{\text{osc}} = \frac{v}{2\pi r} = \frac{3.00 \times 10^7\,\text{m/s}}{2\pi(0.22\,\text{m})} = 2.1 \times 10^7\text{ Hz} .$$

68. The magnetic force on the wire is $F_B = idB$, pointing to the left. Thus $v = at = (F_B/m)t = idBt/m$, to the left (away from the generator).

69. (a) We use Eq. 29-10: $v_d = E/B = (10 \times 10^{-6}\,\text{V}/1.0 \times 10^{-2}\,\text{m})/(1.5\,\text{T}) = 6.7 \times 10^{-4}\,\text{m/s}$.

(b) We rewrite Eq. 29-12 in terms of the electric field:

$$n = \frac{Bi}{V\ell e} = \frac{Bi}{(Ed)\ell e} = \frac{Bi}{EAe}$$

which we use $A = \ell d$. In this experiment, $A = (0.010\,\text{m})(10 \times 10^{-6}\,\text{m}) = 1.0 \times 10^{-7}\,\text{m}^2$. By Eq. 29-10, v_d equals the ratio of the fields (as noted in part (a)), so we are led to

$$\begin{aligned} n &= \frac{Bi}{EAe} = \frac{i}{v_d A e} \\ &= \frac{3.0\,\text{A}}{(6.7 \times 10^{-4}\,\text{m/s})(1.0 \times 10^{-7}\,\text{m}^2)(1.6 \times 10^{-19}\,\text{C})} \\ &= 2.8 \times 10^{29}/\text{m}^3 . \end{aligned}$$

(c) Since a drawing of an inherently 3-D situation can be misleading, we describe it in terms of horizontal *north, south, east, west* and vertical *up* and *down* directions. We assume $\vec{B}$ points up and the conductor's width of 0.010 m is along an east-west line. We take the current going northward. The conduction electrons experience a westward magnetic force (by the right-hand rule), which results in the west side of the conductor being negative and the east side being positive (with reference to the Hall voltage which becomes established).

70. The fact that the fields are uniform, with the feature that the charge moves in a straight line, implies the speed is constant (if it were not, then the magnetic *force* would vary while the electric force could not – causing it to deviate from straight-line motion). This is then the situation leading to Eq. 29-7, and we find

$$|\vec{E}| = v|\vec{B}| = 500 \text{ V/m} .$$

Its direction (so that $\vec{F} = q(\vec{E} + \vec{v} \times \vec{B})$ vanishes) is downward (in "page" coordinates).

71. (a) We use Eq. 29-2 and Eq. 3-30:

$$\begin{aligned}
\vec{F} &= q\vec{v} \times \vec{B} \\
&= (+e)\left((v_y B_z - v_z B_y)\,\hat{\mathrm{i}} + (v_z B_x - v_x B_z)\,\hat{\mathrm{j}} + (v_x B_y - v_y B_x)\,\hat{\mathrm{k}}\right) \\
&= (1.60 \times 10^{-19})\left(((4)(0.008) - (-6)(-0.004))\,\hat{\mathrm{i}} + \right. \\
&\qquad \left. ((-6)(0.002) - (-2)(0.008))\,\hat{\mathrm{j}} + ((-2)(-0.004) - (4)(0.002))\,\hat{\mathrm{k}}\right) \\
&= (1.28 \times 10^{-21})\,\hat{\mathrm{i}} + (6.41 \times 10^{-22})\,\hat{\mathrm{j}}
\end{aligned}$$

with SI units understood.

(b) By definition of the cross product, $\vec{v} \perp \vec{F}$. This is easily verified by taking the dot (scalar) product of $\vec{v}$ with the result of part (a), yielding zero, provided care is taken not to introduce any round-off error.

(c) There are several ways to proceed. It may be worthwhile to note, first, that if B_z were 6.00 mT instead of 8.00 mT then the two vectors would be exactly antiparallel. Hence, the angle θ between $\vec{B}$ and $\vec{v}$ is presumably "close" to 180°. Here, we use Eq. 3-20:

$$\theta = \cos^{-1} \frac{\vec{v} \cdot \vec{B}}{|\vec{v}|\,|\vec{B}|} = \cos^{-1} \frac{-68}{\sqrt{56}\,\sqrt{84}} = 173° .$$

72. (a) From symmetry, we conclude that any x-component of force will vanish (evaluated over the entirety of the bent wire as shown). By the right-hand rule, a field in the $\hat{\mathrm{k}}$ direction produces on each part of the bent wire a y-component of force pointing in the $-\hat{\mathrm{j}}$ direction; each of these components has magnitude

$$|F_y| = i\,\ell\,|\vec{B}| \sin 30° = 8 \text{ N} .$$

Therefore, the the force (in Newtons) on the wire shown in the figure is $-16\,\hat{\mathrm{j}}$.

(b) The force exerted on the left half of the bent wire points in the $-\hat{\mathrm{k}}$ direction, by the right-hand rule, and the force exerted on the right half of the wire points in the $+\hat{\mathrm{k}}$ direction. It is clear that the magnitude of each force is equal, so that the force (evaluated over the entirety of the bent wire as shown) must necessarily vanish.

73. The contribution to the force by the magnetic field ($\vec{B} = B_x\hat{\mathrm{i}} = -0.020\hat{\mathrm{i}}$ T) is given by Eq. 29-2:

$$\begin{aligned}
\vec{F}_B &= q\vec{v} \times \vec{B} \\
&= q\left(\left(17000\hat{\mathrm{i}} \times B_x\hat{\mathrm{i}}\right) + \left(-11000\hat{\mathrm{j}} \times B_x\hat{\mathrm{i}}\right) + \left(7000\hat{\mathrm{k}} \times B_x\hat{\mathrm{i}}\right)\right) \\
&= q\left(-220\hat{\mathrm{k}} - 140\hat{\mathrm{j}}\right)
\end{aligned}$$

in SI units. And the contribution to the force by the electric field ($\vec{E} = E_y\hat{\mathrm{j}} = 300\hat{\mathrm{j}}$ V/m) is given by Eq. 23-1: $\vec{F}_E = qE_y\hat{\mathrm{j}}$. Using $q = 5.0 \times 10^{-6}$ C, the net force (with the unit newton understood) on the particle is

$$\vec{F} = 0.0008\hat{\mathrm{j}} - 0.0011\hat{\mathrm{k}} .$$

74. Letting $B_x = B_y = B_1$ and $B_z = B_2$ and using Eq. 29-2 and Eq. 3-30, we obtain (with SI units understood)

$$\begin{aligned} \vec{F} &= q\vec{v} \times \vec{B} \\ 4\hat{\imath} - 20\hat{\jmath} + 12\hat{k} &= 2\left((4B_2 - 6B_1)\,\hat{\imath} + (6B_1 - 2B_2)\,\hat{\jmath} + (2B_1 - 4B_1)\,\hat{k} \right) . \end{aligned}$$

Equating like components, we find $B_1 = -3$ and $B_2 = -4$. In summary (with the unit Tesla understood), $\vec{B} = -3.0\hat{\imath} - 3.0\hat{\jmath} - 4.0\hat{k}$.

75. (a) We use Eq. 29-16 to calculate r:

$$r = \frac{m_e v}{qB} = \frac{(9.11 \times 10^{-31}\,\text{kg})(0.10)(3.00 \times 10^8\,\text{m/s})}{(1.60 \times 10^{-19}\,\text{C})(0.50\,\text{T})} = 3.4 \times 10^{-4}\ \text{m} .$$

(b) The kinetic energy, computed using the formula from Chapter 7, is

$$K = \frac{1}{2} m_e v^2 = \frac{(9.11 \times 10^{-31}\,\text{kg})(3.0 \times 10^7\,\text{m/s})^2}{2(1.6 \times 10^{-19}\,\text{J/eV})} = 2.6 \times 10^3\ \text{eV} .$$

76. (a) From $m = B^2qx^2/8V$ we have $\Delta m = (B^2q/8V)(2x\Delta x)$. Here $x = \sqrt{8Vm/B^2q}$, which we substitute into the expression for Δm to obtain

$$\Delta m = \left(\frac{B^2 q}{8V}\right) 2\sqrt{\frac{8mV}{B^2 q}}\,\Delta x = B\sqrt{\frac{mq}{2V}}\,\Delta x .$$

(b) The distance between the spots made on the photographic plate is

$$\begin{aligned} \Delta x &= \frac{\Delta m}{B}\sqrt{\frac{2V}{mq}} \\ &= \frac{(37\,\text{u} - 35\,\text{u})(1.66 \times 10^{-27}\,\text{kg/u})}{0.50\,\text{T}} \sqrt{\frac{2(7.3 \times 10^3\,\text{V})}{(36\,\text{u})(1.66 \times 10^{-27}\,\text{kg/u})(1.60 \times 10^{-19}\,\text{C})}} \\ &= 8.2 \times 10^{-3}\ \text{m} . \end{aligned}$$

77. (a) Since $\vec{B}$ is uniform,

$$\vec{F}_B = \int_{\text{wire}} i d\vec{L} \times \vec{B} = i \left(\int_{\text{wire}} d\vec{L} \right) \times \vec{B} = i\vec{L}_{ab} \times \vec{B} ,$$

where we note that $\int_{\text{wire}} d\vec{L} = \vec{L}_{ab}$, with $\vec{L}_{ab}$ being the displacement vector from a to b.

(b) Now $\vec{L}_{ab} = 0$, so $\vec{F}_B = i\vec{L}_{ab} \times \vec{B} = 0$.

78. We use $d\vec{F}_B = i d\vec{L} \times \vec{B}$, where $d\vec{L} = dx\hat{\imath}$ and $\vec{B} = B_x\hat{\imath} + B_y\hat{\jmath}$. Thus,

$$\begin{aligned} \vec{F}_B &= \int i d\vec{L} \times \vec{B} \\ &= \int_{x_i}^{x_f} i\,dx\,\hat{\imath} \times (B_x\hat{\imath} + B_y\hat{\jmath}) = i\int_{x_i}^{x_f} B_y\,dx\hat{k} \\ &= (-5.0\,\text{A}) \left(\int_{1.0}^{3.0} (8.0x^2\,dx)\,(\text{m}\cdot\text{mT}) \right) \hat{k} \\ &= -0.35\ \text{N}\ \hat{k} . \end{aligned}$$

Chapter 30

1. (a) The magnitude of the magnetic field due to the current in the wire, at a point a distance r from the wire, is given by

$$B = \frac{\mu_0 i}{2\pi r} .$$

With $r = 20\,\text{ft} = 6.10\,\text{m}$, we find

$$B = \frac{(4\pi \times 10^{-7}\,\text{T}\cdot\text{m/A})(100\,\text{A})}{2\pi(6.10\,\text{m})} = 3.3 \times 10^{-6}\,\text{T} = 3.3\,\mu\text{T} .$$

(b) This is about one-sixth the magnitude of the Earth's field. It will affect the compass reading.

2. The current i due to the electron flow is $i = ne = (5.6 \times 10^{14}/\text{s})(1.6 \times 10^{-19}\,\text{C}) = 9.0 \times 10^{-5}\,\text{A}$. Thus,

$$B = \frac{\mu_0 i}{2\pi r} = \frac{(4\pi \times 10^{-7})(9.0 \times 10^{-5})}{2\pi(1.5 \times 10^{-3})} = 1.2 \times 10^{-8}\ \text{T} .$$

3. (a) The field due to the wire, at a point 8.0 cm from the wire, must be $39\,\mu\text{T}$ and must be directed due south. Since $B = \mu_0 i/2\pi r$,

$$i = \frac{2\pi r B}{\mu_0} = \frac{2\pi(0.080\,\text{m})(39 \times 10^{-6}\,\text{T})}{4\pi \times 10^{-7}\,\text{T}\cdot\text{m/A}} = 16\ \text{A} .$$

(b) The current must be from west to east to produce a field which is directed southward at points below it.

4. The points must be along a line parallel to the wire and a distance r from it, where r satisfies

$$B_{\text{wire}} = \frac{\mu_0 i}{2\pi r} = B_{\text{ext}} ,$$

or

$$r = \frac{\mu_0 i}{2\pi B_{\text{ext}}} = \frac{(1.26 \times 10^{-6}\,\text{T}\cdot\text{m/A})(100\,\text{A})}{2\pi(5.0 \times 10^{-3}\,\text{T})} = 4.0 \times 10^{-3}\ \text{m} .$$

5. We assume the current flows in the $+x$ direction and the particle is at some distance d in the $+y$ direction (away from the wire). Then, the magnetic field at the location of the charge q is

$$\vec{B} = \frac{\mu_0 i}{2\pi d}\,\hat{\text{k}} .$$

Thus,

$$\vec{F} = q\vec{v} \times \vec{B} = \frac{\mu_0 i q}{2\pi d}\left(\vec{v} \times \hat{\text{k}}\right) .$$

(a) In this situation, $\vec{v} = v(-\hat{\jmath})$ (where v is the speed and is a positive value). Also, the problem specifies $q > 0$. Thus,

$$\vec{F} = \frac{\mu_0 iqv}{2\pi d}\left((-\hat{\jmath}) \times \hat{k}\right) = -\frac{\mu_0 iqv}{2\pi d}(\hat{\imath}) ,$$

which tells us that $\vec{F}_q$ has a magnitude of $\mu_0 iqv/2\pi d$ and is in the direction opposite to that of the current flow.

(b) Now the direction $\vec{v}$ is reversed, and we obtain $\vec{F} = +\mu_0 iqv\hat{\imath}/2\pi d$. The magnitude is identical to that found in part (a), but the direction of the force is now in the same direction as that of the current flow.

6. The straight segment of the wire produces no magnetic field at C (see the *straight sections* discussion in Sample Problem 30-1). Also, the fields from the two semi-circular loops cancel at C (by symmetry). Therefore, $B_C = 0$.

7. Each of the semi-infinite straight wires contributes $\mu_0 i/4\pi R$ (Eq. 30-9) to the field at the center of the circle (both contributions pointing "out of the page"). The current in the arc contributes a term given by Eq. 30-11 pointing into the page, and this is able to produce zero total field at that location if

$$\begin{aligned} B_{\text{arc}} &= 2B_{\text{semi infinite}} \\ \frac{\mu_0 i\phi}{4\pi R} &= 2\left(\frac{\mu_0 i}{4\pi R}\right) \end{aligned}$$

which yields $\phi = 2$ rad.

8. Recalling the *straight sections* discussion in Sample Problem 30-1, we see that the current in segments AH and JD do not contribute to the field at point C. Using Eq. 30-11 (with $\phi = \pi$) and the right-hand rule, we find that the current in the semicircular arc HJ contributes $\mu_0 i/4R_1$ (into the page) to the field at C. Also, arc DA contributes $\mu_0 i/4R_2$ (out of the page) to the field there. Thus, the net field at C is

$$\vec{B} = \frac{\mu_0 i}{4}\left(\frac{1}{R_1} - \frac{1}{R_2}\right) \quad \text{into the page} .$$

9. Recalling the *straight sections* discussion in Sample Problem 30-1, we see that the current in the straight segments colinear with P do not contribute to the field at that point. Using Eq. 30-11 (with $\phi = \theta$) and the right-hand rule, we find that the current in the semicircular arc of radius b contributes $\mu_0 i\theta/4\pi b$ (out of the page) to the field at P. Also, the current in the large radius arc contributes $\mu_0 i\theta/4\pi a$ (into the page) to the field there. Thus, the net field at P is

$$\vec{B} = \frac{\mu_0 i\theta}{4\pi}\left(\frac{1}{b} - \frac{1}{a}\right) \quad \text{out of the page} .$$

10. (a) Recalling the *straight sections* discussion in Sample Problem 30-1, we see that the current in the straight segments colinear with C do not contribute to the field at that point.

(b) Eq. 30-11 (with $\phi = \pi$) indicates that the current in the semicircular arc contributes $\mu_0 i/4R$ to the field at C. The right-hand rule shows that this field is into the page.

(c) The contributions from parts (a) and (b) sum to

$$\vec{B} = \frac{\mu_0 i}{4R} \quad \text{into the page} .$$

11. Our x axis is along the wire with the origin at the midpoint. The current flows in the positive x direction. All segments of the wire produce magnetic fields at P_1 that are out of the page. According to the Biot-Savart law, the magnitude of the field any (infinitesimal) segment produces at P_1 is given by

$$dB = \frac{\mu_0 i}{4\pi}\frac{\sin\theta}{r^2}\,dx$$

where θ (the angle between the segment and a line drawn from the segment to P_1) and r (the length of that line) are functions of x. Replacing r with $\sqrt{x^2+R^2}$ and $\sin\theta$ with $R/r = R/\sqrt{x^2+R^2}$, we integrate from $x=-L/2$ to $x=L/2$. The total field is

$$B = \frac{\mu_0 iR}{4\pi}\int_{-L/2}^{L/2}\frac{dx}{(x^2+R^2)^{3/2}} = \frac{\mu_0 iR}{4\pi}\frac{1}{R^2}\frac{x}{(x^2+R^2)^{1/2}}\bigg|_{-L/2}^{L/2} = \frac{\mu_0 i}{2\pi R}\frac{L}{\sqrt{L^2+4R^2}}\ .$$

If $L \gg R$, then R^2 in the denominator can be ignored and

$$B = \frac{\mu_0 i}{2\pi R}$$

is obtained. This is the field of a long straight wire. For points very close to a finite wire, the field is quite similar to that of an infinitely long wire.

12. The center of a square is a distance $R = a/2$ from the nearest side (each side being of length $L = a$). There are four sides contributing to the field at the center, so the result of problem 11 leads to

$$B_{\text{center}} = 4\left(\frac{\mu_0 i}{2\pi(a/2)}\right)\left(\frac{a}{\sqrt{a^2+4(a/2)^2}}\right) = \frac{2\sqrt{2}\,\mu_0 i}{\pi a}\ .$$

13. Our x axis is along the wire with the origin at the right endpoint, and the current is in the positive x direction. All segments of the wire produce magnetic fields at P_2 that are out of the page. According to the Biot-Savart law, the magnitude of the field any (infinitesimal) segment produces at P_2 is given by

$$dB = \frac{\mu_0 i}{4\pi}\frac{\sin\theta}{r^2}\,dx$$

where θ (the angle between the segment and a line drawn from the segment to P_2) and r (the length of that line) are functions of x. Replacing r with $\sqrt{x^2+R^2}$ and $\sin\theta$ with $R/r = R/\sqrt{x^2+R^2}$, we integrate from $x=-L$ to $x=0$. The total field is

$$B = \frac{\mu_0 iR}{4\pi}\int_{-L}^{0}\frac{dx}{(x^2+R^2)^{3/2}} = \frac{\mu_0 iR}{4\pi}\frac{1}{R^2}\frac{x}{(x^2+R^2)^{1/2}}\bigg|_{-L}^{0} = \frac{\mu_0 i}{4\pi R}\frac{L}{\sqrt{L^2+R^2}}\ .$$

14. We refer to the side of length L as the long side and that of length W as the short side. The center is a distance $W/2$ from the midpoint of each long side, and is a distance $L/2$ from the midpoint of each short side. There are two of each type of side, so the result of problem 11 leads to

$$B = 2\frac{\mu_0 i}{2\pi(W/2)}\frac{L}{\sqrt{L^2+4(W/2)^2}} + 2\frac{\mu_0 i}{2\pi(L/2)}\frac{W}{\sqrt{W^2+4(L/2)^2}}\ .$$

The final form of this expression, shown in the problem statement, derives from finding the common denominator of the above result and adding them, while noting that

$$\frac{L^2+W^2}{\sqrt{W^2+L^2}} = \sqrt{W^2+L^2}\ .$$

15. We imagine the square loop in the yz plane (with its center at the origin) and the evaluation point for the field being along the x axis (as suggested by the notation in the problem). The origin is a distance $a/2$ from each side of the square loop, so the distance from the evaluation point to each side of the square is, by the Pythagorean theorem,

$$R = \sqrt{(a/2)^2+x^2} = \frac{1}{2}\sqrt{a^2+4x^2}\ .$$

Only the x components of the fields (contributed by each side) will contribute to the final result (other components cancel in pairs), so a trigonometric factor of

$$\frac{a/2}{R} = \frac{a}{\sqrt{a^2 + 4x^2}}$$

multiplies the expression of the field given by the result of problem 11 (for each side of length $L = a$). Since there are four sides, we find

$$B(x) = 4\left(\frac{\mu_0 i}{2\pi R}\right)\left(\frac{a}{\sqrt{a^2 + 4R^2}}\right)\left(\frac{a}{\sqrt{a^2 + 4x^2}}\right) = \frac{4\,\mu_0\, i\, a^2}{2\pi\left(\frac{1}{2}\right)\left(\sqrt{a^2 + 4x^2}\right)^2\sqrt{a^2 + 4(a/2)^2 + 4x^2}}$$

which simplifies to the desired result. It is straightforward to set $x = 0$ and see that this reduces to the expression found in problem 12 (noting that $\frac{4}{\sqrt{2}} = 2\sqrt{2}$).

16. Our y axis is along the wire with the origin at the top endpoint, and the current is in the positive y direction. All segments of the wire produce magnetic fields at P that are into the page. According to the Biot-Savart law, the magnitude of the field any (infinitesimal) segment produces at P is given by

$$dB = \frac{\mu_0 i}{4\pi}\frac{\sin\theta}{r^2}\,dy$$

where θ (the angle between the segment and a line drawn from the segment to P) and r (the length of that line) are functions of y. Replacing r with $\sqrt{y^2 + a^2}$ and $\sin\theta$ with $a/r = a/\sqrt{y^2 + a^2}$, we integrate from $y = -a$ to $y = 0$. The total field is

$$B = \frac{\mu_0 i a}{4\pi}\int_{-a}^{0}\frac{dy}{(y^2 + a^2)^{3/2}} = \frac{\mu_0 i a}{4\pi}\frac{1}{a^2}\left.\frac{y}{(y^2 + a^2)^{1/2}}\right|_{-a}^{0} = \frac{\mu_0 i}{4\pi a}\frac{a}{\sqrt{a^2 + a^2}}$$

which simplifies to the desired result (noting that $\frac{1}{4\sqrt{2}} = \frac{\sqrt{2}}{8}$).

17. Using the result of problem 12 and Eq. 30-12, we wish to show that

$$\frac{2\sqrt{2}\,\mu_0 i}{\pi a} > \frac{\mu_0 i}{2R}\ , \quad \text{or} \quad \frac{4\sqrt{2}}{\pi a} > \frac{1}{R}\ ,$$

but to do this we must relate the parameters a and R. If both wires have the same length L then the geometrical relationships $4a = L$ and $2\pi R = L$ provide the necessary connection:

$$4a = 2\pi R \implies a = \frac{\pi R}{2}\ .$$

Thus, our proof consists of the observation that

$$\frac{4\sqrt{2}}{\pi a} = \frac{8\sqrt{2}}{\pi^2 R} > \frac{1}{R}\ ,$$

as one can check numerically (that $8\sqrt{2}/\pi^2 > 1$).

18. Recalling the *straight sections* discussion in Sample Problem 30-1, we see that the current in the straight segments colinear with P do not contribute to the field at that point. We use the result of problem 16 to evaluate the contributions to the field at P, noting that the nearest wire-segments (each of length a) produce magnetism into the page at P and the further wire-segments (each of length $2a$) produce magnetism pointing out of the page at P. Thus, we find (into the page)

$$B_P = 2\left(\frac{\sqrt{2}\mu_0 i}{8\pi a}\right) - 2\left(\frac{\sqrt{2}\mu_0 i}{8\pi(2a)}\right) = \frac{\sqrt{2}\mu_0 i}{8\pi a}\ .$$

19. Consider a section of the ribbon of thickness dx located a distance x away from point P. The current it carries is $di = i\,dx/w$, and its contribution to B_P is

$$dB_P = \frac{\mu_0 di}{2\pi x} = \frac{\mu_0 i dx}{2\pi x w} .$$

Thus,

$$B_P = \int dB_P = \frac{\mu_0 i}{2\pi w}\int_d^{d+w}\frac{dx}{x} = \frac{\mu_0 i}{2\pi w}\ln\left(1+\frac{w}{d}\right) ,$$

and $\vec{B}_P$ points upward.

20. The two small wire-segments, each of length $a/4$, shown in Fig. 30-39 nearest to point P, are labeled 1 and 8 in the figure below.

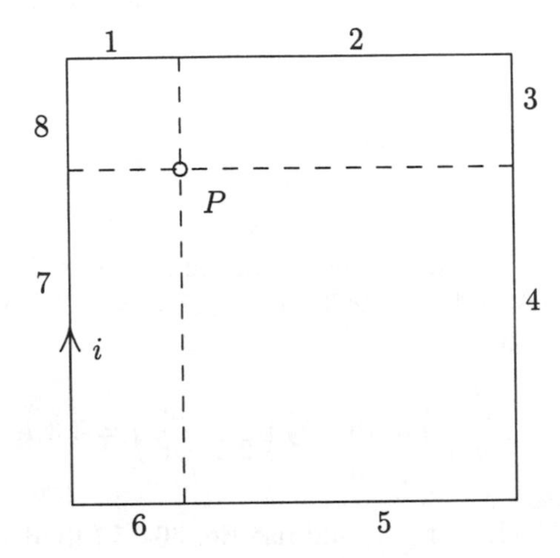

Let $\vec{e}$ be a unit vector pointing into the page. We use the results of problems 13 and 16 to calculate B_{P1} through B_{P8}:

$$\begin{aligned}
B_{P1} &= B_{P8} = \frac{\sqrt{2}\mu_0 i}{8\pi(a/4)} = \frac{\sqrt{2}\mu_0 i}{2\pi a} ,\\
B_{P4} &= B_{P5} = \frac{\sqrt{2}\mu_0 i}{8\pi(3a/4)} = \frac{\sqrt{2}\mu_0 i}{6\pi a} ,\\
B_{P2} &= B_{P7} = \frac{\mu_0 i}{4\pi(a/4)}\cdot\frac{3a/4}{[(3a/4)^2+(a/4)^2]^{1/2}} = \frac{3\mu_0 i}{\sqrt{10}\pi a} ,
\end{aligned}$$

and

$$B_{P3} = B_{P6} = \frac{\mu_0 i}{4\pi(3a/4)}\cdot\frac{a/4}{[(a/4)^2+(3a/4)^2]^{1/2}} = \frac{\mu_0 i}{3\sqrt{10}\pi a} .$$

Finally,

$$\begin{aligned}
\vec{B}_P &= \sum_{n=1}^{8} B_{Pn}\vec{e}\\
&= 2\frac{\mu_0 i}{\pi a}\left(\frac{\sqrt{2}}{2}+\frac{\sqrt{2}}{6}+\frac{3}{\sqrt{10}}+\frac{1}{3\sqrt{10}}\right)\vec{e}\\
&= \frac{2(4\pi\times10^{-7}\,\text{T}\cdot\text{m/A})(10\,\text{A})}{\pi(8.0\times10^{-2}\,\text{m})}\left(\frac{\sqrt{2}}{2}+\frac{\sqrt{2}}{6}+\frac{3}{\sqrt{10}}+\frac{1}{3\sqrt{10}}\right)\vec{e}\\
&= (2.0\times10^{-4}\,\text{T})\,\vec{e} ,
\end{aligned}$$

where $\vec{e}$ is a unit vector pointing into the page.

21. (a) If the currents are parallel, the two fields are in opposite directions in the region between the wires. Since the currents are the same, the total field is zero along the line that runs halfway between the wires. There is no possible current for which the field does not vanish.

    (b) If the currents are antiparallel, the fields are in the same direction in the region between the wires. At a point halfway between they have the same magnitude, $\mu_0 i/2\pi r$. Thus the total field at the midpoint has magnitude $B = \mu_0 i/\pi r$ and

$$i = \frac{\pi r B}{\mu_0} = \frac{\pi(0.040\,\text{m})(300\times10^{-6}\,\text{T})}{4\pi\times10^{-7}\,\text{T}\cdot\text{m/A}} = 30\ \text{A}\ .$$

22. Since they carry current in the same direction, then (by the right-hand rule) the only region in which their fields might cancel is between them. Thus, if the point at which we are evaluating their field is r away from the wire carrying current i and is $d-r$ away from the wire carrying current $3i$, then the canceling of their fields leads to

$$\frac{\mu_0 i}{2\pi r} = \frac{\mu_0(3i)}{2\pi(d-r)} \quad\Longrightarrow\quad r = \frac{d}{4}\ .$$

23. Using the right-hand rule, we see that the current i_2 carried by wire 2 must be out of the page. Now, $B_{P1} = \mu_0 i_1/2\pi r_1$ where $i_1 = 6.5\,\text{A}$ and $r_1 = 0.75\,\text{cm} + 1.5\,\text{cm} = 2.25\,\text{cm}$, and $B_{P2} = \mu_0 i_2/2\pi r_2$ where $r_2 = 1.5\,\text{cm}$. From $B_{P1} = B_{P2}$ we get

$$i_2 = i_1\left(\frac{r_2}{r_1}\right) = (6.5\,\text{A})\left(\frac{1.5\,\text{cm}}{2.25\,\text{cm}}\right) = 4.3\ \text{A}\ .$$

24. We label these wires 1 through 5, left to right, and use Eq. 30-15 (divided by length). Then,

$$\begin{aligned}\vec{F}_1 &= \frac{\mu_0 i^2}{2\pi}\left(\frac{1}{d}+\frac{1}{2d}+\frac{1}{3d}+\frac{1}{4d}\right)\hat{\jmath} = \frac{25\mu_0 i^2}{24\pi d}\hat{\jmath} \\ &= \frac{(13)(4\pi\times10^{-7}\,\text{T}\cdot\text{m/A})(3.00\,\text{A})^2(1.00\,\text{m})\hat{\jmath}}{24\pi(8.00\times10^{-2}\,\text{m})} \\ &= 4.69\times10^{-5}\,\text{N/m}\ \hat{\jmath}\ ;\end{aligned}$$

$$\vec{F}_2 = \frac{\mu_0 i^2}{2\pi}\left(\frac{1}{2d}+\frac{1}{3d}\right)\hat{\jmath} = \frac{5\mu_0 i^2}{12\pi d}\hat{\jmath} = 1.88\times10^{-5}\,\text{N/m}\ \hat{\jmath}\ ;$$

$F_3 = 0$ (because of symmetry); $\vec{F}_4 = -\vec{F}_2$; and $\vec{F}_5 = -\vec{F}_1$.

25. Each wire produces a field with magnitude given by $B = \mu_0 i/2\pi r$, where r is the distance from the corner of the square to the center. According to the Pythagorean theorem, the diagonal of the square has length $\sqrt{2}a$, so $r = a/\sqrt{2}$ and $B = \mu_0 i/\sqrt{2}\pi a$. The fields due to the wires at the upper left and lower right corners both point toward the upper right corner of the square. The fields due to the wires at the upper right and lower left corners both point toward the upper left corner. The horizontal components cancel and the vertical components sum to

$$\begin{aligned}B_{\text{total}} &= 4\frac{\mu_0 i}{\sqrt{2}\pi a}\cos 45^\circ = \frac{2\mu_0 i}{\pi a} \\ &= \frac{2(4\pi\times10^{-7}\,\text{T}\cdot\text{m/A})(20\,\text{A})}{\pi(0.20\,\text{m})} = 8.0\times10^{-5}\ \text{T}\ .\end{aligned}$$

In the calculation $\cos 45^\circ$ was replaced with $1/\sqrt{2}$. The total field points upward.

26. Using Eq. 30-15, the force on, say, wire 1 (the wire at the upper left of the figure) is along the diagonal (pointing towards wire 3 which is at the lower right). Only the forces (or their components) along the diagonal direction contribute. With $\theta = 45°$, we find

$$\begin{aligned} F_1 &= \left|\vec{F}_{12} + \vec{F}_{13} + \vec{F}_{14}\right| \\ &= 2F_{12}\cos\theta + F_{13} \\ &= 2\left(\frac{\mu_0 i^2}{2\pi a}\right)\cos 45° + \frac{\mu_0 i^2}{2\sqrt{2}\pi a} \\ &= 0.338\left(\frac{\mu_0 i^2}{a}\right) . \end{aligned}$$

27. We use Eq. 30-15 and the superposition of forces: $\vec{F}_4 = \vec{F}_{14} + \vec{F}_{24} + \vec{F}_{34}$. With $\theta = 45°$, the situation is as shown below:

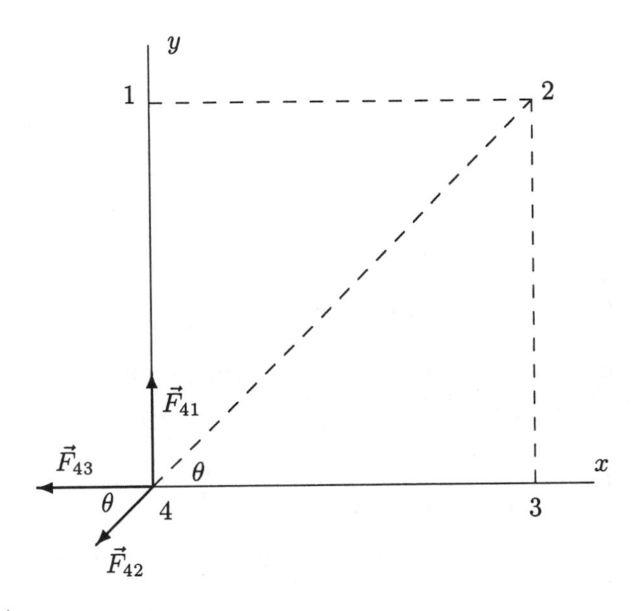

The components of $\vec{F}_4$ are given by

$$\begin{aligned} F_{4x} &= -F_{43} - F_{42}\cos\theta \\ &= -\frac{\mu_0 i^2}{2\pi a} - \frac{\mu_0 i^2 \cos 45°}{2\sqrt{2}\pi a} \\ &= -\frac{3\mu_0 i^2}{4\pi a} \end{aligned}$$

and

$$\begin{aligned} F_{4y} &= F_{41} - F_{42}\sin\theta \\ &= \frac{\mu_0 i^2}{2\pi a} - \frac{\mu_0 i^2 \sin 45°}{2\sqrt{2}\pi a} \\ &= \frac{\mu_0 i^2}{4\pi a} . \end{aligned}$$

Thus,

$$F_4 = (F_{4x}^2 + F_{4y}^2)^{1/2} = \left[\left(-\frac{3\mu_0 i^2}{4\pi a}\right)^2 + \left(\frac{\mu_0 i^2}{4\pi a}\right)^2\right]^{1/2} = \frac{\sqrt{10}\mu_0 i^2}{4\pi a} ,$$

and $\vec{F}_4$ makes an angle ϕ with the positive x axis, where

$$\phi = \tan^{-1}\left(\frac{F_{4y}}{F_{4x}}\right) = \tan^{-1}\left(-\frac{1}{3}\right) = 162° .$$

28. (a) Consider a segment of the projectile between y and $y + dy$. We use Eq. 30-14 to find the magnetic force on the segment, and Eq. 30-9 for the magnetic field of each semi-infinite wire (the top rail referred to as wire 1 and the bottom as wire 2). The current in rail 1 is in the $+\hat{\mathrm{i}}$ direction, and the current in the rail 2 is in the $-\hat{\mathrm{i}}$ direction. The field (in the region between the wires) set up by wire 1 is into the paper (the $-\hat{\mathrm{k}}$ direction) and that set up by wire 2 is also into the paper. The force element (a function of y) acting on the segment of the projectile (in which the current flows in the $-\hat{\mathrm{j}}$ direction) is given below. The coordinate origin is at the bottom of the projectile.

$$\begin{aligned} d\vec{F} &= d\vec{F}_1 + d\vec{F}_2 \\ &= i\,dy(-\hat{\mathrm{j}}) \times \vec{B}_1 + dy(-\hat{\mathrm{j}}) \times \vec{B}_2 \\ &= i[B_1 + B_2]\,\hat{\mathrm{i}}\,dy \\ &= i\left[\frac{\mu_0 i}{4\pi(2R + w - y)} + \frac{\mu_0 i}{4\pi y}\right]\hat{\mathrm{i}}\,dy . \end{aligned}$$

Thus, the force on the projectile is

$$\vec{F} = \int d\vec{F} = \frac{i^2\mu_0}{4\pi}\int_R^{R+w}\left(\frac{1}{2R + w - y} + \frac{1}{y}\right)dy\,\hat{\mathrm{i}} = \frac{\mu_0 i^2}{2\pi}\ln\left(1 + \frac{w}{R}\right)\hat{\mathrm{i}} .$$

(b) Using the work-energy theorem, we have $\Delta K = \frac{1}{2}mv_f^2 = W_{\text{ext}} = \int \vec{F}\cdot d\vec{s} = FL$. Thus, the final speed of the projectile is

$$\begin{aligned} v_f &= \left(\frac{2W_{\text{ext}}}{m}\right)^{1/2} = \left[\frac{2}{m}\frac{\mu_0 i^2}{2\pi}\ln\left(1 + \frac{w}{R}\right)L\right]^{1/2} \\ &= \left[\frac{2(4\pi \times 10^{-7}\,\mathrm{T\cdot m/A})(450 \times 10^3\,\mathrm{A})^2\ln(1 + 1.2\,\mathrm{cm}/6.7\,\mathrm{cm})(4.0\,\mathrm{m})}{2\pi(10 \times 10^{-3}\,\mathrm{kg})}\right]^{1/2} \\ &= 2.3 \times 10^3\ \mathrm{m/s} . \end{aligned}$$

29. The magnitudes of the forces on the sides of the rectangle which are parallel to the long straight wire (with $i_1 = 30$ A) are computed using Eq. 30-15, but the force on each of the sides lying perpendicular to it (along our y axis, with the origin at the top wire and $+y$ downward) would be figured by integrating as follows:

$$F_{\perp\,\text{sides}} = \int_a^{a+b}\frac{i_2\mu_0 i_1}{2\pi y}\,dy .$$

Fortunately, these forces on the two perpendicular sides of length b cancel out. For the remaining two (parallel) sides of length L, we obtain

$$\begin{aligned} F &= \frac{\mu_0 i_1 i_2 L}{2\pi}\left(\frac{1}{a} - \frac{1}{a + d}\right) = \frac{\mu_0 i_1 i_2 b}{2\pi a(a + b)} \\ &= \frac{(4\pi \times 10^{-7}\,\mathrm{T\cdot m/A})(30\,\mathrm{A})(20\,\mathrm{A})(8.0\,\mathrm{cm})(30 \times 10^{-2}\,\mathrm{m})}{2\pi(1.0\,\mathrm{cm} + 8.0\,\mathrm{cm})} \\ &= 3.2 \times 10^{-3}\ \mathrm{N} , \end{aligned}$$

and $\vec{F}$ points toward the wire.

30. A close look at the path reveals that only currents 1, 3, 6 and 7 are enclosed. Thus, noting the different current directions described in the problem, we obtain

$$\oint \vec{B} \cdot d\vec{s} = \mu_0(7i - 6i + 3i + i) \quad = \quad 5\mu_0 i \ .$$

31. (a) Two of the currents are out of the page and one is into the page, so the net current enclosed by the path is 2.0 A, out of the page. Since the path is traversed in the clockwise sense, a current into the page is positive and a current out of the page is negative, as indicated by the right-hand rule associated with Ampere's law. Thus,

$$\oint \vec{B} \cdot d\vec{s} = -\mu_0 i = -(2.0\,\text{A})(4\pi \times 10^{-7}\,\text{T·m/A}) = -2.5 \times 10^{-6}\,\text{T·m} \ .$$

(b) The net current enclosed by the path is zero (two currents are out of the page and two are into the page), so $\oint \vec{B} \cdot d\vec{s} = \mu_0 i_{\text{enc}} = 0$.

32. We use Eq. 30-22 for the B-field inside the wire and Eq. 30-19 for that outside the wire. The plot is shown below (with SI units understood).

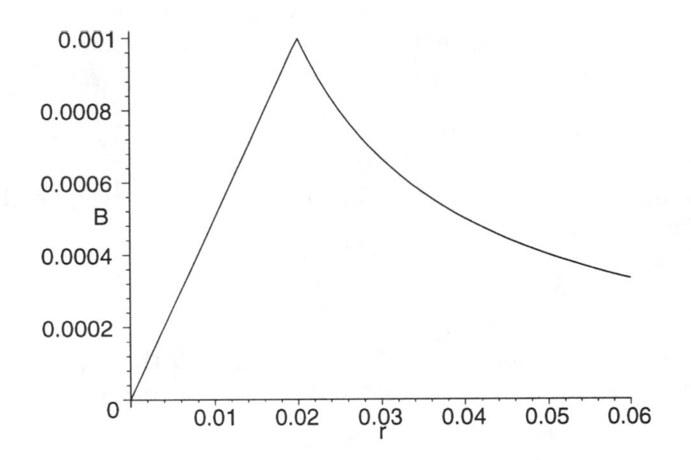

33. We use Ampere's law. For the dotted loop shown on the diagram $i = 0$. The integral $\int \vec{B} \cdot d\vec{s}$ is zero along the bottom, right, and top sides of the loop. Along the right side the field is zero, along the top and bottom sides the field is perpendicular to $d\vec{s}$. If ℓ is the length of the left edge, then direct integration yields $\oint \vec{B} \cdot d\vec{s} = B\ell$, where B is the magnitude of the field at the left side of the loop. Since neither B nor ℓ is zero, Ampere's law is contradicted. We conclude that the geometry shown for the magnetic field lines is in error. The lines actually bulge outward and their density decreases gradually, not discontinuously as suggested by the figure.

34. We use Ampere's law: $\oint \vec{B} \cdot d\vec{s} = \mu_0 i$, where the integral is around a closed loop and i is the net current through the loop. For path 1, the result is

$$\begin{aligned}\oint_1 \vec{B} \cdot d\vec{s} &= \mu_0(-5.0\,\text{A} + 3.0\,\text{A}) = (-2.0\,\text{A})(4\pi \times 10^{-7}\,\text{T·m/A}) \\ &= -2.5 \times 10^{-6}\ \text{T·m} \ .\end{aligned}$$

For path 2, we find

$$\begin{aligned}\oint_2 \vec{B} \cdot d\vec{s} &= \mu_0(-5.0\,\text{A} - 5.0\,\text{A} - 3.0\,\text{A}) = (-13.0\,\text{A})(4\pi \times 10^{-7}\,\text{T·m/A}) \\ &= -1.6 \times 10^{-5}\ \text{T·m} \ .\end{aligned}$$

35. For $r < a$,

$$B(r) = \frac{\mu_0 i_{\text{enc}}}{2\pi r} = \frac{\mu_0}{2\pi r}\int_0^r J(r) 2\pi r\, dr = \frac{\mu_0}{2\pi}\int_0^r J_0\left(\frac{r}{a}\right) 2\pi r\, dr = \frac{\mu_0 J_0 r^2}{3a} .$$

36. (a) Replacing $i/\pi R^2$ with $J = 100\ \text{A/m}^2$, in Eq. 30-22, we have

$$\left|\vec{B}\right| = \left(\frac{\mu_0 J}{2}\right) r = 1.3 \times 10^{-7}\ \text{T}$$

where $r = 0.0020$ m.

(b) Similarly, writing $i = J\pi R^2$ in Eq. 30-19 yields

$$\left|\vec{B}\right| = \frac{\mu_0 J R^2}{2r} = 1.4 \times 10^{-7}\ \text{T}$$

where $r = 0.0040$ m.

37. (a) The the magnetic field at a point within the hole is the sum of the fields due to two current distributions. The first is that of the solid cylinder obtained by filling the hole and has a current density that is the same as that in the original cylinder (with the hole). The second is the solid cylinder that fills the hole. It has a current density with the same magnitude as that of the original cylinder but is in the opposite direction. If these two situations are superposed the total current in the region of the hole is zero. Now, a solid cylinder carrying current i, uniformly distributed over a cross section, produces a magnetic field with magnitude

$$B = \frac{\mu_0 i r}{2\pi R^2}$$

a distance r from its axis, inside the cylinder. Here R is the radius of the cylinder. For the cylinder of this problem the current density is

$$J = \frac{i}{A} = \frac{i}{\pi(a^2 - b^2)} ,$$

where $A = \pi(a^2 - b^2)$ is the cross-sectional area of the cylinder with the hole. The current in the cylinder without the hole is

$$I_1 = JA = \pi J a^2 = \frac{i a^2}{a^2 - b^2}$$

and the magnetic field it produces at a point inside, a distance r_1 from its axis, has magnitude

$$B_1 = \frac{\mu_0 I_1 r_1}{2\pi a^2} = \frac{\mu_0 i r_1 a^2}{2\pi a^2(a^2 - b^2)} = \frac{\mu_0 i r_1}{2\pi(a^2 - b^2)} .$$

The current in the cylinder that fills the hole is

$$I_2 = \pi J b^2 = \frac{i b^2}{a^2 - b^2}$$

and the field it produces at a point inside, a distance r_2 from the its axis, has magnitude

$$B_2 = \frac{\mu_0 I_2 r_2}{2\pi b^2} = \frac{\mu_0 i r_2 b^2}{2\pi b^2(a^2 - b^2)} = \frac{\mu_0 i r_2}{2\pi(a^2 - b^2)} .$$

At the center of the hole, this field is zero and the field there is exactly the same as it would be if the hole were filled. Place $r_1 = d$ in the expression for B_1 and obtain

$$B = \frac{\mu_0 i d}{2\pi(a^2 - b^2)}$$

for the field at the center of the hole. The field points upward in the diagram if the current is out of the page.

(b) If $b = 0$ the formula for the field becomes

$$B = \frac{\mu_0 i d}{2\pi a^2} \ .$$

This correctly gives the field of a solid cylinder carrying a uniform current i, at a point inside the cylinder a distance d from the axis. If $d = 0$ the formula gives $B = 0$. This is correct for the field on the axis of a cylindrical shell carrying a uniform current.

(c) Consider a rectangular path with two long sides (side 1 and 2, each with length L) and two short sides (each of length less than b). If side 1 is directly along the axis of the hole, then side 2 would be also parallel to it and also in the hole. To ensure that the short sides do not contribute significantly to the integral in Ampere's law, we might wish to make L *very* long (perhaps longer than the length of the cylinder), or we might appeal to an argument regarding the angle between $\vec{B}$ and the short sides (which is 90° at the axis of the hole). In any case, the integral in Ampere's law reduces to

$$\begin{aligned}
\oint_{\text{rectangle}} \vec{B} \cdot d\vec{s} &= \mu_0 i_{\text{enclosed}} \\
\int_{\text{side 1}} \vec{B} \cdot d\vec{s} + \int_{\text{side 2}} \vec{B} \cdot d\vec{s} &= \mu_0 i_{\text{in hole}} \\
\left(B_{\text{side 1}} - B_{\text{side 2}}\right) L &= 0
\end{aligned}$$

where $B_{\text{side 1}}$ is the field along the axis found in part (a). This shows that the field at off-axis points (where $B_{\text{side 2}}$ is evaluated) is the same as the field at the center of the hole; therefore, the field in the hole is uniform.

38. The field at the center of the pipe (point C) is due to the wire alone, with a magnitude of

$$B_C = \frac{\mu_0 i_{\text{wire}}}{2\pi(3R)} = \frac{\mu_0 i_{\text{wire}}}{6\pi R} \ .$$

For the wire we have $B_{P,\text{wire}} > B_{C,\text{wire}}$. Thus, for $B_P = B_C = B_{C,\text{wire}}$, i_{wire} must be into the page:

$$B_P = B_{P,\text{wire}} - B_{P,\text{pipe}} = \frac{\mu_0 i_{\text{wire}}}{2\pi R} - \frac{\mu_0 i}{2\pi(2R)} \ .$$

Setting $B_C = -B_P$ we obtain $i_{\text{wire}} = 3i/8$.

39. The "current per unit x-length" may be viewed as current density multiplied by the thickness Δy of the sheet; thus, $\lambda = J\Delta y$. Ampere's law may be (and often is) expressed in terms of the current density vector as follows:

$$\oint \vec{B} \cdot d\vec{s} = \mu_0 \int \vec{J} \cdot d\vec{A}$$

where the area integral is over the region enclosed by the path relevant to the line integral (and $\vec{J}$ is in the $+z$ direction, out of the paper). With J uniform throughout the sheet, then it clear that the right-hand side of this version of Ampere's law should reduce, in this problem, to $\mu_0 JA = \mu_0 J\Delta y\Delta x = \mu_0\lambda\Delta x$.

(a) Figure 30-52 certainly has the horizontal components of $\vec{B}$ drawn correctly at points P and P' (as reference to Fig. 30-4 will confirm [consider the current elements nearest each of those points]), so the question becomes: is it possible for $\vec{B}$ to have vertical components in the figure? Our focus is on point P. Fig. 30-4 suggests that the current element just to the right of the nearest one (the one directly under point P) will contribute a downward component, but by the same reasoning the current element just to the left of the nearest one should contribute an upward component to the field at P. The current elements are all equivalent, as is reflected in the horizontal-translational symmetry built into this problem; therefore, all vertical components should cancel in pairs. The field at P must be purely horizontal, as drawn.

(b) The path used in evaluating $\oint \vec{B} \cdot d\vec{s}$ is rectangular, of horizontal length Δx (the horizontal sides passing through points P and P' respectively) and vertical size $\delta y > \Delta y$. The vertical sides have no contribution to the integral since $\vec{B}$ is purely horizontal (so the scalar dot product produces zero for those sides), and the horizontal sides contribute two equal terms, as shown below. Ampere's law yields

$$2B\Delta x = \mu_0 \lambda \Delta x \implies B = \frac{1}{2}\mu_0 \lambda .$$

40. It is possible (though tedious) to use Eq. 30-28 and evaluate the contributions (with the intent to sum them) of all 1200 loops to the field at, say, the center of the solenoid. This would make use of all the information given in the problem statement, but this is not the method that the student is expected to use here. Instead, Eq. 30-25 for the ideal solenoid (which does not make use of the coil radius) is the preferred method:

$$B = \mu_0 in = \mu_0 i \left(\frac{N}{\ell}\right)$$

where $i = 3.60$ A, $\ell = 0.950$ m and $N = 1200$. This yields $B = 0.00571$ T.

41. It is possible (though tedious) to use Eq. 30-28 and evaluate the contributions (with the intent to sum them) of all 200 loops to the field at, say, the center of the solenoid. This would make use of all the information given in the problem statement, but this is not the method that the student is expected to use here. Instead, Eq. 30-25 for the ideal solenoid (which does not make use of the coil diameter) is the preferred method:

$$B = \mu_0 in = \mu_0 i \left(\frac{N}{\ell}\right)$$

where $i = 0.30$ A, $\ell = 0.25$ m and $N = 200$. This yields $B = 0.0030$ T.

42. We find N, the number of turns of the solenoid, from $B = \mu_0 in = \mu_0 iN/\ell$: $N = B\ell/\mu_0 i$. Thus, the total length of wire used in making the solenoid is

$$2\pi rN = \frac{2\pi rB\ell}{\mu_0 i} = \frac{2\pi(2.60 \times 10^{-2}\,\text{m})(23.0 \times 10^{-3}\,\text{T})(1.30\,\text{m})}{2(4\pi \times 10^{-7}\,\text{T}\cdot\text{m/A})(18.0\,\text{A})} = 108\ \text{m} .$$

43. (a) We use Eq. 30-26. The inner radius is $r = 15.0$ cm, so the field there is

$$B = \frac{\mu_0 iN}{2\pi r} = \frac{(4\pi \times 10^{-7}\,\text{T}\cdot\text{m/A})(0.800\,\text{A})(500)}{2\pi(0.150\,\text{m})} = 5.33 \times 10^{-4}\ \text{T} .$$

(b) The outer radius is $r = 20.0$ cm. The field there is

$$B = \frac{\mu_0 iN}{2\pi r} = \frac{(4\pi \times 10^{-7}\,\text{T}\cdot\text{m/A})(0.800\,\text{A})(500)}{2\pi(0.200\text{m})} = 4.00 \times 10^{-4}\ \text{T} .$$

44. (a) The ideal solenoid is long enough (and we are evaluating the field at a point far enough inside) such that the open ends of the solenoid are "out of sight" and the situation displays a horizontal-translational symmetry (assuming the axis of the cylindrical shape of the solenoid is horizontal). A view of a "slice" of, say, the bottom of the solenoid would therefore appear similar to that shown in Fig. 30-52, where point P is in the interior of the solenoid and point P' is outside the coil. Now, Fig. 30-52 differs in at least one respect from our "slice" view of the solenoid in that the field at P' would be zero instead of what is shown in that figure. The field vanishes there because the top of the solenoid (similar to that shown in Fig. 30-52, in "slice" view, but with the currents and field directions reversed) would contribute an equal and opposite field to any exterior point, thus canceling it. For interior points, the top and bottom "slices" each contribute $\frac{1}{2}\mu_0\lambda$ (in the same direction) [this is shown in the solution to problem 39] and thus produce an interior field equal to $B = \mu_0\lambda$.

(b) Applying Ampere's law to a rectangular path which passes through points P (interior) and P' (exterior) similar to that described in the solution to part (b) of problem 39, we are not surprised to find

$$\oint \vec{B} \cdot d\vec{s} = \left(\vec{B}_P - \vec{B}_{P'}\right) \cdot \hat{\mathrm{i}} \Delta x = \mu_0 \lambda \Delta x$$

just as we found in part (b) of problem 39 (except that we are now taking the $+x$ direction in the same direction as the field at P, to avoid confusion with signs). The difference with the previous solution is that in 39, $\left(\vec{B}_P - \vec{B}_{P'}\right) \cdot \hat{\mathrm{i}}$ was equal to $B - (-B) = 2B$, whereas in this case we have $B - 0 = B$. Although the value of B is different in the two problems, we see that the *change* $\left(\vec{B}_P - \vec{B}_{P'}\right) \cdot \hat{\mathrm{i}}$ is the same: $\mu_0 \lambda$.

45. Consider a circle of radius r, inside the toroid and concentric with it (like either of the loops drawn in Fig. 30-20). The current that passes through the region between this circle and another larger radius circle (well outside the toroid) is Ni, where N is the number of turns and i is the current (note that this region includes a "slice" of the outer rim of the toroid). The current per unit length (of the circle) is $\lambda = Ni/2\pi r$, and $\mu_0 \lambda$ is therefore $\mu_0 Ni/2\pi r$, the magnitude of the magnetic field at the circle (call it B_1). Since the field outside a toroid (call it B_2) is zero, the above result is also the *change* in the magnitude of the field encountered as you move from the circle to the outside (say, to the larger radius circle mentioned above). The equality is not really surprising in light of Ampere's law, particularly if the path used in $\oint \vec{B} \cdot d\vec{s}$ is made to connect the circle in the toroid and the larger radius circle (or portions of each of them, of lengths Δs_1 and Δs_2). The connecting paths (each of size Δr) between the circles can be made perpendicular to the magnetic field lines (so that $\vec{B} \cdot \vec{s} = 0$). In fact, we can keep the connecting paths roughly perpendicular to $\vec{B}$ and manage to have $\Delta s_1 \approx \Delta s_2$ if our Amperian loop is very small (especially if Δr is much smaller than the outer radius of the toroid). Simplifying our notation, the current through the loop is therefore $\Delta s \lambda$, so Ampere's law yields $(B_1 - B_2)\Delta s = \mu_0 \Delta s \lambda$ and $B_2 - B_1 = \mu_0 \lambda$. What this demonstrates is that the change of the magnetic field is $\mu_0 \lambda$ when moving from one point to another (in a direction perpendicular to the field) across a current sheet (as the term is used in problem 39); this principle is useful in any discussion of boundary conditions in electrodynamics applications.

46. The orbital radius for the electron is

$$r = \frac{mv}{eB} = \frac{mv}{e\mu_0 ni}$$

which we solve for i:

$$\begin{aligned} i &= \frac{mv}{e\mu_0 nr} \\ &= \frac{(9.11 \times 10^{-31}\,\mathrm{kg})(0.0460)(3.00 \times 10^8\,\mathrm{m/s})}{(1.60 \times 10^{-19}\,\mathrm{C})(4\pi \times 10^{-7}\,\mathrm{T \cdot m/A})(100/0.0100\,\mathrm{m})(2.30 \times 10^{-2}\,\mathrm{m})} \\ &= 0.272\ \mathrm{A}\ . \end{aligned}$$

47. (a) We denote the $\vec{B}$-fields at point P on the axis due to the solenoid and the wire as $\vec{B}_s$ and $\vec{B}_w$, respectively. Since $\vec{B}_s$ is along the axis of the solenoid and $\vec{B}_w$ is perpendicular to it, $\vec{B}_s \perp \vec{B}_w$, respectively. For the net field $\vec{B}$ to be at 45° with the axis we then must have $B_s = B_w$. Thus,

$$B_s = \mu_0 i_s n = B_w = \frac{\mu_0 i_w}{2\pi d}\ ,$$

which gives the separation d to point P on the axis:

$$d = \frac{i_w}{2\pi i_s n} = \frac{6.00\,\mathrm{A}}{2\pi(20.0 \times 10^{-3}\,\mathrm{A})(10\,\mathrm{turns/cm})} = 4.77\ \mathrm{cm}\ .$$

(b) The magnetic field strength is

$$\begin{aligned} B &= \sqrt{2}B_s \\ &= \sqrt{2}(4\pi \times 10^{-7}\,\mathrm{T\cdot m/A})(20.0 \times 10^{-3}\,\mathrm{A})(10\,\mathrm{turns}/0.0100\,\mathrm{m}) \\ &= 3.55 \times 10^{-5}\,\mathrm{T}\,. \end{aligned}$$

48. (a) We set $z = 0$ in Eq. 30-28 (which is equivalent using to Eq. 30-12 multiplied by the number of loops). Thus, $B(0) \propto i/R$. Since case b has two loops,

$$\frac{B_b}{B_a} = \frac{2i/R_b}{i/R_a} = \frac{2R_a}{R_b} = 4\,.$$

(b) The ratio of their magnetic dipole moments is

$$\frac{\mu_b}{\mu_a} = \frac{2iA_b}{iA_a} = \frac{2R_b^2}{R_a^2} = 2\left(\frac{1}{2}\right)^2 = \frac{1}{2}\,.$$

49. The magnitude of the magnetic dipole moment is given by $\mu = NiA$, where N is the number of turns, i is the current, and A is the area. We use $A = \pi R^2$, where R is the radius. Thus,

$$\mu = (200)(0.30\,\mathrm{A})\pi(0.050\mathrm{m})^2 = 0.47\,\mathrm{A\cdot m^2}\,.$$

50. We use Eq. 30-28 and note that the contributions to $\vec{B}_P$ from the two coils are the same. Thus,

$$B_P = \frac{2\mu_0 iR^2 N}{2[R^2 + (R/2)^2]^{3/2}} = \frac{8\mu_0 Ni}{5\sqrt{5}R}\,.$$

$\vec{B}_P$ is in the positive x direction.

51. (a) The magnitude of the magnetic dipole moment is given by $\mu = NiA$, where N is the number of turns, i is the current, and A is the area. We use $A = \pi R^2$, where R is the radius. Thus,

$$\mu = Ni\pi R^2 = (300)(4.0\,\mathrm{A})\pi(0.025\,\mathrm{m})^2 = 2.4\,\mathrm{A\cdot m^2}\,.$$

(b) The magnetic field on the axis of a magnetic dipole, a distance z away, is given by Eq. 30-29:

$$B = \frac{\mu_0}{2\pi}\frac{\mu}{z^3}\,.$$

We solve for z:

$$z = \left(\frac{\mu_0}{2\pi}\frac{\mu}{B}\right)^{1/3} = \left(\frac{(4\pi \times 10^{-7}\,\mathrm{T\cdot m/A})(2.36\,\mathrm{A\cdot m^2})}{2\pi(5.0 \times 10^{-6}\,\mathrm{T})}\right)^{1/3} = 46\ \mathrm{cm}\,.$$

52. (a) For $x \gg a$, the result of problem 15 reduces to

$$B(x) \approx \frac{4\mu_0 ia^2}{\pi(4x^2)(4x^2)^{1/2}} = \frac{\mu_0(ia^2)}{4\pi x^3}\,,$$

which is indeed the field of a magnetic dipole (see Eq. 30-29).

(b) The magnitude of the magnetic dipole moment is $\mu = ia^2$, by comparison between Eq. 30-29 and the result above.

53. Since the origin is midway between the coils, and the axis is chosen to be x (as opposed to the z used in Eq. 30-28), then the net field of the two coils is

$$B = \frac{\mu_0 N i R^2}{2}\left(\frac{1}{\sqrt{R^2+(R/2-x)^2}} + \frac{1}{\sqrt{R^2+(R/2+x)^2}}\right)$$

where $i = 50$ A, $N = 300$ and $R = 0.050$ m. The graph of this function (using SI units) is shown below.

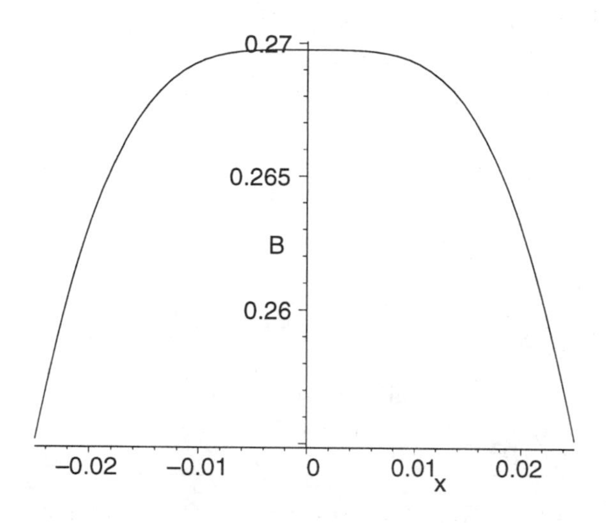

54. (a) By imagining that each of the segments bg and cf (which are shown in the figure as having no current) actually has a pair of currents, where both currents are of the same magnitude (i) but opposite direction (so that the pair effectively cancels in the final sum), one can justify the superposition.

(b) The dipole moment of path $abcdefgha$ is

$$\begin{aligned}\vec{\mu} &= \vec{\mu}_{bcfgb} + \vec{\mu}_{abgha} + \vec{\mu}_{cdefc} = (ia^2)(\hat{\jmath} - \hat{\imath} + \hat{\imath}) = ia^2\hat{\jmath} \\ &= (6.0\,\text{A})(0.10\,\text{m})^2\hat{\jmath} = 6.0 \times 10^{-2}\ \text{A}\cdot\text{m}^2\ \hat{\jmath}\ .\end{aligned}$$

(c) Since both points are far from the cube we can use the dipole approximation. For $(x, y, z) = (0, 5.0\,\text{m}, 0)$

$$\begin{aligned}\vec{B}(0, 5.0\,\text{m}, 0) &\approx \frac{\mu_0}{2\pi}\frac{\vec{\mu}}{y^3} \\ &= \frac{(1.26\times 10^{-6}\,\text{T}\cdot\text{m/A})(6.0\times 10^{-2}\,\text{m}^2\cdot\text{A})\hat{\jmath}}{2\pi(5.0\,\text{m})^3} \\ &= 9.6\times 10^{-11}\ \text{T}\ \hat{\jmath}\ .\end{aligned}$$

For $(x, y, z) = (5.0\,\text{m}, 0, 0)$, note that the line joining the end point of interest and the location of the dipole is perpendicular to the axis of the dipole. You can check easily that if an electric dipole is used, the field would be $E \approx (1/4\pi\varepsilon_0)(p/x^3)$, which is half of the magnitude of E for a point on the y axis the same distance from the dipole. By analogy, in our case B is also half the value or $B(0, 5.0\,\text{m}, 0)$, i.e.,

$$B(5.0\,\text{m}, 0, 0) = \frac{1}{2}B(0, 5.0\,\text{m}, 0) = \frac{1}{2}(9.6\times 10^{-11}\,\text{T}) = 4.8\times 10^{-11}\,\text{T}\ .$$

Just like the electric dipole case, $\vec{B}(5.0\,\text{m}, 0, 0)$ points in the negative y direction.

55. (a) The magnitude of the magnetic field on the axis of a circular loop, a distance z from the loop center, is given by Eq. 30-28:

$$B = \frac{N\mu_0 i R^2}{2(R^2+z^2)^{3/2}} ,$$

where R is the radius of the loop, N is the number of turns, and i is the current. Both of the loops in the problem have the same radius, the same number of turns, and carry the same current. The currents are in the same sense, and the fields they produce are in the same direction in the region between them. We place the origin at the center of the left-hand loop and let x be the coordinate of a point on the axis between the loops. To calculate the field of the left-hand loop, we set $z = x$ in the equation above. The chosen point on the axis is a distance $s - x$ from the center of the right-hand loop. To calculate the field it produces, we put $z = s - x$ in the equation above. The total field at the point is therefore

$$B = \frac{N\mu_0 i R^2}{2}\left[\frac{1}{(R^2+x^2)^{3/2}} + \frac{1}{(R^2+x^2-2sx+s^2)^{3/2}}\right] .$$

Its derivative with respect to x is

$$\frac{dB}{dx} = -\frac{N\mu_0 i R^2}{2}\left[\frac{3x}{(R^2+x^2)^{5/2}} + \frac{3(x-s)}{(R^2+x^2-2sx+s^2)^{5/2}}\right] .$$

When this is evaluated for $x = s/2$ (the midpoint between the loops) the result is

$$\left.\frac{dB}{dx}\right|_{s/2} = -\frac{N\mu_0 i R^2}{2}\left[\frac{3s/2}{(R^2+s^2/4)^{5/2}} - \frac{3s/2}{(R^2+s^2/4-s^2+s^2)^{5/2}}\right] = 0$$

independently of the value of s.

(b) The second derivative is

$$\frac{d^2B}{dx^2} = \frac{N\mu_0 i R^2}{2}\left[-\frac{3}{(R^2+x^2)^{5/2}} + \frac{15x^2}{(R^2+x^2)^{7/2}} - \frac{3}{(R^2+x^2-2sx+s^2)^{5/2}} + \frac{15(x-s)^2}{(R^2+x^2-2sx+s^2)^{7/2}}\right] .$$

At $x = s/2$,

$$\left.\frac{d^2B}{dx^2}\right|_{s/2} = \frac{N\mu_0 i R^2}{2}\left[-\frac{6}{(R^2+s^2/4)^{5/2}} + \frac{30s^2/4}{(R^2+s^2/4)^{7/2}}\right]$$

$$= \frac{N\mu_0 R^2}{2}\left[\frac{-6(R^2+s^2/4)+30s^2/4}{(R^2+s^2/4)^{7/2}}\right] = 3N\mu_0 i R^2\,\frac{s^2-R^2}{(R^2+s^2/4)^{7/2}} .$$

Clearly, this is zero if $s = R$.

56. (a) By the right-hand rule, $\vec{B}$ points into the paper at P (see Fig. 30-6(c)). To find the magnitude of the field, we use Eq. 30-11 for each semicircle ($\phi = \pi$ rad), and use superposition to obtain the result:

$$B = \frac{\mu_0 i\pi}{4\pi a} + \frac{\mu_0 i\pi}{4\pi b} = \frac{\mu_0 i}{4}\left(\frac{1}{a}+\frac{1}{b}\right) .$$

(b) The direction of $\vec{\mu}$ is the same as the $\vec{B}$ found in part (a): into the paper. The enclosed area is $A = (\pi a^2 + \pi b^2)/2$ which means the magnetic dipole moment has magnitude

$$|\vec{\mu}| = \frac{\pi i}{2}\left(a^2+b^2\right) .$$

57. (a) We denote the large loop and small coil with subscripts 1 and 2, respectively.

$$B_1 = \frac{\mu_0 i_1}{2R_1} = \frac{(4\pi \times 10^{-7}\,\mathrm{T\cdot m/A})(15\,\mathrm{A})}{2(0.12\,\mathrm{m})} = 7.9 \times 10^{-5}\ \mathrm{T}\ .$$

(b) The torque has magnitude equal to

$$\begin{aligned}\tau &= \left|\vec{\mu}_2 \times \vec{B}_1\right| = \mu_2 B_1 \sin 90^\circ \\ &= N_2 i_2 A_2 B_1 = \pi N_2 i_2 r_2^2 B_1 \\ &= \pi(50)(1.3\,\mathrm{A})(0.82 \times 10^{-2}\,\mathrm{m})^2(7.9 \times 10^{-5}\,\mathrm{T}) = 1.1 \times 10^{-6}\ \mathrm{N\cdot m}\ .\end{aligned}$$

58. (a) The contribution to B_C from the (infinite) straight segment of the wire is

$$B_{C1} = \frac{\mu_0 i}{2\pi R}\ .$$

The contribution from the circular loop is

$$B_{C2} = \frac{\mu_0 i}{2R}\ .$$

Thus,

$$B_C = B_{C1} + B_{C2} = \frac{\mu_0 i}{2R}\left(1 + \frac{1}{\pi}\right)\ .$$

$\vec{B}_C$ points out of the page.

(b) Now $\vec{B}_{C1} \perp \vec{B}_{C2}$ so

$$B_C = \sqrt{B_{C1}^2 + B_{C2}^2} = \frac{\mu_0 i}{2R}\sqrt{1 + \frac{1}{\pi^2}}\ ,$$

and $\vec{B}_C$ points at an angle (relative to the plane of the paper) equal to

$$\tan^{-1}\left(\frac{B_{C1}}{B_{C2}}\right) = \tan^{-1}\left(\frac{1}{\pi}\right) = 18^\circ\ .$$

59. (a) For the circular path L of radius r concentric with the conductor

$$\oint_L \vec{B}\cdot d\vec{s} = 2\pi r B = \mu_0 i_{\mathrm{enc}} = \mu_0\, i\, \frac{\pi(r^2 - b^2)}{\pi(a^2 - b^2)}\ .$$

Thus,

$$B = \frac{\mu_0 i}{2\pi(a^2 - b^2)}\left(\frac{r^2 - b^2}{r}\right)\ .$$

(b) At $r = a$, the magnetic field strength is

$$\frac{\mu_0 i}{2\pi(a^2 - b^2)}\left(\frac{a^2 - b^2}{a}\right) = \frac{\mu_0 i}{2\pi a}\ .$$

At $r = b$, $B \propto r^2 - b^2 = 0$. Finally, for $b = 0$

$$B = \frac{\mu_0 i}{2\pi a^2}\frac{r^2}{r} = \frac{\mu_0 i r}{2\pi a^2}$$

which agrees with Eq. 30-22.

(c) The field is zero for $r < b$ and is equal to Eq. 30-19 for $r > a$, so this along with the result of part (a) provides a determination of B over the full range of values. The graph (with SI units understood) is shown below.

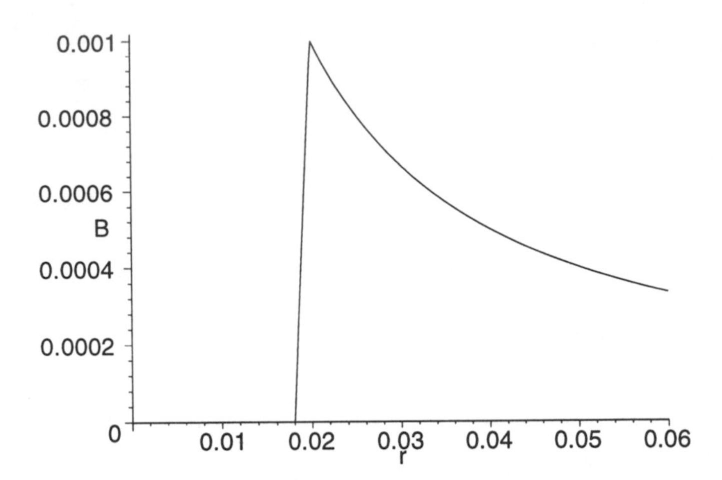

60. (a) Eq. 30-22 applies for $r < c$. Our sign choice is such that i is positive in the smaller cylinder and negative in the larger one.

$$B = \frac{\mu_0 i r}{2\pi c^2} \qquad \text{for} \quad r \le c \ .$$

(b) Eq. 30-19 applies in the region between the conductors.

$$B = \frac{\mu_0 i}{2\pi r} \qquad \text{for} \quad c \le r \le b \ .$$

(c) Within the larger conductor we have a superposition of the field due to the current in the inner conductor (still obeying Eq. 30-19) plus the field due to the (negative) current in the that part of the outer conductor at radius less than r (see part (a) of problem 59 for more details). The result is

$$B = \frac{\mu_0 i}{2\pi r} - \frac{\mu_0 i}{2\pi r}\left(\frac{r^2 - b^2}{a^2 - b^2}\right) \qquad \text{for} \quad b < r \le a \ .$$

If desired, this expression can be simplified to read

$$B = \frac{\mu_0 i}{2\pi r}\left(\frac{a^2 - r^2}{a^2 - b^2}\right) \ .$$

(d) Outside the coaxial cable, the net current enclosed is zero. So $B = 0$ for $r \ge a$.

(e) We test these expressions for one case. If $a \to \infty$ and $b \to \infty$ (such that $a > b$) then we have the situation described on page 696 of the textbook.

(f) Using SI units, the graph of the field is shown below:

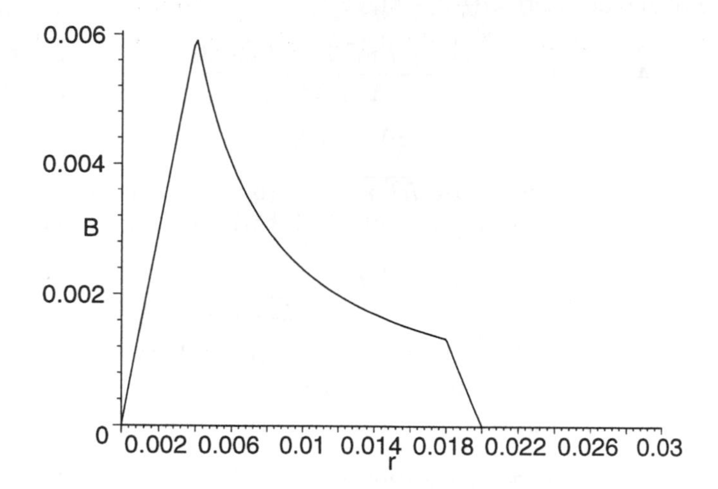

61. (a) We find the field by superposing the results of two semi-infinite wires (Eq. 30-9) and a semicircular arc (Eq. 30-11 with $\phi = \pi$ rad). The direction of $\vec{B}$ is out of the page, as can be checked by referring to Fig. 30-6(c). The magnitude of $\vec{B}$ at point a is therefore

$$B_a = 2\left(\frac{\mu_0 i}{4\pi R}\right) + \frac{\mu_0 i \pi}{4\pi R} = \frac{\mu_0 i}{2R}\left(\frac{1}{\pi} + \frac{1}{2}\right) .$$

With $i = 10$ A and $R = 0.0050$ m, we obtain $B_a = 1.0 \times 10^{-3}$ T. The direction of this field is out of the page, as Fig. 30-6(c) makes clear.

(b) The last remark in the problem statement implies that treating b as a point midway between two infinite wires is a good approximation. Thus, using Eq. 30-6,

$$B_b = 2\left(\frac{\mu_0 i}{2\pi R}\right) = 8.0 \times 10^{-4}\ \text{T} .$$

This field, too, points out of the page.

62. We use $B(x,\, y,\, z) = (\mu_0/4\pi) i\, \Delta\vec{s} \times \vec{r}/r^3$, where $\Delta\vec{s} = \Delta s\hat{\jmath}$ and $\vec{r} = x\hat{\imath} + y\hat{\jmath} + z\vec{k}$. Thus,

$$\vec{B}(x,\, y,\, z) = \left(\frac{\mu_0}{4\pi}\right) \frac{i\,\Delta s\hat{\jmath} \times (x\vec{i} + y\hat{\jmath} + z\vec{k})}{(x^2 + y^2 + z^2)^{3/2}} = \frac{\mu_0 i\,\Delta s\,(z\hat{\imath} - x\hat{k})}{4\pi(x^2 + y^s + z^2)^{3/2}} .$$

(a) The field on the z axis (at $z = 5.0$ m) is

$$\begin{aligned} \vec{B}(0,\, 0,\, 5.0\,\text{m}) &= \frac{(4\pi \times 10^{-7}\,\text{T}\cdot\text{m/A})(2.0\,\text{A})(3.0 \times 10^{-2}\,\text{m})(5.0\,\text{m})\hat{\imath}}{4\pi\,(0^2 + 0^2 + (5.0\,\text{m})^2)^{3/2}} \\ &= 2.4 \times 10^{-10}\ \text{T}\,\hat{\imath} . \end{aligned}$$

(b) $\vec{B}(0,\, 6.0\,\text{m},\, 0)$, since $x = z = 0$.

(c) The field in the xy plane, at $(x, y) = (7, 7)$, is

$$\begin{aligned} \vec{B}(7.0\,\text{m},\, 7.0\,\text{m},\, 0) &= \frac{(4\pi \times 10^{-7}\,\text{T}\cdot\text{m/A})(2.0\,\text{A})(3.0 \times 10^{-2}\,\text{m})(-7.0\,\text{m})\hat{k}}{4\pi\,((7.0\,\text{m})^2 + (7.0\,\text{m})^2 + 0^2)^{3/2}} \\ &= 4.3 \times 10^{-11}\ \text{T}\,\hat{k} . \end{aligned}$$

(d) The field in the xy plane, at $(x, y) = (-3, -4)$, is

$$\begin{aligned} \vec{B}(-3.0\,\text{m}, -4.0\,\text{m}, 0) &= \frac{(4\pi \times 10^{-7}\,\text{T}\cdot\text{m/A})(2.0\,\text{A})(3.0\times 10^{-2}\,\text{m})(3.0\,\text{m})\hat{\text{k}}}{4\pi\left((-3.0\,\text{m})^2 + (-4.0\,\text{m})^2 + 0^2\right)^{3/2}} \\ &= 1.4\times 10^{-10}\ \text{T}\ \hat{\text{k}}\ . \end{aligned}$$

63. (a) Eq. 30-19 applies for each wire, with $r = \sqrt{R^2 + (d/2)^2}$ (by the Pythagorean theorem). The vertical components of the fields cancel, and the two (identical) horizontal components add to yield the final result

$$B = 2\left(\frac{\mu_0 i}{2\pi r}\right)\left(\frac{d/2}{r}\right) = \frac{\mu_0 i d}{2\pi\left(R^2 + (d/2)^2\right)}$$

where $(d/2)/r$ is a trigonometric factor to select the horizontal component. It is clear that this is equivalent to the expression in the problem statement.

(b) Using the right-hand rule, we find both horizontal components point rightward.

64. (a) The difference between this and Sample Problem 6 is that the current in wire 2 is reversed form what is shown in Fig. 30-59(a). Thus, we replace $i \to -i$ in the expression for $B_2(x)$ and add the fields:

$$B_1(x) + B_2(x) = \frac{\mu_0 i}{2\pi(d+x)} + \frac{\mu_0(-i)}{2\pi(d-x)} = -\frac{\mu_0 i x}{\pi\left(d^2 - x^2\right)}$$

which is equivalent to the desired result.

(b) As remarked in that Sample Problem, this expression does not apply within the wires themselves. If we assume the wires have nearly zero thickness, then the expression applies over nearly all of the range $-0.02 < x < 0.02$ (with SI units understood). To be definite about this issue, we have picked a small wire radius (.005 m) and graphed the field over the range $-.0195 \le x \le 0.0195$.

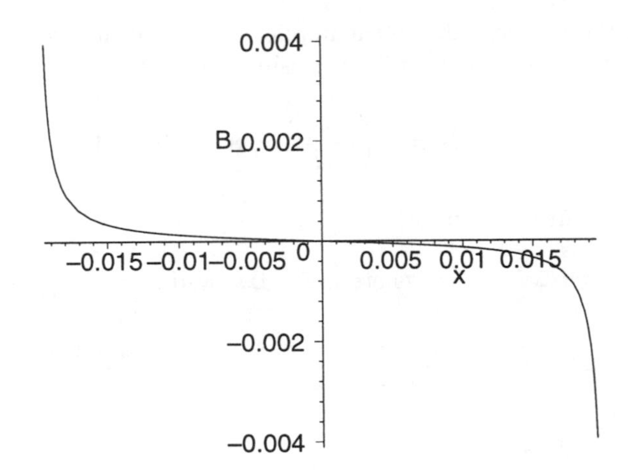

65. (a) All wires carry parallel currents and attract each other; thus, the "top" wire is pulled downward by the other two:

$$|\vec{F}| = \frac{\mu_0\, L(5.0\ \text{A})(3.2\ \text{A})}{2\pi(0.10\ \text{m})} + \frac{\mu_0\, L(5.0\ \text{A})(5.0\ \text{A})}{2\pi(0.20\ \text{m})}$$

where $L = 3.0$ m. Thus, $|\vec{F}| = 1.7\times 10^{-4}$ N.

(b) Now, the "top" wire is pushed upward by the center wire and pulled downward by the bottom wire:

$$|\vec{F}| = \frac{\mu_0\, L(5.0\ \text{A})(3.2\ \text{A})}{2\pi(0.10\ \text{m})} - \frac{\mu_0\, L(5.0\ \text{A})(5.0\ \text{A})}{2\pi(0.20\ \text{m})}$$

so that $|\vec{F}| = 2.1\times 10^{-5}$ N.

66. With cylindrical symmetry, we have, external to the conductors,

$$|\vec{B}| = \frac{\mu_0\, i_{\rm enc}}{2\pi\, r}$$

which produces $i_{\rm enc} = 25$ mA from the given information. Therefore, the thin wire must carry 5 mA in a direction opposite to the 30 mA carried by the thin conducting surface.

67. The area enclosed by the loop L is $A = \frac{1}{2}(4d)(3d) = 6d^2$. Thus

$$\begin{aligned}\oint_c \vec{B}\cdot d\vec{s} &= \mu_0 i = \mu_0 jA \\ &= (4\pi\times10^{-7}\,\mathrm{T\cdot m/A})(15\,\mathrm{A/m^2})(6)(0.20\,\mathrm{m})^2 = 4.5\times10^{-6}\ \mathrm{T{\cdot}m}\ .\end{aligned}$$

68. We refer to the center of the circle (where we are evaluating $\vec{B}$) as C. Recalling the *straight sections* discussion in Sample Problem 30-1, we see that the current in the straight segments which are colinear with C do not contribute to the field there. Eq. 30-11 (with $\phi = \pi/2$ rad) and the right-hand rule indicates that the currents in the two arcs contribute

$$\frac{\mu_0 i(\pi/2)}{4\pi R} - \frac{\mu_0 i(\pi/2)}{4\pi R} = 0$$

to the field at C. Thus, the non-zero contributions come from those straight-segments which are not colinear with C. There are two of these "semi-infinite" segments, one a vertical distance R above C and the other a horizontal distance R to the left of C. Both contribute fields pointing out of the page (see Fig. 30-6(c)). Since the magnitudes of the two contributions (governed by Eq. 30-9) add, then the result is

$$B = 2\left(\frac{\mu_0 i}{4\pi R}\right) = \frac{\mu_0 i}{2\pi R}$$

exactly what one would expect from a single infinite straight wire (see Eq. 30-6). For such a wire to produce such a field (out of the page) with a leftward current requires that the point of evaluating the field be below the wire (again, see Fig. 30-6(c)).

69. Since the radius is $R = 0.0013$ m, then the $i = 50$ A produces

$$B = \frac{\mu_0 i}{2\pi R} = 0.0077\ \mathrm{T}$$

at the edge of the wire. The three equations, Eq. 30-6, Eq. 30-19 and Eq. 30-22, agree at this point.

70. We note that the distance from each wire to P is $r = d/\sqrt{2} = 0.071$ m. In both parts, the current is $i = 100$ A.

(a) With the currents parallel, application of the right-hand rule (to determine each of their contributions to the field at P) reveals that the vertical components cancel and the horizontal components add – yielding the result:

$$B = 2\left(\frac{\mu_0 i}{2\pi r}\right)\cos 45^\circ = 4.0\times10^{-4}\ \mathrm{T}\ .$$

and directed leftward in the figure.

(b) Now, with the currents antiparallel, application of the right-hand rule shows that the horizontal components cancel and the vertical components add. Thus,

$$B = 2\left(\frac{\mu_0 i}{2\pi r}\right)\sin 45^\circ = 4.0\times10^{-4}\ \mathrm{T}\ .$$

and directed upward in the figure.

71. (a) As illustrated in Sample Problem 30-1, the radial segments do not contribute to $\vec{B}_P$ and the arc-segments contribute according to Eq. 30-11 (with angle in radians). If $\hat{k}$ designates the direction "out of the page" then

$$\vec{B} = \frac{\mu_0(0.40\text{ A})(\pi\text{ rad})}{4\pi(0.050\text{ m})}\hat{k} - \frac{\mu_0(0.80\text{ A})\left(\frac{2\pi}{3}\text{ rad}\right)}{4\pi(0.040\text{ m})}\hat{k}$$

which yields $\vec{B} = -1.7 \times 10^{-6}\,\hat{k}$ T.

(b) Now we have

$$\vec{B} = -\frac{\mu_0(0.40\text{ A})(\pi\text{ rad})}{4\pi(0.050\text{ m})}\hat{k} - \frac{\mu_0(0.80\text{ A})\left(\frac{2\pi}{3}\text{ rad}\right)}{4\pi(0.040\text{ m})}\hat{k}$$

which yields $\vec{B} = -6.7 \times 10^{-6}\,\hat{k}$ T.

72. (a) We designate the wire along $y = r_A = 0.100$ m wire A and the wire along $y = r_B = 0.050$ m wire B. Using Eq. 30-6, we have

$$\begin{aligned}\vec{B}_{\text{net}} &= \vec{B}_A + \vec{B}_B \\ &= -\frac{\mu_0\, i_A}{2\pi\, r_A}\hat{k} - \frac{\mu_0\, i_B}{2\pi\, r_B}\hat{k}\end{aligned}$$

which yields $\vec{B}_{\text{net}} = 52.0 \times 10^{-6}\hat{k}$ T.

(b) This will occur for some value $r_B < y < r_A$ such that

$$\frac{\mu_0\, i_A}{2\pi\,(r_A - y)} = \frac{\mu_0\, i_B}{2\pi\,(y - r_B)}\ .$$

Solving, we find $y = 13/160 \approx 0.081$ m.

(c) We eliminate the $y < r_B$ possibility due to wire B carrying the larger current. We expect a solution in the region $y > r_A$ where

$$\frac{\mu_0\, i_A}{2\pi\,(y - r_A)} = \frac{\mu_0\, i_B}{2\pi\,(y - r_B)}\ .$$

Solving, we find $y = 7/40 \approx 0.018$ m.

73. (a) The field in this region is entirely due to the long wire (with, presumably, negligible thickness). Using Eq. 30-19,

$$|\vec{B}| = \frac{\mu_0\, i_w}{2\pi r} = 4.8 \times 10^{-3}\ \text{T}$$

where $i_w = 24$ A and $r = 0.0010$ m.

(b) Now the field consists of two contributions (which are antiparallel) – from the wire (Eq. 30-19) and from a portion of the conductor (Eq. 30-22 modified for annular area):

$$\begin{aligned}|\vec{B}| &= \frac{\mu_0\, i_w}{2\pi r} - \frac{\mu_0\, i_{\text{enc}}}{2\pi r} \\ &= \frac{\mu_0\, i_w}{2\pi r} - \frac{\mu_0\, i_c}{2\pi r}\left(\frac{\pi r^2 - \pi R_i^2}{\pi R_o^2 - \pi R_i^2}\right)\end{aligned}$$

where $r = 0.0030$ m, $R_i = 0.0020$ m, $R_o = 0.0040$ m and $i_c = 24$ A. Thus, we find $|\vec{B}| = 9.3 \times 10^{-4}$ T.

(c) Now, in the external region, the individual fields from the two conductors cancel completely (since $i_c = i_w$): $\vec{B} = 0$.

74. In this case $L = 2\pi r$ is roughly the length of the toroid so

$$B = \mu_0 i_0 \left(\frac{N}{2\pi r}\right) = \mu_0 n i_0 \ .$$

This result is expected, since from the perspective of a point inside the toroid the portion of the toroid in the vicinity of the point resembles part of a long solenoid.

75. We take the current ($i = 50$ A) to flow in the $+x$ direction, and the electron to be at a point P which is $r = 0.050$ m above the wire (where "up" is the $+y$ direction). Thus, the field produced by the current points in the $+z$ direction at P. Then, combining Eq. 30-6 with Eq. 29-2, we obtain $\vec{F}_e = (-e\mu_0 i/2\pi r)(\vec{v} \times \hat{\mathrm{k}})$.

(a) The electron is moving down: $\vec{v} = -v\,\hat{\mathrm{j}}$ (where $v = 1.0 \times 10^7$ m/s is the speed) so

$$\vec{F}_e = \frac{-e\mu_0 iv}{2\pi r}\left(-\hat{\mathrm{i}}\right) = 3.2 \times 10^{-16}\ \mathrm{N}\,\hat{\mathrm{i}}\ .$$

(b) In this case, the electron in the same direction as the current: $\vec{v} = v\,\hat{\mathrm{i}}$ so

$$\vec{F}_e = \frac{-e\mu_0 iv}{2\pi r}\left(-\hat{\mathrm{j}}\right) = 3.2 \times 10^{-16}\ \mathrm{N}\,\hat{\mathrm{j}}\ .$$

(c) Now, $\vec{v} = \pm v\hat{\mathrm{k}}$ so $\vec{F}_e \propto \hat{\mathrm{k}} \times \hat{\mathrm{k}} = 0$.

76. Eq. 30-6 gives

$$i = \frac{2\pi RB}{\mu_0} = \frac{2\pi(0.880\,\mathrm{m})(7.30 \times 10^{-6}\,\mathrm{T})}{4\pi \times 10^{-7}\,\mathrm{T\cdot m/A}} = 32.1\ \mathrm{A}\ .$$

77. For $x > 20$ mm, the field due i_2 is downward and thus subtracts from B_1 and is entirely consistent with the given expression for B_2 (note that it becomes negative when $x > d$). Similarly, for $x < -20$ mm, the field due to i_1 is downward and subtracts from B_2 (which is positive and points upward for all $x < d$). This again is consistent with the expression for B_1 which is seen to become negative for x less than $-d$ (that is, x negative and $|x| > |d|$). We conclude that the given expressions are valid over the whole of the x axis, and their answer (Eq. 30-33) holds for all x (other than at the locations of the wires themselves, where it becomes problematic, as discussed in the Sample Problem).

78. By the right-hand rule, the magnetic field $\vec{B}_1$(evaluated at a) produced by wire 1 (the wire at bottom left) is at $\phi = 150°$ (measured counterclockwise from the $+x$ axis, in the xy plane), and the field produced by wire 2 (the wire at bottom right) is at $\phi = 210°$. By symmetry ($\vec{B}_1 = \vec{B}_2$) we observe that only the x-components survive, yielding

$$\vec{B}_1 + \vec{B}_2 = 2\,\frac{\mu_0\, i}{2\pi\,\ell}\,\cos 150°\,\hat{\mathrm{i}} = -3.46 \times 10^{-5}\,\hat{\mathrm{i}}\ \mathrm{T}$$

where $i = 10$ A, $\ell = 0.10$ m, and Eq. 30-6 has been used. To cancel this, wire b must carry current into the page (that is, the $-\hat{\mathrm{k}}$ direction) of value

$$i_b = \left(3.46 \times 10^{-5}\right)\frac{2\pi\, r}{\mu_0} = 15\ \mathrm{A}$$

where $r = \sqrt{3}\,\ell/2 = 0.087$ m and and Eq. 30-6 has again been used.

79. Using Eq. 30-22 and Eq. 30-19, we have

$$|\vec{B}_1| = \left(\frac{\mu_0\, i}{2\pi\, R^2}\right) r_1$$

$$|\vec{B}_2| = \frac{\mu_0\, i}{2\pi\, r_2}$$

where $r_1 = 0.0040$ m, $|\vec{B}_1| = 2.8 \times 10^{-4}$ T, $r_2 = 0.010$ m and $|\vec{B}_2| = 2.0 \times 10^{-4}$ T. Point 2 is known to be external to the wire since $|\vec{B}_2| < |\vec{B}_1|$. From the second equation, we find $i = 10$ A. Plugging this into the first equation yields $R = 5.3 \times 10^{-3}$ m.

80. Using a magnifying glass, we see that all but i_2 are directed into the page. Wire 3 is therefore attracted to all but wire 2. Letting $d = 0.50$ m, we find the net force (per meter length) using Eq. 30-15, with positive indicated a rightward force:

$$\frac{|\vec{F}|}{\ell} = \frac{\mu_0 \, i_3}{2\pi} \left(-\frac{i_1}{2d} + \frac{i_2}{d} + \frac{i_4}{d} + \frac{i_5}{2d} \right)$$

which yields $|\vec{F}|/\ell = 8.0 \times 10^{-7}$ N/m.

Chapter 31

1. The magnetic field is normal to the plane of the loop and is uniform over the loop. Thus at any instant the magnetic flux through the loop is given by $\Phi_B = AB = \pi r^2 B$, where $A = \pi r^2$ is the area of the loop. According to Faraday's law the magnitude of the emf in the loop is

$$\mathcal{E} = \frac{d\Phi_B}{dt} = \pi r^2 \frac{dB}{dt} = \pi(0.055\,\text{m})^2(0.16\,\text{T/s}) = 1.5 \times 10^{-3}\ \text{V}\ .$$

2. The induced emf is

$$\begin{aligned}
\mathcal{E} &= -\frac{d\Phi_B}{dt} = -\frac{d(BA)}{dt} = -A\frac{dB}{dt} \\
&= -A\frac{d}{dt}(\mu_0 in) = -A\mu_0 n\frac{d}{dt}(i_0 \sin\omega t) \\
&= -A\mu_0 n i_0 \omega \cos\omega t\ .
\end{aligned}$$

3. (a)

$$|\mathcal{E}| = \left|\frac{d\Phi_B}{dt}\right| = \frac{d}{dt}(6.0t^2 + 7.0t) = 12t + 7.0 = 12(2.0) + 7.0 = 31\ \text{mV}\ .$$

   (b) Appealing to Lenz's law (especially Fig. 31-5(a)) we see that the current flow in the loop is clockwise. Thus, the current is from right to left through R.

4. (a) We use $\mathcal{E} = -d\Phi_B/dt = -\pi r^2 dB/dt$. For $0 < t < 2.0\,\text{s}$:

$$\mathcal{E} = -\pi r^2\frac{dB}{dt} = -\pi(0.12\,\text{m})^2\left(\frac{0.5\,\text{T}}{2.0\,\text{s}}\right) = -1.1 \times 10^{-2}\ \text{V}\ .$$

   (b) $2.0\,\text{s} < t < 4.0\,\text{s}:\ \mathcal{E} \propto dB/dt = 0$.

   (c) $4.0\,\text{s} < t < 6.0\,\text{s}:$

$$\mathcal{E} = -\pi r^2\frac{dB}{dt} = -\pi(0.12\,\text{m})^2\left(\frac{-0.5\,\text{T}}{6.0\,\text{s} - 4.0\,\text{s}}\right) = 1.1 \times 10^{-2}\ \text{V}\ .$$

5. (a) Table 27-1 gives the resistivity of copper. Thus,

$$R = \rho\frac{L}{A} = (1.68 \times 10^{-8}\,\Omega\cdot\text{m})\left[\frac{\pi(0.10\,\text{m})}{\pi(2.5 \times 10^{-3})^2/4}\right] = 1.1 \times 10^{-3}\,\Omega\ .$$

   (b) We use $i = |\mathcal{E}|/R = |d\Phi_B/dt|/R = (\pi r^2/R)|dB/dt|$. Thus

$$\left|\frac{dB}{dt}\right| = \frac{iR}{\pi r^2} = \frac{(10\,\text{A})(1.1 \times 10^{-3}\,\Omega)}{\pi(0.05\,\text{m})^2} = 1.4\ \text{T/s}\ .$$

6. (a) Following Sample Problem 31-1, we have

$$\Phi_B = \mu_0 i n A \qquad \text{where} \quad A = \frac{\pi d^2}{4}$$

with $i = 3t + t^2$ (SI units and 2 significant figures understood). The magnitude of the induced emf is therefore

$$\mathcal{E} = N\frac{d\Phi_B}{dt} \approx 0.0012(3 + 2t)$$

where we have used the values specified in Sample Problem 31-1 for all quantities except the current. The plot is shown below.

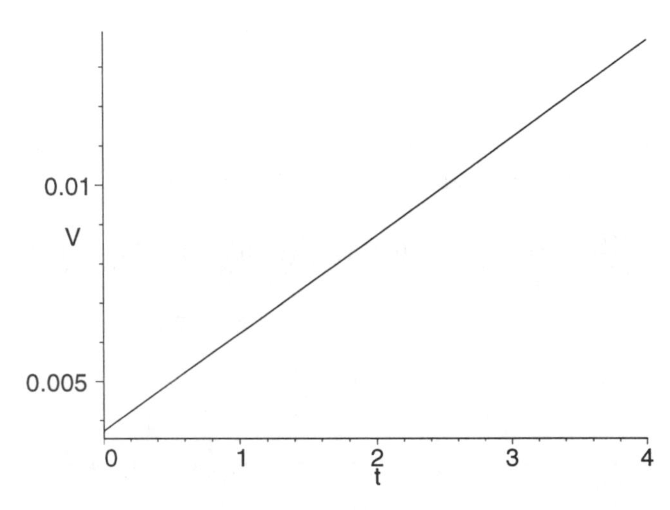

(b) Using Ohm's law, the induced current is

$$i\Big|_{t=2.0\,\text{s}} = \frac{\mathcal{E}\big|_{t=2.0\,\text{s}}}{R} = \frac{0.0087\,\text{V}}{0.15\,\Omega} = 0.058 \text{ A} .$$

7. The primary difference between this and the situation described in Sample Problem 31-1 is in the quantity A. The area through which there is magnetic flux is not the area of the short coil, in this case, but is the area of the solenoid (there is no field outside an ideal solenoid). Actually, because of the current (which we calculate here) in the short coil, there is a very small amount of field outside the solenoid (caused by that current) – but it may be disregarded in this calculation. The values are as indicated in Sample Problem 31-1 except that $A = \pi D^2/4$ (where $D = 0.032$ m) and $N = 120$ for the short coil. Thus, we find $\Phi_{B,i} = 3.3 \times 10^{-5}$ Wb, and the magnitude of the induced emf is 0.16 V. Ohm's law then yields $0.16\,\text{V}/5.3\,\Omega = 0.030$ A.

8. Using Faraday's law, the induced emf is

$$\begin{aligned} \mathcal{E} &= -\frac{d\Phi_B}{dt} = -\frac{d(BA)}{dt} = -B\frac{dA}{dt} = -B\frac{d(\pi r^2)}{dt} = -2\pi r B\frac{dr}{dt} \\ &= -2\pi(0.12\,\text{m})(0.800\,\text{T})(-0.750\,\text{m/s}) = 0.452 \text{ V} . \end{aligned}$$

9. (a) In the region of the smaller loop the magnetic field produced by the larger loop may be taken to be uniform and equal to its value at the center of the smaller loop, on the axis. Eq. 30-29, with $z = x$ (taken to be much greater than R), gives

$$\vec{B} = \frac{\mu_0 i R^2}{2x^3}\,\hat{\text{i}}$$

where the $+x$ direction is upward in Fig. 31-36. The magnetic flux through the smaller loop is, to a good approximation, the product of this field and the area (πr^2) of the smaller loop:

$$\Phi_B = \frac{\pi\mu_0 i r^2 R^2}{2x^3} .$$

(b) The emf is given by Faraday's law:

$$\mathcal{E} = -\frac{d\Phi_B}{dt} = -\left(\frac{\pi\mu_0 i r^2 R^2}{2}\right)\frac{d}{dt}\left(\frac{1}{x^3}\right) = -\left(\frac{\pi\mu_0 i r^2 R^2}{2}\right)\left(-\frac{3}{x^4}\frac{dx}{dt}\right) = \frac{3\pi\mu_0 i r^2 R^2 v}{2x^4} .$$

(c) As the smaller loop moves upward, the flux through it decreases, and we have situation like that shown in Fig. 31-5(b). The induced current will be directed so as to produce a magnetic field that is upward through the smaller loop, in the same direction as the field of the larger loop. It will be counterclockwise as viewed from above, in the same direction as the current in the larger loop.

10. The flux $\Phi_B = BA\cos\theta$ does not change as the loop is rotated. Faraday's law only leads to a nonzero induced emf when the flux is changing, so the result in this instance is 0.

11. (a) Ohm's law combines with Faraday's law to give $i = -\frac{N}{R}\frac{d\Phi_B}{dt}$ where R is the resistance of the coil. In this case, $N = 1$ (it is a single loop), and we integrate to find the charge:

$$\begin{aligned}\int_0^t i\,dt &= -\frac{1}{R}\int_0^t \frac{d\Phi_B}{dt}\,dt \\ q(t) &= -\frac{1}{R}\left(\Phi_B(t) - \Phi_B(0)\right)\end{aligned}$$

which is equivalent to the expression shown in the problem statement. We have used little more than the fundamental theorem of calculus; no particular assumptions have been made about how the integrations should be performed. The result is independent of the way $\vec{B}$ has changed.

(b) If the current is identically zero for over the whole range $0 \to t$ then certainly the left-hand side of our computation, above, gives zero. But the same result can come from the current being in one direction for, say, $0 \to \frac{t}{2}$ and then in the opposite direction for $\frac{t}{2} \to t$ in such a way that $\int_0^t i\,dt = 0$. So a vanishing integral does not necessarily mean the integrand itself is identically zero.

12. (a) Eq. 30-12 gives the field at the center of the large loop with $R = 1.00$ m and current $i(t)$. This is approximately the field throughout the area ($A = 2.00 \times 10^{-4}\ \text{m}^2$) enclosed by the small loop. Thus, with $B = \mu_0 i/2R$ and $i(t) = i_0 + kt$ (where $i_0 = 200$ A and $k = (-200\ \text{A} - 200\ \text{A})/1.00\ \text{s} = -400\ \text{A/s}$), we find

$$\begin{aligned}B\big|_{t=0} &= \frac{\mu_0 i_0}{2R} = \frac{(4\pi\times10^{-7}\ \text{H/m})(200\ \text{A})}{2(1.00\ \text{m})} = 1.26\times10^{-4}\ \text{T} , \\ B\big|_{t=0.500\ \text{s}} &= \frac{(4\pi\times10^{-7}\ \text{H/m})[200\ \text{A} - (400\ \text{A/s})(0.500\ \text{s})]}{2(1.00\ \text{m})} = 0 , \\ B\big|_{t=1.00\ \text{s}} &= \frac{(4\pi\times10^{-7}\ \text{H/m})[200\ \text{A} - (400\ \text{A/s})(1.00\ \text{s})]}{2(1.00\ \text{m})} = -1.26\times10^{-4}\ \text{T} .\end{aligned}$$

(b) Let the area of the small loop be a. Then $\Phi_B = Ba$, and Faraday's law yields

$$\begin{aligned}\mathcal{E} &= -\frac{d\Phi_B}{dt} = -\frac{d(Ba)}{dt} = -a\frac{dB}{dt} = -a\left(\frac{\Delta B}{\Delta t}\right) \\ &= -(2.00\times10^{-4}\ \text{m}^2)\left(\frac{-1.26\times10^{-4}\ \text{T} - 1.26\times10^{-4}\ \text{T}}{1.00\ \text{s}}\right) = 5.04\times10^{-8}\ \text{V} .\end{aligned}$$

13. From the result of the problem 11,

$$q(t) = \frac{1}{R}[\Phi_B(0) - \Phi_B(t)] = \frac{A}{R}[B(0) - B(t)]$$
$$= \frac{1.20 \times 10^{-3}\,\text{m}^2}{13.0\,\Omega}[1.60\,\text{T} - (-1.60\,\text{T})] = 2.95 \times 10^{-2}\ \text{C}\ .$$

14. We note that 1 gauss $= 10^{-4}$ T. Adapting the result of the problem 11,

$$q(t) = \frac{N}{R}[BA\cos 20^\circ - (-BA\cos 20^\circ)] = \frac{2NBA\cos 20^\circ}{R}$$
$$= \frac{2(1000)(0.590 \times 10^{-4}\,\text{T})\pi(0.100\,\text{m})^2(\cos 20^\circ)}{85.0\,\Omega + 140\,\Omega} = 1.55 \times 10^{-5}\ \text{C}\ .$$

Note that the axis of the coil is at 20°, not 70°, from the magnetic field of the Earth.

15. (a) Let L be the length of a side of the square circuit. Then the magnetic flux through the circuit is $\Phi_B = L^2B/2$, and the induced emf is

$$\mathcal{E}_i = -\frac{d\Phi_B}{dt} = -\frac{L^2}{2}\frac{dB}{dt}\ .$$

Now $B = 0.042 - 0.870t$ and $dB/dt = -0.870\,\text{T/s}$. Thus,

$$\mathcal{E}_i = \frac{(2.00\,\text{m})^2}{2}(0.870\,\text{T/s}) = 1.74\ \text{V}\ .$$

The magnetic field is out of the page and decreasing so the induced emf is counterclockwise around the circuit, in the same direction as the emf of the battery. The total emf is $\mathcal{E} + \mathcal{E}_i = 20.0\,\text{V} + 1.74\,\text{V} = 21.7\,\text{V}$.

(b) The current is in the sense of the total emf (counterclockwise).

16. (a) Since $\vec{B} = B\hat{\imath}$ uniformly, then only the area "projected" onto the yz plane will contribute to the flux (due to the scalar [dot] product). This "projected" area corresponds to one-fourth of a circle. Thus, the magnetic flux Φ_B through the loop is

$$\Phi_B = \int \vec{B}\cdot d\vec{A} = \frac{1}{4}\pi r^2 B\ .$$

Thus,

$$|\mathcal{E}| = \left|\frac{d\Phi_B}{dt}\right| = \left|\frac{d}{dt}\left(\frac{1}{4}\pi r^2 B\right)\right| = \frac{\pi r^2}{4}\left|\frac{dB}{dt}\right|$$
$$= \frac{1}{4}\pi(0.10\,\text{m})^2(3.0 \times 10^{-3}\,\text{T/s}) = 2.4 \times 10^{-5}\ \text{V}\ .$$

(b) We have a situation analogous to that shown in Fig. 31-5(a). Thus, the current in segment bc flows from c to b (following Lenz's law).

17. (a) It should be emphasized that the result, given in terms of $\sin(2\pi ft)$, could as easily be given in terms of $\cos(2\pi ft)$ or even $\cos(2\pi ft + \phi)$ where ϕ is a phase constant as discussed in Chapter 16. The angular position θ of the rotating coil is measured from some reference line (or plane), and which line one chooses will affect whether the magnetic flux should be written as $BA\cos\theta$, $BA\sin\theta$ or $BA\cos(\theta + \phi)$. Here our choice is such that $\Phi_B = BA\cos\theta$. Since the coil is rotating steadily, θ increases linearly with time. Thus, $\theta = \omega t$ (equivalent to $\theta = 2\pi ft$) if θ is understood to be in

radians (and ω would be the angular velocity). Since the area of the rectangular coil is $A = ab$, Faraday's law leads to

$$\mathcal{E} = -N\frac{d(BA\cos\theta)}{dt} = -NBA\frac{d\cos(2\pi ft)}{dt} = NBab2\pi f\sin(2\pi ft)$$

which is the desired result, shown in the problem statement. The second way this is written ($\mathcal{E}_0\sin(2\pi ft)$) is meant to emphasize that the voltage output is sinusoidal (in its time dependence) and has an amplitude of $\mathcal{E}_0 = 2\pi fNabB$.

(b) We solve $\mathcal{E}_0 = 150\,\text{V} = 2\pi fNabB$ when $f = 60.0$ rev/s and $B = 0.500$ T. The three unknowns are N, a, and b which occur in a product; thus, we obtain $Nab = 0.796\ \text{m}^2$. This means, for instance, that if we wanted the coil to have a square shape and consist of 50 turns, then the side length of the square would be $a = b = 0.126$ m.

18. (a) The rotational frequency (in revolutions per second) is identical to the time-dependent voltage frequency (in cycles per second, or Hertz). This conclusion should not be considered obvious, and the calculation shown in part (b) should serve to reinforce it.

(b) First, we define angle relative to the plane of Fig. 31-41, such that the semicircular wire is in the $\theta = 0$ position and a quarter of a period (of revolution) later it will be in the $\theta = \pi/2$ position (where its midpoint will reach a distance of a above the plane of the figure). At the moment it is in the $\theta = \pi/2$ position, the area enclosed by the "circuit" will appear to us (as we look down at the figure) to that of a simple rectangle (call this area A_0 which is the area it will again appear to enclose when the wire is in the $\theta = 3\pi/2$ position). Since the area of the semicircle is $\pi a^2/2$ then the area (as it appears to us) enclosed by the circuit, as a function of our angle θ, is

$$A = A_0 + \frac{\pi a^2}{2}\cos\theta$$

where (since θ is increasing at a steady rate) the angle depends linearly on time, which we can write either as $\theta = \omega t$ or $\theta = 2\pi ft$ if we take $t = 0$ to be a moment when the arc is in the $\theta = 0$ position. Since $\vec{B}$ is uniform (in space) and constant (in time), Faraday's law leads to

$$\mathcal{E} = -\frac{d\Phi_B}{dt} = -B\frac{dA}{dt} = -B\frac{d\left(A_0 + \frac{\pi a^2}{2}\cos\theta\right)}{dt} = -B\frac{\pi a^2}{2}\frac{d\cos(2\pi ft)}{dt}$$

which yields $\mathcal{E} = B\pi^2a^2f\sin(2\pi ft)$. This (due to the sinusoidal dependence) reinforces the conclusion in part (a) and also (due to the factors in front of the sine) provides the voltage amplitude: $\mathcal{E}_{\text{max}} = B\pi^2a^2f$.

19. First we write $\Phi_B = BA\cos\theta$. We note that the angular position θ of the rotating coil is measured from some reference line or plane, and we are implicitly making such a choice by writing the magnetic flux as $BA\cos\theta$ (as opposed to, say, $BA\sin\theta$). Since the coil is rotating steadily, θ increases linearly with time. Thus, $\theta = \omega t$ if θ is understood to be in radians (here, $\omega = 2\pi f$ is the angular velocity of the coil in radians per second, and $f = 1000\,\text{rev/min} \approx 16.7$ rev/s is the frequency). Since the area of the rectangular coil is $A = 0.500 \times 0.300 = 0.150\ \text{m}^2$, Faraday's law leads to

$$\mathcal{E} = -N\frac{d(BA\cos\theta)}{dt} = -NBA\frac{d\cos(2\pi ft)}{dt} = NBA2\pi f\sin(2\pi ft)$$

which means it has a voltage amplitude of

$$\mathcal{E}_{\text{max}} = 2\pi fNAB = 2\pi(16.7\,\text{rev/s})(100\,\text{turns})(0.15\,\text{m}^2)(3.5\,\text{T}) = 5.50 \times 10^3\ \text{V}\ .$$

20. The field (due to the current in the straight wire) is out-of-the-page in the upper half of the circle and is into the page in the lower half of the circle, producing zero net flux, at any time. There is no induced current in the circle.

21. Consider a (thin) strip of area of height dy and width $\ell = 0.020$ m. The strip is located at some $0 < y < \ell$. The element of flux through the strip is

$$d\Phi_B = B\,dA = (4t^2y)(\ell\,dy)$$

where SI units (and 2 significant figures) are understood. To find the total flux through the square loop, we integrate:

$$\Phi_B = \int d\Phi_B = \int_0^\ell (4t^2y\ell)\,dy = 2t^2\ell^3 .$$

Thus, Faraday's law yields

$$|\mathcal{E}| = \left|\frac{d\Phi_B}{dt}\right| = 4t\ell^3 .$$

At $t = 2.5$ s, we find the magnitude of the induced emf is 8.0×10^{-5} V. Its "direction" (or "sense") is clockwise, by Lenz's law.

22. (a) First, we observe that a large portion of the figure contributes flux which "cancels out." The field (due to the current in the long straight wire) through the part of the rectangle above the wire is out of the page (by the right-hand rule) and below the wire it is into the page. Thus, since the height of the part above the wire is $b - a$, then a strip below the wire (where the strip borders the long wire, and extends a distance $b - a$ away from it) has exactly the equal-but-opposite flux which cancels the contribution from the part above the wire. Thus, we obtain the non-zero contributions to the flux:

$$\Phi_B = \int B\,dA = \int_{b-a}^{a} \left(\frac{\mu_0 i}{2\pi r}\right)(b\,dr) = \frac{\mu_0 i b}{2\pi}\ln\left(\frac{a}{b-a}\right) .$$

Faraday's law, then, (with SI units and 3 significant figures understood) leads to

$$\begin{aligned}
\mathcal{E} &= -\frac{d\Phi_B}{dt} = -\frac{d}{dt}\left[\frac{\mu_0 i b}{2\pi}\ln\left(\frac{a}{b-a}\right)\right] \\
&= -\frac{\mu_0 b}{2\pi}\ln\left(\frac{a}{b-a}\right)\frac{di}{dt} = -\frac{\mu_0 b}{2\pi}\ln\left(\frac{a}{b-a}\right)\frac{d}{dt}\left(\frac{9}{2}t^2 - 10t\right) \\
&= \frac{-\mu_0 b(9t-10)}{2\pi}\ln\left(\frac{a}{b-a}\right) .
\end{aligned}$$

With $a = 0.120$ m and $b = 0.160$ m, then, at $t = 3.00$ s, the magnitude of the emf induced in the rectangular loop is

$$|\mathcal{E}| = \frac{(4\pi \times 10^{-7})(0.16)(9(3) - 10)}{2\pi}\ln\left(\frac{0.12}{0.16 - 0.12}\right) = 5.98 \times 10^{-7}\text{ V} .$$

(b) We note that $\frac{di}{dt} > 0$ at $t = 3$ s. The situation is roughly analogous to that shown in Fig. 31-5(c). From Lenz's law, then, the induced emf (hence, the induced current) in the loop is counterclockwise.

23. (a) We refer to the (very large) wire length as L and seek to compute the flux per meter: Φ_B/L. Using the right-hand rule discussed in Chapter 30, we see that the net field in the region between the axes of antiparallel currents is the addition of the magnitudes of their individual fields, as given by Eq. 30-19 and Eq. 30-22. There is an evident reflection symmetry in the problem, where the plane of symmetry is midway between the two wires (at what we will call $x = \ell/2$, where $\ell = 20\,\text{mm} = 0.020\,\text{m}$); the net field at any point $0 < x < \ell/2$ is the same at its "mirror image" point $\ell - x$. The central axis of one of the wires passes through the origin, and that of the other passes through $x = \ell$. We make use of the symmetry by integrating over $0 < x < \ell/2$ and then multiplying by 2:

$$\Phi_B = 2\int_0^{\ell/2} B\,dA = 2\int_0^{d/2} B\,(L\,dx) + 2\int_{d/2}^{\ell/2} B\,(L\,dx)$$

where $d = 0.0025$ m is the diameter of each wire. We will use $R = d/2$, and r instead of x in the following steps. Thus, using the equations from Ch. 30 referred to above, we find

$$\begin{aligned}\frac{\Phi_B}{L} &= 2\int_0^R \left(\frac{\mu_0 i}{2\pi R^2}r + \frac{\mu_0 i}{2\pi(\ell - r)}\right) dr + 2\int_R^{\ell/2}\left(\frac{\mu_0 i}{2\pi r} + \frac{\mu_0 i}{2\pi(\ell - r)}\right) dr \\ &= \frac{\mu_0 i}{2\pi}\left(1 - 2\ln\left(\frac{\ell - R}{\ell}\right)\right) + \frac{\mu_0 i}{\pi}\ln\left(\frac{\ell - R}{R}\right) \\ &= 0.23 \times 10^{-5}\ \text{T·m} + 1.08 \times 10^{-5}\ \text{T·m}\end{aligned}$$

which yields $\Phi_B/L = 1.3 \times 10^{-5}$ T·m or 1.3×10^{-5} Wb/m.

(b) The flux (per meter) existing within the regions of space occupied by one or the other wires was computed above to be 0.23×10^{-5} T·m. Thus,

$$\frac{0.23 \times 10^{-5}\ \text{T·m}}{1.3 \times 10^{-5}\ \text{T·m}} = 0.17 = 17\% .$$

(c) What was described in part (a) as a symmetry plane at $x = \ell/2$ is now (in the case of parallel currents) a plane of vanishing field (the fields subtract from each other in the region between them, as the right-hand rule shows). The flux in the $0 < x < \ell/2$ region is now of opposite sign of the flux in the $\ell/2 < x < \ell$ region which causes the total flux (or, in this case, flux per meter) to be zero.

24. (a) We assume the flux is entirely due to the field generated by the long straight wire (which is given by Eq. 30-19). We integrate according to Eq. 31-3, not worrying about the possibility of an overall minus sign since we are asked to find the absolute value of the flux.

$$|\Phi_B| = \int_{r-b/2}^{r+b/2}\left(\frac{\mu_0 i}{2\pi r}\right)(a\,dr) = \frac{\mu_0 i a}{2\pi}\ln\left(\frac{r + \frac{b}{2}}{r - \frac{b}{2}}\right) .$$

(b) Implementing Faraday's law involves taking a derivative of the flux in part (a), and recognizing that $\frac{dr}{dt} = v$. The magnitude of the induced emf divided by the loop resistance then gives the induced current:

$$i_{\text{loop}} = \left|\frac{\mathcal{E}}{R}\right| = -\frac{\mu_0 i a}{2\pi R}\left|\frac{d}{dt}\ln\left(\frac{r + \frac{b}{2}}{r - \frac{b}{2}}\right)\right| = \frac{\mu_0 i a b v}{2\pi R\,(r^2 - (b/2)^2)} .$$

25. Thermal energy is generated at the rate $P = \mathcal{E}^2/R$ (see Eq. 27-23). Using Eq. 27-16, the resistance is given by $R = \rho L/A$, where the resistivity is $1.69 \times 10^{-8}\ \Omega\cdot$m (by Table 27-1) and $A = \pi d^2/4$ is the cross-sectional area of the wire ($d = 0.00100$ m is the wire thickness). The area *enclosed* by the loop is

$$A_{\text{loop}} = \pi r_{\text{loop}}^2 = \pi\left(\frac{L}{2\pi}\right)^2$$

since the length of the wire ($L = 0.500$ m) is the circumference of the loop. This enclosed area is used in Faraday's law (where we ignore minus signs in the interest of finding the magnitudes of the quantities):

$$\mathcal{E} = \frac{d\Phi_B}{dt} = A_{\text{loop}}\frac{dB}{dt} = \frac{L^2}{4\pi}\frac{dB}{dt}$$

where the rate of change of the field is $dB/dt = 0.0100$ T/s. Consequently, we obtain

$$P = \frac{\left(\frac{L^2}{4\pi}\frac{dB}{dt}\right)^2}{4\rho L/\pi d^2} = \frac{d^2 L^3}{64\pi\rho}\left(\frac{dB}{dt}\right)^2 = 3.68 \times 10^{-6}\ \text{W} .$$

26. Noting that $|\Delta B| = B$, we find the thermal energy is

$$P_{\text{thermal}}\Delta t = \frac{\mathcal{E}^2 \Delta t}{R} = \frac{1}{R}\left(-\frac{d\Phi_B}{dt}\right)^2 \Delta t = \frac{1}{R}\left(-A\frac{\Delta B}{\Delta t}\right)^2 \Delta t = \frac{A^2 B^2}{R \Delta t} .$$

27. (a) Eq. 31-10 leads to

$$\mathcal{E} = BLv = (0.350\,\text{T})(0.250\,\text{m})(0.550\,\text{m/s}) = 0.0481 \text{ V} .$$

(b) By Ohm's law, the induced current is $i = 0.0481\,\text{V}/18.0\,\Omega = 0.00267\,\text{A}$. By Lenz's law, the current is clockwise in Fig. 31-46.

(c) Eq. 27-22 leads to $P = i^2 R = 0.000129$ W.

28. Noting that $F_{\text{net}} = BiL - mg = 0$, we solve for the current:

$$i = \frac{mg}{BL} = \frac{|\mathcal{E}|}{R} = \frac{1}{R}\left|\frac{d\Phi_B}{dt}\right| = \frac{B}{R}\left|\frac{dA}{dt}\right| = \frac{Bv_t L}{R} ,$$

which yields $v_t = mgR/B^2L^2$.

29. (a) By Lenz's law, the induced emf is clockwise. In the rod itself, we would say the emf is directed up the page. Eq. 31-10 leads to

$$\mathcal{E} = BLv = (1.2\,\text{T})(0.10\,\text{m})(5.0\,\text{m/s}) = 0.60 \text{ V} .$$

(b) By Ohm's law, the (clockwise) induced current is $i = 0.60\,\text{V}/0.40\,\Omega = 1.5\,\text{A}$.

(c) Eq. 27-22 leads to $P = i^2 R = 0.90$ W.

(d) From Eq. 29-2, we find that the force on the rod associated with the uniform magnetic field is directed rightward and has magnitude

$$F = iLB = (1.5\,\text{A})(0.10\,\text{m})(1.2\,\text{T}) = 0.18 \text{ N} .$$

To keep the rod moving at constant velocity, therefore, a leftward force (due to some external agent) having that same magnitude must be continuously supplied to the rod.

(e) Using Eq. 7-48, we find the power associated with the force being exerted by the external agent: $P = Fv = (0.18\,\text{N})(5.0\,\text{m/s}) = 0.90$ W, which is the same as our result from part (c).

30. (a) The "height" of the triangular area enclosed by the rails and bar is the same as the distance traveled in time v: $d = vt$, where $v = 5.20\,\text{m/s}$. We also note that the "base" of that triangle (the distance between the intersection points of the bar with the rails) is $2d$. Thus, the area of the triangle is

$$A = \frac{1}{2}(\text{base})(\text{height}) = \frac{1}{2}(2vt)(vt) = v^2 t^2 .$$

Since the field is a uniform $B = 0.350\,\text{T}$, then the magnitude of the flux (in SI units) is $\Phi_B = BA = (0.350)(5.20)^2 t^2 = 9.46t^2$. At $t = 3.00$ s, we obtain $\Phi_B = 85.2\,\text{Wb}$.

(b) The magnitude of the emf is the (absolute value of) Faraday's law:

$$\mathcal{E} = \frac{d\Phi_B}{dt} = 9.46\frac{dt^2}{dt} = 18.9t$$

in SI units. At $t = 3.00$ s, this yields $\mathcal{E} = 56.8$ V.

(c) Our calculation in part (b) shows that $n = 1$.

31. (a) Letting x be the distance from the right end of the rails to the rod, we find an expression for the magnetic flux through the area enclosed by the rod and rails. By Eq. 30-19, the field is $B = \mu_0 i/2\pi r$, where r is the distance from the long straight wire. We consider an infinitesimal horizontal strip of length x and width dr, parallel to the wire and a distance r from it; it has area $A = x\,dr$ and the flux $d\Phi_B = (\mu_0 ix/2\pi r)dr$. By Eq. 31-3, the total flux through the area enclosed by the rod and rails is

$$\Phi_B = \frac{\mu_0 ix}{2\pi}\int_a^{a+L}\frac{dr}{r} = \frac{\mu_0 ix}{2\pi}\ln\left(\frac{a+L}{a}\right) .$$

According to Faraday's law the emf induced in the loop is

$$\begin{aligned} \mathcal{E} &= \frac{d\Phi_B}{dt} = \frac{\mu_0 i}{2\pi}\frac{dx}{dt}\ln\left(\frac{a+L}{a}\right) = \frac{\mu_0 iv}{2\pi}\ln\left(\frac{a+L}{a}\right) \\ &= \frac{(4\pi\times10^{-7}\,\mathrm{T\cdot m/A})(100\,\mathrm{A})(5.00\,\mathrm{m/s})}{2\pi}\ln\left(\frac{1.00\,\mathrm{cm}+10.0\,\mathrm{cm}}{1.00\,\mathrm{cm}}\right) \\ &= 2.40\times10^{-4}\ \mathrm{V} . \end{aligned}$$

(b) By Ohm's law, the induced current is $i_\ell = \mathcal{E}/R = (2.40\times10^{-4}\,\mathrm{V})/(0.400\,\Omega) = 6.00\times10^{-4}\,\mathrm{A}$. Since the flux is increasing the magnetic field produced by the induced current must be into the page in the region enclosed by the rod and rails. This means the current is clockwise.

(c) Thermal energy is being generated at the rate $P = i_\ell^2 R = (6.00\times10^{-4}\mathrm{A})^2(0.400\,\Omega) = 1.44\times10^{-7}\mathrm{W}$.

(d) Since the rod moves with constant velocity, the net force on it is zero. The force of the external agent must have the same magnitude as the magnetic force and must be in the opposite direction. The magnitude of the magnetic force on an infinitesimal segment of the rod, with length dr at a distance r from the long straight wire, is $dF_B = i_\ell B\,dr = (\mu_0 i_\ell i/2\pi r)\,dr$. We integrate to find the magnitude of the total magnetic force on the rod:

$$\begin{aligned} F_B &= \frac{\mu_0 i_\ell i}{2\pi}\int_a^{a+L}\frac{dr}{r} = \frac{\mu_0 i_\ell i}{2\pi}\ln\left(\frac{a+L}{a}\right) \\ &= \frac{(4\pi\times10^{-7}\,\mathrm{T\cdot m/A})(6.00\times10^{-4}\,\mathrm{A})(100\,\mathrm{A})}{2\pi}\ln\left(\frac{1.00\,\mathrm{cm}+10.0\,\mathrm{cm}}{1.00\,\mathrm{cm}}\right) \\ &= 2.87\times10^{-8}\ \mathrm{N} . \end{aligned}$$

Since the field is out of the page and the current in the rod is upward in the diagram, the force associated with the magnetic field is toward the right. The external agent must therefore apply a force of 2.87×10^{-8} N, to the left.

(e) By Eq. 7-48, the external agent does work at the rate $P = Fv = (2.87\times10^{-8}\,\mathrm{N})(5.00\,\mathrm{m/s}) = 1.44\times10^{-7}$ W. This is the same as the rate at which thermal energy is generated in the rod. All the energy supplied by the agent is converted to thermal energy.

32.

$$\begin{aligned} \oint_1 \vec{E}\cdot d\vec{s} &= -\frac{d\vec{\Phi}_{B1}}{dt} = \frac{d}{dt}(B_1 A_1) = A_1\frac{dB_1}{dt} = \pi r_1^2\frac{dB_1}{dt} \\ &= \pi(0.200\,\mathrm{m})^2(-8.50\times10^{-3}\,\mathrm{T/s}) = -1.07\times10^{-3}\ \mathrm{V} \end{aligned}$$

$$\begin{aligned} \oint_2 \vec{E}\cdot d\vec{s} &= -\frac{d\vec{\Phi}_{B2}}{dt} = \pi r_2^2\frac{dB_2}{dt} \\ &= \pi(0.300\,\mathrm{m})^2(-8.50\times10^{-3}\,\mathrm{T/s}) = -2.40\times10^{-3}\ \mathrm{V} \end{aligned}$$

$$\oint_3 \vec{E}\cdot d\vec{s} = \oint_1 \vec{E}\cdot d\vec{s} - \oint_2 \vec{E}\cdot d\vec{s} = -1.07\times10^{-3}\,\mathrm{V} - (-2.4\times10^{-3}\,\mathrm{V}) = 1.33\times10^{-3}\ \mathrm{V}$$

33. (a) The point at which we are evaluating the field is inside the solenoid, so Eq. 31-27 applies. The magnitude of the induced electric field is

$$E = \frac{1}{2}\frac{dB}{dt}r = \frac{1}{2}(6.5 \times 10^{-3}\,\mathrm{T/s})(0.0220\,\mathrm{m}) = 7.15 \times 10^{-5}\ \mathrm{V/m}\ .$$

(b) Now point at which we are evaluating the field is outside the solenoid and Eq. 31-29 applies. The magnitude of the induced field is

$$E = \frac{1}{2}\frac{dB}{dt}\frac{R^2}{r} = \frac{1}{2}(6.5 \times 10^{-3}\,\mathrm{T/s})\frac{(0.0600\,\mathrm{m})^2}{(0.0820\,\mathrm{m})} = 1.43 \times 10^{-4}\ \mathrm{V/m}\ .$$

34. The magnetic field B can be expressed as

$$B(t) = B_0 + B_1 \sin(\omega t + \phi_0)\ ,$$

where $B_0 = (30.0\,\mathrm{T} + 29.6\,\mathrm{T})/2 = 29.8\,\mathrm{T}$ and $B_1 = (30.0\,\mathrm{T} - 29.6\,\mathrm{T})/2 = 0.200\,\mathrm{T}$. Then from Eq. 31-27

$$E = \frac{1}{2}\left(\frac{dB}{dt}\right)r = \frac{r}{2}\frac{d}{dt}[B_0 + B_1\sin(\omega t + \phi_0)] = \frac{1}{2}B_1\omega r\cos(\omega t + \phi_0)\ .$$

We note that $\omega = 2\pi f$ and that the factor in front of the cosine is the maximum value of the field. Consequently,

$$E_{\mathrm{max}} = \frac{1}{2}B_1(2\pi f)r = \frac{1}{2}(0.200\,\mathrm{T})(2\pi)(15\,\mathrm{Hz})(1.6 \times 10^{-2}\,\mathrm{m}) = 0.15\ \mathrm{V/m}\ .$$

35. We use Faraday's law in the form $\oint \vec{E}\cdot d\vec{s} = -(d\Phi_B/dt)$, integrating along the dotted path shown in the Figure. At all points on the upper and lower sides the electric field is either perpendicular to the side or else it vanishes. We assume it vanishes at all points on the right side (outside the capacitor). On the left side it is parallel to the side and has constant magnitude. Thus, direct integration yields $\oint \vec{E}\cdot d\vec{s} = EL$, where L is the length of the left side of the rectangle. The magnetic field is zero and remains zero, so $d\Phi_B/dt = 0$. Faraday's law leads to a contradiction: $EL = 0$, but neither E nor L is zero. Therefore, there must be an electric field along the right side of the rectangle.

36. (a) We interpret the question as asking for N multiplied by the flux through one turn:

$$\Phi_{\mathrm{turns}} = N\Phi_B = NBA = NB(\pi r^2) = (30.0)(2.60 \times 10^{-3}\,\mathrm{T})(\pi)(0.100\,\mathrm{m})^2 = 2.45 \times 10^{-3}\ \mathrm{Wb}\ .$$

(b) Eq. 31-35 leads to

$$L = \frac{N\Phi_B}{i} = \frac{2.45 \times 10^{-3}\,\mathrm{Wb}}{3.80\,\mathrm{A}} = 6.45 \times 10^{-4}\ \mathrm{H}\ .$$

37. Since $N\Phi_B = Li$, we obtain

$$\Phi_B = \frac{Li}{N} = \frac{(8.0 \times 10^{-3}\,\mathrm{H})(5.0 \times 10^{-3}\,\mathrm{A})}{400} = 1.0 \times 10^{-7}\ \mathrm{Wb}\ .$$

38. (a) We imagine dividing the one-turn solenoid into N small circular loops placed along the width W of the copper strip. Each loop carries a current $\Delta i = i/N$. Then the magnetic field inside the solenoid is $B = \mu_0 n\Delta i = \mu_0(N/W)(i/N) = \mu_0 i/W$.

(b) Eq. 31-35 leads to

$$L = \frac{\Phi_B}{i} = \frac{\pi R^2 B}{i} = \frac{\pi R^2(\mu_0 i/W)}{i} = \frac{\pi\mu_0 R^2}{W}\ .$$

39. We refer to the (very large) wire length as ℓ and seek to compute the flux per meter: Φ_B/ℓ. Using the right-hand rule discussed in Chapter 30, we see that the net field in the region between the axes of antiparallel currents is the addition of the magnitudes of their individual fields, as given by Eq. 30-19 and Eq. 30-22. There is an evident reflection symmetry in the problem, where the plane of symmetry is midway between the two wires (at $x = d/2$); the net field at any point $0 < x < d/2$ is the same at its "mirror image" point $d - x$. The central axis of one of the wires passes through the origin, and that of the other passes through $x = d$. We make use of the symmetry by integrating over $0 < x < d/2$ and then multiplying by 2:

$$\Phi_B = 2\int_0^{d/2} B\,dA = 2\int_0^a B\,(\ell\,dx) + 2\int_a^{d/2} B\,(\ell\,dx)$$

where $d = 0.0025$ m is diameter of each wire. We will r instead of x in the following steps. Thus, using the equations from Ch. 30 referred to above, we find

$$\begin{aligned}\frac{\Phi_B}{\ell} &= 2\int_0^a \left(\frac{\mu_0 i}{2\pi a^2}r + \frac{\mu_0 i}{2\pi(d-r)}\right)dr + 2\int_a^{d/2}\left(\frac{\mu_0 i}{2\pi r} + \frac{\mu_0 i}{2\pi(d-r)}\right)dr \\ &= \frac{\mu_0 i}{2\pi}\left(1 - 2\ln\left(\frac{d-a}{d}\right)\right) + \frac{\mu_0 i}{\pi}\ln\left(\frac{d-a}{a}\right)\end{aligned}$$

where the first term is the flux within the wires and will be neglected (as the problem suggests). Thus, the flux is approximately $\Phi_B \approx \mu_0 i\ell/\pi \ln((d-a)/a)$. Now, we use Eq. 31-35 (with $N = 1$) to obtain the inductance:

$$L = \frac{\Phi_B}{i} = \frac{\mu_0\ell}{\pi}\ln\left(\frac{d-a}{a}\right).$$

40. (a) Speaking anthropomorphically, the coil wants to fight the changes – so if it wants to push current rightward (when the current is already going rightward) then i must be in the process of decreasing.

(b) From Eq. 31-37 (in absolute value) we get

$$L = \left|\frac{\mathcal{E}}{di/dt}\right| = \frac{17\,\text{V}}{2.5\,\text{kA/s}} = 6.8\times10^{-4}\ \text{H}.$$

41. Since $\mathcal{E} = -L(di/dt)$, we may obtain the desired induced emf by setting

$$\frac{di}{dt} = -\frac{\mathcal{E}}{L} = -\frac{60\,\text{V}}{12\,\text{H}} = -5.0\ \text{A/s}.$$

We might, for example, uniformly reduce the current from 2.0 A to zero in 40 ms.

42. During periods of time when the current is varying linearly with time, Eq. 31-37 (in absolute values) becomes $|\mathcal{E}| = L\left|\frac{\Delta i}{\Delta t}\right|$. For simplicity, we omit the absolute value signs in the following.

(a) For $0 < t < 2$ ms

$$\mathcal{E} = L\frac{\Delta i}{\Delta t} = \frac{(4.6\,\text{H})(7.0\,\text{A} - 0)}{2.0\times10^{-3}\,\text{s}} = 1.6\times10^4\ \text{V}.$$

(b) For 2 ms $< t <$ 5 ms

$$\mathcal{E} = L\frac{\Delta i}{\Delta t} = \frac{(4.6\,\text{H})(5.0\,\text{A} - 7.0\,\text{A})}{(5.0 - 2.0)10^{-3}\,\text{s}} = 3.1\times10^3\ \text{V}.$$

(c) For 5 ms $< t <$ 6 ms

$$\mathcal{E} = L\frac{\Delta i}{\Delta t} = \frac{(4.6\,\text{H})(0 - 5.0\,\text{A})}{(6.0 - 5.0)10^{-3}\,\text{s}} = 2.3\times10^4\ \text{V}.$$

43. (a) Voltage is proportional to inductance (by Eq. 31-37) just as, for resistors, it is proportional to resistance. Since the (independent) voltages for series elements add ($V_1 + V_2$), then inductances in series must *add* just as was the case for resistances.

(b) To ensure the independence of the voltage values, it is important that the inductors not be too close together (the related topic of mutual inductance is treated in §31-12). The requirement is that magnetic field lines from one inductor should not have significant significant presence in any other.

(c) Just as with resistors, $L_{\text{eq}} = \sum_{n=1}^{N} L_n$.

44. (a) Voltage is proportional to inductance (by Eq. 31-37) just as, for resistors, it is proportional to resistance. Now, the (independent) voltages for parallel elements are equal ($V_1 = V_2$), and the currents (which are generally functions of time) add ($i_1(t) + i_2(t) = i(t)$). This leads to the Eq. 28-21 for resistors. We note that this condition on the currents implies

$$\frac{di_1(t)}{dt} + \frac{di_2(t)}{dt} = \frac{di(t)}{dt} .$$

Thus, although the inductance equation Eq. 31-37 involves the rate of change of current, as opposed to current itself, the conditions that led to the parallel resistor formula also applies to inductors. Therefore,

$$\frac{1}{L_{\text{eq}}} = \frac{1}{L_1} + \frac{1}{L_2} .$$

(b) To ensure the independence of the voltage values, it is important that the inductors not be too close together (the related topic of mutual inductance is treated in §31-12). The requirement is that the field of one inductor not have significant influence (or "coupling") in the next.

(c) Just as with resistors, $\frac{1}{L_{\text{eq}}} = \sum_{n=1}^{N} \frac{1}{L_n}$.

45. Starting with zero current at $t = 0$ (the moment the switch is closed) the current in the circuit increases according to

$$i = \frac{\mathcal{E}}{R}\left(1 - e^{-t/\tau_L}\right) ,$$

where $\tau_L = L/R$ is the inductive time constant and $\mathcal{E}$ is the battery emf. To calculate the time at which $i = 0.9990\mathcal{E}/R$, we solve for t:

$$0.9990\frac{\mathcal{E}}{R} = \frac{\mathcal{E}}{R}\left(1 - e^{-t/\tau_L}\right) \implies \ln(0.0010) = -(t/\tau) \implies t = 6.91\tau_L .$$

46. The steady state value of the current is also its maximum value, $\mathcal{E}/R$, which we denote as i_m. We are told that $i = i_m/3$ at $t_0 = 5.00$ s. Eq. 31-43 becomes $i = i_m(1 - e^{-t_0/\tau_L})$, which leads to

$$\tau_L = -\frac{t_0}{\ln(1 - i/i_m)} = -\frac{5.00\,\text{s}}{\ln(1 - 1/3)} = 12.3\ \text{s} .$$

47. The current in the circuit is given by $i = i_0 e^{-t/\tau_L}$, where i_0 is the current at time $t = 0$ and τ_L is the inductive time constant (L/R) . We solve for τ_L. Dividing by i_0 and taking the natural logarithm of both sides, we obtain

$$\ln\left(\frac{i}{i_0}\right) = -\frac{t}{\tau_L} .$$

This yields

$$\tau_L = -\frac{t}{\ln(i/i_0)} = -\frac{1.0\,\text{s}}{\ln((10 \times 10^{-3}\,\text{A})/(1.0\,\text{A}))} = 0.217\ \text{s} .$$

Therefore, $R = L/\tau_L = 10\,\text{H}/0.217\,\text{s} = 46\,\Omega$.

48. (a) Immediately after the switch is closed $\mathcal{E} - \mathcal{E}_L = iR$. But $i = 0$ at this instant, so $\mathcal{E}_L = \mathcal{E}$.

(b) $\mathcal{E}_L(t) = \mathcal{E}e^{-t/\tau_L} = \mathcal{E}e^{-2.0\tau_L/\tau_L} = \mathcal{E}e^{-2.0} = 0.135\mathcal{E}$.

(c) From $\mathcal{E}_L(t) = \mathcal{E}e^{-t/\tau_L}$ we obtain

$$\frac{t}{\tau_L} = \ln\left(\frac{\mathcal{E}}{\mathcal{E}_L}\right) = \ln 2 \implies t = \tau_L \ln 2 \quad = \quad 0.693\tau_L .$$

49. (a) If the battery is switched into the circuit at $t = 0$, then the current at a later time t is given by

$$i = \frac{\mathcal{E}}{R}\left(1 - e^{-t/\tau_L}\right) ,$$

where $\tau_L = L/R$. Our goal is to find the time at which $i = 0.800\mathcal{E}/R$. This means

$$0.800 = 1 - e^{-t/\tau_L} \implies e^{-t/\tau_L} = 0.200 .$$

Taking the natural logarithm of both sides, we obtain $-(t/\tau_L) = \ln(0.200) = -1.609$. Thus

$$t = 1.609\tau_L = \frac{1.609L}{R} = \frac{1.609(6.30 \times 10^{-6}\ \mathrm{H})}{1.20 \times 10^3\ \Omega} = 8.45 \times 10^{-9}\ \mathrm{s} .$$

(b) At $t = 1.0\tau_L$ the current in the circuit is

$$i = \frac{\mathcal{E}}{R}\left(1 - e^{-1.0}\right) = \left(\frac{14.0\ \mathrm{V}}{1.20 \times 10^3\ \Omega}\right)\left(1 - e^{-1.0}\right) = 7.37 \times 10^{-3}\ \mathrm{A} .$$

50. Applying the loop theorem

$$\mathcal{E} - L\left(\frac{di}{dt}\right) = iR ,$$

we solve for the (time-dependent) emf, with SI units understood:

$$\begin{aligned}\mathcal{E} &= L\frac{di}{dt} + iR = L\frac{d}{dt}(3.0 + 5.0t) + (3.0 + 5.0t)R \\ &= (6.0)(5.0) + (3.0 + 5.0t)(4.0) \\ &= (42 + 20t)\end{aligned}$$

in volts if t is in seconds.

51. Taking the time derivative of both sides of Eq. 31-43, we obtain

$$\begin{aligned}\frac{di}{dt} &= \frac{d}{dt}\left[\frac{\mathcal{E}}{R}\left(1 - e^{-Rt/\tau_L}\right)\right] = \frac{\mathcal{E}}{L}e^{-RT/L} \\ &= \left(\frac{45.0\ \mathrm{V}}{50.0 \times 10^{-3}\ \mathrm{H}}\right)e^{-(180\ \Omega)(1.20\times10^{-3}\ \mathrm{s})/50.0\times10^{-3}\ \mathrm{H}} = 12.0\ \mathrm{A/s} .\end{aligned}$$

52. (a) Our notation is as follows: h is the height of the toroid, a its inner radius, and b its outer radius. Since it has a square cross section, $h = b - a = 0.12\ \mathrm{m} - 0.10\ \mathrm{m} = 0.02\ \mathrm{m}$.. We derive the flux using Eq. 30-26 and the self-inductance using Eq. 31-35:

$$\Phi_B = \int_a^b B\,dA = \int_a^b \left(\frac{\mu_0 Ni}{2\pi r}\right) h\,dr = \frac{\mu_0 Nih}{2\pi}\ln\left(\frac{b}{a}\right)$$

and $L = N\Phi_B/i = (\mu_0 N^2 h/2\pi)\ln(b/a)$. We note that the formulas for Φ_B and L can also be found in the Supplement for the chapter, in Sample Problem 31-11. Now, since the inner circumference of the toroid is $l = 2\pi a = 2\pi(10\,\text{cm}) \approx 62.8\,\text{cm}$, the number of turns of the toroid is roughly $N \approx 62.8\,\text{cm}/1.0\,\text{mm} = 628$. Thus

$$\begin{aligned} L &= \frac{\mu_0 N^2 h}{2\pi}\ln\left(\frac{b}{a}\right) \\ &\approx \frac{(4\pi\times10^{-7}\,\text{H/m})\,(628)^2(0.02\,\text{m})}{2\pi}\ln\left(\frac{12}{10}\right) \\ &= 2.9\times10^{-4}\ \text{H}\,. \end{aligned}$$

(b) Noting that the perimeter of a square is four times its sides, the total length ℓ of the wire is $\ell = (628)4(2.0\,\text{cm}) = 50\,\text{m}$, the resistance of the wire is $R = (50\,\text{m})(0.02\,\Omega/\text{m}) = 1.0\,\Omega$. Thus

$$\tau_L = \frac{L}{R} = \frac{2.9\times10^{-4}\,\text{H}}{1.0\,\Omega} = 2.9\times10^{-4}\ \text{s}\,.$$

53. (a) The inductor prevents a fast build-up of the current through it, so immediately after the switch is closed, the current in the inductor is zero. It follows that

$$i_1 = i_2 = \frac{\mathcal{E}}{R_1+R_2} = \frac{100\,\text{V}}{10.0\,\Omega+20.0\,\Omega} = 3.33\ \text{A}\,.$$

(b) After a suitably long time, the current reaches steady state. Then, the emf across the inductor is zero, and we may imagine it replaced by a wire. The current in R_3 is $i_1 - i_2$. Kirchhoff's loop rule gives

$$\mathcal{E} - i_1R_1 - i_2R_2 = 0 \quad\text{and}\quad \mathcal{E} - i_1R_1 - (i_1-i_2)R_3 = 0\,.$$

We solve these simultaneously for i_1 and i_2. The results are

$$\begin{aligned} i_1 &= \frac{\mathcal{E}\,(R_2+R_3)}{R_1R_2+R_1R_3+R_2R_3} \\ &= \frac{(100\,\text{V})(20.0\,\Omega+30.0\,\Omega)}{(10.0\,\Omega)(20.0\,\Omega)+(10.0\,\Omega)(30.0\,\Omega)+(20.0\,\Omega)(30.0\,\Omega)} \\ &= 4.55\ \text{A}\,, \end{aligned}$$

and

$$\begin{aligned} i_2 &= \frac{\mathcal{E}R_3}{R_1R_2+R_1R_3+R_2R_3} \\ &= \frac{(100\,\text{V})(30.0\,\Omega)}{(10.0\,\Omega)(20.0\,\Omega)+(10.0\,\Omega)(30.0\,\Omega)+(20.0\,\Omega)(30.0\,\Omega)} \\ &= 2.73\ \text{A}\,. \end{aligned}$$

(c) The left-hand branch is now broken. We take the current (immediately) as zero in that branch when the switch is opened (that is, $i_1 = 0$). The current in R_3 changes less rapidly because there is an inductor in its branch. In fact, immediately after the switch is opened it has the same value that it had before the switch was opened. That value is $4.55\,\text{A} - 2.73\,\text{A} = 1.82\,\text{A}$. The current in R_2 is the same as that in R_3 (1.82 A).

(d) There are no longer any sources of emf in the circuit, so all currents eventually drop to zero.

54. (a) When switch S is just closed (case I), $V_1 = \mathcal{E}$ and $i_1 = \mathcal{E}/R_1 = 10\,\text{V}/5.0\,\Omega = 2.0\,\text{A}$. After a long time (case II) we still have $V_1 = \mathcal{E}$, so $i_1 = 2.0\,\text{A}$.

(b) Case I: since now $\mathcal{E}_L = \mathcal{E}$, $i_2 = 0$; case II: since now $\mathcal{E}_L = 0$, $i_2 = \mathcal{E}/R_2 = 10\,\text{V}/10\,\Omega = 1.0\,\text{A}$.

(c) Case I: $i = i_1 + i_2 = 2.0\,\text{A} + 0 = 2.0\,\text{A}$; case II: $i = i_1 + i_2 = 2.0\,\text{A} + 1.0\,\text{A} = 3.0\,\text{A}$.

(d) Case I: since $\mathcal{E}_L = \mathcal{E}$, $V_2 = \mathcal{E} - \mathcal{E}_L = 0$; case II: since $\mathcal{E}_L = 0$, $V_2 = \mathcal{E} - \mathcal{E}_L = \mathcal{E} = 10\,\text{V}$.

(e) Case I: $\mathcal{E}_L = \mathcal{E} = 10\,\text{V}$; case II: $\mathcal{E}_L = 0$.

(f) Case I: $di_2/dt = \mathcal{E}_L/L = \mathcal{E}/L = 10\,\text{V}/5.0\,\text{H} = 2.0\,\text{A/s}$; case II: $di_2/dt = \mathcal{E}_L/L = 0$.

55. (a) We assume i is from left to right through the closed switch. We let i_1 be the current in the resistor and take it to be downward. Let i_2 be the current in the inductor, also assumed downward. The junction rule gives $i = i_1 + i_2$ and the loop rule gives $i_1 R - L(di_2/dt) = 0$. According to the junction rule, $(di_1/dt) = -(di_2/dt)$. We substitute into the loop equation to obtain

$$L\frac{di_1}{dt} + i_1 R = 0\,.$$

This equation is similar to Eq. 31-48, and its solution is the function given as Eq. 31-49:

$$i_1 = i_0 e^{-Rt/L}\,,$$

where i_0 is the current through the resistor at $t = 0$, just after the switch is closed. Now just after the switch is closed, the inductor prevents the rapid build-up of current in its branch, so at that moment $i_2 = 0$ and $i_1 = i$. Thus $i_0 = i$, so

$$i_1 = ie^{-Rt/L} \quad \text{and} \quad i_2 = i - i_1 = i\left(1 - e^{-Rt/L}\right)\,.$$

(b) When $i_2 = i_1$,

$$e^{-Rt/L} = 1 - e^{-Rt/L} \implies e^{-Rt/L} = \frac{1}{2}\,.$$

Taking the natural logarithm of both sides (and using $\ln(1/2) = -\ln 2$) we obtain

$$\left(\frac{Rt}{L}\right) = \ln 2 \implies t = \frac{L}{R}\ln 2\,.$$

56. Let $U_B(t) = \frac{1}{2}Li^2(t)$. We require the energy at time t to be half of its final value: $U(t) = \frac{1}{2}U_B(t \to \infty) = \frac{1}{4}Li_f^2$. This gives $i(t) = i_f/\sqrt{2}$. But $i(t) = i_f(1 - e^{-t/\tau_L})$, so

$$1 - e^{-t/\tau_L} = \frac{1}{\sqrt{2}} \implies t = -\tau_L \ln\left(1 - \frac{1}{\sqrt{2}}\right) = 1.23\tau_L\,.$$

57. From Eq. 31-51 and Eq. 31-43, the rate at which the energy is being stored in the inductor is

$$\begin{aligned}\frac{dU_B}{dt} &= \frac{d\left(\frac{1}{2}Li^2\right)}{dt} = Li\frac{di}{dt} \\ &= L\left(\frac{\mathcal{E}}{R}\left(1 - e^{-t/\tau_L}\right)\right)\left(\frac{\mathcal{E}}{R}\frac{1}{\tau_L}e^{-t/\tau_L}\right) \\ &= \frac{\mathcal{E}^2}{R}\left(1 - e^{-t/\tau_L}\right)e^{-t/\tau_L}\end{aligned}$$

where $\tau_L = L/R$ has been used. From Eq. 27-22 and Eq. 31-43, the rate at which the resistor is generating thermal energy is

$$P_{\text{thermal}} = i^2 R = \frac{\mathcal{E}^2}{R^2}\left(1 - e^{-t/\tau_L}\right)^2 R = \frac{\mathcal{E}^2}{R}\left(1 - e^{-t/\tau_L}\right)^2\,.$$

We equate this to dU_B/dt, and solve for the time:

$$\frac{\mathcal{E}^2}{R}\left(1 - e^{-t/\tau_L}\right)^2 = \frac{\mathcal{E}^2}{R}\left(1 - e^{-t/\tau_L}\right)e^{-t/\tau_L} \implies t = \tau_L \ln 2 = (37.0\,\text{ms})\ln 2 = 25.6\,\text{ms}\,.$$

58. (a) From Eq. 31-51 and Eq. 31-43, the rate at which the energy is being stored in the inductor is

$$\begin{aligned}\frac{dU_B}{dt} &= \frac{d\left(\frac{1}{2}Li^2\right)}{dt} = Li\,\frac{di}{dt}\\ &= L\left(\frac{\mathcal{E}}{R}\left(1-e^{-t/\tau_L}\right)\right)\left(\frac{\mathcal{E}}{R}\frac{1}{\tau_L}e^{-t/\tau_L}\right)\\ &= \frac{\mathcal{E}^2}{R}\left(1-e^{-t/\tau_L}\right)e^{-t/\tau_L}\,.\end{aligned}$$

Now, $\tau_L = L/R = 2.0\,\text{H}/10\,\Omega = 0.20\,\text{s}$ and $\mathcal{E} = 100\,\text{V}$, so the above expression yields $dU_B/dt = 2.4\times10^2$ W when $t = 0.10$ s.

(b) From Eq. 27-22 and Eq. 31-43, the rate at which the resistor is generating thermal energy is

$$P_{\text{thermal}} = i^2R = \frac{\mathcal{E}^2}{R^2}\left(1-e^{-t/\tau_L}\right)^2 R = \frac{\mathcal{E}^2}{R}\left(1-e^{-t/\tau_L}\right)^2 .$$

At $t = 0.10$ s, this yields $P_{\text{thermal}} = 1.5\times10^2$ W.

(c) By energy conservation, the rate of energy being supplied to the circuit by the battery is

$$P_{\text{battery}} = P_{\text{thermal}} + \frac{dU_B}{dt} = 3.9\times10^2\ \text{W}\,.$$

We note that this could result could alternatively have been found from Eq. 28-14 (with Eq. 31-43).

59. (a) If the battery is applied at time $t = 0$ the current is given by

$$i = \frac{\mathcal{E}}{R}\left(1-e^{-t/\tau_L}\right),$$

where $\mathcal{E}$ is the emf of the battery, R is the resistance, and τ_L is the inductive time constant (L/R). This leads to

$$e^{-t/\tau_L} = 1-\frac{iR}{\mathcal{E}} \implies -\frac{t}{\tau_L} = \ln\left(1-\frac{iR}{\mathcal{E}}\right).$$

Since

$$\ln\left(1-\frac{iR}{\mathcal{E}}\right) = \ln\left[1-\frac{(2.00\times10^{-3}\,\text{A})(10.0\times10^3\,\Omega)}{50.0\,\text{V}}\right] = -0.5108\,,$$

the inductive time constant is $\tau_L = t/0.5108 = (5.00\times10^{-3}\,\text{s})/0.5108 = 9.79\times10^{-3}\,\text{s}$ and the inductance is

$$L = \tau_L R = (9.79\times10^{-3}\,\text{s})(10.0\times10^3\,\Omega) = 97.9\ \text{H}\,.$$

(b) The energy stored in the coil is

$$U_B = \frac{1}{2}Li^2 = \frac{1}{2}(97.9\,\text{H})(2.00\times10^{-3}\,\text{A})^2 = 1.96\times10^{-4}\ \text{J}\,.$$

60. (a) The energy delivered by the battery is the integral of Eq. 28-14 (where we use Eq. 31-43 for the current):

$$\begin{aligned}\int_0^t P_{\text{battery}}\,dt &= \int_0^t \frac{\mathcal{E}^2}{R}\left(1-e^{-Rt/L}\right)dt = \frac{\mathcal{E}^2}{R}\left[t+\frac{L}{R}\left(e^{-Rt/L}-1\right)\right]\\ &= \frac{(10.0\,\text{V})^2}{6.70\,\Omega}\left[2.00\,\text{s}+\frac{(5.50\,\text{H})\left(e^{-(6.70\,\Omega)(2.00\,\text{s})/5.50\,\text{H}}-1\right)}{6.70\,\Omega}\right]\\ &= 18.7\ \text{J}\,.\end{aligned}$$

(b) The energy stored in the magnetic field is given by Eq. 31-51:

$$\begin{aligned} U_B &= \frac{1}{2}Li^2(t) = \frac{1}{2}L\left(\frac{\mathcal{E}}{R}\right)^2 (1-e^{-Rt/L})^2 \\ &= \frac{1}{2}(5.50\,\mathrm{H})\left(\frac{10.0\,\mathrm{V}}{6.70\,\Omega}\right)^2 \left[1-e^{-(6.70\,\Omega)(2.00\,\mathrm{s})/5.50\,\mathrm{H}}\right]^2 \\ &= 5.10\ \mathrm{J}\ . \end{aligned}$$

(c) The difference of the previous two results gives the amount "lost" in the resistor: $18.7\,\mathrm{J} - 5.10\,\mathrm{J} = 13.6\,\mathrm{J}$.

61. Suppose that the switch had been in position a for a long time so that the current had reached the steady-state value i_0. The energy stored in the inductor is $U_B = \frac{1}{2}Li_0^2$. Now, the switch is thrown to position b at time $t = 0$. Thereafter the current is given by

$$i = i_0 e^{-t/\tau_L}\ ,$$

where τ_L is the inductive time constant, given by $\tau_L = L/R$. The rate at which thermal energy is generated in the resistor is given by

$$P = i^2R = i_0^2 R e^{-2t/\tau_L}\ .$$

Over a long time period the energy dissipated is

$$\int_0^\infty P\,dt = i_0^2 R \int_0^\infty e^{-2t/\tau_L}\,dt = -\frac{1}{2}i_0^2 R\tau_L e^{-2t/\tau_L}\bigg|_0^\infty = \frac{1}{2}i_0^2 R\tau_L\ .$$

Upon substitution of $\tau_L = L/R$ this becomes $\frac{1}{2}Li_0^2$, the same as the total energy originally stored in the inductor.

62. The magnetic energy stored in the toroid is given by $U_B = \frac{1}{2}Li^2$, where L is its inductance and i is the current. By Eq. 31-56, the energy is also given by $U_B = u_B\mathcal{V}$, where u_B is the average energy density and $\mathcal{V}$ is the volume. Thus

$$i = \sqrt{\frac{2u_B\mathcal{V}}{L}} = \sqrt{\frac{2\left(70.0\,\mathrm{J/m^3}\right)(0.0200\,\mathrm{m^3})}{90.0\times10^{-3}\,\mathrm{H}}} = 5.58\ \mathrm{A}\ .$$

63. (a) At any point the magnetic energy density is given by $u_B = B^2/2\mu_0$, where B is the magnitude of the magnetic field at that point. Inside a solenoid $B = \mu_0 n i$, where n, for the solenoid of this problem, is $(950\,\mathrm{turns})/(0.850\,\mathrm{m}) = 1.118\times10^3\,\mathrm{m^{-1}}$. The magnetic energy density is

$$u_B = \frac{1}{2}\mu_0 n^2 i^2 = \frac{1}{2}\left(4\pi\times10^{-7}\,\mathrm{T\cdot m/A}\right)\left(1.118\times10^3\,\mathrm{m^{-1}}\right)^2(6.60\,\mathrm{A})^2 = 34.2\ \mathrm{J/m^3}\ .$$

(b) Since the magnetic field is uniform inside an ideal solenoid, the total energy stored in the field is $U_B = u_B\mathcal{V}$, where $\mathcal{V}$ is the volume of the solenoid. $\mathcal{V}$ is calculated as the product of the cross-sectional area and the length. Thus

$$U_B = (34.2\,\mathrm{J/m^3})(17.0\times10^{-4}\,\mathrm{m^2})(0.850\,\mathrm{m}) = 4.94\times10^{-2}\ \mathrm{J}\ .$$

64. We use $1\,\mathrm{ly} = 9.46\times10^{15}\,\mathrm{m}$, and use the symbol $\mathcal{V}$ for volume.

$$U_B = \mathcal{V}u_B = \frac{\mathcal{V}B^2}{2\mu_0} = \frac{(9.46\times10^{15}\,\mathrm{m})^3(1\times10^{-10}\,\mathrm{T})^2}{2\,(4\pi\times10^{-7}\,\mathrm{H/m})} = 3\times10^{36}\ \mathrm{J}\ .$$

65. We set $u_E = \frac{1}{2}\varepsilon_0 E^2 = u_B = \frac{1}{2}B^2/\mu_0$ and solve for the magnitude of the electric field:

$$E = \frac{B}{\sqrt{\varepsilon_0 \mu_0}} = \frac{0.50\,\text{T}}{\sqrt{(8.85\times10^{-12}\,\text{F/m})\,(4\pi\times10^{-7}\,\text{H/m})}} = 1.5\times10^{8}\ \text{V/m}\ .$$

66. (a) The magnitude of the magnetic field at the center of the loop, using Eq. 30-11, is

$$B = \frac{\mu_0 i}{2R} = \frac{(4\pi\times10^{-7}\,\text{H/m})\,(100\,\text{A})}{2(50\times10^{-3}\,\text{m})} = 1.3\times10^{-3}\ \text{T}\ .$$

(b) The energy per unit volume in the immediate vicinity of the center of the loop is

$$u_B = \frac{B^2}{2\mu_0} = \frac{(1.3\times10^{-3}\,\text{T})^2}{2\,(4\pi\times10^{-7}\,\text{H/m})} = 0.63\ \text{J/m}^3\ .$$

67. (a) The energy per unit volume associated with the magnetic field is

$$u_B = \frac{B^2}{2\mu_0} = \frac{1}{2\mu_0}\left(\frac{\mu_0 i}{2R}\right)^2 = \frac{\mu_0 i^2}{8R^2} = \frac{(4\pi\times10^{-7}\,\text{H/m})\,(10\,\text{A})^2}{8(2.5\times10^{-3}\,\text{m}/2)^2} = 1.0\ \text{J/m}^3\ .$$

(b) The electric energy density is

$$\begin{aligned} u_E &= \frac{1}{2}\varepsilon_0 E^2 = \frac{\epsilon_0}{2}(\rho J)^2 = \frac{\varepsilon_0}{2}\left(\frac{iR}{\ell}\right)^2 \\ &= \frac{1}{2}(8.85\times10^{-12}\,\text{F/m})\left[(10\,\text{A})(3.3\,\Omega/10^3\,\text{m})\right]^2 \\ &= 4.8\times10^{-15}\ \text{J/m}^3\ . \end{aligned}$$

Here we used $J = i/A$ and $R = \rho\ell/A$ to obtain $\rho J = iR/\ell$.

68. (a) The flux in coil 1 is

$$\frac{L_1 i_1}{N_1} = \frac{(25\,\text{mH})(6.0\,\text{mA})}{100} = 1.5\ \mu\text{Wb}\ ,$$

and the magnitude of the self-induced emf is

$$L_1\frac{di_1}{dt} = (25\,\text{mH})(4.0\,\text{A/s}) = 100\ \text{mV}\ .$$

(b) In coil 2, we find

$$\Phi_{21} = \frac{Mi_1}{N_2} = \frac{(3.0\,\text{mH})(6.0\,\text{mA})}{200} = 90\ \text{nWb}\ ,$$

$$\mathcal{E}_{21} = M\frac{di_1}{dt} = (3.0\,\text{mH})(4.0\,\text{A/s}) = 12\ \text{mV}\ .$$

69. (a) Eq. 31-67 yields

$$M = \frac{\mathcal{E}_1}{|di_2/dt|} = \frac{25.0\,\text{mV}}{15.0\,\text{A/s}} = 1.67\ \text{mH}\ .$$

(b) Eq. 31-62 leads to

$$N_2\Phi_{21} = Mi_1 = (1.67\,\text{mH})(3.60\,\text{A}) = 6.00\ \text{mWb}\ .$$

70. We use $\mathcal{E}_2 = -M\,di_1/dt \approx M|\Delta i/\Delta t|$ to find M:

$$M = \left|\frac{\mathcal{E}}{\Delta i_1/\Delta t}\right| = \frac{30\times10^3\,\text{V}}{6.0\,\text{A}/(2.5\times10^{-3}\,\text{s})} = 13\ \text{H}\ .$$

71. (a) We assume the current is changing at (nonzero) rate di/dt and calculate the total emf across both coils. First consider the coil 1. The magnetic field due to the current in that coil points to the right. The magnetic field due to the current in coil 2 also points to the right. When the current increases, both fields increase and both changes in flux contribute emf's in the same direction. Thus, the induced emf's are

$$\mathcal{E}_1 = -(L_1 + M)\frac{di}{dt} \quad \text{and} \quad \mathcal{E}_2 = -(L_2 + M)\frac{di}{dt} .$$

Therefore, the total emf across both coils is

$$\mathcal{E} = \mathcal{E}_1 + \mathcal{E}_2 = -(L_1 + L_2 + 2M)\frac{di}{dt}$$

which is exactly the emf that would be produced if the coils were replaced by a single coil with inductance $L_{\text{eq}} = L_1 + L_2 + 2M$.

(b) We imagine reversing the leads of coil 2 so the current enters at the back of coil rather than the front (as pictured in the diagram). Then the field produced by coil 2 at the site of coil 1 is opposite to the field produced by coil 1 itself. The fluxes have opposite signs. An increasing current in coil 1 tends to increase the flux in that coil, but an increasing current in coil 2 tends to decrease it. The emf across coil 1 is

$$\mathcal{E}_1 = -(L_1 - M)\frac{di}{dt} .$$

Similarly, the emf across coil 2 is

$$\mathcal{E}_2 = -(L_2 - M)\frac{di}{dt} .$$

The total emf across both coils is

$$\mathcal{E} = -(L_1 + L_2 - 2M)\frac{di}{dt} .$$

This the same as the emf that would be produced by a single coil with inductance $L_{\text{eq}} = L_1 + L_2 - 2M$.

72. The coil-solenoid mutual inductance is

$$M = M_{cs} = \frac{N\Phi_{cs}}{i_s} = \frac{N(\mu_0 i_s n\pi R^2)}{i_s} = \mu_0 \pi R^2 nN .$$

As long as the magnetic field of the solenoid is entirely contained within the cross-section of the coil we have $\Phi_{sc} = B_s A_s = B_s \pi R^2$, regardless of the shape, size, or possible lack of close-packing of the coil.

73. Letting the current in solenoid 1 be i, we calculate the flux linkage in solenoid 2. The mutual inductance, then, is this flux linkage divided by i. The magnetic field inside solenoid 1 is parallel to the axis and has uniform magnitude $B = \mu_0 i n_1$, where n_1 is the number of turns per unit length of the solenoid. The cross-sectional area of the solenoid is πR_1^2. Since $\vec{B}$ is normal to the cross section, the flux here is

$$\Phi = AB = \pi R_1^2 \mu_0 n_1 i .$$

Since the magnetic field is zero outside the solenoid, this is also the flux through a cross section of solenoid 2. The number of turns in a length ℓ of solenoid 2 is $N_2 = n_2 \ell$, and the flux linkage is

$$N_2 \Phi = n_2 \ell \pi R_1^2 \mu_0 n_1 i .$$

The mutual inductance is

$$M = \frac{N_2 \Phi}{i} = \pi R_1^2 \ell \mu_0 n_1 n_2 .$$

M does not depend on R_2 because there is no magnetic field in the region between the solenoids. Changing R_2 does not change the flux through solenoid 2, but changing R_1 does.

74. We use the expression for the flux Φ_B over the toroid cross-section derived in our solution to problem 52 obtain the coil-toroid mutual inductance:

$$M_{ct} = \frac{N_c \Phi_{ct}}{i_t} = \frac{N_c}{i_t} \frac{\mu_0 i_t N_t h}{2\pi} \ln\left(\frac{b}{a}\right) = \frac{\mu_0 N_1 N_2 h}{2\pi} \ln\left(\frac{b}{a}\right)$$

where $N_t = N_1$ and $N_c = N_2$. We note that the formula for Φ_B can also be found in the Supplement for the chapter, in Sample Problem 31-11.

75. (a) The flux over the loop cross section due to the current i in the wire is given by

$$\Phi = \int_a^{a+b} B_{\text{wire}} l \, dr = \int_a^{a+b} \frac{\mu_0 i l}{2\pi r} dr = \frac{\mu_0 i l}{2\pi} \ln\left(1 + \frac{b}{a}\right) .$$

Thus,

$$M = \frac{N\Phi}{i} = \frac{N\mu_0 l}{2\pi} \ln\left(1 + \frac{b}{a}\right) .$$

(b) From the formula for M obtained above

$$M = \frac{(100)\,(4\pi \times 10^{-7}\,\text{H/m})\,(0.30\,\text{m})}{2\pi} \ln\left(1 + \frac{8.0}{1.0}\right) = 1.3 \times 10^{-5}\ \text{H} .$$

76. For $t < 0$, no current goes through L_2, so $i_2 = 0$ and $i_1 = \mathcal{E}/R$. As the switch is opened there will be a very brief sparking across the gap. i_1 drops while i_2 increases, both very quickly. The loop rule can be written as

$$\mathcal{E} - i_1 R - L_1 \frac{di_1}{dt} - i_2 R - L_2 \frac{di_2}{dt} = 0 ,$$

where the initial value of i_1 at $t = 0$ is given by $\mathcal{E}/R$ and that of i_2 at $t = 0$ is 0. We consider the situation shortly after $t = 0$. Since the sparking is very brief, we can reasonably assume that both i_1 and i_2 get equalized quickly, before they can change appreciably from their respective initial values. Here, the loop rule requires that $L_1(di_1/dt)$, which is large and negative, must roughly cancel $L_2(di_2/dt)$, which is large and positive:

$$L_1 \frac{di_1}{dt} \approx -L_2 \frac{di_2}{dt} .$$

Let the common value reached by i_1 and i_2 be i, then

$$\frac{di_1}{dt} \approx \frac{\Delta i_1}{\Delta t} = \frac{i - \mathcal{E}/R}{\Delta t}$$

and

$$\frac{di_2}{dt} \approx \frac{\Delta i_2}{\Delta t} = \frac{i - 0}{\Delta t} .$$

The equations above yield

$$L_1\left(i - \frac{\mathcal{E}}{R}\right) = -L_2(i - 0) \implies i = \frac{\mathcal{E} L_1}{L_2 R_1 + L_1 R_2} = \frac{L_1}{L_1 + L_2} \frac{\mathcal{E}}{R} .$$

77. (a) $i_0 = \mathcal{E}/R = 100\,\text{V}/10\,\Omega = 10\,\text{A}$.

(b) $U_B = \frac{1}{2} L i_0^2 = \frac{1}{2}(2.0\,\text{H})(10\,\text{A})^2 = 100\,\text{J}$.

78. We write $i = i_0 e^{-t/\tau_L}$ and note that $i = 10\%\, i_0$. We solve for t:

$$t = \tau_L \ln\left(\frac{i_0}{i}\right) = \frac{L}{R} \ln\left(\frac{i_0}{i}\right) = \frac{2.00\,\text{H}}{3.00\,\Omega} \ln\left(\frac{i_0}{0.100\, i_0}\right) = 1.54\ \text{s} .$$

79. (a) The energy density at any point is given by $u_B = B^2/2\mu_0$, where B is the magnitude of the magnetic field. The magnitude of the field inside a toroid, a distance r from the center, is given by Eq. 30-26: $B = \mu_0 iN/2\pi r$, where N is the number of turns and i is the current. Thus

$$u_B = \frac{1}{2\mu_0}\left(\frac{\mu_0 iN}{2\pi r}\right)^2 = \frac{\mu_0 i^2 N^2}{8\pi^2 r^2} .$$

(b) We evaluate the integral $U_B = \int u_B \, d\mathcal{V}$ over the volume of the toroid. A circular strip with radius r, height h, and thickness dr has volume $d\mathcal{V} = 2\pi rh\, dr$, so

$$U_B = \frac{\mu_0 i^2 N^2}{8\pi^2} 2\pi h \int_a^b \frac{dr}{r} = \frac{\mu_0 i^2 N^2 h}{4\pi} \ln\left(\frac{b}{a}\right) .$$

Substituting in the given values, we find

$$\begin{aligned} U_B &= \frac{(4\pi \times 10^{-7}\,\mathrm{T{\cdot}m/A})(0.500\mathrm{A})^2(1250)^2(13\times 10^{-3}\,\mathrm{m})}{4\pi} \ln\left(\frac{95\,\mathrm{mm}}{52\,\mathrm{mm}}\right) \\ &= 3.06 \times 10^{-4}\ \mathrm{J} . \end{aligned}$$

(c) The inductance is given in Sample Problem 31-11:

$$L = \frac{\mu_0 N^2 h}{2\pi} \ln\left(\frac{b}{a}\right)$$

so the energy is given by

$$U_B = \frac{1}{2} L i^2 = \frac{\mu_0 N^2 i^2 h}{4\pi} \ln\left(\frac{b}{a}\right) .$$

This the exactly the same as the expression found in part (b) and yields the same numerical result.

80. If the solenoid is long and thin, then when it is bent into a toroid $(b-a)/a$ is much less than 1. Therefore,

$$L_{\mathrm{toroid}} = \frac{\mu_0 N^2 h}{2\pi} \ln\left(\frac{b}{a}\right) = \frac{\mu_0 N^2 h}{2\pi} \ln\left(1 + \frac{b-a}{b}\right) \approx \frac{\mu_0 N^2 h(b-a)}{2\pi b} .$$

Since $A = h(b-a)$ is the cross-sectional area and $l = 2\pi b$ is the length of the toroid, we may rewrite this expression for the toroid self-inductance as

$$\frac{L_{\mathrm{toroid}}}{l} \approx \frac{\mu_0 N^2 A}{l^2} = \mu_0 n^2 A ,$$

which indeed reduces to that of a long solenoid. Note that the approximation $\ln(1+x) \approx x$ is used for very small $|x|$.

81. Using Eq. 31-43

$$i = \frac{\mathcal{E}}{R}\left(1 - e^{-t/\tau_L}\right)$$

where $\tau_L = 2.0$ ns, we find

$$t = \tau_L \ln\left(\frac{1}{1 - \frac{iR}{\mathcal{E}}}\right) \approx 1.0\ \mathrm{ns} .$$

82. We note that $n = 100\,\mathrm{turns/cm} = 10000\,\mathrm{turns/m}$. The induced emf is

$$\begin{aligned} \mathcal{E} &= -\frac{d\Phi_B}{dt} = -\frac{d(BA)}{dt} = -A\frac{d}{dt}(\mu_0 n i) = -\mu_0 n \pi r^2 \frac{di}{dt} \\ &= -(4\pi \times 10^{-7}\,\mathrm{T{\cdot}m/A})(10000\,\mathrm{turn/m})(\pi)(25\times 10^{-3}\,\mathrm{m})^2 \left(\frac{0.50\,\mathrm{A} - 1.0\,\mathrm{A}}{10 \times 10^{-3}\,\mathrm{s}}\right) \\ &= 1.2 \times 10^{-3}\ \mathrm{V} . \end{aligned}$$

Note that since $\vec{B}$ only appears inside the solenoid, the area A is be the cross-sectional area of the solenoid, not the (larger) loop.

83. With $\tau_L = L/R = 0.0010$ s, we find the current at $t = 0.0020$ s from Eq. 31-43:

$$i = \frac{\mathcal{E}}{R}\left(1 - e^{-t/\tau_L}\right) = 0.86 \text{ A} .$$

Consequently, the energy stored, from Eq. 31-51, is

$$U_B = \frac{1}{2}Li^2 = 3.7 \times 10^{-3} \text{ J} .$$

84. (a) The magnitude of the average induced emf is

$$\mathcal{E}_{\text{avg}} = \left|\frac{-d\Phi_B}{dt}\right| = \left|\frac{\Delta\Phi_B}{\Delta t}\right| = \frac{BA_i}{t} = \frac{(2.0\,\text{T})(0.20\,\text{m})^2}{0.20\,\text{s}} = -0.40 \text{ V} .$$

(b) The average induced current is

$$i_{\text{avg}} = \frac{\mathcal{E}_{\text{avg}}}{R} = \frac{0.40\,\text{V}}{20 \times 10^{-3}\,\Omega} = 20 \text{ A} .$$

85. (a) As the switch closes at $t = 0$, the current being zero in the inductor serves as an initial condition for the building-up of current in the circuit. Thus, at $t = 0$ any current through the battery is also that through the 20 Ω and 10 Ω resistors. Hence,

$$i = \frac{\mathcal{E}}{30\ \Omega} = 0.40 \text{ A}$$

which results in a voltage drop across the 10 Ω resistor equal to (0.40)(10) = 4.0 V. The inductor must have this same voltage across it $|\mathcal{E}_L|$, and we use (the absolute value of) Eq. 31-37:

$$\frac{di}{dt} = \frac{|\mathcal{E}_L|}{L} = \frac{4.0}{0.010} = 400 \text{ A/s} .$$

(b) Applying the loop rule to the outer loop, we have

$$\mathcal{E} - (0.50 \text{ A})(20\ \Omega) - |\mathcal{E}_L| = 0 .$$

Therefore, $|\mathcal{E}_L| = 2.0$ V, and Eq. 31-37 leads to

$$\frac{di}{dt} = \frac{|\mathcal{E}_L|}{L} = \frac{2.0}{0.010} = 200 \text{ A/s} .$$

(c) As $t \to \infty$, the inductor has $\mathcal{E}_L = 0$ (since the current is no longer changing). Thus, the loop rule (for the outer loop) leads to

$$\mathcal{E} - i\,(20\ \Omega) - |\mathcal{E}_L| = 0 \implies i = 0.60 \text{ A} .$$

86. (a) $L = \Phi/i = 26 \times 10^{-3}\,\text{Wb}/5.5\,\text{A} = 4.7 \times 10^{-3}\,\text{H}$.

(b) We use Eq. 31-43 to solve for t:

$$\begin{aligned} t &= -\tau_L \ln\left(1 - \frac{iR}{\mathcal{E}}\right) = -\frac{L}{R}\ln\left(1 - \frac{iR}{\mathcal{E}}\right) \\ &= -\frac{4.7 \times 10^{-3}\,\text{H}}{0.75\,\Omega}\ln\left[1 - \frac{(2.5\,\text{A})(0.75\,\Omega)}{6.0\,\text{V}}\right] = 2.4 \times 10^{-3} \text{ s} . \end{aligned}$$

87. (a) We use $U_B = \frac{1}{2}Li^2$ to solve for the self-inductance:

$$L = \frac{2U_B}{i^2} = \frac{2(25.0 \times 10^{-3}\,\text{J})}{(60.0 \times 10^{-3}\,\text{A})^2} = 13.9 \text{ H} .$$

(b) Since $U_B \propto i^2$, for U_B to increase by a factor of 4, i must increase by a factor of 2. Therefore, i should be increased to $2(60.0\,\text{mA}) = 120\,\text{mA}$.

88. (a) The self-inductance per meter is

$$\frac{L}{\ell} = \mu_0 n^2 A = (4\pi \times 10^{-7}\,\text{H/m})\,(100\,\text{turns/cm})^2(\pi)(1.6\,\text{cm})^2 = 0.10\ \text{H/m}\ .$$

(b) The induced emf per meter is

$$\frac{\mathcal{E}}{\ell} = \frac{L}{\ell}\frac{di}{dt} = (0.10\,\text{H/m})(13\,\text{A/s}) = 1.3\ \text{V/m}\ .$$

89. (a) The energy needed is

$$U_E = u_E V = \frac{1}{2}\epsilon_0 E^2 V = \frac{1}{2}(8.85 \times 10^{-12}\,\text{F/m})(100\,\text{kV/m})^2(10\,\text{cm})^3 = 4.4 \times 10^{-5}\ \text{J}\ .$$

(b) The energy needed is

$$U_B = u_B V = \frac{1}{2\mu_0}B^2 V = \frac{(1.0\,\text{T})^2}{2\,(4\pi \times 10^{-7}\,\text{H/m})}(10\,\text{cm})^3 = 4.0 \times 10^2\ \text{J}\ .$$

(c) Obviously, since $U_B > U_E$ greater amounts of energy can be stored in the magnetic field.

90. The induced electric field E as a function of r is given by $E(r) = (r/2)(dB/dt)$. So

$$\begin{aligned} a_c &= a_a = \frac{eE}{m} = \frac{er}{2m}\left(\frac{dB}{dt}\right) \\ &= \frac{(1.60 \times 10^{-19}\,\text{C})(5.0 \times 10^{-2}\,\text{m})(10 \times 10^{-3}\,\text{T/s})}{2(9.11 \times 10^{-27}\,\text{kg})} = 4.4 \times 10^7\ \text{m/s}^2\ . \end{aligned}$$

With regard to the directions, $\vec{a}_a$ points to the right and $\vec{a}_c$ points to the left. At point b we have $a_b \propto r_b = 0$.

91. Using Eq. 31-43, we find

$$i = \frac{\mathcal{E}}{R}\left(1 - e^{-t/\tau_L}\right) \implies \tau_L = \frac{t}{\ln\left(\frac{1}{1-\frac{iR}{\mathcal{E}}}\right)} = 22.4\ \text{s}\ .$$

Thus, from Eq. 31-44 (the definition of the time constant), we obtain $L = (22.4\,\text{s})(2.0\,\Omega) = 45$ H.

92. (a) As the switch closes at $t = 0$, the current being zero in the inductors serves as an initial condition for the building-up of current in the circuit. Thus, the current through any element of this circuit is also zero at that instant. Consequently, the loop rule requires the emf ($\mathcal{E}_{L1}$) of the $L_1 = 0.30$ H inductor to cancel that of the battery. We now apply (the absolute value of) Eq. 31-37

$$\frac{di}{dt} = \frac{|\mathcal{E}_{L1}|}{L_1} = \frac{6.0}{0.30} = 20\ \text{A/s}\ .$$

(b) What is being asked for is essentially the current in the battery when the emf's of the inductors vanish (as $t \to \infty$). Applying the loop rule to the outer loop, with $R_1 = 8.0\ \Omega$, we have

$$\mathcal{E} - i R_1 - |\mathcal{E}_{L1}| - |\mathcal{E}_{L2}| = 0 \implies i = \frac{6.0\,\text{V}}{R_1} = 0.75\ \text{A}\ .$$

93. The magnetic flux is

$$\Phi_B = \vec{B}\cdot\vec{A} = BA\cos 57° = \left(4.2\times10^{-6}\,\mathrm{T}\right)\left(2.5\,\mathrm{m}^2\right)\cos 57° = 5.7\times10^{-5}\ \mathrm{Wb}\ .$$

94. From the given information, we find

$$\frac{dB}{dt} = \frac{0.030\ \mathrm{T}}{0.015\ \mathrm{s}} = 2.0\ \mathrm{T/s}\ .$$

Thus, with $N = 1$ and $\cos 30° = \sqrt{3}/2$, and using Faraday's law with Ohm's law, we have

$$i = \frac{|\mathcal{E}|}{R} = \frac{N\pi r^2}{R}\,\frac{\sqrt{3}}{2}\,\frac{dB}{dt} = 0.021\ \mathrm{A}\ .$$

95. Before the fuse blows, the current through the resistor remains zero. We apply the loop theorem to the battery-fuse-inductor loop: $\mathcal{E} - L\,di/dt = 0$. So $i = \mathcal{E}t/L$. As the fuse blows at $t = t_0$, $i = i_0 = 3.0\,\mathrm{A}$. Thus,

$$t_0 = \frac{i_0 L}{\mathcal{E}} = \frac{(3.0\,\mathrm{A})(5.0\,\mathrm{H})}{10\,\mathrm{V}} = 1.5\ \mathrm{s}\ .$$

We do not show the graph here; qualitatively, it would be similar to Fig. 31-14.

96. We write (as functions of time) $V_L(t) = \mathcal{E}e^{-t/\tau_l}$. Considering the first two data points, (V_{L1}, t_1) and (V_{L2}, t_2), satisfying $V_{Li} = \mathcal{E}e^{-t_i/\tau_L}$ $(i = 1, 2)$, we have $V_{L1}/V_{L2} = \mathcal{E}e^{-(t_1-t_2)/\tau_L}$, which gives

$$\tau_L = \frac{t_1 - t_2}{\ln(V_2/V_1)} = \frac{1.0\,\mathrm{ms} - 2.0\,\mathrm{ms}}{\ln(13.8/18.2)} = 3.6\,\mathrm{ms}\ .$$

Therefore, $\mathcal{E} = V_{L1}e^{t_1/\tau_L} = (18.2\,\mathrm{V})e^{1.0\,\mathrm{ms}/3.6\,\mathrm{ms}} = 24\,\mathrm{V}$. One can easily check that the values of τ_L and $\mathcal{E}$ are consistent with the rest of the data points.

97. (a) The energy density is

$$u_B = \frac{B_e^2}{2\mu_0} = \frac{(50\times10^{-6}\,\mathrm{T})^2}{2\,(4\pi\times10^{-7}\,\mathrm{H/m})} = 1.0\times10^{-3}\ \mathrm{J/m}^3\ .$$

(b) The volume of the shell of thickness h is $\mathcal{V} \approx 4\pi R_e^2 h$, where R_e is the radius of the Earth. So

$$U_B \approx \mathcal{V}u_B \approx 4\pi(6.4\times10^6\,\mathrm{m})^2(16\times10^3\,\mathrm{m})(1.0\times10^{-3}\,\mathrm{J/m}^3) = 8.4\times10^{15}\ \mathrm{J}\ .$$

98. (a) $N = 2.0\,\mathrm{m}/2.5\,\mathrm{mm} = 800$.

(b) $L/l = \mu_0 n^2 A = \left(4\pi\times10^{-7}\,\mathrm{H/m}\right)(800/2.0\,\mathrm{m})^2(\pi)(0.040\,\mathrm{m})^2/4 = 2.5\times10^{-4}\,\mathrm{H}$.

99. The self-inductance and resistance of the coil may be treated as a "pure" inductor in series with a "pure" resistor, in which case the situation described in the problem is equivalent to that analyzed in §31-9 with solution Eq. 31-43. The derivative of that solution is

$$\frac{di}{dt} = \frac{\mathcal{E}}{R\,\tau_L}\,e^{-t/\tau_L} = \frac{\mathcal{E}}{L}\,e^{-t/\tau_L}\ .$$

With $\tau_L = 0.28$ ms (by Eq. 31-44), $L = 0.050$ H and $\mathcal{E} = 45$ V, we obtain $di/dt = 12$ A/s when $t = 1.2$ ms.

100. (a) We apply Newton's second law to the rod

$$m\frac{dv}{dt} = iBL\ ,$$

and integrate to obtain

$$v = \frac{iBLt}{m}\ .$$

The velocity $\vec{v}$ points away from the generator G.

(b) When the current i in the rod becomes zero, the rod will no longer be accelerated by a force $F = iBL$ and will therefore reach a constant terminal velocity. This occurs when $|\mathcal{E}_{\text{induced}}| = \mathcal{E}$. Specifically,

$$|\mathcal{E}_{\text{induced}}| = \left|\frac{d\Phi_B}{dt}\right| = \left|\frac{d(BA)}{dt}\right| = B\left|\frac{dA}{dt}\right| = BvL = \mathcal{E} \ .$$

Thus, $\vec{v} = \mathcal{E}/BL$, leftward.

(c) In case (a) electric energy is supplied by the generator and is transferred into the kinetic energy of the rod. In the case considered now the battery initially supplies electric energy to the rod, causing its kinetic energy to increase to a maximum value of $\frac{1}{2}mv^2 = \frac{1}{2}(\mathcal{E}/BL)^2$. Afterwards, there is no further energy transfer from the battery to the rod, and the kinetic energy of the rod remains constant.

101. (a) At $t = 0.50$ s and $t = 1.5$ s, the magnetic field is decreasing at a rate of 3/2 mT/s, leading to

$$i = \frac{|\mathcal{E}|}{R} = \frac{A\,|dB/dt|}{R} = \frac{(3.0)(3/2)}{9.0} = 0.50 \text{ mA}$$

with a counterclockwise sense (by Lenz's law).

(b) See the results of part (a).

(c) and (d) For $t > 2.0$ s, there is no change in flux and therefore no induced current.

102. The magnetic flux is

$$\begin{aligned}\Phi_B &= BA = \left(\frac{\mu_0 i_0 N}{2\pi r}\right) A \\ &= \frac{(4\pi \times 10^{-7}\,\text{H/m})\,(0.800\,\text{A})(500)(5.00\times 10^{-2}\,\text{m})^2}{2\pi(0.150\,\text{m} + 0.0500\,\text{m}/2)} \\ &= 1.15 \times 10^{-6} \text{ Wb} \ .\end{aligned}$$

103. (a) As the switch closes at $t = 0$, the current being zero in the inductor serves as an initial condition for the building-up of current in the circuit. Thus, at $t = 0$ the current through the battery is also zero.

(b) With no current anywhere in the circuit at $t = 0$, the loop rule requires the emf of the inductor $\mathcal{E}_L$ to cancel that of the battery ($\mathcal{E} = 40$ V). Thus, the absolute value of Eq. 31-37 yields

$$\frac{di}{dt} = \frac{|\mathcal{E}_L|}{L} = \frac{40}{0.050} = 800 \text{ A/s} \ .$$

(c) This circuit becomes equivalent to that analyzed in §31-9 when we replace the parallel set of 20000 Ω resistors with $R = 10000$ Ω. Now, with $\tau_L = L/R = 5 \times 10^{-6}$ s, we have $t/\tau_L = 3/5$, and we apply Eq. 31-43:

$$i = \frac{\mathcal{E}}{R}\left(1 - e^{-3/5}\right) \approx 1.8 \times 10^{-3} \text{ A} \ .$$

(d) The rate of change of the current is figured from the loop rule (and Eq. 31-37):

$$\mathcal{E} - iR - |\mathcal{E}_L| = 0 \ .$$

Using the values from part (c), we obtain $|\mathcal{E}_L| \approx 22$ V. Then,

$$\frac{di}{dt} = \frac{|\mathcal{E}_L|}{L} = \frac{22}{0.050} \approx 440 \text{ A/s} \ .$$

(e) and (f) As $t \to \infty$, the circuit reaches a steady state condition, so that $di/dt = 0$ and $\mathcal{E}_L = 0$. The loop rule then leads to

$$\mathcal{E} - iR - |\mathcal{E}_L| = 0 \implies i = \frac{40}{10000} = 4.0 \times 10^{-3} \text{ A} .$$

104. The magnetic flux Φ_B through the loop is given by $\Phi_B = 2B(\pi r^2/2)(\cos 45°) = \pi r^2 B/\sqrt{2}$. Thus

$$\begin{aligned} \mathcal{E} &= -\frac{d\Phi_B}{dt} = -\frac{d}{dt}\left(\frac{\pi r^2 B}{\sqrt{2}}\right) = -\frac{\pi r^2}{\sqrt{2}}\left(\frac{\Delta B}{\Delta t}\right) \\ &= -\frac{\pi(3.7 \times 10^{-2}\,\text{m})^2}{\sqrt{2}}\left(\frac{0 - 76 \times 10^{-3}\,\text{T}}{4.5 \times 10^{-3}\,\text{s}}\right) \\ &= 5.1 \times 10^{-2} \text{ V} . \end{aligned}$$

The direction of the induced current is clockwise when viewed along the direction of $\vec{B}$.

105. The area enclosed by any turn of the coil is πr^2 where $r = 0.15$ m, and the coil has $N = 50$ turns. Thus, the magnitude of the induced emf, using Eq. 31-7, is

$$|\mathcal{E}| = N\pi r^2 \left|\frac{dB}{dt}\right| = (3.53 \text{ m}^2)\left|\frac{dB}{dt}\right|$$

where $\left|\frac{dB}{dt}\right| = (0.0126 \text{ T/s})|\cos \omega t|$. Thus, using Ohm's law, we have

$$i = \frac{|\mathcal{E}|}{R} = \frac{(3.53)(0.0126)}{4.0}|\cos \omega t| .$$

When $t = 0.020$ s, this yields $i = 0.011$ A.

106. (First problem of **Cluster**)
Combining Ohm's and Faraday's laws, the current magnitude is

$$i = \frac{|\mathcal{E}|}{R} = \frac{BLv}{R}$$

for this "one-loop" circuit, where the version of Faraday's law expressed in Eq. 31-10 (often called "motional emf") has been used. Here, $B = |\vec{B}| = 0.200$ T, $L = 0.300$ m and $v = 12.0$ m/s. Reasoning with Lenz's law, the sense of the induced current is *counterclockwise* (to produce field in its interior out of the page, "fighting" the increasing inward pointed flux due to the applied field).

(a) With $R = 5.00\ \Omega$, this yields $i = 0.144$ A.

(b) With $R = 7.00\ \Omega$, we obtain $i = 0.103$ A.

107. (Second problem of **Cluster**)

(a) With $L = 0.50$ m and $R = 5.00\ \Omega$, we combine Ohm's and Faraday's laws, so that the current magnitude is

$$i = \frac{|\mathcal{E}|}{R} = \frac{BLv}{R} = 0.240 \text{ A} .$$

The direction is counterclockwise, as explained in the solution to the previous problem.

(b) The area in the loop is $A = \frac{1}{2}(L_0 + L)x$ where $x = vt$ and $L_0 = 0.300$ m. But the value of L depends on the distance from the resistor x:

$$\begin{aligned} L &= 30 \text{ cm} + \left(\frac{20 \text{ cm}}{1 \text{ m}}\right)x \\ &= L_0 + 0.200(vt) \end{aligned}$$

where $x = vt$ has been used. Therefore, the area becomes

$$A = L_0\, vt + 0.100\, v^2 t^2 \quad .$$

The induced emf is, from Faraday's law,

$$\mathcal{E} = \frac{d\Phi}{dt} = B\,\frac{dA}{dt} = B\left(L_0\, v + 2(0.100)v^2 t\right)$$

and the induced current is

$$i = \frac{\mathcal{E}}{R} = 0.144 + 1.152t$$

in SI units and is counterclockwise (for reasons given in previous solution).

108. (Third problem of **Cluster**)

(a) , (b) and (c) The area enclosed by the loop is that of a rectangle with one side (x) expanding. With $B_0 = 0.200$ T and $\xi = 0.050$ T/s (the rate of field increase), we have

$$\begin{aligned} \Phi &= BA = (B_0 + \xi t)\,(Lx) \\ &= B_0 Lvt + \xi Lvt^2 \end{aligned}$$

where $x = vt$ has been used. Thus, from Faraday's and Ohm's laws, the induced current is

$$i = \frac{\mathcal{E}}{R} = \frac{B_0 Lv}{R} + 2\frac{\xi Lv}{R}\,t$$

and is counterclockwise (to produce field in the loop's interior pointing out of the page, "fighting" the increasing inward pointed flux due to the applied field). Therefore, the current at $t = 0$ is $B_0 Lv/R = 0.144$ A. And its value at $t = 1.00$ s is $(B_0 + 2\xi)Lv/R = 0.216$ A.

Chapter 32

1. (a) Since the field lines of a bar magnet point towards its South pole, then the $\vec{B}$ arrows in one's sketch should point generally towards the left and also towards the central axis.

   (b) The sign of $\vec{B}\cdot d\vec{A}$ for every $d\vec{A}$ on the side of the paper cylinder is negative.

   (c) No, because Gauss' law for magnetism applies to an *enclosed* surface only. In fact, if we include the top and bottom of the cylinder to form an enclosed surface S then $\oint_s \vec{B}\cdot d\vec{A} = 0$ will be valid, as the flux through the open end of the cylinder near the magnet is positive.

2. We use $\sum_{n=1}^{6} \Phi_{Bn} = 0$ to obtain

$$\Phi_{B6} = -\sum_{n=1}^{5} \Phi_{Bn} = -(-1\,\mathrm{Wb} + 2\,\mathrm{Wb} - 3\,\mathrm{Wb} + 4\,\mathrm{Wb} - 5\,\mathrm{Wb}) = +3\ \mathrm{Wb}\ .$$

3. We use Gauss' law for magnetism: $\oint \vec{B}\cdot d\vec{A} = 0$. Now, $\oint \vec{B}\cdot d\vec{A} = \Phi_1 + \Phi_2 + \Phi_C$, where Φ_1 is the magnetic flux through the first end mentioned, Φ_2 is the magnetic flux through the second end mentioned, and Φ_C is the magnetic flux through the curved surface. Over the first end the magnetic field is inward, so the flux is $\Phi_1 = -25.0\,\mu\mathrm{Wb}$. Over the second end the magnetic field is uniform, normal to the surface, and outward, so the flux is $\Phi_2 = AB = \pi r^2 B$, where A is the area of the end and r is the radius of the cylinder. It value is

$$\Phi_2 = \pi(0.120\,\mathrm{m})^2(1.60\times 10^{-3}\,\mathrm{T}) = +7.24\times 10^{-5}\,\mathrm{Wb} = +72.4\,\mu\mathrm{Wb}\ .$$

   Since the three fluxes must sum to zero,

$$\Phi_C = -\Phi_1 - \Phi_2 = 25.0\,\mu\mathrm{Wb} - 72.4\,\mu\mathrm{Wb} = -47.4\,\mu\mathrm{Wb}\ .$$

   The minus sign indicates that the flux is inward through the curved surface.

4. The flux through Arizona is

$$\Phi = -B_r A = -(43\times 10^{-6}\,\mathrm{T})(295{,}000\,\mathrm{km}^2)(10^3\,\mathrm{m/km})^2 = -1.3\times 10^7\ \mathrm{Wb}\ ,$$

   inward. By Gauss' law this is equal to the negative value of the flux Φ' through the rest of the surface of the Earth. So $\Phi' = 1.3\times 10^7$ Wb, outward.

5. The horizontal component of the Earth's magnetic field is given by $B_h = B\cos\phi_i$, where B is the magnitude of the field and ϕ_i is the inclination angle. Thus

$$B = \frac{B_h}{\cos\phi_i} = \frac{16\,\mu\mathrm{T}}{\cos 73^\circ} = 55\,\mu\mathrm{T}\ .$$

6. (a) The Pythagorean theorem leads to

$$\begin{aligned} B &= \sqrt{B_h^2 + B_v^2} = \sqrt{\left(\frac{\mu_0\mu}{4\pi r^3}\cos\lambda_m\right)^2 + \left(\frac{\mu_0\mu}{2\pi r^3}\sin\lambda_m\right)^2} \\ &= \frac{\mu_0\mu}{4\pi r^3}\sqrt{\cos^2\lambda_m + 4\sin^2\lambda_m} = \frac{\mu_0\mu}{4\pi r^3}\sqrt{1+3\sin^2\lambda_m}\ , \end{aligned}$$

where $\cos^2\lambda_m + \sin^2\lambda_m = 1$ was used.

(b) We use Eq. 3-6:

$$\tan\phi_i = \frac{B_v}{B_h} = \frac{(\mu_0\mu/2\pi r^3)\sin\lambda_m}{(\mu_0\mu/4\pi r^3)\cos\lambda_m} = 2\tan\lambda_m\ .$$

7. (a) At the magnetic equator ($\lambda_m = 0$), the field is

$$B = \frac{\mu_0\mu}{4\pi r^3} = \frac{(4\pi\times 10^{-7}\,\text{T}\cdot\text{m/A})\,(8.00\times 10^{22}\,\text{A}\cdot\text{m}^2)}{4\pi(6.37\times 10^6\,\text{m})^3} = 3.10\times 10^{-5}\,\text{T}\ ,$$

and $\phi_i = \tan^{-1}(2\tan\lambda_m) = \tan^{-1}(0) = 0$.

(b) At $\lambda_m = 60°$, we find

$$B = \frac{\mu_0\mu}{4\pi r^3}\sqrt{1+3\sin^2\lambda_m} = (3.10\times 10^{-5})\sqrt{1+3\sin^2 60°} = 5.6\times 10^{-5}\,\text{T}\ ,$$

and $\phi_i = \tan^{-1}(2\tan 60°) = 74°$.

(c) At the north magnetic pole ($\lambda_m = 90.0°$), we obtain

$$B = \frac{\mu_0\mu}{4\pi r^3}\sqrt{1+3\sin^2\lambda_m} = (3.1\times 10^{-5})\sqrt{1+3(1.00)^2} = 6.20\times 10^{-5}\,\text{T}\ ,$$

and $\phi_i = \tan^{-1}(2\tan 90°) = 90°$.

8. (a) At a distance r from the center of the Earth, the magnitude of the magnetic field is given by

$$B = \frac{\mu_0\mu}{4\pi r^3}\sqrt{1+3\sin^2\lambda_m}\ ,$$

where μ is the Earth's dipole moment and λ_m is the magnetic latitude. The ratio of the field magnitudes for two different distances at the same latitude is

$$\frac{B_2}{B_1} = \frac{r_1^3}{r_2^3}\ .$$

With B_1 being the value at the surface and B_2 being half of B_1, we set r_1 equal to the radius R_e of the Earth and r_2 equal to $R_e + h$, where h is altitude at which B is half its value at the surface. Thus,

$$\frac{1}{2} = \frac{R_e^3}{(R_e+h)^3}\ .$$

Taking the cube root of both sides and solving for h, we get

$$h = \left(2^{1/3}-1\right)R_e = \left(2^{1/3}-1\right)(6370\,\text{km}) = 1660\,\text{km}\ .$$

(b) We use the expression for B obtained in problem 6, part (a). For maximum B, we set $\sin\lambda_m = 1$. Also, $r = 6370\,\text{km} - 2900\,\text{km} = 3470\,\text{km}$. Thus,

$$\begin{aligned} B_{\text{max}} &= \frac{\mu_0\mu}{4\pi r^3}\sqrt{1+3\sin^2\lambda_m} \\ &= \frac{(4\pi\times 10^{-7}\,\text{T}\cdot\text{m/A})\,(8.00\times 10^{22}\,\text{A}\cdot\text{m}^2)}{4\pi(3.47\times 10^6\,\text{m})^3}\sqrt{1+3(1)^2} = 3.83\times 10^{-4}\,\text{T}\ . \end{aligned}$$

(c) The angle between the magnetic axis and the rotational axis of the Earth is 11.5°, so $\lambda_m = 90.0° - 11.5° = 78.5°$ at Earth's geographic north pole. Also $r = R_e = 6370\,\text{km}$. Thus,

$$\begin{aligned} B &= \frac{\mu_0 \mu}{4\pi R_E^3}\sqrt{1 + 3\sin^2\lambda_m} \\ &= \frac{(4\pi \times 10^{-7}\,\text{T}\cdot\text{m/A})\,(8.0 \times 10^{22}\,\text{J/T})\sqrt{1 + 3\sin^2 78.5°}}{4\pi (6.37 \times 10^6\,\text{m})^3} = 6.11 \times 10^{-5}\ \text{T}\ , \end{aligned}$$

and, using the result of part (b) of problem 6,

$$\phi_i = \tan^{-1}(2\tan 78.5°) = 84.2°\ .$$

A plausible explanation to the discrepancy between the calculated and measured values of the Earth's magnetic field is that the formulas we obtained in problem 6 are based on dipole approximation, which does not accurately represent the Earth's actual magnetic field distribution on or near its surface. (Incidentally, the dipole approximation becomes more reliable when we calculate the Earth's magnetic field far from its center.)

9. We use Eq. 32-11: $\mu_{\text{orb},z} = -m_l \mu_B$.

   (a) For $m_l = 1$, $\mu_{\text{orb},z} = -(1)\,(9.27 \times 10^{-24}\,\text{J/T}) = -9.27 \times 10^{-24}\,\text{J/T}$.

   (b) For $m_l = -2$, $\mu_{\text{orb},z} = -(-2)\,(9.27 \times 10^{-24}\,\text{J/T}) = 1.85 \times 10^{-23}\,\text{J/T}$.

10. We use Eq. 32-7 to obtain $\Delta U = -\Delta(\mu_{s,z} B) = -B\Delta\mu_{s,z}$, where $\mu_{s,z} = \pm eh/4\pi m_e = \pm\mu_B$ (see Eqs. 32-4 and 32-5). Thus,

$$\Delta U = -B[\mu_B - (-\mu_B)] = 2\mu_B B = 2\,(9.27 \times 10^{-24}\,\text{J/T})\,(0.25\,\text{T}) = 4.6 \times 10^{-24}\ \text{J}\ .$$

11. (a) Since $m_l = 0$, $L_{\text{orb},z} = m_l h/2\pi = 0$.

    (b) Since $m_l = 0$, $\mu_{\text{orb},z} = -m_l \mu_B = 0$.

    (c) Since $m_l = 0$, then from Eq. 32-12, $U = -\mu_{\text{orb},z} B_{\text{ext}} = -m_l \mu_B B_{\text{ext}} = 0$.

    (d) Regardless of the value of m_l, we find for the spin part

$$U = -\mu_{s,z} B = \pm\mu_B B = \pm\,(9.27 \times 10^{-24}\,\text{J/T})\,(35\,\text{mT}) = \pm 3.2 \times 10^{-25}\ \text{J}\ .$$

    (e) Now $m_l = -3$, so

$$L_{\text{orb},z} = \frac{m_l h}{2\pi} = \frac{(-3)\,(6.63 \times 10^{-27}\,\text{J}\cdot\text{s})}{2\pi} = -3.16 \times 10^{-34}\ \text{J}\cdot\text{s}$$

and

$$\mu_{\text{orb},z} = -m_l \mu_B = -(-3)\,(9.27 \times 10^{-24}\,\text{J/T}) = 2.78 \times 10^{-23}\ \text{J/T}\ .$$

The potential energy associated with the electron's orbital magnetic moment is now

$$U = -\mu_{\text{orb},z} B_{\text{ext}} = -(2.78 \times 10^{-23}\,\text{J/T})(35 \times 10^{-3}\,\text{T}) = -9.73 \times 10^{-25}\ \text{J}\ ;$$

while the potential energy associated with the electron spin, being independent of m_l, remains the same: $\pm 3.2 \times 10^{-25}$ J.

12. (a) A sketch of the field lines (due to the presence of the bar magnet) in the vicinity of the loop is shown below:

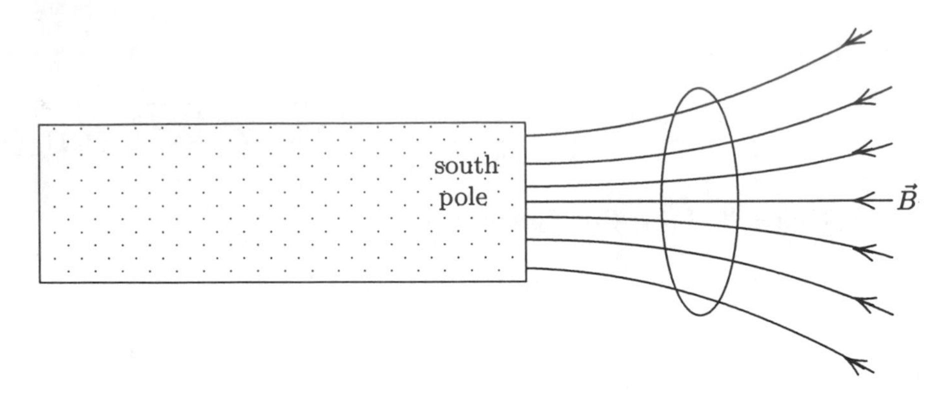

(b) The primary conclusion of §32-6 is two-fold: $\vec{\mu}$ is opposite to $\vec{B}$, and the effect of $\vec{F}$ is to move the material towards regions of smaller $|\vec{B}|$ values. The direction of the magnetic moment vector (of our loop) is toward the left in our sketch.

(c) See our comments in part (b). Since the size of $|\vec{B}|$ relates to the "crowdedness" of the field lines, we see that $\vec{F}$ is towards the right in our sketch.

13. An electric field with circular field lines is induced as the magnetic field is turned on. Suppose the magnetic field increases linearly from zero to B in time t. According to Eq. 31-27, the magnitude of the electric field at the orbit is given by

$$E = \left(\frac{r}{2}\right)\frac{dB}{dt} = \left(\frac{r}{2}\right)\frac{B}{t},$$

where r is the radius of the orbit. The induced electric field is tangent to the orbit and changes the speed of the electron, the change in speed being given by

$$\Delta v = at = \frac{eE}{m_e}t = \left(\frac{e}{m_e}\right)\left(\frac{r}{2}\right)\left(\frac{B}{t}\right)t = \frac{erB}{2m_e}.$$

The average current associated with the circulating electron is $i = ev/2\pi r$ and the dipole moment is

$$\mu = Ai = \left(\pi r^2\right)\left(\frac{ev}{2\pi r}\right) = \frac{1}{2}evr.$$

The change in the dipole moment is

$$\Delta\mu = \frac{1}{2}er\,\Delta v = \frac{1}{2}er\left(\frac{erB}{2m_e}\right) = \frac{e^2r^2B}{4m_e}.$$

14. Reviewing Sample Problem 32-1 before doing this exercise is helpful. Let

$$K = \frac{3}{2}kT = \left|\vec{\mu}\cdot\vec{B} - (-\vec{\mu}\cdot\vec{B})\right| = 2\mu B$$

which leads to

$$T = \frac{4\mu B}{3k} = \frac{4(1.0\times10^{-23}\,\text{J/T})(0.50\,\text{T})}{3(1.38\times10^{-23}\,\text{J/K})} = 0.48\text{ K}.$$

15. The magnetization is the dipole moment per unit volume, so the dipole moment is given by $\mu = M\mathcal{V}$, where M is the magnetization and $\mathcal{V}$ is the volume of the cylinder ($\mathcal{V} = \pi r^2 L$, where r is the radius of the cylinder and L is its length). Thus,

$$\mu = M\pi r^2 L = (5.30\times10^3\,\text{A/m})\pi(0.500\times10^{-2}\,\text{m})^2(5.00\times10^{-2}\,\text{m}) = 2.08\times10^{-2}\text{ J/T}.$$

16. (a) A sketch of the field lines (due to the presence of the bar magnet) in the vicinity of the loop is shown below:

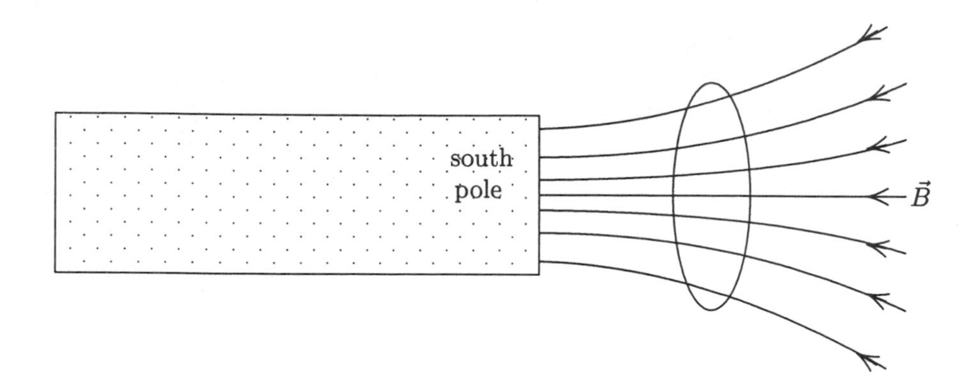

(b) The textbook, in §32-7, makes it clear that $\vec{\mu}$ is in the same direction as $\vec{B}$, and the effect of $\vec{F}$ is to move the material towards regions of larger $|\vec{B}|$ values. The direction of the magnetic moment vector (of our loop) is toward the right in our sketch.

(c) See our comments in part (b). Since the size of $|\vec{B}|$ relates to the "crowdedness" of the field lines, we see that $\vec{F}$ is towards the left in our sketch.

17. For the measurements carried out, the largest ratio of the magnetic field to the temperature is $(0.50\,\text{T})/(10\,\text{K}) = 0.050\,\text{T/K}$. Look at Fig. 32-9 to see if this is in the region where the magnetization is a linear function of the ratio. It is quite close to the origin, so we conclude that the magnetization obeys Curie's law.

18. (a) From Fig. 32-9 we estimate a slope of $B/T = 0.50\,\text{T/K}$ when $M/M_{\text{max}} = 50\%$. So $B = 0.50\,\text{T} = (0.50\,\text{T/K})(300\,\text{K}) = 150\,\text{T}$.

(b) Similarly, now $B/T \approx 2$ so $B = (2)(300) = 600\,\text{T}$.

(c) Except for very short times and in very small volumes, these values are not attainable in the lab.

19. (a) A charge e traveling with uniform speed v around a circular path of radius r takes time $T = 2\pi r/v$ to complete one orbit, so the average current is

$$i = \frac{e}{T} = \frac{ev}{2\pi r} \;.$$

The magnitude of the dipole moment is this multiplied by the area of the orbit:

$$\mu = \frac{ev}{2\pi r}\,\pi r^2 = \frac{evr}{2} \;.$$

Since the magnetic force of with magnitude evB is centripetal, Newton's law yields $evB = m_e v^2/r$, so

$$r = \frac{m_e v}{eB} \;.$$

Thus,

$$\mu = \frac{1}{2}(ev)\left(\frac{m_e v}{eB}\right) = \left(\frac{1}{B}\right)\left(\frac{1}{2}m_e v^2\right) = \frac{K_e}{B} \;.$$

The magnetic force $-e\vec{v} \times \vec{B}$ must point toward the center of the circular path. If the magnetic field is directed into the page, for example, the electron will travel clockwise around the circle. Since the electron is negative, the current is in the opposite direction, counterclockwise and, by the right-hand rule for dipole moments, the dipole moment is out of the page. That is, the dipole moment is directed opposite to the magnetic field vector.

(b) We note that the charge canceled in the derivation of $\mu = K_e/B$. Thus, the relation $\mu = K_i/B$ holds for a positive ion. If the magnetic field is directed into the page, the ion travels counterclockwise around a circular orbit and the current is in the same direction. Therefore, the dipole moment is again out of the page, opposite to the magnetic field.

(c) The magnetization is given by $M = \mu_e n_e + \mu_i n_i$, where μ_e is the dipole moment of an electron, n_e is the electron concentration, μ_i is the dipole moment of an ion, and n_i is the ion concentration. Since $n_e = n_i$, we may write n for both concentrations. We substitute $\mu_e = K_e/B$ and $\mu_i = K_i/B$ to obtain

$$M = \frac{n}{B}(K_e + K_i) = \frac{5.3 \times 10^{21}\,\text{m}^{-3}}{1.2\,\text{T}}\left(6.2 \times 10^{-20}\,\text{J} + 7.6 \times 10^{-21}\,\text{J}\right) = 310\ \text{A/m}\ .$$

20. The Curie temperature for iron is 770°C. If x is the depth at which the temperature has this value, then $10°\text{C} + (30°\text{C/km})x = 770°\text{C}$. Therefore,

$$x = \frac{770°\text{C} - 10°\text{C}}{30°\text{C/km}} = 25\ \text{km}\ .$$

21. (a) The field of a dipole along its axis is given by Eq. 30-29:

$$B = \frac{\mu_0}{2\pi}\frac{\mu}{z^3}\ ,$$

where μ is the dipole moment and z is the distance from the dipole. Thus,

$$B = \frac{(4\pi \times 10^{-7}\,\text{T}\cdot\text{m}/A)(1.5 \times 10^{-23}\,\text{J/T})}{2\pi(10 \times 10^{-9}\,\text{m})} = 3.0 \times 10^{-6}\ \text{T}\ .$$

(b) The energy of a magnetic dipole $\vec{\mu}$ in a magnetic field $\vec{B}$ is given by $U = -\vec{\mu}\cdot\vec{B} = -\mu B\cos\phi$, where ϕ is the angle between the dipole moment and the field. The energy required to turn it end-for-end (from $\phi = 0°$ to $\phi = 180°$) is

$$\Delta U = 2\mu B = 2(1.5 \times 10^{-23}\,\text{J/T})(3.0 \times 10^{-6}\,\text{T}) = 9.0 \times 10^{-29}\ \text{J} = 5.6 \times 10^{-10}\ \text{eV}\ .$$

The mean kinetic energy of translation at room temperature is about 0.04 eV. Thus, if dipole-dipole interactions were responsible for aligning dipoles, collisions would easily randomize the directions of the moments and they would not remain aligned.

22. (a) The number of iron atoms in the iron bar is

$$N = \frac{\left(7.9\,\text{g/cm}^3\right)(5.0\,\text{cm})\left(1.0\,\text{cm}^2\right)}{(55.847\,\text{g/mol})\,/\,(6.022 \times 10^{23}/\text{mol})} = 4.3 \times 10^{23}\ .$$

Thus the dipole moment of the iron bar is

$$\mu = \left(2.1 \times 10^{-23}\,\text{J/T}\right)\left(4.3 \times 10^{23}\right) = 8.9\ \text{A}\cdot\text{m}^2\ .$$

(b) $\tau = \mu B\sin 90° = (8.9\,\text{A}\cdot\text{m}^2)(1.57\,\text{T}) = 13\,\text{N}\cdot\text{m}$.

23. The saturation magnetization corresponds to complete alignment of all atomic dipoles and is given by $M_{\text{sat}} = \mu n$, where n is the number of atoms per unit volume and μ is the magnetic dipole moment of an atom. The number of nickel atoms per unit volume is $n = \rho/m$, where ρ is the density of nickel. The mass of a single nickel atom is calculated using $m = M/N_A$, where M is the atomic mass of nickel and N_A is Avogadro's constant. Thus,

$$\begin{aligned} n &= \frac{\rho N_A}{M} = \frac{(8.90\,\text{g/cm}^3)(6.02 \times 10^{23}\,\text{atoms/mol})}{58.71\,\text{g/mol}} \\ &= 9.126 \times 10^{22}\,\text{atoms/cm}^3 = 9.126 \times 10^{28}\ \text{atoms/m}^3\ . \end{aligned}$$

The dipole moment of a single atom of nickel is

$$\mu = \frac{M_{\text{sat}}}{n} = \frac{4.70 \times 10^5\ \text{A/m}}{9.126 \times 10^{28}\ \text{m}^3} = 5.15 \times 10^{-24}\ \text{A}\cdot\text{m}^2\ .$$

24. From the way the wire is wound it is clear that P_2 is the magnetic north pole while P_1 is the south pole.

    (a) The deflection will be toward P_1 (away from the magnetic north pole).

    (b) As the electromagnet is turned on, the magnetic flux Φ_B through the aluminum changes abruptly, causing a strong induced current which produces a magnetic field opposite to that of the electromagnet. As a result, the aluminum sample will be pushed toward P_1, away from the magnetic north pole of the bar magnet. As Φ_B reaches a constant value, however, the induced current disappears and the aluminum sample, being paramagnetic, will move slightly toward P_2, the magnetic north pole of the electromagnet.

    (c) A magnetic north pole will now be induced on the side of the sample closer to P_1, and a magnetic south pole will appear on the other side. If the magnitude of the field of the electromagnet is larger near P_1, then the sample will move toward P_1.

25. (a) If the magnetization of the sphere is saturated, the total dipole moment is $\mu_{\text{total}} = N\mu$, where N is the number of iron atoms in the sphere and μ is the dipole moment of an iron atom. We wish to find the radius of an iron sphere with N iron atoms. The mass of such a sphere is Nm, where m is the mass of an iron atom. It is also given by $4\pi\rho R^3/3$, where ρ is the density of iron and R is the radius of the sphere. Thus $Nm = 4\pi\rho R^3/3$ and

$$N = \frac{4\pi\rho R^3}{3m}\ .$$

We substitute this into $\mu_{\text{total}} = N\mu$ to obtain

$$\mu_{\text{total}} = \frac{4\pi\rho R^3\mu}{3m}\ .$$

We solve for R and obtain

$$R = \left(\frac{3m\mu_{\text{total}}}{4\pi\rho\mu}\right)^{1/3}\ .$$

The mass of an iron atom is

$$m = 56\,\text{u} = (56\,\text{u})(1.66 \times 10^{-27}\,\text{kg/u}) = 9.30 \times 10^{-26}\ \text{kg}\ .$$

Therefore,

$$R = \left[\frac{3(9.30 \times 10^{-26}\,\text{kg})(8.0 \times 10^{22}\,\text{J/T})}{4\pi(14 \times 10^3\,\text{kg/m}^3)(2.1 \times 10^{-23}\,\text{J/T})}\right]^{1/3} = 1.8 \times 10^5\ \text{m}\ .$$

(b) The volume of the sphere is

$$V_s = \frac{4\pi}{3}R^3 = \frac{4\pi}{3}(1.82 \times 10^5\,\text{m})^3 = 2.53 \times 10^{16}\,\text{m}^3$$

and the volume of the Earth is

$$V_e = \frac{4\pi}{3}(6.37 \times 10^6\,\text{m})^3 = 1.08 \times 10^{21}\ \text{m}^3\ ,$$

so the fraction of the Earth's volume that is occupied by the sphere is

$$\frac{2.53 \times 10^{16}\,\text{m}^3}{1.08 \times 10^{21}\,\text{m}^3} = 2.3 \times 10^{-5}\ .$$

26. Let R be the radius of a capacitor plate and r be the distance from axis of the capacitor. For points with $r \le R$, the magnitude of the magnetic field is given by

$$B = \frac{\mu_0 \varepsilon_0 r}{2} \frac{dE}{dt} ,$$

and for $r \ge R$, it is

$$B = \frac{\mu_0 \varepsilon_0 R^2}{2r} \frac{dE}{dt} .$$

The maximum magnetic field occurs at points for which $r = R$, and its value is given by either of the formulas above:

$$B_{\rm max} = \frac{\mu_0 \varepsilon_0 R}{2} \frac{dE}{dt} .$$

There are two values of r for which $B = B_{\rm max}/2$: one less than R and one greater. To find the one that is less than R, we solve

$$\frac{\mu_0 \varepsilon_0 r}{2} \frac{dE}{dt} = \frac{\mu_0 \varepsilon_0 R}{4} \frac{dE}{dt}$$

for r. The result is $r = R/2 = (55.0\,{\rm mm})/2 = 27.5\,{\rm mm}$. To find the one that is greater than R, we solve

$$\frac{\mu_0 \varepsilon_0 R^2}{2r} \frac{dE}{dt} = \frac{\mu_0 \varepsilon_0 R}{4} \frac{dE}{dt}$$

for r. The result is $r = 2R = 2(55.0\,{\rm mm}) = 110\,{\rm mm}$.

27. We use the result of part (b) in Sample Problem 32-3:

$$B = \frac{\mu_0 \varepsilon_0 R^2}{2r} \frac{dE}{dt} \qquad \text{(for } r \ge R)$$

to solve for dE/dt:

$$\begin{aligned} \frac{dE}{dt} &= \frac{2Br}{\mu_0 \varepsilon_0 R^2} \\ &= \frac{2(2.0 \times 10^{-7}\,T)(6.0 \times 10^{-3}\,{\rm m})}{(4\pi \times 10^{-7}\,{\rm T{\cdot}m/A})\,(8.85 \times 10^{-12}\,\frac{\rm C^2}{\rm N{\cdot}m^2})\,(3.0 \times 10^{-3}\,{\rm m})^2} = 2.4 \times 10^{13}\,\frac{\rm V}{\rm m{\cdot}s} . \end{aligned}$$

28. (a) Noting that the magnitude of the electric field (assumed uniform) is given by $E = V/d$ (where $d = 5.0$ mm), we use the result of part (a) in Sample Problem 32-3

$$B = \frac{\mu_0 \varepsilon_0 r}{2} \frac{dE}{dt} = \frac{\mu_0 \varepsilon_0 r}{2d} \frac{dV}{dt} \qquad \text{(for } r \le R) .$$

We also use the fact that the time derivative of $\sin(\omega t)$ (where $\omega = 2\pi f = 2\pi(60) \approx 377/{\rm s}$ in this problem) is $\omega \cos(\omega t)$. Thus, we find the magnetic field as a function of r (for $r \le R$; note that this neglects "fringing" and related effects at the edges):

$$B = \frac{\mu_0 \varepsilon_0 r}{2d} V_{\rm max} \omega \cos(\omega t) \implies B_{\rm max} = \frac{\mu_0 \varepsilon_0 r V_{\rm max} \omega}{2d}$$

where $V_{\rm max} = 150$ V. This grows with r until reaching its highest value at $r = R = 30$ mm:

$$\begin{aligned} B_{\rm max}\Big|_{r=R} &= \frac{\mu_0 \varepsilon_0 R V_{\rm max} \omega}{2d} \\ &= \frac{(4\pi \times 10^{-7}\,{\rm H/m})\,(8.85 \times 10^{-12}\,{\rm F/m})\,(30 \times 10^{-3}\,{\rm m})\,(150\,{\rm V})(377/{\rm s})}{2(5.0 \times 10^{-3}\,{\rm m})} \\ &= 1.9 \times 10^{-12}\ {\rm T} . \end{aligned}$$

(b) For $r \le 0.03\,\text{m}$, we use the $B_{\text{max}} = \frac{\mu_0\varepsilon_0 r V_{\text{max}}\omega}{2d}$ expression found in part (a) (note the $B \propto r$ dependence), and for $r \ge 0.03\,\text{m}$ we perform a similar calculation starting with the result of part (b)in Sample Problem 32-3:

$$\begin{aligned} B_{\text{max}} &= \left(\frac{\mu_0\varepsilon_0 R^2}{2r}\frac{dE}{dt}\right)_{\text{max}} \\ &= \left(\frac{\mu_0\varepsilon_0 R^2}{2rd}\frac{dV}{dt}\right)_{\text{max}} \\ &= \left(\frac{\mu_0\varepsilon_0 R^2}{2rd} V_{\text{max}}\omega\cos(\omega t)\right)_{\text{max}} \\ &= \frac{\mu_0\varepsilon_0 R^2 V_{\text{max}}\omega}{2rd} \qquad (\text{for } r \ge R) \end{aligned}$$

(note the $B \propto r^{-1}$ dependence – See also Eqs. 32-40 and 32-41). The plot (with SI units understood) is shown below.

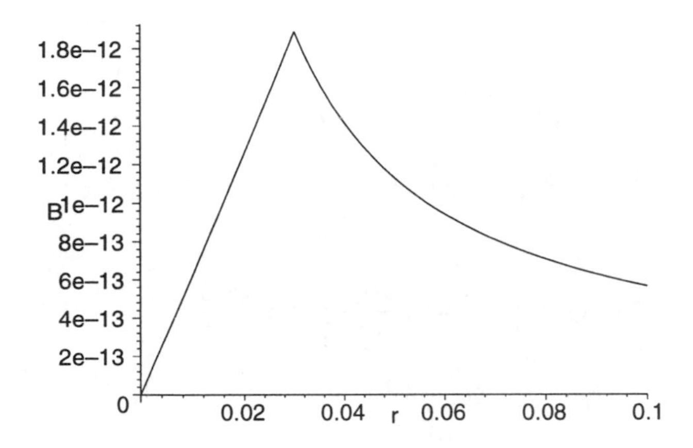

29. The displacement current is given by

$$i_d = \varepsilon_0 A\frac{dE}{dt} ,$$

where A is the area of a plate and E is the magnitude of the electric field between the plates. The field between the plates is uniform, so $E = V/d$, where V is the potential difference across the plates and d is the plate separation. Thus

$$i_d = \frac{\varepsilon_0 A}{d}\frac{dV}{dt} .$$

Now, $\varepsilon_0 A/d$ is the capacitance C of a parallel-plate capacitor (not filled with a dielectric), so

$$i_d = C\frac{dV}{dt} .$$

30. Let the area plate be A and the plate separation be d. We use Eq. 32-34:

$$i_d = \varepsilon_0\frac{d\Phi_E}{dt} = \varepsilon_0\frac{d}{dt}(AE) = \varepsilon_0 A\frac{d}{dt}\left(\frac{V}{d}\right) = \frac{\varepsilon_0 A}{d}\left(\frac{dV}{dt}\right) ,$$

or

$$\frac{dV}{dt} = \frac{i_d d}{\varepsilon_0 A} = \frac{i_d}{C} = \frac{1.5\text{ A}}{2.0\times10^{-6}\text{ F}} = 7.5\times10^5\text{ V/s} .$$

Therefore, we need to change the voltage difference across the capacitor at the rate of 7.5×10^5 V/s.

31. Consider an area A, normal to a uniform electric field $\vec{E}$. The displacement current density is uniform and normal to the area. Its magnitude is given by $J_d = i_d/A$. For this situation

$$i_d = \varepsilon_0 A \frac{dE}{dt} ,$$

so

$$J_d = \frac{1}{A} \varepsilon_0 A \frac{dE}{dt} = \varepsilon_0 \frac{dE}{dt} .$$

32. We use Eq. 32-38:

$$i_d = \varepsilon_0 A \frac{dE}{dt} .$$

Note that, in this situation, A is the area over which a changing electric field is present. In this case $r > R$, so $A = \pi R^2$. Thus,

$$\frac{dE}{dt} = \frac{i_d}{\varepsilon_0 A} = \frac{i_d}{\varepsilon_0 \pi R^2} = \frac{2.0\,\text{A}}{\pi\left(8.85 \times 10^{-12}\,\frac{\text{C}^2}{\text{N}\cdot\text{m}^2}\right)(0.10\,\text{m})^2} = 7.2 \times 10^{12}\,\frac{\text{V}}{\text{m}\cdot\text{s}} .$$

33. (a) We use $\oint \vec{B}\cdot d\vec{s} = \mu_0 I_{\text{enclosed}}$ to find

$$\begin{aligned} B &= \frac{\mu_0 I_{\text{enclosed}}}{2\pi r} = \frac{\mu_0 (J_d \pi r^2)}{2\pi r} = \frac{1}{2}\mu_0 J_d r \\ &= \frac{1}{2}(1.26 \times 10^{-6}\,\text{H/m})(20\,\text{A/m}^2)(50 \times 10^{-3}\,\text{m}) = 6.3 \times 10^{-7}\,\text{T} . \end{aligned}$$

(b) From

$$i_d = J_d \pi r^2 = \varepsilon_0 \frac{d\Phi_E}{dt} = \varepsilon_0 \pi r^2 \frac{dE}{dt}$$

we get

$$\frac{dE}{dt} = \frac{J_d}{\varepsilon_0} = \frac{20\,\text{A/m}^2}{8.85 \times 10^{-12}\,\text{F/m}} = 2.3 \times 10^{12}\,\frac{\text{V}}{\text{m}\cdot\text{s}} .$$

34. (a) From Eq. 32-34,

$$\begin{aligned} i_d &= \varepsilon_0 \frac{d\Phi_E}{dt} = \varepsilon_0 A \frac{dE}{dt} = \varepsilon_0 A \frac{d}{dt}\left[(4.0 \times 10^5) - (6.0 \times 10^4 t)\right] \\ &= -\varepsilon_0 A (6.0 \times 10^4\,\text{V/m}\cdot\text{s}) \\ &= -\left(8.85 \times 10^{-12}\,\frac{\text{C}^2}{\text{N}\cdot\text{m}^2}\right)(4.0 \times 10^{-2}\,\text{m}^2)(6.0 \times 10^4\,\text{V/m}\cdot\text{s}) \\ &= -2.1 \times 10^{-8}\,\text{A} . \end{aligned}$$

(b) If one draws a counterclockwise circular loop s around the plates, then according to Eq. 32-42

$$\oint_s \vec{B}\cdot d\vec{s} = \mu_0 i_d < 0 ,$$

which means that $\vec{B}\cdot d\vec{s} < 0$. Thus $\vec{B}$ must be clockwise.

35. (a) In region a of the graph,

$$\begin{aligned} |i_d| &= \varepsilon_0 \left|\frac{d\Phi_E}{dt}\right| = \varepsilon_0 A \left|\frac{dE}{dt}\right| \\ &= (8.85 \times 10^{-12}\,\text{F/m})(1.6\,\text{m}^2)\left|\frac{4.5 \times 10^5\,\text{N/C} - 6.0 \times 10^5\,\text{N/C}}{4.0 \times 10^{-6}\,\text{s}}\right| = 0.71\,\text{A}. \end{aligned}$$

(b) $i_d \propto dE/dt = 0$.

(c) In region c of the graph,

$$|i_d| = \varepsilon_0 A \left|\frac{dE}{dt}\right| = (8.85\times10^{-12}\,\text{F/m})(1.6\,\text{m}^2)\left|\frac{-4.0\times10^5\,\text{N/C}}{15\times10^{-6}\,\text{s} - 10\times10^{-6}\,\text{s}}\right| = 1.1\ \text{A}\ .$$

36. Using Eq. 32-38, we have

$$\frac{d\left|\vec{E}\right|}{dt} = \frac{i_d}{\varepsilon_0\,A} = 7.2\times10^{12}$$

where $A = \pi(0.10)^2$ (fringing is being neglected in §32-10) and SI units are understood.

37. (a) At any instant the displacement current i_d in the gap between the plates equals the conduction current i in the wires. Thus $i_d = i = 2.0$ A.

(b) The rate of change of the electric field is

$$\frac{dE}{dt} = \frac{1}{\varepsilon_0 A}\left(\varepsilon_0\frac{d\Phi_E}{dt}\right) = \frac{i_d}{\varepsilon_0 A} = \frac{2.0\,\text{A}}{(8.85\times10^{-12}\,\text{F/m})(1.0\,\text{m})^2} = 2.3\times10^{11}\ \frac{\text{V}}{\text{m}\cdot\text{s}}\ .$$

(c) The displacement current through the indicated path is

$$i'_d = i_d \times \left(\frac{\text{area enclosed by the path}}{\text{area of each plate}}\right) = (2.0\,\text{A})\left(\frac{0.50\,\text{m}}{1.0\,\text{m}}\right)^2 = 0.50\ \text{A}\ .$$

(d) The integral of the field around the indicated path is

$$\oint \vec{B}\cdot d\vec{s} = \mu_0 i'_d = (1.26\times10^{-6}\,\text{H/m})(0.50\,\text{A}) = 6.3\times10^{-7}\ \text{T}\cdot\text{m}\ .$$

38. (a) Since $i = i_d$ (Eq. 32-39) then the portion of displacement current enclosed is

$$i_{d,\text{enc}} = i\,\frac{\pi\left(\frac{R}{3}\right)^2}{\pi R^2} = i\,\frac{1}{9} = 1.33\ \text{A}\ .$$

(b) We see from Sample Problems 32-3 and 32-4 that the maximum field is at $r = R$ and that (in the interior) the field is simply proportional to r. Therefore,

$$\frac{B}{B_{\text{max}}} = \frac{3.00\ \text{mT}}{12.0\ \text{mT}} = \frac{r}{R}$$

which yields $r = R/4$ as a solution. We now look for a solution in the exterior region, where the field is inversely proportional to r (by Eq. 32-41):

$$\frac{B}{B_{\text{max}}} = \frac{3.00\ \text{mT}}{12.0\ \text{mT}} = \frac{R}{r}$$

which yields $r = 4R$ as a solution.

39. (a) Using Eq. 27-10, we find

$$E = \rho J = \frac{\rho i}{A} = \frac{(1.62\times10^{-8}\,\Omega\cdot\text{m})(100\,\text{A})}{5.00\times10^{-6}\,\text{m}^2} = 0.324\ \text{V/m}\ .$$

(b) The displacement current is

$$\begin{aligned} i_d &= \varepsilon_0\frac{d\Phi_E}{dt} = \varepsilon_0 A\frac{dE}{dt} = \varepsilon_0 A\frac{d}{dt}\left(\frac{\rho i}{A}\right) = \varepsilon_0\rho\frac{di}{dt} \\ &= (8.85\times10^{-12}\,\text{F})(1.62\times10^{-8}\,\Omega)(2000\,\text{A/s}) = 2.87\times10^{-16}\ \text{A}\ . \end{aligned}$$

(c) The ratio of fields is

$$\frac{B(\text{ due to } i_d)}{B(\text{ due to } i)} = \frac{\mu_0 i_d/2\pi r}{\mu_0 i/2\pi r} = \frac{i_d}{i} = \frac{2.87 \times 10^{-16}\text{ A}}{100\text{ A}} = 2.87 \times 10^{-18} .$$

40. (a) From Sample Problem 32-3 we know that $B \propto r$ for $r \leq R$ and $B \propto r^{-1}$ for $r \geq R$. So the maximum value of B occurs at $r = R$, and there are two possible values of r at which the magnetic field is 75% of B_{max}. We denote these two values as r_1 and r_2, where $r_1 < R$ and $r_2 > R$. Then $0.75B_{\text{max}}/B_{\text{max}} = r_1/R$, or $r_1 = 0.75R$; and $0.75B_{\text{max}}/B_{\text{max}} = (r_2/R)^{-1}$, or $r_2 = R/0.75 = 1.3R$.

(b) From Eqs. 32-39 and 32-41,

$$B_{\text{max}} = \frac{\mu_0 i_d}{2\pi R} = \frac{\mu_0 i}{2\pi R} = \frac{(4\pi \times 10^{-7}\,\text{T}\cdot\text{m/A})\,(6.0\,\text{A})}{2\pi(0.040\,\text{m})} = 3.0 \times 10^{-5}\text{ T} .$$

41. (a) At any instant the displacement current i_d in the gap between the plates equals the conduction current i in the wires. Thus $i_{\text{max}} = i_{d\text{ max}} = 7.60\,\mu\text{A}$.

(b) Since $i_d = \varepsilon_0\,(d\Phi_E/dt)$,

$$\left(\frac{d\Phi_E}{dt}\right)_{\text{max}} = \frac{i_{d\text{ max}}}{\varepsilon_0} = \frac{7.60 \times 10^{-6}\text{ A}}{8.85 \times 10^{-12}\text{ F/m}} = 8.59 \times 10^5\text{V}\cdot\text{m/s} .$$

(c) According to problem 29,

$$i_d = C\frac{dV}{dt} = \frac{\varepsilon_0 A}{d}\frac{dV}{dt} .$$

Now the potential difference across the capacitor is the same in magnitude as the emf of the generator, so $V = \mathcal{E}_{\text{m}} \sin\omega t$ and $dV/dt = \omega\mathcal{E}_{\text{m}}\cos\omega t$. Thus,

$$i_d = \frac{\varepsilon_0 A\omega\mathcal{E}_{\text{m}}}{d}\cos\omega t$$

and

$$i_{d\text{ max}} = \frac{\varepsilon_0 A\omega\mathcal{E}_{\text{m}}}{d} .$$

This means

$$\begin{aligned} d &= \frac{\varepsilon_0 A\omega\mathcal{E}_{\text{m}}}{i_{d\,\text{max}}} = \frac{(8.85 \times 10^{-12}\,\text{F/m})\pi(0.180\text{m})^2(130\,\text{rad/s})(220\text{V})}{7.60 \times 10^{-6}\text{ A}} \\ &= 3.39 \times 10^{-3}\,\text{m} , \end{aligned}$$

where $A = \pi R^2$ was used.

(d) We use the Ampere-Maxwell law in the form $\oint \vec{B} \cdot d\vec{s} = \mu_0 I_d$, where the path of integration is a circle of radius r between the plates and parallel to them. I_d is the displacement current through the area bounded by the path of integration. Since the displacement current density is uniform between the plates $I_d = (r^2/R^2)i_d$, where i_d is the total displacement current between the plates and R is the plate radius. The field lines are circles centered on the axis of the plates, so $\vec{B}$ is parallel to $d\vec{s}$. The field has constant magnitude around the circular path, so $\oint \vec{B}\cdot d\vec{s} = 2\pi r B$. Thus,

$$2\pi r B = \mu_0\left(\frac{r^2}{R^2}\right)i_d$$

and

$$B = \frac{\mu_0 i_d r}{2\pi R^2} .$$

The maximum magnetic field is given by

$$B_{\text{max}} = \frac{\mu_0 i_{d\text{ max}}\, r}{2\pi R^2} = \frac{(4\pi \times 10^{-7}\text{T}\cdot\text{m/A})(7.6 \times 10^{-6}\text{ A})(0.110\text{m})}{2\pi(0.180\text{m})^2} = 5.16 \times 10^{-12}\text{ T} .$$

42. From Gauss' law for magnetism, the flux through S_1 is equal to that through S_2, the portion of the xz plane that lies within the cylinder. Here the normal direction of S_2 is $+y$. Therefore,

$$\begin{aligned}\Phi_B(S_1) &= \Phi_B(S_2) = \int_{-r}^{r} B(x)L\,dx \\ &= 2\int_{-r}^{r} B_{\text{left}}(x)L\,dx \\ &= 2\int_{-r}^{r} \frac{\mu_0 i}{2\pi}\frac{1}{2r-x}L\,dx \quad = \quad \frac{\mu_0 iL}{\pi}\ln 3\,.\end{aligned}$$

43. (a) Again from Fig. 32-9, for $M/M_{\text{max}} = 50\%$ we have $B/T = 0.50$. So $T = B/0.50 = 2/0.50 = 4\,\text{K}$.

(b) Now $B/T = 2.0$, so $T = 2/2.0 = 1\,\text{K}$.

44. (a) For a given value of l, m_l varies from $-l$ to $+l$. Thus, in our case $l = 3$, and the number of different m_l's is $2l + 1 = 2(3) + 1 = 7$. Thus, since $L_{\text{orb},z} \propto m_l$, there are a total of seven different values of $L_{\text{orb},z}$.

(b) Similarly, since $\mu_{\text{orb},z} \propto m_l$, there are also a total of seven different values of $\mu_{\text{orb},z}$.

(c) Since $L_{\text{orb},z} = m_l h/2\pi$, the greatest allowed value of $L_{\text{orb},z}$ is given by $|m_l|_{\text{max}}h/2\pi = 3h/2\pi$; while the least allowed value is given by $|m_l|_{\text{min}}h/2\pi = 0$.

(d) Similar to part (c), since $\mu_{\text{orb},z} = -m_l\mu_B$, the greatest allowed value of $\mu_{\text{orb},z}$ is given by $|m_l|_{\text{max}}\mu_B = 3eh/4\pi m_e$; while the least allowed value is given by $|m_l|_{\text{min}}\mu_B = 0$.

(e) From Eqs. 32-3 and 32-9 the z component of the net angular momentum of the electron is given by

$$L_{\text{net},z} = L_{\text{orb},z} + L_{s,z} = \frac{m_l h}{2\pi} + \frac{m_s h}{2\pi}\,.$$

For the maximum value of $L_{\text{net},z}$ let $m_l = [m_l]_{\text{max}} = 3$ and $m_s = \frac{1}{2}$. Thus

$$[L_{\text{net},z}]_{\text{max}} = \left(3 + \frac{1}{2}\right)\frac{h}{2\pi} = \frac{3.5h}{2\pi}\,.$$

(f) Since the maximum value of $L_{\text{net},z}$ is given by $[m_J]_{\text{max}}h/2\pi$ with $[m_J]_{\text{max}} = 3.5$ (see the last part above), the number of allowed values for the z component of $L_{\text{net},z}$ is given by $2[m_J]_{\text{max}} + 1 = 2(3.5) + 1 = 8$.

45. (a) We use the result of part (a) in Sample Problem 32-3:

$$B = \frac{\mu_0\varepsilon_0 r}{2}\frac{dE}{dt} \qquad (\text{for } r \le R)\,,$$

where $r = 0.80R$ and

$$\frac{dE}{dt} = \frac{d}{dt}\left(\frac{V}{d}\right) = \frac{1}{d}\frac{d}{dt}\left(V_0 e^{-t/\tau}\right) = -\frac{V_0}{\tau d}e^{-t/\tau}\,.$$

Here $V_0 = 100\,\text{V}$. Thus

$$\begin{aligned}B(t) &= \left(\frac{\mu_0\varepsilon_0 r}{2}\right)\left(-\frac{V_0}{\tau d}e^{-t/\tau}\right) = -\frac{\mu_0\varepsilon_0 V_0 r}{2\tau d}e^{-t/\tau} \\ &= -\frac{(4\pi\times10^{-7}\,\text{T}\cdot\text{m/A})\left(8.85\times10^{-12}\,\frac{\text{C}^2}{\text{N}\cdot\text{m}^2}\right)(100\,\text{V})(0.80)(16\,\text{mm})}{2(12\times10^{-3}\,\text{s})(5.0\,\text{mm})}e^{-t/12\,\text{ms}} \\ &= -(1.2\times10^{-13}\,\text{T})\,e^{-t/12\,\text{ms}}\,.\end{aligned}$$

The minus sign here is insignificant.

(b) At time $t = 3\tau$, $B(t) = -(1.2 \times 10^{-13}\,\mathrm{T})e^{-3\tau/\tau} = -5.9 \times 10^{-15}\,\mathrm{T}$.

46. The given value 7.0 mW should be 7.0 mWb. From Eq. 32-1, we have

$$\begin{aligned}(\Phi_B)_{\rm in} &= (\Phi_B)_{\rm out} \\ &= 0.0070\,\mathrm{Wb} + (0.40\,\mathrm{T})(\pi r^2) \\ &= 9.2 \times 10^{-3}\ \mathrm{Wb}\ .\end{aligned}$$

Thus, the magnetic flux at the sides is inward with absolute-value equal to 9.2 mWb.

47. The definition of displacement current is Eq. 32-34, and the formula of greatest convenience here is Eq. 32-41:

$$i_d = \frac{2\pi\, r\, B}{\mu_0} = \frac{2\pi(0.0300\,\mathrm{m})\,(2.00 \times 10^{-6}\,\mathrm{T})}{4\pi \times 10^{-7}\,\mathrm{T\cdot m/A}} = 0.30\ \mathrm{A}\ .$$

48. Ignoring points where the determination of the slope is problematic, we find the interval of largest $\Delta|\vec{E}|/\Delta t$ is $6\,\mu\mathrm{s} < t < 7\,\mu\mathrm{s}$. During that time, we have, from Eq. 32-38,

$$i_d = \varepsilon_0\, A\, \frac{\Delta|\vec{E}|}{\Delta t} = \varepsilon_0\,(2.0\,\mathrm{m}^2)\,(2.0 \times 10^6\,\mathrm{V/m})$$

which yields $i_d = 3.5 \times 10^{-5}$ A.

49. (a) We use the notation $P(\mu)$ for the probability of a dipole being parallel to $\vec{B}$, and $P(-\mu)$ for the probability of a dipole being antiparallel to the field. The magnetization may be thought of as a "weighted average" in terms of these probabilities:

$$M = \frac{N\mu P(\mu) - N\mu P(-\mu)}{P(\mu) + P(-\mu)} = \frac{N\mu\left(e^{\mu B/KT} - e^{-\mu B/KT}\right)}{e^{\mu B/KT} + e^{-\mu B/KT}} = N\mu \tanh\left(\frac{\mu B}{kT}\right)\ .$$

(b) For $\mu B \ll kT$ (that is, $\mu B/kT \ll 1$) we have $e^{\pm\mu B/kT} \approx 1 \pm \mu B/kT$, so

$$M = N\mu \tanh\left(\frac{\mu B}{kT}\right) \approx \frac{N\mu[(1 + \mu B/kT) - (1 - \mu B/kT)]}{(1 + \mu B/kT) + (1 - \mu B/kT)} = \frac{N\mu^2 B}{kT}\ .$$

(c) For $\mu B \gg kT$ we have $\tanh(\mu B/kT) \approx 1$, so

$$M = N\mu \tanh\left(\frac{\mu B}{kT}\right) \approx N\mu\ .$$

(d) One can easily plot the tanh function using, for instance, a graphical calculator. One can then note the resemblance between such a plot and Fig. 32-9. By adjusting the parameters used in one's plot, the curve in Fig. 32-9 can reliably be fit with a tanh function.

50. (a) From Eq. 22-3,

$$E = \frac{e}{4\pi\varepsilon_0 r^2} = \frac{(1.60 \times 10^{-19}\,\mathrm{C})(8.99 \times 10^9\,\mathrm{N\cdot m^2/C^2})}{(5.2 \times 10^{-11}\,\mathrm{m})^2} = 5.3 \times 10^{11}\ \mathrm{N/C}\ .$$

(b) We use Eq. 30-28:

$$B = \frac{\mu_0\, \mu_p}{2\pi\, r^3} = \frac{(4\pi \times 10^{-7}\,\mathrm{T\cdot m/A})\,(1.4 \times 10^{-26}\,\mathrm{J/T})}{2\pi(5.2 \times 10^{-11}\,\mathrm{m})^3} = 2.0 \times 10^{-2}\ \mathrm{T}\ .$$

(c) From Eq. 32-10,

$$\frac{\mu_{\rm orb}}{\mu_p} = \frac{eh/4\pi m_e}{\mu_p} = \frac{\mu_B}{\mu_p} = \frac{9.27 \times 10^{-24}\,\mathrm{J/T}}{1.4 \times 10^{-26}\,\mathrm{J/T}} = 6.6 \times 10^2\ .$$

51. The interacting potential energy between the magnetic dipole of the compass and the Earth's magnetic field is $U = -\vec{\mu}\cdot\vec{B}_e = -\mu B_e \cos\theta$, where θ is the angle between $\vec{\mu}$ and $\vec{B}_e$. For small angle θ

$$U(\theta) = -\mu B_e \cos\theta \quad \approx \quad -\mu B_e\left(1 - \frac{\theta^2}{2}\right) = \frac{1}{2}\kappa\theta^2 - \mu B_e$$

where $\kappa = \mu B_e$. Conservation of energy for the compass then gives

$$\frac{1}{2}I\left(\frac{d\theta}{dt}\right)^2 + \frac{1}{2}\kappa\theta^2 = \text{const.}\ .$$

This is to be compared with the following expression for the mechanical energy of a spring-mass system:

$$\frac{1}{2}m\left(\frac{dx}{dt}\right)^2 + \frac{1}{2}kx^2 = \text{const.}\ ,$$

which yields $\omega = \sqrt{k/m}$. So by analogy, in our case

$$\omega = \sqrt{\frac{\kappa}{I}} = \sqrt{\frac{\mu B_e}{I}} = \sqrt{\frac{\mu B_e}{ml^2/12}}\ ,$$

which leads to

$$\mu = \frac{ml^2\omega^2}{12B_e} = \frac{(0.050\,\text{kg})(4.0\times10^{-2}\,\text{m})^2(45\,\text{rad/s})^2}{12(16\times10^{-6}\,\text{T})} = 8.4\times10^2\ \text{J/T}\ .$$

52. Let the area of each circular plate be A and that of the central circular section be a, then

$$\frac{A}{a} = \frac{\pi R^2}{\pi(R/2)^2} = 4\ .$$

Thus, from Eqs. 32-38 and 32-39 the total discharge current is given by $i = i_d = 4(2.0\,\text{A}) = 8.0\,\text{A}$.

53. (a) Using Eq. 32-11, we find $\mu_{\text{orb},z} = -3\mu_B = -2.78\times10^{-23}$ J/T (that these are acceptable units for magnetic moment is seen from Eq. 32-12 or Eq. 32-7; they are equivalent to $\text{A}\cdot\text{m}^2$).

(b) Similarly, for $m_\ell = -4$ we obtain $\mu_{\text{orb},z} = 3.71\times10^{-23}$ J/T.

54. (a) Since the field is decreasing, the displacement current (by Eq. 32-38) is downward, which produces (by the right-hand rule) a clockwise sense for the induced magnetic field.

(b) See the solution for part (a).

(c) and (d) We write $\vec{E} = E_z\hat{\text{k}} = (E_0 - \xi t)\hat{\text{k}}$ where $\xi = 60000(\text{V/m})/\text{s}$. From Eq. 32-36 (treated in absolute value)

$$i_d = \varepsilon_0\, A\left|\frac{dE_z}{dt}\right| = \varepsilon_0\, A\,\xi$$

which yields $i_d = 2.1\times10^{-8}$ A for all values of t.

55. (a) From $\mu = iA = i\pi R_e^2$ we get

$$i = \frac{\mu}{\pi R_e^2} = \frac{8.0\times10^{22}\,\text{J/T}}{\pi(6.37\times10^6\,\text{m})^2} = 6.3\times10^8\ \text{A}\ .$$

(b) Yes, because far away from the Earth the fields of both the Earth itself and the current loop are dipole fields. If these two dipoles cancel each other out, then the net field will be zero.

(c) No, because the field of the current loop is not that of a magnetic dipole in the region close to the loop.

56. (a) The period of rotation is $T = 2\pi/\omega$ and in this time all the charge passes any fixed point near the ring. The average current is $i = q/T = q\omega/2\pi$ and the magnitude of the magnetic dipole moment is

$$\mu = iA = \frac{q\omega}{2\pi}\pi r^2 = \frac{1}{2}q\omega r^2 .$$

(b) We curl the fingers of our right hand in the direction of rotation. Since the charge is positive, the thumb points in the direction of the dipole moment. It is the same as the direction of the angular momentum vector of the ring.

57. (a) The potential energy of the atom in association with the presence an external magnetic field $\vec{B}_{\text{ext}}$ is given by Eqs. 32-11 and 32-12:

$$U = -\mu_{\text{orb}} \cdot \vec{B}_{\text{ext}} = -\mu_{\text{orb},z} B_{\text{ext}} = -m_l \mu_B B_{\text{ext}} .$$

For level E_1 there is no change in energy as a result of the introduction of $\vec{B}_{\text{ext}}$, so $U \propto m_l = 0$, meaning that $m_l = 0$ for this level. For level E_2 the single level splits into a triplet (i.e., three separate ones) in the presence of $\vec{B}_{\text{ext}}$, meaning that there are three different values of m_l. The middle one in the triplet is unshifted from the original value of E_2 so its m_l must be equal to 0.

(b) The other two in the triplet then correspond to $m_l = -1$ and $m_1 = +1$, respectively.

(c) For any pair of adjacent levels in the triplet $|\Delta m_l| = 1$. Thus, the spacing is given by

$$\begin{aligned} \Delta U &= |\Delta(-m_l \mu_B B)| = |\Delta m_l| \mu_B B = \mu_B B \\ &= (9.27 \times 10^{-24}\,\text{J/T})\,(0.50\,\text{T}) = 4.6 \times 10^{-24}\,\text{J} \end{aligned}$$

which is equivalent to 2.9×10^{-5} eV.

58. (a) The magnitude of the toroidal field is given by $B_0 = \mu_0 n i_p$, where n is the number of turns per unit length of toroid and i_p is the current required to produce the field (in the absence of the ferromagnetic material). We use the average radius ($r_{\text{avg}} = 5.5$cm) to calculate n:

$$n = \frac{N}{2\pi r_{\text{avg}}} = \frac{400\,\text{turns}}{2\pi(5.5 \times 10^{-2}\,\text{m})} = 1.16 \times 10^3\,\text{turns/m} .$$

Thus,

$$i_p = \frac{B_0}{\mu_0 n} = \frac{0.20 \times 10^{-3}\,\text{T}}{(4\pi \times 10^{-7}\,\text{T}\cdot\text{m/A})(1.16 \times 10^3/\text{m})} = 0.14\,\text{A} .$$

(b) If Φ is the magnetic flux through the secondary coil, then the magnitude of the emf induced in that coil is $\mathcal{E} = N(d\Phi/dt)$ and the current in the secondary is $i_s = \mathcal{E}/R$, where R is the resistance of the coil. Thus

$$i_s = \left(\frac{N}{R}\right)\frac{d\Phi}{dt} .$$

The charge that passes through the secondary when the primary current is turned on is

$$q = \int i_s\,dt = \frac{N}{R}\int \frac{d\Phi}{dt}\,dt = \frac{N}{R}\int_0^{\Phi} d\Phi = \frac{N\Phi}{R} .$$

The magnetic field through the secondary coil has magnitude $B = B_0 + B_M = 801 B_0$, where B_M is the field of the magnetic dipoles in the magnetic material. The total field is perpendicular to the plane of the secondary coil, so the magnetic flux is $\Phi = AB$, where A is the area of the Rowland ring (the field is inside the ring, not in the region between the ring and coil). If r is the radius of the ring's cross section, then $A = \pi r^2$. Thus

$$\Phi = 801\pi r^2 B_0 \ .$$

The radius r is $(6.0\,\text{cm} - 5.0\,\text{cm})/2 = 0.50\,\text{cm}$ and

$$\Phi = 801\pi(0.50 \times 10^{-2}\,\text{m})^2(0.20 \times 10^{-3}\,\text{T}) = 1.26 \times 10^{-5}\,\text{Wb} \ .$$

Consequently,

$$q = \frac{50(1.26 \times 10^{-5}\,\text{Wb})}{8.0\,\Omega} = 7.9 \times 10^{-5}\ \text{C} \ .$$

59. Combining Eq. 32-7 with Eq. 32-2 and Eq. Eq. 32-3, we obtain

$$\Delta U = 2\,\mu_B\,B$$

where μ_B is the Bohr magneton (evaluated in Eq. 32-5). Thus, with $\Delta U = 6.0 \times 10^{-25}$ J, we find $B = |\vec{B}| = 0.032$ T.

60. (a) Using Eq. 32-37 but noting that the capacitor is being *discharged*, we have

$$\frac{d|\vec{E}|}{dt} = -\frac{i}{\varepsilon_0\,A} = -8.8 \times 10^{15}$$

where $A = (0.0080)^2$ and SI units are understood.

(b) Assuming a perfectly uniform field, even so near to an edge (which is consistent with the fact that fringing is neglected in §32-10), we follow part (a) of Sample Problem 32-4 and relate the (absolute value of the) line integral to the portion of displacement current enclosed.

$$\begin{aligned} \left|\oint \vec{B} \cdot d\vec{s}\right| &= \mu_0\, i_{d,\text{enc}} \\ &= \mu_0\,\frac{W\,H}{L^2}\,i \\ &= 5.9 \times 10^{-7}\ \text{Wb/m} \ . \end{aligned}$$

Chapter 33

1. We find the capacitance from $U = \frac{1}{2}Q^2/C$:

$$C = \frac{Q^2}{2U} = \frac{(1.60 \times 10^{-6}\,\mathrm{C})^2}{2(140 \times 10^{-6}\,\mathrm{J})} = 9.14 \times 10^{-9}\ \mathrm{F}\ .$$

2. According to $U = \frac{1}{2}LI^2 = \frac{1}{2}Q^2/C$, the current amplitude is

$$I = \frac{Q}{\sqrt{LC}} = \frac{3.00 \times 10^{-6}\,\mathrm{C}}{\sqrt{(1.10 \times 10^{-3}\,\mathrm{H})(4.00 \times 10^{-6}\,\mathrm{F})}} = 4.52 \times 10^{-2}\ \mathrm{A}\ .$$

3. (a) All the energy in the circuit resides in the capacitor when it has its maximum charge. The current is then zero. If Q is the maximum charge on the capacitor, then the total energy is

$$U = \frac{Q^2}{2C} = \frac{(2.90 \times 10^{-6}\,\mathrm{C})^2}{2(3.60 \times 10^{-6}\,\mathrm{F})} = 1.17 \times 10^{-6}\ \mathrm{J}\ .$$

(b) When the capacitor is fully discharged, the current is a maximum and all the energy resides in the inductor. If I is the maximum current, then $U = LI^2/2$ leads to

$$I = \sqrt{\frac{2U}{L}} = \sqrt{\frac{2(1.168 \times 10^{-6}\,\mathrm{J})}{75 \times 10^{-3}\,\mathrm{H}}} = 5.58 \times 10^{-3}\ \mathrm{A}\ .$$

4. (a) The period is $T = 4(1.50\,\mu\mathrm{s}) = 6.00\,\mu\mathrm{s}$.

(b) The frequency is the reciprocal of the period:

$$f = \frac{1}{T} = \frac{1}{6.00\,\mu\mathrm{s}} = 1.67 \times 10^5\ \mathrm{Hz}\ .$$

(c) The magnetic energy does not depend on the direction of the current (since $U_B \propto i^2$), so this will occur after one-half of a period, or $3.00\,\mu\mathrm{s}$.

5. (a) We recall the fact that the period is the reciprocal of the frequency. It is helpful to refer also to Fig. 33-1. The values of t when plate A will again have maximum positive charge are multiples of the period:

$$t_A = nT = \frac{n}{f} = \frac{n}{2.00 \times 10^3\,\mathrm{Hz}} = n(5.00\,\mu\mathrm{s})\ ,$$

where $n = 1, 2, 3, 4, \cdots$.

(b) We note that it takes $t = \frac{1}{2}T$ for the charge on the other plate to reach its maximum positive value for the first time (compare steps a and e in Fig. 33-1). This is when plate A acquires its most negative charge. From that time onward, this situation will repeat once every period. Consequently,

$$t = \frac{1}{2}T + nT = \frac{1}{2}(2n+1)T = \frac{(2n+1)}{2f} = \frac{(2n+1)}{2(2 \times 10^3\,\mathrm{Hz})} = (2n+1)(2.50\,\mu\mathrm{s})\ ,$$

where $n = 0, 1, 2, 3, 4, \cdots$.

(c) At $t = \frac{1}{4}T$, the current and the magnetic field in the inductor reach maximum values for the first time (compare steps a and c in Fig. 33-1). Later this will repeat every half-period (compare steps c and g in Fig. 33-1). Therefore,

$$t_L = \frac{T}{4} + \frac{nT}{2} = (1+2n)\frac{T}{4} = (2n+1)(1.25\,\mu\text{s}) ,$$

where $n = 0, 1, 2, 3, 4, \cdots$.

6. (a) The angular frequency is

$$\omega = \sqrt{\frac{k}{m}} = \sqrt{\frac{F/x}{m}} = \sqrt{\frac{8.0\,\text{N}}{(2.0 \times 10^{-3}\,\text{m})(0.50\,\text{kg})}} = 89 \text{ rad/s} .$$

(b) The period is $1/f$ and $f = \omega/2\pi$. Therefore,

$$T = \frac{2\pi}{\omega} = \frac{2\pi}{89\,\text{rad/s}} = 7.0 \times 10^{-2} \text{ s} .$$

(c) From $\omega = (LC)^{-1/2}$, we obtain

$$C = \frac{1}{\omega^2 L} = \frac{1}{(89\,\text{rad/s})^2(5.0\,\text{H})} = 2.5 \times 10^{-5} \text{ F} .$$

7. (a) The mass m corresponds to the inductance, so $m = 1.25\,\text{kg}$.

(b) The spring constant k corresponds to the reciprocal of the capacitance. Since the total energy is given by $U = Q^2/2C$, where Q is the maximum charge on the capacitor and C is the capacitance,

$$C = \frac{Q^2}{2U} = \frac{(175 \times 10^{-6}\,\text{C})^2}{2(5.70 \times 10^{-6}\,\text{J})} = 2.69 \times 10^{-3}\,\text{F}$$

and

$$k = \frac{1}{2.69 \times 10^{-3}\,\text{m/N}} = 372 \text{ N/m} .$$

(c) The maximum displacement corresponds to the maximum charge, so $x_{\text{max}} = 175 \times 10^{-6}\,\text{m}$.

(d) The maximum speed v_{max} corresponds to the maximum current. The maximum current is

$$I = Q\omega = \frac{Q}{\sqrt{LC}} = \frac{175 \times 10^{-6}\,\text{C}}{\sqrt{(1.25\,\text{H})(2.69 \times 10^{-3}\,\text{F})}} = 3.02 \times 10^{-3} \text{ A} .$$

Consequently, $v_{\text{max}} = 3.02 \times 10^{-3}\,\text{m/s}$.

8. We find the inductance from $f = \omega/2\pi = (2\pi\sqrt{LC})^{-1}$.

$$L = \frac{1}{4\pi^2 f^2 C} = \frac{1}{4\pi^2(10 \times 10^3\,\text{Hz})^2(6.7 \times 10^{-6}\,\text{F})} = 3.8 \times 10^{-5} \text{ H} .$$

9. The time required is $t = T/4$, where the period is given by $T = 2\pi/\omega = 2\pi\sqrt{LC}$. Consequently,

$$t = \frac{T}{4} = \frac{2\pi\sqrt{LC}}{4} = \frac{2\pi\sqrt{(0.050\,\text{H})(4.0 \times 10^{-6}\,\text{F})}}{4} = 7.0 \times 10^{-4} \text{ s} .$$

10. We apply the loop rule to the entire circuit:

$$\begin{aligned}\mathcal{E}_{\text{total}} &= \mathcal{E}_{L_1} + \mathcal{E}_{C_1} + \mathcal{E}_{R_1} + \cdots \\ &= \sum_j \left(\mathcal{E}_{L_j} + \mathcal{E}_{C_j} + \mathcal{E}_{R_j}\right) \\ &= \sum_j \left(L_j \frac{di}{dt} + \frac{q}{C_j} + iR_j\right) \\ &= L\frac{di}{dt} + \frac{q}{C} + iR \qquad \text{where } L = \sum_j L_j\,,\ \frac{1}{C} = \sum_j \frac{1}{C_j}\,,\ R = \sum_j R_j\end{aligned}$$

where we require $\mathcal{E}_{\text{total}} = 0$. This is equivalent to the simple LRC circuit shown in Fig. 33-22(b).

11. (a) $Q = CV_{\max} = (1.0 \times 10^{-9}\,\text{F})(3.0\,\text{V}) = 3.0 \times 10^{-9}\,\text{C}$.

(b) From $U = \frac{1}{2}LI^2 = \frac{1}{2}Q^2/C$ we get

$$I = \frac{Q}{\sqrt{LC}} = \frac{3.0 \times 10^{-9}\,\text{C}}{\sqrt{(3.0 \times 10^{-3}\,\text{H})(1.0 \times 10^{-9}\,\text{F})}} = 1.7 \times 10^{-3}\ \text{A}\,.$$

(c) When the current is at a maximum, the magnetic field is at maximum:

$$U_{B,\max} = \frac{1}{2}LI^2 = \frac{1}{2}(3.0 \times 10^{-3}\,\text{H})(1.7 \times 10^{-3}\,\text{A})^2 = 4.5 \times 10^{-9}\ \text{J}\,.$$

12. (a) We use $U = \frac{1}{2}LI^2 = \frac{1}{2}Q^2/C$ to solve for L:

$$\begin{aligned}L &= \frac{1}{C}\left(\frac{Q}{I}\right)^2 = \frac{1}{C}\left(\frac{CV_{\max}}{I}\right)^2 \\ &= C\left(\frac{V_{\max}}{I}\right)^2 \\ &= (4.00 \times 10^{-6}\,\text{F})\left(\frac{1.50\,\text{V}}{50.0 \times 10^{-3}\,\text{A}}\right)^2 \\ &= 3.60 \times 10^{-3}\ \text{H}\,.\end{aligned}$$

(b) Since $f = \omega/2\pi$, the frequency is

$$f = \frac{1}{2\pi\sqrt{LC}} = \frac{1}{2\pi\sqrt{(3.60 \times 10^{-3}\,\text{H})(4.00 \times 10^{-6}\,\text{F})}} = 1.33 \times 10^3\ \text{Hz}\,.$$

(c) Referring to Fig. 33-1, we see that the required time is one-fourth of a period (where the period is the reciprocal of the frequency). Consequently,

$$t = \frac{1}{4}T = \frac{1}{4f} = \frac{1}{4(1.33 \times 10^3\,\text{Hz})} = 1.88 \times 10^{-4}\ \text{s}\,.$$

13. (a) After the switch is thrown to position b the circuit is an LC circuit. The angular frequency of oscillation is $\omega = 1/\sqrt{LC}$. Consequently,

$$f = \frac{\omega}{2\pi} = \frac{1}{2\pi\sqrt{LC}} = \frac{1}{2\pi\sqrt{(54.0 \times 10^{-3}\,\text{H})(6.20 \times 10^{-6}\,\text{F})}} = 275\ \text{Hz}\,.$$

(b) When the switch is thrown, the capacitor is charged to $V = 34.0\,\text{V}$ and the current is zero. Thus, the maximum charge on the capacitor is $Q = VC = (34.0\,\text{V})(6.20\times10^{-6}\,\text{F}) = 2.11\times10^{-4}\,\text{C}$. The current amplitude is

$$I = \omega Q = 2\pi f Q = 2\pi(275\,\text{Hz})(2.11\times10^{-4}\,\text{C}) = 0.365\ \text{A}\ .$$

14. The capacitors C_1 and C_2 can be used in four different ways: (1) C_1 only; (2) C_2 only; (3) C_1 and C_2 in parallel; and (4) C_1 and C_2 in series. The corresponding oscillation frequencies are:

$$f_1 = \frac{1}{2\pi\sqrt{LC_1}} = \frac{1}{2\pi\sqrt{(1.0\times10^{-2}\,\text{H})(5.0\times10^{-6}\,\text{F})}} = 7.1\times10^2\ \text{Hz}$$

$$f_2 = \frac{1}{2\pi\sqrt{LC_2}} = \frac{1}{2\pi\sqrt{(1.0\times10^{-2}\,\text{H})(2.0\times10^{-6}\,\text{F})}} = 1.1\times10^3\ \text{Hz}$$

$$f_3 = \frac{1}{2\pi\sqrt{L(C_1+C_2)}} = \frac{1}{2\pi\sqrt{(1.0\times10^{-2}\,\text{H})(2.0\times10^{-6}\,\text{F}+5.0\times10^{-6}\,\text{F})}} = 6.0\times10^2\ \text{Hz}$$

$$\begin{aligned} f_4 &= \frac{1}{2\pi\sqrt{LC_1C_2/(C_1+C_2)}} = \frac{1}{2\pi}\sqrt{\frac{2.0\times10^{-6}\,\text{F}+5.0\times10^{-6}\,\text{F}}{(1.0\times10^{-2}\,\text{H})(2.0\times10^{-6}\,\text{F})(5.0\times10^{-6}\,\text{F})}} \\ &= 1.3\times10^3\ \text{Hz} \end{aligned}$$

15. (a) Since the frequency of oscillation f is related to the inductance L and capacitance C by $f = 1/2\pi\sqrt{LC}$, the smaller value of C gives the larger value of f. Consequently, $f_{\max} = 1/2\pi\sqrt{LC_{\min}}$, $f_{\min} = 1/2\pi\sqrt{LC_{\max}}$, and

$$\frac{f_{\max}}{f_{\min}} = \frac{\sqrt{C_{\max}}}{\sqrt{C_{\min}}} = \frac{\sqrt{365\,\text{pF}}}{\sqrt{10\,\text{pF}}} = 6.0\ .$$

(b) An additional capacitance C is chosen so the ratio of the frequencies is

$$r = \frac{1.60\,\text{MHz}}{0.54\,\text{MHz}} = 2.96\ .$$

Since the additional capacitor is in parallel with the tuning capacitor, its capacitance adds to that of the tuning capacitor. If C is in picofarads, then

$$\frac{\sqrt{C+365\,\text{pF}}}{\sqrt{C+10\,\text{pF}}} = 2.96\ .$$

The solution for C is

$$C = \frac{(365\,\text{pF}) - (2.96)^2(10\,\text{pF})}{(2.96)^2 - 1} = 36\ \text{pF}\ .$$

We solve $f = 1/2\pi\sqrt{LC}$ for L. For the minimum frequency $C = 365\,\text{pF} + 36\,\text{pF} = 401\,\text{pF}$ and $f = 0.54\,\text{MHz}$. Thus

$$L = \frac{1}{(2\pi)^2Cf^2} = \frac{1}{(2\pi)^2(401\times10^{-12}\,\text{F})(0.54\times10^6\,\text{Hz})^2} = 2.2\times10^{-4}\ \text{H}\ .$$

16. (a) Since the percentage of energy stored in the electric field of the capacitor is $(1 - 75.0\%) = 25.0\%$, then

$$\frac{U_E}{U} = \frac{q^2/2C}{Q^2/2C} = 25.0\%$$

which leads to $q = \sqrt{0.250}\,Q = 0.500Q$.

(b) From

$$\frac{U_B}{U} = \frac{Li^2/2}{LI^2/2} = 75.0\% \,,$$

we find $i = \sqrt{0.750}\, I = 0.866I$.

17. (a) The total energy U is the sum of the energies in the inductor and capacitor:

$$\begin{aligned} U &= U_E + U_B = \frac{q^2}{2C} + \frac{i^2 L}{2} \\ &= \frac{(3.80 \times 10^{-6}\,\text{C})^2}{2(7.80 \times 10^{-6}\,\text{F})} + \frac{(9.20 \times 10^{-3}\,\text{A})^2 (25.0 \times 10^{-3}\,\text{H})}{2} = 1.98 \times 10^{-6}\ \text{J} \,. \end{aligned}$$

(b) We solve $U = Q^2/2C$ for the maximum charge:

$$Q = \sqrt{2CU} = \sqrt{2(7.80 \times 10^{-6}\,\text{F})(1.98 \times 10^{-6}\,\text{J})} = 5.56 \times 10^{-6}\ \text{C} \,.$$

(c) From $U = I^2 L/2$, we find the maximum current:

$$I = \sqrt{\frac{2U}{L}} = \sqrt{\frac{2(1.98 \times 10^{-6}\,\text{J})}{25.0 \times 10^{-3}\,\text{H}}} = 1.26 \times 10^{-2}\ \text{A} \,.$$

(d) If q_0 is the charge on the capacitor at time $t = 0$, then $q_0 = Q \cos\phi$ and

$$\phi = \cos^{-1}\left(\frac{q}{Q}\right) = \cos^{-1}\left(\frac{3.80 \times 10^{-6}\,\text{C}}{5.56 \times 10^{-6}\,\text{C}}\right) = \pm 46.9^\circ \,.$$

For $\phi = +46.9^\circ$ the charge on the capacitor is decreasing, for $\phi = -46.9^\circ$ it is increasing. To check this, we calculate the derivative of q with respect to time, evaluated for $t = 0$. We obtain $-\omega Q \sin\phi$, which we wish to be positive. Since $\sin(+46.9^\circ)$ is positive and $\sin(-46.9^\circ)$ is negative, the correct value for increasing charge is $\phi = -46.9^\circ$.

(e) Now we want the derivative to be negative and $\sin\phi$ to be positive. Thus, we take $\phi = +46.9^\circ$.

18. The linear relationship between θ (the knob angle in degrees) and frequency f is

$$f = f_0 \left(1 + \frac{\theta}{180^\circ}\right) \quad \Longrightarrow \quad \theta = 180^\circ \left(\frac{f}{f_0} - 1\right)$$

where $f_0 = 2 \times 10^5$ Hz. Since $f = \omega/2\pi = 1/2\pi\sqrt{LC}$, we are able to solve for C in terms of θ:

$$C = \frac{1}{4\pi^2 L f_0^2 \left(1 + \frac{\theta}{180^\circ}\right)^2} = \frac{81}{400000\pi^2 (180^\circ + \theta)^2}$$

with SI units understood. After multiplying by 10^{12} (to convert to picofarads), this is plotted, below.

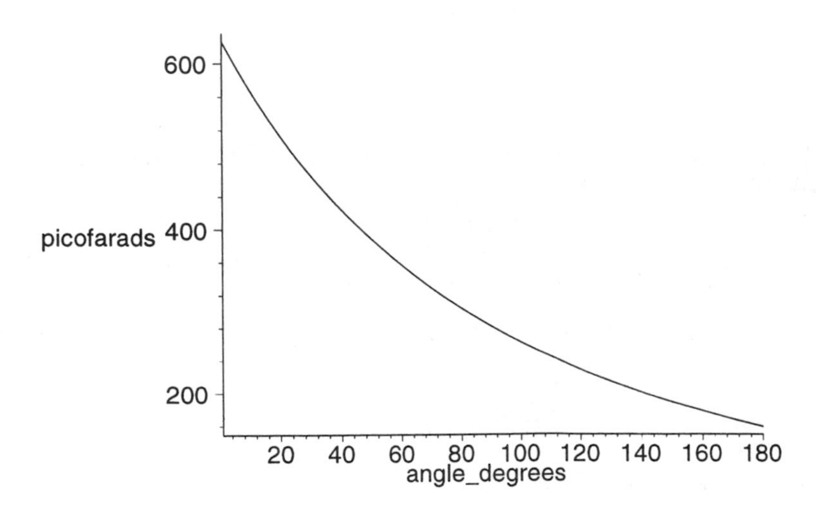

19. (a) The charge (as a function of time) is given by $q = Q \sin \omega t$, where Q is the maximum charge on the capacitor and ω is the angular frequency of oscillation. A sine function was chosen so that $q = 0$ at time $t = 0$. The current (as a function of time) is

$$i = \frac{dq}{dt} = \omega Q \cos \omega t ,$$

and at $t = 0$, it is $I = \omega Q$. Since $\omega = 1/\sqrt{LC}$,

$$Q = I\sqrt{LC} = (2.00\,\mathrm{A})\sqrt{(3.00 \times 10^{-3}\,\mathrm{H})(2.70 \times 10^{-6}\,\mathrm{F})} = 1.80 \times 10^{-4}\ \mathrm{C} .$$

(b) The energy stored in the capacitor is given by

$$U_E = \frac{q^2}{2C} = \frac{Q^2 \sin^2 \omega t}{2C}$$

and its rate of change is

$$\frac{dU_E}{dt} = \frac{Q^2 \omega \sin \omega t \cos \omega t}{C} .$$

We use the trigonometric identity $\cos \omega t \sin \omega t = \frac{1}{2} \sin(2\omega t)$ to write this as

$$\frac{dU_E}{dt} = \frac{\omega Q^2}{2C} \sin(2\omega t) .$$

The greatest rate of change occurs when $\sin(2\omega t) = 1$ or $2\omega t = \pi/2\,\mathrm{rad}$. This means

$$t = \frac{\pi}{4\omega} = \frac{\pi T}{4(2\pi)} = \frac{T}{8}$$

where T is the period of oscillation. The relationship $\omega = 2\pi/T$ was used.

(c) Substituting $\omega = 2\pi/T$ and $\sin(2\omega t) = 1$ into $dU_E/dt = (\omega Q^2/2C)\sin(2\omega t)$, we obtain

$$\left(\frac{dU_E}{dt}\right)_{\max} = \frac{2\pi Q^2}{2TC} = \frac{\pi Q^2}{TC} .$$

Now $T = 2\pi\sqrt{LC} = 2\pi\sqrt{(3.00 \times 10^{-3}\,\mathrm{H})(2.70 \times 10^{-6}\,\mathrm{F})} = 5.655 \times 10^{-4}\,\mathrm{s}$, so

$$\left(\frac{dU_E}{dt}\right)_{\max} = \frac{\pi(1.80 \times 10^{-4}\,\mathrm{C})^2}{(5.655 \times 10^{-4}\,\mathrm{s})(2.70 \times 10^{-6}\,\mathrm{F})} = 66.7\ \mathrm{W} .$$

We note that this is a positive result, indicating that the energy in the capacitor is indeed increasing at $t = T/8$.

20. For the first circuit $\omega = (L_1C_1)^{-1/2}$, and for the second one $\omega = (L_2C_2)^{-1/2}$. When the two circuits are connected in series, the new frequency is

$$\begin{aligned} \omega' &= \frac{1}{\sqrt{L_{\mathrm{eq}}C_{\mathrm{eq}}}} \\ &= \frac{1}{\sqrt{(L_1 + L_2)C_1C_2/(C_1 + C_2)}} = \frac{1}{\sqrt{(L_1C_1C_2 + L_2C_2C_1)/(C_1 + C_2)}} \\ &= \frac{1}{\sqrt{L_1C_1}} \frac{1}{\sqrt{(C_1 + C_2)/(C_1 + C_2)}} = \omega , \end{aligned}$$

where we use $\omega^{-1} = \sqrt{L_1C_1} = \sqrt{L_2C_2}$.

21. (a) We compare this expression for the current with $i = I\sin(\omega t+\phi_0)$. Setting $(\omega t+\phi) = 2500t+0.680 = \pi/2$, we obtain $t = 3.56 \times 10^{-4}$ s.

(b) Since $\omega = 2500\,\text{rad/s} = (LC)^{-1/2}$,

$$L = \frac{1}{\omega^2 C} = \frac{1}{(2500\,\text{rad/s})^2(64.0\times10^{-6}\,\text{F})} = 2.50\times10^{-3}\ \text{H}\ .$$

(c) The energy is

$$U = \frac{1}{2}LI^2 = \frac{1}{2}(2.50\times10^{-3}\,\text{H})(1.60\,\text{A})^2 = 3.20\times10^{-3}\ \text{J}\ .$$

22. (a) The figure implies that the the instantaneous current through the leftmost inductor is the same as that through the rightmost one, which means there is no current through the middle inductor (at any instant). Applying the loop rule to the outer loop (including the rightmost and leftmost inductors), with the current suitably related to the rate of change of charge, we find

$$2L\frac{d^2q}{dt^2} + \frac{2}{C}q = 0 \implies \omega = \frac{1}{\sqrt{(2L)(C/2)}} = \frac{1}{\sqrt{LC}}\ .$$

(b) In this case, we see that the middle inductor must have current $2i(t)$ flowing downward, and application of the loop rule to, say, the left loop leads to

$$L\frac{d^2q}{dt^2} + L\left(2\frac{d^2q}{dt^2}\right) + \frac{1}{C}q = 0 \implies \omega = \frac{1}{\sqrt{(3L)(C)}} = \frac{1}{\sqrt{3LC}}\ .$$

23. The energy needed to charge the $100\,\mu$F capacitor to 300 V is

$$\frac{1}{2}C_2V^2 = \frac{1}{2}(100\times10^{-6}\,\text{F})(300\,\text{V})^2 = 4.50\ \text{J}\ .$$

The energy initially in the $900\,\mu$F capacitor is

$$\frac{1}{2}C_1V^2 = \frac{1}{2}(900\times10^{-6}\,\text{F})(100\,\text{V})^2 = 4.50\ \text{J}\ .$$

All the energy originally in the $900\,\mu$F capacitor must be transferred to the $100\,\mu$F capacitor. The plan is to store it temporarily in the inductor. We do this by leaving switch S_1 open and closing switch S_2. We wait until the $900\,\mu$F capacitor is completely discharged and the current in the circuit is at maximum (this occurs at $t = T_1/4$, one quarter of the relevant period). Since

$$T_1 = 2\pi\sqrt{LC_1} = 2\pi\sqrt{(10.0\,\text{H})(900\times10^{-6}\,\text{F})} = 0.596\ \text{s}\ ,$$

we wait until $t = (0.596\,\text{s})/4 = 0.149\,\text{s}$. Now, we close switch S_1 while simultaneously opening switch S_2. Next, we wait for one-fourth of the T_2 period to elapse and open switch S_1. The $100\,\mu$F capacitor then has maximum charge, and all the energy resides in it. Since

$$T_2 = 2\pi\sqrt{LC_2} = 2\pi\sqrt{(10.0\,\text{H})(100\times10^{-6}\,\text{F})} = 0.199\ \text{s}\ ,$$

we must keep S_1 closed for $(0.199\,\text{s})/4 = 0.0497\,\text{s}$. It is helpful to refer to Figure 23-1 to appreciate the emphasis on "quarter-periods" in this solution.

24. (a) Since $T = 2\pi/\omega = 2\pi\sqrt{LC}$, we may rewrite the power on the exponential factor as

$$-\pi R\sqrt{\frac{C}{L}}\frac{t}{T} = -\pi R\sqrt{\frac{C}{L}}\frac{t}{2\pi\sqrt{LC}} = -\frac{Rt}{2L}\ .$$

Thus $e^{-Rt/2L} = e^{-\pi R\sqrt{C/L}(t/T)}$.

(b) Since $-\pi R\sqrt{C/L}(t/T)$ must be unitless (as is t/T), $R\sqrt{C/L}$ must also be unitless. Thus, the SI unit of $\sqrt{C/L}$ must be Ω^{-1}. In other words, the SI unit for $\sqrt{L/C}$ is Ω.

(c) Since the amplitude of oscillation reduces by a factor of $e^{-\pi R\sqrt{C/L}(T/T)} = e^{-\pi R\sqrt{C/L}}$ after each cycle, the condition is equivalent to $\pi R\sqrt{C/L} \ll 1$, or $R \ll \sqrt{L/C}$.

25. Since $\omega \approx \omega'$, we may write $T = 2\pi/\omega$ as the period and $\omega = 1/\sqrt{LC}$ as the angular frequency. The time required for 50 cycles (with 3 significant figures understood) is

$$\begin{aligned} t &= 50T = 50\left(\frac{2\pi}{\omega}\right) = 50\left(2\pi\sqrt{LC}\right) \\ &= 50\left(2\pi\sqrt{(220\times10^{-3}\,\text{H})\,(12.0\times10^{-6}\,\text{F})}\right) = 0.5104\ \text{s}\ . \end{aligned}$$

The maximum charge on the capacitor decays according to

$$q_{\max} = Qe^{-Rt/2L}$$

(this is called the *exponentially decaying amplitude* in §33-5), where Q is the charge at time $t = 0$ (if we take $\phi = 0$ in Eq. 33-25). Dividing by Q and taking the natural logarithm of both sides, we obtain

$$\ln\left(\frac{q_{\max}}{Q}\right) = -\frac{Rt}{2L}$$

which leads to

$$R = -\frac{2L}{t}\ln\left(\frac{q_{\max}}{Q}\right) = -\frac{2(220\times10^{-3}\,\text{H})}{0.5104\,\text{s}}\ln(0.99) = 8.66\times10^{-3}\ \Omega\ .$$

26. The charge q after N cycles is obtained by substituting $t = NT = 2\pi N/\omega'$ into Eq. 33-25:

$$\begin{aligned} q &= Qe^{-Rt/2L}\cos(\omega' t+\phi) = Qe^{-RNT/2L}\cos(\omega'(2\pi N/\omega')+\phi) \\ &= Qe^{-RN(2\pi\sqrt{L/C})/2L}\cos(2\pi N+\phi) \\ &= Qe^{-N\pi R\sqrt{C/L}}\cos(\phi). \end{aligned}$$

We note that the initial charge (setting $N = 0$ in the above expression) is $q_0 = Q\cos\phi$, where $q_0 = 6.2\,\mu\text{C}$ is given (with 3 significant figures understood). Consequently, we write the above result as $q_N = q_0 e^{-N\pi R\sqrt{C/L}}$ and obtain

$$\begin{aligned} q_5 &= (6.2\,\mu\text{C})e^{-5\pi(7.2\,\Omega)\sqrt{0.0000032\,\text{F}/12\,\text{H}}} = 5.85\,\mu\text{C} \\ q_{10} &= (6.2\,\mu\text{C})e^{-10\pi(7.2\,\Omega)\sqrt{0.0000032\,\text{F}/12\,\text{H}}} = 5.52\,\mu\text{C} \\ q_{100} &= (6.2\,\mu\text{C})e^{-100\pi(7.2\,\Omega)\sqrt{0.0000032\,\text{F}/12\,\text{H}}} = 1.93\,\mu\text{C}\ . \end{aligned}$$

27. The assumption stated at the end of the problem is equivalent to setting $\phi = 0$ in Eq. 33-25. Since the maximum energy in the capacitor (each cycle) is given by $q_{\max}^2/2C$, where $q_{\max}$ is the maximum charge (during a given cycle), then we seek the time for which

$$\frac{q_{\max}^2}{2C} = \frac{1}{2}\frac{Q^2}{2C} \implies q_{\max} = \frac{Q}{\sqrt{2}}\ .$$

Now $q_{\max}$ (referred to as the *exponentially decaying amplitude* in §33-5) is related to Q (and the other parameters of the circuit) by

$$q_{\max} = Qe^{-Rt/2L} \implies \ln\left(\frac{q_{\max}}{Q}\right) = -\frac{Rt}{2L}\ .$$

Setting $q_{\max} = Q/\sqrt{2}$, we solve for t:

$$t = -\frac{2L}{R}\ln\left(\frac{q_{\max}}{Q}\right) = -\frac{2L}{R}\ln\left(\frac{1}{\sqrt{2}}\right) = \frac{L}{R}\ln 2 \, .$$

The identities $\ln(1/\sqrt{2}) = -\ln\sqrt{2} = -\frac{1}{2}\ln 2$ were used to obtain the final form of the result.

28. (a) In Eq. 33-25, we set $q = 0$ and $t = 0$ to obtain $0 = Q\cos\phi$. This gives $\phi = \pm\pi/2$ (assuming $Q \neq 0$). It should be noted that other roots are possible (for instance, $\cos(3\pi/2) = 0$) but the $\pm\pi/2$ choices for the phase constant are in some sense the "simplest." We choose $\phi = -\pi/2$ to make the manipulation of signs in the expressions below easier to follow. To simplify the work in part (b), we note that $\cos(\omega' t - \pi/2) = \sin(\omega' t)$.

(b) First, we calculate the time-dependent current $i(t)$ from Eq. 33-25:

$$\begin{aligned} i(t) &= \frac{dq}{dt} = \frac{d}{dt}\left(Qe^{-Rt/2L}\sin(\omega' t)\right) \\ &= -\frac{QR}{2L}e^{-Rt/2L}\sin(\omega' t) + Q\omega' e^{-Rt/2L}\cos(\omega' t) \\ &= Qe^{-Rt/2L}\left(-\frac{R\sin(\omega' t)}{2L} + \omega'\cos(\omega' t)\right) , \end{aligned}$$

which we evaluate at $t = 0$: $i(0) = Q\omega'$. If we denote $i(0) = I$ as suggested in the problem, then $Q = I/\omega'$. Returning this to Eq. 33-25 leads to

$$q = Qe^{-Rt/2L}\cos(\omega' t + \phi) = \left(\frac{I}{\omega'}\right)e^{-Rt/2L}\cos\left(\omega' t - \frac{\pi}{2}\right) = Ie^{-Rt/2L}\frac{\sin(\omega' t)}{\omega'}$$

which answers the question if we interpret "current amplitude" as I. If one, instead, interprets an (exponentially decaying) "current amplitude" to be more appropriately defined as $i_{\max} = i(t)/\cos(\cdots)$ (that is, the current after dividing out its oscillatory behavior), then another step is needed in the $i(t)$ manipulations, above. Using the identity $x\cos\alpha - y\sin\alpha = r\cos(\alpha+\beta)$ where $r = \sqrt{x^2+y^2}$ and $\tan\beta = y/x$, we can write the current as

$$i(t) = Qe^{-Rt/2L}\left(-\frac{R\sin(\omega' t)}{2L} + \omega'\cos(\omega' t)\right) = Q\sqrt{\omega'^2 + \left(\frac{R}{2L}\right)^2}\, e^{-Rt/2L}\cos(\omega' t + \theta)$$

where $\theta = \tan^{-1}(R/2L\omega')$. Thus, the current amplitude defined in this second way becomes (using Eq. 33-26 for ω')

$$i_{\max} = Q\sqrt{\omega'^2 + \left(\frac{R}{2L}\right)^2}\, e^{-Rt/2L} = Q\omega e^{-Rt/2L} \, .$$

In terms of $i_{\max}$ the expression for charge becomes

$$q = Qe^{-Rt/2L}\sin(\omega' t) = \left(\frac{i_{\max}}{\omega}\right)\sin(\omega' t)$$

which is remarkably similar to our previous "result" in terms of I, except for the fact that ω' in the denominator has now been replaced with ω (and, of course, the exponential has been absorbed into the definition of $i_{\max}$).

29. Let t be a time at which the capacitor is fully charged in some cycle and let $q_{\max 1}$ be the charge on the capacitor then. The energy in the capacitor at that time is

$$U(t) = \frac{q_{\max 1}^2}{2C} = \frac{Q^2}{2C}e^{-Rt/L}$$

where

$$q_{\max 1} = Q\, e^{-Rt/2L}$$

(see the discussion of the *exponentially decaying amplitude* in §33-5). One period later the charge on the fully charged capacitor is

$$q_{\max 2} = Q\, e^{-R(t+T)/2L} \qquad \text{where} \quad T = \frac{2\pi}{\omega'} ,$$

and the energy is

$$U(t+T) = \frac{q^2_{\max 2}}{2C} = \frac{Q^2}{2C}\, e^{-R(t+T)/L} .$$

The fractional loss in energy is

$$\frac{|\Delta U|}{U} = \frac{U(t) - U(t+T)}{U(t)} = \frac{e^{-Rt/L} - e^{-R(t+T)/L}}{e^{-Rt/L}} = 1 - e^{-RT/L} .$$

Assuming that RT/L is very small compared to 1 (which would be the case if the resistance is small), we expand the exponential (see Appendix E). The first few terms are:

$$e^{-RT/L} \approx 1 - \frac{RT}{L} + \frac{R^2T^2}{2L^2} + \cdots .$$

If we approximate $\omega \approx \omega'$, then we can write T as $2\pi/\omega$. As a result, we obtain

$$\frac{|\Delta U|}{U} \approx 1 - \left(1 - \frac{RT}{L} + \cdots\right) \approx \frac{RT}{L} = \frac{2\pi R}{\omega L} .$$

30. (a) We use $I = \mathcal{E}/X_c = \omega_d C\mathcal{E}$:

$$I = \omega_d C\mathcal{E}_m = 2\pi f_d C\mathcal{E}_m = 2\pi(1.00 \times 10^3\,\text{Hz})(1.50 \times 10^{-6}\,\text{F})(30.0\,\text{V}) = 0.283\ \text{A} .$$

(b) $I = 2\pi(8.00 \times 10^3\,\text{Hz})(1.50 \times 10^{-6}\,\text{F})(30.0\,\text{V}) = 2.26\,\text{A}$.

31. (a) The current amplitude I is given by $I = V_L/X_L$, where $X_L = \omega_d L = 2\pi f_d L$. Since the circuit contains only the inductor and a sinusoidal generator, $V_L = \mathcal{E}_m$. Therefore,

$$I = \frac{V_L}{X_L} = \frac{\mathcal{E}_m}{2\pi f_d L} = \frac{30.0\,\text{V}}{2\pi(1.00 \times 10^3\,\text{Hz})(50.0 \times 10^{-3}\,\text{H})} = 0.0955\ \text{A} .$$

(b) The frequency is now eight times larger than in part (a), so the inductive reactance X_L is eight times larger and the current is one-eighth as much. The current is now $(0.0955\,\text{A})/8 = 0.0119\,\text{A}$.

32. (a) and (b) Regardless of the frequency of the generator, the current through the resistor is

$$I = \frac{\mathcal{E}_m}{R} = \frac{30.0\,\text{V}}{50\,\Omega} = 0.60\ \text{A} .$$

33. (a) The inductive reactance for angular frequency ω_d is given by $X_L = \omega_d L$, and the capacitive reactance is given by $X_C = 1/\omega_d C$. The two reactances are equal if $\omega_d L = 1/\omega_d C$, or $\omega_d = 1/\sqrt{LC}$. The frequency is

$$f_d = \frac{\omega_d}{2\pi} = \frac{1}{2\pi\sqrt{LC}} = \frac{1}{2\pi\sqrt{(6.0 \times 10^{-3}\,\text{H})(10 \times 10^{-6}\,\text{F})}} = 650\ \text{Hz} .$$

(b) The inductive reactance is $X_L = \omega_d L = 2\pi f_d L = 2\pi(650\,\text{Hz})(6.0 \times 10^{-3}\,\text{H}) = 24\,\Omega$. The capacitive reactance has the same value at this frequency.

(c) The natural frequency for free LC oscillations is $f = \omega/2\pi = 1/2\pi\sqrt{LC}$, the same as we found in part (a).

34. (a) The circuit consists of one generator across one inductor; therefore, $\mathcal{E}_m = V_L$. The current amplitude is

$$I = \frac{\mathcal{E}_m}{X_L} = \frac{\mathcal{E}_m}{\omega_d L} = \frac{25.0\,\text{V}}{(377\,\text{rad/s})(12.7\,\text{H})} = 5.22 \times 10^{-3}\ \text{A}\ .$$

(b) When the current is at a maximum, its derivative is zero. Thus, Eq. 31-37 gives $\mathcal{E}_L = 0$ at that instant. Stated another way, since $\mathcal{E}(t)$ and $i(t)$ have a 90° phase difference, then $\mathcal{E}(t)$ must be zero when $i(t) = I$. The fact that $\phi = 90° = \pi/2$ rad is used in part (c).

(c) Consider Eq. 32-28 with $\mathcal{E} = -\frac{1}{2}\mathcal{E}_m$. In order to satisfy this equation, we require $\sin(\omega_d t) = -1/2$. Now we note that the problem states that $\mathcal{E}$ is increasing *in magnitude*, which (since it is already negative) means that it is becoming more negative. Thus, differentiating Eq. 32-28 with respect to time (and demanding the result be negative) we must also require $\cos(\omega_d t) < 0$. These conditions imply that ωt must equal $(2n\pi - 5\pi/6)$ [n = integer]. Consequently, Eq. 33-29 yields (for all values of n)

$$i = I\ \sin\left(2n\pi - \frac{5\pi}{6} - \frac{\pi}{2}\right) = (5.22 \times 10^{-3}\ \text{A})\left(\frac{\sqrt{3}}{2}\right) = 4.51 \times 10^{-3}\ \text{A}\ .$$

35. (a) The generator emf is a maximum when $\sin(\omega_d t - \pi/4) = 1$ or $\omega_d t - \pi/4 = (\pi/2) \pm 2n\pi$ [n = integer]. The first time this occurs after $t = 0$ is when $\omega_d t - \pi/4 = \pi/2$ (that is, $n = 0$). Therefore,

$$t = \frac{3\pi}{4\omega_d} = \frac{3\pi}{4(350\,\text{rad/s})} = 6.73 \times 10^{-3}\ \text{s}\ .$$

(b) The current is a maximum when $\sin(\omega_d t - 3\pi/4) = 1$, or $\omega_d t - 3\pi/4 = (\pi/2) \pm 2n\pi$ [n = integer]. The first time this occurs after $t = 0$ is when $\omega_d t - 3\pi/4 = \pi/2$ (as in part (a), $n = 0$). Therefore,

$$t = \frac{5\pi}{4\omega_d} = \frac{5\pi}{4(350\,\text{rad/s})} = 1.12 \times 10^{-2}\ \text{s}\ .$$

(c) The current lags the emf by $+\frac{\pi}{2}$ rad, so the circuit element must be an inductor.

(d) The current amplitude I is related to the voltage amplitude V_L by $V_L = IX_L$, where X_L is the inductive reactance, given by $X_L = \omega_d L$. Furthermore, since there is only one element in the circuit, the amplitude of the potential difference across the element must be the same as the amplitude of the generator emf: $V_L = \mathcal{E}_m$. Thus, $\mathcal{E}_m = I\omega_d L$ and

$$L = \frac{\mathcal{E}_m}{I\omega_d} = \frac{30.0\,\text{V}}{(620 \times 10^{-3}\,\text{A})(350\,\text{rad/s})} = 0.138\ \text{H}\ .$$

36. (a) The circuit consists of one generator across one capacitor; therefore, $\mathcal{E}_m = V_C$. Consequently, the current amplitude is

$$I = \frac{\mathcal{E}_m}{X_C} = \omega C\mathcal{E}_m = (377\,\text{rad/s})(4.15 \times 10^{-6}\,\text{F})(25.0\,\text{V}) = 3.91 \times 10^{-2}\,\text{A}\ .$$

(b) When the current is at a maximum, the charge on the capacitor is changing at its largest rate. This happens not when it is fully charged ($\pm q_{\text{max}}$), but rather as it passes through the (momentary) states of being uncharged ($q = 0$). Since $q = CV$, then the voltage across the capacitor (and at the generator, by the loop rule) is zero when the current is at a maximum. Stated more precisely, the time-dependent emf $\mathcal{E}(t)$ and current $i(t)$ have a $\phi = -90°$ phase relation, implying $\mathcal{E}(t) = 0$ when $i(t) = I$. The fact that $\phi = -90° = -\pi/2$ rad is used in part (c).

(c) Consider Eq. 32-28 with $\mathcal{E} = -\frac{1}{2}\mathcal{E}_m$. In order to satisfy this equation, we require $\sin(\omega_d t) = -1/2$. Now we note that the problem states that $\mathcal{E}$ is increasing *in magnitude*, which (since it is already negative) means that it is becoming more negative. Thus, differentiating Eq. 32-28 with respect to time (and demanding the result be negative) we must also require $\cos(\omega_d t) < 0$. These conditions imply that ωt must equal $(2n\pi - 5\pi/6)$ [$n =$ integer]. Consequently, Eq. 33-29 yields (for all values of n)

$$i = I\,\sin\left(2n\pi - \frac{5\pi}{6} + \frac{\pi}{2}\right) = (3.91\times10^{-3}\ \text{A})\left(-\frac{\sqrt{3}}{2}\right) = -3.38\times10^{-2}\ \text{A}\ .$$

37. (a) Now $X_C = 0$, while $R = 160\,\Omega$ and $X_L = 86.7\,\Omega$ remain unchanged. Therefore, the impedance is

$$Z = \sqrt{R^2 + X_L^2} = \sqrt{(160\,\Omega)^2 + (86.7\,\Omega)^2} = 182\ \Omega\ .$$

The current amplitude is now found to be

$$I = \frac{\mathcal{E}_m}{Z} = \frac{36.0\ \text{V}}{182\,\Omega} = 0.198\ \text{A}\ .$$

The phase angle is, from Eq. 33-65,

$$\phi = \tan^{-1}\left(\frac{X_L - X_C}{R}\right) = \tan^{-1}\left(\frac{86.7\,\Omega - 0}{160\,\Omega}\right) = 28.5°\ .$$

(b) We first find the voltage amplitudes across the circuit elements:

$$\begin{aligned} V_R &= IR = (0.198\ \text{A})(160\,\Omega) \approx 32\ \text{V} \\ V_L &= IX_L = (0.216\ \text{A})(86.7\,\Omega) \approx 17\ \text{V} \end{aligned}$$

This is an inductive circuit, so $\mathcal{E}_m$ leads I. The phasor diagram is drawn to scale below.

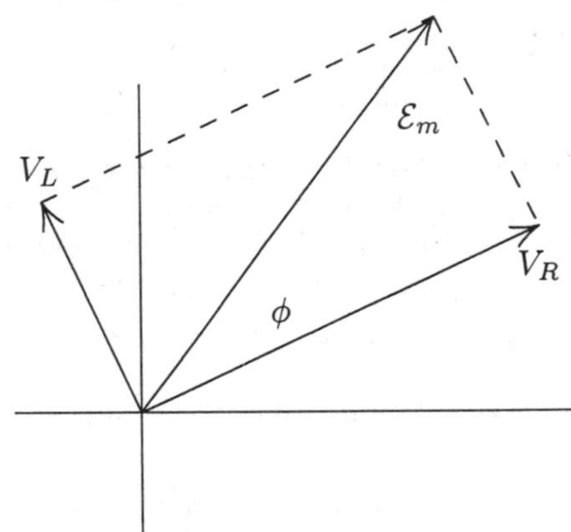

38. (a) Now $X_L = 0$, while $R = 160\,\Omega$ and $X_C = 177\,\Omega$ remain as shown in the Sample Problem. Therefore, the impedance, current amplitude and phase angle are

$$\begin{aligned} Z &= \sqrt{R^2 + X_C^2} = \sqrt{(160\,\Omega)^2 + (177\,\Omega)^2} = 239\ \Omega\ , \\ I &= \frac{\mathcal{E}_m}{Z} = \frac{36.0\ \text{V}}{239\,\Omega} = 0.151\ \text{A}\ , \\ \phi &= \tan^{-1}\left(\frac{X_L - X_C}{R}\right) = \tan^{-1}\left(\frac{0 - 177\,\Omega}{160\,\Omega}\right) = -47.9°\ . \end{aligned}$$

(b) We first find the voltage amplitudes across the circuit elements:

$$V_R = IR = (0.151\,\text{A})(160\,\Omega) \approx 24\text{ V}$$
$$V_C = IX_C = (0.151\,\text{A})(177\,\Omega) \approx 27\text{ V}$$

The circuit is capacitive, so I leads $\mathcal{E}_m$. The phasor diagram is drawn to scale below.

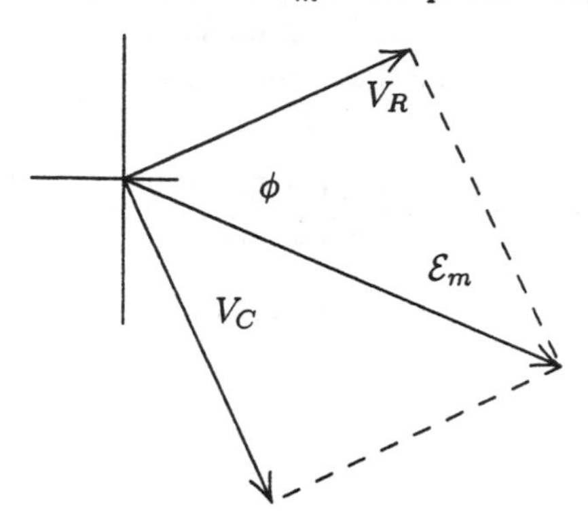

39. (a) The capacitive reactance is

$$X_C = \frac{1}{\omega_d C} = \frac{1}{2\pi f_d C} = \frac{1}{2\pi(60.0\,\text{Hz})(70.0\times10^{-6}\,\text{F})} = 37.9\ \Omega\ .$$

The inductive reactance $86.7\,\Omega$ is unchanged. The new impedance is

$$Z = \sqrt{R^2 + (X_L - X_C)^2} = \sqrt{(160\,\Omega)^2 + (37.9\,\Omega - 86.7\,\Omega)^2} = 167\ \Omega\ .$$

The current amplitude is

$$I = \frac{\mathcal{E}_m}{Z} = \frac{36.0\,\text{V}}{167\,\Omega} = 0.216\text{ A}\ .$$

The phase angle is

$$\phi = \tan^{-1}\left(\frac{X_L - X_C}{R}\right) = \tan^{-1}\left(\frac{86.7\,\Omega - 37.9\,\Omega}{160\,\Omega}\right) = 17.0^\circ\ .$$

(b) We first find the voltage amplitudes across the circuit elements:

$$V_R = IR = (0.216\,\text{A})(160\,\Omega) = 34.6\text{ V}$$
$$V_L = IX_L = (0.216\,\text{A})(86.7\,\Omega) = 18.7\text{ V}$$
$$V_C = IX_C = (0.216\,\text{A})(37.9\,\Omega) = 8.19\text{ V}$$

Note that $X_L > X_C$, so that $\mathcal{E}_m$ leads I. The phasor diagram is drawn to scale below.

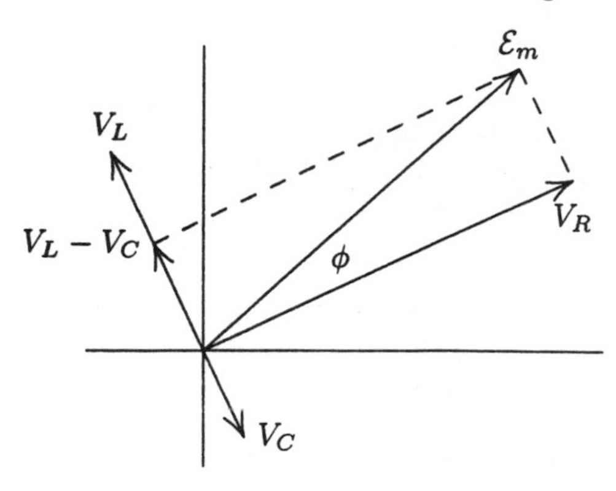

40. (a) The resonance frequency f_0 of the circuit is about $(1.50\,\text{kHz} + 1.30\,\text{kHz})/2 = 1.40\,\text{kHz}$. Thus, from $2\pi f_0 = (LC)^{-1/2}$ we get

$$L = \frac{1}{4\pi^2 f_0^2 C} = \frac{1}{4\pi^2(1.40 \times 10^3\,\text{Hz})^2(5.50 \times 10^{-6}\,\text{F})} = 2.35 \times 10^{-3}\ \text{H} .$$

(b) From the resonance curves shown in the textbook, we see that as R increases the resonance curve gets more spread out, so the two frequencies at which the amplitude is at half-maximum level will move away from each other.

41. The amplitude of the voltage across the inductor in an RLC series circuit is given by $V_L = IX_L = I\omega_d L$. At resonance, the driving angular frequency equals the natural angular frequency: $\omega_d = \omega = 1/\sqrt{LC}$. For the given circuit

$$X_L = \frac{L}{\sqrt{LC}} = \frac{1.0\,\text{H}}{\sqrt{(1.0\,\text{H})(1.0 \times 10^{-6}\,\text{F})}} = 1000\ \Omega .$$

At resonance the capacitive reactance has this same value, and the impedance reduces simply: $Z = R$. Consequently,

$$I = \left.\frac{\mathcal{E}_m}{Z}\right|_{\text{resonance}} = \frac{\mathcal{E}_m}{R} = \frac{10\,\text{V}}{10\,\Omega} = 1.0\ \text{A} .$$

The voltage amplitude across the inductor is therefore

$$V_L = IX_L = (1.0\,\text{A})(1000\,\Omega) = 1000\ \text{V}$$

which is much larger than the amplitude of the generator emf.

42. (a) We note that we obtain the maximum value in Eq. 33-28 when we set

$$t = \frac{\pi}{2\omega_d} = \frac{1}{4f} = \frac{1}{4(60)} = 0.00417\ \text{s}$$

or 4.17 ms. The result is $\mathcal{E}_m \sin(\pi/2) = \mathcal{E}_m \sin(90°) = 36.0\,\text{V}$. We note, for reference in the subsequent parts, that at $t = 4.17\,\text{ms}$, the current is

$$i = I\sin(\omega_d t - \phi) = I\sin(90° - (-29.4°)) = (0.196\,\text{A})\cos(29.4°) = 0.171\ \text{A}$$

using Eq. 33-29 and the results of the Sample Problem.

(b) At $t = 4.17$ ms, Ohm's law directly gives

$$v_R = iR = (I\cos(29.4°))\,R(0.171\,\text{A})(160\,\Omega) = 27.3\ \text{V} .$$

(c) The capacitor voltage phasor is 90° less than that of the current. Thus, at $t = 4.17$ ms, we obtain

$$v_C = I\sin(90° - (-29.4°) - 90°)X_C = IX_C\sin(29.4°) = (0.196\,\text{A})(177\,\Omega)\sin(29.4°) = 17.0\ \text{V} .$$

(d) The inductor voltage phasor is 90° more than that of the current. Therefore, at $t = 4.17$ ms, we find

$$v_L = I\sin(90° - (-29.4°) + 90°)X_L = -IX_L\sin(29.4°) = -(0.196\,\text{A})(86.7\,\Omega)\sin(29.4°) = -8.3\ \text{V} .$$

(e) Our results for parts (b), (c) and (d) add to give 36.0 V, the same as the answer for part (a).

43. The resistance of the coil is related to the reactances and the phase constant by Eq. 33-65. Thus,

$$\frac{X_L - X_C}{R} = \frac{\omega_d L - 1/\omega_d C}{R} = \tan\phi ,$$

which we solve for R:

$$\begin{aligned} R &= \frac{1}{\tan\phi}\left(\omega_d L - \frac{1}{\omega_d C}\right) \\ &= \frac{1}{\tan 75^\circ}\left[(2\pi)(930\,\text{Hz})(8.8\times 10^{-2}\,\text{H}) - \frac{1}{(2\pi)(930\,\text{Hz})(0.94\times 10^{-6}\,\text{F})}\right] \\ &= 89\,\Omega\,. \end{aligned}$$

44. (a) The capacitive reactance is

$$X_C = \frac{1}{2\pi f C} = \frac{1}{2\pi(400\,\text{Hz})(24.0\times 10^{-6}\,\text{F})} = 16.6\,\Omega\,.$$

(b) The impedance is

$$\begin{aligned} Z &= \sqrt{R^2 + (X_L - X_C)^2} = \sqrt{R^2 + (2\pi f L - X_C)^2} \\ &= \sqrt{(220\,\Omega)^2 + [2\pi(400\,\text{Hz})(150\times 10^{-3}\,\text{H}) - 16.6\,\Omega]^2} = 422\,\Omega\,. \end{aligned}$$

(c) The current amplitude is

$$I = \frac{\mathcal{E}_m}{Z} = \frac{220\,\text{V}}{422\,\Omega} = 0.521\text{ A}\,.$$

(d) Now $X_C \propto C_{\text{eq}}^{-1}$. Thus, X_C increases as C_{eq} decreases.

(e) Now $C_{\text{eq}} = C/2$, and the new impedance is

$$Z = \sqrt{(220\,\Omega)^2 + [2\pi(400\,\text{Hz})(150\times 10^{-3}\,\text{H}) - 2(16.6\,\Omega)]^2} = 408\,\Omega < 422\,\Omega\,.$$

Therefore, the impedance decreases.

(f) Since $I \propto Z^{-1}$, it increases.

45. (a) For a given amplitude $(E)_m$ of the generator emf, the current amplitude is given by

$$I = \frac{(E)_m}{Z} = \frac{(E)_m}{\sqrt{R^2 + (\omega_d L - 1/\omega_d C)^2}}\,.$$

We find the maximum by setting the derivative with respect to ω_d equal to zero:

$$\frac{dI}{d\omega_d} = -(E)_m\left[R^2 + (\omega_d L - 1/\omega_d C)^2\right]^{-3/2}\left[\omega_d L - \frac{1}{\omega_d C}\right]\left[L + \frac{1}{\omega_d^2 C}\right]\,.$$

The only factor that can equal zero is $\omega_d L - (1/\omega_d C)$; it does so for $\omega_d = 1/\sqrt{LC} = \omega$. For this circuit,

$$\omega_d = \frac{1}{\sqrt{LC}} = \frac{1}{\sqrt{(1.00\,\text{H})(20.0\times 10^{-6}\,\text{F})}} = 224\text{ rad/s}\,.$$

(b) When $\omega_d = \omega$, the impedance is $Z = R$, and the current amplitude is

$$I = \frac{(E)_m}{R} = \frac{30.0\,\text{V}}{5.00\,\Omega} = 6.00\text{ A}\,.$$

(c) We want to find the (positive) values of ω_d for which $I = \frac{(E)_m}{2R}$:

$$\frac{(E)_m}{\sqrt{R^2 + (\omega_d L - 1/\omega_d C)^2}} = \frac{(E)_m}{2R}\,.$$

This may be rearranged to yield

$$\left(\omega_d L - \frac{1}{\omega_d C}\right)^2 = 3R^2 .$$

Taking the square root of both sides (acknowledging the two $\pm$ roots) and multiplying by $\omega_d C$, we obtain

$$\omega_d^2 (LC) \pm \omega_d \left(\sqrt{3}CR\right) - 1 = 0 .$$

Using the quadratic formula, we find the smallest positive solution

$$\begin{aligned}\omega_2 &= \frac{-\sqrt{3}CR + \sqrt{3C^2R^2 + 4LC}}{2LC} \\ &= \frac{-\sqrt{3}(20.0 \times 10^{-6}\,\text{F})(5.00\,\Omega)}{2(1.00\,\text{H})(20.0 \times 10^{-6}\,\text{F})} \\ &\quad + \frac{\sqrt{3(20.0 \times 10^{-6}\,\text{F})^2(5.00\,\Omega)^2 + 4(1.00\,\text{H})(20.0 \times 10^{-6}\,\text{F})}}{2(1.00\,\text{H})(20.0 \times 10^{-6}\,\text{F})} \\ &= 219\ \text{rad/s} ,\end{aligned}$$

and the largest positive solution

$$\begin{aligned}\omega_1 &= \frac{+\sqrt{3}CR + \sqrt{3C^2R^2 + 4LC}}{2LC} \\ &= \frac{+\sqrt{3}(20.0 \times 10^{-6}\,\text{F})(5.00\,\Omega)}{2(1.00\,\text{H})(20.0 \times 10^{-6}\,\text{F})} \\ &\quad + \frac{\sqrt{3(20.0 \times 10^{-6}\,\text{F})^2(5.00\,\Omega)^2 + 4(1.00\,\text{H})(20.0 \times 10^{-6}\,\text{F})}}{2(1.00\,\text{H})(20.0 \times 10^{-6}\,\text{F})} \\ &= 228\ \text{rad/s} .\end{aligned}$$

(d) The fractional width is

$$\frac{\omega_1 - \omega_2}{\omega_0} = \frac{228\,\text{rad/s} - 219\,\text{rad/s}}{224\,\text{rad/s}} = 0.04 .$$

46. Four possibilities exist: (1) $C_1 = 4.00\,\mu\text{F}$ is used alone; (2) $C_2 = 6.00\,\mu\text{F}$ is used alone; (3) C_1 and C_2 are connected in series; and (4) C_1 and C_2 are connected in parallel. The corresponding resonant frequencies are

$$\begin{aligned}f_1 &= \frac{1}{2\pi\sqrt{LC_1}} = \frac{1}{2\pi\sqrt{(2.00 \times 10^{-3}\,\text{H})(4.00 \times 10^{-6}\,\text{F})}} = 1.78 \times 10^3\ \text{Hz} \\ f_2 &= \frac{1}{2\pi\sqrt{LC_2}} = \frac{1}{2\pi\sqrt{(2.00 \times 10^{-3}\,\text{H})(6.00 \times 10^{-6}\,\text{F})}} = 1.45 \times 10^3\ \text{Hz} \\ f_3 &= \frac{1}{2\pi\sqrt{LC_1C_2/(C_1 + C_2)}} = 2.30 \times 10^3\ \text{Hz} \\ f_4 &= \frac{1}{2\pi\sqrt{L(C_1 + C_2)}} = 1.13 \times 10^3\ \text{Hz} .\end{aligned}$$

47. We use the expressions found in Problem 45:

$$\begin{aligned}\omega_1 &= \frac{+\sqrt{3}CR + \sqrt{3C^2R^2 + 4LC}}{2LC} \\ \omega_2 &= \frac{-\sqrt{3}CR + \sqrt{3C^2R^2 + 4LC}}{2LC} .\end{aligned}$$

We also use Eq. 33-4. Thus,

$$\frac{\Delta\omega_d}{\omega} = \frac{\omega_1 - \omega_2}{\omega} = \frac{2\sqrt{3}CR\sqrt{LC}}{2LC} = R\sqrt{\frac{3C}{L}} \,.$$

For the data of Problem 45,

$$\frac{\Delta\omega_d}{\omega} = (5.00\,\Omega)\sqrt{\frac{3(20.0\times 10^{-6}\,\text{F})}{1.00\,\text{H}}} = 3.87\times 10^{-2} \,.$$

This is in agreement with the result of Problem 45. The method of Problem 45, however, gives only one significant figure since two numbers close in value are subtracted ($\omega_1 - \omega_2$). Here the subtraction is done algebraically, and three significant figures are obtained.

48. (a) Since $L_{\text{eq}} = L_1 + L_2$ and $C_{\text{eq}} = C_1 + C_2 + C_3$ for the circuit, the resonant frequency is

$$\begin{aligned}
\omega &= \frac{1}{2\pi\sqrt{L_{\text{eq}}C_{\text{eq}}}} = \frac{1}{2\pi\sqrt{(L_1+L_2)(C_1+C_2+C_3)}} \\
&= \frac{1}{2\pi\sqrt{(1.70\times 10^{-3}\,\text{H} + 2.30\times 10^{-3}\,\text{H})(4.00\times 10^{-6}\,\text{F} + 2.50\times 10^{-6}\,\text{F} + 3.50\times 10^{-6}\,\text{F})}} \\
&= 796\ \text{Hz} \,.
\end{aligned}$$

(b) The resonant frequency does not depend on R so it will not change as R increases.

(c) Since $\omega \propto (L_1 + L_2)^{-1/2}$, it will decrease as L_1 increases.

(d) Since $\omega \propto C_{\text{eq}}^{-1/2}$ and C_{eq} decreases as C_3 is removed, ω will increase.

49. The average power dissipated in resistance R when the current is alternating is given by $P_{\text{avg}} = I_{\text{rms}}^2 R$, where I_{rms} is the root-mean-square current. Since $I_{\text{rms}} = I/\sqrt{2}$, where I is the current amplitude, this can be written $P_{\text{avg}} = I^2R/2$. The power dissipated in the same resistor when the current i_d is direct is given by $P = i_d^2 R$. Setting the two powers equal to each other and solving, we obtain

$$i_d = \frac{I}{\sqrt{2}} = \frac{2.60\ \text{A}}{\sqrt{2}} = 1.84\ \text{A} \,.$$

50. Since the impedance of the voltmeter is large, it will not affect the impedance of the circuit when connected in parallel with the circuit. So the reading will be 100 V in all three cases.

51. The amplitude (peak) value is

$$V_{\text{max}} = \sqrt{2}\,V_{\text{rms}} = \sqrt{2}(100\,\text{V}) = 141\ \text{V} \,.$$

52. (a) We refer to problem 34, part (c). The power delivered by the generator at this instant is $P = \mathcal{E}(t)i(t) = \mathcal{E}_m \sin(2n\pi - \pi/6) I \sin(\pi/3) = -\mathcal{E}_m I \sin(\pi/6)\sin(\pi/3)$. This is less than zero, so it is taking energy from the rest of the circuit.

(b) We refer to problem 36, part (c). The power delivered by the generator at this instant is $P = \mathcal{E}(t)i(t) = \mathcal{E}_m \sin(2n\pi - \pi/6) I \sin(-2\pi/3) = \mathcal{E}_m I \sin(\pi/6)\sin(2\pi/3)$. Since this is positive, it is supplying energy to the rest of the system.

53. We use $P_{\text{avg}} = I_{\text{rms}}^2 R = \frac{1}{2}I^2R$.

- $P_{\text{avg}} = 0$, since $R = 0$.
- $P_{\text{avg}} = \frac{1}{2}I^2R = \frac{1}{2}(0.600\,\text{A})^2(50\,\Omega) = 9.0\,\text{W}$.
- $P_{\text{avg}} = \frac{1}{2}I^2R = \frac{1}{2}(0.198\,\text{A})^2(160\,\Omega) = 3.14\,\text{W}$.
- $P_{\text{avg}} = \frac{1}{2}I^2R = \frac{1}{2}(0.151\,\text{A})^2(160\,\Omega) = 1.82\,\text{W}$.

54. We start with Eq. 33-76:

$$P_{\text{avg}} = \mathcal{E}_{\text{rms}} I_{\text{rms}} \cos\phi = \mathcal{E}_{\text{rms}} \left(\frac{\mathcal{E}_{\text{rms}}}{Z}\right)\left(\frac{R}{Z}\right) = \frac{\mathcal{E}_{\text{rms}}^2 R}{Z^2} .$$

For a purely resistive circuit, $Z = R$, and this result reduces to Eq. 27-23 (with V replaced with $\mathcal{E}_{\text{rms}}$). This is also the case for a series RLC circuit at resonance. The average rate for dissipating energy is, of course, zero if $R = 0$, as would be the case for a purely inductive circuit.

55. (a) Using Eq. 33-61, the impedance is

$$Z = \sqrt{(12.0\,\Omega)^2 + (1.30\,\Omega - 0)^2} = 12.1\ \Omega .$$

(b) We use the result of problem 54:

$$P_{\text{avg}} = \frac{\mathcal{E}_{\text{rms}}^2 R}{Z^2} = \frac{(120\,\text{V})^2(12.0\,\Omega)}{(12.1\,\Omega)^2} = 1.18 \times 10^3\ \text{W} .$$

56. The current in the circuit satisfies $i(t) = I\sin(\omega_d t - \phi)$, where

$$\begin{aligned} I &= \frac{\mathcal{E}_m}{Z} = \frac{\mathcal{E}_m}{\sqrt{R^2 + (\omega_d L - 1/\omega_d C)^2}} \\ &= \frac{45.0\,\text{V}}{\sqrt{(16.0\,\Omega)^2 + \{(3000\,\text{rad/s})(9.20\,\text{mH}) - 1/[(3000\,\text{rad/s})(31.2\,\mu\text{F})]\}^2}} \\ &= 1.93\ \text{A} \end{aligned}$$

and

$$\begin{aligned} \phi &= \tan^{-1}\left(\frac{X_L - X_C}{R}\right) = \tan^{-1}\left(\frac{\omega_d L - 1/\omega_d C}{R}\right) \\ &= \tan^{-1}\left[\frac{(3000\,\text{rad/s})(9.20\,\text{mH})}{16.0\,\Omega} - \frac{1}{(3000\,\text{rad/s})(16.0\,\Omega)(31.2\,\mu\text{F})}\right] \\ &= 46.5^\circ . \end{aligned}$$

(a) The power supplied by the generator is

$$\begin{aligned} P_g &= i(t)\mathcal{E}(t) = I\sin(\omega_d t - \phi)\mathcal{E}_m \sin\omega_d t \\ &= (1.93\,\text{A})(45.0\,\text{V})\sin[(3000\,\text{rad/s})(0.442\,\text{ms})]\sin[(3000\,\text{rad/s})(0.442\,\text{ms}) - 46.5^\circ] \\ &= 41.4\ \text{W} . \end{aligned}$$

(b) The rate at which the energy in the capacitor changes is

$$\begin{aligned} P_c &= -\frac{d}{dt}\left(\frac{q^2}{2C}\right) = -i\frac{q}{C} = -iV_c \\ &= -I\sin(\omega_d t - \phi)\left(\frac{I}{\omega_d C}\right)\cos(\omega_d t - \phi) = -\frac{I^2}{2\omega_d C}\sin[2(\omega_d t - \phi)] \\ &= -\frac{(1.93\,\text{A})^2}{2(3000\,\text{rad/s})(31.2 \times 10^{-6}\,\text{F})}\sin[2(3000\,\text{rad/s})(0.442\,\text{ms}) - 2(46.5^\circ)] \\ &= -17.0\ \text{W} . \end{aligned}$$

(c) The rate at which the energy in the inductor changes is

$$\begin{aligned}P_i &= \frac{d}{dt}\left(\frac{1}{2}Li^2\right) = Li\frac{di}{dt} = LI\sin(\omega_d t - \phi)\frac{d}{dt}[I\sin(\omega_d t - \phi)] \\ &= \frac{1}{2}\omega_d LI^2 \sin[2(\omega_d t - \phi)] \\ &= \frac{1}{2}(3000\,\text{rad/s})(1.93\,\text{A})^2(9.20\,\text{mH})\sin[2(3000\,\text{rad/s})(0.442\,\text{ms}) - 2(46.5°)] \\ &= 44.1\text{ W} .\end{aligned}$$

(d) The rate at which energy is being dissipated by the resistor is

$$\begin{aligned}P_r &= i^2R = I^2R\sin^2(\omega_d t - \phi) \\ &= (1.93\,\text{A})^2(16.0\,\Omega)\sin^2\left[(3000\,\text{rad/s})(0.442\,\text{ms}) - 46.5°\right] \\ &= 14.4\,\text{W} .\end{aligned}$$

(e) The negative result for P_i means that energy is being taken away from the inductor at this particular time.

(f) $P_i + P_r + P_c = 44.1\text{W} - 17.0\,\text{W} + 14.4\,\text{W} = 41.5\,\text{W} = P_g$.

57. (a) The power factor is $\cos\phi$, where ϕ is the phase constant defined by the expression $i = I\sin(\omega t - \phi)$. Thus, $\phi = -42.0°$ and $\cos\phi = \cos(-42.0°) = 0.743$.

(b) Since $\phi < 0$, $\omega t - \phi > \omega t$. The current leads the emf.

(c) The phase constant is related to the reactance difference by $\tan\phi = (X_L - X_C)/R$. We have $\tan\phi = \tan(-42.0°) = -0.900$, a negative number. Therefore, $X_L - X_C$ is negative, which leads to $X_C > X_L$. The circuit in the box is predominantly capacitive.

(d) If the circuit were in resonance X_L would be the same as X_C, $\tan\phi$ would be zero, and ϕ would be zero. Since ϕ is not zero, we conclude the circuit is not in resonance.

(e) Since $\tan\phi$ is negative and finite, neither the capacitive reactance nor the resistance are zero. This means the box must contain a capacitor and a resistor. The inductive reactance may be zero, so there need not be an inductor. If there is an inductor its reactance must be less than that of the capacitor at the operating frequency.

(f) The average power is

$$P_{\text{avg}} = \frac{1}{2}\mathcal{E}_m I\cos\phi = \frac{1}{2}(75.0\,\text{V})(1.20\,\text{A})(0.743) = 33.4\text{ W} .$$

(g) The answers above depend on the frequency only through the phase constant ϕ, which is given. If values were given for R, L and C then the value of the frequency would also be needed to compute the power factor.

58. This circuit contains no reactances, so $\mathcal{E}_{\text{rms}} = I_{\text{rms}}R_{\text{total}}$. Using Eq. 33-71, we find the average dissipated power in resistor R is

$$P_R = I_{\text{rms}}^2 R = \left(\frac{\mathcal{E}_m}{r+R}\right)^2 R .$$

In order to maximize P_R we set the derivative equal to zero:

$$\frac{dP_R}{dR} = \frac{\mathcal{E}_m^2[(r+R)^2 - 2(r+R)R]}{(r+R)^4} = \frac{\mathcal{E}_m^2(r-R)}{(r+R)^3} = 0 \implies R = r$$

59. We use the result of problem 54:

$$P_{\text{avg}} = \frac{(E)_m^2 R}{2Z^2} = \frac{(E)_m^2 R}{2\left[R^2 + (\omega_d L - 1/\omega_d C)^2\right]} .$$

We use the expression $Z = \sqrt{R^2 + (\omega_d L - 1/\omega_d C)^2}$ for the impedance in terms of the angular frequency.

(a) Considered as a function of C, P_{avg} has its largest value when the factor $R^2 + (\omega_d L - 1/\omega_d C)^2$ has the smallest possible value. This occurs for $\omega_d L = 1/\omega_d C$, or

$$C = \frac{1}{\omega_d^2 L} = \frac{1}{(2\pi)^2(60.0\,\text{Hz})^2(60.0 \times 10^{-3}\,\text{H})} = 1.17 \times 10^{-4}\ \text{F} .$$

The circuit is then at resonance.

(b) In this case, we want Z^2 to be as large as possible. The impedance becomes large without bound as C becomes very small. Thus, the smallest average power occurs for $C = 0$ (which is not very different from a simple open switch).

(c) When $\omega_d L = 1/\omega_d C$, the expression for the average power becomes

$$P_{\text{avg}} = \frac{(E)_m^2}{2R} ,$$

so the maximum average power is in the resonant case and is equal to

$$P_{\text{avg}} = \frac{(30.0\,\text{V})^2}{2(5.00\,\Omega)} = 90.0\ \text{W} .$$

On the other hand, the minimum average power is $P_{\text{avg}} = 0$ (as it would be for an open switch).

(d) At maximum power, the reactances are equal: $X_L = X_C$. The phase angle ϕ in this case may be found from

$$\tan\phi = \frac{X_L - X_C}{R} = 0 ,$$

which implies $\phi = 0$. On the other hand, at minimum power $X_C \propto 1/C$ is infinite, which leads us to set $\tan\phi = -\infty$. In this case, we conclude that $\phi = -90°$.

(e) At maximum power, the power factor is $\cos\phi = \cos 0° = 1$, and at minimum power, it is $\cos\phi = \cos(-90°) = 0$.

60. (a) The power consumed by the light bulb is $P = I^2R/2$. So we must let $P_{\text{max}}/P_{\text{min}} = (I/I_{\text{min}})^2 = 5$, or

$$\left(\frac{I}{I_{\text{min}}}\right)^2 = \left(\frac{\mathcal{E}_m/Z_{\text{min}}}{\mathcal{E}_m/Z_{\text{max}}}\right)^2 = \left(\frac{Z_{\text{max}}}{Z_{\text{min}}}\right)^2 = \left(\frac{\sqrt{R^2 + (\omega L_{\text{max}})^2}}{R}\right)^2 = 5 .$$

We solve for L_{max}:

$$L_{\text{max}} = \frac{2R}{\omega} = \frac{2(120\,\text{V})^2/1000\,\text{W}}{2\pi(60.0\,\text{Hz})} = 7.64 \times 10^{-2}\ \text{H} .$$

(b) Now we must let

$$\left(\frac{R_{\text{max}} + R_{\text{bulb}}}{R_{\text{bulb}}}\right)^2 = 5 ,$$

or

$$R_{\text{max}} = (\sqrt{5} - 1)R_{\text{bulb}} = (\sqrt{5} - 1)\frac{(120\,\text{V})^2}{1000\,\text{W}} = 17.8\ \Omega .$$

This is not done because the resistors would consume, rather than temporarily store, electromagnetic energy.

61. (a) The rms current is

$$\begin{aligned} I_{\rm rms} &= \frac{\mathcal{E}_{\rm rms}}{Z} = \frac{\mathcal{E}_{\rm rms}}{\sqrt{R^2 + (2\pi f L - 1/2\pi f C)^2}} \\ &= \frac{75.0\,\text{V}}{\sqrt{(15.0\,\Omega)^2 + \{2\pi(550\,\text{Hz})(25.0\,\text{mH}) - 1/[2\pi(550\,\text{Hz})(4.70\mu\text{F})]\}^2}} \\ &= 2.59\ \text{A}\ . \end{aligned}$$

(b) The various rms voltages are:

$$\begin{aligned} V_{ab} &= I_{\rm rms} R = (2.59\,\text{A})(15.0\,\Omega) = 38.8\ \text{V} \\ V_{bc} &= I_{\rm rms} X_C = \frac{I_{\rm rms}}{2\pi f C} = \frac{2.59\,\text{A}}{2\pi(550\,\text{Hz})(4.70\,\mu\text{F})} = 159\ \text{V} \\ V_{cd} &= I_{\rm rms} X_L = 2\pi I_{\rm rms} f L = 2\pi(2.59\,\text{A})(550\,\text{Hz})(25.0\,\text{mH}) = 224\ \text{V} \\ V_{bd} &= |V_{bc} - V_{cd}| = |159.5\,\text{V} - 223.7\,\text{V}| = 64.2\ \text{V} \\ V_{ad} &= \sqrt{V_{ab}^2 + V_{bd}^2} = \sqrt{(38.8\,\text{V})^2 + (64.2\,\text{V})^2} = 75.0\ \text{V} \end{aligned}$$

(c) For L and C, the rate is zero since they do not dissipate energy. For R,

$$P_R = \frac{V_{ab}^2}{R} = \frac{(38.8\,\text{V})^2}{15.0\,\Omega} = 100\ \text{W}\ .$$

62. We use Eq. 33-79 to find

$$V_s = V_p \left(\frac{N_s}{N_p}\right) = (100\,\text{V})\left(\frac{500}{50}\right) = 1.00 \times 10^3\ \text{V}\ .$$

63. (a) The stepped-down voltage is

$$V_s = V_p \left(\frac{N_s}{N_p}\right) = (120\,\text{V})\left(\frac{10}{500}\right) = 2.4\ \text{V}\ .$$

(b) By Ohm's law, the current in the secondary is

$$I_s = \frac{V_s}{R_s} = \frac{2.4\,\text{V}}{15\,\Omega} = 0.16\ \text{A}\ .$$

We find the primary current from Eq. 33-80:

$$I_p = I_s \left(\frac{N_s}{N_p}\right) = (0.16\,\text{A})\left(\frac{10}{500}\right) = 3.2 \times 10^{-3}\ \text{A}\ .$$

64. Step up:

- We use T_1T_2 as primary and T_1T_3 as secondary coil: $V_{13}/V_{12} = (800 + 200)/200 = 5.00$.
- We use T_1T_2 as primary and T_2T_3 as secondary coil: $V_{23}/V_{13} = 800/200 = 4.00$.
- We use T_2T_3 as primary and T_1T_3 as secondary coil: $V_{13}/V_{23} = (800 + 200)/800 = 1.25$.

Step down: By exchanging the primary and secondary coils in each of the three cases above we get the following possible ratios:

- $1/5.00 = 0.200$
- $1/4.00 = 0.250$

- $1/1.25 = 0.800$

65. The amplifier is connected across the primary windings of a transformer and the resistor R is connected across the secondary windings. If I_s is the rms current in the secondary coil then the average power delivered to R is $P_{\text{avg}} = I_s^2 R$. Using $I_s = (N_p/N_s)I_p$, we obtain

$$P_{\text{avg}} = \left(\frac{I_p N_p}{N_s}\right)^2 R .$$

Next, we find the current in the primary circuit. This is effectively a circuit consisting of a generator and two resistors in series. One resistance is that of the amplifier (r), and the other is the equivalent resistance R_{eq} of the secondary circuit. Therefore,

$$I_p = \frac{\mathcal{E}_{\text{rms}}}{r + R_{\text{eq}}} = \frac{\mathcal{E}_{\text{rms}}}{r + (N_p/N_s)^2 R}$$

where Eq. 33-82 is used for R_{eq}. Consequently,

$$P_{\text{avg}} = \frac{\mathcal{E}^2 (N_p/N_s)^2 R}{\left[r + (N_p/N_s)^2 R\right]^2} .$$

Now, we wish to find the value of N_p/N_s such that P_{avg} is a maximum. For brevity, let $x = (N_p/N_s)^2$. Then

$$P_{\text{avg}} = \frac{\mathcal{E}^2 R x}{(r + xR)^2} ,$$

and the derivative with respect to x is

$$\frac{dP_{\text{avg}}}{dx} = \frac{\mathcal{E}^2 R(r - xR)}{(r + xR)^3} .$$

This is zero for $x = r/R = (1000\,\Omega)/(10\,\Omega) = 100$. We note that for small x, P_{avg} increases linearly with x, and for large x it decreases in proportion to $1/x$. Thus $x = r/R$ is indeed a maximum, not a minimum. Recalling $x = (N_p/N_s)^2$, we conclude that the maximum power is achieved for $N_p/N_s = \sqrt{x} = 10$. The diagram below is a schematic of a transformer with a ten to one turns ratio. An actual transformer would have many more turns in both the primary and secondary coils.

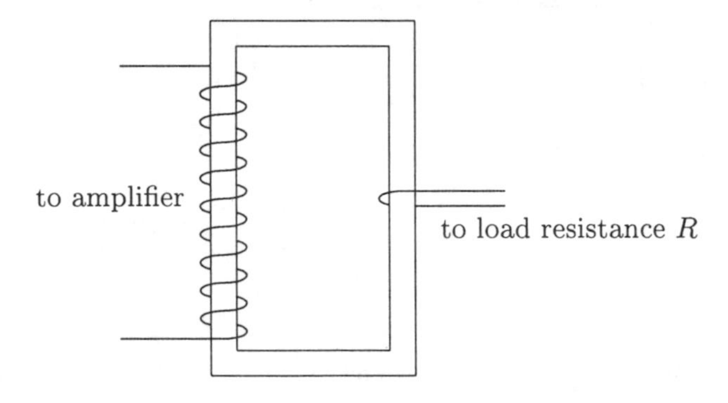

66. The effective resistance R_{eff} satisfies $I_{\text{rms}}^2 R_{\text{eff}} = P_{\text{mechanical}}$, or

$$R_{\text{eff}} = \frac{P_{\text{mechanical}}}{I_{\text{rms}}^2} = \frac{(0.100\,\text{hp})(746\,\text{W/hp})}{(0.650\,\text{A})^2} = 177\,\Omega .$$

This is not the same as the resistance R of its coils, but just the effective resistance for power transfer from electrical to mechanical form. In fact $I_{\text{rms}}^2 R$ would not give $P_{\text{mechanical}}$ but rather the rate of energy loss due to thermal dissipation.

67. The rms current in the motor is

$$I_{\rm rms} = \frac{\mathcal{E}_{\rm rms}}{Z} = \frac{\mathcal{E}_{\rm rms}}{\sqrt{R^2 + X_L^2}} = \frac{420\ {\rm V}}{\sqrt{(45.0\ \Omega)^2 + (32.0\ \Omega)^2}} = 7.61\ {\rm A}\ .$$

68. We use $nT/2$ to represent the integer number of half-periods specified in the problem. Note that $T = 2\pi/\omega$. We use the calculus-based definition of an average of a function:

$$\begin{aligned}
\left[\sin^2(\omega t - \phi)\right]_{\rm avg} &= \frac{1}{nT/2}\int_0^{\frac{nT}{2}} \sin^2(\omega t - \phi)\, dt \\
&= \frac{2}{nT}\int_0^{\frac{nT}{2}} \frac{1 - \cos(2\omega t - 2\phi)}{2}\, dt \\
&= \frac{2}{nT}\left[\frac{t}{2} - \frac{1}{4\omega}\sin(2\omega t - 2\phi)\right]\Bigg|_0^{\frac{nT}{2}} \\
&= \frac{1}{2} - \frac{1}{2nT\omega}\left[\sin(n\omega T - 2\phi) + \sin 2\phi\right]\ .
\end{aligned}$$

Since $n\omega T = n\omega(2\pi/\omega) = 2n\pi$, we have $\sin(n\omega T - 2\phi) = \sin(2n\pi - 2\phi) = -\sin 2\phi$ so $[\sin(n\omega T - 2\phi) + \sin 2\phi] = 0$. Thus,

$$\left[\sin^2(\omega t - \phi)\right]_{\rm avg} = \frac{1}{2}\ .$$

69. (a) The energy stored in the capacitor is given by $U_E = q^2/2C$. Since q is a periodic function of t with period T, so must be U_E. Consequently, U_E will not be changed over one complete cycle. Actually, U_E has period $T/2$, which does not alter our conclusion.

(b) Similarly, the energy stored in the inductor is $U_B = \frac{1}{2}i^2L$. Since i is a periodic function of t with period T, so must be U_B.

(c) The energy supplied by the generator is

$$P_{\rm avg}T = (I_{\rm rms}\mathcal{E}_{\rm rms}\cos\phi)T = \left(\frac{1}{2}T\right)\mathcal{E}_m I \cos\phi$$

where we substitute $I_{\rm rms} = I/\sqrt{2}$ and $\mathcal{E}_{\rm rms} = \mathcal{E}_m/\sqrt{2}$.

(d) The energy dissipated by the resistor is

$$P_{\rm avg,resistor}\,T = (I_{\rm rms}V_R)T = I_{\rm rms}(I_{\rm rms}R)T = \left(\frac{1}{2}T\right)I^2R\ .$$

(e) Since $\mathcal{E}_m I\cos\phi = \mathcal{E}_m I(V_R/\mathcal{E}_m) = \mathcal{E}_m I(IR/\mathcal{E}_m) = I^2R$, the two quantities are indeed the same.

70. (a) The rms current in the cable is $I_{\rm rms} = P/V_t = 250 \times 10^3\ {\rm W}/(80 \times 10^3\ {\rm V}) = 3.125\ {\rm A}$. The rms voltage drop is then $\Delta V = I_{\rm rms}R = (3.125\ {\rm A})(2)(0.30\ \Omega) = 1.9\ {\rm V}$, and the rate of energy dissipation is $P_d = I_{\rm rms}^2R = (3.125\ {\rm A})(2)(0.60\ \Omega) = 5.9\ {\rm W}$.

(b) Now $I_{\rm rms} = 250 \times 10^3\ {\rm W}/(8.0 \times 10^3\ {\rm V}) = 31.25\ {\rm A}$, so $\Delta V = (31.25\ {\rm A})(0.60\ \Omega) = 19\ {\rm V}$ and $P_d = (3.125\ {\rm A})^2(0.60\ \Omega) = 5.9 \times 10^2\ {\rm W}$.

(c) Now $I_{\rm rms} = 250 \times 10^3\ {\rm W}/(0.80 \times 10^3\ {\rm V}) = 312.5\ {\rm A}$, so $\Delta V = (312.5\ {\rm A})(0.60\ \Omega) = 1.9 \times 10^2\ {\rm V}$ and $P_d = (312.5\ {\rm A})^2(0.60\ \Omega) = 5.9 \times 10^4\ {\rm W}$. Both the rate of energy dissipation and the voltage drop increase as V_t decreases. Therefore, to minimize these effects the best choice among the three V_t's above is $V_t = 80\ {\rm kV}$.

71. (a) The impedance is

$$Z = \frac{\mathcal{E}_{\rm m}}{I} = \frac{125\ {\rm V}}{3.20\ {\rm A}} = 39.1\ \Omega\ .$$

(b) From $V_R = IR = \mathcal{E}_m \cos\phi$, we get

$$R = \frac{\mathcal{E}_m \cos\phi}{I} = \frac{(125\text{ V})\cos(0.982\text{ rad})}{3.20\text{ A}} = 21.7\ \Omega\ .$$

(c) Since $X_L - X_C \propto \sin\phi = \sin(-0.982\text{ rad})$, we conclude that $X_L < X_C$. The circuit is predominantly capacitive.

72. (a) The phase constant is given by

$$\phi = \tan^{-1}\left(\frac{V_L - V_C}{R}\right) = \tan^{-1}\left(\frac{V_L - V_L/2.00}{V_L/2.00}\right) = \tan^{-1}(1.00) = 45.0^\circ\ .$$

(b) We solve R from $\mathcal{E}_m \cos\phi = IR$:

$$R = \frac{\mathcal{E}_m \cos\phi}{I} = \frac{(30.0\text{ V})(\cos 45^\circ)}{300\times 10^{-3}\text{ A}} = 70.7\ \Omega\ .$$

73. (a) We solve L from Eq. 33-4, using the fact that $\omega = 2\pi f$:

$$L = \frac{1}{4\pi^2 f^2 C} = \frac{1}{4\pi^2(10.4\times 10^3\text{ Hz})^2(340\times 10^{-6}\text{ F})} = 6.89\times 10^{-7}\text{ H}\ .$$

(b) The total energy may be figured from the inductor (when the current is at maximum):

$$U = \frac{1}{2}LI^2 = \frac{1}{2}(6.89\times 10^{-7}\text{ H})(7.20\times 10^{-3}\text{ A})^2 = 1.79\times 10^{-11}\text{ J}\ .$$

(c) We solve for Q from $U = \frac{1}{2}Q^2/C$:

$$Q = \sqrt{2CU} = \sqrt{2(340\times 10^{-6}\text{ F})(1.79\times 10^{-11}\text{ J})} = 1.10\times 10^{-7}\text{ C}\ .$$

74. (a) Let $\omega t - \pi/4 = \pi/2$ to obtain $t = 3\pi/4\omega = 3\pi/[4(350\text{ rad/s})] = 6.73\times 10^{-3}$ s.

(b) Let $\omega t + \pi/4 = \pi/2$ to obtain $t = \pi/4\omega = \pi/[4(350\text{ rad/s})] = 2.24\times 10^{-3}$ s.

(c) Since i leads $\mathcal{E}$ in phase by $\pi/2$, the element must be a capacitor.

(d) We solve C from $X_C = (\omega C)^{-1} = \mathcal{E}_m/I$:

$$C = \frac{I}{\mathcal{E}_m \omega} = \frac{6.20\times 10^{-3}\text{ A}}{(30.0\text{ V})(350\text{ rad/s})} = 5.90\times 10^{-5}\text{ F}\ .$$

75. From the problem statement $2\pi f_0 = (LC)^{-1/2} = 6000$ Hz, $Z = \sqrt{R^2 + (2\pi f_1 L - 1/2\pi f_1 C)^2} = 1000\ \Omega$ where $f_1 = 8000$ Hz, and $\cos\phi = R/Z = \cos 45^\circ$. We solve these equations for the unknowns.

(a) $R = Z\cos\phi = (1000\ \Omega)\cos 45^\circ = 707\ \Omega$

(b) The self-inductance is

$$L = \frac{\sqrt{Z^2 - R^2}}{2\pi(f_1 - f_0^2/f_1)} = \frac{\sqrt{(1000\ \Omega)^2 - (707\ \Omega)^2}}{2\pi[8000\text{ Hz} - (6000\text{ Hz})^2/8000\text{ Hz}]} = 3.22\times 10^{-2}\text{ H}\ .$$

(c) The capacitance is

$$C = \frac{1}{4\pi^2 f_0^2 L} = \frac{1}{4\pi^2(6000\text{ Hz})^2(3.22\times 10^{-2}\text{ H})} = 2.19\times 10^{-8}\text{ F}\ .$$

76. (a) From Eq. 33-65, we have

$$\phi = \tan^{-1}\left(\frac{V_L - V_C}{V_R}\right) = \tan^{-1}\left(\frac{V_L - (V_L/1.50)}{(V_L/2.00)}\right)$$

which becomes $\tan^{-1} 2/3 = 33.7°$ or 0.588 rad.

(b) Since $\phi > 0$, it is inductive ($X_L > X_C$).

(c) We have $V_R = IR = 9.98$ V, so that $V_L = 2.00V_R = 20.0$ V and $V_C = V_L/1.50 = 13.3$ V. Therefore, from Eq. 33-60,

$$\mathcal{E}_m = \sqrt{V_R^2 + (V_L - V_C)^2}$$

we find $\mathcal{E}_m = 12.0$ V.

77. (a) With f understood to be in Hertz, the capacitive reactance is $X_C = \left[(2\pi)(45 \times 10^{-6}\,\text{F})f\right]^{-1}$.

(b) The resistance, reactance and impedance are plotted over the range $10 \leq f \leq 70$ Hz. The horizontal line is R, and the curve that crosses that line is X_C. SI units are understood.

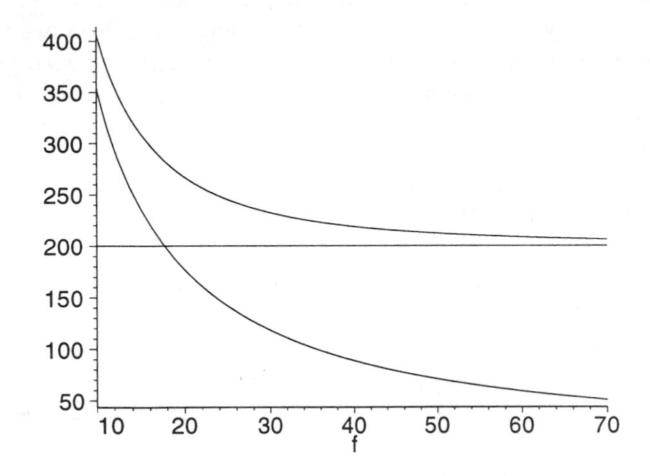

(c) From the graph, we estimate the crossing point to be at about 18 Hz. More careful considerations lead to $f = 17.7$ Hz as the frequency where $X_C = R$.

78. (a) The voltage amplitude for the source is $V_s = 100\,\text{V} = IZ = I\sqrt{R^2 + X_C^2}$, from which we can determine the current at each frequency (the explicit dependence of X_C on frequency is stated in the solution to part (a) of problem 77). This leads to the voltage amplitude across the resistor $V_R = IR$ and the voltage amplitude across the capacitor

$$V_C = IX_C = \left(\frac{V_s}{\sqrt{R^2 + X_C^2}}\right) X_C \quad \text{where} \quad X_C = \frac{1}{2\pi C f}$$

using the values $R = 200\,\Omega$ and $C = 45 \times 10^{-6}$ F given in problem 77. We show, below, the graphs of V_s, V_R and V_C over the range $0 < f < 100$ Hz. The falling curve is V_C and the rising curve is V_R.

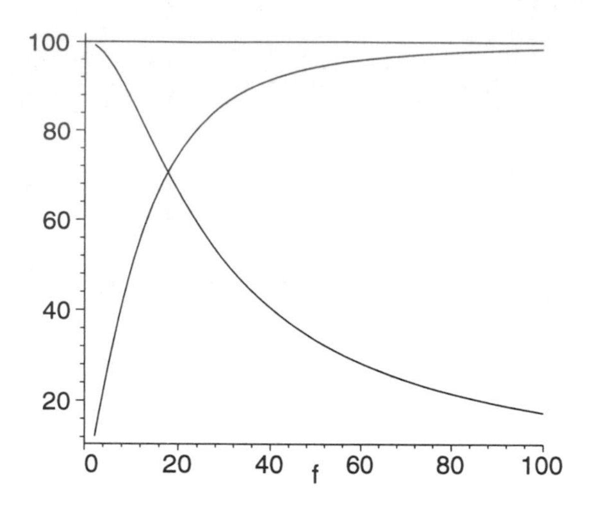

(b) The graph indicates that V_C and V_R are equal at roughly 18 Hz. More careful considerations lead to $f = 17.7$ Hz as the frequency for which $V_C = V_R$.

79. When switch S_1 is closed and the others are open, the inductor is essentially out of the circuit and what remains is an RC circuit. The time constant is $\tau_C = RC$. When switch S_2 is closed and the others are open, the capacitor is essentially out of the circuit. In this case, what we have is an LR circuit with time constant $\tau_L = L/R$. Finally, when switch S_3 is closed and the others are open, the resistor is essentially out of the circuit and what remains is an LC circuit that oscillates with period $T = 2\pi\sqrt{LC}$. Substituting $L = R\tau_L$ and $C = \tau_C/R$, we obtain $T = 2\pi\sqrt{\tau_C\tau_L}$.

80. (a) From Eq. 33-25,

$$\frac{dq}{dt} = \frac{d}{dt}\left[Qe^{-Rt/2L}\cos(\omega' t + \phi)\right] = -\frac{RQ}{2L}e^{-Rt/2L}\cos(\omega' t + \phi) - \omega' Q e^{-Rt/2L}\sin(\omega' t + \phi)$$

and

$$\begin{aligned}\frac{d^2q}{dt^2} &= \left(\frac{R}{2L}\right)e^{-Rt/2L}\left[\left(\frac{RQ}{2L}\right)\cos(\omega' t + \phi) - \omega' Q\sin(\omega' t + \phi)\right] \\ &\quad + e^{-Rt/2L}\left[\frac{RQ\omega'}{2L}\sin(\omega' t + \phi) - \omega'^2 Q\cos(\omega' t + \phi)\right] .\end{aligned}$$

Substituting these expressions, and Eq. 33-25 itself, into Eq. 33-24, we obtain

$$Qe^{-Rt/2L}\left[-\omega'^2 L - \left(\frac{R}{2L}\right)^2 + \frac{1}{c}\right]\cos(\omega' t + \phi) = 0 .$$

Since this equation is valid at any time t, we must have

$$-\omega'^2 L - \left(\frac{R}{2L}\right)^2 + \frac{1}{C} = 0 \implies \omega' = \sqrt{\frac{1}{LC} - \left(\frac{R}{2L}\right)^2} = \sqrt{\omega^2 - \left(\frac{R}{2L}\right)^2} .$$

(b) The fractional shift in frequency is

$$\begin{aligned}\frac{\Delta f}{f} = \frac{\Delta\omega}{\omega} &= \frac{\omega - \omega'}{\omega} = 1 - \frac{\sqrt{(1/LC) - (R/2L)^2}}{\sqrt{1/LC}} = 1 - \sqrt{1 - \frac{R^2C}{4L}} \\ &= 1 - \sqrt{1 - \frac{(100\,\Omega)^2(7.30 \times 10^{-6}\,\text{F})}{4(4.40\,\text{H})}} = 0.00210 = 0.210\% .\end{aligned}$$

81. (a) We find L from $X_L = \omega L = 2\pi f L$:

$$f = \frac{X_L}{2\pi L} = \frac{1.30 \times 10^3 \Omega}{2\pi(45.0 \times 10^{-3}\,\text{H})} = 4.60 \times 10^3 \text{ Hz} .$$

(b) The capacitance is found from $X_C = (\omega C)^{-1} = (2\pi f C)^{-1}$:

$$C = \frac{1}{2\pi f X_C} = \frac{1}{2\pi(4.60 \times 10^3\,\text{Hz})(1.30 \times 10^3\,\Omega)} = 2.66 \times 10^{-8} \text{ F} .$$

(c) Noting that $X_L \propto f$ and $X_C \propto f^{-1}$, we conclude that when f is doubled, X_L doubles and X_C reduces by half. Thus, $X_L = 2(1.30 \times 10^3\,\Omega) = 2.60 \times 10^3\,\Omega$ and $X_C = 1.30 \times 10^3\,\Omega/2 = 6.50 \times 10^2\,\Omega$.

82. (a) We consider the following combinations: $\Delta V_{12} = V_1 - V_2$, $\Delta V_{13} = V_1 - V_3$, and $\Delta V_{23} = V_2 - V_3$. For ΔV_{12},

$$\Delta V_{12} = A\sin(\omega_d t) - A\sin(\omega_d t - 120°) = 2A\sin\left(\frac{120°}{2}\right)\cos\left(\frac{2\omega_d t - 120°}{2}\right) = \sqrt{3}\,A\cos(\omega_d t - 60°)$$

where we use $\sin\alpha - \sin\beta = 2\sin[(\alpha-\beta)/2]\cos[(\alpha+\beta)/2]$ and $\sin 60° = \sqrt{3}/2$. Similarly,

$$\Delta V_{13} = A\sin(\omega_d t) - A\sin(\omega_d t - 240°) = 2A\sin\left(\frac{240°}{2}\right)\cos\left(\frac{2\omega_d t - 240°}{2}\right) = \sqrt{3}\,A\cos(\omega_d t - 120°)$$

and

$$\Delta V_{23} = A\sin(\omega_d t - 120°) - A\sin(\omega_d t - 240°) = 2A\sin\left(\frac{120°}{2}\right)\cos\left(\frac{2\omega_d t - 360°}{2}\right) = \sqrt{3}\,A\cos(\omega_d t - 180°)$$

All three expressions are sinusoidal functions of t with angular frequency ω_d.

(b) We note that each of the above expressions has an amplitude of $\sqrt{3}A$.

83. When the switch is open, we have a series LRC circuit involving just the one capacitor near the upper right corner. Eq. 33-65 leads to

$$\frac{\omega_d L - \frac{1}{\omega_d C}}{R} = \tan\phi_o = \tan(-20°) = -\tan 20° .$$

Now, when the switch is in position 1, the equivalent capacitance in the circuit is $2C$. In this case, we have

$$\frac{\omega_d L - \frac{1}{2\omega_d C}}{R} = \tan\phi_1 = \tan 10.0° .$$

Finally, with the switch in position 2, the circuit is simply an LC circuit with current amplitude

$$I_2 = \frac{\mathcal{E}_m}{Z_{LC}} = \frac{\mathcal{E}_m}{\sqrt{\left(\omega_d L - \frac{1}{\omega_d C}\right)^2}} = \frac{\mathcal{E}_m}{\frac{1}{\omega_d C} - \omega_d L}$$

where we use the fact that $\frac{1}{\omega_d C} > \omega_d L$ in simplifying the square root (this fact is evident from the description of the first situation, when the switch was open). We solve for L, R and C from the three equations above:

$$R = \frac{-\mathcal{E}_m}{I_2\tan\phi_o} = \frac{120\,\text{V}}{(2.00\,\text{A})\tan 20.0°} = 165\,\Omega$$

$$C = \frac{I_2}{2\omega_d\mathcal{E}_m\left(1 - \frac{\tan\phi_1}{\tan\phi_o}\right)} = \frac{2.00\,\text{A}}{2(2\pi)(60.0\,\text{Hz})(120\,\text{V})\left(1 + \frac{\tan 10.0°}{\tan 20.0°}\right)} = 1.49 \times 10^{-5}\text{ F}$$

$$L = \frac{\mathcal{E}_m}{\omega_d I_2}\left(1 - 2\frac{\tan\phi_1}{\tan\phi_o}\right) = \frac{120\,\text{V}}{2\pi(60.0\,\text{Hz})(2.00\,\text{A})}\left(1 + 2\frac{\tan 10.0°}{\tan 20.0°}\right) 0.313\text{ H}$$

84. (a) Using $X_C = 1/\omega C$ and $V_C = I_C X_C$, we find

$$\omega = \frac{I_C}{CV_C} = 5.77 \times 10^5 \text{ rad/s} .$$

This value is used in the subsequent parts. The period is $T = 2\pi/\omega = 1.09 \times 10^{-5}$ s.

(b) Adapting Eq. 26-22 to the notation of this chapter,

$$U_{E,\text{max}} = \frac{1}{2}CV_C^2 = 4.5 \times 10{-}9 \text{ J} .$$

(c) The discussion in §33-4 shows that $U_{E,\text{max}} = U_{B,\text{max}}$.

(d) We return to Eq. 31-37 (though other, equivalent, approaches could be explored):

$$\frac{di}{dt} = \frac{-\mathcal{E}_L}{L}$$

By the loop rule, $\mathcal{E}_L$ is at its most negative value when the capacitor voltage is at its most positive (V_C). Using this plus the frequency relationship between L and C (Eq. 33-4) leads to

$$\left|\frac{di}{dt}\right|_{\text{max}} = \omega^2 CV_C = 998 \text{ A/s} .$$

(e) Differentiating Eq. 31-51, we have

$$\frac{dU_B}{dt} = Li\frac{di}{dt} .$$

As in the previous part, we use $L = 1/\omega^2 C$. We also use a simple sinusoidal form for the current, $i = I \sin \omega t$:

$$\frac{dU_B}{dt} = \frac{1}{\omega^2 C} I^2 \, \omega \, \sin \omega t \, \cos \omega t$$

where this I is equivalent to the I_C used in part (a). Using a well-known trig identity, we obtain

$$\left(\frac{dU_B}{dt}\right)_{\text{max}} = \frac{I^2}{2\omega^2 C} (\sin 2\omega t)_{\text{max}} = \frac{I^2}{2\omega^2 C}$$

which yields a (maximum) time rate of change (for U_B) equal to 2.60×10^{-3} J/s.

85. (a) At any time, the total energy U in the circuit is the sum of the energy U_E in the capacitor and the energy U_B in the inductor. When $U_E = 0.500 U_B$ (at time t), then $U_B = 2.00 U_E$ and $U = U_E + U_B = 3.00 U_E$. Now, U_E is given by $q^2/2C$, where q is the charge on the capacitor at time t. The total energy U is given by $Q^2/2C$, where Q is the maximum charge on the capacitor. Thus, $Q^2/2C = 3.00 q^2/2C$ or $q = Q/\sqrt{3.00} = 0.577Q$.

(b) If the capacitor is fully charged at time $t = 0$, then the time-dependent charge on the capacitor is given by $q = Q \cos \omega t$. This implies that the condition $q = 0.577Q$ is satisfied when $\cos \omega t = 0.557$, or $\omega t = 0.955$ rad. Since $\omega = 2\pi/T$ (where T is the period of oscillation), $t = 0.955T/2\pi = 0.152T$.

86. (a) Eqs. 33-4 and 33-14 lead to

$$Q = \frac{I}{\omega} = I\sqrt{LC} = 1.27 \times 10^{-6} \text{ C} .$$

(b) We choose the phase constant in Eq. 33-12 to be $\phi = -\pi/2$, so that $i_0 = I$ in Eq. 33-15). Thus, the energy in the capacitor is

$$U_E = \frac{q^2}{2C} = \frac{Q^2}{2C} (\sin \omega t)^2 .$$

Differentiating and using the fact that $2\sin\theta\cos\theta = \sin 2\theta$, we obtain

$$\frac{dU_E}{dt} = \frac{Q^2}{2C}\,\omega\,\sin 2\omega t\ .$$

We find the maximum value occurs whenever $\sin 2\omega t = 1$, which leads (with $n =$ *odd* integer) to

$$t = \frac{1}{2\omega}\,\frac{n\pi}{2} = \frac{n\pi}{4\omega} = \frac{n\pi}{4}\sqrt{LC} = 8.31\times 10^{-5}\ \text{s},\ 2.49\times 10^{-4}\ \text{s},\ \ldots\ .$$

(c) Returning to the above expression for dU_E/dt with the requirement that $\sin 2\omega t = 1$, we obtain

$$\left(\frac{dU_E}{dt}\right)_{\text{max}} = \frac{Q^2}{2C}\,\omega = \frac{\left(I\sqrt{LC}\right)^2}{2C}\,\frac{1}{\sqrt{LC}} = \frac{I^2}{2}\sqrt{\frac{L}{C}} = 5.44\times 10^{-3}\ \text{J/s}\ .$$

87. (a) We observe that $\omega = 6597$ rad/s, and, consequently, $X_L = 594\ \Omega$ and $X_C = 303\ \Omega$. Since $X_L > X_C$, the phase angle is positive: $\phi = +60°$.

(b) From Eq. 33-65, we obtain

$$R = \frac{X_L - X_C}{\tan\phi} = 168\ \Omega\ .$$

(c) Since we are already on the "high side" of resonance, increasing f will only decrease the current further, but *decreasing* f brings us closer to resonance and, consequently, large values of I.

(d) Increasing L increases X_L, but we already have $X_L > X_C$. Thus, if we wish to move closer to resonance (where X_L must equal X_C), we need to *decrease* the value of L.

(e) To change the present condition of $X_C < X_L$ to something closer to $X_C = X_L$ (resonance, large current), we can increase X_C. Since X_C depends inversely on C, this means *decreasing* C.

88. (a) We observe that $\omega_d = 12566$ rad/s. Consequently, $X_L = 754\ \Omega$ and $X_C = 199\ \Omega$. Hence, Eq. 33-65 gives

$$\phi = \tan^{-1}\left(\frac{X_L - X_C}{R}\right) = 1.22\ \text{rad}\ .$$

(b) We find the current amplitude from Eq. 33-60:

$$I = \frac{\mathcal{E}_m}{\sqrt{R^2 + (X_L - X_C)^2}} = 0.288\ \text{A}\ .$$

89. From Eq. 33-60, we have

$$\left(\frac{220\ \text{V}}{3.00\ \text{A}}\right)^2 = R^2 + X_L^2 \implies X_L = 69.3\ \Omega\ .$$

90. (a) We observe that $\omega = 7540$ rad/s, and, consequently, $X_L = 377\ \Omega$ and $X_C = 15.3\ \Omega$. Therefore, Eq. 33-64 leads to

$$I_{\text{rms}} = \frac{112\ \text{V}}{\sqrt{(35\ \Omega)^2 + (377\ \Omega - 15\ \Omega)^2}} = 0.308\ \text{A}\ .$$

(b) (c) (d) (e) (f) and (g) For the individual elements, we have:

$$\begin{aligned} V_{R,\text{rms}} &= I_{\text{rms}}R = 10.8\ \text{V} \\ V_{C,\text{rms}} &= I_{\text{rms}}X_C = 4.73\ \text{V} \\ V_{L,\text{rms}} &= I_{\text{rms}}X_L = 116\ \text{V} \end{aligned}$$

The capacitor and inductor are not dissipative elements; the only power dissipated (by definition) is in the resistor. If a coil, perhaps referred to as an inductor in building a circuit, is found to have an internal resistance, then the coil (for purposes of analysis) is taken to be an inductor plus a resistor. The power dissipated in the resistive element is $P_{\text{avg}} = (0.308\ \text{A})^2(35\ \Omega) = 3.33$ W.

91. From Eq. 33-4, with $\omega = 2\pi f = 4.49 \times 10^3$ rad/s, we obtain

$$L = \frac{1}{\omega^2 C} = 7.08 \times 10^{-3} \text{ H} .$$

92. (a) From Eq. 33-4, with $\omega = 2\pi f$, we have

$$f = \frac{1}{2\pi\sqrt{LC}} = 7.08 \times 10^{-3} \text{ H} .$$

(b) The maximum current in the oscillator is

$$i_{\text{max}} = I_C = \frac{V_C}{X_C} = \omega C v_{\text{max}} = 4.00 \times 10^{-3} \text{ A} .$$

(c) Using Eq. 31-51, we find the maximum magnetic energy:

$$U_{B,\text{max}} = \frac{1}{2} L i_{\text{max}}^2 = 1.6 \times 10^{-8} \text{ J} .$$

(d) Adapting Eq. 31-37 to the notation of this chapter,

$$v_{\text{max}} = L \left| \frac{di}{dt} \right|_{\text{max}}$$

which yields a (maximum) time rate of change (for i) equal to 2000 A/s.

Chapter 34

1. The time for light to travel a distance d in free space is $t = d/c$, where c is the speed of light (3.00×10^8 m/s).

   (a) We take d to be $150\,\text{km} = 150 \times 10^3$ m. Then,

   $$t = \frac{d}{c} = \frac{150 \times 10^3\,\text{m}}{3.00 \times 10^8\,\text{m/s}} = 5.00 \times 10^{-4}\ \text{s}\ .$$

   (b) At full moon, the Moon and Sun are on opposite sides of Earth, so the distance traveled by the light is $d = (1.5 \times 10^8\,\text{km}) + 2(3.8 \times 10^5\,\text{km}) = 1.51 \times 10^8\,\text{km} = 1.51 \times 10^{11}$ m. The time taken by light to travel this distance is

   $$t = \frac{d}{c} = \frac{1.51 \times 10^{11}\,\text{m}}{3.00 \times 10^8\,\text{m/s}} = 500\,\text{s} = 8.4\ \text{min}\ .$$

   (c) We take d to be $2(1.3 \times 10^9\,\text{km}) = 2.6 \times 10^{12}$ m. Then,

   $$t = \frac{d}{c} = \frac{2.6 \times 10^{12}\,\text{m}}{3.00 \times 10^8\,\text{m/s}} = 8.7 \times 10^3\,\text{s} = 2.4\ \text{h}\ .$$

   (d) We take d to be 6500 ly and the speed of light to be 1.00 ly/y. Then,

   $$t = \frac{d}{c} = \frac{6500\,\text{ly}}{1.00\,\text{ly/y}} = 6500\,\text{y}\ .$$

   The explosion took place in the year $1054 - 6500 = -5446$ or 5446 BCE.

2. (a)

   $$f = \frac{c}{\lambda} = \frac{3.0 \times 10^8\,\text{m/s}}{(1.0 \times 10^5)(6.4 \times 10^6\,\text{m})} = 4.7 \times 10^{-3}\ \text{Hz}\ .$$

   (b)

   $$T = \frac{1}{f} = \frac{1}{4.7 \times 10^{-3}\,\text{Hz}} = 212\,\text{s} = 3\,\text{min}\ 32\,\text{s}\ .$$

3. (a) From Fig. 34-2 we find the wavelengths in question to be about 515 nm and 610 nm.

   (b) Again from Fig. 34-2 the wavelength is about 555 nm. Therefore,

   $$f = \frac{c}{\lambda} = \frac{3.00 \times 10^8\,\text{m/s}}{555\,\text{nm}} = 5.41 \times 10^{14}\ \text{Hz}\ ,$$

   and the period is $(5.41 \times 10^{14}\,\text{Hz})^{-1} = 1.85 \times 10^{-15}$ s.

4. Since $\Delta\lambda \ll \lambda$, we find Δf is equal to

$$\left|\Delta\left(\frac{c}{\lambda}\right)\right| \approx \frac{c\Delta\lambda}{\lambda^2} = \frac{(3.0\times 10^8\,\text{m/s})(0.0100\times 10^{-9}\,\text{m})}{(632.8\times 10^{-9}\,\text{m})^2} = 7.49\times 10^9\ \text{Hz}\ .$$

5. (a) Suppose that at time t_1, the moon is starting a revolution (on the verge of going behind Jupiter, say) and that at this instant, the distance between Jupiter and Earth is ℓ_1. The time of the start of the revolution as seen on Earth is $t_1^* = t_1 + \ell_1/c$. Suppose the moon starts the next revolution at time t_2 and at that instant, the Earth-Jupiter distance is ℓ_2. The start of the revolution as seen on Earth is $t_2^* = t_2 + \ell_2/c$. Now, the actual period of the moon is given by $T = t_2 - t_1$ and the period as measured on Earth is

$$T^* = t_2^* - t_1^* = t_2 - t_1 + \frac{\ell_2}{c} - \frac{\ell_1}{c} = T + \frac{\ell_2 - \ell_1}{c}\ .$$

The period as measured on Earth is longer than the actual period. This is due to the fact that Earth moves during a revolution, and light takes a finite time to travel from Jupiter to Earth. For the situation depicted in Fig. 34-38, light emitted at the end of a revolution travels a longer distance to get to Earth than light emitted at the beginning. Suppose the position of Earth is given by the angle θ, measured from x. Let R be the radius of Earth's orbit and d be the distance from the Sun to Jupiter. The law of cosines, applied to the triangle with the Sun, Earth, and Jupiter at the vertices, yields $\ell^2 = d^2 + R^2 - 2dR\cos\theta$. This expression can be used to calculate ℓ_1 and ℓ_2. Since Earth does not move very far during one revolution of the moon, we may approximate $\ell_2 - \ell_1$ by $(d\ell/dt)T$ and T^* by $T + (d\ell/dt)(T/c)$. Now

$$\frac{d\ell}{dt} = \frac{2Rd\sin\theta}{\sqrt{d^2+R^2-2dR\cos\theta}}\,\frac{d\theta}{dt} = \frac{2vd\sin\theta}{\sqrt{d^2+R^2-2dR\cos\theta}}\ ,$$

where $v = R\,(d\theta/dt)$ is the speed of Earth in its orbit. For $\theta = 0$, $(d\ell/dt) = 0$ and $T^* = T$. Since Earth is then moving perpendicularly to the line from the Sun to Jupiter, its distance from the planet does not change appreciably during one revolution of the moon. On the other hand, when $\theta = 90°$, $d\ell/dt = vd/\sqrt{d^2+R^2}$ and

$$T^* = T\left(1 + \frac{vd}{c\sqrt{d^2+R^2}}\right)\ .$$

The Earth is now moving parallel to the line from the Sun to Jupiter, and its distance from the planet changes during a revolution of the moon.

(b) Our notation is as follows: t is the actual time for the moon to make N revolutions, and t^* is the time for N revolutions to be observed on Earth. Then,

$$t^* = t + \frac{\ell_2 - \ell_1}{c}\ ,$$

where ℓ_1 is the Earth-Jupiter distance at the beginning of the interval and ℓ_2 is the Earth-Jupiter distance at the end. Suppose Earth is at position x at the beginning of the interval, and at y at the end. Then, $\ell_1 = d - R$ and $\ell_2 = \sqrt{d^2+R^2}$. Thus,

$$t^* = t + \frac{\sqrt{d^2+R^2} - (d-R)}{c}\ .$$

A value can be found for t by measuring the observed period of revolution when Earth is at x and multiplying by N. We note that the observed period is the true period when Earth is at x. The time interval as Earth moves from x to y is t^*. The difference is

$$t^* - t = \frac{\sqrt{d^2+R^2} - (d-R)}{c}\ .$$

If the radii of the orbits of Jupiter and Earth are known, the value for $t^* - t$ can be used to compute c. Since Jupiter is much further from the Sun than Earth, $\sqrt{d^2 + R^2}$ may be approximated by d and $t^* - t$ may be approximated by R/c. In this approximation, only the radius of Earth's orbit need be known.

6. The emitted wavelength is

$$\begin{aligned}\lambda &= \frac{c}{f} = 2\pi c\sqrt{LC} \\ &= 2\pi(2.998 \times 10^8\,\text{m/s})\sqrt{(0.253 \times 10^{-6}\,\text{H})(25.0 \times 10^{-12}\,\text{F})} = 4.74\ \text{m}\ .\end{aligned}$$

7. If f is the frequency and λ is the wavelength of an electromagnetic wave, then $f\lambda = c$. The frequency is the same as the frequency of oscillation of the current in the LC circuit of the generator. That is, $f = 1/2\pi\sqrt{LC}$, where C is the capacitance and L is the inductance. Thus

$$\frac{\lambda}{2\pi\sqrt{LC}} = c\ .$$

The solution for L is

$$L = \frac{\lambda^2}{4\pi^2 Cc^2} = \frac{(550 \times 10^{-9}\,\text{m})^2}{4\pi^2(17 \times 10^{-12}\,\text{F})(2.998 \times 10^8\,\text{m/s})^2} = 5.00 \times 10^{-21}\ \text{H}\ .$$

This is exceedingly small.

8. The amplitude of the magnetic field in the wave is

$$B_m = \frac{E_m}{c} = \frac{3.20 \times 10^{-4}\,\text{V/m}}{2.998 \times 10^8\,\text{m/s}} = 1.07 \times 10^{-12}\ \text{T}\ .$$

9. Since the $\vec{E}$-wave oscillates in the z direction and travels in the x direction, we have $B_x = B_z = 0$. With SI units understood, we find

$$\begin{aligned}B_y &= B_{\text{m}} \cos\left[\pi \times 10^{15}\left(t - \frac{x}{c}\right)\right] = \frac{2.0\cos[10^{15}\pi(t - x/c)]}{3.0 \times 10^8} \\ &= (6.7 \times 10^{-9})\cos\left[10^{15}\pi\left(t - \frac{x}{c}\right)\right]\end{aligned}$$

10. Using $\vec{S} = (1/\mu_0)\vec{E} \times \vec{B}$, we see that (on the right hand) letting the thumb be in the $\vec{E}$ direction and the index finger be in the $\vec{B}$ direction means that the middle finger (held perpendicular to the other two, making a "triad" of the thumb and two fingers) points in the direction of wave propagation (the direction of $\vec{S}$). Holding the right hand in this manner can facilitate checking the directions in the Figures. A more algebraic approach is to note that $\hat{\text{j}} \times \hat{\text{k}} = \hat{\text{i}}$. This is especially useful for checking Figures 34-6 and 34-7.

11. If P is the power and Δt is the time interval of one pulse, then the energy in a pulse is

$$E = P\,\Delta t = (100 \times 10^{12}\,\text{W})(1.0 \times 10^{-9}\,\text{s}) = 1.0 \times 10^5\ \text{J}\ .$$

12. The intensity of the signal at Proxima Centauri is

$$I = \frac{P}{4\pi r^2} = \frac{1.0 \times 10^6\,\text{W}}{4\pi[(4.3\,\text{ly})(9.46 \times 10^{15}\,\text{m/ly})]^2} = 4.8 \times 10^{-29}\ \text{W/m}^2\ .$$

13. The region illuminated on the Moon is a circle with radius $R = r\theta/2$, where r is the Earth-Moon distance (3.82×10^8 m) and θ is the full-angle beam divergence in radians. The area A illuminated is

$$A = \pi R^2 = \frac{\pi r^2 \theta^2}{4} = \frac{\pi (3.82 \times 10^8\,\text{m})^2 (0.880 \times 10^{-6}\,\text{rad})^2}{4} = 8.88 \times 10^4\ \text{m}^2 .$$

14. The intensity is the average of the Poynting vector:

$$I = S_{\text{avg}} = \frac{cB_m^2}{2\mu_0} = \frac{(3.0 \times 10^8\,\text{m/s})(1.0 \times 10^{-4}\,\text{T})^2}{2(1.26 \times 10^{-6}\,\text{H/m})^2} = 1.2 \times 10^6\ \text{W/m}^2 .$$

15. (a) The amplitude of the magnetic field in the wave is

$$B_m = \frac{E_m}{c} = \frac{5.00\,\text{V/m}}{2.998 \times 10^8\,\text{m/s}} = 1.67 \times 10^{-8}\ \text{T} .$$

(b) The intensity is the average of the Poynting vector:

$$I = S_{\text{avg}} = \frac{E_m^2}{2\mu_0 c} = \frac{(5.00\,\text{V/m})^2}{2(4\pi \times 10^{-7}\,\text{T}\cdot\text{m/A})(2.998 \times 10^8\,\text{m/s})} = 3.31 \times 10^{-2}\ \text{W/m}^2 .$$

16. We use $I = E_m^2/2\mu_0 c$ to calculate E_m:

$$\begin{aligned} E_m &= \sqrt{2\mu_0 I c} = \sqrt{2(4\pi \times 10^{-7}\,\text{T}\cdot\text{m/A})(1.40 \times 10^3\,\text{W/m}^2)(2.998 \times 10^8\,\text{m/s})} \\ &= 1.03 \times 10^3\ \text{V/m} . \end{aligned}$$

The magnetic field amplitude is therefore

$$B_m = \frac{E_m}{c} = \frac{1.03 \times 10^4\,\text{V/m}}{2.998 \times 10^8\,\text{m/s}} = 3.43 \times 10^{-6}\ \text{T} .$$

17. (a) The magnetic field amplitude of the wave is

$$B_m = \frac{E_m}{c} = \frac{2.0\,\text{V/m}}{2.998 \times 10^8\,\text{m/s}} = 6.7 \times 10^{-9}\ \text{T} .$$

(b) The intensity is

$$I = \frac{E_m^2}{2\mu_0 c} = \frac{(2.0\,\text{V/m})^2}{2(4\pi \times 10^{-7}\,\text{T}\cdot\text{m/A})(2.998 \times 10^8\,\text{m/s})} = 5.3 \times 10^{-3}\ \text{W/m}^2 .$$

(c) The power of the source is

$$P = 4\pi r^2 I_{\text{avg}} = 4\pi (10\,\text{m})^2 (5.3 \times 10^{-3}\,\text{W/m}^2) = 6.7\ \text{W} .$$

18. (a) The power received is

$$P_r = (1.0 \times 10^{-12}\,\text{W}) \frac{\pi[(1000\,\text{ft})(0.3048\,\text{m/ft})]^2/4}{4\pi (6.37 \times 10^6\,\text{m})^2} = 1.4 \times 10^{-22}\ \text{W} .$$

(b) The power of the source would be

$$P = 4\pi r^2 I = 4\pi [(2.2 \times 10^4\,\text{ly})(9.46 \times 10^{15}\,\text{m/ly})]^2 \left[\frac{1.0 \times 10^{-12}\,\text{W}}{4\pi (6.37 \times 10^6\,\text{m})^2}\right] = 1.1 \times 10^{15}\ \text{W} .$$

19. (a) The average rate of energy flow per unit area, or intensity, is related to the electric field amplitude E_m by $I = E_m^2/2\mu_0 c$, so

$$\begin{aligned} E_m &= \sqrt{2\mu_0 c I} = \sqrt{2(4\pi \times 10^{-7}\,\text{H/m})(2.998 \times 10^8\,\text{m/s})(10 \times 10^{-6}\,\text{W/m}^2)} \\ &= 8.7 \times 10^{-2}\ \text{V/m}\ . \end{aligned}$$

(b) The amplitude of the magnetic field is given by

$$B_m = \frac{E_m}{c} = \frac{8.7 \times 10^{-2}\,\text{V/m}}{2.998 \times 10^8\,\text{m/s}} = 2.9 \times 10^{-10}\ \text{T}\ .$$

(c) At a distance r from the transmitter, the intensity is $I = P/4\pi r^2$, where P is the power of the transmitter. Thus

$$P = 4\pi r^2 I = 4\pi(10 \times 10^3\,\text{m})^2(10 \times 10^{-6}\,\text{W/m}^2) = 1.3 \times 10^4\ \text{W}\ .$$

20. The radiation pressure is

$$p_r = \frac{I}{c} = \frac{10\,\text{W/m}^2}{2.998 \times 10^8\,\text{m/s}} = 3.3 \times 10^{-8}\ \text{Pa}\ .$$

21. The plasma completely reflects all the energy incident on it, so the radiation pressure is given by $p_r = 2I/c$, where I is the intensity. The intensity is $I = P/A$, where P is the power and A is the area intercepted by the radiation. Thus

$$p_r = \frac{2P}{Ac} = \frac{2(1.5 \times 10^9\,\text{W})}{(1.00 \times 10^{-6}\,\text{m}^2)(2.998 \times 10^8\,\text{m/s})} = 1.0 \times 10^7\,\text{Pa} = 10\ \text{MPa}\ .$$

22. (a) The radiation pressure produces a force equal to

$$\begin{aligned} F_r &= p_r(\pi R_e^2) = \left(\frac{I}{c}\right)(\pi R_e^2) \\ &= \frac{\pi(1.4 \times 10^3\,\text{W/m}^2)(6.37 \times 10^6\,\text{m})^2}{2.998 \times 10^8\,\text{m/s}} = 6.0 \times 10^8\ \text{N}\ . \end{aligned}$$

(b) The gravitational pull of the Sun on Earth is

$$\begin{aligned} F_{\text{grav}} &= \frac{GM_sM_e}{d_{es}^2} \\ &= \frac{(6.67 \times 10^{-11}\,\text{N}\cdot\text{m}^2/\text{kg}^2)(2.0 \times 10^{30}\,\text{kg})(5.98 \times 10^{24}\,\text{kg})}{(1.5 \times 10^{11}\,\text{m})^2} \\ &= 3.6 \times 10^{22}\ \text{N}\ , \end{aligned}$$

which is much greater than F_r.

23. Since the surface is perfectly absorbing, the radiation pressure is given by $p_r = I/c$, where I is the intensity. Since the bulb radiates uniformly in all directions, the intensity a distance r from it is given by $I = P/4\pi r^2$, where P is the power of the bulb. Thus

$$p_r = \frac{P}{4\pi r^2 c} = \frac{500\,\text{W}}{4\pi(1.5\,\text{m})^2(2.998 \times 10^8\,\text{m/s})} = 5.9 \times 10^{-8}\ \text{Pa}\ .$$

24. (a) We note that the cross section area of the beam is $\pi d^2/4$, where d is the diameter of the spot ($d = 2.00\lambda$). The beam intensity is

$$I = \frac{P}{\pi d^2/4} = \frac{5.00 \times 10^{-3}\,\text{W}}{\pi[(2.00)(633 \times 10^{-9}\,\text{m})]^2/4} = 3.97 \times 10^9\ \text{W/m}^2 \ .$$

(b) The radiation pressure is

$$p_r = \frac{I}{c} = \frac{3.97 \times 10^9\,\text{W/m}^2}{2.998 \times 10^8\,\text{m/s}} = 13.2\ \text{Pa} \ .$$

(c) In computing the corresponding force, we can use the power and intensity to eliminate the area (mentioned in part (a)). We obtain

$$F_r = \left(\frac{\pi d^2}{4}\right) p_r = \left(\frac{P}{I}\right) p_r = \frac{(5.00 \times 10^{-3}\,\text{W})(13.2\,\text{Pa})}{3.97 \times 10^9\,\text{W/m}^2} = 1.67 \times 10^{-11}\ \text{N} \ .$$

(d) The acceleration of the sphere is

$$\begin{aligned} a &= \frac{F_r}{m} = \frac{F_r}{\rho(\pi d^3/6)} = \frac{6(1.67 \times 10^{-11}\,\text{N})}{\pi(5.00 \times 10^3\,\text{kg/m}^3)[(2.00)(633 \times 10^{-9}\,\text{m})]^3} \\ &= 3.14 \times 10^3\ \text{m/s}^2 \ . \end{aligned}$$

25. (a) Since $c = \lambda f$, where λ is the wavelength and f is the frequency of the wave,

$$f = \frac{c}{\lambda} = \frac{2.998 \times 10^8\,\text{m/s}}{3.0\,\text{m}} = 1.0 \times 10^8\ \text{Hz} \ .$$

(b) The magnetic field amplitude is

$$B_m = \frac{E_m}{c} = \frac{300\,\text{V/m}}{2.998 \times 10^8\,\text{m/s}} = 1.00 \times 10^{-6}\ \text{T} \ .$$

$\vec{B}$ must be in the positive z direction when $\vec{E}$ is in the positive y direction in order for $\vec{E} \times \vec{B}$ to be in the positive x direction (the direction of propagation).

(c) The angular wave number is

$$k = \frac{2\pi}{\lambda} = \frac{2\pi}{3.0\,\text{m}} = 2.1\,\text{rad/m} \ .$$

The angular frequency is

$$\omega = 2\pi f = 2\pi(1.0 \times 10^8\,\text{Hz}) = 6.3 \times 10^8\ \text{rad/s} \ .$$

(d) The intensity of the wave is

$$I = \frac{E_m^2}{2\mu_0 c} = \frac{(300\,\text{V/m})^2}{2(4\pi \times 10^{-7}\,\text{H/m})(2.998 \times 10^8\,\text{m/s})} = 119\ \text{W/m}^2 \ .$$

(e) Since the sheet is perfectly absorbing, the rate per unit area with which momentum is delivered to it is I/c, so

$$\frac{dp}{dt} = \frac{IA}{c} = \frac{(119\,\text{W/m}^2)(2.0\,\text{m}^2)}{2.998 \times 10^8\,\text{m/s}} = 8.0 \times 10^{-7}\,\text{N} \ .$$

The radiation pressure is

$$p_r = \frac{dp/dt}{A} = \frac{8.0 \times 10^{-7}\,\text{N}}{2.0\,\text{m}^2} = 4.0 \times 10^{-7}\ \text{Pa} \ .$$

26. The mass of the cylinder is $m = \rho(\pi d_1^2/4)H$, where d_1 is the diameter of the cylinder. Since it is in equilibrium

$$F_{\text{net}} = mg - F_r = \frac{\pi H d_1^2 g\rho}{4} - \left(\frac{\pi d_1^2}{4}\right)\left(\frac{2I}{c}\right) = 0 \, .$$

We solve for H:

$$\begin{aligned} H &= \frac{2I}{gc\rho} = \left(\frac{2P}{\pi d^2/4}\right)\frac{1}{gc\rho} \\ &= \frac{8(4.60\,\text{W})}{\pi(2.60\times10^{-3}\,\text{m})^2(9.8\,\text{m/s}^2)(3.0\times10^8\,\text{m/s})(1.20\times10^3\,\text{kg/m}^3)} \\ &= 4.91\times10^{-7}\,\text{m} \, . \end{aligned}$$

27. Let f be the fraction of the incident beam intensity that is reflected. The fraction absorbed is $1-f$. The reflected portion exerts a radiation pressure of

$$p_r = \frac{2fI_0}{c}$$

and the absorbed portion exerts a radiation pressure of

$$p_a = \frac{(1-f)I_0}{c} \, ,$$

where I_0 is the incident intensity. The factor 2 enters the first expression because the momentum of the reflected portion is reversed. The total radiation pressure is the sum of the two contributions:

$$p_{\text{total}} = p_r + p_a = \frac{2fI_0 + (1-f)I_0}{c} = \frac{(1+f)I_0}{c} \, .$$

To relate the intensity and energy density, we consider a tube with length ℓ and cross-sectional area A, lying with its axis along the propagation direction of an electromagnetic wave. The electromagnetic energy inside is $U = uA\ell$, where u is the energy density. All this energy passes through the end in time $t = \ell/c$, so the intensity is

$$I = \frac{U}{At} = \frac{uA\ell c}{A\ell} = uc \, .$$

Thus $u = I/c$. The intensity and energy density are positive, regardless of the propagation direction. For the partially reflected and partially absorbed wave, the intensity just outside the surface is $I = I_0 + fI_0 = (1+f)I_0$, where the first term is associated with the incident beam and the second is associated with the reflected beam. Consequently, the energy density is

$$u = \frac{I}{c} = \frac{(1+f)I_0}{c} \, ,$$

the same as radiation pressure.

28. We imagine the bullets (of mass m and speed v each) which will strike a surface of area A of the plane within time t to $t+\Delta t$ to be contained in a cylindrical volume at time t. Since the number of bullets contained in the cylinder is $N = n(Av\Delta t)$ and each bullet changes its momentum by $\Delta p_b = mv$, the rate of change of the total momentum for the bullets that strike the area is

$$F = \frac{\Delta P_{\text{total}}}{\Delta t} = N\frac{p_b}{\Delta t} = \frac{(Av\Delta t)nmv}{\Delta t} = Anmv^2$$

where n is the number density of the bullets (bullets per unit volume). The pressure is then

$$p_r = \frac{F}{A} = nmv^2 = 2nK \, ,$$

where $K = \frac{1}{2}mv^2$. Note that nK is the kinetic energy density. Also note that the relation between energy and momentum for a bullet is quite different from the relation between those quantities for an electromagnetic wave.

29. If the beam carries energy U away from the spaceship, then it also carries momentum $p = U/c$ away. Since the total momentum of the spaceship and light is conserved, this is the magnitude of the momentum acquired by the spaceship. If P is the power of the laser, then the energy carried away in time t is $U = Pt$. We note that there are 86400 seconds in a day. Thus, $p = Pt/c$ and, if m is mass of the spaceship, its speed is

$$v = \frac{p}{m} = \frac{Pt}{mc} = \frac{(10 \times 10^3\,\text{W})(86400\,\text{s})}{(1.5 \times 10^3\,\text{kg})(2.998 \times 10^8\,\text{m/s})} = 1.9 \times 10^{-3}\ \text{m/s}\ .$$

30. We require $F_{\text{grav}} = F_r$ or

$$G\frac{mM_s}{d_{es}^2} = \frac{2IA}{c}\ ,$$

and solve for the area A:

$$\begin{aligned} A &= \frac{cGmM_s}{2Id_{es}^2} = \frac{(6.67 \times 10^{-11}\,\text{N}\cdot\text{m}^2/\text{kg}^2)(1500\,\text{kg})(1.99 \times 10^{30}\,\text{kg})(2.998 \times 10^8\,\text{m/s})}{2(1.40 \times 10^3\,\text{W/m}^2)(1.50 \times 10^{11}\,\text{m})^2} \\ &= 9.5 \times 10^5\,\text{m}^2 = 0.95\ \text{km}^2\ . \end{aligned}$$

31. (a) Let r be the radius and ρ be the density of the particle. Since its volume is $(4\pi/3)r^3$, its mass is $m = (4\pi/3)\rho r^3$. Let R be the distance from the Sun to the particle and let M be the mass of the Sun. Then, the gravitational force of attraction of the Sun on the particle has magnitude

$$F_g = \frac{GMm}{R^2} = \frac{4\pi GM\rho r^3}{3R^2}\ .$$

If P is the power output of the Sun, then at the position of the particle, the radiation intensity is $I = P/4\pi R^2$, and since the particle is perfectly absorbing, the radiation pressure on it is

$$p_r = \frac{I}{c} = \frac{P}{4\pi R^2 c}\ .$$

All of the radiation that passes through a circle of radius r and area $A = \pi r^2$, perpendicular to the direction of propagation, is absorbed by the particle, so the force of the radiation on the particle has magnitude

$$F_r = p_r A = \frac{\pi P r^2}{4\pi R^2 c} = \frac{Pr^2}{4R^2 c}\ .$$

The force is radially outward from the Sun. Notice that both the force of gravity and the force of the radiation are inversely proportional to R^2. If one of these forces is larger than the other at some distance from the Sun, then that force is larger at all distances. The two forces depend on the particle radius r differently: F_g is proportional to r^3 and F_r is proportional to r^2. We expect a small radius particle to be blown away by the radiation pressure and a large radius particle with the same density to be pulled inward toward the Sun. The critical value for the radius is the value for which the two forces are equal. Equating the expressions for F_g and F_r, we solve for r:

$$r = \frac{3P}{16\pi GM\rho c}\ .$$

(b) According to Appendix C, $M = 1.99 \times 10^{30}\,\text{kg}$ and $P = 3.90 \times 10^{26}\,\text{W}$. Thus,

$$\begin{aligned} r &= \frac{3(3.90 \times 10^{26}\,\text{W})}{16\pi(6.67 \times 10^{-11}\,\text{N}\cdot\text{m}^2/\text{kg}^2)(1.99 \times 10^{30}\,\text{kg})(1.0 \times 10^3\,\text{kg/m}^3)(3.00 \times 10^8\,\text{m/s})} \\ &= 5.8 \times 10^{-7}\ \text{m}\ . \end{aligned}$$

32. (a) The discussion in §17-5 regarding the argument of the sine function $(kx + \omega t)$ makes it clear that the wave is traveling in the negative y direction. Thus, $\vec{S}$ points in the $-\hat{\jmath}$ direction.

(b) Since $\vec{E} \times \vec{B} \propto \vec{S}$ and $\vec{B}$ points in the $\hat{\text{i}}$ direction, then we may conclude that $\vec{E}$ points in the $-\hat{\text{k}}$ direction (recall that $\hat{\text{k}} \times \hat{\text{i}} = \hat{\text{j}}$). Therefore, $E_x = E_y = 0$ and $E_z = -cB\sin(kx + \omega t)$.

(c) Since $E_x = E_y = 0$, the wave is polarized along the z axis.

33. (a) Since the incident light is unpolarized, half the intensity is transmitted and half is absorbed. Thus the transmitted intensity is $I = 5.0\,\text{mW/m}^2$. The intensity and the electric field amplitude are related by $I = E_m^2/2\mu_0 c$, so

$$\begin{aligned} E_m &= \sqrt{2\mu_0 cI} = \sqrt{2(4\pi \times 10^{-7}\,\text{H/m})(3.00 \times 10^8\,\text{m/s})(5.0 \times 10^{-3}\,\text{W/m}^2)} \\ &= 1.9\ \text{V/m}\ . \end{aligned}$$

(b) The radiation pressure is $p_r = I_a/c$, where I_a is the absorbed intensity. Thus

$$p_r = \frac{5.0 \times 10^{-3}\,\text{W/m}^2}{3.00 \times 10^8\,\text{m/s}} = 1.7 \times 10^{-11}\ \text{Pa}\ .$$

34. After passing through the first polarizer the initial intensity I_0 reduces by a factor of 1/2. After passing through the second one it is further reduced by a factor of $\cos^2(\pi - \theta_1 - \theta_2) = \cos^2(\theta_1 + \theta_2)$. Finally, after passing through the third one it is again reduced by a factor of $\cos^2(\pi - \theta_2 - \theta_3) = \cos^2(\theta_2 + \theta_3)$. Therefore,

$$\begin{aligned} \frac{I_f}{I_0} &= \frac{1}{2}\cos^2(\theta_1 + \theta_2)\cos^2(\theta_2 + \theta_3) \\ &= \frac{1}{2}\cos^2(50° + 50°)\cos^2(50° + 50°) = 4.5 \times 10^{-4}\ . \end{aligned}$$

35. Let I_0 be the intensity of the unpolarized light that is incident on the first polarizing sheet. The transmitted intensity is $I_1 = \frac{1}{2}I_0$, and the direction of polarization of the transmitted light is $\theta_1 = 40°$ counterclockwise from the y axis in the diagram. The polarizing direction of the second sheet is $\theta_2 = 20°$ clockwise from the y axis, so the angle between the direction of polarization that is incident on that sheet and the polarizing direction of the sheet is $40° + 20° = 60°$. The transmitted intensity is

$$I_2 = I_1\cos^2 60° = \frac{1}{2}I_0\cos^2 60°\ ,$$

and the direction of polarization of the transmitted light is 20° clockwise from the y axis. The polarizing direction of the third sheet is $\theta_3 = 40°$ counterclockwise from the y axis. Consequently, the angle between the direction of polarization of the light incident on that sheet and the polarizing direction of the sheet is $20° + 40° = 60°$. The transmitted intensity is

$$I_3 = I_2\cos^2 60° = \frac{1}{2}I_0\cos^4 60° = 3.1 \times 10^{-2}\ .$$

Thus, 3.1% of the light's initial intensity is transmitted.

36. As the unpolarized beam of intensity I_0 passes the first polarizer, its intensity is reduced to $\frac{1}{2}I_0$. After passing through the second polarizer, for which the direction of polarization is at an angle θ from that of the first one, the intensity is $I = \frac{1}{2}I_0\cos^2\theta = \frac{1}{3}I_0$. Thus, $\cos^2\theta = 2/3$, which leads to $\theta = 35°$.

37. The angle between the direction of polarization of the light incident on the first polarizing sheet and the polarizing direction of that sheet is $\theta_1 = 70°$. If I_0 is the intensity of the incident light, then the intensity of the light transmitted through the first sheet is

$$I_1 = I_0\cos^2\theta_1 = (43\,\text{W/m}^2)\cos^2 70° = 5.03\ \text{W/m}^2\ .$$

The direction of polarization of the transmitted light makes an angle of 70° with the vertical and an angle of $\theta_2 = 20°$ with the horizontal. θ_2 is the angle it makes with the polarizing direction of the second polarizing sheet. Consequently, the transmitted intensity is

$$I_2 = I_1\cos^2\theta_2 = (5.03\,\text{W/m}^2)\cos^2 20° = 4.4\ \text{W/m}^2\ .$$

38. In this case, we replace $I_0 \cos^2 70°$ by $\frac{1}{2}I_0$ as the intensity of the light after passing through the first polarizer. Therefore,

$$I_f = \frac{1}{2} I_0 \cos^2(90° - 70°) = \frac{1}{2}(43\,\mathrm{W/m^2})(\cos^2 20°) = 19\ \mathrm{W/m^2}\ .$$

39. Let I_0 be the intensity of the incident beam and f be the fraction that is polarized. Thus, the intensity of the polarized portion is fI_0. After transmission, this portion contributes $fI_0 \cos^2\theta$ to the intensity of the transmitted beam. Here θ is the angle between the direction of polarization of the radiation and the polarizing direction of the filter. The intensity of the unpolarized portion of the incident beam is $(1-f)I_0$ and after transmission, this portion contributes $(1-f)I_0/2$ to the transmitted intensity. Consequently, the transmitted intensity is

$$I = fI_0 \cos^2\theta + \frac{1}{2}(1-f)I_0\ .$$

As the filter is rotated, $\cos^2\theta$ varies from a minimum of 0 to a maximum of 1, so the transmitted intensity varies from a minimum of

$$I_{\min} = \frac{1}{2}(1-f)I_0$$

to a maximum of

$$I_{\max} = fI_0 + \frac{1}{2}(1-f)I_0 = \frac{1}{2}(1+f)I_0\ .$$

The ratio of $I_{\max}$ to $I_{\min}$ is

$$\frac{I_{\max}}{I_{\min}} = \frac{1+f}{1-f}\ .$$

Setting the ratio equal to 5.0 and solving for f, we get $f = 0.67$.

40. (a) The fraction of light which is transmitted by the glasses is

$$\frac{I_f}{I_0} = \frac{E_f^2}{E_0^2} = \frac{E_v^2}{E_v^2 + E_h^2} = \frac{E_v^2}{E_v^2 + (2.3E_v)^2} = 0.16\ .$$

(b) Since now the horizontal component of $\vec{E}$ will pass through the glasses,

$$\frac{I_f}{I_0} = \frac{E_h^2}{E_v^2 + E_h^2} = \frac{(2.3E_v)^2}{E_v^2 + (2.3E_v)^2} = 0.84\ .$$

41. (a) The rotation cannot be done with a single sheet. If a sheet is placed with its polarizing direction at an angle of 90° to the direction of polarization of the incident radiation, no radiation is transmitted. It can be done with two sheets. We place the first sheet with its polarizing direction at some angle θ, between 0 and 90°, to the direction of polarization of the incident radiation. Place the second sheet with its polarizing direction at 90° to the polarization direction of the incident radiation. The transmitted radiation is then polarized at 90° to the incident polarization direction. The intensity is $I_0 \cos^2\theta \cos^2(90° - \theta) = I_0 \cos^2\theta \sin^2\theta$, where I_0 is the incident radiation. If θ is not 0 or 90°, the transmitted intensity is not zero.

(b) Consider n sheets, with the polarizing direction of the first sheet making an angle of $\theta = 90°/n$ relative to the direction of polarization of the incident radiation. The polarizing direction of each successive sheet is rotated $90°/n$ in the same sense from the polarizing direction of the previous sheet. The transmitted radiation is polarized, with its direction of polarization making an angle of 90° with the direction of polarization of the incident radiation. The intensity is $I = I_0 \cos^{2n}(90°/n)$. We want the smallest integer value of n for which this is greater than $0.60I_0$. We start with $n = 2$ and calculate $\cos^{2n}(90°/n)$. If the result is greater than 0.60, we have obtained the solution. If it is less, increase n by 1 and try again. We repeat this process, increasing n by 1 each time, until we have a value for which $\cos^{2n}(90°/n)$ is greater than 0.60. The first one will be $n = 5$.

42. The angle of incidence for the light ray on mirror B is $90° - \theta$. So the outgoing ray r' makes an angle $90° - (90° - \theta) = \theta$ with the vertical direction, and is antiparallel to the incoming one. The angle between i and r' is therefore $180°$.

43. The law of refraction states

$$n_1 \sin\theta_1 = n_2 \sin\theta_2 \ .$$

We take medium 1 to be the vacuum, with $n_1 = 1$ and $\theta_1 = 32.0°$. Medium 2 is the glass, with $\theta_2 = 21.0°$. We solve for n_2:

$$n_2 = n_1 \frac{\sin\theta_1}{\sin\theta_2} = (1.00)\left(\frac{\sin 32.0°}{\sin 21.0°}\right) = 1.48 \ .$$

44. (a) The law of refraction requires that $\sin\theta_1/\sin\theta_2 = n_{\text{water}} = \text{const}$. We can check that this is indeed valid for any given pair of θ_1 and θ_2. For example $\sin 10°/\sin 8° = 1.3$, and $\sin 20°/\sin 15°30' = 1.3$, etc.

(b) $n_{\text{water}} = 1.3$, as shown in part (a).

45. Note that the normal to the refracting surface is vertical in the diagram. The angle of refraction is $\theta_2 = 90°$ and the angle of incidence is given by $\tan\theta_1 = w/h$, where h is the height of the tank and w is its width. Thus

$$\theta_1 = \tan^{-1}\left(\frac{w}{h}\right) = \tan^{-1}\left(\frac{1.10\,\text{m}}{0.850\,\text{m}}\right) = 52.31° \ .$$

The law of refraction yields

$$n_1 = n_2 \frac{\sin\theta_2}{\sin\theta_1} = (1.00)\left(\frac{\sin 90°}{\sin 52.31°}\right) = 1.26 \ ,$$

where the index of refraction of air was taken to be unity.

46. (a) Approximating $n = 1$ for air, we have

$$n_1 \sin\theta_1 = (1)\sin\theta_5 \implies 56.9° = \theta_5$$

and with the more accurate value for n_{air} in Table 34-1, we obtain 56.8°.

(b) Eq. 34-44 leads to

$$n_1 \sin\theta_1 = n_2 \sin\theta_2 = n_3 \sin\theta_3 = n_4 \sin\theta_4$$

so that

$$\theta_4 = sin^{-1}\left(\frac{n_1}{n_4}\sin\theta_1\right) = 35.3° \ .$$

47. Consider a ray that grazes the top of the pole, as shown in the diagram below. Here $\theta_1 = 35°$, $\ell_1 = 0.50\,\text{m}$, and $\ell_2 = 1.50\,\text{m}$. The length of the shadow is $x + L$. x is given by $x = \ell_1 \tan\theta_1 = (0.50\,\text{m})\tan 35° = 0.35\,\text{m}$. According to the law of refraction, $n_2 \sin\theta_2 = n_1 \sin\theta_1$. We take $n_1 = 1$ and $n_2 = 1.33$ (from Table 34–1). Then,

$$\theta_2 = \sin^{-1}\left(\frac{\sin\theta_1}{n_2}\right) = \sin^{-1}\left(\frac{\sin 35.0°}{1.33}\right) = 25.55° \ .$$

L is given by

$$L = \ell_2 \tan\theta_2 = (1.50\,\text{m})\tan 25.55° = 0.72\,\text{m} \ .$$

The length of the shadow is $0.35\,\mathrm{m} + 0.72\,\mathrm{m} = 1.07\,\mathrm{m}$.

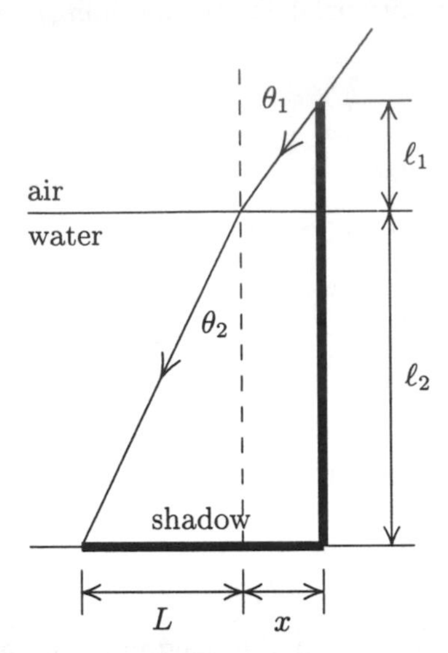

48. We use the law of refraction (assuming $n_{\mathrm{air}} = 1$) and the law of sines to determine the paths of various light rays. The index of refraction for fused quartz can be found in Fig. 34-19. We estimate $n_{\mathrm{blue}} = 1.463$, $n_{\mathrm{y\,g}} = 1.459$, and $n_{\mathrm{red}} = 1.456$. The light rays as they leave the prism (from the right side of the prism shown below) are very close together; on the scale we used below, the individual rays are difficult to resolve. Measured from the surface of the prism (at the face from which they emerge from the prism) their angles are $\theta_{\mathrm{blue}} = 28.51°, \theta_{\mathrm{y\,g}} = 28.95°$, and $\theta_{\mathrm{red}} = 29.29°$. The angle between the incident rays (on the left side of the picture) and the dashed line (the axis normal to the left face of the prism) is 35°.

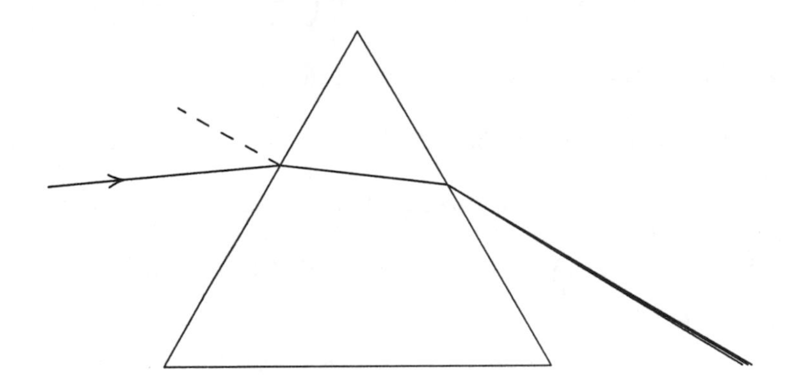

49. Let θ be the angle of incidence and θ_2 be the angle of refraction at the left face of the plate. Let n be the index of refraction of the glass. Then, the law of refraction yields $\sin\theta = n\sin\theta_2$. The angle of incidence at the right face is also θ_2. If θ_3 is the angle of emergence there, then $n\sin\theta_2 = \sin\theta_3$. Thus $\sin\theta_3 = \sin\theta$ and $\theta_3 = \theta$. The emerging ray is parallel to the incident ray. We wish to derive an expression for x in terms of θ. If D is the length of the ray in the glass, then $D\cos\theta_2 = t$ and $D = t/\cos\theta_2$. The angle α in the diagram equals $\theta - \theta_2$ and $x = D\sin\alpha = D\sin(\theta - \theta_2)$. Thus

$$x = \frac{t\sin(\theta - \theta_2)}{\cos\theta_2} \,.$$

If all the angles θ, θ_2, θ_3, and $\theta - \theta_2$ are small and measured in radians, then $\sin\theta \approx \theta$, $\sin\theta_2 \approx \theta_2$, $\sin(\theta - \theta_2) \approx \theta - \theta_2$, and $\cos\theta_2 \approx 1$. Thus $x \approx t(\theta - \theta_2)$. The law of refraction applied to the point of incidence at the left face of the plate is now $\theta \approx n\theta_2$, so $\theta_2 \approx \theta/n$ and

$$x \approx t\left(\theta - \frac{\theta}{n}\right) = \frac{(n-1)t\theta}{n} .$$

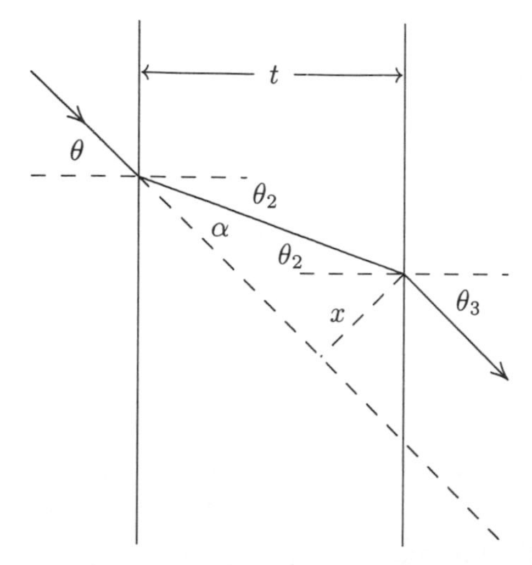

50. (a) An incident ray which is normal to the water surface is not refracted, so the angle at which it strikes the first mirror is $\theta_1 = 45°$. According to the law of reflection, the angle of reflection is also 45°. This means the ray is horizontal as it leaves the first mirror, and the angle of incidence at the second mirror is $\theta_2 = 45°$. Since the angle of reflection at the second mirror is also 45° the ray leaves that mirror normal again to the water surface. There is no refraction at the water surface, and the emerging ray is parallel to the incident ray.

    (b) We imagine that the incident ray makes an angle θ_1 with the normal to the water surface. The angle of refraction θ_2 is found from $\sin\theta_1 = n\sin\theta_2$, where n is the index of refraction of the water. The normal to the water surface and the normal to the first mirror make an angle of 45°. If the normal to the water surface is continued downward until it meets the normal to the first mirror, the triangle formed has an interior angle of $180° - 45° = 135°$ at the vertex formed by the normal. Since the interior angles of a triangle must sum to 180°, the angle of incidence at the first mirror satisfies $\theta_3 + \theta_2 + 135° = 180°$, so $\theta_3 = 45° - \theta_2$. Using the law of reflection, the angle of reflection at the first mirror is also $45° - \theta_2$. We note that the triangle formed by the ray and the normals to the two mirrors is a right triangle. Consequently, $\theta_3 + \theta_4 + 90° = 180°$ and $\theta_4 = 90° - \theta_3 = 90° - 45° + \theta_2 = 45° + \theta_2$. The angle of reflection at the second mirror is also $45° + \theta_2$. Now, we continue the normal to the water surface downward from the exit point of the ray to the second mirror. It makes an angle of 45° with the mirror. Consider the triangle formed by the second mirror, the ray, and the normal to the water surface. The angle at the intersection of the normal and the mirror is $180° - 45° = 135°$. The angle at the intersection of the ray and the mirror is $90° - \theta_4 = 90° - (45° + \theta_2) = 45° - \theta_2$. The angle at the intersection of the ray and the water surface is θ_5. These three angles must sum to 180°, so $135° + 45° - \theta_2 + \theta_5 = 180°$. This means $\theta_5 = \theta_2$. Finally, we use the law of refraction to find θ_6:

$$\sin\theta_6 = n\sin\theta_5 \implies \sin\theta_6 = n\sin\theta_2 ,$$

    since $\theta_5 = \theta_2$. Finally, since $\sin\theta_1 = n\sin\theta_2$, we conclude that $\sin\theta_6 = \sin\theta_1$ and $\theta_6 = \theta_1$. The exiting ray is parallel to the incident ray.

51. We label the light ray's point of entry A, the vertex of the prism B, and the light ray's exit point C. Also, the point in Fig. 34-49 where ψ is defined (at the point of intersection of the extrapolations of the incident and emergent rays) is denoted D. The angle indicated by ADC is the supplement of ψ, so we

denote it $\psi_s = 180° - \psi$. The angle of refraction in the glass is $\theta_2 = \frac{1}{n}\sin\theta$. The angles between the interior ray and the nearby surfaces is the complement of θ_2, so we denote it $\theta_{2c} = 90° - \theta_2$. Now, the angles in the triangle ABC must add to 180°:

$$180° = 2\theta_{2c} + \phi \implies \theta_2 = \frac{\phi}{2} \ .$$

Also, the angles in the triangle ADC must add to 180°:

$$180° = 2(\theta - \theta_2) + \psi_s \implies \theta = 90° + \theta_2 - \frac{1}{2}\psi_s$$

which simplifies to $\theta = \theta_2 + \frac{1}{2}\psi$. Combining this with our previous result, we find $\theta = \frac{1}{2}(\phi + \psi)$. Thus, the law of refraction yields

$$n = \frac{\sin(\theta)}{\sin(\theta_2)} = \frac{\sin\left(\frac{1}{2}(\phi + \psi)\right)}{\sin\left(\frac{1}{2}\phi\right)} \ .$$

52. The critical angle is

$$\theta_c = \sin^{-1}\left(\frac{1}{n}\right) = \sin^{-1}\left(\frac{1}{1.8}\right) = 34° \ .$$

53. Let $\theta_1 = 45°$ be the angle of incidence at the first surface and θ_2 be the angle of refraction there. Let θ_3 be the angle of incidence at the second surface. The condition for total internal reflection at the second surface is $n\sin\theta_3 \geq 1$. We want to find the smallest value of the index of refraction n for which this inequality holds. The law of refraction, applied to the first surface, yields $n\sin\theta_2 = \sin\theta_1$. Consideration of the triangle formed by the surface of the slab and the ray in the slab tells us that $\theta_3 = 90° - \theta_2$. Thus, the condition for total internal reflection becomes $1 \leq n\sin(90° - \theta_2) = n\cos\theta_2$. Squaring this equation and using $\sin^2\theta_2 + \cos^2\theta_2 = 1$, we obtain $1 \leq n^2(1 - \sin^2\theta_2)$. Substituting $\sin\theta_2 = (1/n)\sin\theta_1$ now leads to

$$1 \leq n^2\left(1 - \frac{\sin^2\theta_1}{n^2}\right) = n^2 - \sin^2\theta_1 \ .$$

The largest value of n for which this equation is true is the value for which $1 = n^2 - \sin^2\theta_1$. We solve for n:

$$n = \sqrt{1 + \sin^2\theta_1} = \sqrt{1 + \sin^2 45°} = 1.22 \ .$$

54. Reference to Fig. 34-24 may help in the visualization of why there appears to be a "circle of light" (consider revolving that picture about a vertical axis). The depth and the radius of that circle (which is from point a to point f in that figure) is related to the tangent of the angle of incidence. Thus, the diameter D of the circle in question is

$$D = 2h\tan\theta_c = 2h\tan\left[\sin^{-1}\left(\frac{1}{n_w}\right)\right] = 2(80.0\,\text{cm})\tan\left[\sin^{-1}\left(\frac{1}{1.33}\right)\right] = 182\ \text{cm} \ .$$

55. (a) No refraction occurs at the surface ab, so the angle of incidence at surface ac is $90° - \phi$. For total internal reflection at the second surface, $n_g\sin(90° - \phi)$ must be greater than n_a. Here n_g is the index of refraction for the glass and n_a is the index of refraction for air. Since $\sin(90° - \phi) = \cos\phi$, we want the largest value of ϕ for which $n_g\cos\phi \geq n_a$. Recall that $\cos\phi$ decreases as ϕ increases from zero. When ϕ has the largest value for which total internal reflection occurs, then $n_g\cos\phi = n_a$, or

$$\phi = \cos^{-1}\left(\frac{n_a}{n_g}\right) = \cos^{-1}\left(\frac{1}{1.52}\right) = 48.9° \ .$$

The index of refraction for air is taken to be unity.

(b) We now replace the air with water. If $n_w = 1.33$ is the index of refraction for water, then the largest value of ϕ for which total internal reflection occurs is

$$\phi = \cos^{-1}\left(\frac{n_w}{n_g}\right) = \cos^{-1}\left(\frac{1.33}{1.52}\right) = 29.0^\circ .$$

56. (a) (b) and (c) The index of refraction n for fused quartz is slightly higher on the bluish side of the visible light spectrum (with shorter wavelength). We estimate $n = 1.463$ for blue and $n = 1.456$ for red. Since $\sin\theta_c = 1/n$, the critical angle is slightly smaller for blue than it is for red: $\theta_c = 43.12^\circ$ for blue and $\theta_c = 43.38^\circ$ for red. Thus, at an angle of incidence of, say, $\theta = 43.29^\circ$, the refracted beam would be depleted of blue (and would appear to an outside observer as reddish), and the reflected beam would consequently appear to be bluish (to someone able to observe that beam, the operational details of which are not discussed here).

57. (a) The diagram below shows a cross section, through the center of the cube and parallel to a face. L is the length of a cube edge and S labels the spot. A portion of a ray from the source to a cube face is also shown. Light leaving the source at a small angle θ is refracted at the face and leaves the cube; light leaving at a sufficiently large angle is totally reflected. The light that passes through the cube face forms a circle, the radius r being associated with the critical angle for total internal reflection. If θ_c is that angle, then

$$\sin\theta_c = \frac{1}{n}$$

where n is the index of refraction for the glass. As the diagram shows, the radius of the circle is given by $r = (L/2)\tan\theta_c$. Now,

$$\tan\theta_c = \frac{\sin\theta_c}{\cos\theta_c} = \frac{\sin\theta_c}{\sqrt{1-\sin^2\theta_c}} = \frac{1/n}{\sqrt{1-(1/n)^2}} = \frac{1}{\sqrt{n^2-1}}$$

and the radius of the circle is

$$r = \frac{L}{2\sqrt{n^2-1}} = \frac{10\,\text{mm}}{2\sqrt{(1.5)^2-1}} = 4.47\,\text{mm} .$$

If an opaque circular disk with this radius is pasted at the center of each cube face, the spot will not be seen (provided internally reflected light can be ignored).

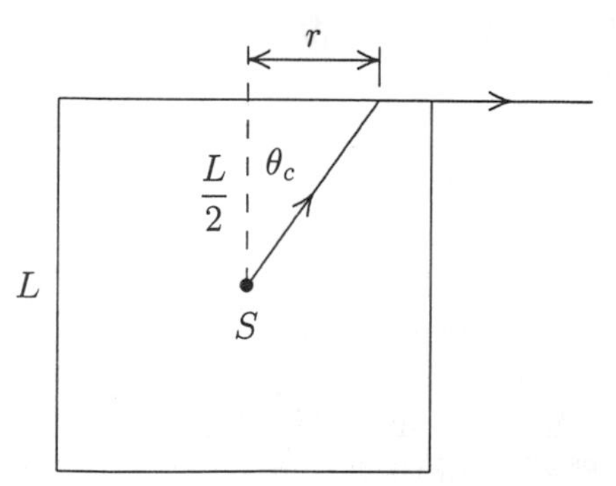

(b) There must be six opaque disks, one for each face. The total area covered by disks is $6\pi r^2$ and the total surface area of the cube is $6L^2$. The fraction of the surface area that must be covered by disks is

$$f = \frac{6\pi r^2}{6L^2} = \frac{\pi r^2}{L^2} = \frac{\pi(4.47\,\text{mm})^2}{(10\,\text{mm})^2} = 0.63 .$$

58. (a) We refer to the entry point for the original incident ray as point A (which we take to be on the left side of the prism, as in Fig. 34-49), the prism vertex as point B, and the point where the interior

ray strikes the right surface of the prism as point C. The angle between line AB and the interior ray is β (the complement of the angle of refraction at the first surface), and the angle between the line BC and the interior ray is α (the complement of its angle of incidence when it strikes the second surface). When the incident ray is at the minimum angle for which light is able to exit the prism, the light exits along the second face. That is, the angle of refraction at the second face is 90°, and the angle of incidence there for the interior ray is the critical angle for total internal reflection. Let θ_1 be the angle of incidence for the original incident ray and θ_2 be the angle of refraction at the first face, and let θ_3 be the angle of incidence at the second face. The law of refraction, applied to point C, yields $n \sin\theta_3 = 1$, so $\sin\theta_3 = 1/n = 1/1.60 = 0.625$ and $\theta_3 = 38.68°$. The interior angles of the triangle ABC must sum to 180°, so $\alpha + \beta = 120°$. Now, $\alpha = 90° - \theta_3 = 51.32°$, so $\beta = 120° - 51.32° = 69.68°$. Thus, $\theta_2 = 90° - \beta = 21.32°$. The law of refraction, applied to point A, yields $\sin\theta_1 = n \sin\theta_2 = 1.60 \sin 21.32° = 0.5817$. Thus $\theta_1 = 35.6°$.

(b) We apply the law of refraction to point C. Since the angle of refraction there is the same as the angle of incidence at A, $n \sin\theta_3 = \sin\theta_1$. Now, $\alpha + \beta = 120°$, $\alpha = 90° - \theta_3$, and $\beta = 90° - \theta_2$, as before. This means $\theta_2 + \theta_3 = 60°$. Thus, the law of refraction leads to

$$\sin\theta_1 = n\sin(60° - \theta_2) \implies \sin\theta_1 = n\sin 60° \cos\theta_2 - n\cos 60° \sin\theta_2$$

where the trigonometric identity $\sin(A - B) = \sin A \cos B - \cos A \sin B$ is used. Next, we apply the law of refraction to point A:

$$\sin\theta_1 = n\sin\theta_2 \implies \sin\theta_2 = (1/n)\sin\theta_1$$

which yields $\cos\theta_2 = \sqrt{1 - \sin^2\theta_2} = \sqrt{1 - (1/n^2)\sin^2\theta_1}$. Thus,

$$\sin\theta_1 = n\sin 60° \sqrt{1 - (1/n)^2 \sin^2\theta_1} - \cos 60° \sin\theta_1$$

or

$$(1 + \cos 60°)\sin\theta_1 = \sin 60° \sqrt{n^2 - \sin^2\theta_1}\ .$$

Squaring both sides and solving for $\sin\theta_1$, we obtain

$$\sin\theta_1 = \frac{n\sin 60°}{\sqrt{(1+\cos 60°)^2 + \sin^2 60°}} = \frac{1.60\sin 60°}{\sqrt{(1+\cos 60°)^2 + \sin^2 60°}} = 0.80$$

and $\theta_1 = 53.1°$.

59. (a) A ray diagram is shown below. Let θ_1 be the angle of incidence and θ_2 be the angle of refraction at the first surface. Let θ_3 be the angle of incidence at the second surface. The angle of refraction there is $\theta_4 = 90°$. The law of refraction, applied to the second surface, yields $n \sin\theta_3 = \sin\theta_4 = 1$. As shown in the diagram, the normals to the surfaces at P and Q are perpendicular to each other. The interior angles of the triangle formed by the ray and the two normals must sum to 180°, so $\theta_3 = 90° - \theta_2$ and $\sin\theta_3 = \sin(90° - \theta_2) = \cos\theta_2 = \sqrt{1 - \sin^2\theta_2}$. According to the law of refraction, applied at Q, $n\sqrt{1 - \sin^2\theta_2} = 1$. The law of refraction, applied to point P, yields $\sin\theta_1 = n\sin\theta_2$, so $\sin\theta_2 = (\sin\theta_1)/n$ and

$$n\sqrt{1 - \frac{\sin^2\theta_1}{n^2}} = 1\ .$$

Squaring both sides and solving for n, we get

$$n = \sqrt{1 + \sin^2\theta_1}\ .$$

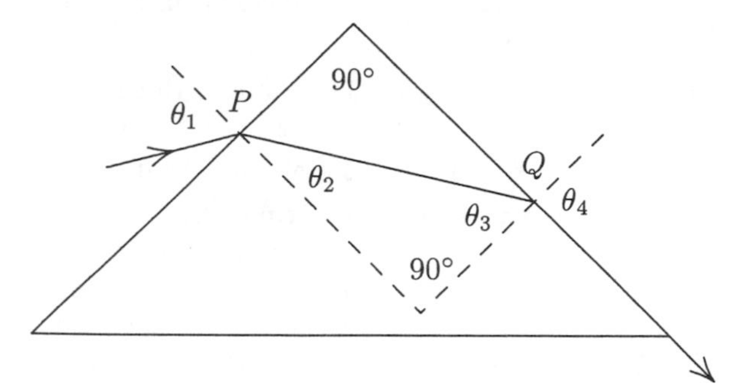

(b) The greatest possible value of $\sin^2\theta_1$ is 1, so the greatest possible value of n is $n_{\text{max}} = \sqrt{2} = 1.41$.

(c) For a given value of n, if the angle of incidence at the first surface is greater than θ_1, the angle of refraction there is greater than θ_2 and the angle of incidence at the second face is less than θ_3 $(= 90° - \theta_2)$. That is, it is less than the critical angle for total internal reflection, so light leaves the second surface and emerges into the air.

(d) If the angle of incidence at the first surface is less than θ_1, the angle of refraction there is less than θ_2 and the angle of incidence at the second surface is greater than θ_3. This is greater than the critical angle for total internal reflection, so all the light is reflected at Q.

60. (a) We use Eq. 34-49: $\theta_{\text{B}} = \tan^{-1} n_w = \tan^{-1}(1.33) = 53.1°$.

(b) Yes, since n_w depends on the wavelength of the light.

61. The angle of incidence θ_B for which reflected light is fully polarized is given by Eq. 34–48 of the text. If n_1 is the index of refraction for the medium of incidence and n_2 is the index of refraction for the second medium, then $\theta_B = \tan^{-1}(n_2/n_1) = \tan^{-1}(1.53/1.33) = 63.8°$.

62. From Fig. 34-19 we find $n_{\text{max}} = 1.470$ for $\lambda = 400\,\text{nm}$ and $n_{\text{min}} = 1.456$ for $\lambda = 700\,\text{nm}$. The corresponding Brewster's angles are $\theta_{\text{B,max}} = \tan^{-1} n_{\text{max}} = \tan^{-1}(1.470) = 55.77°$ and $\theta_{\text{B,min}} = \tan^{-1}(1.456) = 55.52°$.

63. (a) The Sun is far enough away that we approximate its rays as "parallel" in this Figure. That is, if the sunray makes angle θ from horizontal when the bird is in one position, then it makes the same angle θ when the bird is any other position. Therefore, its shadow on the ground moves as the bird moves: at 15 m/s.

(b) If the bird is in a position, a distance $x > 0$ from the wall, such that its shadow is on the wall at a distance $0 \geq y \geq h$ from the top of the wall, then it is clear from the Figure that $tan\theta = y/x$. Thus,

$$\frac{dy}{dt} = \frac{dx}{dt}\tan\theta = (-15\,\text{m/s})\tan 30° = -8.7\ \text{m/s}\ ,$$

which means that the distance y (which was measured as a positive number downward from the top of the wall) is shrinking at the rate of 8.7 m/s.

(c) Since $\tan\theta$ grows as $0 \leq \theta < 90°$ increases, then a larger value of $|dy/dt|$ implies a larger value of θ. The Sun is higher in the sky when the hawk glides by.

(d) With $|dy/dt| = 45$ m/s, we find

$$v_{\text{hawk}} = \left|\frac{dx}{dt}\right| = \frac{\left|\frac{dy}{dt}\right|}{\tan\theta}$$

so that we obtain $\theta = 72°$ if we assume $v_{\text{hawk}} = 15$ m/s.

64. (a) The 63.00 ns arrival times are consistent with the top of the tomb being 31.50 ns (pulse travel time) away from the surface. Since the pulses travel at 10.0 cm/ns in the soil, this travel time corresponds to a distance equal to 315 cm = 3.15 m.

(b) We are told that the locations in Fig. 34-54 are 2.0 m apart. Return pulses are registered at stations 2 through 7, but the returns from stations 2 and 7 are not "robust." The tomb's horizontal length is therefore at least 9 m long, and very probably less than 12 m in length.

(c) As demonstrated in part (a), we divide the travel times by 2 to infer depth. Thus, at station 3: the top of the tomb is 31.50 ns (pulse travel time in soil) from the surface; the top stone slab is 1.885 ns thick (pulse travel time in stone); the interior of the tomb is 8.00 ns high (pulse travel time in air); and the bottom stone slab is 1.885 ns thick (pulse travel time in stone). Since the pulse travels at 30 cm/s in the air, the interior of the tomb under station 3 (at the west end of the tomb) is 240 cm = 2.40 m high. At the east end (under, say, station 5), the corresponding time difference is

$$\frac{74.77\,\text{ns} - 66.77\,\text{ns}}{2} = 4.00\ \text{ns}$$

which corresponds to an interior height equal to (4.00 ns)(30 cm/s) = 120 cm/s = 1.20 m.

65. Since the layers are parallel, the angle of refraction regarding the first surface is the same as the angle of incidence regarding the second surface (as is suggested by the notation in Fig. 34-55). We recall that as part of the derivation of Eq. 34-49 (Brewster's angle), the textbook shows that the refracted angle is the complement of the incident angle:

$$\theta_2 = (\theta_1)_c = 90^\circ - \theta_1 \ .$$

We apply Eq. 34-49 to both refractions, setting up a product:

$$\begin{aligned} \left(\frac{n_2}{n_1}\right)\left(\frac{n_3}{n_2}\right) &= (\tan\theta_{B\,1\to 2})\,(\tan\theta_{B\,2\to 3}) \\ \frac{n_3}{n_1} &= (\tan\theta_1)\,(\tan\theta_2) \ . \end{aligned}$$

Now, since θ_2 is the complement of θ_1 we have

$$\tan\theta_2 = \tan(\theta_1)_c = \frac{1}{\tan\theta_1} \ .$$

Therefore, the product of tangents cancel and we obtain $n_3/n_1 = 1$. Consequently, the third medium is air: $n_3 = 1.0$.

66. In air, light travels at roughly $c = 3.0 \times 10^8$ m/s. Therefore, for $t = 1.0$ ns, we have a distance of

$$d = ct = \left(3.0 \times 10^8\ \text{m/s}\right)\left(1.0 \times 10^{-9}\ \text{s}\right) = 0.30\ \text{m} \ .$$

67. (a) The first contribution to the overall deviation is at the first refraction: $\delta\theta_1 = \theta_i - \theta_r$. The next contribution to the overall deviation is the reflection. Noting that the angle between the ray right before reflection and the axis normal to the back surface of the sphere is equal to θ_r, and recalling the law of reflection, we conclude that the angle by which the ray turns (comparing the direction of propagation before and after the reflection) is $\delta\theta_2 = 180^\circ - 2\theta_r$. The final contribution is the refraction suffered by the ray upon leaving the sphere: $\delta\theta_3 = \theta_i - \theta_r$ again. Therefore,

$$\theta_{\text{dev}} = \delta\theta_1 + \delta\theta_2 + \delta\theta_3 = 180^\circ + 2\theta_i - 4\theta_r \ .$$

(b) We substitute $\theta_r = \sin^{-1}(\frac{1}{n}\sin\theta_i)$ into the expression derived in part (a), using the two given values for n. The higher curve is for the blue light.

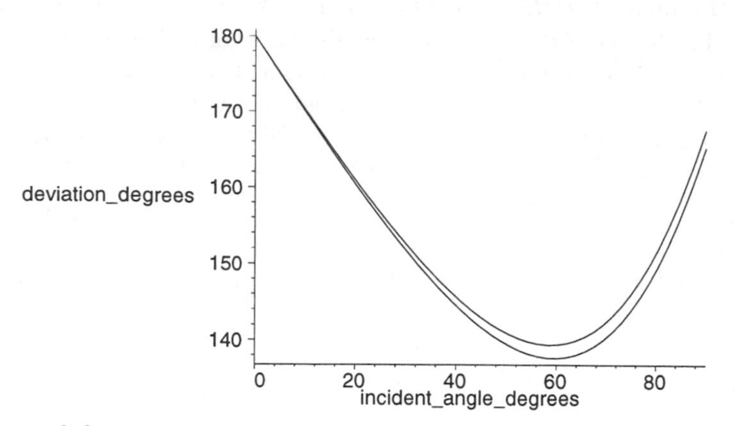

(c) We can expand the graph and try to estimate the minimum, or search for it with a more sophisticated numerical procedure. We find that the θ_{dev} minimum for red light is 137.63°, and this occurs at $\theta_i = 59.52°$.

(d) For blue light, we find that the θ_{dev} minimum is 139.35°, and this occurs at $\theta_i = 59.52°$.

(e) The difference in θ_{dev} in the previous two parts is 1.72°.

68. (a) The first contribution to the overall deviation is at the first refraction: $\delta\theta_1 = \theta_i - \theta_r$. The next contribution(s) to the overall deviation is (are) the reflection(s). Noting that the angle between the ray right before reflection and the axis normal to the back surface of the sphere is equal to θ_r, and recalling the law of reflection, we conclude that the angle by which the ray turns (comparing the direction of propagation before and after [each] reflection) is $\delta\theta_r = 180° - 2\theta_r$. Thus, for k reflections, we have $\delta\theta_2 = k\theta_r$ to account for these contributions. The final contribution is the refraction suffered by the ray upon leaving the sphere: $\delta\theta_3 = \theta_i - \theta_r$ again. Therefore,

$$\theta_{\text{dev}} = \delta\theta_1 + \delta\theta_2 + \delta\theta_3 = 2(\theta_i - \theta_r) + k(180° - 2\theta_r) = k(180°) + 2\theta_i - 2(k+1)\theta_r \ .$$

(b) For $k = 2$ and $n = 1.331$ (given in problem 67), we search for the second-order rainbow angle numerically. We find that the θ_{dev} minimum for red light is 230.37°, and this occurs at $\theta_i = 71.90°$.

(c) Similarly, we find that the second-order θ_{dev} minimum for blue light (for which $n = 1.343$) is 233.48°, and this occurs at $\theta_i = 71.52°$.

(d) The difference in θ_{dev} in the previous two parts is 3.11°.

(e) Setting $k = 3$, we search for the third-order rainbow angle numerically. We find that the θ_{dev} minimum for red light is 317.53°, and this occurs at $\theta_i = 76.88°$.

(f) Similarly, we find that the third-order θ_{dev} minimum for blue light is 321.89°, and this occurs at $\theta_i = 76.62°$.

(g) The difference in θ_{dev} in the previous two parts is 4.37°.

69. Reference to Fig. 34-24 may help in the visualization of why there appears to be a "circle of light" (consider revolving that picture about a vertical axis). The depth and the radius of that circle (which is from point a to point f in that figure) is related to the tangent of the angle of incidence. The diameter of the circle in question is given by $d = 2h\tan\theta_c$. For water $n = 1.33$, so Eq. 34-47 gives $\sin\theta_c = 1/1.33$, or $\theta_c = 48.75°$. Thus,

$$d = 2h\tan\theta_c = 2(2.00\,\text{m})(\tan 48.75°) = 4.56\ \text{m} \ .$$

70. We apply Eq. 34-42 (twice) to obtain

$$I = I_0 \cos^2\theta_1 \cos^2\theta_2$$

where $\theta_1 = 20°$ and $\theta_2 = (20° + \theta)$. Since $I/I_0 = 0.200$, we find $\cos\theta_2 = \sqrt{0.2265}$ which leads to $\theta_2 = 62°$ and consequently to $\theta = 42°$.

71. (a) The electric field amplitude is $E_m = \sqrt{2}\,E_{\text{rms}} = 70.7$ V/m, so that the magnetic field amplitude is $B_m = 2.36 \times 10^{-7}$ T by Eq. 34-5. Since the direction of propagation, $\vec{E}$, and $\vec{B}$ are mutually perpendicular, we infer that the only non-zero component of $\vec{B}$ is B_x, and note that the direction of propagation being along the $-z$ axis means the spatial and temporal parts of the wave function argument are of like sign (see §17-5). Also, from $\lambda = 250$ nm, we find that $f = c/\lambda = 1.20 \times 10^{15}$ Hz, which leads to $\omega = 2\pi f = 7.53 \times 10^{15}$ rad/s. Also, we note that $k = 2\pi/\lambda = 2.51 \times 10^7$ m^{-1}. Thus, assuming some "initial condition" (that, say the field is zero, with its derivative positive, at $z = 0$ when $t = 0$), we have

$$B_x = 2.36 \times 10^{-7} \;\sin\left(\left(2.51 \times 10^7\right) z + \left(7.53 \times 10^{15}\right) t\right)$$

in SI units.

(b) The exposed area of the triangular chip is $A = \sqrt{3}\,\ell^2/8$, where $\ell = 2.00 \times 10^{-6}$ m. The intensity of the wave is

$$I = \frac{1}{c\mu_0}\,E_{\text{rms}}^2 = 6.64 \text{ W/m}^2 .$$

Thus, Eq. 34-33 leads to

$$F = \frac{2IA}{c} = 3.83 \times 10^{-20} \text{ N} .$$

72. We follow Sample Problem 34-2 in computing the sunlight intensity at the sail's location.

$$I = \frac{P_S}{4\pi r^2} = \frac{3.9 \times 10^{26} \text{ W}}{4\pi \left(3.0 \times 10^{11} \text{ m}\right)^2} = 345 \text{ W/m}^2$$

With $A = (2.0 \text{ m})^2$, we use Eq. 34-33 to obtain the radiation force:

$$F = \frac{2IA}{c} = 9.2 \times 10^{-6} \text{ N} .$$

73. (a) Eq. 34-5 gives $E = cB$, which relates the field values at any instant – and so relates rms values to rms values, and amplitude values to amplitude values, as the case may be. Thus, $E_{\text{rms}} = cB_{\text{rms}} = 16.8$ V/m. Multiplying by $\sqrt{2}$ yields the electric field amplitude $E_m = 23.7$ V/m.

(b) We use Eq. 34-26:

$$I = \frac{1}{\mu_0\, c}\,E_{\text{rms}}^2 = 0.748 \text{ W/m}^2 .$$

74. Consider two wavelengths, λ_1 and λ_2, whose corresponding frequencies are f_1 and f_2. Then $\lambda_1 = C/f_1$ and $\lambda_2 = C/f_2$. If $\lambda_1/\lambda_2 = 10$, then

$$\frac{\lambda_1}{\lambda_2} = \frac{C/f_1}{C/f_2} = \frac{f_2}{f_1} = 10 .$$

The spaces are the same on both scales.

75. We take the derivative with respect to x of both sides of Eq. 34-11:

$$\frac{\partial}{\partial x}\left(\frac{\partial E}{\partial x}\right) = \frac{\partial^2 E}{\partial x^2} = \frac{\partial}{\partial x}\left(-\frac{\partial B}{\partial t}\right) = -\frac{\partial^2 B}{\partial x \partial t} .$$

Now we differentiate both sides of Eq. 34-18 with respect to t:

$$\frac{\partial}{\partial t}\left(-\frac{\partial B}{\partial x}\right) = -\frac{\partial^2 B}{\partial x \partial t} = \frac{\partial}{\partial t}\left(\varepsilon_0\mu_0\frac{\partial E}{\partial t}\right) = \varepsilon_0\mu_0\frac{\partial^2 E}{\partial t^2} .$$

Substituting $\partial^2 E/\partial x^2 = -\partial^2 B/\partial x\partial t$ from the first equation above into the second one, we get

$$\varepsilon_0\mu_0\frac{\partial^2 E}{\partial t^2} = \frac{\partial^2 E}{\partial x^2} ,$$

or

$$\frac{\partial^2 E}{\partial t^2} = \frac{1}{\varepsilon_0\mu_0}\frac{\partial^2 E}{\partial x^2} = c^2\frac{\partial^2 E}{\partial x^2} .$$

Similarly, we differentiate both sides of Eq. 34-11 with respect to t

$$\frac{\partial^2 E}{\partial x\partial t} = -\frac{\partial^2 B}{\partial t^2} ,$$

and differentiate both sides of Eq. 34-18 with respect to x

$$-\frac{\partial^2 B}{\partial x^2} = \varepsilon_0\mu_0\frac{\partial^2 E}{\partial x\partial t} .$$

Combining these two equations, we get

$$\frac{\partial^2 B}{\partial t^2} = \frac{1}{\varepsilon_0\mu_0}\frac{\partial^2 B}{\partial x^2} = c^2\frac{\partial^2 B}{\partial x^2} .$$

76. The energy density of an electromagnetic wave is given by $u = u_E + u_B$. From the discussion in §34-4, $u_E = u_B = \frac{1}{2}\varepsilon_0 E^2$, so $u = 2u_E = \varepsilon_0 E^2$. Upon averaging over time this becomes

$$u_{\text{avg}} = \varepsilon_0\overline{E^2} = \varepsilon_0 E_{\text{rms}}^2 .$$

Combining this equation with Eq. 34-26 in the textbook, we obtain

$$I = \frac{1}{c\mu_0}E_{\text{rms}}^2 = \frac{1}{c\mu_0}\frac{u_{\text{avg}}}{\varepsilon_0} = \frac{c^2 u_{\text{avg}}}{c} = cu_{\text{avg}}$$

where $c^2 = 1/\varepsilon_0\mu_0$ is used.

77. (a) Assuming complete absorption, the radiation pressure is

$$p_r = \frac{I}{c} = \frac{1.4\times 10^3\ \text{W/m}^2}{3.0\times 10^8\ \text{m/s}} = 4.7\times 10^{-6}\ \text{N/m}^2 .$$

(b) We compare values by setting up a ratio:

$$\frac{p_r}{p_0} = \frac{4.7\times 10^{-6}\ \text{N/m}^2}{1.0\times 10^5\ \text{N/m}^2} = 4.7\times 10^{-11} .$$

78. (a) Suppose there are a total of N transparent layers ($N = 5$ in our case). We label these layers from left to right with indices 1, 2, ..., N. Let the index of refraction of the air be n_0. We denote the initial angle of incidence of the light ray upon the air-layer boundary as θ_i and the angle of the emerging light ray as θ_f. We note that, since all the boundaries are parallel to each other, the angle of incidence θ_j at the boundary between the j-th and the $(j+1)$-th layers is the same as the angle between the transmitted light ray and the normal in the j-th layer. Thus, for the first boundary (the one between the air and the first layer)

$$\frac{n_1}{n_0} = \frac{\sin\theta_i}{\sin\theta_1} ,$$

for the second boundary

$$\frac{n_2}{n_1} = \frac{\sin\theta_1}{\sin\theta_2} ,$$

and so on. Finally, for the last boundary

$$\frac{n_0}{n_N} = \frac{\sin\theta_N}{\sin\theta_f} .$$

Multiplying these equations, we obtain

$$\left(\frac{n_1}{n_0}\right)\left(\frac{n_2}{n_1}\right)\left(\frac{n_3}{n_2}\right)\cdots\left(\frac{n_0}{n_N}\right) = \left(\frac{\sin\theta_i}{\sin\theta_1}\right)\left(\frac{\sin\theta_1}{\sin\theta_2}\right)\left(\frac{\sin\theta_2}{\sin\theta_3}\right)\cdots\left(\frac{\sin\theta_N}{\sin\theta_f}\right) .$$

We see that the L.H.S. of the equation above can be reduced to n_0/n_0 while the R.H.S. is equal to $\sin\theta_i/\sin\theta_f$. Equating these two expressions, we find

$$\sin\theta_f = \left(\frac{n_0}{n_0}\right)\sin\theta_i = \sin\theta_i ,$$

which gives $\theta_i = \theta_f$. So for the two light rays in the problem statement, the angle of the emerging light rays are both the same as their respective incident angles. Thus, $\theta_f = 0$ for ray a and $\theta_f = 20°$ for ray b.

(b) In this case, all we need to do is to change the value of n_0 from 1.0 (for air) to 1.5 (for glass). This does not change the result above. Note that the result of this problem is fairly general. It is independent of the number of layers and the thickness and index of refraction of each layer.

79. We use the result of the problem 51 to solve for ψ. Note that $\phi = 60.0°$ in our case. Thus, from

$$n = \frac{\sin\frac{1}{2}(\psi + \phi)}{\sin\frac{1}{2}\phi} ,$$

we get

$$\sin\frac{1}{2}(\psi + \phi) = n\sin\frac{1}{2}\phi = (1.31)\sin\left(\frac{60.0°}{2}\right) = 0.655 ,$$

which gives $\frac{1}{2}(\psi + \phi) = \sin^{-1}(0.655) = 40.9°$. Thus, $\psi = 2(40.9°) - \phi = 2(40.9°) - 60.0° = 21.8°$.

80. (a) The light that passes through the surface of the lake is within a cone of apex angle $2\theta_c$ making a "circle of light" there; reference to Fig. 34-24 may help in visualizing this (consider revolving that picture about a vertical axis). Since the source is point-like, its energy spreads out with perfect spherical symmetry, until reaching the surface and other boundaries of the lake. The problem asks us to assume there are no partial reflections at the surface, only the total reflections outside the "circle of light." Thus, of the full sphere of light (of area $A_s = 4\pi R^2$) emitted by the source, only a fraction of it – coinciding with the cone of apex angle $2\theta_c$ – enters the air above. If we label the area of that portion of the sphere which reaches the air above as A, then the fraction of the total energy emitted that passes through the surface is

$$frac = \frac{A}{4\pi R^2} \qquad \text{where} \qquad R = \frac{h}{\cos\theta_c}$$

is the distance from the point-source to the edge of the "circle of light." Now, the area A of the spherical cap of height H bounded by that circle is

$$A = 2\pi RH = 2\pi R(R - h)$$

may be looked up in a number of references, or can be derived from $A = 2\pi R^2 \int_0^{\theta_c} \sin\theta\, d\theta$. Consequently,

$$frac = \frac{2\pi R(R-h)}{4\pi R^2} = \frac{1}{2}\left(1 - \frac{h}{R}\right) = \frac{1}{2}(1 - \cos\theta_c) .$$

The critical angle is given by $\sin\theta_c = 1/n$, which implies $\cos\theta_c = \sqrt{1-\sin^2\theta_c} = \sqrt{1-1/n^2}$. When this expression is substituted into our result above, we obtain

$$frac = \frac{1}{2}\left(1-\sqrt{1-\frac{1}{n^2}}\right) .$$

(b) For $n = 1.33$,

$$frac = \frac{1}{2}\left(1-\sqrt{1-\frac{1}{(1.33)^2}}\right) = 0.170 .$$

81. We apply Eq. 34-40 (once) and Eq. 34-42 (twice) to obtain

$$I = \frac{1}{2} I_0 \cos^2\theta_1 \cos^2\theta_2 .$$

With $\theta_1 = 60° - 20° = 40°$ and $\theta_2 = 40° + 30° = 70°$, this yields $I/I_0 = 0.034$.

82. (a) From $kc = \omega$ where $k = 1.00 \times 10^6$ m^{-1}, we obtain $\omega = 3.00 \times 10^{14}$ rad/s. The magnetic field amplitude is, from Eq. 34-5, $B = (5.00 \text{ V/m})/c = 1.67 \times 10^{-8}$ T. From the fact that $-\hat{k}$ (the direction of propagation), $\vec{E} = E_y\hat{j}$, and $\vec{B}$ are mutually perpendicular, we conclude that the only non-zero component of $\vec{B}$ is B_x, so that we have (in SI units)

$$B_x = 1.67 \times 10^{-8} \sin\left(\left(1.00 \times 10^6\right) z + \left(3.00 \times 10^{14}\right) t\right) .$$

(b) The wavelength is $\lambda = 2\pi/k = 6.28 \times 10^{-6}$ m.

(c) The period is $T = 2\pi/\omega = 2.09 \times 10^{-14}$ s.

(d) The intensity is

$$I = \frac{1}{c\mu_0}\left(\frac{5.00 \text{ V/m}}{\sqrt{2}}\right)^2 = 0.0332 \text{ W/m}^2 .$$

(e) As noted in part (a), the only nonzero component of $\vec{B}$ is B_x. The magnetic field oscillates along the x axis.

(f) The wavelength found in part (b) places this in the infrared portion of the spectrum.

83. We write $m = \rho\mathcal{V}$ where $\mathcal{V} = 4\pi R^3/3$ is the volume. Plugging this into $F = ma$ and then into Eq. 34-32 (with $A = \pi R^2$, assuming the light is in the form of plane waves), we find

$$\rho\frac{4\pi R^3}{3} a = \frac{I\pi R^2}{c} .$$

This simplifies to

$$a = \frac{3I}{4\rho c R}$$

which yields $a = 1.5 \times 10^{-9}$ m/s^2.

84. Since intensity is power divided by area (and the area is spherical in the isotropic case), then the intensity at a distance of $r = 20$ m from the source is

$$I = \frac{P}{4\pi r^2} = 0.040 \text{ W/m}^2 .$$

as illustrated in Sample Problem 34-2. Now, in Eq. 34-32 for a totally absorbing area A, we note that the exposed area of the small sphere is that on a flat circle $A = \pi(0.020\,\text{m})^2 = 0.0013$ m^2. Therefore,

$$F = \frac{IA}{c} = \frac{(0.040)(0.0013)}{3 \times 10^8} = 1.7 \times 10^{-13} \text{ N} .$$

95. From Eq. 34-26, we have $E_{\rm rms} = \sqrt{\mu_0 c I} = 1941$ V/m, which implies (using Eq. 34-5) that $B_{\rm rms} = 1941/c = 6.47 \times 10^{-6}$ T. Multiplying by $\sqrt{2}$ yields the magnetic field amplitude $B_m = 9.16 \times 10^{-6}$ T.

96. (a) The frequency is

$$f = \frac{c}{\lambda} = \frac{3.0 \times 10^8 \text{ m/s}}{0.067 \times 10^{-15} \text{ m}} = 4.5 \times 10^{24} \text{ Hz} .$$

(b) In this case, the (very long) wavelength is

$$\lambda = \frac{c}{f} = \frac{3.0 \times 10^8 \text{ m/s}}{30 \text{ Hz}} = 1.0 \times 10^7 \text{ m}$$

which is about 1.6 Earth radii.

97. The fraction is

$$\frac{\pi R_e^2}{4\pi d_{es}^2} = \frac{1}{4}\left(\frac{6.37 \times 10^6 \text{ m}}{1.50 \times 10^{11} \text{ m}}\right)^2 = 4.51 \times 10^{-10} .$$

98. (a) When examining Fig. 34-73, it is important to note that the angle (measured from the central axis) for the light ray in air, θ, is not the angle for the ray in the glass core, which we denote θ'. The law of refraction leads to

$$\sin\theta' = \frac{1}{n_1}\sin\theta \qquad \text{assuming} \quad n_{\rm air} = 1 .$$

The angle of incidence for the light ray striking the coating is the complement of θ', which we denote as $\theta'_{\rm comp}$ and recall that

$$\sin\theta'_{\rm comp} = \cos\theta' = \sqrt{1 - \sin^2\theta'} .$$

In the critical case, $\theta'_{\rm comp}$ must equal θ_c specified by Eq. 34-47. Therefore,

$$\frac{n_2}{n_1} = \sin\theta'_{\rm comp} = \sqrt{1 - \sin^2\theta'} = \sqrt{1 - \left(\frac{1}{n_1}\sin\theta\right)^2}$$

which leads to the result: $\sin\theta = \sqrt{n_1^2 - n_2^2}$.

(b) With $n_1 = 1.58$ and $n_2 = 1.53$, we obtain

$$\theta = \sin^{-1}\left(1.58^2 - 1.53^2\right) = 23.2° .$$

99. (a) In our solution here, we assume the reader has looked at our solution for problem 98. A light ray traveling directly along the central axis reaches the end in time

$$t_{\rm direct} = \frac{L}{v_1} = \frac{n_1 L}{c} .$$

For the ray taking the critical zig-zag path, only its velocity component along the core axis direction contributes to reaching the other end of the fiber. That component is $v_1\cos\theta'$, so the time of travel for this ray is

$$t_{\rm zig\,zag} = \frac{L}{v_1\cos\theta'} = \frac{n_1 L}{c\sqrt{1 - \left(\frac{1}{n_1}\sin\theta\right)^2}}$$

using results from the previous solution. Plugging in $\sin\theta = \sqrt{n_1^2 - n_2^2}$ and simplifying, we obtain

$$t_{\rm zig\,zag} = \frac{n_1 L}{c(n_2/n_1)} = \frac{n_1^2 L}{n_2 c} .$$

The difference $t_{\rm zig\,zag} - t_{\rm direct}$ readily yields the result shown in the problem statement.

(b) With $n_1 = 1.58$, $n_2 = 1.53$ and $L = 300$ m, we obtain $\Delta t = 52$ ns.

100. (a) The condition (in Eq. 34-44) required in the critical angle calculation is $\theta_3 = 90°$. Thus (with $\theta_2 = \theta_c$, which we don't compute here),

$$n_1 \sin\theta_1 = n_2 \sin\theta_2 = n_3 \sin\theta_3$$

leads to $\theta_1 = \theta = \sin^{-1} n_3/n_1 = 54.3°$.

(b) Reducing θ leads to a reduction of θ_2 so that it becomes less than the critical angle; therefore, there will be some transmission of light into material 3.

101. (a) We note that the complement of the angle of refraction (in material 2) is the critical angle. Thus,

$$n_1 \sin\theta = n_2 \cos\theta_c = n_2 \sqrt{1 - \left(\frac{n_3}{n_2}\right)^2} = \sqrt{n_2^2 - n_3^2}$$

leads to $\theta = 51.1°$.

(b) Reducing θ leads to an increase of the angle with which the light strikes the interface between materials 2 and 3, so it becomes greater than the critical angle. Therefore, there will be no transmission of light into material 3.

Chapter 35

1. The image is 10 cm behind the mirror and you are 30 cm in front of the mirror. You must focus your eyes for a distance of 10 cm + 30 cm = 40 cm.

2. The bird is a distance d_2 in front of the mirror; the plane of its image is that same distance d_2 behind the mirror. The lateral distance between you and the bird is $d_3 = 5.00$ m. We denote the distance from the camera to the mirror as d_1, and we construct a right triangle out of d_3 and the distance between the camera and the image plane $(d_1 + d_2)$. Thus, the focus distance is

$$\begin{aligned} d &= \sqrt{(d_1 + d_2)^2 + d_3^2} \\ &= \sqrt{(4.30\,\text{m} + 3.30\,\text{m})^2 + (5.00\,\text{m})^2} \\ &= 9.10\ \text{m}\ . \end{aligned}$$

3. (a) There are three images. Two are formed by single reflections from each of the mirrors and the third is formed by successive reflections from both mirrors.

   (b) The positions of the images are shown on the two diagrams below. The diagram on the left below shows the image I_1, formed by reflections from the left-hand mirror. It is the same distance behind the mirror as the object O is in front, and lies on the line perpendicular to the mirror and through the object. Image I_2 is formed by light that is reflected from both mirrors. We may consider I_2 to be the image of I_1 formed by the right-hand mirror, extended. I_2 is the same distance behind the line of the right-hand mirror as I_1 is in front and it is on the line that is perpendicular to the line of the mirror. The diagram on the right, below, shows image I_3, formed by reflections from the right-hand mirror. It is the same distance behind the mirror as the object is in front, and lies on the line perpendicular to the mirror and through the object. As the diagram shows, light that is first reflected from the right-hand mirror and then from the left-hand mirror forms an image at I_2.

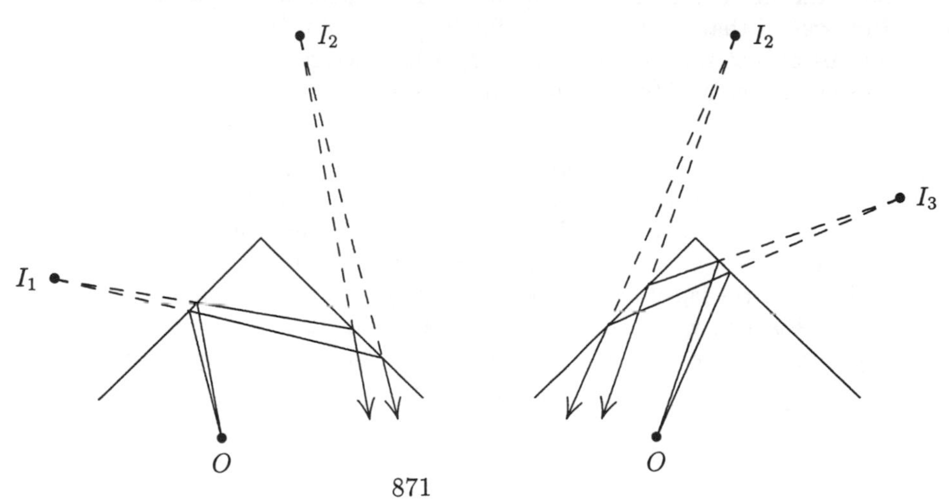

4. In each case there is an object and its "first" image in the mirror closest to it (this image is the same distance behind the mirror as the object is in front of it and might be referred to as the object's "twin"). The rest of the "figuring" consists of drawing perpendiculars from these (or imagining doing so) to the mirror-planes and constructing further images.

   (a) For $\theta = 45°$, we have two images in the second mirror caused by the object and its "first" image, and from these one can construct two new images I and I' behind the first mirror plane. Extending the second mirror plane, we can find two further images of I and I' which are on equal sides of the extension of the first mirror plane. This circumstance implies there are no further images, since these final images are each other's "twins." We show this construction in the figure below. Summarizing, we find $1 + 2 + 2 + 2 = 7$ images in this case.

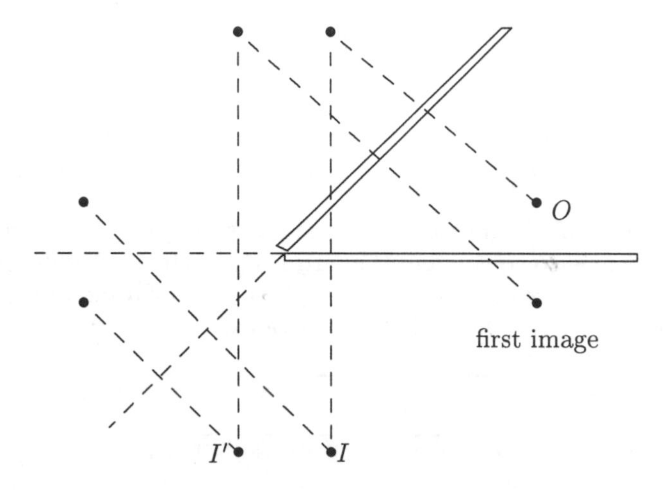

   (b) For $\theta = 60°$, we have two images in the second mirror caused by the object and its "first" image, and from these one can construct two new images I and I' behind the first mirror plane. The images I and I' are each other's "twins" in the sense that they are each other's reflections about the extension of the second mirror plane; there are no further images. Summarizing, we find $1+2+2 = 5$ images in this case.

   (c) For $\theta = 120°$, we have two images I_1' and I_2 behind the extension of the second mirror plane, caused by the object and its "first" image (which we refer to here as I_1). No further images can be constructed from I_1' and I_2, since the method indicated above would place any further possibilities in front of the mirrors. This construction has the disadvantage of deemphasizing the actual ray-tracing, and thus any dependence on where the observer of these images is actually placing his or her eyes. It turns out in this case that the number of images that can be seen ranges from 1 to 3, depending on the locations of both the object and the observer. As an example, if the observer's eye is collinear with I_1 and I_1', then the observer can only see one image (I_1 and not the one behind it). Another observer, close to the second mirror would probably be able to see only I_1 and I_2. However, if that observer moves further back from the vertex of the two mirrors he or she should also be able to see the third image, I_1', which is essentially the "twin" image formed from I_1 relative to the extension of the second mirror plane.

5. Consider a single ray from the source to the mirror and let θ be the angle of incidence. The angle of reflection is also θ and the reflected ray makes an angle of 2θ with the incident ray. Now we rotate the mirror through the angle α so that the angle of incidence increases to $\theta + \alpha$. The reflected ray now makes an angle of $2(\theta + \alpha)$ with the incident ray. The reflected ray has been rotated through an angle of 2α. If the mirror is rotated so the angle of incidence is decreased by α, then the reflected ray makes an angle of $2(\theta - \alpha)$ with the incident ray. Again it has been rotated through 2α. The diagrams below show the situation for $\alpha = 45°$. The ray from the object to the mirror is the same in both cases and the

reflected rays are 90° apart.

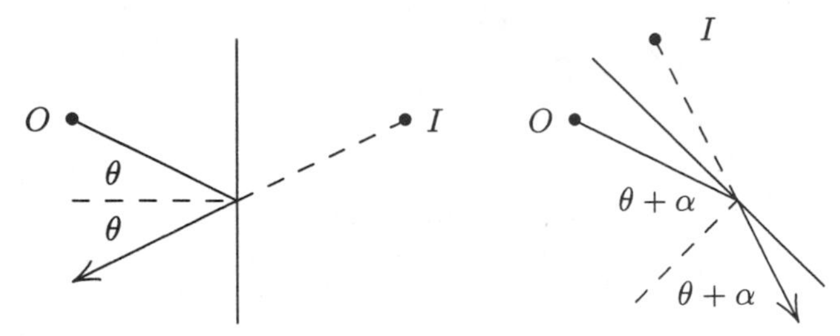

6. When S is barely able to see B the light rays from B must reflect to S off the edge of the mirror. The angle of reflection in this case is 45°, since a line drawn from S to the mirror's edge makes a 45° angle relative to the wall. By the law of reflection, we find

$$\frac{x}{d/2} = \tan 45^\circ \quad \Longrightarrow \quad x = \frac{d}{2} = \frac{3.0\,\text{m}}{2} = 1.5\ \text{m}\ .$$

7. The intensity of light from a point source varies as the inverse of the square of the distance from the source. Before the mirror is in place, the intensity at the center of the screen is given by $I_0 = A/d^2$, where A is a constant of proportionality. After the mirror is in place, the light that goes directly to the screen contributes intensity I_0, as before. Reflected light also reaches the screen. This light appears to come from the image of the source, a distance d behind the mirror and a distance $3d$ from the screen. Its contribution to the intensity at the center of the screen is

$$I_r = \frac{A}{(3d)^2} = \frac{A}{9d^2} = \frac{I_0}{9}\ .$$

The total intensity at the center of the screen is

$$I = I_0 + I_r = I_0 + \frac{I_0}{9} = \frac{10}{9} I_0\ .$$

The ratio of the new intensity to the original intensity is $I/I_0 = 10/9$.

8. We apply the law of refraction, assuming all angles are in radians:

$$\frac{\sin\theta}{\sin\theta'} = \frac{n_w}{n_{\text{air}}}\ ,$$

which in our case reduces to $\theta' \approx \theta/n_w$ (since both θ and θ' are small, and $n_{\text{air}} \approx 1$). We refer to our figure, below. The object O is a vertical distance h_1 above the water, and the water surface is a vertical distance h_2 above the mirror. We are looking for a distance d (treated as a positive number) below the mirror where the image I of the object is formed. In the triangle OAB

$$|AB| = h_1 \tan\theta \approx h_1\theta\ ,$$

and in the triangle CBD

$$|BC| = 2h_2 \tan\theta' \approx 2h_2\theta' \approx \frac{2h_2\theta}{n_w}\ .$$

Finally, in the triangle ACI, we have $|AI| = d + h_2$. Therefore,

$$\begin{aligned} d &= |AI| - h_2 = \frac{|AC|}{\tan\theta} - h_2 \\ &\approx \frac{|AB| + |BC|}{\theta} - h_2 \\ &= \left(\frac{h_1}{\theta} + \frac{2h_2\theta}{n_w}\right)\frac{1}{\theta} - h_2 = h_1 + \frac{2h_2}{n_w} - h_2 \\ &= 250\,\text{cm} + \frac{2(200\,\text{cm})}{1.33} - 200\,\text{cm} = 351\ \text{cm}\ . \end{aligned}$$

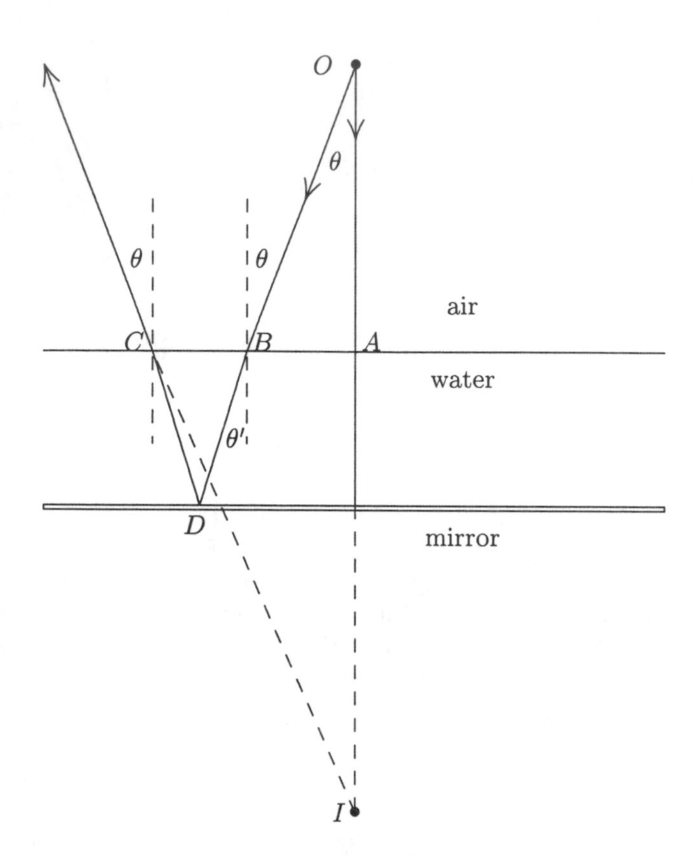

9. We use Eqs. 35-3 and 35-4, and note that $m = -i/p$. Thus,

$$\frac{1}{p} - \frac{1}{pm} = \frac{1}{f} = \frac{2}{r} \, .$$

We solve for p:

$$p = \frac{r}{2}\left(1 - \frac{1}{m}\right) = \frac{35.0\,\text{cm}}{2}\left(1 - \frac{1}{2.50}\right) = 10.5 \text{ cm} \, .$$

10. (a) $f = +20\,\text{cm}$ (positive, because the mirror is concave); $r = 2f = 2(+20\,\text{cm}) = +40\,\text{cm}$; $i = (1/f - 1/p)^{-1} = (1/20\,\text{cm} - 1/10\,\text{cm})^{-1} = -20\,\text{cm}$; $m = -i/p = -(-20\,\text{cm}/10\,\text{cm}) = +2.0$. The image is virtual and upright. The ray diagram would be similar to Fig. 35-8(a) in the textbook.

(b) The fact that the magnification is 1 and the image is virtual means that the mirror is flat (plane). Flat mirrors (and flat "lenses" such as a window pane) have $f = \infty$ (or $f = -\infty$ since the sign does not matter in this extreme case), and consequently $r = \infty$ (or $r = -\infty$) by Eq. 35-3. Eq. 35-4 readily yields $i = -10\,\text{cm}$. The magnification being positive implies the image is upright; the answer is "no" (it's not inverted). The ray diagram would be similar to Fig. 35-6(a) in the textbook.

(c) Since $f > 0$, the mirror is concave. Using Eq. 35-3, we obtain $r = 2f = +40\,\text{cm}$. Eq. 35-4 readily yields $i = +60\,\text{cm}$. Substituting this (and the given object distance) into Eq. 35-6 gives $m = -2.0$. Since $i > 0$, the answer is "yes" (the image is real). Since $m < 0$, our answer is "yes" (the image is inverted). The ray diagram would be similar to Fig. 35-8(c) in the textbook.

(d) Since $m < 0$, the image is inverted. With that in mind, we examine the various possibilities in Figs. 35-6, 35-8 and 35-9, and note that an inverted image (for reflections from a single mirror) can only occur if the mirror is concave (and if $p > f$). Next, we find i from Eq. 35-6 (which yields $i = 30\,\text{cm}$) and then use this value (and Eq. 35-4) to compute the focal length; we obtain $f = +20\,\text{cm}$. Then, Eq. 35-3 gives $r = +40\,\text{cm}$. As already noted, $i = +30\,\text{cm}$. Yes, the image is real (since $i > 0$). Yes, the image is inverted (as already noted). The ray diagram would be similar to Figs. 35-9(a) and (b) in the textbook.

(e) Since $r < 0$ then (by Eq. 35-3) $f < 0$, which means the mirror is convex. The focal length is $f = r/2 = -20\,\text{cm}$. Eq. 35-4 leads to $p = +20\,\text{cm}$, and Eq. 35-6 gives $m = +0.50$. No, the image is virtual. No, the image is upright. The ray diagram would be similar to Figs. 35-9(c) and (d) in the textbook.

(f) Since $0 < m < 1$, the image is upright but smaller than the object. With that in mind, we examine the various possibilities in Figs. 35-6, 35-8 and 35-9, and note that such an image (for reflections from a single mirror) can only occur if the mirror is convex. Thus, we must put a minus sign in front of the "20" value given for f. Eq. 35-3 then gives $r = -40\,\text{cm}$. To solve for i and p we must set up Eq. 35-4 and Eq. 35-6 as a simultaneous set and solve for the two unknowns. The results are $i = -18\,\text{cm}$ and $p = +180\,\text{cm}$. No, the image is virtual (since $i < 0$). No, the image is upright (as already noted). The ray diagram would be similar to Figs. 35-9(c) and (d) in the textbook.

(g) Knowing the mirror is convex means we must put a minus sign in front of the "40" value given for r. Then, Eq. 35-3 yields $f = r/2 = -20\,\text{cm}$. The fact that the mirror is convex also means that we need to insert a minus sign in front of the "4.0" value given for i, since the image in this case must be virtual (see Figs. 35-6, 35-8 and 35-9). Eq. 35-4 leads to $p = +5.0\,\text{cm}$, and Eq. 35-6 gives $m = +0.8$. No, the image is virtual. No, the image is upright. The ray diagram would be similar to Figs. 35-9(c) and (d) in the textbook.

(h) Since the image is inverted, we can scan Figs. 35-6, 35-8 and 35-9 in the textbook and find that the mirror must be concave. This also implies that we must put a minus sign in front of the "0.50" value given for m. To solve for f, we first find $i = +12\,\text{cm}$ from Eq. 35-6 and plug into Eq. 35-4; the result is $f = +8\,\text{cm}$. Thus, $r = 2f = +16\,\text{cm}$. Yes, the image is real (since $i > 0$). The ray diagram would be similar to Figs. 35-9(a) and (b) in the textbook.

11. (a) Suppose one end of the object is a distance p from the mirror and the other end is a distance $p + L$. The position i_1 of the image of the first end is given by

$$\frac{1}{p} + \frac{1}{i_1} = \frac{1}{f}$$

where f is the focal length of the mirror. Thus,

$$i_1 = \frac{fp}{p - f} .$$

The image of the other end is located at

$$i_2 = \frac{f(p + L)}{p + L - f} ,$$

so the length of the image is

$$L' = i_1 - i_2 = \frac{fp}{p - f} - \frac{f(p + L)}{p + L - f} = \frac{f^2 L}{(p - f)(p + L - f)} .$$

Since the object is short compared to $p - f$, we may neglect the L in the denominator and write

$$L' = L\left(\frac{f}{p - f}\right)^2 .$$

(b) The lateral magnification is $m = -i/p$ and since $i = fp/(p-f)$, this can be written $m = -f/(p-f)$. The longitudinal magnification is

$$m' = \frac{L'}{L} = \left(\frac{f}{p-f}\right)^2 = m^2 .$$

12. (a) From Eqs. 35-3 and 35-4, we obtain $i = pf/(p-f) = pr/(2p-r)$. Differentiating both sides with respect to time and using $v_O = -dp/dt$, we find

$$v_I = \frac{di}{dt} = \frac{d}{dt}\left(\frac{pr}{2p-r}\right) = \frac{-rv_O(2p-r) + 2v_O pr}{(2p-r)^2} = \left(\frac{r}{2p-r}\right)^2 v_O .$$

(b) If $p = 30\,\text{cm}$, we obtain

$$v_I = \left[\frac{15\,\text{cm}}{2(30\,\text{cm}) - 15\,\text{cm}}\right]^2 (5.0\,\text{cm/s}) = 0.56\ \text{cm/s} .$$

(c) If $p = 8.0\,\text{cm}$, we obtain

$$v_I = \left[\frac{15\,\text{cm}}{2(8.0\,\text{cm}) - 15\,\text{cm}}\right]^2 (5.0\,\text{cm/s}) = 1.1 \times 10^3\ \text{cm/s} .$$

(d) If $p = 1.0\,\text{cm}$, we obtain

$$v_I = \left[\frac{15\,\text{cm}}{2(1.0\,\text{cm}) - 15\,\text{cm}}\right]^2 (5.0\,\text{cm/s}) = 6.7\ \text{cm/s} .$$

13. (a) We use Eq. 35-8 and note that $n_1 = n_{\text{air}} = 1.00$, $n_2 = n$, $p = \infty$, and $i = 2r$:

$$\frac{1.00}{\infty} + \frac{n}{2r} = \frac{n-1}{r} .$$

We solve for the unknown index: $n = 2.00$.

(b) Now $i = r$ so Eq. 35-8 becomes

$$\frac{n}{r} = \frac{n-1}{r} ,$$

which is not valid unless $n \to \infty$ or $r \to \infty$. It is impossible to focus at the center of the sphere.

14. We remark that the sign convention for r (for these refracting surfaces) is the opposite of what was used for mirrors. This point is discussed in §35-5.

(a) We use Eq. 35-8:

$$i = n_2\left(\frac{n_2 - n_1}{r} - \frac{n_1}{p}\right)^{-1} = 1.5\left(\frac{1.5 - 1.0}{30\,\text{cm}} - \frac{1.0}{10\,\text{cm}}\right)^{-1} = -18\ \text{cm} .$$

The image is virtual and upright. The ray diagram would be similar to Fig. 35-10(c) in the textbook.

(b) We manipulate Eq. 35-8 to find r:

$$r = (n_2 - n_1)\left(\frac{n_1}{p} + \frac{n_2}{i}\right)^{-1} = (1.5 - 1.0)\left(\frac{1.0}{10} + \frac{1.5}{-13}\right)^{-1} = -32.5\ \text{cm}$$

which should be rounded to two significant figures. The image is virtual and upright. The ray diagram would be similar to Fig. 35-10(e) in the textbook, but with the object and the image placed closer to the surface.

(c) We manipulate Eq. 35-8 to find p:

$$p = \frac{n_1}{\frac{n_2-n_1}{r} - \frac{n_2}{i}} = \frac{1.0}{\frac{1.5-1.0}{30} - \frac{1.5}{600}} = 71 \text{ cm} .$$

The image is real and inverted. The ray diagram would be similar to Fig. 35-10(a) in the textbook.

(d) We manipulate Eq. 35-8 to separate the indices:

$$\begin{aligned} n_2\left(\frac{1}{r} - \frac{1}{i}\right) &= \left(\frac{n_1}{p} + \frac{n_1}{r}\right) \\ n_2\left(\frac{1}{-20} - \frac{1}{-20}\right) &= \left(\frac{1.0}{20} + \frac{1.0}{-20}\right) \\ n_2(0) &= 0 \end{aligned}$$

which is identically satisfied for any choice of n_2. The ray diagram would be similar to Fig. 35-10(d) in the textbook, but with C, O and I together at the same point. The image is virtual and upright.

(e) We manipulate Eq. 35-8 to find r:

$$r = (n_2 - n_1)\left(\frac{n_1}{p} + \frac{n_2}{i}\right)^{-1} = (1.0 - 1.5)\left(\frac{1.5}{10} + \frac{1.0}{-6.0}\right)^{-1} = 30 \text{ cm} .$$

The image is virtual and upright. The ray diagram would be similar to Fig. 35-10(f) in the textbook, but with the object and the image located closer to the surface.

(f) We manipulate Eq. 35-8 to find p:

$$p = \frac{n_1}{\frac{n_2-n_1}{r} - \frac{n_2}{i}} = \frac{1.5}{\frac{1.0-1.5}{-30} - \frac{1.0}{-7.5}} = 10 \text{ cm} .$$

The image is virtual and upright. The ray diagram would be similar to Fig. 35-10(d) in the textbook.

(g) We manipulate Eq. 35-8 to find the image distance:

$$i = n_2\left(\frac{n_2 - n_1}{r} - \frac{n_1}{p}\right)^{-1} = 1.0\left(\frac{1.0 - 1.5}{30\,\text{cm}} - \frac{1.5}{70\,\text{cm}}\right)^{-1} = -26 \text{ cm} .$$

The image is virtual and upright. The ray diagram would be similar to Fig. 35-10(f) in the textbook.

(h) We manipulate Eq. 35-8 to separate the indices:

$$\begin{aligned} n_2\left(\frac{1}{r} - \frac{1}{i}\right) &= \left(\frac{n_1}{p} + \frac{n_1}{r}\right) \\ n_2\left(\frac{1}{-30} - \frac{1}{600}\right) &= \left(\frac{1.5}{100} + \frac{1.5}{-30}\right) \\ n_2(-0.035) &= -0.035 \end{aligned}$$

which implies $n_2 = 1.0$. The ray diagram would be similar to Fig. 35-10(b) in the textbook, but with C, O and I together at the same point. The image is real and inverted.

15. The water is medium 1, so $n_1 = n_w$ which we simply write as n. The air is medium 2, for which $n_2 \approx 1$. We refer points where the light rays strike the water surface as A (on the left side of Fig. 35-32) and B (on the right side of the picture). The point midway between A and B (the center point in the picture) is C. The penny P is directly below C, and the location of the "apparent" or Virtual penny is V. We note that the angle $\angle CVB$ (the same as $\angle CVA$) is equal to θ_2, and the angle $\angle CPB$ (the same as $\angle CPA$) is

equal to θ_1. The triangles CVB and CPB share a common side, the horizontal distance from C to B (which we refer to as x). Therefore,

$$\tan\theta_2 = \frac{x}{d_a} \quad \text{and} \quad \tan\theta_1 = \frac{x}{d} \ .$$

Using the small angle approximation (so a ratio of tangents is nearly equal to a ratio of sines) and the law of refraction, we obtain

$$\begin{aligned} \frac{\tan\theta_2}{\tan\theta_1} &\approx \frac{\sin\theta_2}{\sin\theta_1} \\ \frac{\frac{x}{d_a}}{\frac{x}{d}} &\approx \frac{n_1}{n_2} \\ \frac{d}{d_a} &\approx n \end{aligned}$$

which yields the desired relation: $d_a = d/n$.

16. First, we note that – *relative to the water* – the index of refraction of the carbon tetrachloride should be thought of as $n = 1.46/1.33 = 1.1$ (this notation is chosen to be consistent with problem 15). Now, if the observer were in the water, directly above the 40 mm deep carbon tetrachloride layer, then the apparent depth of the penny as measured below the surface of the carbon tetrachloride is $d_a = 40\,\text{mm}/1.1 = 36.4\,\text{mm}$. This "apparent penny" serves as an "object" for the rays propagating upward through the 20 mm layer of water, where this "object" should be thought of as being $20\,\text{mm} + 36.4\,\text{mm} = 56.4\,\text{mm}$ from the top surface. Using the result of problem 15 again, we find the perceived location of the penny, for a person at the normal viewing position above the water, to be $56.4\,\text{mm}/1.33 = 42\,\text{mm}$ below the water surface.

17. We solve Eq. 35-9 for the image distance i: $i = pf/(p - f)$. The lens is diverging, so its focal length is $f = -30\,\text{cm}$. The object distance is $p = 20\,\text{cm}$. Thus,

$$i = \frac{(20\,\text{cm})(-30\,\text{cm})}{(20\,\text{cm}) - (-30\,\text{cm})} = -12\,\text{cm} \ .$$

The negative sign indicates that the image is virtual and is on the same side of the lens as the object. The ray diagram, drawn to scale, is shown on the right.

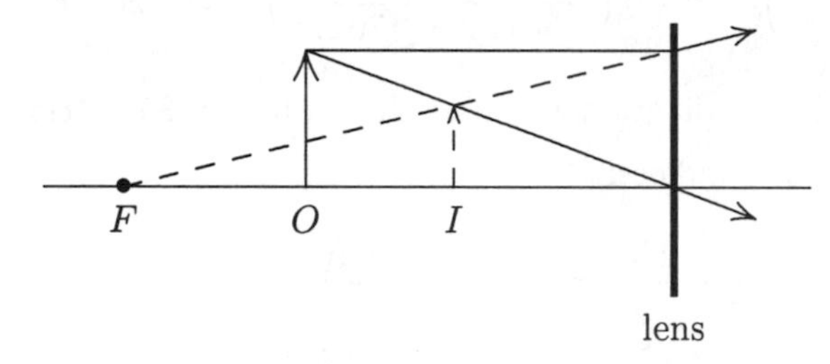

18. Let the diameter of the Sun be d_s and that of the image be d_i. Then, Eq. 35-5 leads to

$$\begin{aligned} d_i &= |m|d_s = \left(\frac{i}{p}\right) d_s \approx \left(\frac{f}{p}\right) d_s \\ &= \frac{(20.0 \times 10^{-2}\,\text{m})(2)(6.96 \times 10^{8}\,\text{m})}{1.50 \times 10^{11}\,\text{m}} \\ &= 1.86 \times 10^{-3}\,\text{m} = 1.86\,\text{mm} \ . \end{aligned}$$

19. We use the lens maker's equation, Eq. 35–10:

$$\frac{1}{f} = (n-1)\left(\frac{1}{r_1} - \frac{1}{r_2}\right)$$

where f is the focal length, n is the index of refraction, r_1 is the radius of curvature of the first surface encountered by the light and r_2 is the radius of curvature of the second surface. Since one surface has twice the radius of the other and since one surface is convex to the incoming light while the other is concave, set $r_2 = -2r_1$ to obtain

$$\frac{1}{f} = (n-1)\left(\frac{1}{r_1} + \frac{1}{2r_1}\right) = \frac{3(n-1)}{2r_1} .$$

We solve for r_1:

$$r_1 = \frac{3(n-1)f}{2} = \frac{3(1.5-1)(60\,\text{mm})}{2} = 45\,\text{mm} .$$

The radii are 45 mm and 90 mm.

20. (a) We use Eq. 35-10:

$$f = \left[(n-1)\left(\frac{1}{r_1} - \frac{1}{r_2}\right)\right]^{-1} = \left[(1.5-1)\left(\frac{1}{\infty} - \frac{1}{-20\,\text{cm}}\right)\right]^{-1} = +40\text{ cm} .$$

(b) From Eq. 35-9,

$$i = \left(\frac{1}{f} - \frac{1}{p}\right)^{-1} = \left(\frac{1}{40\,\text{cm}} - \frac{1}{40\,\text{cm}}\right)^{-1} = \infty .$$

21. For a thin lens, $(1/p) + (1/i) = (1/f)$, where p is the object distance, i is the image distance, and f is the focal length. We solve for i:

$$i = \frac{fp}{p-f} .$$

Let $p = f + x$, where x is positive if the object is outside the focal point and negative if it is inside. Then,

$$i = \frac{f(f+x)}{x} .$$

Now let $i = f + x'$, where x' is positive if the image is outside the focal point and negative if it is inside. Then,

$$x' = i - f = \frac{f(f+x)}{x} - f = \frac{f^2}{x}$$

and $xx' = f^2$.

22. We solve Eq. 35-9 for the image distance:

$$i = \left(\frac{1}{f} - \frac{1}{p}\right)^{-1} = \frac{fp}{p-f} .$$

The height of the image is thus

$$h_i = mh_p = \left(\frac{i}{p}\right)h_p = \frac{fh_p}{p-f} = \frac{(75\,\text{mm})(1.80\,\text{m})}{27\,\text{m} - 0.075\,\text{m}} = 5.0\text{ mm} .$$

23. Using Eq. 35-9 and noting that $p + i = d = 44\,\text{cm}$, we obtain $p^2 - dp + df = 0$. Therefore,

$$p = \frac{1}{2}(d \pm \sqrt{d^2 - 4df}) = 22\,\text{cm} \pm \frac{1}{2}\sqrt{(44\,\text{cm})^2 - 4(44\,\text{cm})(11\,\text{cm})} = 22\text{ cm} .$$

24. (a) Since this is a converging lens ("C") then $f > 0$, so we should put a plus sign in front of the "10" value given for the focal length. There is not enough information to determine r_1 and r_2. Eq. 35-9 gives

$$i = \frac{1}{\frac{1}{f} - \frac{1}{p}} = \frac{1}{\frac{1}{10} - \frac{1}{20}} = +20 \text{ cm} .$$

There is insufficient information for the determination of n. From Eq. 35-6, $m = -20/20 = -1.0$. The image is real (since $i > 0$) and inverted (since $m < 0$). The ray diagram would be similar to Fig. 35-14(a) in the textbook.

(b) Since $f > 0$, this is a converging lens ("C"). There is not enough information to determine r_1 and r_2. Eq. 35-9 gives

$$i = \frac{1}{\frac{1}{f} - \frac{1}{p}} = \frac{1}{\frac{1}{10} - \frac{1}{5}} = -10 \text{ cm} .$$

There is insufficient information for the determination of n. From Eq. 35-6, $m = -(-10)/5 = +2.0$. The image is virtual (since $i < 0$) and upright (since $m > 0$). The ray diagram would be similar to Fig. 35-14(b) in the textbook.

(c) We are told the magnification is positive and greater than 1. Scanning the single-lens-image figures in the textbook (Figs. 35-13, 35-14 and 35-16), we see that such a magnification (which implies an upright image larger than the object) is only possible if the lens is of the converging ("C") type (and if $p < f$). Thus, we should put a plus sign in front of the "10" value given for the focal length. Eq. 35-9 gives

$$i = \frac{1}{\frac{1}{f} - \frac{1}{p}} = \frac{1}{\frac{1}{10} - \frac{1}{5}} = -10 \text{ cm} ,$$

which implies the image is virtual. There is insufficient information for the determinations of n, r_1 and r_2. The ray diagram would be similar to Fig. 35-14(b) in the textbook.

(d) We are told the magnification is less than 1, and we note that $p < |f|$). Scanning Figs. 35-13, 35-14 and 35-16, we see that such a magnification (which implies an image smaller than the object) and object position (being fairly close to the lens) are simultaneously possible only if the lens is of the diverging ("D") type. Thus, we should put a minus sign in front of the "10" value given for the focal length. Eq. 35-9 gives

$$i = \frac{1}{\frac{1}{f} - \frac{1}{p}} = \frac{1}{\frac{1}{-10} - \frac{1}{5}} = -3.3 \text{ cm} ,$$

which implies the image is virtual (and upright). There is insufficient information for the determinations of n, r_1 and r_2. The ray diagram would be similar to Fig. 35-14(c) in the textbook.

(e) Eq. 35-10 yields $f = \frac{1}{n-1}(1/r_1 - 1/r_2)^{-1} = +30$ cm. Since $f > 0$, this must be a converging ("C") lens. From Eq. 35-9, we obtain

$$i = \frac{1}{\frac{1}{f} - \frac{1}{p}} = \frac{1}{\frac{1}{30} - \frac{1}{10}} = -15 \text{ cm} .$$

Eq. 35-6 yields $m = -(-15)/10 = +1.5$. Therefore, the image is virtual ($i < 0$) and upright ($m > 0$). The ray diagram would be similar to Fig. 35-14(b) in the textbook.

(f) Eq. 35-10 yields $f = \frac{1}{n-1}(1/r_1 - 1/r_2)^{-1} = -30$ cm. Since $f < 0$, this must be a diverging ("D") lens. From Eq. 35-9, we obtain

$$i = \frac{1}{\frac{1}{f} - \frac{1}{p}} = \frac{1}{\frac{1}{-30} - \frac{1}{10}} = -7.5 \text{ cm} .$$

Eq. 35-6 yields $m = -(-7.5)/10 = +0.75$. Therefore, the image is virtual ($i < 0$) and upright ($m > 0$). The ray diagram would be similar to Fig. 35-14(c) in the textbook.

(g) Eq. 35-10 yields $f = \frac{1}{n-1}(1/r_1 - 1/r_2)^{-1} = -120$ cm. Since $f < 0$, this must be a diverging ("D") lens. From Eq. 35-9, we obtain

$$i = \frac{1}{\frac{1}{f} - \frac{1}{p}} = \frac{1}{\frac{1}{-120} - \frac{1}{10}} = -9.2 \text{ cm} .$$

Eq. 35-6 yields $m = -(-9.2)/10 = +0.92$. Therefore, the image is virtual ($i < 0$) and upright ($m > 0$). The ray diagram would be similar to Fig. 35-14(c) in the textbook.

(h) We are told the absolute value of the magnification is 0.5 and that the image was upright. Thus, $m = +0.5$. Using Eq. 35-6 and the given value of p, we find $i = -5.0$ cm; it is a virtual image. Eq. 35-9 then yields the focal length: $f = -10$ cm. Therefore, the lens is of the diverging ("D") type. The ray diagram would be similar to Fig. 35-14(c) in the textbook. There is insufficient information for the determinations of n, r_1 and r_2.

(i) Using Eq. 35-6 (which implies the image is inverted) and the given value of p, we find $i = -mp = +5.0$ cm; it is a real image. Eq. 35-9 then yields the focal length: $f = +3.3$ cm. Therefore, the lens is of the converging ("C") type. The ray diagram would be similar to Fig. 35-14(a) in the textbook. There is insufficient information for the determinations of n, r_1 and r_2.

25. For an object in front of a thin lens, the object distance p and the image distance i are related by $(1/p) + (1/i) = (1/f)$, where f is the focal length of the lens. For the situation described by the problem, all quantities are positive, so the distance x between the object and image is $x = p + i$. We substitute $i = x - p$ into the thin lens equation and solve for x:

$$x = \frac{p^2}{p - f} .$$

To find the minimum value of x, we set $dx/dp = 0$ and solve for p. Since

$$\frac{dx}{dp} = \frac{p(p - 2f)}{(p - f)^2} ,$$

the result is $p = 2f$. The minimum distance is

$$x_{\text{min}} = \frac{p^2}{p - f} = \frac{(2f)^2}{2f - f} = 4f .$$

This is a minimum, rather than a maximum, since the image distance i becomes large without bound as the object approaches the focal point.

26. (a) (b) (c) and (d) Our first step is to form the image from the first lens. With $p_1 = 10$ cm and $f_1 = -15$ cm, Eq. 35-9 leads to

$$\frac{1}{p_1} + \frac{1}{i_1} = \frac{1}{f_1} \implies i_1 = -6 \text{ cm} .$$

The corresponding magnification is $m_1 = -i_1/p_1 = 0.6$. This image serves the role of "object" for the second lens, with $p_2 = 12 + 6 = 18$ cm, and $f_2 = 12$ cm. Now, Eq. 35-9 leads to

$$\frac{1}{p_2} + \frac{1}{i_2} = \frac{1}{f_2} \implies i_2 = 36 \text{ cm}$$

with a corresponding magnification of $m_2 = -i_2/p_2 = -2$, resulting in a net magnification of $m = m_1 m_2 = -1.2$. The fact that m is positive means that the orientation of the final image is inverted with respect to the (original) object. The height of the final image is (in absolute value) $(1.2)(1.0\,\text{cm}) = 1.2$ cm. The fact that i_2 is positive means that the final image is real.

27. Without the diverging lens (lens 2), the real image formed by the converging lens (lens 1) is located at a distance

$$i_1 = \left(\frac{1}{f_1} - \frac{1}{p_1}\right)^{-1} = \left(\frac{1}{20\,\text{cm}} - \frac{1}{40\,\text{cm}}\right)^{-1} = 40\ \text{cm}$$

to the right of lens 1. This image now serves as an object for lens 2, with $p_2 = -(40\,\text{cm} - 10\,\text{cm}) = -30\,\text{cm}$. So

$$i_2 = \left(\frac{1}{f_2} - \frac{1}{p_2}\right)^{-1} = \left(\frac{1}{-15\,\text{cm}} - \frac{1}{-30\,\text{cm}}\right)^{-1} = -30\ \text{cm}\ .$$

Thus, the image formed by lens 2 is located 30 cm to the left of lens 2. It is virtual (since $i_2 < 0$). The magnification is $m = (-i_1/p_1) \times (-i_2/p_2) = +1$, so the image has the same size and orientation as the object.

28. (a) For the image formed by the first lens

$$i_1 = \left(\frac{1}{f_1} - \frac{1}{p_1}\right)^{-1} = \left(\frac{1}{10\,\text{cm}} - \frac{1}{20\,\text{cm}}\right)^{-1} = 20\ \text{cm}\ .$$

For the subsequent image formed by the second lens $p_2 = 30\,\text{cm} - 20\,\text{cm} = 10\,\text{cm}$, so

$$i_2 = \left(\frac{1}{f_2} - \frac{1}{p_2}\right)^{-1} = \left(\frac{1}{12.5\,\text{cm}} - \frac{1}{10\,\text{cm}}\right)^{-1} = -50\ \text{cm}\ .$$

Thus, the final image is 50 cm to the left of the second lens, which means that it coincides with the object. The magnification is

$$m = \left(\frac{i_1}{p_1}\right)\left(\frac{i_2}{p_2}\right) = \left(\frac{20\,\text{cm}}{20\,\text{cm}}\right)\left(\frac{-50\,\text{cm}}{10\,\text{cm}}\right) = -5.0\ ,$$

which means that the final image is five times larger than the original object.

(b) The ray diagram would be very similar to Fig. 35-17 in the textbook, except that the final image would be directly underneath the original object.

(c) and (d) It is virtual and inverted.

29. We place an object far away from the composite lens and find the image distance i. Since the image is at a focal point, $i = f$, where f equals the effective focal length of the composite. The final image is produced by two lenses, with the image of the first lens being the object for the second. For the first lens, $(1/p_1) + (1/i_1) = (1/f_1)$, where f_1 is the focal length of this lens and i_1 is the image distance for the image it forms. Since $p_1 = \infty$, $i_1 = f_1$. The thin lens equation, applied to the second lens, is $(1/p_2) + (1/i_2) = (1/f_2)$, where p_2 is the object distance, i_2 is the image distance, and f_2 is the focal length. If the thicknesses of the lenses can be ignored, the object distance for the second lens is $p_2 = -i_1$. The negative sign must be used since the image formed by the first lens is beyond the second lens if i_1 is positive. This means the object for the second lens is virtual and the object distance is negative. If i_1 is negative, the image formed by the first lens is in front of the second lens and p_2 is positive. In the thin lens equation, we replace p_2 with $-f_1$ and i_2 with f to obtain

$$-\frac{1}{f_1} + \frac{1}{f} = \frac{1}{f_2}$$

or

$$\frac{1}{f} = \frac{1}{f_1} + \frac{1}{f_2} = \frac{f_1 + f_2}{f_1 f_2}\ .$$

Thus,

$$f = \frac{f_1 f_2}{f_1 + f_2}\ .$$

30. (a) A convex (converging) lens, since a real image is formed.

(b) Since $i = d - p$ and $i/p = 1/2$,

$$p = \frac{2d}{3} = \frac{2(40.0\,\text{cm})}{3} = 26.7 \text{ cm} .$$

(c) The focal length is

$$f = \left(\frac{1}{i} + \frac{1}{p}\right)^{-1} = \left(\frac{1}{d/3} + \frac{1}{2d/3}\right)^{-1} = \frac{2d}{9} = \frac{2(40.0\,\text{cm})}{9} = 8.89\,\text{cm} .$$

31. (a) If the object distance is x, then the image distance is $D - x$ and the thin lens equation becomes

$$\frac{1}{x} + \frac{1}{D-x} = \frac{1}{f} .$$

We multiply each term in the equation by $fx(D-x)$ and obtain $x^2 - Dx + Df = 0$. Solving for x, we find that the two object distances for which images are formed on the screen are

$$x_1 = \frac{D - \sqrt{D(D-4f)}}{2} \quad \text{and} \quad x_2 = \frac{D + \sqrt{D(D-4f)}}{2} .$$

The distance between the two object positions is

$$d = x_2 - x_1 = \sqrt{D(D-4f)} .$$

(b) The ratio of the image sizes is the same as the ratio of the lateral magnifications. If the object is at $p = x_1$, the magnitude of the lateral magnification is

$$|m_1| = \frac{i_1}{p_1} = \frac{D - x_1}{x_1} .$$

Now $x_1 = \frac{1}{2}(D - d)$, where $d = \sqrt{D(D-f)}$, so

$$|m_1| = \frac{D - (D-d)/2}{(D-d)/2} = \frac{D+d}{D-d} .$$

Similarly, when the object is at x_2, the magnitude of the lateral magnification is

$$|m_2| = \frac{I_2}{p_2} = \frac{D - x_2}{x_2} = \frac{D - (D+d)/2}{(D+d)/2} = \frac{D-d}{D+d} .$$

The ratio of the magnifications is

$$\frac{m_2}{m_1} = \frac{(D-d)/(D+d)}{(D+d)/(D-d)} = \left(\frac{D-d}{D+d}\right)^2 .$$

32. The minimum diameter of the eyepiece is given by

$$d_{ey} = \frac{d_{ob}}{m_\theta} = \frac{75\,\text{mm}}{36} = 2.1 \text{ mm} .$$

33. (a) If L is the distance between the lenses, then according to Fig. 35-17, the tube length is $s = L - f_{ob} - f_{ey} = 25.0\,\text{cm} - 4.00\,\text{cm} - 8.00\,\text{cm} = 13.0\,\text{cm}$.

(b) We solve $(1/p) + (1/i) = (1/f_{ob})$ for p. The image distance is $i = f_{ob} + s = 4.00\,\text{cm} + 13.0\,\text{cm} = 17.0\,\text{cm}$, so

$$p = \frac{i f_{ob}}{i - f_{ob}} = \frac{(17.0\,\text{cm})(4.00\,\text{cm})}{17.0\,\text{cm} - 4.00\,\text{cm}} = 5.23 \text{ cm} .$$

(c) The magnification of the objective is

$$m = -\frac{i}{p} = -\frac{17.0\,\text{cm}}{5.23\,\text{cm}} = -3.25\ .$$

(d) The angular magnification of the eyepiece is

$$m_\theta = \frac{25\,\text{cm}}{f_{\text{ey}}} = \frac{25\,\text{cm}}{8.00\,\text{cm}} = 3.13\ .$$

(e) The overall magnification of the microscope is

$$M = mm_\theta = (-3.25)(3.13) = -10.2\ .$$

34. (a) Without the magnifier, $\theta = h/P_n$ (see Fig. 35-16). With the magnifier, letting $p = P_n$ and $i = -|i| = -P_n$, we obtain

$$\frac{1}{p} = \frac{1}{f} - \frac{1}{i} = \frac{1}{f} + \frac{1}{|i|} = \frac{1}{f} + \frac{1}{P_n}\ .$$

Consequently,

$$m_\theta = \frac{\theta'}{\theta} = \frac{h/p}{h/P_n} = \frac{1/f + 1/P_n}{1/P_n} = 1 + \frac{P_n}{f} = 1 + \frac{25\,\text{cm}}{f}\ .$$

(b) Now $i = -|i| \to -\infty$, so $1/p + 1/i = 1/p = 1/f$ and

$$m_\theta = \frac{\theta'}{\theta} = \frac{h/p}{h/P_n} = \frac{1/f}{1/P_n} = \frac{P_n}{f} = \frac{25\,\text{cm}}{f}\ .$$

(c) For $f = 10\,\text{cm}$,

$$m_\theta = 1 + \frac{25\,\text{cm}}{10\,\text{cm}} = 3.5\ (\text{case (a)}) \quad \text{and} \quad \frac{25\,\text{cm}}{10\,\text{cm}} = 2.5\ (\text{case (b)})\ .$$

35. (a) When the eye is relaxed, its lens focuses far-away objects on the retina, a distance i behind the lens. We set $p = \infty$ in the thin lens equation to obtain $1/i = 1/f$, where f is the focal length of the relaxed effective lens. Thus, $i = f = 2.50\,\text{cm}$. When the eye focuses on closer objects, the image distance i remains the same but the object distance and focal length change. If p is the new object distance and f' is the new focal length, then

$$\frac{1}{p} + \frac{1}{i} = \frac{1}{f'}\ .$$

We substitute $i = f$ and solve for f':

$$f' = \frac{pf}{f + p} = \frac{(40.0\,\text{cm})(2.50\,\text{cm})}{40.0\,\text{cm} + 2.50\,\text{cm}} = 2.35\ \text{cm}\ .$$

(b) Consider the lensmaker's equation

$$\frac{1}{f} = (n-1)\left(\frac{1}{r_1} - \frac{1}{r_2}\right)$$

where r_1 and r_2 are the radii of curvature of the two surfaces of the lens and n is the index of refraction of the lens material. For the lens pictured in Fig. 35-34, r_1 and r_2 have about the same magnitude, r_1 is positive, and r_2 is negative. Since the focal length decreases, the combination $(1/r_1) - (1/r_2)$ must increase. This can be accomplished by decreasing the magnitudes of both radii.

36. Refer to Fig. 35-17. For the intermediate image $p = 10\,\text{mm}$ and $i = (f_{\text{ob}}+s+f_{\text{ey}})-f_{\text{ey}} = 300\,\text{m}-50\,\text{mm} = 250\,\text{mm}$, so

$$\frac{1}{f_{\text{ob}}} = \frac{1}{i} + \frac{1}{p} = \frac{1}{250\,\text{mm}} + \frac{1}{10\,\text{mm}} \implies f_{\text{ob}} = 9.62\ \text{mm}\ ,$$

and $s = (f_{\text{ob}} + s + f_{\text{ey}}) - f_{\text{ob}} - f_{\text{ey}} = 300\,\text{mm} - 9.62\,\text{mm} - 50\,\text{mm} = 240\,\text{mm}$. Then from Eq. 35-14,

$$M = -\frac{s}{f_{\text{ob}}}\frac{25\,\text{cm}}{f_{\text{ey}}} = -\left(\frac{240\,\text{mm}}{9.62\,\text{mm}}\right)\left(\frac{150\,\text{mm}}{50\,\text{mm}}\right) = -125\ .$$

37. (a) Now, the lens-film distance is

$$i = \left(\frac{1}{f} - \frac{1}{p}\right)^{-1} = \left(\frac{1}{5.0\,\text{cm}} - \frac{1}{100\,\text{cm}}\right)^{-1} = 5.3\ \text{cm}\ .$$

(b) The change in the lens-film distance is $5.3\,\text{cm} - 5.0\,\text{cm} = 0.30\,\text{cm}$.

38. We combine Eq. 35-4 and Eq. 35-6 and arrive at

$$m = -\frac{pf/(p-f)}{p} = \frac{1}{1-r} \qquad \text{where} \quad r = \frac{p}{f}$$

We emphasize that this r (for ratio) is not the radius of curvature. The magnification as a function of r is graphed below:

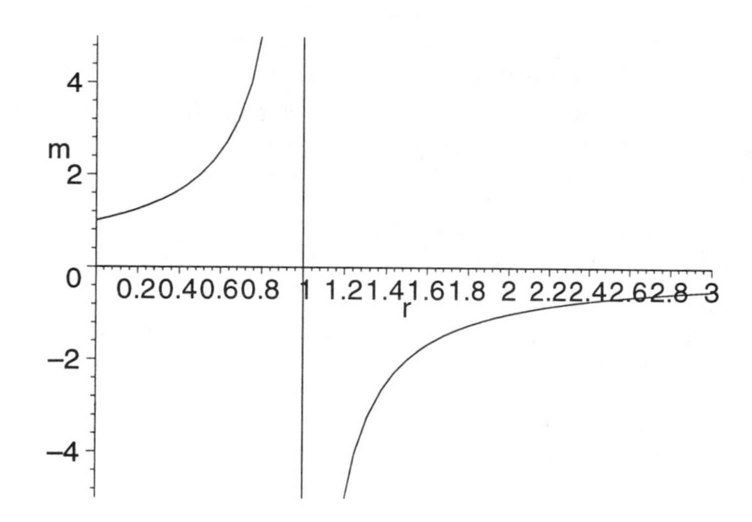

39. (a) The discussion in the textbook of the refracting telescope (a subsection of §35-7) applies to the Newtonian arrangement if we replace the objective lens of Fig. 35-18 with an objective mirror (with the light incident on it from the right). This might suggest that the incident light would be blocked by the person's head in Fig. 35-18, which is why Newton added the mirror M' in his design (to move the head and eyepiece out of the way of the incoming light). The beauty of the idea of characterizing both lenses and mirrors by focal lengths is that it is easy, in a case like this, to simply carry over the results of the objective-lens telescope to the objective-mirror telescope, so long as we replace a positive f device with another positive f device. Thus, the converging lens serving as the objective of Fig. 35-18 must be replaced (as Newton has done in Fig. 35-11) with a concave mirror. With this change of language, the discussion in the textbook leading up to Eq. 35-15 applies equally as well to the Newtonian telescope: $m_\theta = -f_{\text{ob}}/f_{\text{ey}}$.

(b) A meter stick (held perpendicular to the line of sight) at a distance of 2000 m subtends an angle of

$$\theta_{\text{stick}} \approx \frac{1\,\text{m}}{2000\,\text{m}} = 0.0005 \text{ rad} .$$

Multiplying this by the mirror focal length gives (16.8 m)(0.0005) = 8.4 mm for the size of the image.

(c) With $r = 10$ m, Eq. 35-3 gives $f_{\text{ob}} = 5$ m. Plugging this into (the absolute value of) Eq. 35-15 leads to $f_{\text{ey}} = 5/200 = 2.5$ cm.

40. (a) The "object" for the mirror which results in that box-image is equally in front of the mirror (4 cm). This object is actually the first image formed by the system (produced by the first transmission through the lens); in those terms, it corresponds to $i_1 = 10 - 4 = 6$ cm. Thus, with $f_1 = 2$ cm, Eq. 35-9 leads to

$$\frac{1}{p_1} + \frac{1}{i_1} = \frac{1}{f_1} \implies p_1 = 3.00 \text{ cm} .$$

(b) The previously mentioned box-image (4 cm behind the mirror) serves as an "object" (at $p_3 = 14$ cm) for the return trip of light through the lens ($f_3 = f_1 = 2$ cm). This time, Eq. 35-9 leads to

$$\frac{1}{p_3} + \frac{1}{i_3} = \frac{1}{f_3} \implies i_3 = 2.33 \text{ cm} .$$

41. (a) In this case $m > +1$ and we know we are dealing with a converging lens (producing a virtual image), so that our result for focal length should be positive. Since $|p + i| = 20$ cm and $i = -2p$, we find $p = 20$ cm and $i = -40$ cm. Substituting these into Eq. 35-9,

$$\frac{1}{p} + \frac{1}{i} = \frac{1}{f}$$

leads to $f = +40$ cm, which is positive as we expected.

(b) In this case $0 < m < 1$ and we know we are dealing with a diverging lens (producing a virtual image), so that our result for focal length should be negative. Since $|p + i| = 20$ cm and $i = -p/2$, we find $p = 40$ cm and $i = -20$ cm. Substituting these into Eq. 35-9 leads to $f = -40$ cm, which is negative as we expected.

42. (a) The first image is figured using Eq. 35-8, with $n_1 = 1$ (using the rounded-off value for air) and $n_2 = 8/5$.

$$\frac{1}{p} + \frac{8}{5i} = \frac{1.6 - 1}{r}$$

For a "flat lens" $r = \infty$, so we obtain $i = -8p/5 = -64/5$ (with the unit cm understood) for that object at $p = 10$ cm. Relative to the second surface, this image is at a distance of $3 + 64/5 = 79/5$. This serves as an object in order to find the final image, using Eq. 35-8 again (and $r = \infty$) but with $n_1 = 8/5$ and $n_2 = 4/3$.

$$\frac{8}{5p'} + \frac{4}{3i'} = 0$$

which produces (for $p' = 79/5$) $i' = -5p/6 = -79/6 \approx -13.2$. This means the observer appears $13.2 + 6.8 = 20$ cm from the fish.

(b) It is straightforward to "reverse" the above reasoning, the result being that the final fish-image is 7.0 cm to the right of the air-wall interface, and thus 15 cm from the observer.

43. (a) (b) and (c) Since $m = +0.250$, we have $i = -0.25p$ which indicates that the image is virtual (as well as being diminished in size). We conclude from this that the mirror is convex and that $f < 0$; in fact, $f = -2.00$ cm. Substituting $i = -p/4$ into Eq. 35-4 produces

$$\frac{1}{p} - \frac{4}{p} = -\frac{3}{p} = \frac{1}{f}$$

Therefore, we find $p = 6.00$ cm and $i = -1.50$ cm.

44. (a) A parallel ray of light focuses at the focal point behind the lens. In the case of farsightedness we need to bring the focal point closer. That is, we need to reduce the focal length. From problem 29, we know that we need to use a converging lens of certain focal length $f_1 > 0$ which, when combined with the eye of focal length f_2, gives $f = f_1 f_2/(f_1 + f_2) < f_2$. Similarly, we see that in the case of nearsightness we need to do a similar computation but with a diverging ($f_1 < 0$) lens.

(b) In this case, the unaided eyes are able to accommodate rays of light coming from distant (and medium-range) sources, but not from close ones. The person (not wearing glasses) is able to see far (not near), so the person is farsighted.

(c) The bifocal glasses can provide suitable corrections for different types of visual defects that prove a hindrance in different situations, such as reading (difficult for the farsighted individual) and viewing a distant object (difficult for a nearsighted individual).

45. (a) We use Eq. 35-10, with the conventions for signs discussed in §35-5 and §35-6.

(b) For the bi-convex (or double convex) case, we have

$$f = \left[(n-1)\left(\frac{1}{r_1} - \frac{1}{r_2}\right)\right]^{-1} = \left[(1.5-1)\left(\frac{1}{40\text{ cm}} - \frac{1}{-40\text{ cm}}\right)\right]^{-1} = 40\text{ cm} .$$

Since $f > 0$ the lens forms a real image of the Sun.

(c) For the planar convex lens, we find

$$f = \left[(1.5-1)\left(\frac{1}{\infty} - \frac{1}{-40\text{ cm}}\right)\right]^{-1} = 80\text{ cm} ,$$

and the image formed is real (since $f > 0$).

(d) Now

$$f = \left[(1.5-1)\left(\frac{1}{40\text{ cm}} - \frac{1}{60\text{ cm}}\right)\right]^{-1} = 240\text{ cm} ,$$

and the image formed is real (since $f > 0$).

(e) For the bi-concave lens, the focal length is

$$f = \left[(1.5-1)\left(\frac{1}{-40\text{ cm}} - \frac{1}{40\text{ cm}}\right)\right]^{-1} = -40\text{ cm} ,$$

and the image formed is virtual (since $f < 0$).

(f) In this case,

$$f = \left[(1.5-1)\left(\frac{1}{\infty} - \frac{1}{40\text{ cm}}\right)\right]^{-1} = -80\text{ cm} ,$$

and the image formed is virtual (since $f < 0$).

(g) Now

$$f = \left[(1.5-1)\left(\frac{1}{60\text{ cm}} - \frac{1}{40\text{ cm}}\right)\right]^{-1} = -240\text{ cm} ,$$

and the image formed is virtual (since $f < 0$).

46. Of course, the shortest possible path between A and B is the straight line path which does not go to the mirror at all. In this problem, we are concerned with only those paths which do strike the mirror. The problem statement suggests that we turn our attention to the mirror-image point of A (call it A') and requests that we construct a proof without calculus. We can see that the length of any line segment AP drawn from A to the mirror (at point P on the mirror surface) is the same as the length of its "mirror segment" $A'P$ drawn from A' to that point P. Thus, the total length of the light path from A to P

to B is the same as the total length of segments drawn from A' to P to B. Now, we dismissed (in the first sentence of this solution) the possibility of a straight line path directly from A to B because it does not strike the mirror. However, we *can* construct a straight line path from A' to B which does intersect the mirror surface! Any other pair of segments ($A'P$ and PB) would give greater total length than the straight path (with $A'P$ and PB collinear), so if the straight path $A'B$ obeys the law of reflection, then we have our proof. Now, since $A'P$ is the mirror-twin of AP, then they both approach the mirror surface with the same angle α (one from the front side and the other from the back side). And since $A'P$ is collinear with PB, then PB also makes the same angle α with respect to the mirror surface (by vertex angles). If AP and PB are each α degrees away from the front of the mirror, then they are each θ degrees (where θ is the complement of α) measured from the normal axis. Thus, the law of reflection is consistent with the concept of the shortest light path.

47. (a) (b) and (c) Our first step is to form the image from the first lens. With $p_1 = 4$ cm and $f_1 = -4$ cm, Eq. 35-9 leads to

$$\frac{1}{p_1} + \frac{1}{i_1} = \frac{1}{f_1} \implies i_1 = -2 \text{ cm} .$$

The corresponding magnification is $m_1 = -i_1/p_1 = 1/2$. This image serves the role of "object" for the second lens, with $p_2 = 10 + 2 = 12$ cm, and $f_2 = -4$ cm. Now, Eq. 35-9 leads to

$$\frac{1}{p_2} + \frac{1}{i_2} = \frac{1}{f_2} \implies i_2 = -3.00 \text{ cm}$$

with a corresponding magnification of $m_2 = -i_2/p_2 = 1/4$, resulting in a net magnification of $m = m_1 m_2 = 1/8$. The fact that m is positive means that the orientation of the final image is the same as the (original) object. The fact that i_2 is negative means that the final image is virtual.

48. (a) (b) (c) and (d) Our first step is to form the image from the first lens. With $p_1 = 3$ cm and $f_1 = +4$ cm, Eq. 35-9 leads to

$$\frac{1}{p_1} + \frac{1}{i_1} = \frac{1}{f_1} \implies i_1 = -12 \text{ cm} .$$

The corresponding magnification is $m_1 = -i_1/p_1 = 4$. This image serves the role of "object" for the second lens, with $p_2 = 8 + 12 = 20$ cm, and $f_2 = -4$ cm. Now, Eq. 35-9 leads to

$$\frac{1}{p_2} + \frac{1}{i_2} = \frac{1}{f_2} \implies i_2 = -3.33 \text{ cm}$$

with a corresponding magnification of $m_2 = -i_2/p_2 = 1/6$, resulting in a net magnification of $m = m_1 m_2 = 2/3$. The fact that m is positive means that the orientation of the final image is the same as the (original) object. The fact that i_2 is negative means that the final image is virtual (and therefore to the left of the second lens).

49. Since $0 < m < 1$, we conclude the lens is of the diverging type (so $f = -40$ cm). Thus, substituting $i = -3p/10$ into Eq. 35-9 produces

$$\frac{1}{p} - \frac{10}{3p} = -\frac{7}{3p} = \frac{1}{f} .$$

Therefore, we find $p = 93.3$ cm and $i = -28.0$ cm.

50. (a) We use Eq. 35-8 (and Fig. 35-10(b) is useful), with $n_1 = 1$ (using the rounded-off value for air) and $n_2 = 1.5$.

$$\frac{1}{p} + \frac{1.5}{i} = \frac{1.5 - 1}{r}$$

Using the sign convention for r stated in the paragraph following Eq. 35-8 (so that $r = +6.0$ cm), we obtain $i = -90$ cm for objects at $p = 10$ cm. Thus, the object and image are 80 cm apart.

(b) The image distance i is negative with increasing magnitude as p increases from very small values to some value p_0 at which point $i \to -\infty$. Since $1/(-\infty) = 0$, the above equation yields

$$\frac{1}{p_0} = \frac{1.5 - 1}{r} \quad \Longrightarrow \quad p_0 = 2r .$$

Thus, the range for producing virtual images is $0 < p \leq 12$ cm.

51. (a) Since $m = +0.200$, we have $i = -0.2p$ which indicates that the image is virtual (as well as being diminished in size). We conclude from this that the mirror is convex (and that $f = -40.0$ cm).

(b) Substituting $i = -p/5$ into Eq. 35-4 produces

$$\frac{1}{p} - \frac{5}{p} = -\frac{4}{p} = \frac{1}{f} .$$

Therefore, we find $p = 160$ cm.

52. (a) First, the lens forms a real image of the object located at a distance

$$i_1 = \left(\frac{1}{f_1} - \frac{1}{p_1}\right)^{-1} = \left(\frac{1}{f_1} - \frac{1}{2f_1}\right)^{-1} = 2f_1$$

to the right of the lens, or at $p_2 = 2(f_1 + f_2) - 2f_1 = 2f_2$ in front of the mirror. The subsequent image formed by the mirror is located at a distance

$$i_2 = \left(\frac{1}{f_2} - \frac{1}{p_2}\right)^{-1} = \left(\frac{1}{f_2} - \frac{1}{2f_2}\right)^{-1} = 2f_2$$

to the left of the mirror, or at $p_1' = 2(f_1 + f_2) - 2f_2 = 2f_1$ to the right of the lens. The final image formed by the lens is that at a distance i_1' to the left of the lens, where

$$i_1' = \left(\frac{1}{f_1} - \frac{1}{p_1'}\right)^{-1} = \left(\frac{1}{f_1} - \frac{1}{2f_1}\right)^{-1} = 2f_1 .$$

This turns out to be the same as the location of the original object. The final image is real and inverted. The lateral magnification is

$$m = \left(-\frac{i_1}{p_1}\right)\left(-\frac{i_2}{p_2}\right)\left(-\frac{i_1'}{p_1'}\right) = \left(-\frac{2f_1}{2f_1}\right)\left(-\frac{2f_2}{2f_2}\right)\left(-\frac{2f_1}{2f_1}\right) = -1.0 .$$

(b) The ray diagram is shown below. We set the ratio $f_2/f_1 = 1/2$ for the purposes of this sketch. The intermediate images are not shown explicitly, but they are both located on the plane indicated by the dashed line.

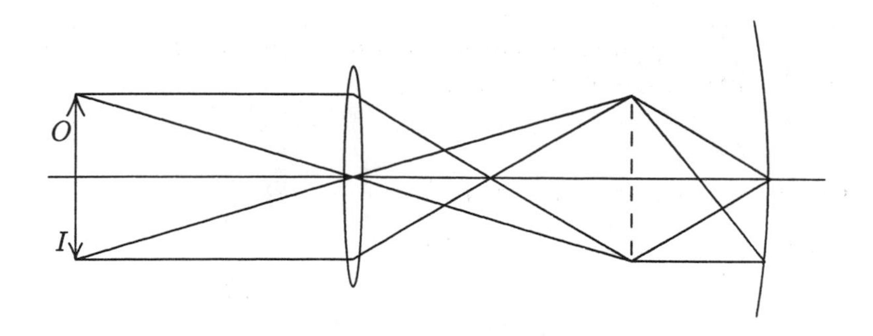

53. From Eq. 35-10, if

$$f \propto \left(\frac{1}{r_1} - \frac{1}{r_2}\right)^{-1} = \frac{r_1 r_2}{r_2 - r_1}$$

is positive (that is, if $r_2 > r_1$), then the lens is converging. Otherwise it is diverging.

(a) Converging, since $r_2 \to \infty$ and r_1 is finite (so $r_2 > r_1$).

(b) Diverging, since $r_1 \to \infty$ and r_2 is finite (so $r_2 < r_1$).

(c) Converging, since $r_2 > r_1$.

(d) Diverging, since $r_2 < r_1$.

54. We refer to Fig. 35-2 in the textbook. Consider the two light rays, r and r', which are closest to and on either side of the normal ray (the ray that reverses when it reflects). Each of these rays has an angle of incidence equal to θ when they reach the mirror. Consider that these two rays reach the top and bottom edges of the pupil after they have reflected. If ray r strikes the mirror at point A and ray r' strikes the mirror at B, the distance between A and B (call it x) is

$$x = 2d_o \tan\theta$$

where d_o is the distance from the mirror to the object. We can construct a right triangle starting with the image point of the object (a distance d_o behind the mirror; see I in Fig. 35-2). One side of the triangle follows the extended normal axis (which would reach from I to the middle of the pupil), and the hypotenuse is along the extension of ray r (after reflection). The distance from the pupil to I is $d_{\text{ey}} + d_o$, and the small angle in this triangle is again θ. Thus,

$$\tan\theta = \frac{R}{d_{\text{ey}} + d_o}$$

where R is the pupil radius (2.5 mm). Combining these relations, we find

$$x = 2d_o \frac{R}{d_{\text{ey}} + d_o} = 2(100\,\text{mm})\frac{2.5\,\text{mm}}{300\,\text{mm} + 100\,\text{mm}}$$

which yields $x = 1.67$ mm. Now, x serves as the diameter of a circular area A on the mirror, in which all rays that reflect will reach the eye. Therefore,

$$A = \frac{1}{4}\pi x^2 = \frac{\pi}{4}(1.67\,\text{mm})^2 = 2.2\ \text{mm}^2\ .$$

55. The sphere (of radius 0.35 m) is a convex mirror with focal length $f = -0.175$ m. We adopt the approximation that the rays are close enough to the central axis for Eq. 35-4 to be applicable. We also take the "1.0 m in front of ... [the] sphere" to mean $p = 1.0$ m (measured from the front surface as opposed to being measured from the center-point of the sphere).

(a) The equation $1/p + 1/i = 1/f$ yields $i = -0.15$ m, which means the image is 15 cm from the front surface, appearing to be *inside* the sphere.

(b) and (c) The lateral magnification is $m = -i/p$ which yields $m = 0.15$. Therefore, the image distance is $(0.15)(2.0\,\text{m}) = 0.30$ m; that this is a positive value implies the image is erect (upright).

56. (a) The mirror has focal length $f = 12$ cm. With $m = +3$, we have $i = -3p$. We substitute this into Eq. 35-4:

$$\begin{aligned} \frac{1}{p} + \frac{1}{i} &= \frac{1}{f} \\ \frac{1}{p} + \frac{1}{-3p} &= \frac{1}{12} \\ \frac{2}{3p} &= \frac{1}{12} \end{aligned}$$

with the unit cm understood. Consequently, we find $p = 2(12)/3 = 8.0$ cm.

(b) With $m = -3$, we have $i = +3p$, which we substitute into Eq. 35-4:

$$\begin{aligned} \frac{1}{p} + \frac{1}{i} &= \frac{1}{f} \\ \frac{1}{p} + \frac{1}{3p} &= \frac{1}{12} \\ \frac{4}{3p} &= \frac{1}{12} \end{aligned}$$

with the unit cm understood. Consequently, we find $p = 4(12)/3 = 16$ cm.

(c) With $m = -1/3$, we have $i = p/3$. Thus, Eq. 35-4 leads to

$$\begin{aligned} \frac{1}{p} + \frac{1}{i} &= \frac{1}{f} \\ \frac{1}{p} + \frac{3}{p} &= \frac{1}{12} \\ \frac{4}{p} &= \frac{1}{12} \end{aligned}$$

with the unit cm understood. Consequently, we find $p = 4(12) = 48$ cm.

57. Since $m = -2$ and $p = 4$ cm, then $i = 8$ cm (and is real). Eq. 35-9 is

$$\frac{1}{p} + \frac{1}{i} = \frac{1}{f}$$

and leads to $f = 2.67$ cm (which is positive, as it must be for a converging lens).

58. We use Eq. 35-8 (and Fig. 35-10(d) is useful), with $n_1 = 1.6$ and $n_2 = 1$ (using the rounded-off value for air).

$$\frac{1.6}{p} + \frac{1}{i} = \frac{1 - 1.6}{r}$$

Using the sign convention for r stated in the paragraph following Eq. 35-8 (so that $r = -5.0$ cm), we obtain $i = -2.4$ cm for objects at $p = 3.0$ cm. Returning to Fig. 35-52 (and noting the location of the observer), we conclude that the tabletop seems 7.4 cm away.

59. The fact that it is inverted implies $m < 0$. Therefore, with $m = -1/2$, we have $i = p/2$, which we substitute into Eq. 35-4:

$$\begin{aligned} \frac{1}{p} + \frac{1}{i} &= \frac{1}{f} \\ \frac{1}{p} + \frac{2}{p} &= \frac{1}{f} \\ \frac{3}{30.0} &= \frac{1}{f} \end{aligned}$$

with the unit cm understood. Consequently, we find $f = 30/3 = 10.0$ cm. The fact that $f > 0$ implies the mirror is concave.

60. (a) Suppose that the lens is placed to the left of the mirror. The image formed by the converging lens is located at a distance

$$i = \left(\frac{1}{f} - \frac{1}{p}\right)^{-1} = \left(\frac{1}{0.50\,\text{m}} - \frac{1}{1.0\,\text{m}}\right)^{-1} = 1.0\ \text{m}$$

to the right of the lens, or $2.0\,\text{m} - 1.0\,\text{m} = 1.0\,\text{m}$ in front of the mirror. The image formed by the mirror for this real image is then at 1.0 m to the right of the the mirror, or $2.0\,\text{m} + 1.0\,\text{m} = 3.0\,\text{m}$

to the right of the lens. This image then results in another image formed by the lens, located at a distance

$$i' = \left(\frac{1}{f} - \frac{1}{p'}\right)^{-1} = \left(\frac{1}{0.50\,\text{m}} - \frac{1}{3.0\,\text{m}}\right)^{-1} = 6.0\text{ m}$$

to the left of the lens (that is, 2.6 cm from the mirror).

(b) The final image is real since $i' > 0$.

(c) It also has the same orientation as the object, as one can verify by drawing a ray diagram or finding the product of the magnifications (see the next part, which shows $m > 0$).

(d) The lateral magnification is

$$m = \left(-\frac{i}{p}\right)\left(-\frac{i'}{p'}\right) = \left(-\frac{1.0\,\text{m}}{1.0\,\text{m}}\right)\left(-\frac{0.60\,\text{m}}{3.0\,\text{m}}\right) = +0.20\,.$$

61. (a) Parallel rays are bent by positive-f lenses to their focal points F_1, and rays that come from the focal point positions F_2 in front of positive-f lenses are made to emerge parallel. The key, then, to this type of beam expander is to have the rear focal point F_1 of the first lens coincide with the front focal point F_2 of the second lens. Since the triangles that meet at the coincident focal point are similar (they share the same angle; they are vertex angles), then $W_2/f_2 = W_1/f_1$ follows immediately.

(b) The previous argument can be adapted to the first lens in the expanding pair being of the diverging type, by ensuring that the front focal point of the first lens coincides with the front focal point of the second lens. The distance between the lenses in this case is $f_2 - |f_1|$ (where we assume $f_2 > |f_1|$), which we can write as $f_2 + f_1$ just as in part (a).

62. The area is proportional to W^2, so the result of problem 61 plus the definition of intensity (power P divided by area) leads to

$$\frac{I_2}{I_1} = \frac{P/W_2^2}{P/W_1^2} = \frac{W_1^2}{W_2^2} = \frac{f_1^2}{f_2^2}\,.$$

63. (a) Virtual, since the image is formed by plane mirrors.

(b) Same. One can easily verify this by locating, for example, the images of two points, one at the head of the penguin and the other at its feet.

(c) Same, since the image formed by any plane mirror retains the original shape and size of an object.

(d) The image of the penguin formed by the top mirror is located a distance D above the top mirror, or $L + D$ above the bottom one. Therefore, the final image of the penguin, formed by the bottom mirror, is a distance $L + D$ from the bottom mirror.

64. In the closest mirror, the "first" image I_1 is 10 cm behind the mirror and therefore 20 cm from the object O. There are images from both O and I_1 in the more distant mirror: an image I_2 which is 30 cm behind that mirror (since O is 30 cm in front of it), and an image I_3 which is 50 cm behind the mirror (since I_1 is 50 cm in front of it). We note that I_2 is 60 cm from O, and I_3 is 80 cm from O. Returning to the closer mirror, we find images of I_2 and I_3, as follows: an image I_4 which is 70 cm behind the mirror (since I_2 is 70 cm in front of it) and an image I_5 which is 90 cm behind the mirror (since I_3 is 90 cm in front of it). The distances (measured from O) for I_4 and I_5 are 80 cm and 100 cm, respectively.

Chapter 36

1. (a) The frequency of yellow sodium light is

$$f = \frac{c}{\lambda} = \frac{2.998 \times 10^8\,\text{m/s}}{589 \times 10^{-9}\,\text{m}} = 5.09 \times 10^{14}\ \text{Hz}\ .$$

   (b) When traveling through the glass, its wavelength is

$$\lambda_n = \frac{\lambda}{n} = \frac{589\,\text{nm}}{1.52} = 388\ \text{nm}\ .$$

   (c) The light speed when traveling through the glass is

$$v = f\lambda_n = (5.09 \times 10^{14}\,\text{Hz})(388 \times 10^{-9}\,\text{m}) = 1.97 \times 10^8\ \text{m/s}\ .$$

2. Comparing the light speeds in sapphire and diamond, we obtain

$$\begin{aligned} \Delta v &= v_s - v_d = c\left(\frac{1}{n_s} - \frac{1}{n_d}\right) \\ &= (2.998 \times 10^8\,\text{m/s})\left(\frac{1}{1.77} - \frac{1}{2.42}\right) = 4.55 \times 10^7\ \text{m/s}\ . \end{aligned}$$

3. The index of refraction is found from Eq. 36-3:

$$n = \frac{c}{v} = \frac{2.998 \times 10^8\,\text{m/s}}{1.92 \times 10^8\,\text{m/s}} = 1.56\ .$$

4. The index of refraction of fused quartz at $\lambda = 550\,\text{nm}$ is about 1.459, obtained from Fig. 34-19. Thus, from Eq. 36-3, we find

$$v = \frac{c}{n} = \frac{2.998 \times 10^8\,\text{m/s}}{1.459} = 2.06 \times 10^8\ \text{m/s}\ .$$

5. Applying the law of refraction, we obtain $\sin\theta/\sin 30^\circ = v_s/v_d$. Consequently,

$$\theta = \sin^{-1}\left(\frac{v_s \sin 30^\circ}{v_d}\right) = \sin^{-1}\left[\frac{(3.0\,\text{m/s})\sin 30^\circ}{4.0\,\text{m/s}}\right] = 22^\circ\ .$$

   The angle of incidence is gradually reduced due to refraction, such as shown in the calculation above (from 30° to 22°). Eventually after many refractions, θ will be virtually zero. This is why most waves come in normal to a shore.

6. (a) The time t_2 it takes for pulse 2 to travel through the plastic is

$$t_2 = \frac{L}{c/1.55} + \frac{L}{c/1.70} + \frac{L}{c/1.60} + \frac{L}{c/1.45} = \frac{6.30L}{c}\ .$$

Similarly for pulse 1:

$$t_1 = \frac{2L}{c/1.59} + \frac{L}{c/1.65} + \frac{L}{c/1.50} = \frac{6.33L}{c} .$$

Thus, pulse 2 travels through the plastic in less time.

(b) The time difference (as a multiple of L/c) is

$$\Delta t = t_2 - t_1 = \frac{6.33L}{c} - \frac{6.30L}{c} = \frac{0.03L}{c} .$$

7. (a) We take the phases of both waves to be zero at the front surfaces of the layers. The phase of the first wave at the back surface of the glass is given by $\phi_1 = k_1 L - \omega t$, where k_1 $(= 2\pi/\lambda_1)$ is the angular wave number and λ_1 is the wavelength in glass. Similarly, the phase of the second wave at the back surface of the plastic is given by $\phi_2 = k_2 L - \omega t$, where k_2 $(= 2\pi/\lambda_2)$ is the angular wave number and λ_2 is the wavelength in plastic. The angular frequencies are the same since the waves have the same wavelength in air and the frequency of a wave does not change when the wave enters another medium. The phase difference is

$$\phi_1 - \phi_2 = (k_1 - k_2)L = 2\pi\left(\frac{1}{\lambda_1} - \frac{1}{\lambda_2}\right)L .$$

Now, $\lambda_1 = \lambda_{\text{air}}/n_1$, where λ_{air} is the wavelength in air and n_1 is the index of refraction of the glass. Similarly, $\lambda_2 = \lambda_{\text{air}}/n_2$, where n_2 is the index of refraction of the plastic. This means that the phase difference is $\phi_1 - \phi_2 = (2\pi/\lambda_{\text{air}})(n_1 - n_2)L$. The value of L that makes this 5.65 rad is

$$L = \frac{(\phi_1 - \phi_2)\lambda_{\text{air}}}{2\pi(n_1 - n_2)} = \frac{5.65(400 \times 10^{-9}\,\text{m})}{2\pi(1.60 - 1.50)} = 3.60 \times 10^{-6}\ \text{m} .$$

(b) 5.65 rad is less than 2π rad $= 6.28$ rad, the phase difference for completely constructive interference, and greater than π rad ($= 3.14$ rad), the phase difference for completely destructive interference. The interference is, therefore, intermediate, neither completely constructive nor completely destructive. It is, however, closer to completely constructive than to completely destructive.

8. (a) Eq. 36-11 (in absolute value) yields

$$\frac{L}{\lambda}\,|n_2 - n_1| = \frac{(8.50 \times 10^{-6}\,\text{m})}{500 \times 10^{-9}\,\text{m}}(1.60 - 1.50) = 1.70 .$$

(b) Similarly,

$$\frac{L}{\lambda}\,|n_2 - n_1| = \frac{(8.50 \times 10^{-6}\,\text{m})}{500 \times 10^{-9}\,\text{m}}(1.72 - 1.62) = 1.70 .$$

(c) In this case, we obtain

$$\frac{L}{\lambda}\,|n_2 - n_1| = \frac{(3.25 \times 10^{-6}\,\text{m})}{500 \times 10^{-9}\,\text{m}}(1.79 - 1.59) = 1.30 .$$

(d) Since their phase differences were identical, the brightness should be the same for (a) and (b). Now, the phase difference in (c) differs from an integer by 0.30, which is also true for (a) and (b). Thus, their effective phase differences are equal, and the brightness in case (c) should be the same as that in (a) and (b).

9. (a) We choose a horizontal x axis with its origin at the left edge of the plastic. Between $x = 0$ and $x = L_2$ the phase difference is that given by Eq. 36-11 (with L in that equation replaced with L_2). Between $x = L_2$ and $x = L_1$ the phase difference is given by an expression similar to Eq. 36-11 but with L replaced with $L_1 - L_2$ and n_2 replaced with 1 (since the top ray in Fig. 36-28 is now traveling through air, which has index of refraction approximately equal to 1). Thus, combining these phase differences and letting all lengths be in μm (so $\lambda = 0.600$), we have

$$\frac{L_2}{\lambda}(n_2 - n_1) + \frac{L_1 - L_2}{\lambda}(1 - n_1) = \frac{3.50}{0.600}(1.60 - 1.40) + \frac{4.00 - 3.50}{0.600}(1 - 1.40) = 0.833 .$$

(b) Since the answer in part (a) is closer to an integer than to a half-integer, then the interference is more nearly constructive than destructive.

10. (a) We wish to set Eq. 36-11 equal to $\frac{1}{2}$, since a half-wavelength phase difference is equivalent to a π radians difference. Thus,

$$L_{\text{min}} = \frac{\lambda}{2\,(n_2 - n_1)} = \frac{620\,\text{nm}}{2(1.65 - 1.45)} = 1550\,\text{nm} = 1.55\ \mu\text{m}\ .$$

(b) Since a phase difference of $\frac{3}{2}$ (wavelengths) is effectively the same as what we required in part (a), then

$$L = \frac{3\lambda}{2\,(n_2 - n_1)} = 3L_{\text{min}} = 3(1.55\mu\text{m}) = 4.65\ \mu\text{m}\ .$$

11. (a) We use Eq. 36-14 with $m = 3$:

$$\theta = \sin^{-1}\left(\frac{m\lambda}{d}\right) = \sin^{-1}\left[\frac{2(550 \times 10^{-9}\,\text{m})}{7.70 \times 10^{-6}\,\text{m}}\right] = 0.216\ \text{rad}\ .$$

(b) $\theta = (0.216)(180°/\pi) = 12.4°$.

12. Here we refer to phase difference in radians (as opposed to wavelengths or degrees). For the first dark fringe $\phi_1 = \pm\pi$, and for the second one $\phi_2 = \pm 3\pi$, etc. For the mth one $\phi_m = \pm(2m+1)\pi$.

13. The condition for a maximum in the two-slit interference pattern is $d\sin\theta = m\lambda$, where d is the slit separation, λ is the wavelength, m is an integer, and θ is the angle made by the interfering rays with the forward direction. If θ is small, $\sin\theta$ may be approximated by θ in radians. Then, $\theta = m\lambda/d$, and the angular separation of adjacent maxima, one associated with the integer m and the other associated with the integer $m+1$, is given by $\Delta\theta = \lambda/d$. The separation on a screen a distance D away is given by $\Delta y = D\,\Delta\theta = \lambda D/d$. Thus,

$$\Delta y = \frac{(500 \times 10^{-9}\,\text{m})(5.40\,\text{m})}{1.20 \times 10^{-3}\,\text{m}} = 2.25 \times 10^{-3}\,\text{m} = 2.25\ \text{mm}\ .$$

14. (a) For the maximum adjacent to the central one, we set $m = 1$ in Eq. 36-14 and obtain

$$\theta_1 = \sin^{-1}\left(\frac{m\lambda}{d}\right)\bigg|_{m=1} = \sin^{-1}\left[\frac{(1)(\lambda)}{100\lambda}\right] = 0.010\ \text{rad}\ .$$

(b) Since $y_1 = D\tan\theta_1$ (see Fig. 36-8(a)), we obtain $y_1 = (500\,\text{mm})\tan(0.010\,\text{rad}) = 5.0\,\text{mm}$. The separation is $\Delta y = y_1 - y_0 = y_1 - 0 = 5.0\,\text{mm}$.

15. The angular positions of the maxima of a two-slit interference pattern are given by $d\sin\theta = m\lambda$, where d is the slit separation, λ is the wavelength, and m is an integer. If θ is small, $\sin\theta$ may be approximated by θ in radians. Then, $\theta = m\lambda/d$ to good approximation. The angular separation of two adjacent maxima is $\Delta\theta = \lambda/d$. Let λ' be the wavelength for which the angular separation is 10.0% greater. Then, $1.10\lambda/d = \lambda'/d$ or $\lambda' = 1.10\lambda = 1.10(589\,\text{nm}) = 648\,\text{nm}$.

16. In Sample Problem 36-2, an experimentally useful relation is derived: $\Delta y = \lambda D/d$. Dividing both sides by D, this becomes $\Delta\theta = \lambda/d$ with θ in radians. In the steps that follow, however, we will end up with an expression where degrees may be directly used. Thus, in the present case,

$$\Delta\theta_n = \frac{\lambda_n}{d} = \frac{\lambda}{nd} = \frac{\Delta\theta}{n} = \frac{0.20°}{1.33} = 0.15°\ .$$

17. Interference maxima occur at angles θ such that $d\sin\theta = m\lambda$, where m is an integer. Since $d = 2.0\,\text{m}$ and $\lambda = 0.50\,\text{m}$, this means that $\sin\theta = 0.25m$. We want all values of m (positive and negative) for which $|0.25m| \le 1$. These are -4, -3, -2, -1, 0, $+1$, $+2$, $+3$, and $+4$. For each of these except -4 and $+4$, there are two different values for θ. A single value of θ ($-90°$) is associated with $m = -4$ and a single value ($+90°$) is associated with $m = +4$. There are sixteen different angles in all and, therefore, sixteen maxima.

18. Initially, source A leads source B by 90°, which is equivalent to 1/4 wavelength. However, source A also lags behind source B since r_A is longer than r_B by 100 m, which is $100\,\text{m}/400\,\text{m} = 1/4$ wavelength. So the net phase difference between A and B at the detector is zero.

19. The maxima of a two-slit interference pattern are at angles θ given by $d\sin\theta = m\lambda$, where d is the slit separation, λ is the wavelength, and m is an integer. If θ is small, $\sin\theta$ may be replaced by θ in radians. Then, $d\theta = m\lambda$. The angular separation of two maxima associated with different wavelengths but the same value of m is $\Delta\theta = (m/d)(\lambda_2 - \lambda_1)$, and their separation on a screen a distance D away is

$$\begin{aligned}\Delta y &= D\tan\Delta\theta \approx D\,\Delta\theta = \left[\frac{mD}{d}\right](\lambda_2 - \lambda_1)\\ &= \left[\frac{3(1.0\,\text{m})}{5.0\times 10^{-3}\,\text{m}}\right](600\times 10^{-9}\,\text{m} - 480\times 10^{-9}\,\text{m}) = 7.2\times 10^{-5}\,\text{m}\,.\end{aligned}$$

The small angle approximation $\tan\Delta\theta \approx \Delta\theta$ (in radians) is made.

20. Let the distance in question be x. The path difference (between rays originating from S_1 and S_2 and arriving at points on the $x > 0$ axis) is

$$\sqrt{d^2 + x^2} - x = \left(m + \frac{1}{2}\right)\lambda\,,$$

where we are requiring destructive interference (half-integer wavelength phase differences) and $m = 0, 1, 2, \cdots$. After some algebraic steps, we solve for the distance in terms of m:

$$x = \frac{d^2}{(2m+1)\lambda} - \frac{(2m+1)\lambda}{4}\,.$$

To obtain the largest value of x, we set $m = 0$:

$$x_0 = \frac{d^2}{\lambda} - \frac{\lambda}{4} = \frac{(3.00\lambda)^2}{\lambda} - \frac{\lambda}{4} = 8.75\lambda\,.$$

21. Consider the two waves, one from each slit, that produce the seventh bright fringe in the absence of the mica. They are in phase at the slits and travel different distances to the seventh bright fringe, where they have a phase difference of $2\pi m = 14\pi$. Now a piece of mica with thickness x is placed in front of one of the slits, and an additional phase difference between the waves develops. Specifically, their phases at the slits differ by

$$\frac{2\pi x}{\lambda_m} - \frac{2\pi x}{\lambda} = \frac{2\pi x}{\lambda}(n-1)$$

where λ_m is the wavelength in the mica and n is the index of refraction of the mica. The relationship $\lambda_m = \lambda/n$ is used to substitute for λ_m. Since the waves are now in phase at the screen,

$$\frac{2\pi x}{\lambda}(n-1) = 14\pi$$

or

$$x = \frac{7\lambda}{n-1} = \frac{7(550\times 10^{-9}\,\text{m})}{1.58 - 1} = 6.64\times 10^{-6}\,\text{m}\,.$$

22. (a) We use $\Delta y = D\lambda/d$ (see Sample Problem 36-2). Because of the placement of the mirror in the problem $D = 2(20.0\,\text{m}) = 40.0\,\text{m}$, which we express in millimeters in the calculation below:

$$d = \frac{D\lambda}{\Delta y} = \frac{(4.00 \times 10^4\,\text{mm})(632.8 \times 10^{-6}\,\text{mm})}{100\,\text{mm}} = 0.253\ \text{mm}\ .$$

(b) In this case the interference pattern will be shifted. At the location of the original central maximum, the effective phase difference is now $\frac{1}{2}$ wavelength, so there is now a minimum instead of a maximum.

23. The phasor diagram is shown below. Here $E_1 = 1.00$, $E_2 = 2.00$, and $\phi = 60°$. The resultant amplitude E_m is given by the trigonometric law of cosines:

$$E_m^2 = E_1^2 + E_2^2 - 2E_1E_2\cos(180° - \phi)\ .$$

Thus,

$$E_m = \sqrt{(1.00)^2 + (2.00)^2 - 2(1.00)(2.00)\cos 120°} = 2.65\ .$$

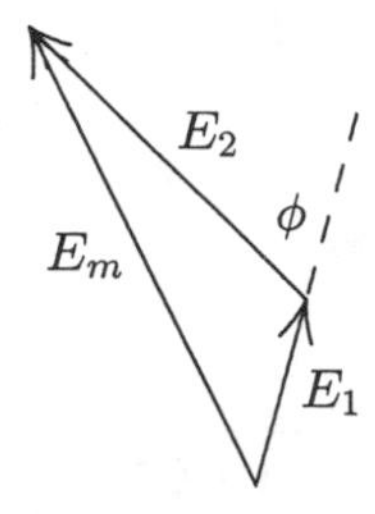

24. In adding these with the phasor method (as opposed to, say, trig identities), we may set $t = 0$ (see Sample Problem 36-3) and add them as vectors:

$$\begin{aligned} y_h &= 10\cos 0° + 8.0\cos 30° = 16.9 \\ y_v &= 10\sin 0° + 8.0\sin 30° = 4.0 \end{aligned}$$

so that

$$\begin{aligned} y_R &= \sqrt{y_h^2 + y_v^2} = 17.4 \\ \beta &= \tan^{-1}\left(\frac{y_v}{y_h}\right) = 13.3°\ . \end{aligned}$$

Thus, $y = y_1 + y_2 = y_R\sin(\omega t + \beta) = 17.4\sin(\omega t + 13.3°)$.

25. In adding these with the phasor method (as opposed to, say, trig identities), we may set $t = 0$ (see Sample Problem 36-3) and add them as vectors:

$$\begin{aligned} y_h &= 10\cos 0° + 15\cos 30° + 5.0\cos(-45°) = 26.5 \\ y_v &= 10\sin 0° + 15\sin 30° + 5.0\sin(-45°) = 4.0 \end{aligned}$$

so that

$$\begin{aligned} y_R &= \sqrt{y_h^2 + y_v^2} = 26.8 \\ \beta &= \tan^{-1}\left(\frac{y_v}{y_h}\right) = 8.5°\ . \end{aligned}$$

Thus, $y = y_1 + y_2 + y_3 = y_R\sin(\omega t + \beta) = 26.8\sin(\omega t + 8.5°)$.

26. Fig. 36-9 in the textbook is plotted versus the phase difference (in radians), whereas this problem requests that we plot the intensity versus the physical angle θ (defined in Fig. 36-8). The values given in the problem imply $d\lambda = 1000$. Combining this with Eq. 36-22 and Eq. 36-21, we solve for the (normalized) intensity:

$$\frac{I}{4I_0} = \cos^2(1000\pi \sin\theta) \ .$$

This is plotted over $0 \le \theta \le 0.0040$ rad:

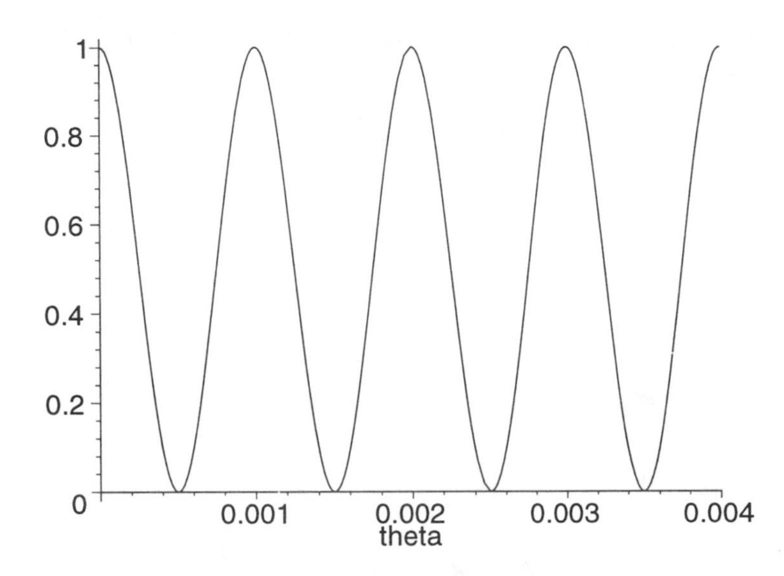

27. (a) To get to the detector, the wave from S_1 travels a distance x and the wave from S_2 travels a distance $\sqrt{d^2 + x^2}$. The phase difference (in terms of wavelengths) between the two waves is

$$\sqrt{d^2 + x^2} - x = m\lambda \qquad m = 0, 1, 2, \ldots$$

where we are requiring constructive interference. The solution is

$$x = \frac{d^2 - m^2\lambda^2}{2m\lambda} \ .$$

The largest value of m that produces a positive value for x is $m = 3$. This corresponds to the maximum that is nearest S_1, at

$$x = \frac{(4.00\,\text{m})^2 - 9(1.00\,\text{m})^2}{(2)(3)(1.00\,\text{m})} = 1.17\ \text{m} \ .$$

For the next maximum, $m = 2$ and $x = 3.00\,\text{m}$. For the third maximum, $m = 1$ and $x = 7.50\,\text{m}$.

(b) Minima in intensity occur where the phase difference is π rad; the intensity at a minimum, however, is not zero because the amplitudes of the waves are different. Although the amplitudes are the same at the sources, the waves travel different distances to get to the points of minimum intensity and each amplitude decreases in inverse proportion to the distance traveled.

28. Setting $I = 2I_0$ in Eq. 36-21 and solving for the smallest (in absolute value) two roots for $\phi/2$, we find

$$\phi = 2\cos^{-1}\left(\frac{1}{\sqrt{2}}\right) = \pm\frac{\pi}{2}\ \text{rad} \ .$$

Now, for small θ in radians, Eq. 36-22 becomes $\phi = 2\pi d\theta/\lambda$. This leads to two corresponding angle values:

$$\theta = \pm\frac{\lambda}{4d} \ .$$

The difference between these two values is $\Delta\theta = \frac{\lambda}{4d} - \left(-\frac{\lambda}{4d}\right) = \frac{\lambda}{2d}$.

29. We take the electric field of one wave, at the screen, to be

$$E_1 = E_0 \sin(\omega t)$$

and the electric field of the other to be

$$E_2 = 2E_0 \sin(\omega t + \phi) \ ,$$

where the phase difference is given by

$$\phi = \left(\frac{2\pi d}{\lambda}\right) \sin\theta \ .$$

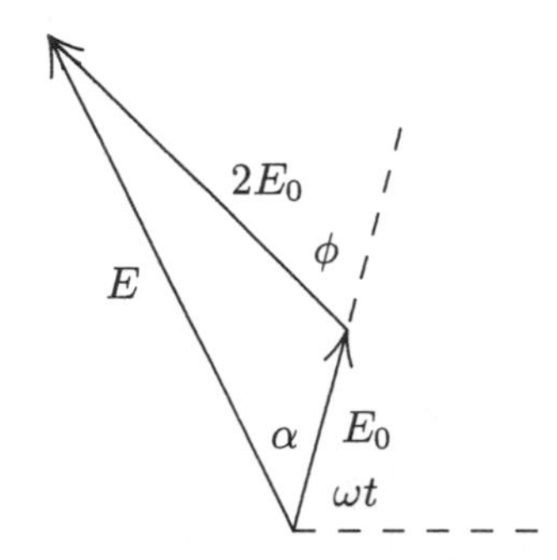

Here d is the center-to-center slit separation and λ is the wavelength. The resultant wave can be written $E = E_1 + E_2 = E\sin(\omega t + \alpha)$, where α is a phase constant. The phasor diagram is shown above. The resultant amplitude E is given by the trigonometric law of cosines:

$$E^2 = E_0^2 + (2E_0)^2 - 4E_0^2\cos(180° - \phi) = E_0^2(5 + 4\cos\phi) \ .$$

The intensity is given by $I = I_0(5 + 4\cos\phi)$, where I_0 is the intensity that would be produced by the first wave if the second were not present. Since $\cos\phi = 2\cos^2(\phi/2) - 1$, this may also be written $I = I_0\left[1 + 8\cos^2(\phi/2)\right]$.

30. The fact that wave W_2 reflects two additional times has no substantive effect on the calculations, since two reflections amount to a $2(\lambda/2) = \lambda$ phase difference, which is effectively not a phase difference at all. The substantive difference between W_2 and W_1 is the extra distance $2L$ traveled by W_2.

   (a) For wave W_2 to be a half-wavelength "behind" wave W_1, we require $2L = \lambda/2$, or $L = \lambda/4 = 155$ nm using the wavelength value given in the problem.

   (b) Destructive interference will again appear if W_2 is $\frac{3}{2}\lambda$ "behind" the other wave. In this case, $2L' = 3\lambda/2$, and the difference is

$$L' - L = \frac{3\lambda}{4} - \frac{\lambda}{4} = \frac{\lambda}{2} = 310 \text{ nm} \ .$$

31. The wave reflected from the front surface suffers a phase change of π rad since it is incident in air on a medium of higher index of refraction. The phase of the wave reflected from the back surface does not change on reflection since the medium beyond the soap film is air and has a lower index of refraction than the film. If L is the thickness of the film, this wave travels a distance $2L$ farther than the wave reflected from the front surface. The phase difference of the two waves is $2L(2\pi/\lambda_f) - \pi$, where λ_f is the wavelength in the film. If λ is the wavelength in vacuum and n is the index of refraction of the soap film, then $\lambda_f = \lambda/n$ and the phase difference is

$$2nL\left(\frac{2\pi}{\lambda}\right) - \pi = 2(1.33)(1.21 \times 10^{-6}\text{ m})\left(\frac{2\pi}{585 \times 10^{-9}\text{ m}}\right) - \pi = 10\pi \text{ rad} \ .$$

Since the phase difference is an even multiple of π, the interference is completely constructive.

32. In contrast to the initial conditions of problem 30, we now consider waves W_2 and W_1 with an initial effective phase difference (in wavelengths) equal to $\frac{1}{2}$, and seek positions of the sliver which cause the wave to constructively interfere (which corresponds to an integer-valued phase difference in wavelengths). Thus, the extra distance $2L$ traveled by W_2 must amount to $\frac{1}{2}\lambda$, $\frac{3}{2}\lambda$, and so on. We may write this requirement succinctly as

$$L = \frac{2m+1}{4}\lambda \qquad \text{where } m = 0, 1, 2, \ldots \ .$$

33. For constructive interference, we use Eq. 36-34: $2n_2L = (m + 1/2)\lambda$. For the two smallest values of L, let $m = 0$ and 1:

$$\begin{aligned} L_0 &= \frac{\lambda/2}{2n_2} = \frac{624\,\text{nm}}{4(1.33)} = 117\,\text{nm} = 0.117\,\mu\text{m} \\ L_1 &= \frac{(1+1/2)\lambda}{2n_2} = \frac{3\lambda}{2n_2} = 3L_0 = 3(0.1173\mu\text{m}) = 0.352\,\mu\text{m} \ . \end{aligned}$$

34. We use the formula obtained in Sample Problem 36-5:

$$L_{\text{min}} = \frac{\lambda}{4n_2} = \frac{\lambda}{4(1.25)} = 0.200\lambda \ .$$

35. Light reflected from the front surface of the coating suffers a phase change of π rad while light reflected from the back surface does not change phase. If L is the thickness of the coating, light reflected from the back surface travels a distance $2L$ farther than light reflected from the front surface. The difference in phase of the two waves is $2L(2\pi/\lambda_c) - \pi$, where λ_c is the wavelength in the coating. If λ is the wavelength in vacuum, then $\lambda_c = \lambda/n$, where n is the index of refraction of the coating. Thus, the phase difference is $2nL(2\pi/\lambda) - \pi$. For fully constructive interference, this should be a multiple of 2π. We solve

$$2nL\left(\frac{2\pi}{\lambda}\right) - \pi = 2m\pi$$

for L. Here m is an integer. The solution is

$$L = \frac{(2m+1)\lambda}{4n} \ .$$

To find the smallest coating thickness, we take $m = 0$. Then,

$$L = \frac{\lambda}{4n} = \frac{560 \times 10^{-9}\,\text{m}}{4(2.00)} = 7.00 \times 10^{-8}\ \text{m} \ .$$

36. Let the thicknesses (which appear in Fig. 36-31 as different heights h) of the structure be $h = kL$, where k is a pure number. In section (b), for example, $k = 2$. Using Eq. 36-34, the condition for constructive interference becomes

$$2h = 2(kL) = \frac{(m+1/2)\lambda}{n_2} \qquad \text{where } m = 0, 1, 2, \ldots$$

which leads to

$$k = \frac{(m+1/2)\lambda}{2n_2L} = \frac{(m+1/2)(600\,\text{nm})}{2(1.50)(4.00\times 10^3\,\text{nm})} = \frac{2m+1}{40} \ ,$$

or $40k - 1 = 2m$. This means that $40k - 1$ would have to be an even integer. One can check that none of the given values of k (1, 2, $\frac{1}{2}$, 3, $\frac{1}{10}$) will satisfy this condition. Therefore, none of the sections provides the right thickness for constructive interference.

37. For complete destructive interference, we want the waves reflected from the front and back of the coating to differ in phase by an odd multiple of π rad. Each wave is incident on a medium of higher index of refraction from a medium of lower index, so both suffer phase changes of π rad on reflection. If L is the thickness of the coating, the wave reflected from the back surface travels a distance $2L$ farther than the wave reflected from the front. The phase difference is $2L(2\pi/\lambda_c)$, where λ_c is the wavelength in the coating. If n is the index of refraction of the coating, $\lambda_c = \lambda/n$, where λ is the wavelength in vacuum, and the phase difference is $2nL(2\pi/\lambda)$. We solve

$$2nL\left(\frac{2\pi}{\lambda}\right) = (2m+1)\pi$$

for L. Here m is an integer. The result is

$$L = \frac{(2m+1)\lambda}{4n} .$$

To find the least thickness for which destructive interference occurs, we take $m = 0$. Then,

$$L = \frac{\lambda}{4n} = \frac{600 \times 10^{-9}\,\text{m}}{4(1.25)} = 1.2 \times 10^{-7}\ \text{m} .$$

38. Eqs. 36-14 and 36-16 treat the interference of reflections, and here we are concerned with interference of the transmitted light. Maxima in the reflections should, reasonably enough, correspond to minima in the transmissions, and vice versa. So we might expect to apply those equations to this case if we switch the designations "maxima" and "minima," *if* we are careful with the phase shifts that occur at the points of reflection (which depend on the relative values of n). Now, if the expression $2L = m\lambda/n_2$ is to give the condition for constructive interference for the transmitted light, then the situation should be similar to that which led in the textbook to Eqs. 36-14 and 36-16; namely, the thin film should be surrounded by two higher-index or two lower-index media. Such is the case for Fig. 36-32(a) and Fig. 36-32(c), but not for the others.

39. The situation is analogous to that treated in Sample Problem 36-5, in the sense that the incident light is in a low index medium, the thin film has somewhat higher $n = n_2$, and the last layer has the highest refractive index. To see very little or no reflection, according to the Sample Problem, the condition

$$2L = \left(m + \frac{1}{2}\right)\frac{\lambda}{n_2} \qquad \text{where} \quad m = 0, 1, 2, \ldots$$

must hold. The value of L which corresponds to no reflection corresponds, reasonably enough, to the value which gives maximum transmission of light (into the highest index medium – which in this problem is the water).

(a) If $2L = \left(m + \frac{1}{2}\right)\frac{\lambda}{n_2}$ (Eq. 36-34) gives zero reflection in this type of system, then we might reasonably expect that its counterpart, Eq. 36-35, gives maximum reflection here. A more careful analysis such as that given in §36-7 bears this out. We disregard the $m = 0$ value (corresponding to $L = 0$) since there is *some* oil on the water. Thus, for $m = 1, 2, \ldots$ maximum reflection occurs for wavelengths

$$\lambda = \frac{2n_2L}{m} = \frac{2(1.20)(460\,\text{nm})}{m} = 1104\,\text{nm}\,,\ 552\,\text{nm}\,,\ 368\,\text{nm}\ \ldots$$

We note that only the 552 nm wavelength falls within the visible light range.

(b) As remarked above, maximum transmission into the water occurs for wavelengths given by

$$2L = \left(m + \frac{1}{2}\right)\frac{\lambda}{n_2} \implies \lambda = \frac{4n_2L}{2m+1}$$

which yields $\lambda = 2208\,\text{nm}\,,\ 736\,\text{nm}\,,\ 442\,\text{nm}\ \ldots$ for the different values of m. We note that only the 442 nm wavelength (blue) is in the visible range, though we might expect some red contribution since the 736 nm is very close to the visible range.

40. The situation is analogous to that treated in Sample Problem 36-5, in the sense that the incident light is in a low index medium, the thin film of oil has somewhat higher $n = n_2$, and the last layer (the glass plate) has the highest refractive index. To see very little or no reflection, according to the Sample Problem, the condition

$$2L = \left(m + \frac{1}{2}\right)\frac{\lambda}{n_2} \qquad \text{where} \quad m = 0, 1, 2, \ldots$$

must hold. With $\lambda = 500$ nm and $n_2 = 1.30$, the possible answers for L are

$$L = 96\,\text{nm}\,,\ 288\,\text{nm}\,,\ 481\,\text{nm}\,,\ 673\,\text{nm}\,,\ 865\,\text{nm}\,,\ \ldots$$

And, with $\lambda = 700$ nm and the same value of n_2, the possible answers for L are

$$L = 135\,\text{nm}\,,\ 404\,\text{nm}\,,\ 673\,\text{nm}\,,\ 942\,\text{nm}\,,\ \ldots$$

The lowest number these lists have in common is $L = 673$ nm.

41. Light reflected from the upper oil surface (in contact with air) changes phase by π rad. Light reflected from the lower surface (in contact with glass) changes phase by π rad if the index of refraction of the oil is less than that of the glass and does not change phase if the index of refraction of the oil is greater than that of the glass.

- First, suppose the index of refraction of the oil is greater than the index of refraction of the glass. The condition for fully destructive interference is $2n_o d = m\lambda$, where d is the thickness of the oil film, n_o is the index of refraction of the oil, λ is the wavelength in vacuum, and m is an integer. For the shorter wavelength, $2n_o d = m_1\lambda_1$ and for the longer, $2n_o d = m_2\lambda_2$. Since λ_1 is less than λ_2, m_1 is greater than m_2, and since fully destructive interference does not occur for any wavelengths between, $m_1 = m_2 + 1$. Solving $(m_2 + 1)\lambda_1 = m_2\lambda_2$ for m_2, we obtain

$$m_2 = \frac{\lambda_1}{\lambda_2 - \lambda_1} = \frac{500\,\text{nm}}{700\,\text{nm} - 500\,\text{nm}} = 2.50\,.$$

  Since m_2 must be an integer, the oil cannot have an index of refraction that is greater than that of the glass.

- Now suppose the index of refraction of the oil is less than that of the glass. The condition for fully destructive interference is then $2n_o d = (2m+1)\lambda$. For the shorter wavelength, $2m_o d = (2m_1+1)\lambda_1$, and for the longer, $2n_o d = (2m_2 + 1)\lambda_2$. Again, $m_1 = m_2 + 1$, so $(2m_2 + 3)\lambda_1 = (2m_2 + 1)\lambda_2$. This means the value of m_2 is

$$m_2 = \frac{3\lambda_1 - \lambda_2}{2(\lambda_2 - \lambda_1)} = \frac{3(500\,\text{nm}) - 700\,\text{nm}}{2(700\,\text{nm} - 500\,\text{nm})} = 2.00\,.$$

  This is an integer. Thus, the index of refraction of the oil is less than that of the glass.

42. We solve Eq. 36-34 with $n_2 = 1.33$ and $\lambda = 600$ nm for $m = 1, 2, 3, \ldots$:

$$L = 113\,\text{nm}\,,\ 338\,\text{nm}\,,\ 564\,\text{nm}\,,\ 789\,\text{nm}\,,\ \ldots$$

And, we similarly solve Eq. 36-35 with the same n_2 and $\lambda = 450$ nm:

$$L = 0\,,\ 169\,\text{nm}\,,\ 338\,\text{nm}\,,\ 508\,\text{nm}\,,\ 677\,\text{nm}\,,\ \ldots$$

The lowest number these lists have in common is $L = 338$ nm.

43. Consider the interference of waves reflected from the top and bottom surfaces of the air film. The wave reflected from the upper surface does not change phase on reflection but the wave reflected from the bottom surface changes phase by π rad. At a place where the thickness of the air film is L, the condition

for fully constructive interference is $2L = (m + \frac{1}{2})\lambda$, where λ (= 683 nm) is the wavelength and m is an integer. This is satisfied for $m = 140$:

$$L = \frac{(m + \frac{1}{2})\lambda}{2} = \frac{(140.5)(683 \times 10^{-9}\,\text{m})}{2} = 4.80 \times 10^{-5}\,\text{m} = 0.048\ \text{mm}\ .$$

At the thin end of the air film, there is a bright fringe. It is associated with $m = 0$. There are, therefore, 140 bright fringes in all.

44. (a) At the left end, the plates touch, so $L = 0$ there, which is clearly consistent with Eq. 36-35 (the destructive interference or "dark fringe" equation) for $m = 0$.

    (b) Eq. 36-35 shows a simple proportionality between L and λ. So as we slowly increase L (from zero – its value in part (a)), the smallest nonzero value of L for which the equation (which specifies destructive interference) is satisfied occurs for the lowest possible value of λ. Wavelengths for blue light are the shortest of the visible portion of the spectrum.

45. Assume the wedge-shaped film is in air, so the wave reflected from one surface undergoes a phase change of π rad while the wave reflected from the other surface does not. At a place where the film thickness is L, the condition for fully constructive interference is $2nL = (m + \frac{1}{2})\lambda$, where n is the index of refraction of the film, λ is the wavelength in vacuum, and m is an integer. The ends of the film are bright. Suppose the end where the film is narrow has thickness L_1 and the bright fringe there corresponds to $m = m_1$. Suppose the end where the film is thick has thickness L_2 and the bright fringe there corresponds to $m = m_2$. Since there are ten bright fringes, $m_2 = m_1 + 9$. Subtract $2nL_1 = (m_1 + \frac{1}{2})\lambda$ from $2nL_2 = (m_1 + 9 + \frac{1}{2})\lambda$ to obtain $2n\,\Delta L = 9\lambda$, where $\Delta L = L_2 - L_1$ is the change in the film thickness over its length. Thus,

$$\Delta L = \frac{9\lambda}{2n} = \frac{9(630 \times 10^{-9}\,\text{m})}{2(1.50)} = 1.89 \times 10^{-6}\ \text{m}\ .$$

46. The situation is analogous to that treated in Sample Problem 36-5, in the sense that the incident light is in a low index medium, the thin film of acetone has somewhat higher $n = n_2$, and the last layer (the glass plate) has the highest refractive index. To see very little or no reflection, according to the Sample Problem, the condition

$$2L = \left(m + \frac{1}{2}\right)\frac{\lambda}{n_2} \qquad \text{where} \quad m = 0, 1, 2, \ldots$$

must hold. This is the same as Eq. 36-34 which was developed for the opposite situation (constructive interference) regarding a thin film surrounded on both sides by air (a very different context than the one in this problem). By analogy, we expect Eq. 36-35 to apply in this problem to reflection *maxima*. A more careful analysis such as that given in §36-7 bears this out. Thus, using Eq. 36-35 with $n_2 = 1.25$ and $\lambda = 700$ nm yields

$$L = 0,\ 280\,\text{nm},\ 560\,\text{nm},\ 840\,\text{nm},\ 1120\,\text{nm},\ \ldots$$

for the first several m values. And the equation shown above (equivalent to Eq. 36-34) gives, with $\lambda = 600$ nm,

$$L = 120\,\text{nm},\ 360\,\text{nm},\ 600\,\text{nm},\ 840\,\text{nm},\ 1080\,\text{nm},\ \ldots$$

for the first several m values. The lowest number these lists have in common is $L = 840$ nm.

47. We use Eq. 36-34:

$$\begin{aligned} L_{16} &= \left(16 + \frac{1}{2}\right)\frac{\lambda}{2n_2} \\ L_6 &= \left(6 + \frac{1}{2}\right)\frac{\lambda}{2n_2} \end{aligned}$$

The difference between these, using the fact that $n_2 = n_{\text{air}} = 1.0$, is

$$L_{16} - L_6 = (10)\frac{480\,\text{nm}}{2(1.0)} = 2400\ \text{nm}\ .$$

48. We apply Eq. 36-25 to both scenarios: $m = 4001$ and $n_2 = n_{\text{air}}$, and $m = 4000$ and $n_2 = n_{\text{vacuum}} = 1.00000$:

$$2L = (4001)\frac{\lambda}{n_{\text{air}}} \qquad \text{and} \qquad 2L = (4000)\frac{\lambda}{1.00000}\ .$$

Since the $2L$ factor is the same in both cases, we set the right hand sides of these expressions equal to each other and cancel the wavelength. Finally, we obtain

$$n_{\text{air}} = (1.00000)\frac{4001}{4000} = 1.00025\ .$$

We remark that this same result can be obtained starting with Eq. 36-41 (which is developed in the textbook for a somewhat different situation) and using Eq. 36-40 to eliminate the $2L/\lambda$ term.

49. Consider the interference pattern formed by waves reflected from the upper and lower surfaces of the air wedge. The wave reflected from the lower surface undergoes a π rad phase change while the wave reflected from the upper surface does not. At a place where the thickness of the wedge is d, the condition for a maximum in intensity is $2d = (m + \frac{1}{2})\lambda$, where λ is the wavelength in air and m is an integer. Thus, $d = (2m+1)\lambda/4$. As the geometry of Fig. 36-34 shows, $d = R - \sqrt{R^2 - r^2}$, where R is the radius of curvature of the lens and r is the radius of a Newton's ring. Thus, $(2m+1)\lambda/4 = R - \sqrt{R^2 - r^2}$. First, we rearrange the terms so the equation becomes

$$\sqrt{R^2 - r^2} = R - \frac{(2m+1)\lambda}{4}\ .$$

Next, we square both sides, rearrange to solve for r^2, then take the square root. We get

$$r = \sqrt{\frac{(2m+1)R\lambda}{2} - \frac{(2m+1)^2\lambda^2}{16}}\ .$$

If R is much larger than a wavelength, the first term dominates the second and

$$r = \sqrt{\frac{(2m+1)R\lambda}{2}}\ .$$

50. (a) We find m from the last formula obtained in problem 49:

$$m = \frac{r^2}{R\lambda} - \frac{1}{2} = \frac{(10 \times 10^{-3}\,\text{m})^2}{(5.0\,\text{m})(589 \times 10^{-9}\,\text{m})} - \frac{1}{2}$$

which (rounding down) yields $m = 33$. Since the first bright fringe corresponds to $m = 0$, $m = 33$ corresponds to the thirty-fourth bright fringe.

(b) We now replace λ by $\lambda_n = \lambda/n_w$. Thus,

$$m_n = \frac{r^2}{R\lambda_n} - \frac{1}{2} = \frac{n_w r^2}{R\lambda} - \frac{1}{2} = \frac{(1.33)(10 \times 10^{-3}\,\text{m})^2}{(5.0\,\text{m})(589 \times 10^{-9}\,\text{m})} - \frac{1}{2} = 45\ .$$

This corresponds to the forty-sixth bright fringe (see remark at the end of our solution in part (a)).

51. We solve for m using the formula $r = \sqrt{(2m+1)R\lambda/2}$ obtained in problem 49 and find $m = r^2/R\lambda - 1/2$. Now, when m is changed to $m + 20$, r becomes r', so $m + 20 = r'^2/R\lambda - 1/2$. Taking the difference between the two equations above, we eliminate m and find

$$R = \frac{r'^2 - r^2}{20\lambda} = \frac{(0.368\,\text{cm})^2 - (0.162\,\text{cm})^2}{20(546 \times 10^{-7}\,\text{cm})} = 100\ \text{cm}\ .$$

52. (a) The binomial theorem (Appendix E) allows us to write

$$\sqrt{k(1+x)} = \sqrt{k}\left(1 + \frac{x}{2} + \frac{x^2}{8} + \frac{3x^3}{48} + \cdots\right) \approx \sqrt{k} + \frac{x}{2}\sqrt{k}$$

for $x \ll 1$. Thus, the end result from the solution of problem 49 yields

$$r_m = \sqrt{R\lambda m\left(1 + \frac{1}{2m}\right)} \approx \sqrt{R\lambda m} + \frac{1}{4m}\sqrt{R\lambda m}$$

and

$$r_{m+1} = \sqrt{R\lambda m\left(1 + \frac{3}{2m}\right)} \approx \sqrt{R\lambda m} + \frac{3}{4m}\sqrt{R\lambda m}$$

for very large values of m. Subtracting these, we obtain

$$\Delta r = \frac{3}{4m}\sqrt{R\lambda m} - \frac{1}{4m}\sqrt{R\lambda m} = \frac{1}{2}\sqrt{\frac{R\lambda}{m}} \, .$$

(b) We take the differential of the area: $dA = d(\pi r^2) = 2\pi r\,dr$, and replace dr with Δr in anticipation of using the result from part (a). Thus, the area between adjacent rings for large values of m is

$$2\pi r_m(\Delta r) \approx 2\pi\left(\sqrt{R\lambda m} + \frac{1}{4m}\sqrt{R\lambda m}\right)\left(\frac{1}{2}\sqrt{\frac{R\lambda}{m}}\right) \approx 2\pi\left(\sqrt{R\lambda m}\right)\left(\frac{1}{2}\sqrt{\frac{R\lambda}{m}}\right)$$

which simplifies to the desired result ($\pi\lambda R$).

53. The wave that goes directly to the receiver travels a distance L_1 and the reflected wave travels a distance L_2. Since the index of refraction of water is greater than that of air this last wave suffers a phase change on reflection of half a wavelength. To obtain constructive interference at the receiver, the difference $L_2 - L_1$ must be an odd multiple of a half wavelength. Consider the diagram below. The right triangle on the left, formed by the vertical line from the water to the transmitter T, the ray incident on the water, and the water line, gives $D_a = a/\tan\theta$. The right triangle on the right, formed by the vertical line from the water to the receiver R, the reflected ray, and the water line leads to $D_b = x/\tan\theta$. Since $D_a + D_b = D$,

$$\tan\theta = \frac{a+x}{D} \, .$$

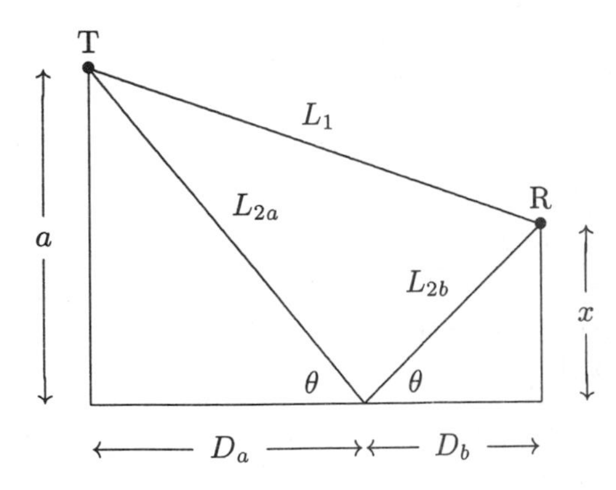

We use the identity $\sin^2\theta = \tan^2\theta/(1+\tan^2\theta)$ to show that $\sin\theta = (a+x)/\sqrt{D^2+(a+x)^2}$. This means

$$L_{2a} = \frac{a}{\sin\theta} = \frac{a\sqrt{D^2+(a+x)^2}}{a+x}$$

and

$$L_{2b} = \frac{x}{\sin\theta} = \frac{x\sqrt{D^2 + (a+x)^2}}{a+x} .$$

Therefore,

$$L_2 = L_{2a} + L_{2b} = \frac{(a+x)\sqrt{D^2+(a+x)^2}}{a+x} = \sqrt{D^2+(a+x)^2} .$$

Using the binomial theorem, with D^2 large and $a^2 + x^2$ small, we approximate this expression: $L_2 \approx D + (a+x)^2/2D$. The distance traveled by the direct wave is $L_1 = \sqrt{D^2 + (a-x)^2}$. Using the binomial theorem, we approximate this expression: $L_1 \approx D + (a-x)^2/2D$. Thus,

$$L_2 - L_1 \approx D + \frac{a^2 + 2ax + x^2}{2D} - D - \frac{a^2 - 2ax + x^2}{2D} = \frac{2ax}{D} .$$

Setting this equal to $(m + \frac{1}{2})\lambda$, where m is zero or a positive integer, we find $x = (m + \frac{1}{2})(D/2a)\lambda$.

54. According to Eq. 36-41, the number of fringes shifted (ΔN) due to the insertion of the film of thickness L is $\Delta N = (2L/\lambda)(n-1)$. Therefore,

$$L = \frac{\lambda \Delta N}{2(n-1)} = \frac{(589\,\text{nm})(7.0)}{2(1.40-1)} = 5.2\,\mu\text{m} .$$

55. A shift of one fringe corresponds to a change in the optical path length of one wavelength. When the mirror moves a distance d the path length changes by $2d$ since the light traverses the mirror arm twice. Let N be the number of fringes shifted. Then, $2d = N\lambda$ and

$$\lambda = \frac{2d}{N} = \frac{2(0.233 \times 10^{-3}\,\text{m})}{792} = 5.88 \times 10^{-7}\,\text{m} = 588\,\text{nm} .$$

56. We denote the two wavelengths as λ and λ', respectively. We apply Eq. 36-40 to both wavelengths and take the difference:

$$N' - N = \frac{2L}{\lambda'} - \frac{2L}{\lambda} = 2L\left(\frac{1}{\lambda'} - \frac{1}{\lambda}\right) .$$

We now require $N' - N = 1$ and solve for L:

$$\begin{aligned} L &= \frac{1}{2}\left(\frac{1}{\lambda} - \frac{1}{\lambda'}\right)^{-1} \\ &= \frac{1}{2}\left(\frac{1}{589.10\,\text{nm}} - \frac{1}{589.59\,\text{nm}}\right)^{-1} \\ &= 3.54 \times 10^5\,\text{nm} = 354\,\mu\text{m} . \end{aligned}$$

57. Let ϕ_1 be the phase difference of the waves in the two arms when the tube has air in it, and let ϕ_2 be the phase difference when the tube is evacuated. These are different because the wavelength in air is different from the wavelength in vacuum. If λ is the wavelength in vacuum, then the wavelength in air is λ/n, where n is the index of refraction of air. This means

$$\phi_1 - \phi_2 = 2L\left[\frac{2\pi n}{\lambda} - \frac{2\pi}{\lambda}\right] = \frac{4\pi(n-1)L}{\lambda}$$

where L is the length of the tube. The factor 2 arises because the light traverses the tube twice, once on the way to a mirror and once after reflection from the mirror. Each shift by one fringe corresponds to a change in phase of 2π rad, so if the interference pattern shifts by N fringes as the tube is evacuated,

$$\frac{4\pi(n-1)L}{\lambda} = 2N\pi$$

and

$$n = 1 + \frac{N\lambda}{2L} = 1 + \frac{60(500 \times 10^{-9}\,\text{m})}{2(5.0 \times 10^{-2}\,\text{m})} = 1.00030 .$$

58. Let the position of the mirror measured from the point at which $d_1 = d_2$ be x. We assume the beam-splitting mechanism is such that the two waves interfere constructively for $x = 0$ (with some beam-splitters, this would not be the case). We can adapt Eq. 36-22 to this situation by incorporating a factor of 2 (since the interferometer utilizes directly reflected light in contrast to the double-slit experiment) and eliminating the $\sin\theta$ factor. Thus, the phase difference between the two light paths is $\Delta\phi = 2(2\pi x/\lambda) = 4\pi x/\lambda$. Then from Eq. 36-21 (writing $4I_0$ as I_m) we find

$$I = I_m \cos^2\left(\frac{\Delta\phi}{2}\right) = I_m \cos^2\left(\frac{2\pi x}{\lambda}\right) .$$

59. (a) To get to the detector, the wave from S_1 travels a distance x and the wave from S_2 travels a distance $\sqrt{d^2 + x^2}$. The phase difference (in terms of wavelengths) between the two waves is

$$\sqrt{d^2 + x^2} - x = m\lambda \qquad m = 0, 1, 2, \ldots$$

where we are requiring constructive interference. The solution is

$$x = \frac{d^2 - m^2\lambda^2}{2m\lambda} .$$

We see that setting $m = 0$ in this expression produces $x = \infty$; hence, the phase difference between the waves when P is very far away is 0.

(b) The result of part (a) implies that the waves constructively interfere at P.

(c) As is particularly evident from our results in part (d), the phase difference increases as x decreases.

(d) We can use our formula from part (a) for the 0.5λ, 1.50λ, etc differences by allowing m in our formula to take on half-integer values. The half- integer values, though, correspond to destructive interference. Using the values $\lambda = 0.500\,\mu$m and $d = 2.00\,\mu$m, we find $x = 7.88\,\mu$m for $m = \frac{1}{2}$, $x = 3.75\,\mu$m for $m = 1$, $x = 2.29\,\mu$m for $m = \frac{3}{2}$, $x = 1.50\,\mu$m for $m = 2$, and $x = 0.975\,\mu$m for $m = \frac{5}{2}$.

60. (a) In a reference frame fixed on Earth, the ether travels leftward with speed v. Thus, the speed of the light beam in this reference frame is $c - v$ as the beam travels rightward from M to M_1, and $c + v$ as it travels back from M_1 to M. The total time for the round trip is therefore given by

$$t_1 = \frac{d_1}{c - v} + \frac{d_1}{c + v} = \frac{2cd_1}{c^2 - v^2} .$$

(b) In a reference frame fixed on the ether, the mirrors travel rightward with speed v, while the speed of the light beam remains c. Thus, in this reference frame, the total distance the beam has to travel is given by

$$d_2' = 2\sqrt{d_2^2 + \left[v\left(\frac{t_2}{2}\right)\right]^2}$$

[see Fig. 36-37(h)-(j)]. Thus,

$$t_2 = \frac{d_2'}{c} = \frac{2}{c}\sqrt{d_2^2 + \left[v\left(\frac{t_2}{2}\right)\right]^2} ,$$

which we solve for t_2:

$$t_2 = \frac{2d_2}{\sqrt{c^2 - v^2}} .$$

(c) We use the binomial expansion (Appendix E)

$$(1+x)^n = 1 + nx + \cdots \approx 1 + nx \qquad (|x| \ll 1) .$$

In our case let $x = v/c \ll 1$, then

$$L_1 = \frac{2c^2 d_1}{c^2 - v^2} = 2d_1 \left[1 - \left(\frac{v}{c}\right)^2\right]^{-1} \approx 2d_1 \left[1 + \left(\frac{v}{c}\right)^2\right] ,$$

and

$$L_2 = \frac{2cd_2}{\sqrt{c^2 - v^2}} = 2d_2 \left[1 - \left(\frac{v}{c}\right)^2\right]^{-1/2} \approx 2d_2 \left[1 + \frac{1}{2}\left(\frac{v}{c}\right)^2\right] .$$

Thus, if $d_1 = d_2 = d$ then

$$\Delta L = L_1 - L_2 \approx 2d \left[1 + \left(\frac{v}{c}\right)^2\right] - 2d \left[1 + \frac{1}{2}\left(\frac{v}{c}\right)^2\right] = \frac{dv^2}{c^2} .$$

(d) In terms of the wavelength, the phase difference is given by

$$\frac{\Delta L}{\lambda} = \frac{dv^2}{\lambda c^2} .$$

(e) We now must reverse the indices 1 and 2, so the new phase difference is

$$\frac{-\Delta L}{\lambda} = -\frac{dv^2}{\lambda c^2} .$$

The shift in phase difference between these two cases is

$$\text{shift} = \left(\frac{\Delta L}{\lambda}\right) - \left(-\frac{\Delta L}{\lambda}\right) = \frac{2dv^2}{\lambda c^2} .$$

(f) Assume that v is about the same as the orbital speed of the Earth, so that $v \approx 29.8\,\text{km/s}$ (see Appendix C). Thus,

$$\text{shift} = \frac{2dv^2}{\lambda c^2} = \frac{2(10\,\text{m})(29.8 \times 10^3\,\text{m/s})^2}{(500 \times 10^{-9}\,\text{m})(2.998 \times 10^8\,\text{m/s})^2} = 0.40 .$$

61. (a) Every time one more destructive (constructive) fringe appears the increase in thickness of the air gap is $\lambda/2$. Now that there are 6 more destructive fringes in addition to the one at point A, the thickness at B is $t_B = 6(\lambda/2) = 3(600\,\text{nm}) = 1.80\,\mu\text{m}$.

(b) We must now replace λ by $\lambda' = \lambda/n_w$. Since t_B is unchanged $t_B = N(\lambda'/2) = N(\lambda/2n_w)$, or

$$N = \frac{2t_B n_w}{\lambda} = \frac{2(3\lambda)n_w}{\lambda} = 6n_w = 6(1.33) = 8 .$$

62. We adapt Eq. 36-21 to the non-reflective coating on a glass lens: $I = I_{\max}\cos^2(\phi/2)$, where $\phi = (2\pi/\lambda)(2n_2 L) + \pi$. At $\lambda = 450\,\text{nm}$

$$\begin{aligned}\frac{I}{I_{\max}} &= \cos^2\left(\frac{\phi}{2}\right) = \cos^2\left(\frac{2\pi n_2 L}{\lambda} + \frac{\pi}{2}\right) \\ &= \cos^2\left[\frac{2\pi(1.38)(99.6\,\text{nm})}{450\,\text{nm}} + \frac{\pi}{2}\right] = 0.883 ,\end{aligned}$$

and at $\lambda = 650\,\text{nm}$

$$\frac{I}{I_{\max}} = \cos^2\left[\frac{2\pi(1.38)(99.6\,\text{nm})}{650\,\text{nm}} + \frac{\pi}{2}\right] = 0.942 .$$

63. For the fifth maximum $y_5 = D\sin\theta_5 = D(5\lambda/d)$, and for the seventh minimum $y_7' = D\sin\theta_7' = D[(6 + 1/2)\lambda/d]$. Thus,

$$\begin{aligned}\Delta y &= y_7' - y_5 = D\left[\frac{(6+1/2)\lambda}{d}\right] - D\left(\frac{5\lambda}{d}\right) = \frac{3\lambda D}{2d} \\ &= \frac{3(546\times10^{-9}\,\text{m})(20\times10^{-2}\,\text{m})}{2(0.10\times10^{-3}\,\text{m})} \\ &= 1.6\times10^{-3}\,\text{m} = 1.6\ \text{mm}\ .\end{aligned}$$

64. Let the $m = 10$ bright fringe on the screen be a distance y from the central maximum. Then from Fig. 36-8(a)

$$r_1 - r_2 = \sqrt{(y+d/2)^2 + D^2} - \sqrt{(y-d/2)^2 + D^2} = 10\lambda\ ,$$

from which we may solve for y. To the order of $(d/D)^2$ we find

$$y = y_0 + \frac{y(y^2 + d^2/4)}{2D^2}\ ,$$

where $y_0 = 10D\lambda/d$. Thus, we find the percent error as follows:

$$\frac{y_0(y_0^2 + d^2/4)}{2y_0D^2} = \frac{1}{2}\left(\frac{10\lambda}{D}\right)^2 + \frac{1}{8}\left(\frac{d}{D}\right)^2 = \frac{1}{2}\left(\frac{5.89\,\mu\text{m}}{2000\,\mu\text{m}}\right)^2 + \frac{1}{8}\left(\frac{2.0\,\text{mm}}{40\,\text{mm}}\right)^2$$

which yields 0.03%.

65. $v_{\min} = c/n = (2.998\times10^8\,\text{m/s})/1.54 = 1.95\times10^8\,\text{m/s}$.

66. With phasor techniques, this amounts to a vector addition problem $\vec{R} = \vec{A}+\vec{B}+\vec{C}$ where (in magnitude-angle notation) $\vec{A} = (10\ \angle\ 0°)$, $\vec{B} = (5\ \angle\ 45°)$, and $\vec{C} = (5\ \angle{-45°})$, where the magnitudes are understood to be in μV/m. We obtain the resultant (especially efficient on a vector capable calculator in polar mode):

$$\vec{R} = (10\ \angle\ 0°) + (5\ \angle\ 45°) + (5\ \angle - 45°) = (17.1\ \angle\ 0°)$$

which leads to

$$E_R = (17.1\ \mu\text{V/m})\sin(\omega t)$$

where $\omega = 2.0\times10^{14}$ rad/s.

67. (a) and (b) Dividing Eq. 36-12 by the wavelength, we obtain

$$N = \frac{\Delta L}{\lambda} = \frac{d}{\lambda}\sin\theta = 39.6$$

wavelengths. This is close to a half-integer value (destructive interference), so that the correct response is "intermediate illumination but closer to darkness."

68. To explore one quadrant of the circle, we look for angles where Eq. 36-14 is satisfied.

$$\theta = \sin^{-1}\frac{m\lambda}{d} \qquad \text{for } m = 0, 1, 2...$$

where $m\lambda/d$ cannot exceed unity. For $m = 1...7$ we have solutions that are "mirrored" in every other quadrant; so there are $4\times7 = 28$ of these. The solutions at $m = 0$ and $m = 8$ are "special" in that they have twins (at 180° and 270°, respectively) and their multiplicity is 2, not 4. Thus, we have $28+2(2) = 32$ points of maxima.

69. In this case the path traveled by ray no. 2 is longer than that of ray no. 1 by $2L/\cos\theta_r$, instead of $2L$. Here $\sin\theta_i/\sin\theta_r = n_2$, or $\theta_r = \sin^{-1}(\sin\theta_i/n_2)$. So if we replace $2L$ by $2L/\cos\theta_r$ in Eqs. 36-34 and 36-35, we obtain

$$\frac{2n_2L}{\cos\theta_r} = \left(m + \frac{1}{2}\right)\lambda \qquad m = 0,\, 1,\, 2,\cdots$$

for the maxima, and

$$\frac{2n_2L}{cos\theta_r} = m\lambda \qquad m = 0,\, 1,\, 2,\cdots$$

for the minima.

70. (a) and (b) Straightforward application of Eq. 36-3 and $v = \Delta x/\Delta t$ yields the result: pistol 1 with a time equal to 42.03×10^{-12} s; pistol 2 with a time equal to 42.3×10^{-12} s; pistol 3 with a time equal to 43.2×10^{-12} s; and, pistol 4 with a time equal to 41.96×10^{-12} s. We see that the blast from pistol 1 arrives first.

71. We use Eq. 36-34 for constructive interference: $2n_2L = (m + 1/2)\lambda$, or

$$\lambda = \frac{2n_2L}{m+1/2} = \frac{2(1.50)(410\,\text{nm})}{m+1/2} = \frac{1230\,\text{nm}}{m+1/2},$$

where $m = 0, 1, 2, \cdots$. The only value of m which, when substituted into the equation above, would yield a wavelength which falls within the visible light range is $m = 1$. Therefore,

$$\lambda = \frac{1230\,\text{nm}}{1+1/2} = 492 \text{ nm} .$$

72. For the first maximum $m = 0$ and for the tenth one $m = 9$. The separation is $\Delta y = (D\lambda/d)\Delta m = 9D\lambda/d$. We solve for the wavelength:

$$\lambda = \frac{d\Delta y}{9D} = \frac{(0.15\times10^{-3}\,\text{m})(18\times10^{-3}\,\text{m})}{9(50\times10^{-2}\,\text{m})} = 6.0\times10^{-7}\,\text{m} = 600 \text{ nm} .$$

73. In the case of a distant screen the angle θ is close to zero so $\sin\theta \approx \theta$. Thus from Eq. 36-14,

$$\Delta\theta \approx \Delta\sin\theta = \Delta\left(\frac{m\lambda}{d}\right) = \frac{\lambda}{d}\Delta m = \frac{\lambda}{d},$$

or $d \approx \lambda/\Delta\theta = 589\times10^{-9}\,\text{m}/0.018\,\text{rad} = 3.3\times10^{-5}\,\text{m} = 33\,\mu\text{m}$.

74. Using the relations of §36-7, we find that the (vertical) change between the center of one dark band and the next is

$$\Delta y = \lambda 2 = 2.5\times10^{-4} \text{ mm} .$$

Thus, with the (horizontal) separation of dark bands given by $\Delta x = 1.2$ mm, we have

$$\theta \approx \tan\theta = \frac{\Delta y}{\Delta x} = 2.08\times10^{-4} \text{ rad} .$$

Converting this angle into degrees, we arrive at $\theta = 0.012°$.

75. (a) A path length difference of $\lambda/2$ produces the first dark band, of $3\lambda/2$ produces the second dark band, and so on. Therefore, the fourth dark band corresponds to a path length difference of $7\lambda/2 = 1750$ nm.

(b) In the small angle approximation (which we assume holds here), the fringes are equally spaced, so that if Δy denotes the distance from one maximum to the next, then the distance from the middle of the pattern to the fourth dark band must be $16.8\,\text{mm} = 3.5\Delta y$. Therefore, we obtain $\Delta y = 16.8/3.5 = 4.8$ mm.

76. (a) With $\lambda = 0.5\,\mu$m, Eq. 36-14 leads to

$$\theta = \sin^{-1}\frac{(3)(0.5\,\mu\text{m})}{2.00\,\mu\text{m}} = 48.6^\circ\ .$$

(b) Decreasing the frequency means increasing the wavelength – which implies y increases. Qualitatively, this is easily seen with Eq. 36-17. One should exercise caution in appealing to Eq. 36-17 here, due to the fact the small angle approximation is not justified in this problem. The new wavelength is $0.5/0.9 = 0.556\,\mu$m, which produces a new angle of

$$\theta = \sin^{-1}\frac{(3)(0.556\,\mu\text{m})}{2.00\,\mu\text{m}} = 56.4^\circ\ .$$

Using $y = D\tan\theta$ for the old and new angles, and subtracting, we find

$$\Delta y = D\,(\tan 56.4^\circ - \tan 48.6^\circ) = 1.49\ \text{m}\ .$$

77. (a) Following Sample Problem 36-1, we have

$$N_2 - N_1 = \frac{L}{\lambda}(n_2 - n_1) = 1.87$$

which represents a meaningful difference of 0.87 wavelength.

(b) The result in part (a) is closer to 1 wavelength (constructive interference) than it is to $\frac{1}{2}$ wavelength (destructive interference) so the latter choice applies.

(c) This would insert a $\pm\frac{1}{2}$ wavelength into the previous result – resulting in a meaningful difference (between the two rays) equal to $0.87 - 0.50 = 0.37$ wavelength, which is closer to the destructive interference condition. Thus, there is intermediate illumination but closer to darkness.

78. (a) Straightforward application of Eq. 36-3 and $v = \Delta x/\Delta t$ yields the result: film 1 with a traversal time equal to 4.0×10^{-15} s.

(b) Use of Eq. 36-9 leads to the number of wavelengths:

$$N = \frac{L_1 n_1 + L_2 n_2 + L_3 n_3}{\lambda} = 7.5\ .$$

79. (a) In this case, we are dealing with the situation that leads in the textbook to Eq. 36-35 for minima in reflected light from a thin film. The smallest non-zero answer, then, is for $m = 1$: $L = \lambda/2n_2$.

(b) Now, we are dealing with a situation exactly like that treated in Sample Problem 36-5, where the relation $L = \lambda/4n_2$ is derived.

(c) The indices bear the same relation here as in part (b), but we are looking now for the "opposite" result (maximum reflection instead of maximum transmission). We adapt the treatment in Sample Problem 36-5 by requiring $2L = m\lambda/n_2$ instead of $(m + \frac{1}{2})\lambda/2$. The smallest nonzero result in this case is for $m = 1$: $L = \lambda/2n_2$.

80. (a) Since $n_2 > n_3$, this case has no π-phase shift, and the condition for constructive interference is $m\lambda = 2Ln_2$. We solve for L:

$$L = \frac{m\lambda}{2n_2} = \frac{m(525\,\text{nm})}{2(1.55)} = (169\,\text{nm})m\ .$$

For the minimum value of L, let $m = 1$ to obtain $L_{\text{min}} = 169\,$nm.

(b) The light of wavelength λ (other than 525 nm) that would also be preferentially transmitted satisfies $m'\lambda = 2n_2L$, or

$$\lambda = \frac{2n_2L}{m'} = \frac{2(1.55)(169\,\text{nm})}{m'} = \frac{525\,\text{nm}}{m'} \ .$$

Here $m' = 2, 3, 4, \ldots$ (note that $m' = 1$ corresponds to the $\lambda = 525\,\text{nm}$ light, so it should not be included here). Since the minimum value of m' is 2, one can easily verify that no m' will give a value of λ which falls into the visible light range. So no other parts of the visible spectrum will be preferentially transmitted. They are, in fact, reflected.

(c) For a sharp reduction of transmission let

$$\lambda = \frac{2n_2L}{m' + 1/2} = \frac{525\,\text{nm}}{m' + 1/2} \ ,$$

where $m' = 0, 1, 2, 3, \cdots$. In the visible light range $m' = 1$ and $\lambda = 350\,\text{nm}$. This corresponds to the blue-violet light.

81. We adapt the result of problem 21. Now, the phase difference in radians is

$$\frac{2\pi t}{\lambda}(n_2 - n_1) = 2m\pi \ .$$

The problem implies $m = 5$, so the thickness is

$$t = \frac{m\lambda}{n_2 - n_1} = \frac{5(480\,\text{nm})}{1.7 - 1.4} = 8.0 \times 10^3\,\text{nm} = 8.0\,\mu\text{m} \ .$$

82. In Sample Problem 36-2, the relation $\Delta y = \lambda D/d$ is derived. Thus, to prevent Δy from changing, then (since $\Delta y \propto D/d$) we need to double D if d is doubled.

83. (a) In this case, the film has a smaller index material on one side (air) and a larger index material on the other (glass), and we are dealing (in part (a)) with strongly transmitted light, so the condition is given by Eq. 36-35 (which would give dark *reflection* in this scenario)

$$L = \frac{\lambda}{2n_2}\left(m + \frac{1}{2}\right) = 110 \text{ nm}$$

for $n_2 = 1.25$ and $m = 0$.

(b) Now, we are dealing with strongly reflected light, so the condition is given by Eq. 36-34 (which would give no *transmission* in this scenario)

$$L = \frac{m\lambda}{2n_2} = 220 \text{ nm}$$

for $n_2 = 1.25$ and $m = 1$ (the $m = 0$ option is excluded in the problem statement).

84. We infer from Sample Problem 36-2, that (with angle in radians)

$$\Delta\theta = \frac{\lambda}{d}$$

for adjacent fringes. With the wavelength change ($\lambda' = \lambda/n$ by Eq. 36-8), this equation becomes

$$\Delta\theta' = \frac{\lambda'}{d} \ .$$

Dividing one equation by the other, the requirement of *radians* can now be relaxed and we obtain

$$\frac{\Delta\theta'}{\Delta\theta} = \frac{\lambda'}{\lambda} = \frac{1}{n} \ .$$

Therefore, with $n = 1.33$ and $\Delta\theta = 0.30°$, we find $\Delta\theta' = 0.23°$.

85. Using Eq. 36-16 with the small-angle approximation (illustrated in Sample Problem 36-2), we arrive at

$$y = \frac{\left(m + \frac{1}{2}\right)\lambda D}{d}$$

for the position of the $(m+1)^{\text{th}}$ dark band (a simple way to get this is by averaging the expressions in Eq. 36-17 and Eq. 36-18). Thus, with $m = 1$, $y = 0.012$ m and $d = 800\lambda$, we find $D = 6.4$ m.

86. (a) The path length difference between Rays 1 and 2 is $7d - 2d = 5d$. For this to correspond to a half-wavelength requires $5d = \lambda/2$, so that $d = 50.0$ nm.

    (b) The above requirement becomes $5d = \lambda/2n$ in the presence of the solution, with $n = 1.38$. Therefore, $d = 36.2$ nm.

87. (a) The path length difference is $0.5\,\mu$m $= 500$ nm, which is represents $500/400 = 1.25$ wavelengths – that is, a meaningful difference of 0.25 wavelengths. In angular measure, this corresponds to a phase difference of $(0.25)2\pi = \pi/2$ radians.

    (b) When a difference of index of refraction is involved, the approach used in Eq. 36-9 is quite useful. In this approach, we count the wavelengths between S_1 and the origin

$$N_1 = \frac{Ln}{\lambda} + \frac{L'n'}{\lambda}$$

where $n = 1$ (rounding off the index of air), $L = 5.0\,\mu$m, $n' = 1.5$ and $L' = 1.5\,\mu$m. This yields $N_1 = 18.125$ wavelengths. The number of wavelengths between S_2 and the origin is (with $L_2 = 6.0\,\mu$m) given by

$$N_2 = \frac{L_2 n}{\lambda} = 15.000\ .$$

Thus, $N_1 - N_2 = 3.125$ wavelengths, which gives us a meaningful difference of 0.125 wavelength and which "converts" to a phase of $\pi/4$ radian.

Chapter 37

1. The condition for a minimum of a single-slit diffraction pattern is

$$a \sin\theta = m\lambda$$

where a is the slit width, λ is the wavelength, and m is an integer. The angle θ is measured from the forward direction, so for the situation described in the problem, it is 0.60° for $m = 1$. Thus

$$a = \frac{m\lambda}{\sin\theta} = \frac{633 \times 10^{-9}\,\text{m}}{\sin 0.60°} = 6.04 \times 10^{-5}\ \text{m}\ .$$

2. (a) $\theta = \sin^{-1}(1.50\,\text{cm}/2.00\,\text{m}) = 0.430°$.

   (b) For the mth diffraction minimum $a \sin\theta = m\lambda$. We solve for the slit width:

$$a = \frac{m\lambda}{\sin\theta} = \frac{2(441\,\text{nm})}{\sin 0.430°} = 0.118\ \text{mm}\ .$$

3. (a) The condition for a minimum in a single-slit diffraction pattern is given by $a \sin\theta = m\lambda$, where a is the slit width, λ is the wavelength, and m is an integer. For $\lambda = \lambda_a$ and $m = 1$, the angle θ is the same as for $\lambda = \lambda_b$ and $m = 2$. Thus $\lambda_a = 2\lambda_b$.

   (b) Let m_a be the integer associated with a minimum in the pattern produced by light with wavelength λ_a, and let m_b be the integer associated with a minimum in the pattern produced by light with wavelength λ_b. A minimum in one pattern coincides with a minimum in the other if they occur at the same angle. This means $m_a\lambda_a = m_b\lambda_b$. Since $\lambda_a = 2\lambda_b$, the minima coincide if $2m_a = m_b$. Consequently, every other minimum of the λ_b pattern coincides with a minimum of the λ_a pattern.

4. (a) We use Eq. 37-3 to calculate the separation between the first ($m_1 = 1$) and fifth ($m_2 = 5$) minima:

$$\Delta y = D\Delta \sin\theta = D\Delta\left(\frac{m\lambda}{a}\right) = \frac{D\lambda}{a}\Delta m = \frac{D\lambda}{a}(m_2 - m_1)\ .$$

   Solving for the slit width, we obtain

$$a = \frac{D\lambda(m_2 - m_1)}{\Delta y} = \frac{(400\,\text{mm})(550 \times 10^{-6}\,\text{mm})(5 - 1)}{0.35\,\text{mm}} = 2.5\ \text{mm}\ .$$

   (b) For $m = 1$,

$$\sin\theta = \frac{m\lambda}{a} = \frac{(1)(550 \times 10^{-6}\,\text{mm})}{2.5\,\text{mm}} = 2.2 \times 10^{-4}\ .$$

   The angle is $\theta = \sin^{-1}(2.2 \times 10^{-4}) = 2.2 \times 10^{-4}\,\text{rad}$.

5. (a) A plane wave is incident on the lens so it is brought to focus in the focal plane of the lens, a distance of 70 cm from the lens.

(b) Waves leaving the lens at an angle θ to the forward direction interfere to produce an intensity minimum if $a\sin\theta = m\lambda$, where a is the slit width, λ is the wavelength, and m is an integer. The distance on the screen from the center of the pattern to the minimum is given by $y = D\tan\theta$, where D is the distance from the lens to the screen. For the conditions of this problem,

$$\sin\theta = \frac{m\lambda}{a} = \frac{(1)(590\times10^{-9}\,\text{m})}{0.40\times10^{-3}\,\text{m}} = 1.475\times10^{-3}\ .$$

This means $\theta = 1.475\times10^{-3}$ rad and $y = (70\times10^{-2}\,\text{m})\tan(1.475\times10^{-3}\,\text{rad}) = 1.03\times10^{-3}$ m.

6. Let the first minimum be a distance y from the central axis which is perpendicular to the speaker. Then $\sin\theta = y/(D^2+y^2)^{1/2} = m\lambda/a = \lambda/a$ (for $m=1$). Therefore,

$$\begin{aligned} y &= \frac{D}{\sqrt{(a/\lambda)^2-1}} = \frac{D}{\sqrt{(af/v_s)^2-1}} \\ &= \frac{100\,\text{m}}{\sqrt{[(0.300\,\text{m})(3000\,\text{Hz})/(343\,\text{m/s})]^2-1}} = 41.2\ \text{m}\ . \end{aligned}$$

7. The condition for a minimum of intensity in a single-slit diffraction pattern is $a\sin\theta = m\lambda$, where a is the slit width, λ is the wavelength, and m is an integer. To find the angular position of the first minimum to one side of the central maximum, we set $m=1$:

$$\theta_1 = \sin^{-1}\left(\frac{\lambda}{a}\right) = \sin^{-1}\left(\frac{589\times10^{-9}\,\text{m}}{1.00\times10^{-3}\,\text{m}}\right) = 5.89\times10^{-4}\ \text{rad}\ .$$

If D is the distance from the slit to the screen, the distance on the screen from the center of the pattern to the minimum is

$$y_1 = D\tan\theta_1 = (3.00\,\text{m})\tan(5.89\times10^{-4}\,\text{rad}) = 1.767\times10^{-3}\ \text{m}\ .$$

To find the second minimum, we set $m=2$:

$$\theta_2 = \sin^{-1}\left(\frac{2(589\times10^{-9}\,\text{m})}{1.00\times10^{-3}\,\text{m}}\right) = 1.178\times10^{-3}\ \text{rad}\ .$$

The distance from the center of the pattern to this second minimum is $y_2 = D\tan\theta_2 = (3.00\,\text{m})\tan(1.178\times10^{-3}\,\text{rad}) = 3.534\times10^{-3}\,\text{m}$. The separation of the two minima is $\Delta y = y_2 - y_1 = 3.534\,\text{mm} - 1.767\,\text{mm} = 1.77\,\text{mm}$.

8. We note that $\text{nm} = 10^{-9}\,\text{m} = 10^{-6}\,\text{mm}$. From Eq. 37-4,

$$\Delta\phi = \left(\frac{2\pi}{\lambda}\right)(\Delta x\sin\theta) = \left(\frac{2\pi}{589\times10^{-6}\,\text{mm}}\right)\left(\frac{0.10\,\text{mm}}{2}\right)\sin 30^\circ = 266.7\ \text{rad}\ .$$

This is equivalent to $266.7 - 84\pi = 2.8\,\text{rad} = 160^\circ$.

9. We imagine dividing the original slit into N strips and represent the light from each strip, when it reaches the screen, by a phasor. Then, at the central maximum in the diffraction pattern, we would add the N phasors, all in the same direction and each with the same amplitude. We would find that the intensity there is proportional to N^2. If we double the slit width, we need $2N$ phasors if they are each to have the amplitude of the phasors we used for the narrow slit. The intensity at the central maximum is proportional to $(2N)^2$ and is, therefore, four times the intensity for the narrow slit. The energy reaching the screen per unit time, however, is only twice the energy reaching it per unit time when the narrow slit is in place. The energy is simply redistributed. For example, the central peak is now half as wide and the integral of the intensity over the peak is only twice the analogous integral for the narrow slit.

10. (a) $\theta = \sin^{-1}(0.011\,\text{cm}/3.5\,\text{m}) = 0.18^\circ$.

(b) We use Eq. 37-6:

$$\alpha = \left(\frac{\pi a}{\lambda}\right)\sin\theta = \frac{\pi(0.025\,\text{mm})\sin 0.18^\circ}{538\times 10^{-6}\,\text{mm}} = 0.46\ \text{rad}\ .$$

(c) Making sure our calculator is in radian mode, Eq. 37-5 yields

$$\frac{I(\theta)}{I_m} = \left(\frac{\sin\alpha}{\alpha}\right)^2 = 0.93\ .$$

11. (a) The intensity for a single-slit diffraction pattern is given by

$$I = I_m\,\frac{\sin^2\alpha}{\alpha^2}$$

where $\alpha = (\pi a/\lambda)\sin\theta$, a is the slit width and λ is the wavelength. The angle θ is measured from the forward direction. We require $I = I_m/2$, so

$$\sin^2\alpha = \frac{1}{2}\alpha^2\ .$$

(b) We evaluate $\sin^2\alpha$ and $\alpha^2/2$ for $\alpha = 1.39\,\text{rad}$ and compare the results. To be sure that 1.39 rad is closer to the correct value for α than any other value with three significant digits, we could also try 1.385 rad and 1.395 rad.

(c) Since $\alpha = (\pi a/\lambda)\sin\theta$,

$$\theta = \sin^{-1}\left(\frac{\alpha\lambda}{\pi a}\right)\ .$$

Now $\alpha/\pi = 1.39/\pi = 0.442$, so

$$\theta = \sin^{-1}\left(\frac{0.442\lambda}{a}\right)\ .$$

The angular separation of the two points of half intensity, one on either side of the center of the diffraction pattern, is

$$\Delta\theta = 2\theta = 2\sin^{-1}\left(\frac{0.442\lambda}{a}\right)\ .$$

(d) For $a/\lambda = 1.0$,

$$\Delta\theta = 2\sin^{-1}(0.442/1.0) = 0.916\,\text{rad} = 52.5^\circ\ ,$$

for $a/\lambda = 5.0$,

$$\Delta\theta = 2\sin^{-1}(0.442/5.0) = 0.177\,\text{rad} = 10.1^\circ\ ,$$

and for $a/\lambda = 10$,

$$\Delta\theta = 2\sin^{-1}(0.442/10) = 0.0884\,\text{rad} = 5.06^\circ\ .$$

12. Consider Huygens' explanation of diffraction phenomena. When A is in place only the Huygens' wavelets that pass through the hole get to point P. Suppose they produce a resultant electric field E_A. When B is in place, the light that was blocked by A gets to P and the light that passed through the hole in A is blocked. Suppose the electric field at P is now $\vec{E}_B$. The sum $\vec{E}_A + \vec{E}_B$ is the resultant of all waves that get to P when neither A nor B are present. Since P is in the geometric shadow, this is zero. Thus $\vec{E}_A = -\vec{E}_B$, and since the intensity is proportional to the square of the electric field, the intensity at P is the same when A is present as when B is present.

13. (a) The intensity for a single-slit diffraction pattern is given by

$$I = I_m\,\frac{\sin^2\alpha}{\alpha^2}$$

where α is described in the text (see Eq. 37-6). To locate the extrema, we set the derivative of I with respect to α equal to zero and solve for α. The derivative is

$$\frac{dI}{d\alpha} = 2I_m \frac{\sin\alpha}{\alpha^3} (\alpha\cos\alpha - \sin\alpha) \ .$$

The derivative vanishes if $\alpha \neq 0$ but $\sin\alpha = 0$. This yields $\alpha = m\pi$, where m is a nonzero integer. These are the intensity minima: $I = 0$ for $\alpha = m\pi$. The derivative also vanishes for $\alpha\cos\alpha - \sin\alpha = 0$. This condition can be written $\tan\alpha = \alpha$. These implicitly locate the maxima.

(b) The values of α that satisfy $\tan\alpha = \alpha$ can be found by trial and error on a pocket calculator or computer. Each of them is slightly less than one of the values $(m + \frac{1}{2})\pi$ rad, so we start with these values. The first few are 0, 4.4934, 7.7252, 10.9041, 14.0662, and 17.2207. They can also be found graphically. As in the diagram below, we plot $y = \tan\alpha$ and $y = \alpha$ on the same graph. The intersections of the line with the $\tan\alpha$ curves are the solutions. The first two solutions listed above are shown on the diagram.

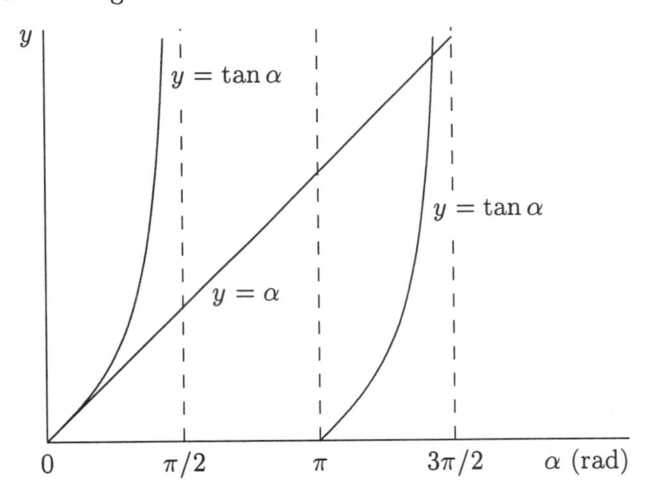

(c) We write $\alpha = (m + \frac{1}{2})\pi$ for the maxima. For the central maximum, $\alpha = 0$ and $m = -\frac{1}{2}$. For the next, $\alpha = 4.4934$ and $m = 0.930$. For the next, $\alpha = 7.7252$ and $m = 1.959$.

14. We use Eq. 37-12 with $\theta = 2.5°/2 = 1.25°$. Thus,

$$d = \frac{1.22\lambda}{\sin\theta} = \frac{1.22(550\,\text{nm})}{\sin 1.25°} = 31\ \mu\text{m} \ .$$

15. (a) We use the Rayleigh criteria. Thus, the angular separation (in radians) of the sources must be at least $\theta_R = 1.22\lambda/d$, where λ is the wavelength and d is the diameter of the aperture. For the headlights of this problem,

$$\theta_R = \frac{1.22(550 \times 10^{-9}\,\text{m})}{5.0 \times 10^{-3}\,\text{m}} = 1.34 \times 10^{-4}\ \text{rad} \ .$$

(b) If L is the distance from the headlights to the eye when the headlights are just resolvable and D is the separation of the headlights, then $D = L\theta_R$, where the small angle approximation is made. This is valid for θ_R in radians. Thus,

$$L = \frac{D}{\theta_R} = \frac{1.4\,\text{m}}{1.34 \times 10^{-4}\,\text{rad}} = 1.0 \times 10^4\ \text{m} = 10\ \text{km} \ .$$

16. (a) We use Eq. 37-14:

$$\theta_R = 1.22\frac{\lambda}{d} = \frac{(1.22)(540 \times 10^{-6}\,\text{mm})}{5.0\,\text{mm}} = 1.3 \times 10^{-4}\ \text{rad} \ .$$

(b) The linear separation is $D = L\theta_R = (160 \times 10^3\,\text{m})(1.3 \times 10^{-4}\,\text{rad}) = 21\,\text{m}$.

17. Using the notation of Sample Problem 37-6 (which is in the textbook supplement), the minimum separation is

$$D = L\theta_R = L\left(1.22\frac{\lambda}{d}\right) = (3.82 \times 10^8\,\text{m})\frac{(1.22)(550 \times 10^{-9}\,\text{m})}{5.1\,\text{m}} = 50\,\text{m} .$$

18. Using the notation of Sample Problem 37-6 (which is in the textbook supplement), the maximum distance is

$$L = \frac{D}{\theta_R} = \frac{D}{1.22\lambda/d} = \frac{(5.0 \times 10^{-3}\,\text{m})(4.0 \times 10^{-3}\,\text{m})}{1.22(550 \times 10^{-9}\,\text{m})} = 30\,\text{m} .$$

19. (a) We use the Rayleigh criteria. If L is the distance from the observer to the objects, then the smallest separation D they can have and still be resolvable is $D = L\theta_R$, where θ_R is measured in radians. The small angle approximation is made. Thus,

$$D = \frac{1.22L\lambda}{d} = \frac{1.22(8.0 \times 10^{10}\,\text{m})(550 \times 10^{-9}\,\text{m})}{5.0 \times 10^{-3}\,\text{m}} = 1.1 \times 10^7\,\text{m} = 1.1 \times 10^4\,\text{km} .$$

This distance is greater than the diameter of Mars; therefore, one part of the planet's surface cannot be resolved from another part.

(b) Now $d = 5.1\,\text{m}$ and

$$D = \frac{1.22(8.0 \times 10^{10}\,\text{m})(550 \times 10^{-9}\,\text{m})}{5.1\,\text{m}} = 1.1 \times 10^4\,\text{m} = 11\,\text{km} .$$

20. Using the notation of Sample Problem 37-6 (which is in the textbook supplement), the minimum separation is

$$D = L\theta_R = L\left(\frac{1.22\lambda}{d}\right) = \frac{(6.2 \times 10^3\,\text{m})(1.22)(1.6 \times 10^{-2}\,\text{m})}{2.3\,\text{m}} = 53\,\text{m} .$$

21. Eq. 37-14 gives $\theta_R = 1.22\lambda/d$, where in our case $\theta_R \approx D/L$, with $D = 60\,\mu\text{m}$ being the size of the object your eyes must resolve, and L being the maximum viewing distance in question. If $d = 3.00\,\text{mm} = 3000\,\mu\text{m}$ is the diameter of your pupil, then

$$L = \frac{Dd}{1.22\lambda} = \frac{(60\,\mu\text{m})(3000\,\mu\text{m})}{1.22(0.55\,\mu\text{m})} = 2.7 \times 10^5\,\mu\text{m} = 27\,\text{cm} .$$

22. Since we are considering the *diameter* of the central diffraction maximum, then we are working with *twice* the Rayleigh angle. Using notation similar to that in Sample Problem 37-6 (which is in the textbook supplement), we have $2(1.22\lambda/d) = D/L$. Therefore,

$$d = 2\frac{1.22\lambda L}{D} = 2\frac{(1.22)(500 \times 10^{-9}\,\text{m})(3.54 \times 10^5\,\text{m})}{9.1\,\text{m}} = 0.047\,\text{m} .$$

23. (a) The first minimum in the diffraction pattern is at an angular position θ, measured from the center of the pattern, such that $\sin\theta = 1.22\lambda/d$, where λ is the wavelength and d is the diameter of the antenna. If f is the frequency, then the wavelength is

$$\lambda = \frac{c}{f} = \frac{3.00 \times 10^8\,\text{m/s}}{220 \times 10^9\,\text{Hz}} = 1.36 \times 10^{-3}\,\text{m} .$$

Thus

$$\theta = \sin^{-1}\left(\frac{1.22\lambda}{d}\right) = \sin^{-1}\left(\frac{1.22(1.36 \times 10^{-3}\,\text{m})}{55.0 \times 10^{-2}\,\text{m}}\right) = 3.02 \times 10^{-3}\,\text{rad} .$$

The angular width of the central maximum is twice this, or $6.04 \times 10^{-3}\,\text{rad}$ $(0.346°)$.

(b) Now $\lambda = 1.6\,\text{cm}$ and $d = 2.3\,\text{m}$, so

$$\theta = \sin^{-1}\left(\frac{1.22(1.6\times10^{-2}\,\text{m})}{2.3\,\text{m}}\right) = 8.5\times10^{-3}\ \text{rad}\ .$$

The angular width of the central maximum is $1.7\times10^{-2}\,\text{rad}$ ($0.97°$).

24. (a) Since $\theta = 1.22\lambda/d$, the larger the wavelength the larger the radius of the first minimum (and second maximum, etc). Therefore, the white pattern is outlined by red lights (with longer wavelength than blue lights).

(b) The diameter of a water drop is

$$d = \frac{1.22\lambda}{\theta} \approx \frac{1.22(7\times10^{-7}\,\text{m})}{1.5(0.50°)(\pi/180°)/2} = 1.3\times10^{-4}\ \text{m}\ .$$

25. (a) Using Eq. 37-14, the angular separation is

$$\theta_\text{R} = \frac{1.22\lambda}{d} = \frac{(1.22)(550\times10^{-9}\,\text{m})}{0.76\,\text{m}} = 8.8\times10^{-7}\ \text{rad}\ .$$

(b) Using the notation of Sample Problem 37-6 (which is in the textbook supplement), the distance between the stars is

$$D = L\theta_\text{R} = \frac{(10\,\text{ly})(9.46\times10^{12}\,\text{km/ly})(0.18)\pi}{(3600)(180)} = 8.4\times10^{7}\ \text{km}\ .$$

(c) The diameter of the first dark ring is

$$d = 2\theta_\text{R}L = \frac{2(0.18)(\pi)(14\,\text{m})}{(3600)(180)} = 2.5\times10^{-5}\,\text{m} = 0.025\ \text{mm}\ .$$

26. We denote the Earth-Moon separation as L. The energy of the beam of light which is projected onto the moon is concentrated in a circular spot of diameter d_1, where $d_1/L = 2\theta_\text{R} = 2(1.22\lambda/d_0)$, with d_0 the diameter of the mirror on Earth. The fraction of energy picked up by the reflector of diameter d_2 on the Moon is then $\eta' = (d_2/d_1)^2$. This reflected light, upon reaching the Earth, has a circular cross section of diameter d_3 satisfying $d_3/L = 2\theta_\text{R} = 2(1.22\lambda/d_2)$. The fraction of the reflected energy that is picked up by the telescope is then $\eta'' = (d_0/d_3)^2$. Consequently, the fraction of the original energy picked up by the detector is

$$\begin{aligned}\eta &= \eta'\eta'' = \left(\frac{d_0}{d_3}\right)^2\left(\frac{d_2}{d_1}\right)^2 = \left[\frac{d_0d_2}{(2.44\lambda d_{em}/d_0)(2.44\lambda d_{em}/d_2)}\right]^2 = \left(\frac{d_0d_2}{2.44\lambda d_{em}}\right)^4 \\ &= \left[\frac{(2.6\,\text{m})(0.10\,\text{m})}{2.44(0.69\times10^{-6}\,\text{m})(3.82\times10^{8}\,\text{m})}\right]^4 \approx 4\times10^{-13}\ .\end{aligned}$$

27. Bright interference fringes occur at angles θ given by $d\sin\theta = m\lambda$, where m is an integer. For the slits of this problem, $d = 11a/2$, so $a\sin\theta = 2m\lambda/11$ (see Sample Problem 37-4). The first minimum of the diffraction pattern occurs at the angle θ_1 given by $a\sin\theta_1 = \lambda$, and the second occurs at the angle θ_2 given by $a\sin\theta_2 = 2\lambda$, where a is the slit width. We should count the values of m for which $\theta_1 < \theta < \theta_2$, or, equivalently, the values of m for which $\sin\theta_1 < \sin\theta < \sin\theta_2$. This means $1 < (2m/11) < 2$. The values are $m = 6$, 7, 8, 9, and 10. There are five bright fringes in all.

28. In a manner similar to that discussed in Sample Problem 37-4, we find the number is $2(d/a) - 1 = 2(2a/a) - 1 = 3$.

29. (a) In a manner similar to that discussed in Sample Problem 37-4, we find the ratio should be $d/a = 4$. Our reasoning is, briefly, as follows: we let the location of the fourth bright fringe coincide with the first minimum of diffraction pattern, and then set $\sin\theta = 4\lambda/d = \lambda/a$ (so $d = 4a$).

(b) Any bright fringe which happens to be at the same location with a diffraction minimum will vanish. Thus, if we let $\sin\theta = m_1\lambda/d = m_2\lambda/a = m_1\lambda/4a = m_2\lambda/a$, or $m_1 = 4m_2$ where $m_2 = 1, 2, 3, \cdots$. The fringes missing are the 4th, 8th, 12th, and so on. Hence, every fourth fringe is missing.

30. The angular location of the mth bright fringe is given by $d\sin\theta = m\lambda$, so the linear separation between two adjacent fringes is

$$\Delta y = \Delta(D\sin\theta) = \Delta\left(\frac{D_m\lambda}{d}\right) = \frac{D\lambda}{d}\Delta m = \frac{D\lambda}{d}\,.$$

31. (a) The angular positions θ of the bright interference fringes are given by $d\sin\theta = m\lambda$, where d is the slit separation, λ is the wavelength, and m is an integer. The first diffraction minimum occurs at the angle θ_1 given by $a\sin\theta_1 = \lambda$, where a is the slit width. The diffraction peak extends from $-\theta_1$ to $+\theta_1$, so we should count the number of values of m for which $-\theta_1 < \theta < +\theta_1$, or, equivalently, the number of values of m for which $-\sin\theta_1 < \sin\theta < +\sin\theta_1$. This means $-1/a < m/d < 1/a$ or $-d/a < m < +d/a$. Now $d/a = (0.150 \times 10^{-3}\,\text{m})/(30.0 \times 10^{-6}\,\text{m}) = 5.00$, so the values of m are $m = -4, -3, -2, -1, 0, +1, +2, +3$, and $+4$. There are nine fringes.

(b) The intensity at the screen is given by

$$I = I_m\left(\cos^2\beta\right)\left(\frac{\sin\alpha}{\alpha}\right)^2$$

where $\alpha = (\pi a/\lambda)\sin\theta$, $\beta = (\pi d/\lambda)\sin\theta$, and I_m is the intensity at the center of the pattern. For the third bright interference fringe, $d\sin\theta = 3\lambda$, so $\beta = 3\pi$ rad and $\cos^2\beta = 1$. Similarly, $\alpha = 3\pi a/d = 3\pi/5.00 = 0.600\pi$ rad and

$$\left(\frac{\sin\alpha}{\alpha}\right)^2 = \left(\frac{\sin 0.600\pi}{0.600\pi}\right)^2 = 0.255\,.$$

The intensity ratio is $I/I_m = 0.255$.

32. (a) The first minimum of the diffraction pattern is at 5.00°, so

$$a = \frac{\lambda}{\sin\theta} = \frac{0.440\,\mu\text{m}}{\sin 5.00°} = 5.05\,\mu\text{m}\,.$$

(b) Since the fourth bright fringe is missing, $d = 4a = 4(5.05\,\mu\text{m}) = 20.2\,\mu\text{m}$.

(c) For the $m = 1$ bright fringe,

$$\alpha = \frac{\pi a\sin\theta}{\lambda} = \frac{\pi(5.05\,\mu\text{m})\sin 1.25°}{0.440\,\mu\text{m}} = 0.787\text{ rad}\,.$$

Consequently, the intensity of the $m = 1$ fringe is

$$I = I_m\left(\frac{\sin\alpha}{\alpha}\right)^2 = (7.0\,\text{mW/cm}^2)\left(\frac{\sin 0.787\,\text{rad}}{0.787}\right)^2 = 5.7\text{ mW/cm}^2\,,$$

which agrees with Fig. 37-36. Similarly for $m = 2$, the intensity is $I = 2.9\,\text{mW/cm}^2$, also in agreement with Fig. 37-36.

33. (a) $d = 20.0\,\text{mm}/6000 = 0.00333\,\text{mm} = 3.33\,\mu\text{m}$.

(b) Let $d\sin\theta = m\lambda\,(m = 0, \pm1, \pm2, \cdots)$. We find $\theta = 0$ for $m = 0$, and

$$\theta = \sin^{-1}(\pm\lambda/d) = \sin^{-1}\left(\pm\frac{0.589\,\mu\text{m}}{3.30\,\mu\text{m}}\right) = \pm10.2°$$

for $m = \pm1$. Similarly, we find $\pm20.7°$ for $m = \pm2$, $\pm32.2°$ for $m = \pm3$, $\pm45°$ for $m = \pm4$, and $\pm62.2°$ for $m = \pm5$. Since $|m|\lambda/d > 1$ for $|m| \geq 6$, these are all the maxima.

34. The angular location of the mth order diffraction maximum is given by $m\lambda = d\sin\theta$. To be able to observe the fifth-order maximum, we must let $\sin\theta|_{m=5} = 5\lambda/d < 1$, or

$$\lambda < \frac{d}{5} = \frac{1.00\,\text{nm}/315}{5} = 635\text{ nm}\ .$$

Therefore, all wavelengths shorter than 635 nm can be used.

35. The ruling separation is $d = 1/(400\,\text{mm}^{-1}) = 2.5\times10^{-3}\,\text{mm}$. Diffraction lines occur at angles θ such that $d\sin\theta = m\lambda$, where λ is the wavelength and m is an integer. Notice that for a given order, the line associated with a long wavelength is produced at a greater angle than the line associated with a shorter wavelength. We take λ to be the longest wavelength in the visible spectrum (700 nm) and find the greatest integer value of m such that θ is less than 90°. That is, find the greatest integer value of m for which $m\lambda < d$. Since $d/\lambda = (2.5\times10^{-6}\,\text{m})/(700\times10^{-9}\,\text{m}) = 3.57$, that value is $m = 3$. There are three complete orders on each side of the $m = 0$ order. The second and third orders overlap.

36. We use Eq. 37-22 for diffraction maxima: $d\sin\theta = m\lambda$. In our case, since the angle between the $m = 1$ and $m = -1$ maxima is 26°, the angle θ corresponding to $m = 1$ is $\theta = 26°/2 = 13°$. We solve for the grating spacing:

$$d = \frac{m\lambda}{\sin\theta} = \frac{(1)(550\,\text{nm})}{\sin 13°} = 2.4\,\mu\text{m}\ .$$

37. (a) Maxima of a diffraction grating pattern occur at angles θ given by $d\sin\theta = m\lambda$, where d is the slit separation, λ is the wavelength, and m is an integer. The two lines are adjacent, so their order numbers differ by unity. Let m be the order number for the line with $\sin\theta = 0.2$ and $m+1$ be the order number for the line with $\sin\theta = 0.3$. Then, $0.2d = m\lambda$ and $0.3d = (m+1)\lambda$. We subtract the first equation from the second to obtain $0.1d = \lambda$, or $d = \lambda/0.1 = (600\times10^{-9}\,\text{m})/0.1 = 6.0\times10^{-6}\,\text{m}$.

(b) Minima of the single-slit diffraction pattern occur at angles θ given by $a\sin\theta = m\lambda$, where a is the slit width. Since the fourth-order interference maximum is missing, it must fall at one of these angles. If a is the smallest slit width for which this order is missing, the angle must be given by $a\sin\theta = \lambda$. It is also given by $d\sin\theta = 4\lambda$, so $a = d/4 = (6.0\times10^{-6}\,\text{m})/4 = 1.5\times10^{-6}\,\text{m}$.

(c) First, we set $\theta = 90°$ and find the largest value of m for which $m\lambda < d\sin\theta$. This is the highest order that is diffracted toward the screen. The condition is the same as $m < d/\lambda$ and since $d/\lambda = (6.0\times10^{-6}\,\text{m})/(600\times10^{-9}\,\text{m}) = 10.0$, the highest order seen is the $m = 9$ order. The fourth and eighth orders are missing, so the observable orders are $m = 0$, 1, 2, 3, 5, 6, 7, and 9.

38. (a) For the maximum with the greatest value of $m\,(= M)$ we have $M\lambda = a\sin\theta < d$, so $M < d/\lambda = 900\,\text{nm}/600\,\text{nm} = 1.5$, or $M = 1$. Thus three maxima can be seen, with $m = 0, \pm1$.

(b) From Eq. 37-25

$$\begin{aligned}\Delta\theta_{\text{hw}} &= \frac{\lambda}{Nd\cos\theta} = \frac{d\sin\theta}{Nd\cos\theta} = \frac{\tan\theta}{N} = \frac{1}{N}\tan\left[\sin^{-1}\left(\frac{\lambda}{d}\right)\right] \\ &= \frac{1}{1000}\tan\left[\sin^{-1}\left(\frac{600\,\text{nm}}{900\,\text{nm}}\right)\right] = 0.051°\ .\end{aligned}$$

39. The angular positions of the first-order diffraction lines are given by $d\sin\theta = \lambda$. Let λ_1 be the shorter wavelength (430 nm) and θ be the angular position of the line associated with it. Let λ_2 be the longer wavelength (680 nm), and let $\theta+\Delta\theta$ be the angular position of the line associated with it. Here $\Delta\theta = 20°$. Then, $d\sin\theta = \lambda_1$ and $d\sin(\theta+\Delta\theta) = \lambda_2$. We write $\sin(\theta+\Delta\theta)$ as $\sin\theta\cos\Delta\theta + \cos\theta\sin\Delta\theta$, then use the equation for the first line to replace $\sin\theta$ with λ_1/d, and $\cos\theta$ with $\sqrt{1-\lambda_1^2/d^2}$. After multiplying by d, we obtain

$$\lambda_1\cos\Delta\theta + \sqrt{d^2-\lambda_1^2}\sin\Delta\theta = \lambda_2 \ .$$

Solving for d, we find

$$\begin{aligned} d &= \sqrt{\frac{(\lambda_2-\lambda_1\cos\Delta\theta)^2 + (\lambda_1\sin\Delta\theta)^2}{\sin^2\Delta\theta}} \\ &= \sqrt{\frac{[(680\,\text{nm})-(430\,\text{nm})\cos 20°]^2 + [(430\,\text{nm})\sin 20°]^2}{\sin^2 20°}} \\ &= 914\,\text{nm} = 9.14\times10^{-4}\ \text{mm} \ . \end{aligned}$$

There are $1/d = 1/(9.14\times10^{-4}\,\text{mm}) = 1090$ rulings per mm.

40. We use Eq. 37-22. For $m = \pm1$

$$\lambda = \frac{d\sin\theta}{m} = \frac{(1.73\mu\text{m})\sin(\pm17.6°)}{\pm1} = 523\ \text{nm} \ ,$$

and for $m = \pm2$

$$\lambda = \frac{(1.73\mu\text{m})\sin(\pm37.3°)}{\pm2} = 524\,\text{nm} \ .$$

Similarly, we may compute the values of λ corresponding to the angles for $m = \pm3$. The average value of these λ's is 523 nm.

41. Consider two of the rays shown in Fig. 37-37, one just above the other. The extra distance traveled by the lower one may be found by drawing perpendiculars from where the top ray changes direction (point P) to the incident and diffracted paths of the lower one. Where these perpendiculars intersect the lower ray's paths are here referred to as points A and C. Where the bottom ray changes direction is point B. We note that angle $\angle APB$ is the same as ψ, and angle BPC is the same as θ (see Fig. 37-37). The difference in path lengths between the two adjacent light rays is $\Delta x = |AB| + |BC| = d\sin\psi + d\sin\theta$. The condition for bright fringes to occur is therefore

$$\Delta x = d\,(\sin\psi + \sin\theta) = m\lambda$$

where $m = 0, 1, 2, \cdots$. If we set $\psi = 0$ then this reduces to Eq. 37-22.

42. Referring to problem 41, we note that the angular deviation of a diffracted ray (the angle between the forward extrapolation of the incident ray and its diffracted ray) is $\psi + \theta$. For $m = 1$, this becomes

$$\psi + \theta = \psi + \sin^{-1}\left(\frac{\lambda}{d} - \sin\psi\right)$$

where the ratio $\lambda/d = 0.40$ using the values given in the problem statement. The graph of this is shown below (with radians used along both axes).

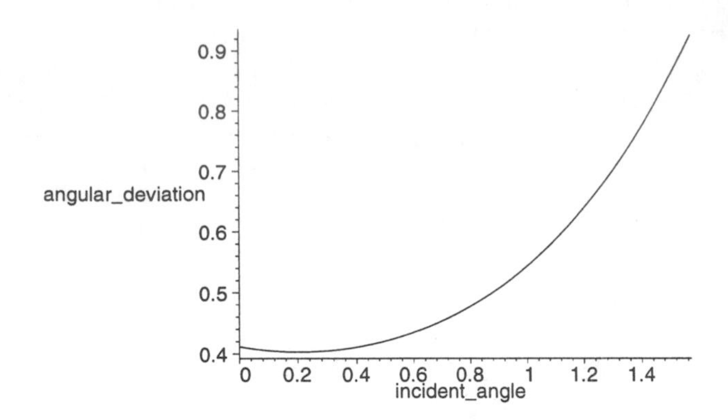

43. The derivation is similar to that used to obtain Eq. 37-24. At the first minimum beyond the mth principal maximum, two waves from adjacent slits have a phase difference of $\Delta\phi = 2\pi m + (2\pi/N)$, where N is the number of slits. This implies a difference in path length of $\Delta L = (\Delta\phi/2\pi)\lambda = m\lambda + (\lambda/N)$. If θ_m is the angular position of the mth maximum, then the difference in path length is also given by $\Delta L = d\sin(\theta_m + \Delta\theta)$. Thus $d\sin(\theta_m + \Delta\theta) = m\lambda + (\lambda/N)$. We use the trigonometric identity $\sin(\theta_m + \Delta\theta) = \sin\theta_m\cos\Delta\theta + \cos\theta_m\sin\Delta\theta$. Since $\Delta\theta$ is small, we may approximate $\sin\Delta\theta$ by $\Delta\theta$ in radians and $\cos\Delta\theta$ by unity. Thus $d\sin\theta_m + d\,\Delta\theta\cos\theta_m = m\lambda + (\lambda/N)$. We use the condition $d\sin\theta_m = m\lambda$ to obtain $d\,\Delta\theta\cos\theta_m = \lambda/N$ and

$$\Delta\theta = \frac{\lambda}{Nd\cos\theta_m} .$$

44. At the point on the screen where we find the inner edge of the hole, we have $\tan\theta = 5.0\,\text{cm}/30\,\text{cm}$, which gives $\theta = 9.46°$. We note that d for the grating is equal to $1.0\,\text{mm}/350 = 1.0\times10^6\,\text{nm}/350$. From $m\lambda = d\sin\theta$, we find

$$m = \frac{d\sin\theta}{\lambda} = \frac{\left(\frac{1.0\times10^6\,\text{nm}}{350}\right)(0.1644)}{\lambda} = \frac{470\,\text{nm}}{\lambda} .$$

Since for white light $\lambda > 400\,\text{nm}$, the only integer m allowed here is $m = 1$. Thus, at one edge of the hole, $\lambda = 470\,\text{nm}$. However, at the other edge, we have $\tan\theta' = 6.0\,\text{cm}/30\,\text{cm}$, which gives $\theta' = 11.31°$. This leads to

$$\lambda' = d\sin\theta' = \left(\frac{1.0\times10^6\,\text{nm}}{350}\right)\sin 11.31° = 560\text{ nm} .$$

Consequently, the range of wavelength is from 470 to 560 nm.

45. Since the slit width is much less than the wavelength of the light, the central peak of the single-slit diffraction pattern is spread across the screen and the diffraction envelope can be ignored. Consider three waves, one from each slit. Since the slits are evenly spaced, the phase difference for waves from the first and second slits is the same as the phase difference for waves from the second and third slits. The electric fields of the waves at the screen can be written $E_1 = E_0\sin(\omega t)$, $E_2 = E_0\sin(\omega t + \phi)$, and $E_3 = E_0\sin(\omega t + 2\phi)$, where $\phi = (2\pi d/\lambda)\sin\theta$. Here d is the separation of adjacent slits and λ is the wavelength. The phasor diagram is shown below. It yields

$$E = E_0\cos\phi + E_0 + E_0\cos\phi = E_0(1 + 2\cos\phi)$$

for the amplitude of the resultant wave. Since the intensity of a wave is proportional to the square of the electric field, we may write $I = AE_0^2(1 + 2\cos\phi)^2$, where A is a constant of proportionality. If I_m is the intensity at the center of the pattern, for which $\phi = 0$, then $I_m = 9AE_0^2$. We take A to be $I_m/9E_0^2$ and obtain

$$I = \frac{I_m}{9}(1 + 2\cos\phi)^2 = \frac{I_m}{9}\left(1 + 4\cos\phi + 4\cos^2\phi\right) .$$

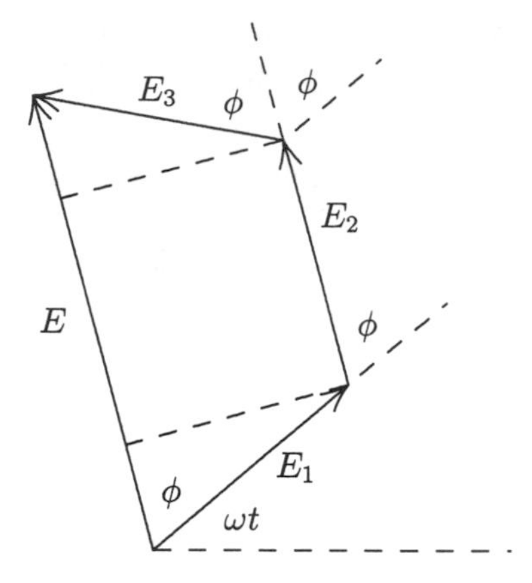

46. Letting $R = \lambda/\Delta\lambda = Nm$, we solve for N:

$$N = \frac{\lambda}{m\Delta\lambda} = \frac{(589.6\,\text{nm} + 589.0\,\text{nm})/2}{2(589.6\,\text{nm} - 589.0\,\text{nm})} = 491 \ .$$

47. If a grating just resolves two wavelengths whose average is λ_{avg} and whose separation is $\Delta\lambda$, then its resolving power is defined by $R = \lambda_{\text{avg}}/\Delta\lambda$. The text shows this is Nm, where N is the number of rulings in the grating and m is the order of the lines. Thus $\lambda_{\text{avg}}/\Delta\lambda = Nm$ and

$$N = \frac{\lambda_{\text{avg}}}{m\,\Delta\lambda} = \frac{656.3\,\text{nm}}{(1)(0.18\,\text{nm})} = 3650\,\text{rulings} \ .$$

48. (a) We find $\Delta\lambda$ from $R = \lambda/\Delta\lambda = Nm$:

$$\Delta\lambda = \frac{\lambda}{Nm} = \frac{500\,\text{nm}}{(600/\text{mm})(5.0\,\text{mm})(3)} = 0.056\,\text{nm} = 56\ \text{pm} \ .$$

(b) Since $\sin\theta = m_{\text{max}}\lambda/d < 1$,

$$m_{\text{max}} < \frac{d}{\lambda} = \frac{1}{(600/\text{mm})(500 \times 10^{-6}\,\text{mm})} = 3.3 \ .$$

Therefore, $m_{\text{max}} = 3$. No higher orders of maxima can be seen.

49. The dispersion of a grating is given by $D = d\theta/d\lambda$, where θ is the angular position of a line associated with wavelength λ. The angular position and wavelength are related by $\mathbf{d}\sin\theta = m\lambda$, where $\mathbf{d}$ is the slit separation (which we made boldfaced in order not to confuse it with the d used in the derivative, below) and m is an integer. We differentiate this expression with respect to θ to obtain

$$\frac{d\theta}{d\lambda}\,\mathbf{d}\,\cos\theta = m \ ,$$

or

$$D = \frac{d\theta}{d\lambda} = \frac{m}{\mathbf{d}\cos\theta} \ .$$

Now $m = (\mathbf{d}/\lambda)\sin\theta$, so

$$D = \frac{\mathbf{d}\sin\theta}{\mathbf{d}\lambda\cos\theta} = \frac{\tan\theta}{\lambda} \ .$$

50. (a) From $d\sin\theta = m\lambda$ we find

$$d = \frac{m\lambda_{\text{avg}}}{\sin\theta} = \frac{3(589.3\,\text{nm})}{\sin 10^\circ} = 1.0 \times 10^4\,\text{nm} = 10\,\mu\text{m} \ .$$

(b) The total width of the ruling is

$$L = Nd = \left(\frac{R}{m}\right)d = \frac{\lambda_{\text{avg}}d}{m\Delta\lambda} = \frac{(589.3\,\text{nm})(10\,\mu\text{m})}{3(589.59\,\text{nm} - 589.00\,\text{nm})} = 3.3\times10^3\mu\text{m} = 3.3\ \text{mm}\ .$$

51. (a) Since the resolving power of a grating is given by $R = \lambda/\Delta\lambda$ and by Nm, the range of wavelengths that can just be resolved in order m is $\Delta\lambda = \lambda/Nm$. Here N is the number of rulings in the grating and λ is the average wavelength. The frequency f is related to the wavelength by $f\lambda = c$, where c is the speed of light. This means $f\,\Delta\lambda + \lambda\,\Delta f = 0$, so

$$\Delta\lambda = -\frac{\lambda}{f}\,\Delta f = -\frac{\lambda^2}{c}\,\Delta f$$

where $f = c/\lambda$ is used. The negative sign means that an increase in frequency corresponds to a decrease in wavelength. We may interpret Δf as the range of frequencies that can be resolved and take it to be positive. Then,

$$\frac{\lambda^2}{c}\,\Delta f = \frac{\lambda}{Nm}$$

and

$$\Delta f = \frac{c}{Nm\lambda}\ .$$

(b) The difference in travel time for waves traveling along the two extreme rays is $\Delta t = \Delta L/c$, where ΔL is the difference in path length. The waves originate at slits that are separated by $(N-1)d$, where d is the slit separation and N is the number of slits, so the path difference is $\Delta L = (N-1)d\sin\theta$ and the time difference is

$$\Delta t = \frac{(N-1)d\sin\theta}{c}\ .$$

If N is large, this may be approximated by $\Delta t = (Nd/c)\sin\theta$. The lens does not affect the travel time.

(c) Substituting the expressions we derived for Δt and Δf, we obtain

$$\Delta f\,\Delta t = \left(\frac{c}{Nm\lambda}\right)\left(\frac{Nd\sin\theta}{c}\right) = \frac{d\sin\theta}{m\lambda} = 1\ .$$

The condition $d\sin\theta = m\lambda$ for a diffraction line is used to obtain the last result.

52. (a) From the expression for the half-width $\Delta\theta_{\text{hw}}$ (given by Eq. 37-25) and that for the resolving power R (given by Eq. 37-29), we find the product of $\Delta\theta_{\text{hw}}$ and R to be

$$\Delta\theta_{\text{hw}}R = \left(\frac{\lambda}{Nd\cos\theta}\right)Nm = \frac{m\lambda}{d\cos\theta} = \frac{d\sin\theta}{d\cos\theta} = \tan\theta\,,$$

where we used $m\lambda = d\sin\theta$ (see Eq. 37-22).

(b) For first order $m = 1$, so the corresponding angle θ_1 satisfies $d\sin\theta_1 = m\lambda = \lambda$. Thus the product in question is given by

$$\begin{aligned}\tan\theta_1 &= \frac{\sin\theta_1}{\cos\theta_1} = \frac{\sin\theta_1}{\sqrt{1-\sin^2\theta_1}}\\ &= \frac{1}{\sqrt{(1/\sin\theta_1)^2-1}} = \frac{1}{\sqrt{(d/\lambda)^2-1}}\\ &= \frac{1}{\sqrt{(900\,\text{nm}/600\,\text{nm})^2-1}} = 0.89\ .\end{aligned}$$

53. Bragg's law gives the condition for a diffraction maximum:

$$2d\sin\theta = m\lambda$$

where d is the spacing of the crystal planes and λ is the wavelength. The angle θ is measured from the surfaces of the planes. For a second-order reflection $m = 2$, so

$$d = \frac{m\lambda}{2\sin\theta} = \frac{2(0.12\times10^{-9}\,\text{m})}{2\sin 28^\circ} = 2.56\times10^{-10}\,\text{m} = 256\ \text{pm}\ .$$

54. We use Eq. 37-31. From the peak on the left at angle 0.75° (estimated from Fig. 37-38), we have

$$\lambda_1 = 2d\sin\theta_1 = 2(0.94\,\text{nm})\sin(0.75^\circ) = 0.025\,\text{nm} = 25\ \text{pm}\ .$$

This estimation should be viewed as reliable to within $\pm 2\,\text{pm}$. We now consider the next peak:

$$\lambda_2 = 2d\sin\theta_2 = 2(0.94\,\text{nm})\sin 1.15^\circ = 0.038\,\text{nm} = 38\ \text{pm}\ .$$

One can check that the third peak from the left is the second-order one for λ_1.

55. The x ray wavelength is $\lambda = 2d\sin\theta = 2(39.8\,\text{pm})\sin 30.0^\circ = 39.8\,\text{pm}$.

56. (a) For the first beam $2d\sin\theta_1 = \lambda_A$ and for the second one $2d\sin\theta_2 = 3\lambda_B$. The values of d and λ_A can then be determined:

$$d = \frac{3\lambda_B}{2\sin\theta_2} = \frac{3(97\,\text{pm})}{2\sin 60^\circ} = 1.7\times10^2\ \text{pm}\ .$$

(b)

$$\lambda_A = 2d\sin\theta_1 = 2(1.7\times10^2\,\text{pm})(\sin 23^\circ) = 1.3\times10^2\ \text{pm}\ .$$

57. There are two unknowns, the x-ray wavelength λ and the plane separation d, so data for scattering at two angles from the same planes should suffice. The observations obey Bragg's law, so

$$2d\sin\theta_1 = m_1\lambda$$

and

$$2d\sin\theta_2 = m_2\lambda\ .$$

However, these cannot be solved for the unknowns. For example, we can use the first equation to eliminate λ from the second. We obtain

$$m_2\sin\theta_1 = m_1\sin\theta_2\ ,$$

an equation that does not contain either of the unknowns.

58. The angle of incidence on the reflection planes is $\theta = 63.8^\circ - 45.0^\circ = 18.8^\circ$, and the plane-plane separation is $d = a_0/\sqrt{2}$. Thus, using $2d\sin\theta = \lambda$, we get

$$a_0 = \sqrt{2}d = \frac{\sqrt{2}\lambda}{2\sin\theta} = \frac{0.260\,\text{nm}}{\sqrt{2}\sin 18.8^\circ} = 0.570\ \text{nm}\ .$$

59\. (a) The sets of planes with the next five smaller interplanar spacings (after a_0) are shown in the diagram below.

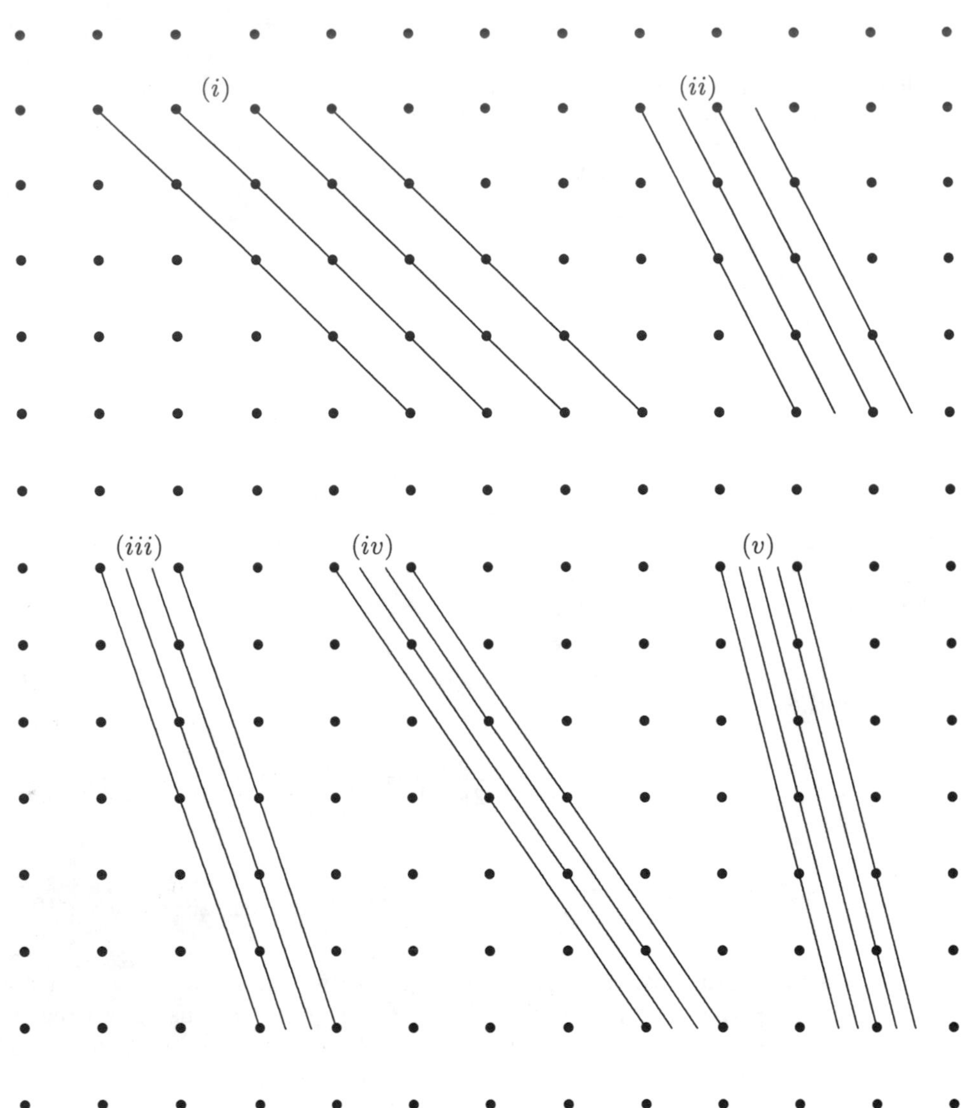

In terms of a_0, the spacings are:

$$\begin{aligned}
(i):&\quad a_0/\sqrt{2} = 0.7071a_0\\
(ii):&\quad a_0/\sqrt{5} = 0.4472a_0\\
(iii):&\quad a_0/\sqrt{10} = 0.3162a_0\\
(iv):&\quad a_0/\sqrt{13} = 0.2774a_0\\
(v):&\quad a_0/\sqrt{17} = 0.2425a_0
\end{aligned}$$

(b) Since a crystal plane passes through lattice points, its slope can be written as the ratio of two integers. Consider a set of planes with slope m/n, as shown in the diagram below. The first and last planes shown pass through adjacent lattice points along a horizontal line and there are $m - 1$ planes between. If h is the separation of the first and last planes, then the interplanar spacing is $d = h/m$. If the planes make the angle θ with the horizontal, then the normal to the planes (shown dotted) makes the angle $\phi = 90° - \theta$. The distance h is given by $h = a_0 \cos\phi$

and the interplanar spacing is $d = h/m = (a_0/m)\cos\phi$. Since $\tan\theta = m/n$, $\tan\phi = n/m$ and $\cos\phi = 1/\sqrt{1+\tan^2\phi} = m/\sqrt{n^2+m^2}$. Thus,

$$d = \frac{h}{m} = \frac{a_0\cos\phi}{m} = \frac{a_0}{\sqrt{n^2+m^2}} \; .$$

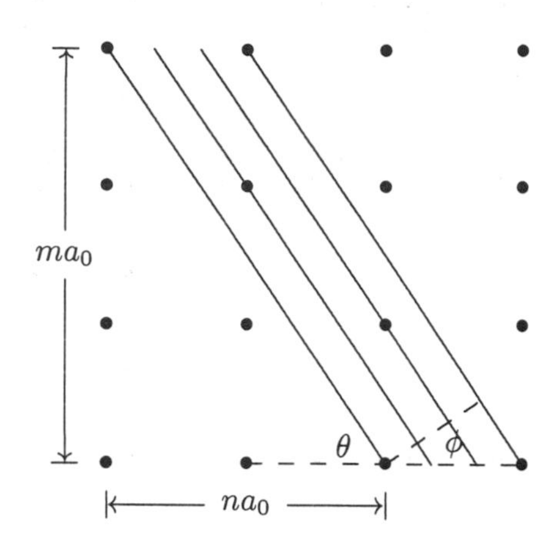

60. The wavelengths satisfy $m\lambda = 2d\sin\theta = 2(275\,\text{pm})(\sin 45°) = 389\,\text{pm}$. In the range of wavelengths given, the allowed values of m are $m = 3$, 4, with the corresponding wavelengths being $389\,\text{pm}/3 = 130\,\text{pm}$ and $389\,\text{pm}/4 = 97.2\,\text{pm}$, respectively.

61. We want the reflections to obey the Bragg condition $2d\sin\theta = m\lambda$, where θ is the angle between the incoming rays and the reflecting planes, λ is the wavelength, and m is an integer. We solve for θ:

$$\theta = \sin^{-1}\left(\frac{m\lambda}{2d}\right) = \sin^{-1}\left(\frac{(0.125\times10^{-9}\,\text{m})m}{2(0.252\times10^{-9}\,\text{m})}\right) = 0.2480m \; .$$

For $m = 1$ this gives $\theta = 14.4°$. The crystal should be turned $45° - 14.4° = 30.6°$ clockwise. For $m = 2$ it gives $\theta = 29.7°$. The crystal should be turned $45° - 29.7° = 15.3°$ clockwise. For $m = 3$ it gives $\theta = 48.1°$. The crystal should be turned $48.1° - 45° = 3.1°$ counterclockwise. For $m = 4$ it gives $\theta = 82.8°$. The crystal should be turned $82.8° - 45° = 37.8°$ counterclockwise. There are no intensity maxima for $m > 4$ as one can verify by noting that $m\lambda/2d$ is greater than 1 for m greater than 4.

62. (a) Eq. 37-3 and Eq. 37-12 imply smaller angles for diffraction for smaller wavelengths. This suggests that diffraction effects in general would decrease.

(b) Using Eq. 37-3 with $m = 1$ and solving for 2θ (the angular width of the central diffraction maximum), we find

$$2\theta = 2\sin^{-1}\left(\frac{\lambda}{a}\right) = 2\sin^{-1}\left(\frac{0.50\,\text{m}}{5.0\,\text{m}}\right) = 11° \; .$$

(c) A similar calculation yields 0.23° for $\lambda = 0.010$ m.

63. (a) Using the notation of Sample Problem 37-6 (which is in the textbook supplement), the minimum separation is

$$D = L\theta_\text{R} = L\left(\frac{1.22\lambda}{d}\right) = \frac{(400\times10^3\,\text{m})(1.22)(550\times10^{-9}\,\text{m})}{(0.005\,\text{m})} \approx 50\ \text{m} \; .$$

(b) The Rayleigh criterion suggests that the astronaut will not be able to discern the Great Wall (see the result of part (a)).

(c) The signs of intelligent life would probably be, at most, ambiguous on the sunlit half of the planet. However, while passing over the half of the planet on the opposite side from the Sun, the astronaut would be able to notice the effects of artificial lighting.

64. Consider two light rays crossing each other at the middle of the lens (see Fig. 37-42(c)). The rays come from opposite sides of the circular dot of diameter D, a distance L from the eyes, so we are using the same notation found in Sample Problem 37-6 (which is in the textbook supplement). Those two rays reach the retina a distance L' behind the lens, striking two points there which are a distance D' apart. Therefore,

$$\frac{D}{L} = \frac{D'}{L'}$$

where $D = 2\,\text{mm}$ and $L' = 20\,\text{mm}$. If we estimate $L \approx 450\,\text{mm}$, we find $D' \approx 0.09\,\text{mm}$. Turning our attention to Fig. 37-42(d), we see

$$\theta = \tan^{-1}\left(\frac{\frac{1}{2}D'}{x}\right)$$

which we wish to set equal to the angle in Eq. 37-12. We could use the small angle approximation $\sin\theta \approx \tan\theta$ to relate these directly, or we could be "exact" – as we show below:

$$\text{If} \quad \tan\phi = \frac{b}{a}\ , \qquad \text{then} \quad \sin\phi = \frac{b}{\sqrt{a^2+b^2}} \quad .$$

Therefore, this "exact" use of Eq. 37-12 leads to

$$1.22\frac{\lambda}{d} = \sin\theta = \frac{\frac{1}{2}D'}{\sqrt{x^2+(D'/2)^2}}$$

where $\lambda = 550 \times 10^{-6}\,\text{mm}$ and $1\,\text{mm} \le x \le 15\,\text{mm}$. Using the value of D' found above, this leads to a range of d values: $0.015\,\text{mm} \le d \le 0.23\,\text{mm}$.

65. Using the same notation found in Sample Problem 37-6,

$$\frac{D}{L} = \theta_R = 1.22\frac{\lambda}{d}$$

where we will assume a "typical" wavelength for visible light: $\lambda \approx 550 \times 10^{-9}$ m.

(a) With $L = 400 \times 10^3$ m and $D = 0.85$ m, the above relation leads to $d = 0.32$ m.

(b) Now with $D = 0.10$ m, the above relation leads to $d = 2.7$ m.

(c) The military satellites do not use Hubble Telescope-sized apertures. A great deal of very sophisticated optical filtering and digital signal processing techniques go into the final product, for which there is not space for us to describe here.

66. Assuming all $N = 2000$ lines are uniformly illuminated, we have

$$\frac{\lambda_{av}}{\Delta\lambda} = Nm$$

from Eq. 37-28 and Eq. 37-29. With $\lambda_{av} = 600$ nm and $m = 2$, we find $\Delta\lambda = 0.15$ nm.

67. The central diffraction envelope spans the range $-\theta_1 < \theta < +\theta_1$ where

$$\theta_1 = \sin^{-1}\frac{\lambda}{a}\ .$$

The maxima in the double-slit pattern are located at

$$\theta_m = \sin^{-1}\frac{m\lambda}{d}\ ,$$

so that our range specification becomes

$$-\sin^{-1}\frac{\lambda}{a} \ < \ \sin^{-1}\frac{m\lambda}{d} \ < \ +\sin^{-1}\frac{\lambda}{a}\ ,$$

which we change (since sine is a monotonically increasing function in the fourth and first quadrants, where all these angles lie) to

$$-\frac{\lambda}{a} < \frac{m\lambda}{d} < +\frac{\lambda}{a} .$$

Rewriting this as $-d/a < m < +d/a$, we find $-6 < m < +6$, or, since m is an integer, $-5 \le m \le +5$. Thus, we find eleven values of m that satisfy this requirement.

68. Employing Eq. 37-3, we find (with $m = 3$ and all lengths in μm)

$$\theta = \sin^{-1}\frac{m\lambda}{a} = \sin^{-1}\frac{(3)(0.5)}{2}$$

which yields $\theta = 48.6°$. Now, we use the experimental geometry ($\tan\theta = y/D$ where y locates the minimum relative to the middle of the pattern) to find

$$y = D\tan\theta = 2.27 \text{ m} .$$

69. (a) From $R = \lambda/\Delta\lambda = Nm$ we find

$$N = \frac{\lambda}{m\Delta\lambda} = \frac{(415.496\,\text{nm} + 415.487\,\text{nm})/2}{2(415.96\,\text{nm} - 415.487\,\text{nm})} = 23100 .$$

(b) We note that $d = (4.0 \times 10^7\,\text{nm})/23100 = 1732\,\text{nm}$. The maxima are found at

$$\theta = \sin^{-1}\left(\frac{m\lambda}{d}\right) = \sin^{-1}\left[\frac{(2)(415.5\,\text{nm})}{1732\,\text{nm}}\right] = 28.7° .$$

70. We use Eq. 37-31. For smallest value of θ, we let $m = 1$. Thus,

$$\theta_{\text{min}} = \sin^{-1}\left(\frac{m\lambda}{2d}\right) = \sin^{-1}\left[\frac{(1)(30\,\text{pm})}{2(0.30 \times 10^3\,\text{pm})}\right] = 2.9° .$$

71. (a) We use Eq. 37-12:

$$\begin{aligned}\theta &= \sin^{-1}\left(\frac{1.22\lambda}{d}\right) = \sin^{-1}\left[\frac{1.22(v_s/f)}{d}\right] \\ &= \sin^{-1}\left[\frac{(1.22)(1450\,\text{m/s})}{(25 \times 10^3\,\text{Hz})(0.60\,\text{m})}\right] = 6.8° .\end{aligned}$$

(b) Now $f = 1.0 \times 10^3$ Hz so

$$\frac{1.22\lambda}{d} = \frac{(1.22)(1450\,\text{m/s})}{(1.0 \times 10^3\,\text{Hz})(0.60\,\text{m})} = 2.9 > 1 .$$

Since $\sin\theta$ cannot exceed 1 there is no minimum.

72. From Eq. 37-3,

$$\frac{a}{\lambda} = \frac{m}{\sin\theta} = \frac{1}{\sin 45.0°} = 1.41 .$$

73. (a) Use of Eq. 37-22 for the limit-wavelengths ($\lambda_1 = 700$ nm and $\lambda_2 = 550$ nm) leads to the condition

$$m_1\lambda_1 \ge m_2\lambda_2$$

for $m_1 + 1 = m_2$ (the low end of a high-order spectrum is what is overlapping with the high end of the next-lower-order spectrum). Assuming equality in the above equation, we can solve for "m_1" (realizing it might not be an integer) and obtain $m_1 \approx 4$ where we have rounded *up*. It is the fourth order spectrum that is the lowest-order spectrum to overlap with the next higher spectrum.

(b) The problem specifies $d = 1/200$ using the mm unit, and we note there are no refraction angles greater than 90°. We concentrate on the largest wavelength $\lambda = 700$ nm $= 7 \times 10^{-4}$ mm and solve Eq. 37-22 for "m_{max}" (realizing it might not be an integer):

$$m_{\text{max}} = \frac{d \sin 90^\circ}{\lambda} = \frac{1}{(200)(7 \times 10^{-4})} \approx 7$$

where we have rounded down. There are no values of m (for the appearance of the full spectrum) greater than $m = 7$.

74. The central diffraction envelope spans the range $-\theta_1 < \theta < +\theta_1$ where

$$\theta_1 = \sin^{-1} \frac{\lambda}{a} .$$

The maxima in the double-slit pattern are at

$$\theta_m = \sin^{-1} \frac{m\lambda}{d} ,$$

so that our range specification becomes

$$-\sin^{-1} \frac{\lambda}{a} < \sin^{-1} \frac{m\lambda}{d} < +\sin^{-1} \frac{\lambda}{a} ,$$

which we change (since sine is a monotonically increasing function in the fourth and first quadrants, where all these angles lie) to

$$-\frac{\lambda}{a} < \frac{m\lambda}{d} < +\frac{\lambda}{a} .$$

Rewriting this as $-d/a < m < +d/a$ we arrive at the result $m_{\text{max}} < d/a \le m_{\text{max}} + 1$. Due to the symmetry of the pattern, the multiplicity of the m values is $2m_{\text{max}} + 1 = 17$ so that $m_{\text{max}} = 8$, and the result becomes

$$8 < \frac{d}{a} \le 9$$

where these numbers are as accurate as the experiment allows (that is, "9" means "9.000" if our measurements are that good).

75. As a slit is narrowed, the pattern spreads outward, so the question about "minimum width" suggests that we are looking at the lowest possible values of m (the label for the minimum produced by light $\lambda = 600$ nm) and m' (the label for the minimum produced by light $\lambda' = 500$ nm). Since the angles are the same, then Eq. 37-3 leads to

$$m\lambda = m'\lambda'$$

which leads to the choices $m = 5$ and $m' = 6$. We find the slit width from Eq. 37-3:

$$a = \frac{m\lambda}{\sin\theta} \approx \frac{m\lambda}{\theta}$$

which yields $a = 3.0$ mm.

76. (a) We note that $d = (76 \times 10^6\,\text{nm})/40000 = 1900\,\text{nm}$. For the first order maxima $\lambda = d\sin\theta$, which leads to

$$\theta = \sin^{-1}\left(\frac{\lambda}{d}\right) = \sin^{-1}\left(\frac{589\,\text{nm}}{1900\,\text{nm}}\right) = 18^\circ .$$

Now, substituting $m = d\sin\theta/\lambda$ into Eq. 37-27 leads to $D = \tan\theta/\lambda = \tan 18^\circ/589\,\text{nm} = 5.5 \times 10^{-4}\,\text{rad/nm} = 0.032^\circ/\text{nm}$. Similarly for $m = 2$ and $m = 3$, we have $\theta = 38^\circ$ and 68°, and the corresponding values of dispersion are $0.076^\circ/\text{nm}$ and $0.24^\circ/\text{nm}$, respectively.

(b) $R = Nm = 40000\,m = 40000$ (for $m = 1$); 80000 (for $m = 2$); and, 120,000 (for $m = 3$).

77. Letting $d\sin\theta = (L/N)\sin\theta = m\lambda$, we get

$$\lambda = \frac{(L/N)\sin\theta}{m} = \frac{(1.0\times10^7\,\text{nm})(\sin 30^\circ)}{(1)(10000)} = 500\text{ nm}\ .$$

78. (a) Using the notation of Sample Problem 37-6,

$$L = \frac{D}{1.22\lambda/d} = \frac{2(50\times10^{-6}\,\text{m})(1.5\times10^{-3}\,\text{m})}{1.22(650\times10^{-9}\,\text{m})} = 0.19\text{ m}\ .$$

(b) The wavelength of the blue light is shorter so $L_{\text{max}} \propto \lambda^{-1}$ will be larger.

79. From $y = m\lambda D/a$ we get

$$\Delta y = \Delta\left(\frac{m\lambda D}{a}\right) = \frac{\lambda D}{a}\Delta m = \frac{(632.8\,\text{nm})(2.60)}{1.37\,\text{mm}}[10-(-10)] = 24.0\text{ mm}\ .$$

80. For $\lambda = 0.10$ nm, we have scattering for order m, and for $\lambda' = 0.075$ nm, we have scattering for order m'. From Eq. 37-31, we see that we must require

$$m\lambda = m'\lambda'$$

which suggests (looking for the smallest integer solutions) that $m = 3$ and $m' = 4$. Returning with this result and with $d = 0.25$ nm to Eq. 37-31, we obtain

$$\theta = \sin^{-1}\frac{m\lambda}{2d} = 37^\circ\ .$$

Studying Figure 37-26, we conclude that the angle between incident and scattered beams is $180^\circ - 2\theta = 106^\circ$.

81. (a) We express all lengths in mm, and since $1/d = 180$, we write Eq. 37-22 as

$$\theta = \sin^{-1}\left(\frac{1}{d}m\lambda\right) = \sin^{-1}(180)(2)\lambda$$

where $\lambda_1 = 4\times10^{-4}$ and $\lambda_2 = 5\times10^{-4}$ (in mm). Thus, $\Delta\theta = \theta_2 - \theta_1 = 2.1^\circ$.

(b) Use of Eq. 37-22 for each wavelength leads to the condition

$$m_1\lambda_1 = m_2\lambda_2$$

for which the smallest possible choices are $m_1 = 5$ and $m_2 = 4$. Returning to Eq. 37-22, then, we find

$$\theta = \sin^{-1}\left(\frac{1}{d}m_1\lambda_1\right) = 21^\circ\ .$$

(c) There are no refraction angles greater than 90°, so we can solve for "m_{max}" (realizing it might not be an integer):

$$m_{\text{max}} = \frac{d\sin 90^\circ}{\lambda_2} = 11$$

where we have rounded down. There are no values of m (for light of wavelength λ_2) greater than $m = 11$.

82. Following Sample Problem 37-6, we use Eq. 37-35:

$$L = \frac{Dd}{1.22\lambda} = 164\text{ m}\ .$$

83. (a) Employing Eq. 37-3 with the small angle approximation ($\sin\theta \approx \tan\theta = y/D$ where y locates the minimum relative to the middle of the pattern), we find (with $m = 1$ and all lengths in mm)

$$D = \frac{ya}{m\lambda} = \frac{(0.9)(0.4)}{4.5 \times 10^{-4}} = 800$$

which places the screen 80 cm away from the slit.

(b) The above equation gives for the value of y (for $m = 3$)

$$y = \frac{(3)\lambda D}{a} = 2.7 \text{ mm} .$$

Subtracting this from the first minimum position $y = 0.9$ mm, we find the result $\Delta y = 1.8$ mm.

84. (a) We require that $\sin\theta = m\lambda_{1,2}/d \leq \sin 30°$, where $m = 1, 2$ and $\lambda_1 = 500\,\text{nm}$. This gives

$$d \geq \frac{2\lambda_s}{\sin 30°} = \frac{2(600\,\text{nm})}{\sin 30°} = 2400\,\text{nm} .$$

For a grating of given total width L we have $N = L/d \propto d^{-1}$, so we need to minimize d to maximize $R = mN \propto d^{-1}$. Thus we choose $d = 2400\,\text{nm}$.

(b) Let the third-order maximum for $\lambda_2 = 600\,\text{nm}$ be the first minimum for the single-slit diffraction profile. This requires that $d\sin\theta = 3\lambda_2 = a\sin\theta$, or $a = d/3 = 2400\,\text{nm}/3 = 800\,\text{nm}$.

(c) Letting $\sin\theta = m_{\text{max}}\lambda_2/d \leq 1$, we obtain

$$m_{\text{max}} \leq \frac{d}{\lambda_2} = \frac{2400\,\text{nm}}{800\,\text{nm}} = 3 .$$

Since the third order is missing the only maxima present are the ones with $m = 0$, 1 and 2.

85. (a) Letting $d\sin\theta = m\lambda$, we solve for λ:

$$\lambda = \frac{d\sin\theta}{m} = \frac{(1.0\,\text{mm}/200)(\sin 30°)}{m} = \frac{2500\,\text{nm}}{m}$$

where $m = 1, 2, 3\cdots$. In the visible light range m can assume the following values: $m_1 = 4$, $m_2 = 5$ and $m_3 = 6$. The corresponding wavelengths are $\lambda_1 = 2500\,\text{nm}/4 = 625\,\text{nm}$, $\lambda_2 = 2500\,\text{nm}/5 = 500\,\text{nm}$, and $\lambda_3 = 2500\,\text{nm}/6 = 416\,\text{nm}$.

(b) The colors are orange (for $\lambda_1 = 625\,\text{nm}$), blue-green (for $\lambda_2 = 500\,\text{nm}$), and violet (for $\lambda_3 = 416\,\text{nm}$).

86. Using the notation of Sample Problem 37-6,

$$L = \frac{D}{\theta_R} = \frac{D}{1.22\lambda/d} = \frac{(5.0 \times 10^{-2}\,\text{m})(4.0 \times 10^{-3}\,\text{m})}{1.22(0.10 \times 10^{-9}\,\text{m})} = 1.6 \times 10^6\,\text{m} = 1600 \text{ km} .$$

87. The condition for a minimum in a single-slit diffraction pattern is given by Eq. 37-3, which we solve for the wavelength:

$$\lambda = \frac{a\sin\theta}{m} = \frac{(0.022\,\text{mm})\sin 1.8°}{1} = 6.9 \times 10^{-4}\,\text{mm} = 690 \text{ nm} .$$

Chapter 38

1. (a) The time an electron with a horizontal component of velocity v takes to travel a horizontal distance L is

$$t = \frac{L}{v} = \frac{20 \times 10^{-2}\,\text{m}}{(0.992)(2.998 \times 10^{8}\,\text{m/s})} = 6.72 \times 10^{-10}\ \text{s}\ .$$

(b) During this time, it falls a vertical distance

$$y = \frac{1}{2}gt^2 = \frac{1}{2}(9.8\,\text{m/s}^2)(6.72 \times 10^{-10}\,\text{s})^2 = 2.2 \times 10^{-18}\ \text{m}\ .$$

This distance is much less than the radius of a proton. We can conclude that for particles traveling near the speed of light in a laboratory, Earth may be considered an approximately inertial frame.

2. (a) The speed parameter β is v/c. Thus,

$$\beta = \frac{(3\,\text{cm/y})(0.01\,\text{m/cm})(1\,\text{y}/3.15 \times 10^{7}\,\text{s})}{3.0 \times 10^{8}\,\text{m/s}} = 3 \times 10^{-18}\ .$$

(b) For the highway speed limit, we find

$$\beta = \frac{(90\,\text{km/h})(1000\,\text{m/km})(1\,\text{h}/3600\,\text{s})}{3.0 \times 10^{8}\,\text{m/s}} = 8.3 \times 10^{-8}\ .$$

(c) Mach 2.5 corresponds to

$$\beta = \frac{(1200\,\text{km/h})(1000\,\text{m/km})(1\,\text{h}/3600\,\text{s})}{3.0 \times 10^{8}\,\text{m/s}} = 1.1 \times 10^{-6}\ .$$

(d) We refer to Table 14-2:

$$\beta = \frac{(11.2\,\text{km/s})(1000\,\text{m/km})}{3.0 \times 10^{8}\,\text{m/s}} = 3.7 \times 10^{-5}\ .$$

(e) For the quasar recession speed, we obtain

$$\beta = \frac{(3.0 \times 10^{4}\,\text{km/s})(1000\,\text{m/km})}{3.0 \times 10^{8}\,\text{m/s}} = 0.10\ .$$

3. From the time dilation equation $\Delta t = \gamma \Delta t_0$ (where Δt_0 is the proper time interval, $\gamma = 1/\sqrt{1-\beta^2}$, and $\beta = v/c$), we obtain

$$\beta = \sqrt{1 - \left(\frac{\Delta t_0}{\Delta t}\right)^2}\ .$$

The proper time interval is measured by a clock at rest relative to the muon. Specifically, $\Delta t_0 = 2.2\,\mu s$. We are also told that Earth observers (measuring the decays of moving muons) find $\Delta t = 16\,\mu s$. Therefore,

$$\beta = \sqrt{1 - \left(\frac{2.2\,\mu s}{16\,\mu s}\right)^2} = 0.9905 .$$

The muon speed is $v = \beta c = 0.9905(2.998 \times 10^8\,\text{m/s}) = 2.97 \times 10^8\,\text{m/s}$.

4. (a) We find β from $\gamma = 1/\sqrt{1-\beta^2}$:

$$\beta = \sqrt{1 - \frac{1}{\gamma^2}} = \sqrt{1 - \frac{1}{(1.01)^2}} = 0.140371 \approx 0.140 .$$

(b) Similarly, $\beta = \sqrt{1 - (10.0)^{-2}} = 0.994987 \approx 0.9950$.

(c) In this case, $\beta = \sqrt{1 - (100)^{-2}} = 0.999\,950$.

(d) This last case might prove problematic for some calculators. The result is $\beta = \sqrt{1 - (1000)^{-2}} = 0.999\,999\,50$. The discussion in Sample Problem 38-7 dealing with large γ values may prove helpful for those whose calculators do not yield this answer.

5. In the laboratory, it travels a distance $d = 0.00105\,\text{m} = vt$, where $v = 0.992c$ and t is the time measured on the laboratory clocks. We can use Eq. 38-7 to relate t to the proper lifetime of the particle t_0:

$$t = \frac{t_0}{\sqrt{1 - (v/c)^2}} \implies t_0 = t\sqrt{1 - \left(\frac{v}{c}\right)^2} = \frac{d}{0.992c}\sqrt{1 - 0.992^2}$$

which yields $t_0 = 4.46 \times 10^{-13}$ s.

6. (a) The round-trip (discounting the time needed to "turn around") should be one year according to the clock you are carrying (this is your proper time interval Δt_0) and 1000 years according to the clocks on Earth which measure Δt. We solve Eq. 38-7 for v and then plug in:

$$\begin{aligned} v &= c\sqrt{1 - \left(\frac{\Delta t_0}{\Delta t}\right)^2} \\ &= (299792458\,\text{m/s})\sqrt{1 - \left(\frac{1\,\text{y}}{1000\,\text{y}}\right)^2} \\ &= 299792308\,\text{m/s} \end{aligned}$$

which may also be expressed as $v = c\sqrt{1 - (1000)^{-2}} = 0.999\,999\,50c$. The discussion in Sample Problem 38-7 dealing with these sorts of values may prove helpful for those whose calculators do not yield this answer.

(b) The equations do not show a dependence on acceleration (or on the direction of the velocity vector), which suggests that a circular journey (with its constant magnitude centripetal acceleration) would give the same result (if the speed is the same) as the one described in the problem. A more careful argument can be given to support this, but it should be admitted that this is a fairly subtle question which has occasionally precipitated debates among professional physicists.

7. The length L of the rod, as measured in a frame in which it is moving with speed v parallel to its length, is related to its rest length L_0 by $L = L_0/\gamma$, where $\gamma = 1/\sqrt{1-\beta^2}$ and $\beta = v/c$. Since γ must be greater than 1, L is less than L_0. For this problem, $L_0 = 1.70\,\text{m}$ and $\beta = 0.630$, so $L = (1.70\,\text{m})\sqrt{1 - (0.630)^2} = 1.32\,\text{m}$.

8. The contracted length of the tube would be

$$L = L_0\sqrt{1-\beta^2} = (3.00\,\text{m})\sqrt{1-0.999987^2} = 0.0153\ \text{m}\ .$$

9. Only the "component" of the length in the x direction contracts, so its y component stays

$$\ell'_y = \ell_y = \ell \sin 30° = 0.5000\ \text{m}$$

while its x component becomes

$$\ell'_x = \ell_x\sqrt{1-\beta^2} = \ell \cos 30°\sqrt{1-0.90^2} = 0.3775\ \text{m}\ .$$

Therefore, using the Pythagorean theorem, the length measured from S' is

$$\ell' = \sqrt{(\ell'_x)^2 + (\ell'_y)^2} = 0.626\ \text{m}\ .$$

10. (a) We solve Eq. 38-13 for v and then plug in:

$$\begin{aligned} v &= c\sqrt{1-\left(\frac{L}{L_0}\right)^2} \\ &= (299792458\,\text{m/s})\sqrt{1-\left(\frac{1}{2}\right)^2} \\ &= 259627884\ \text{m/s} \end{aligned}$$

which may also be expressed as $v = 0.8660254c$.

(b) The Lorentz factor in this case is $\gamma = \frac{1}{\sqrt{1-(v/c)^2}} = 2$ "exactly."

11. (a) The rest length $L_0 = 130\,\text{m}$ of the spaceship and its length L as measured by the timing station are related by Eq. 38-13. Therefore, $L = (130\,\text{m})\sqrt{1-(0.740)^2} = 87.4\,\text{m}$.

(b) The time interval for the passage of the spaceship is

$$\Delta t = \frac{L}{v} = \frac{87.4\,\text{m}}{(0.740)(3.00\times10^8\,\text{m/s})} = 3.94\times10^{-7}\ \text{s}\ .$$

12. (a) According solely to the principles of Special Relativity, yes. If the person moves fast enough, then the time dilation argument will allow for his proper travel time to be much less than that measured from the Earth. Stated differently, length contraction can make that travel distance seem much shorter to the traveler than to our Earth-based estimations. This does not include important considerations such as fuel requirements, stresses to the human body (due to the accelerations, primarily), and so on.

(b) Let $d = 23000\,\text{ly} = 23000\,c\,\text{y}$, which would give the distance in meters if we included a conversion factor for years $\rightarrow$ seconds. With $\Delta t_0 = 30\,\text{y}$ and $\Delta t = d/v$ (see Eq. 38-10), we wish to solve for v from Eq. 38-7. Our first step is as follows:

$$\begin{aligned} \Delta t &= \frac{\Delta t_0}{\sqrt{1-(v/c)^2}} \\ \frac{d}{v} &= \frac{\Delta t_0}{\sqrt{1-(v/c)^2}} \\ \frac{23000\,c\,\text{y}}{v} &= \frac{30\,\text{y}}{\sqrt{1-(v/c)^2}}\ , \end{aligned}$$

at which point we can cancel the unit year and manipulate the equation to solve for the speed. After a couple of algebraic steps, we obtain

$$\begin{aligned} v &= \frac{c}{\sqrt{1+\left(\frac{30}{23000}\right)^2}} \\ &= \frac{299792458\,\text{m/s}}{\sqrt{1+0.000017013}} \\ &= 299792203\ \text{m/s} \end{aligned}$$

which may also be expressed as $v = 0.999\,999\,15c$. The discussion in Sample Problem 38-7 dealing with these sorts of values may prove helpful for those whose calculators do not yield this answer.

13. (a) The speed of the traveler is $v = 0.99c$, which may be equivalently expressed as $0.99\,\text{ly/y}$. Let d be the distance traveled. Then, the time for the trip as measured in the frame of Earth is $\Delta t = d/v = (26\,\text{ly})/(0.99\,\text{ly/y}) = 26.3\,\text{y}$.

    (b) The signal, presumed to be a radio wave, travels with speed c and so takes 26.0 y to reach Earth. The total time elapsed, in the frame of Earth, is $26.3\,\text{y} + 26.0\,\text{y} = 52.3\,\text{y}$.

    (c) The proper time interval is measured by a clock in the spaceship, so $\Delta t_0 = \Delta t/\gamma$. Now $\gamma = 1/\sqrt{1-\beta^2} = 1/\sqrt{1-(0.99)^2} = 7.09$. Thus, $\Delta t_0 = (26.3\,\text{y})/(7.09) = 3.7\,\text{y}$.

14. The "coincidence" of $x = x' = 0$ at $t = t' = 0$ is important for Eq. 38-20 to apply without additional terms. In part (a), we apply these equations directly with $v = +0.400c = 1.199 \times 10^8\,\text{m/s}$, and in part (b) we simply change $v \to -v$ and recalculate the primed values.

    (a) The position coordinate measured in the S' frame is

$$\begin{aligned} x' &= \gamma(x - vt) = \frac{x - vt}{\sqrt{1-\beta^2}} \\ &= \frac{3.00 \times 10^8\,\text{m} - (1.199 \times 10^8\,\text{m/s})(2.50\,\text{s})}{\sqrt{1-(0.400)^2}} \\ &= 2.7 \times 10^5\ \text{m/s} \approx 0\ , \end{aligned}$$

where we conclude that the numerical result (2.7×10^5 or 2.3×10^5 depending on how precise a value of v is used) is not meaningful (in the significant figures sense) and should be set equal to zero (that is, it is "consistent with zero" in view of the statistical uncertainties involved). The time coordinate measured in the S' frame is

$$\begin{aligned} t' &= \gamma\left(t - \frac{vx}{c^2}\right) = \frac{t - \frac{\beta x}{c}}{\sqrt{1-\beta^2}} \\ &= \frac{2.50\,\text{s} - \frac{(0.400)(3.00\times10^3\,\text{m})}{2.998\times10^8\,\text{m/s}}}{\sqrt{1-(0.400)^2}} \\ &= 2.29\ \text{s}\ . \end{aligned}$$

    (b) Now, we obtain

$$x' = \frac{x + vt}{\sqrt{1-\beta^2}} = \frac{3.00 \times 10^8\,\text{m} + (1.199 \times 10^8\,\text{m/s})(2.50\,\text{s})}{\sqrt{1-(0.400)^2}} = 6.54 \times 10^8\ \text{m}\ ,$$

and

$$t' = \gamma\left(t + \frac{vx}{c^2}\right) = \frac{2.50\,\text{s} + \frac{(0.400)(3.00\times10^8\,\text{m})}{2.998\times10^8\,\text{m/s}}}{\sqrt{1-(0.400)^2}} = 3.16\ \text{s}\ .$$

15. The proper time is not measured by clocks in either frame S or frame S' since a single clock at rest in either frame cannot be present at the origin and at the event. The full Lorentz transformation must be used:

$$x' = \gamma(x - vt) \qquad \text{and} \qquad t' = \gamma(t - \beta x/c)$$

where $\beta = v/c = 0.950$ and $\gamma = 1/\sqrt{1-\beta^2} = 1/\sqrt{1-(0.950)^2} = 3.20256$. Thus,

$$\begin{aligned} x' &= (3.20256)\left(100 \times 10^3\,\text{m} - (0.950)(2.998 \times 10^8\,\text{m/s})(200 \times 10^{-6}\,\text{s})\right) \\ &= 1.38 \times 10^5\,\text{m} = 138\,\text{km} \end{aligned}$$

and

$$t' = (3.20256)\left[200 \times 10^{-6}\,\text{s} - \frac{(0.950)(100 \times 10^3\,\text{m})}{2.998 \times 10^8\,\text{m/s}}\right] = -3.74 \times 10^{-4}\,\text{s} = -374\,\mu\text{s} .$$

16. The "coincidence" of $x = x' = 0$ at $t = t' = 0$ is important for Eq. 38-20 to apply without additional terms. We label the event coordinates with subscripts: $(x_1, t_1) = (0, 0)$ and $(x_2, t_2) = (3000, 4.0 \times 10^{-6})$ with SI units understood. Of course, we expect $(x'_1, t'_1) = (0, 0)$, and this may be verified using Eq. 38-20. We now compute (x'_2, t'_2), assuming $v = +0.60c = +1.799 \times 10^8$ m/s (the sign of v is not made clear in the problem statement, but the Figure referred to, Fig. 38-9, shows the motion in the positive x direction).

$$\begin{aligned} x'_2 &= \frac{x - vt}{\sqrt{1-\beta^2}} = \frac{3000 - (1.799 \times 10^8)(4.0 \times 10^{-6})}{\sqrt{1-(0.60)^2}} = 2.85 \times 10^3 \\ t'_2 &= \frac{t - \beta x/c}{\sqrt{1-\beta^2}} = \frac{4.0 \times 10^{-6} - (0.60)(3000)/(2.998 \times 10^8)}{\sqrt{1-(0.60)^2}} = -2.5 \times 10^{-6} \end{aligned}$$

The two events in frame S occur in the order: first 1, then 2. However, in frame S' where $t'_2 < 0$, they occur in the reverse order: first 2, then 1. We note that the distances $x_2 - x_1$ and $x'_2 - x'_1$ are larger than how far light can travel during the respective times ($c(t_2 - t_1) = 1.2$ km and $c|t'_2 - t'_1| \approx 750$ m), so that no inconsistencies arise as a result of the order reversal (that is, no signal from event 1 could arrive at event 2 or vice versa).

17. (a) We take the flashbulbs to be at rest in frame S, and let frame S' be the rest frame of the second observer. Clocks in neither frame measure the proper time interval between the flashes, so the full Lorentz transformation (Eq. 38-20) must be used. Let t_s be the time and x_s be the coordinate of the small flash, as measured in frame S. Then, the time of the small flash, as measured in frame S', is

$$t'_s = \gamma\left(t_s - \frac{\beta x_s}{c}\right)$$

where $\beta = v/c = 0.250$ and $\gamma = 1/\sqrt{1-\beta^2} = 1/\sqrt{1-(0.250)^2} = 1.0328$. Similarly, let t_b be the time and x_b be the coordinate of the big flash, as measured in frame S. Then, the time of the big flash, as measured in frame S', is

$$t'_b = \gamma\left(t_b - \frac{\beta x_b}{c}\right) .$$

Subtracting the second Lorentz transformation equation from the first and recognizing that $t_s = t_b$ (since the flashes are simultaneous in S), we find

$$\Delta t' = -\frac{\gamma\beta(x_s - x_b)}{c} = -\frac{(1.0328)(0.250)(30 \times 10^3\,\text{m})}{3.00 \times 10^8\,\text{m/s}} = -2.58 \times 10^{-5}\,\text{s}$$

where $\Delta t' = t'_s - t'_b$.

(b) Since $\Delta t'$ is negative, t'_b is greater than t'_s. The small flash occurs first in S'.

18. (a) In frame S, our coordinates are such that $x_1 = +1200$ m for the big flash, and $x_2 = 1200 - 720 = 480\,\text{m}$ for the small flash (which occurred later). Thus, $\Delta x = x_2 - x_1 = -720\,\text{m}$. If we set $\Delta x' = 0$ in Eq. 38-24, we find

$$0 = \gamma\,(\Delta x - v\Delta t) = \gamma\,(-720\,\text{m} - v(5.00 \times 10^{-6}\,\text{s}))$$

which yields $v = -1.44 \times 10^8$ m/s. Therefore, frame S' must be moving in the $-x$ direction with a speed of $0.480c$.

(b) Eq. 38-27 leads to

$$\Delta t' = \gamma\left(\Delta t - \frac{v\Delta x}{c^2}\right) = \gamma\left(5.00 \times 10^{-6}\,\text{s} - \frac{(-1.44 \times 10^8\,\text{m/s})(-720\,\text{m})}{(2.998 \times 10^8\,\text{m/s})^2}\right)$$

which turns out to be positive (regardless of the specific value of γ). Thus, the order of the flashes is the same in the S' frame as it is in the S frame (where Δt is also positive). Thus, the big flash occurs first, and the small flash occurs later.

(c) Finishing the computation begun in part (b), we obtain

$$\Delta t' = \frac{5.00 \times 10^{-6}\,\text{s} - \frac{(-1.44\times 10^8\,\text{m/s})(-720\,\text{m})}{(2.998\times 10^8\,\text{m/s})^2}}{\sqrt{1 - 0.480^2}} = 4.39 \times 10^{-6}\ \text{s}\ .$$

19. (a) The Lorentz factor is

$$\gamma = \frac{1}{\sqrt{1-\beta^2}} = \frac{1}{\sqrt{1-(0.600)^2}} = 1.25\ .$$

(b) In the unprimed frame, the time for the clock to travel from the origin to $x = 180\,\text{m}$ is

$$t = \frac{x}{v} = \frac{180\,\text{m}}{(0.600)(3.00 \times 10^8\,\text{m/s})} = 1.00 \times 10^{-6}\ \text{s}\ .$$

The proper time interval between the two events (at the origin and at $x = 180\,\text{m}$) is measured by the clock itself. The reading on the clock at the beginning of the interval is zero, so the reading at the end is

$$t' = \frac{t}{\gamma} = \frac{1.00 \times 10^{-6}\,\text{s}}{1.25} = 8.00 \times 10^{-7}\ \text{s}\ .$$

20. We refer to the solution of problem 18. We wish to adjust Δt so that

$$\Delta x' = 0 = \gamma\,(-720\,\text{m} - v\Delta t)$$

in the limiting case of $|v| \to c$. Thus,

$$\Delta t = \frac{720\,\text{m}}{2.998 \times 10^8\,\text{m/s}} = 2.40 \times 10^{-6}\ \text{s}\ .$$

21. We assume S' is moving in the $+x$ direction. With $u' = +0.40c$ and $v = +0.60c$, Eq. 38-28 yields

$$u = \frac{u' + v}{1 + u'v/c^2} = \frac{0.40c + 0.60c}{1 + (0.40c)(+0.60c)/c^2} = 0.81c\ .$$

22. (a) We use Eq. 38-28:

$$v = \frac{v' + u}{1 + uv'/c^2} = \frac{0.47c + 0.62c}{1 + (0.47)(0.62)} = 0.84c\ ,$$

in the direction of increasing x (since $v > 0$). The classical theory predicts that $v = 0.47c + 0.62c = 1.1c > c$.

(b) Now $v' = -0.47c$ so

$$v = \frac{v' + u}{1 + uv'/c^2} = \frac{-0.47c + 0.62c}{1 + (-0.47)(0.62)} = 0.21c ,$$

again in the direction of increasing x. By contrast, the classical prediction is $v = 0.62c - 0.47c = 0.15c$.

23. (a) One thing Einstein's relativity has in common with the more familiar (Galilean) relativity is the reciprocity of relative velocity. If Joe sees Fred moving at 20 m/s eastward away from him (Joe), then Fred should see Joe moving at 20 m/s westward away from him (Fred). Similarly, if we see Galaxy A moving away from us at $0.35c$ then an observer in Galaxy A should see our galaxy move away from him at $0.35c$.

(b) We take the positive axis to be in the direction of motion of Galaxy A, as seen by us. Using the notation of Eq. 38-28, the problem indicates $v = +0.35c$ (velocity of Galaxy A relative to Earth) and $u = -0.35c$ (velocity of Galaxy B relative to Earth). We solve for the velocity of B relative to A:

$$u' = \frac{u - v}{1 - uv/c^2} = \frac{(-0.35c) - 0.35c}{1 - (-0.35)(0.35)} = -0.62c$$

or $u' = -1.87 \times 10^8$ m/s.

24. Using the notation of Eq. 38-28 and taking "away" (from us) as the positive direction, the problem indicates $v = +0.4c$ and $u = +0.8c$ (with 3 significant figures understood). We solve for the velocity of Q_2 relative to Q_1:

$$u' = \frac{u - v}{1 - uv/c^2} = \frac{0.8c - 0.4c}{1 - (0.8)(0.4)} = 0.588c$$

or $u' = 1.76 \times 10^8$ m/s in a direction away from Earth.

25. Using the notation of Eq. 38-28 and taking the micrometeorite motion as the positive direction, the problem indicates $v = -0.82c$ (spaceship velocity) and $u = +0.82c$ (micrometeorite velocity). We solve for the velocity of the micrometeorite relative to the spaceship:

$$u' = \frac{u - v}{1 - uv/c^2} = \frac{0.82c - (-0.82c)}{1 - (0.82)(-0.82)} = 0.98c$$

or 2.94×10^8 m/s. Using Eq. 38-10, we conclude that observers on the ship measure a transit time for the micrometeorite (as it passes along the length of the ship) equal to

$$\Delta t = \frac{d}{u'} = \frac{350\,\text{m}}{2.94 \times 10^8\,\text{m/s}} = 1.2 \times 10^{-6}\ \text{s} .$$

26. (a) In the messenger's rest system (called S_m), the velocity of the armada is

$$v' = \frac{v - v_m}{1 - vv_m/c^2} = \frac{0.80c - 0.95c}{1 - (0.80c)(0.95c)/c^2} = -0.625c .$$

The length of the armada as measured in S_m is

$$L_1 = \frac{L_0}{\gamma_{v'}} = (1.0\,\text{ly})\sqrt{1 - (-0.625)^2} = 0.781\ \text{ly} .$$

Thus, the length of the trip is

$$t' = \frac{L'}{|v'|} = \frac{0.781\,\text{ly}}{0.625c} = 1.25\ \text{y} .$$

(b) In the armada's rest frame (called S_a), the velocity of the messenger is

$$v' = \frac{v - v_a}{1 - vv_a/c^2} = \frac{0.95c - 0.80c}{1 - (0.95c)(0.80c)/c^2} = 0.625c \ .$$

Now, the length of the trip is

$$t' = \frac{L_0}{v'} = \frac{1.0\,\text{ly}}{0.625c} = 1.6 \text{ y} \ .$$

(c) Measured in system S, the length of the armada is

$$L = \frac{L_0}{\gamma} = 1.0\,\text{ly}\sqrt{1 - (0.80)^2} = 0.60 \text{ ly} \ ,$$

so the length of the trip is

$$t = \frac{L}{v_m - v_a} = \frac{0.60\,\text{ly}}{0.95c - 0.80c} = 4.0 \text{ y} \ .$$

27. The spaceship is moving away from Earth, so the frequency received is given directly by Eq. 38-30. Thus,

$$f = f_0\sqrt{\frac{1-\beta}{1+\beta}} = (100\,\text{MHz})\sqrt{\frac{1-0.9000}{1+0.9000}} = 22.9 \text{ MHz} \ .$$

28. (a) Eq. 38-33 leads to

$$v = \frac{\Delta\lambda}{\lambda}c = \frac{12\,\text{nm}}{513\,\text{nm}}\left(2.998\times 10^8\,\text{m/s}\right) = 7.0\times 10^6 \text{ m/s} \ .$$

(b) The line is shifted to a larger wavelength, which means shorter frequency. Recalling Eq. 38-30 and the discussion that follows it, this means galaxy NGC is moving away from Earth.

29. Eq. 38-33 leads to a recessional speed of

$$v = \frac{\Delta\lambda}{\lambda}c = (0.004)\left(3.0\times 10^8\,\text{m/s}\right) = 1\times 10^6 \text{ m/s} \ .$$

30. We obtain

$$v = \frac{\Delta\lambda}{\lambda}c = \left(\frac{620-540}{620}\right)c = 0.13c = 3.9\times 10^6 \text{ m/s} \ .$$

31. The frequency received is given by

$$f = f_0\sqrt{\frac{1-\beta}{1+\beta}}$$

$$\frac{c}{\lambda} = \frac{c}{\lambda_0}\sqrt{\frac{1-0.20}{1+0.20}}$$

which implies

$$\lambda = (450\,\text{nm})\sqrt{\frac{1+0.20}{1-0.20}} = 550 \text{ nm} \ .$$

This is in the yellow-green portion of the visible spectrum.

32. (a) The work-kinetic energy theorem applies as well to Einsteinian physics as to Newtonian; the only difference is the specific formula for kinetic energy. Thus, we use $W = \Delta K = m_ec^2(\gamma - 1)$ (Eq. 38-49) and $m_ec^2 = 511\,\text{keV} = 0.511\,\text{MeV}$ (Table 38-3), and obtain

$$W = m_ec^2\left(\frac{1}{\sqrt{1-\beta^2}} - 1\right) = (511\,\text{keV})\left[\frac{1}{\sqrt{1-(0.50)^2}} - 1\right] = 79 \text{ keV} \ .$$

(b)

$$W = (0.511\,\text{MeV})\left(\frac{1}{\sqrt{1-(0.990)^2}} - 1\right) = 3.11\ \text{MeV}\ .$$

(c)

$$W = (0.511\,\text{MeV})\left(\frac{1}{\sqrt{1-(0.9990)^2}} - 1\right) = 10.9\ \text{MeV}\ .$$

33. (a) Using $K = m_e c^2(\gamma - 1)$ (Eq. 38-49) and $m_e c^2 = 511\,\text{keV} = 0.511\,\text{MeV}$ (Table 38-3), we obtain

$$\gamma = \frac{K}{m_e c^2} + 1 = \frac{1.00\,\text{keV}}{511\,\text{keV}} + 1 = 1.00196\ .$$

Therefore, the speed parameter is

$$\beta = \sqrt{1 - \frac{1}{\gamma^2}} = \sqrt{1 - \frac{1}{1.00196^2}} = 0.0625\ .$$

(b) We could first find β and then find γ, as illustrated here: With $K = 1.00\,\text{MeV}$, we find

$$\beta = \sqrt{1 - \left(\frac{1.00\,\text{MeV}}{0.511\,\text{MeV}} + 1\right)^{-2}} = 0.941$$

and $\gamma = 1/\sqrt{1-\beta^2} = 2.96$.

(c) Finally, $K = 1000\,\text{MeV}$, so

$$\beta = \sqrt{1 - \left(\frac{1000\,\text{MeV}}{0.511\,\text{MeV}} + 1\right)^{-2}} = 0.999\,999\,87$$

and $\gamma = 1000\,\text{MeV}/0.511\,\text{MeV} + 1 = 1.96 \times 10^3$. The discussion in Sample Problem 38-7 dealing with these sorts of values may prove helpful for those whose calculators do not yield these answers.

34. From Eq. 38-49, $\gamma = (K/mc^2) + 1$, and from Eq. 38-8, the speed parameter is $\beta = \sqrt{1 - (1/\gamma)^2}$.

(a) Table 38-3 gives $m_e c^2 = 511\,\text{keV} = 0.511\,\text{MeV}$, so the Lorentz factor is

$$\gamma = \frac{10.0\,\text{MeV}}{0.511\,\text{MeV}} + 1 = 20.57\ ,$$

and the speed parameter is

$$\beta = \sqrt{1 - \frac{1}{(20.57)^2}} = 0.9988\ .$$

(b) Table 38-3 gives $m_p c^2 = 938\,\text{MeV}$, so the Lorentz factor is $\gamma = 1 + 10.0\,\text{MeV}/938\,\text{MeV} = 1.01$, and the speed parameter is

$$\beta = \sqrt{1 - \frac{1}{1.01^2}} = 0.145\ .$$

(c) If we refer to the data shown in problem 36, we find $m_\alpha = 4.0026$ u, which (using Eq. 38-43) implies $m_\alpha c^2 = 3728$ MeV. This leads to $\gamma = 10/3728 + 1 = 1.0027$. And, being careful not to do any unnecessary rounding off in the intermediate steps, we find $\beta = 0.073$. We remark that the mass value used in our solution is not exactly the alpha particle mass (it's the helium-4 atomic mass), but this slight difference does not introduce significant error in this computation.

35. From Eq. 38-49, $\gamma = (K/mc^2) + 1$, and from Eq. 38-8, the speed parameter is $\beta = \sqrt{1 - (1/\gamma)^2}$. Table 38-3 gives $m_e c^2 = 511\,\text{keV} = 0.511\,\text{MeV}$, so the Lorentz factor is

$$\gamma = \frac{100\,\text{MeV}}{0.511\,\text{MeV}} + 1 = 197 \ ,$$

and the speed parameter is

$$\beta = \sqrt{1 - \frac{1}{(197)^2}} = 0.999987 \ .$$

Thus, the speed of the electron is $0.999987c$, or 99.9987% of the speed of light. The discussion in Sample Problem 38-7 dealing with these sorts of values may prove helpful for those whose calculators do not yield this answer.

36. The mass change is

$$\Delta M = (4.002603\,\text{u} + 15.994915\,\text{u}) - (1.007825\,\text{u} + 18.998405\,\text{u}) = -0.008712\ \text{u} \ .$$

Using Eq. 38-47 and Eq. 38-43, this leads to

$$Q = -\Delta M\,c^2 = -(-0.008712\,\text{u})(931.5\,\text{MeV/u}) = 8.12\ \text{MeV} \ .$$

37. Since the rest energy E_0 and the mass m of the quasar are related by $E_0 = mc^2$, the rate P of energy radiation and the rate of mass loss are related by $P = dE_0/dt = (dm/dt)c^2$. Thus,

$$\frac{dm}{dt} = \frac{P}{c^2} = \frac{1 \times 10^{41}\ \text{W}}{(2.998 \times 10^8\,\text{m/s})^2} = 1.11 \times 10^{24}\ \text{kg/s} \ .$$

Since a solar mass is 2.0×10^{30} kg and a year is 3.156×10^7 s,

$$\frac{dm}{dt} = (1.11 \times 10^{24}\,\text{kg/s})\left(\frac{3.156 \times 10^7\ \text{s/y}}{2.0 \times 10^{30}\,\text{kg/smu}}\right) \approx 18\ \text{smu/y} \ .$$

38. (a) The work-kinetic energy theorem applies as well to Einsteinian physics as to Newtonian; the only difference is the specific formula for kinetic energy. Thus, we use $W = \Delta K$ where $K = m_e c^2(\gamma - 1)$ (Eq. 38-49), and $m_e c^2 = 511\,\text{keV} = 0.511\,\text{MeV}$ (Table 38-3). Noting that $\Delta K = m_e c^2(\gamma_f - \gamma_i)$, we obtain

$$W = m_e c^2\left(\frac{1}{\sqrt{1-\beta_f^2}} - \frac{1}{\sqrt{1-\beta_i^2}}\right) = (511\,\text{keV})\left(\frac{1}{\sqrt{1-(0.19)^2}} - \frac{1}{\sqrt{1-(0.18)^2}}\right) = 0.996\ \text{keV} \ .$$

(b) Similarly,

$$W = (511\,\text{keV})\left(\frac{1}{\sqrt{1-(0.99)^2}} - \frac{1}{\sqrt{1-(0.98)^2}}\right) = 1055\ \text{keV} \ .$$

We see the dramatic increase in difficulty in trying to accelerate a particle when its initial speed is very close to the speed of light.

39. (a) We set Eq. 38-38 equal to mc, as required by the problem, and solve for the speed. Thus,

$$\frac{mv}{\sqrt{1 - v^2/c^2}} = mc$$

leads to $v = c/\sqrt{2} = 0.707c$.

(b) Substituting $v = \sqrt{2}c$ into the definition of γ, we obtain

$$\gamma = \frac{1}{\sqrt{1 - v^2/c^2}} = \frac{1}{\sqrt{1 - (1/2)}} = \sqrt{2} \approx 1.41 .$$

(c) The kinetic energy is

$$K = (\gamma - 1)mc^2 = (\sqrt{2} - 1)mc^2 = 0.414mc^2 .$$

40. (a) We set Eq. 38-49 equal to $2mc^2$, as required by the problem, and solve for the speed. Thus,

$$mc^2 \left(\frac{1}{\sqrt{1 - \left(\frac{v}{c}\right)^2}} - 1 \right) = 2mc^2$$

leads to $v = \frac{2\sqrt{2}}{3} c \approx 0.943c$.

(b) We now set Eq. 38-45 equal to $2mc^2$ and solve for the speed. In this case,

$$\frac{mc^2}{\sqrt{1 - \left(\frac{v}{c}\right)^2}} = 2mc^2$$

leads to $v = \frac{\sqrt{3}}{2} c \approx 0.866c$.

41. We set Eq. 38-52 equal to $(3mc^2)^2$, as required by the problem, and solve for the speed. Thus,

$$(pc)^2 + (mc^2)^2 = 9(mc^2)^2$$

leads to $p = mc\sqrt{8}$.

42. (a) Squaring Eq. 38-44 gives

$$E^2 = \left(mc^2\right)^2 + 2mc^2K + K^2$$

which we set equal to Eq. 38-52. Thus,

$$\left(mc^2\right)^2 + 2mc^2K + K^2 = (pc)^2 + \left(mc^2\right)^2 \implies m = \frac{(pc)^2 - K^2}{2Kc^2} .$$

(b) At low speeds, the pre-Einsteinian expressions $p = mv$ and $K = \frac{1}{2}mv^2$ apply. We note that $pc \gg K$ at low speeds since $c \gg v$ in this regime. Thus,

$$m \to \frac{(mvc)^2 - \left(\frac{1}{2}mv^2\right)^2}{2\left(\frac{1}{2}mv^2\right)c^2} \approx \frac{(mvc)^2}{2\left(\frac{1}{2}mv^2\right)c^2} = m .$$

(c) Here, $pc = 121\,\text{MeV}$, so

$$m = \frac{121^2 - 55^2}{2(55)c^2} = 105.6\ \text{MeV}/c^2 .$$

Now, the mass of the electron (see Table 38-3) is $m_e = 0.511\,\text{MeV}/c^2$, so our result is roughly 207 times bigger than an electron mass.

43. The energy equivalent of one tablet is $mc^2 = (320 \times 10^{-6}\,\text{kg})(3.00 \times 10^8\,\text{m/s})^2 = 2.88 \times 10^{13}\,\text{J}$. This provides the same energy as $(2.88 \times 10^{13}\,\text{J})/(3.65 \times 10^7\,\text{J/L}) = 7.89 \times 10^5\,\text{L}$ of gasoline. The distance the car can go is $d = (7.89 \times 10^5\,\text{L})(12.75\,\text{km/L}) = 1.01 \times 10^7\,\text{km}$. This is roughly 250 times larger than the circumference of Earth (see Appendix C).

44. (a) The proper lifetime Δt_0 is 2.20 μs, and the lifetime measured by clocks in the laboratory (through which the muon is moving at high speed) is $\Delta t = 6.90\,\mu$s. We use Eq. 38-7 to solve for the speed:

$$v = c\sqrt{1 - \left(\frac{\Delta t_0}{\Delta t}\right)^2} = 0.9478c$$

or $v = 2.84 \times 10^8$ m/s.

(b) From the answer to part (a), we find $\gamma = 3.136$. Thus, with $m_\mu c^2 = 207 m_e c^2 = 105.8$ MeV (see Table 38-3), Eq. 38-49 yields

$$K = m_\mu c^2(\gamma - 1) = 226 \text{ MeV} .$$

(c) We write $m_\mu c = 105.8$ MeV/c and apply Eq. 38-38:

$$p = \gamma m_\mu v = \gamma m_\mu c\beta = (3.136)(105.8\,\text{MeV}/c)(0.9478) = 314 \text{ MeV}/c$$

which can also be expressed in SI units ($p = 1.7 \times 10^{-19}$ kg·m/s).

45. The distance traveled by the pion in the frame of Earth is (using Eq. 38-12) $d = v\,\Delta t$. The proper lifetime Δt_0 is related to Δt by the time-dilation formula: $\Delta t = \gamma \Delta t_0$. To use this equation, we must first find the Lorentz factor γ (using Eq. 38-45). Since the total energy of the pion is given by $E = 1.35 \times 10^5$ MeV and its mc^2 value is 139.6 MeV, then

$$\gamma = \frac{E}{mc^2} = \frac{1.35 \times 10^5\,\text{MeV}}{139.6\,\text{MeV}} = 967.05 .$$

Therefore, the lifetime of the moving pion as measured by Earth observers is

$$\Delta t = \gamma \Delta t_0 = (967.1)(35.0 \times 10^{-9}\,\text{s}) = 3.385 \times 10^{-5} \text{ s} ,$$

and the distance it travels is

$$d \approx c\Delta t = (2.998 \times 10^8\,\text{m/s})(3.385 \times 10^{-5}\,\text{s}) = 1.015 \times 10^4\,\text{m} = 10.15 \text{ km}$$

where we have approximated its speed as c (note: its speed can be found by solving Eq. 38-8, which gives $v = 0.9999995c$; this more precise value for v would not significantly alter our final result). Thus, the altitude at which the pion decays is 120 km − 10.15 km = 110 km.

46. The q in the denominator is to be interpreted as $|q|$ (so that the orbital radius r is a positive number). We interpret the given 10.0 MeV to be the kinetic energy of the electron. In order to make use of the mc^2 value for the electron given in Table 38-3 (511 keV = 0.511 MeV) we write the classical kinetic energy formula as

$$K_{\text{classical}} = \frac{1}{2}mv^2 = \frac{1}{2}\left(mc^2\right)\left(\frac{v^2}{c^2}\right) = \frac{1}{2}\left(mc^2\right)\beta^2 .$$

(a) If $K_{\text{classical}} = 10.0$ MeV, then

$$\beta = \sqrt{\frac{2K_{\text{classical}}}{mc^2}} = \sqrt{\frac{2(10.0\,\text{MeV})}{0.511\,\text{MeV}}} = 6.256 ,$$

which, of course, is impossible (see the Ultimate Speed subsection of §38-2). If we use this value anyway, then the classical orbital radius formula yields

$$\begin{aligned} r &= \frac{mv}{|q|B} = \frac{m\beta c}{eB} \\ &= \frac{(9.11 \times 10^{-31}\,\text{kg})\,(6.256)\,(2.998 \times 10^8\,\text{m/s})}{(1.6 \times 10^{-19}\,\text{C})\,(2.20\,\text{T})} \\ &= 4.85 \times 10^{-3} \text{ m} . \end{aligned}$$

If, however, we use the correct value for β (calculated in the next part) then the classical radius formula would give about 0.77 mm.

(b) Before using the relativistically correct orbital radius formula, we must compute β in a relativistically correct way:

$$K = mc^2(\gamma - 1) \implies \gamma = \frac{10.0\,\text{MeV}}{0.511\,\text{MeV}} + 1 = 20.57$$

which implies (from Eq. 38-8)

$$\beta = \sqrt{1 - \frac{1}{\gamma^2}} = 0.99882\ .$$

Therefore,

$$\begin{aligned} r &= \frac{\gamma m v}{|q|B} = \frac{\gamma m \beta c}{eB} \\ &= \frac{(20.57)\,(9.11 \times 10^{-31}\,\text{kg})\,(0.99882)\,(2.998 \times 10^{8}\,\text{m/s})}{(1.6 \times 10^{-19}\,\text{C})\,(2.20\,\text{T})} \\ &= 1.59 \times 10^{-2}\ \text{m}\ . \end{aligned}$$

(c) The period is

$$T = \frac{2\pi r}{\beta c} = \frac{2\pi(0.0159\,\text{m})}{(0.99882)\,(2.998 \times 10^{8}\,\text{m/s})} = 3.34 \times 10^{-10}\ \text{s}\ .$$

Whereas the purely classical result gives a period which is independent of speed, this is no longer true in the relativistic case (due to the γ factor in the equation).

47. The radius r of the path is given in problem 46 as $r = \gamma m v q B$. Thus,

$$\begin{aligned} m &= \frac{qBr\sqrt{1-\beta^2}}{v} \\ &= \frac{2(1.60 \times 10^{-19}\,\text{C})(1.00\,\text{T})(6.28\,\text{m})\sqrt{1-(0.710)^2}}{(0.710)(3.00 \times 10^{8}\,\text{m/s})} \\ &= 6.64 \times 10^{-27}\ \text{kg}\ . \end{aligned}$$

Since $1.00\,\text{u} = 1.66 \times 10^{-27}\,\text{kg}$, the mass is $m = 4.00\,\text{u}$. The nuclear particle contains four nucleons. Since there must be two protons to provide the charge $2e$, the nuclear particle is a helium nucleus (usually referred to as an alpha particle) with two protons and two neutrons.

48. We interpret the given $10\,\text{GeV} = 10000\,\text{MeV}$ to be the kinetic energy of the proton. Using Table 38-3 and Eq. 38-49, we find

$$\gamma = \frac{K}{m_p c^2} + 1 = \frac{10000\,\text{MeV}}{938\,\text{MeV}} + 1 = 11.66\ ,$$

and (from Eq. 38-8)

$$\beta = \sqrt{1 - \frac{1}{\gamma^2}} = 0.9963\ .$$

Therefore, using the equation introduced in problem 46, we obtain

$$\begin{aligned} r &= \frac{\gamma m v}{qB} = \frac{\gamma m_p \beta c}{eB} \\ &= \frac{(11.66)\,(1.67 \times 10^{-27}\,\text{kg})\,(0.9963)\,(2.998 \times 10^{8}\,\text{m/s})}{(1.6 \times 10^{-19}\,\text{C})\,(55 \times 10^{-6}\,\text{T})} \\ &= 6.6 \times 10^{5}\ \text{m}\ . \end{aligned}$$

49. We interpret the given 2.50 MeV = 2500 keV to be the kinetic energy of the electron. Using Table 38-3 and Eq. 38-49, we find

$$\gamma = \frac{K}{m_e c^2} + 1 = \frac{2500\,\text{keV}}{511\,\text{keV}} + 1 = 5.892 \; ,$$

and (from Eq. 38-8)

$$\beta = \sqrt{1 - \frac{1}{\gamma^2}} = 0.9855 \; .$$

Therefore, using the equation introduced in problem 46 (with "q" interpreted as $|q|$), we obtain

$$\begin{aligned} B &= \frac{\gamma m_e v}{|q|\, r} = \frac{\gamma m_e \beta c}{er} \\ &= \frac{(5.892)\,(9.11 \times 10^{-31}\,\text{kg})\,(0.9855)\,(2.998 \times 10^8\,\text{m/s})}{(1.6 \times 10^{-19}\,\text{C})\,(0.030\,\text{m})} \\ &= 0.33\ \text{T} \; . \end{aligned}$$

50. (a) Using Table 38-3 and Eq. 38-49 (or, to be more precise, the value given at the end of the problem statement), we find

$$\gamma = \frac{K}{m_p c^2} + 1 = \frac{500 \times 10^3\,\text{MeV}}{938.3\,\text{MeV}} + 1 = 533.88 \; .$$

(b) From Eq. 38-8, we obtain

$$\beta = \sqrt{1 - \frac{1}{\gamma^2}} = 0.99999825 \; .$$

The discussion in Sample Problem 38-7 dealing with large γ values may prove helpful for those whose calculators do not yield this answer.

(c) To make use of the precise $m_p c^2$ value given here, we rewrite the expression introduced in problem 46 (as applied to the proton) as follows:

$$r = \frac{\gamma m v}{qB} = \frac{\gamma\,(mc^2)\,\left(\frac{v}{c^2}\right)}{eB} = \frac{\gamma\,(mc^2)\,\beta}{ecB} \; .$$

Therefore, the magnitude of the magnetic field is

$$\begin{aligned} B &= \frac{\gamma\,(mc^2)\,\beta}{ecr} \\ &= \frac{(533.88)(938.3\,\text{MeV})(0.99999825)}{ec(750\,\text{m})} \\ &= \frac{667.92 \times 10^6\,\text{V/m}}{c} \end{aligned}$$

where we note the cancellation of the "e" in MeV with the e in the denominator. After substituting $c = 2.998 \times 10^8$ m/s, we obtain $B = 2.23$ T.

51. (a) Before looking at our solution to part (a) (which uses momentum conservation), it might be advisable to look at our solution (and accompanying remarks) for part (b) (where a very different approach is used). Since momentum is a vector, its conservation involves two equations (along the original direction of alpha particle motion, the x direction, as well as along the final proton direction of motion, the y direction). The problem states that all speeds are much less than the speed of light, which allows us to use the classical formulas for kinetic energy and momentum ($K = \frac{1}{2}mv^2$ and

$\vec{p} = m\vec{v}$, respectively). Along the x and y axes, momentum conservation gives (for the components of $\vec{v}_{\rm oxy}$):

$$\begin{aligned} m_\alpha v_\alpha &= m_{\rm oxy} v_{{\rm oxy},x} &\Longrightarrow\ & v_{{\rm oxy},x} = \frac{m_\alpha}{m_{\rm oxy}} v_\alpha \approx \frac{4}{17} v_\alpha \\ 0 &= m_{\rm oxy} v_{{\rm oxy},y} + m_p v_p &\Longrightarrow\ & v_{{\rm oxy},y} = -\frac{m_p}{m_{\rm oxy}} v_p \approx -\frac{1}{17} v_p \ . \end{aligned}$$

To complete these determinations, we need values (inferred from the kinetic energies given in the problem) for the initial speed of the alpha particle (v_α) and the final speed of the proton (v_p). One way to do this is to rewrite the classical kinetic energy expression as $K = \frac{1}{2}(mc^2)\beta^2$ and solve for β (using Table 38-3 and/or Eq. 38-43). Thus, for the proton, we obtain

$$\beta_p = \sqrt{\frac{2K_p}{m_p c^2}} = \sqrt{\frac{2(4.44\,{\rm MeV})}{938\,{\rm MeV}}} = 0.0973 \ .$$

This is almost 10% the speed of light, so one might worry that the relativistic expression (Eq. 38-49) should be used. If one does so, one finds $\beta_p = 0.969$, which is reasonably close to our previous result based on the classical formula. For the alpha particle, we write $m_\alpha c^2 = (4.0026\,{\rm u})(931.5\,{\rm MeV/u}) =$ 3728 MeV (which is actually an overestimate due to the use of the "atomic mass" value in our calculation, but this does not cause significant error in our result), and obtain

$$\beta_\alpha = \sqrt{\frac{2K_\alpha}{m_\alpha c^2}} = \sqrt{\frac{2(7.70\,{\rm MeV})}{3728\,{\rm MeV}}} = 0.064 \ .$$

Returning to our oxygen nucleus velocity components, we are now able to conclude:

$$\begin{aligned} v_{{\rm oxy},x} &\approx \frac{4}{17} v_\alpha &\Longrightarrow\ & \beta_{{\rm oxy},x} \approx \frac{4}{17}\beta_\alpha = \frac{4}{17}(0.064) = 0.015 \\ |v_{{\rm oxy},y}| &\approx \frac{1}{17} v_p &\Longrightarrow\ & \beta_{{\rm oxy},y} \approx \frac{1}{17}\beta_p = \frac{1}{17}(0.097) = 0.0057 \end{aligned}$$

Consequently, with $m_{\rm oxy}c^2 \approx (17\,{\rm u})(931.5\,{\rm MeV/u}) = 1.58 \times 10^4$ MeV, we obtain

$$K_{\rm oxy} = \frac{1}{2}\left(m_{\rm oxy}c^2\right)\left(\beta^2_{{\rm oxy},x} + \beta^2_{{\rm oxy},y}\right) = \frac{1}{2}\left(1.58 \times 10^4\,{\rm MeV}\right)\left(0.015^2 + 0.0057^2\right) \approx 2.0\ {\rm MeV} \ .$$

(b) Using Eq. 38-47 and Eq. 38-43,

$$Q = -(1.007825\,{\rm u} + 16.99914\,{\rm u} - 4.00260\,{\rm u} - 14.00307\,{\rm u})c^2 = -(0.001295\,{\rm u})(931.5\,{\rm MeV/u})$$

which yields $Q = -1.206$ MeV. Incidentally, this provides an alternate way to obtain the answer (and a more accurate one at that!) to part (a). Eq. 38-46 leads to

$$K_{\rm oxy} = K_\alpha + Q - K_p = 7.70\,{\rm MeV} - 1.206\,{\rm MeV} - 4.44\,{\rm MeV} = 2.05\ {\rm MeV} \ .$$

This approach to finding $K_{\rm oxy}$ avoids the many computational steps and approximations made in part (a).

52. (a) From the length contraction equation, the length L'_c of the car according to Garageman is

$$L'_c = \frac{L_c}{\gamma} = L_c\sqrt{1-\beta^2} = (30.5\,{\rm m})\sqrt{1-(0.9980)^2} = 1.93\ {\rm m} \ .$$

(b) Since the x_g axis is fixed to the garage $x_{g2} = L_g = 6.00\,{\rm m}$. As for t_{g2}, note from Fig. 38-21(b) that, at $t_g = t_{g1} = 0$ the coordinate of the front bumper of the limo in the x_g frame is L'_c, meaning that the front of the limo is still a distance $L_g - L'_c$ from the back door of the garage. Since the limo travels at a speed v, the time it takes for the front of the limo to reach the back door of the garage is given by

$$\Delta t_g = t_{g2} - t_{g1} = \frac{L_g - L'_c}{v} = \frac{6.00\,\text{m} - 1.93\,\text{m}}{0.9980(2.998 \times 10^8\,\text{m/s})} = 1.36 \times 10^{-8}\ \text{s}\,.$$

Thus $t_{g2} = t_{g1} + \Delta t_g = 0 + 1.36 \times 10^{-8}\,\text{s} = 1.36 \times 10^{-8}\,\text{s}$.

(c) The limo is inside the garage between times t_{g1} and t_{g2}, so the time duration is $t_{g2} - t_{g1} = 1.36 \times 10^{-8}\,\text{s}$.

(d) Again from Eq. 38-13, the length L'_g of the garage according to Carman is

$$L'_g = \frac{L_g}{\gamma} = L_g\sqrt{1-\beta^2} = (6.00\,\text{m})\,\sqrt{1-(0.9980)^2} = 0.379\ \text{m}\,.$$

(e) Again, since the x_c axis is fixed to the limo $x_{c2} = L_c = 30.5\,\text{m}$. Now, from the two diagrams described in part (h) below, we know that at $t_c = t_{c2}$ (when event 2 takes place), the distance between the rear bumper of the limo and the back door of the garage is given by $L_c - L'_g$. Since the garage travels at a speed v, the front door of the garage will reach the rear bumper of the limo a time Δt_c later, where Δt_c satisfies

$$\Delta t_c = t_{c1} - t_{c2} = \frac{L_c - L'_g}{v} = \frac{30.5\,\text{m} - 0.379\,\text{m}}{0.9980(2.998 \times 10^8\,\text{m/s})} = 1.01 \times 10^{-7}\ \text{s}\,.$$

Thus $t_{c2} = t_{c1} - \Delta t_c = 0 - 1.01 \times 10^{-7}\,\text{s} = -1.01 \times 10^{-7}\,\text{s}$.

(f) From Carman's point of view, the answer is clearly no.

(g) Event 2 occurs first according to Carman, since $t_{c2} < t_{c1}$.

(h) We describe the essential features of the two pictures. For event 2, the front of the limo coincides with the back door, and the garage itself seems very short (perhaps failing to reach as far as the front window of the limo). For event 1, the rear of the car coincides with the front door and the front of the limo has traveled a significant distance beyond the back door. In this picture, as in the other, the garage seems very short compared to the limo.

(i) Both Carman and Garageman are correct in their respective reference frames. But, in a sense, Carman should lose the bet since he dropped his physics course before reaching the Theory of Special Relativity!

53. (a) The spatial separation between the two bursts is vt. We project this length onto the direction perpendicular to the light rays headed to Earth and obtain $D_{\text{app}} = vt \sin\theta$.

(b) Burst 1 is emitted a time t ahead of burst 2. Also, burst 1 has to travel an extra distance L more than burst 2 before reaching the Earth, where $L = vt\cos\theta$ (see Fig. 38-22); this requires an additional time $t' = L/c$. Thus, the apparent time is given by

$$T_{\text{app}} = t - t' = t - \frac{vt\cos\theta}{c} = t\left[1 - \left(\frac{v}{c}\right)\cos\theta\right]\,.$$

(c) We obtain

$$V_{\text{app}} = \frac{D_{\text{app}}}{T_{\text{app}}} = \left[\frac{(v/c)\sin\theta}{1-(v/c)\cos\theta}\right]c = \left[\frac{(0.980)\sin 30.0^\circ}{1-(0.980)\cos 30.0^\circ}\right]c = 3.24\,c\,.$$

54. (a) The strategy is to find the γ factor from $E = 14.24 \times 10^{-9}$ J and $m_pc^2 = 1.5033 \times 10^{-10}$ J and from that find the contracted length. From the energy relation (Eq. 38-45), we obtain

$$\gamma = \frac{E}{mc^2} = 94.73\,.$$

Consequently, Eq. 38-13 yields

$$L = \frac{L_0}{\gamma} = 0.222\,\text{cm} = 2.22 \times 10^{-3}\ \text{m}\ .$$

(b) and (c) From the γ factor, we find the speed:

$$v = c\sqrt{1 - \left(\frac{1}{\gamma}\right)^2} = 0.99994c\ .$$

Therefore, the trip (according to the proton) took $\Delta t_0 = 2.22 \times 10^{-3}/0.99994c = 7.40 \times 10^{-12}$ s. Finally, the time dilation formula (Eq. 38-7) leads to

$$\Delta t = \gamma \Delta t_0 = 7.01 \times 10^{-10}\ \text{s}$$

which can be checked using $\Delta t = L_0/v$ in our frame of reference.

55. Since it has two protons, its kinetic energy is 600 MeV. With the given value $mc^2 = 3727$ MeV, we use Eq. 38-37:

$$pc = \sqrt{K^2 + 2Kmc^2} = \sqrt{600^2 + 2(600)(3727)}$$

which yields $p = 2198\ \text{MeV}/c$.

56. For the purposes of using Eq. 38-28, we choose our frame to be the primed frame and note that, as a consequence, $v = -0.800c\,\hat{\imath}$ for the velocity of us relative to Bullwinkle.

$$u = \frac{u' + v}{1 + u'v/c^2} = \frac{0.990c\hat{\imath} - 0.800c\hat{\imath}}{1 - (0.990)(0.800)} = 0.913c\,\hat{\imath}\ .$$

57. (a) We compute

$$\gamma = \frac{1}{\sqrt{1 - (0.9990)^2}} = 22.4$$

Now, the length contraction formula (Eq. 38-13) yields

$$L = \frac{2.50\ \text{m}}{\gamma} = 0.112\ \text{m}\ .$$

(b) (c) and (d) We assume our spacetime coordinate origins coincide and use the Lorentz transformations (Eq. 38-20, but with primes and non-primes swapped, and $v \to -v$). Lengths are in meters and time is in nanoseconds (so that $c = 0.2998$ in these units).

$$\begin{aligned}
x_\alpha &= \gamma\,(4.0 + (0.9990c)(40)) = 357 \\
t_\alpha &= \gamma\left(40 + (0.9990c)(4.0)/c^2\right) = 1193 \\
x_\beta &= \gamma\,(-4.0 + (0.9990c)(80)) = 446 \\
t_\beta &= \gamma\left(80 + (0.9990c)(-4.0)/c^2\right) = 1491
\end{aligned}$$

Thus, our reckoning of the distance between events is $x_\beta - x_\alpha = 89.0$ m. We note that event alpha took place first (smallest value of t) and that the time-separation is $t_\alpha - t_\beta = 298$ ns.

58. Using Eq. 38-10,

$$v = \frac{d}{t} = \frac{6.0\,\text{ly}}{2.0\,\text{y} + 6.0\,\text{y}} = \frac{(6.0c)(1.0\,\text{y})}{2.0\,\text{y} + 6.0\,\text{y}} = 0.75c.$$

59. To illustrate the technique, we derive Eq. 1′ from Eqs. 1 and 2 (in Table 38-2). We multiply Eq. 2 by speed v and subtract it from Eq. 1:

$$\Delta x - v\Delta t = \gamma\left(\Delta x' + v\Delta t'\right) - v\gamma\left(\Delta t' + \frac{v\Delta x'}{c^2}\right) = \gamma\Delta x'\left(1 - \frac{v^2}{c^2}\right)$$

We note that $\gamma\left(1 - v^2/c^2\right) = 1/\gamma$ (using Eq. 38-8), so that if we multiply the above equation by γ we obtain Eq. 1′:

$$\gamma\left(\Delta x - v\Delta t\right) = \gamma\left(\gamma\Delta x'\left(1 - \frac{v^2}{c^2}\right)\right) = \Delta x'$$

60. (a) $v_r = 2v = 2(27000\,\text{km/h}) = 54000\,\text{km/h}$.

(b) We can express c in these units by multiplying by 3.6: $c = 1.08 \times 10^9$ km/h. The correct formula for v_r is $v_r = 2v/(1 + v^2/c^2)$, so the fractional error is

$$1 - \frac{1}{1 + v^2/c^2} = 1 - \frac{1}{1 + [(27000\,\text{km/h})/(1.08 \times 10^9\,\text{km/h})]^2} = 6.3 \times 10^{-10}\ .$$

The discussion in Sample Problem 38-7 dealing with numerical considerations may prove helpful for those whose calculators do not yield this answer.

61. (a) We assume the electron starts from rest. The classical formula for kinetic energy is Eq. 38-48, so if $v = c$ then this (for an electron) would be $\frac{1}{2}mc^2 = \frac{1}{2}(511\,\text{keV}) = 255.5$ keV (using Table 38-3). Setting this equal to the potential energy loss (which is responsible for its acceleration), we find (using Eq. 25-7)

$$V = \frac{255.5\,\text{keV}}{|q|} = \frac{255\,\text{keV}}{e} = 255.5\ \text{kV}\ .$$

(b) Setting this amount of potential energy loss ($|\Delta U| = 255.5$ keV) equal to the correct relativistic kinetic energy, we obtain (using Eq. 38-49)

$$mc^2\left(\frac{1}{\sqrt{1 - (v/c)^2}} - 1\right) = |\Delta U| \implies v = c\sqrt{1 + \left(\frac{1}{1 - \Delta U/mc^2}\right)^2}$$

which yields $v = 0.745c = 2.23 \times 10^8$ m/s.

62. (a) $\Delta E = \Delta mc^2 = (3.0\,\text{kg})(0.0010)(2.998 \times 10^8\,\text{m/s})^2 = 2.7 \times 10^{14}$ J.

(b) The mass of TNT is

$$m_{\text{TNT}} = \frac{(2.7 \times 10^{14}\,\text{J})(0.227\,\text{kg/mol})}{3.4 \times 10^6\,\text{J}} = 1.8 \times 10^7\ \text{kg}\ .$$

(c) The fraction of mass converted in the TNT case is

$$\frac{\Delta m_{\text{TNT}}}{m_{\text{TNT}}} = \frac{(3.0\,\text{kg})(0.0010)}{1.8 \times 10^7\,\text{kg}} = 1.6 \times 10^{-9}\ ,$$

Therefore, the fraction is $0.0010/1.6 \times 10^{-9} = 6.0 \times 10^6$.

63. (a) Eq. 38-33 yields

$$v = \frac{\Delta\lambda}{\lambda}c = \left(\frac{462 - 434}{434}\right)c = 0.065c$$

or $v = 1.93 \times 10^7$ m/s.

(b) Since it is shifted "towards the red" (towards longer wavelengths) then the galaxy is moving away from us (receding).

64. When $\beta = 0.9860$, we have $\gamma = 5.9972$, and when $\beta = 0.9850$, we have $\gamma = 5.7953$. Thus, $\Delta\gamma = 0.202$ and the change in kinetic energy (equal to the work) becomes (using Eq. 38-49)

$$W = \Delta K = mc^2 \Delta\gamma = 189 \text{ MeV}$$

where $mc^2 = 938$ MeV has been used (see Table 38-3).

65. Using $m_p = 1.672623 \times 10^{-27}$ kg in Eq. 38-45 yields

$$\gamma = \frac{E}{m_p c^2} = \frac{14.242 \times 10^{-9} \text{ J}}{1.50328 \times 10^{-10} \text{ J}} = 94.740 \ .$$

Solving for the speed , we obtain

$$v = c\sqrt{1 - \left(\frac{1}{\gamma}\right)^2} = 0.99994c \ .$$

66. (a) According to ship observers, the duration of proton flight is $\Delta t' = (760\,\text{m})/0.980c = 2.59\ \mu\text{s}$ (assuming it travels the entire length of the ship).

(b) To transform to our point of view, we use Eq. 2 in Table 38-2. Thus, with $\Delta x' = -750$ m, we have

$$\Delta t = \gamma\left(\Delta t' + (0.950c)\Delta x'/c^2\right) = 0.57\ \mu\text{s} \ .$$

(c) and (d) For the ship observers, firing the proton from back to front makes no difference, and $\Delta t' = 2.59\ \mu\text{s}$ as before. For us, the fact that now $\Delta x' = +750$ m is a significant change.

$$\Delta t = \gamma\left(\Delta t' + (0.950c)\Delta x'/c^2\right) = 16.0\ \mu\text{s} \ .$$

67. (a) Our lab-based measurement of its lifetime is figured simply from $t = L/v = 7.99 \times 10^{-13}$ s. Use of the time-dilation relation (Eq. 38-7) leads to

$$\Delta t_0 = \left(7.99 \times 10^{-13}\ \text{s}\right)\sqrt{1 - (0.960)^2} = 2.24 \times 10^{-13}\ \text{s} \ .$$

(b) The length contraction formula can be used, or we can use the simple speed-distance relation (from the point of view of the particle, who watches the lab and all its meter sticks rushing past him at $0.960c$ until he expires): $L = v\Delta t_0 = 6.44 \times 10^{-5}$ m.

68. Using Appendix C, we find that the contraction is

$$\begin{aligned} |\Delta L| &= L_0 - L = L_0\left(1 - \frac{1}{\gamma}\right) = L_0(1 - \sqrt{1-\beta^2}) \\ &= 2(6.370 \times 10^6\ \text{m})\left(1 - \sqrt{1 - \left(\frac{3.0 \times 10^4\ \text{m/s}}{2.998 \times 10^8\ \text{m/s}}\right)^2}\right) \\ &= 0.064\ \text{m} \ . \end{aligned}$$

The discussion in Sample Problem 38-7 dealing with numerical considerations may prove helpful for those whose calculators do not yield this answer.

69. The speed of the spaceship after the first increment is $v_1 = 0.5c$. After the second one, it becomes

$$v_2 = \frac{v' + v_1}{1 + v'v_1/c^2} = \frac{0.50c + 0.50c}{1 + (0.50c)^2/c^2} = 0.80c \ ,$$

and after the third one, the speed is

$$v_3 = \frac{v' + v_2}{1 + v'v_2/c^2} = \frac{0.50c + 0.50c}{1 + (0.50c)(0.80c)/c^2} = 0.929c \ .$$

Continuing with this process, we get $v_4 = 0.976c$, $v_5 = 0.992c$, $v_6 = 0.997c$ and $v_7 = 0.999c$. Thus, seven increments are needed.

70. We use the transverse Doppler shift formula, Eq. 38-34: $f = f_0\sqrt{1-\beta^2}$, or

$$\frac{1}{\lambda} = \frac{1}{\lambda_0}\sqrt{1-\beta^2} .$$

We solve for $\lambda - \lambda_0$:

$$\lambda - \lambda_0 = \lambda_0\left(\frac{1}{\sqrt{1-\beta^2}} - 1\right) = (589.00\,\text{mm})\left[\frac{1}{\sqrt{1-(0.100)^2}} - 1\right] = +2.97 \text{ nm} .$$

71. The mean lifetime of a pion measured by observers on the Earth is $\Delta t = \gamma\Delta t_0$, so the distance it can travel (using Eq. 38-12) is

$$d = v\Delta t = \gamma v \Delta t_0 = \frac{(0.99)(2.998\times10^8\,\text{m/s})(26\times10^{-9}\,\text{s})}{\sqrt{1-(0.99)^2}} = 55 \text{ m} .$$

72. (a) For a proton (using Table 38-3), our results are:

$$E = \gamma m_p c^2 = \frac{938\,\text{MeV}}{\sqrt{1-(0.990)^2}} = 6.65 \text{ GeV}$$

$$K = E - m_p c^2 = 6.65\,\text{GeV} - 938\,\text{MeV} = 5.71 \text{ GeV}$$

$$p = \gamma m_p v = \gamma(m_p c^2)\beta/c = \frac{(938\,\text{MeV})(0.990)/c}{\sqrt{1-(0.990)^2}} = 6.59 \text{ GeV}/c$$

(b) For an electron:

$$E = \gamma m_e c^2 = \frac{0.511\,\text{MeV}}{\sqrt{1-(0.990)^2}} = 3.62 \text{ MeV}$$

$$K = E - m_e c^2 = 3.625\,\text{MeV} - 0.511\,\text{MeV} = 3.11 \text{ MeV}$$

$$p = \gamma m_e v = \gamma(m_e c^2)\beta/c = \frac{(0.511\,\text{MeV})(0.990)/c}{\sqrt{1-(0.990)^2}} = 3.59 \text{ MeV}/c$$

73. The strategy is to find the speed from $E = 1533$ MeV and $mc^2 = 0.511$ MeV (see Table 38-3) and from that find the time. From the energy relation (Eq. 38-45), we obtain

$$v = c\sqrt{1-\left(\frac{mc^2}{E}\right)^2} = 0.99999994c \approx c$$

so that we conclude it took the electron 26 y to reach us. In order to transform to its own "clock" it's useful to compute γ directly from Eq. 38-45:

$$\gamma = \frac{E}{mc^2} = 3000$$

though if one is careful one can also get this result from $\gamma = 1/\sqrt{1-(v/c)^2}$. Then, Eq. 38-7 leads to

$$\Delta t_0 = \frac{26 \text{ y}}{\gamma} = 0.0087\,\text{y}$$

so that the electron "concludes" the distance he traveled is 0.0087 light-years (stated differently, the Earth, which is rushing towards him at very nearly the speed of light, seemed to start its journey from a distance of 0.0087 light-years away).

74. (a) Using Eq. 38-7, we expect the dilated time intervals to be

$$\tau = \gamma\tau_0 = \frac{\tau_0}{\sqrt{1-(v/c)^2}} .$$

(b) We rewrite Eq. 38-30 using the fact that period is the reciprocal of frequency ($f_R = \tau_R^{-1}$ and $f_0 = \tau_0^{-1}$):

$$\tau_R = \frac{1}{f_R} = \left(f_0\sqrt{\frac{1-\beta}{1+\beta}}\right)^{-1} = \tau_0\sqrt{\frac{1+\beta}{1-\beta}} = \tau_0\sqrt{\frac{c+v}{c-v}} .$$

(c) The Doppler shift combines two physical effects: the time dilation of the moving source *and* the travel-time differences involved in periodic emission (like a sine wave or a series of pulses) from a traveling source to a "stationary" receiver). To isolate the purely time-dilation effect, it's useful to consider "local" measurements (say, comparing the readings on a moving clock to those of two of your clocks, spaced some distance apart, such that the moving clock and each of your clocks can make a close-comparison of readings at the moment of passage).

75. We use the relative velocity formula (Eq. 38-28) with the primed measurements being those of the scout ship. We note that $v = -0.900c$ since the velocity of the scout ship relative to the cruiser is opposite to that of the cruiser relative to the scout ship.

$$u = \frac{u'+v}{1+u'v/c^2} = \frac{0.980c - 0.900c}{1-(0.980)(0.900)} = 0.678c .$$

76. We solve the time dilation equation for the time elapsed (as measured by Earth observers):

$$\Delta t = \frac{\Delta t_0}{\sqrt{1-(0.9990)^2}}$$

where $\Delta t_0 = 120$ y. This yields $\Delta t = 2684$ y.

77. (a) The relative contraction is

$$\begin{aligned}
\frac{|\Delta L|}{L_0} &= \frac{L_0(1-\gamma^{-1})}{L_0} = 1-\sqrt{1-\beta^2} \\
&\approx 1-\left(1-\frac{1}{2}\beta^2\right) = \frac{1}{2}\beta^2 \\
&= \frac{1}{2}\left(\frac{630\,\text{m/s}}{3.00\times10^8\,\text{m/s}}\right)^2 \\
&= 2.21\times10^{-12} .
\end{aligned}$$

(b) Letting $|\Delta t - \Delta t_0| = \Delta t_0(\gamma - 1) = \tau = 1.00\,\mu$s, we solve for Δt_0:

$$\begin{aligned}
\Delta t_0 &= \frac{\tau}{\gamma-1} = \frac{\tau}{(1-\beta^2)^{-1/2}-1} \approx \frac{\tau}{1+\frac{1}{2}\beta^2-1} = \frac{2\tau}{\beta^2} \\
&= \frac{2(1.00\times10^{-6}\,\text{s})(1\,\text{d}/86400\,\text{s})}{[(630\,\text{m/s})/(2.998\times10^8\,\text{m/s})]^2} \\
&= 5.25\ \text{d} .
\end{aligned}$$

78. Let the reference frame be S in which the particle (approaching the South Pole) is at rest, and let the frame that is fixed on Earth be S'. Then $v = 0.60c$ and $u' = 0.80c$ (calling "downwards" [in the sense of Fig. 38-31] positive). The relative speed is now the speed of the other particle as measured in S:

$$u = \frac{u'+v}{1+u'v/c^2} = \frac{0.80c+0.60c}{1+(0.80c)(0.60c)/c^2} = 0.95c .$$

79. We refer to the particle in the first sentence of the problem statement as particle 2. Since the total momentum of the two particles is zero in S', it must be that the velocities of these two particles are equal in magnitude and opposite in direction in S'. Letting the velocity of the S' frame be v relative to S, then the particle which is at rest in S must have a velocity of $u'_1 = -v$ as measured in S', while the velocity of the other particle is given by solving Eq. 38-28 for u':

$$u'_2 = \frac{u_2 - v}{1 - u_2 v/c^2} = \frac{\left(\frac{c}{2}\right) - v}{1 - \left(\frac{c}{2}\right)\left(\frac{v}{c^2}\right)} .$$

Letting $u'_2 = -u'_1 = v$, we obtain

$$\frac{\left(\frac{c}{2}\right) - v}{1 - \left(\frac{c}{2}\right)\left(\frac{v}{c^2}\right)} = v \implies v = c(2 \pm \sqrt{3}) \approx 0.27c$$

where the quadratic formula has been used (with the smaller of the two roots chosen so that $v \le c$).

80. From Eq. 28-37, we have

$$Q = -\Delta M c^2 = -\left(3(4.00151\,\mathrm{u}) - 11.99671\,\mathrm{u}\right)c^2 = -(0.00782\,\mathrm{u})(931.5\,\mathrm{MeV/u}) = -7.28\ \mathrm{MeV} .$$

Thus, it takes a minimum of 7.28 MeV supplied to the system to cause this reaction. We note that the masses given in this problem are strictly for the nuclei involved; they are not the "atomic" masses which are quoted in several of the other problems in this chapter.

81. We use Eq. 38-51 with $mc^2 = 0.511$ MeV (see Table 38-3):

$$pc = \sqrt{K^2 + 2Kmc^2} = \sqrt{(2.00)^2 + 2(2.00)(0.511)}$$

This readily yields $p = 2.46\ \mathrm{MeV}/c$.

Chapter 39

1. Eq. 39-3 gives $h = 4.14 \times 10^{-15}\,\text{eV}\cdot\text{s}$, but the metric prefix which stands for 10^{-15} is femto (f). Thus, $h = 4.14\,\text{eV}\cdot\text{fs}$.

2. Let $E = 1240\,\text{eV}\cdot\text{nm}/\lambda_{\text{min}} = 0.6\,\text{eV}$ to get $\lambda = 2.1 \times 10^3\,\text{nm} = 2.1\,\mu\text{m}$. It is in the infrared region.

3. The energy of a photon is given by $E = hf$, where h is the Planck constant and f is the frequency. The wavelength λ is related to the frequency by $\lambda f = c$, so $E = hc/\lambda$. Since $h = 6.626 \times 10^{-34}\,\text{J}\cdot\text{s}$ and $c = 2.998 \times 10^8\,\text{m/s}$,

$$hc = \frac{(6.626 \times 10^{-34}\,\text{J}\cdot\text{s})(2.998 \times 10^8\,\text{m/s})}{(1.602 \times 10^{-19}\,\text{J/eV})(10^{-9}\,\text{m/nm})} = 1240\,\text{eV}\cdot\text{nm} .$$

   Thus,

$$E = \frac{1240\,\text{eV}\cdot\text{nm}}{\lambda} .$$

4. From the result of problem 3,

$$E = \frac{hc}{\lambda} = \frac{1240\,\text{eV}\cdot\text{nm}}{589\,\text{nm}} = 2.11\ \text{eV} .$$

5. Let R be the rate of photon emission (number of photons emitted per unit time) of the Sun and let E be the energy of a single photon. Then the power output of the Sun is given by $P = RE$. Now $E = hf = hc/\lambda$, where h is the Planck constant, f is the frequency of the light emitted, and λ is the wavelength. Thus $P = Rhc/\lambda$ and

$$R = \frac{\lambda P}{hc} = \frac{(550\,\text{nm})(3.9 \times 10^{26}\,\text{W})}{(6.63 \times 10^{-34}\,\text{J}\cdot\text{s})(2.998 \times 10^8\,\text{m/s})} = 1.0 \times 10^{45}\ \text{photons/s} .$$

6. We denote the diameter of the laser beam as d. The cross-sectional area of the beam is $A = \pi d^2/4$. From the formula obtained in problem 5, the rate is given by

$$\begin{aligned} \frac{R}{A} &= \frac{\lambda P}{hc(\pi d^2/4)} \\ &= \frac{4(633\,\text{nm})(5.0 \times 10^{-3}\,\text{W})}{\pi(6.63 \times 10^{-34}\,\text{J}\cdot\text{s})(2.998 \times 10^8\,\text{m/s})(3.5 \times 10^{-3}\,\text{m})^2} \\ &= 1.7 \times 10^{21}\,\frac{\text{photons}}{\text{m}^2 \cdot \text{s}} . \end{aligned}$$

7. Using the result of problem 3,

$$E = \frac{hc}{\lambda} = \frac{1240\,\text{eV}\cdot\text{nm}}{21 \times 10^7\,\text{nm}} = 5.9 \times 10^{-6}\,\text{eV} = 5.9\,\mu\text{eV} .$$

8. Let
$$\frac{1}{2}m_e v^2 = E_{\text{photon}} = \frac{hc}{\lambda}$$
and solve for v:
$$\begin{aligned} v &= \sqrt{\frac{2hc}{\lambda m_e}} = \sqrt{\frac{2hc}{\lambda m_e c^2}\,c^2} = c\sqrt{\frac{2hc}{\lambda (m_e c^2)}} \\ &= (2.998 \times 10^8\ \text{m/s})\sqrt{\frac{2(1240\,\text{eV}\cdot\text{nm})}{(590\,\text{nm})(511 \times 10^3\,\text{eV})}} = 8.6 \times 10^5\ \text{m/s}\ . \end{aligned}$$
Since $v \ll c$, the non-relativistic formula $K = \frac{1}{2}mv^2$ may be used. The result of problem 3 and the $m_e c^2$ value of Table 38-3 are used in our calculation.

9. Since $\lambda = (1{,}650{,}763.73)^{-1}\,\text{m} = 6.0578021 \times 10^{-7}\,\text{m} = 605.78021\,\text{nm}$, the energy is (using the result of problem 3)
$$E = \frac{hc}{\lambda} = \frac{1240\,\text{eV}\cdot\text{nm}}{605.78021\,\text{nm}} = 2.047\ \text{eV}\ .$$

10. Following Sample Problem 39-1, we have
$$P = \frac{Rhc}{\lambda} = \frac{(100/\text{s})\,(6.63 \times 10^{-34}\,\text{J}\cdot\text{s})\,(2.998 \times 10^8\,\text{m/s})}{550 \times 10^{-9}\,\text{m}} = 3.6 \times 10^{-17}\ \text{W}\ .$$

11. The total energy emitted by the bulb is $E = 0.93Pt$, where $P = 60$ W and $t = 730\,\text{h} = (730\,\text{h})(3600\,\text{s/h}) = 2.628 \times 10^6\,\text{s}$. The energy of each photon emitted is $E_{\text{ph}} = hc/\lambda$. Therefore, the number of photons emitted is
$$N = \frac{E}{E_{\text{ph}}} = \frac{0.93Pt}{hc/\lambda} = \frac{(0.93)(60\,\text{W})(2.628 \times 10^6\,\text{s})}{(6.63 \times 10^{-34}\,\text{J}\cdot\text{s})\,(2.998 \times 10^8\,\text{m/s})\,/(630 \times 10^{-9}\,\text{m})} = 4.7 \times 10^{26}\ .$$

12. The rate at which photons are emitted from the argon laser source is given by $R = P/E_{\text{ph}}$, where $P = 1.5$ W is the power of the laser beam and $E_{\text{ph}} = hc/\lambda$ is the energy of each photon of wavelength λ. Since $\alpha = 84\%$ of the energy of the laser beam falls within the central disk, the rate of photon absorption of the central disk is
$$\begin{aligned} R' &= \alpha R = \frac{\alpha P}{hc/\lambda} = \frac{(0.84)(1.5\,\text{W})}{(6.63 \times 10^{-34}\,\text{J}\cdot\text{s})\,(2.998 \times 10^8\,\text{m/s})\,/(515 \times 10^{-9}\,\text{m})} \\ &= 3.3 \times 10^{18}\ \text{photons/s}\ . \end{aligned}$$

13. (a) Let R be the rate of photon emission (number of photons emitted per unit time) and let E be the energy of a single photon. Then, the power output of a lamp is given by $P = RE$ if all the power goes into photon production. Now, $E = hf = hc/\lambda$, where h is the Planck constant, f is the frequency of the light emitted, and λ is the wavelength. Thus $P = Rhc/\lambda$ and $R = \lambda P/hc$. The lamp emitting light with the longer wavelength (the 700 nm lamp) emits more photons per unit time. The energy of each photon is less, so it must emit photons at a greater rate.

(b) Let R be the rate of photon production for the 700 nm lamp. Then,
$$R = \frac{\lambda P}{hc} = \frac{(700\,\text{nm})(400\,\text{J/s})}{(1.60 \times 10^{-19}\,\text{J/eV})(1240\,\text{eV}\cdot\text{nm})} = 1.41 \times 10^{21}\ \text{photon/s}\ .$$
The result $hc = 1240\,\text{eV}\cdot\text{nm}$ developed in Exercise 3 is used.

14. (a) The rate at which solar energy strikes the panel is
$$P = (1.39\,\text{kW/m}^2)(2.60\,\text{m}^2) = 3.61\ \text{kW}\ .$$

(b) The rate at which solar photons are absorbed by the panel is

$$R = \frac{P}{E_{ph}} = \frac{3.61 \times 10^3 \text{ W}}{(6.63 \times 10^{-34}\,\text{J}\cdot\text{s})\,(2.998 \times 10^8\,\text{m/s})\,/(550 \times 10^{-9}\,\text{m})} = 1.00 \times 10^{22}/\text{s}\ .$$

(c) The time in question is given by

$$t = \frac{N_A}{R} = \frac{6.02 \times 10^{23}}{1.00 \times 10^{22}/\text{s}} = 60.2\ \text{s}\ .$$

15. (a) We assume all the power results in photon production at the wavelength $\lambda = 589\,\text{nm}$. Let R be the rate of photon production and E be the energy of a single photon. Then, $P = RE = Rhc/\lambda$, where $E = hf$ and $f = c/\lambda$ are used. Here h is the Planck constant, f is the frequency of the emitted light, and λ is its wavelength. Thus,

$$R = \frac{\lambda P}{hc} = \frac{(589 \times 10^{-9}\,\text{m})(100\,\text{W})}{(6.63 \times 10^{-34}\,\text{J}\cdot\text{s})(3.00 \times 10^8\,\text{m/s})} = 2.96 \times 10^{20}\ \text{photon/s}\ .$$

(b) Let I be the photon flux a distance r from the source. Since photons are emitted uniformly in all directions, $R = 4\pi r^2 I$ and

$$r = \sqrt{\frac{R}{4\pi I}} = \sqrt{\frac{2.96 \times 10^{20}\,\text{photon/s}}{4\pi(1.00 \times 10^4\,\text{photon/m}^2\cdot\text{s})}} = 4.85 \times 10^7\ \text{m}\ .$$

(c) The photon flux is

$$I = \frac{R}{4\pi r^2} = \frac{2.96 \times 10^{20}\,\text{photon/s}}{4\pi(2.00\,\text{m})^2} = 5.89 \times 10^{18}\ \frac{\text{photon}}{\text{m}^2\cdot\text{s}}\ .$$

16. (a) Since $E_{ph} = h/\lambda = 1240\,\text{eV}\cdot\text{nm}/680\,\text{nm} = 1.82\,\text{eV} < \Phi = 2.28\,\text{eV}$, there is no photoelectric emission. The result of problem 3 is used in our calculation.

(b) The cutoff wavelength is the longest wavelength of photons which will cause photoelectric emission. In sodium, this is given by $E_{ph} = hc/\lambda_{max} = \Phi$, or $\lambda_{max} = hc/\Phi = (1240\,\text{eV}\cdot\text{nm})/2.28\,\text{eV} = 544\,\text{nm}$. This corresponds to the color green.

17. The energy of the most energetic photon in the visible light range (with wavelength of about 400 nm) is about $E = (1240\,\text{eV}\cdot\text{nm}/400\,\text{nm}) = 3.1\,\text{eV}$ (using the result of problem 3). Consequently, barium and lithium can be used, since their work functions are both lower than 3.1 eV.

18. (a) For $\lambda = 565\,\text{nm}$

$$hf = \frac{hc}{\lambda} = \frac{1240\,\text{eV}\cdot\text{nm}}{565\,\text{nm}} = 2.20\ \text{eV}\ .$$

Since $\Phi_{potassium} > hf > \Phi_{cesium}$, the photoelectric effect can occur in cesium but not in potassium at this wavelength. The result of problem 3 is used in our calculation.

(b) Now $\lambda = 518\,\text{nm}$ so

$$hf = \frac{hc}{\lambda} = \frac{1240\,\text{eV}\cdot\text{nm}}{518\,\text{m}} = 2.40\,\text{eV}\ .$$

This is greater than both Φ_{cesium} and $\Phi_{potassium}$, so the photoelectric effect can now occur for both metals.

19. The energy of an incident photon is $E = hf = hc/\lambda$, where h is the Planck constant, f is the frequency of the electromagnetic radiation, and λ is its wavelength. The kinetic energy of the most energetic electron emitted is $K_m = E - \Phi = (hc/\lambda) - \Phi$, where Φ is the work function for sodium. The stopping potential V_0 is related to the maximum kinetic energy by $eV_0 = K_m$, so $eV_0 = (hc/\lambda) - \Phi$ and

$$\lambda = \frac{hc}{eV_0 + \Phi} = \frac{1240\,\text{eV}\cdot\text{nm}}{5.0\,\text{eV} + 2.2\,\text{eV}} = 170\ \text{nm}\ .$$

Here $eV_0 = 5.0\,\text{eV}$ and $hc = 1240\,\text{eV}\cdot\text{nm}$ are used. See problem 3.

20. We use Eq. 39-5 to find the maximum kinetic energy of the ejected electrons:

$$K_{\text{max}} = hf - \Phi = (4.14 \times 10^{-15}\,\text{eV·s})(3.0 \times 10^{15}\,\text{Hz}) - 2.3\,\text{eV} = 10\ \text{eV}\ .$$

21. The speed v of the electron satisfies $K_{\text{max}} = \frac{1}{2}m_e v^2 = \frac{1}{2}(m_e c^2)(v/c)^2 = E_{\text{photon}} - \Phi$. Using Table 38-3, we find

$$v = c\sqrt{\frac{2(E_{\text{photon}} - \Phi)}{m_e c^2}} = (2.998 \times 10^8\ \text{m/s})\sqrt{\frac{2(5.80\,\text{eV} - 4.50\,\text{eV})}{511 \times 10^3\,\text{eV}}} = 6.76 \times 10^5\ \text{m/s}\ .$$

22. (a) We use Eq. 39-6:

$$V_{\text{stop}} = \frac{hf - \Phi}{e} = \frac{hc/\lambda - \Phi}{e} = \frac{(1240\,\text{eV·nm}/400\,\text{nm}) - 1.8\,\text{eV}}{e} = 1.3\ \text{V}\ .$$

(b) We use the formula obtained in the solution of problem 21:

$$\begin{aligned} v &= \sqrt{\frac{2(E_{\text{photon}} - \Phi)}{m_e}} = \sqrt{\frac{2eV_{\text{stop}}}{m_e}} = c\sqrt{\frac{2eV_{\text{stop}}}{m_e c^2}} \\ &= (2.998 \times 10^8\ \text{m/s})\sqrt{\frac{2e(1.3\,\text{V})}{511 \times 10^3\,\text{eV}}} \\ &= 6.8 \times 10^5\ \text{m/s}\ . \end{aligned}$$

23. (a) The kinetic energy K_m of the fastest electron emitted is given by $K_m = hf - \Phi = (hc/\lambda) - \Phi$, where Φ is the work function of aluminum, f is the frequency of the incident radiation, and λ is its wavelength. The relationship $f = c/\lambda$ was used to obtain the second form. Thus,

$$K_m = \frac{1240\,\text{eV} \cdot \text{nm}}{200\,\text{nm}} - 4.20\,\text{eV} = 2.00\ \text{eV}$$

where the result of Exercise 3 is used.

(b) The slowest electron just breaks free of the surface and so has zero kinetic energy.

(c) The stopping potential V_0 is given by $K_m = eV_0$, so $V_0 = K_m/e = (2.00\,\text{eV})/e = 2.00\,\text{V}$.

(d) The value of the cutoff wavelength is such that $K_m = 0$. Thus $hc/\lambda = \Phi$ or $\lambda = hc/\Phi = (1240\,\text{eV·nm})/(4.2\,\text{eV}) = 295\,\text{nm}$. If the wavelength is longer, the photon energy is less and a photon does not have sufficient energy to knock even the most energetic electron out of the aluminum sample.

24. We use Eq. 39-6 and the result of problem 3:

$$K_{\text{max}} = E_{\text{photon}} - \Phi = \frac{hc}{\lambda} - \frac{hc}{\lambda_{\text{max}}} = \frac{1240\,\text{eV·nm}}{254\,\text{nm}} - \frac{1240\,\text{eV·nm}}{325\,\text{nm}} = 1.07\ \text{eV}\ .$$

25. To find the longest possible wavelength λ_{max} (corresponding to the lowest possible energy) of a photon which can produce a photoelectric effect in platinum, we set $K_{\text{max}} = 0$ in Eq. 39-5 and use $hf = hc/\lambda$. Thus $hc/\lambda_{\text{max}} = \Phi$. We solve for λ_{max}:

$$\lambda_{\text{max}} = \frac{hc}{\Phi} = \frac{1240\,\text{eV·nm}}{5.32\,\text{eV}} = 233\ \text{nm}\ .$$

26. (a) For the first and second case (labeled 1 and 2) we have $eV_{01} = hc/\lambda_1 - \Phi$ and $eV_{02} = hc/\lambda_2 - \Phi$, from which h and Φ can be determined. Thus,

$$\begin{aligned} h &= \frac{e(V_1 - V_2)}{c(\lambda_1^{-1} - \lambda_2^{-1})} = \frac{1.85\,\text{eV} - 0.820\,\text{eV}}{(3.00 \times 10^{17}\,\text{nm/s})[(300\,\text{nm})^{-1} - (400\,\text{nm})^{-1}]} \\ &= 4.12 \times 10^{-15}\ \text{eV·s}\ . \end{aligned}$$

(b) The work function is

$$\Phi = \frac{3(V_2\lambda_2 - V_1\lambda_1)}{\lambda_1 - \lambda_2} = \frac{(0.820\,\text{eV})(400\,\text{nm}) - (1.85\,\text{eV})(300\,\text{nm})}{300\,\text{nm} - 400\,\text{nm}} = 2.27\ \text{eV}\ .$$

(c) Let $\Phi = hc/\lambda_{\text{max}}$ to obtain

$$\lambda_{\text{max}} = \frac{hc}{\Phi} = \frac{1240\,\text{eV}\cdot\text{nm}}{2.27\,\text{eV}} = 545\ \text{nm}\ .$$

27. (a) We use the photoelectric effect equation (Eq. 39-5) in the form $hc/\lambda = \Phi + K_m$. The work function depends only on the material and the condition of the surface, and not on the wavelength of the incident light. Let λ_1 be the first wavelength described and λ_2 be the second. Let $K_{m1} = 0.710\,\text{eV}$ be the maximum kinetic energy of electrons ejected by light with the first wavelength, and $K_{m2} = 1.43\,\text{eV}$ be the maximum kinetic energy of electrons ejected by light with the second wavelength. Then,

$$\frac{hc}{\lambda_1} = \Phi + K_{m1} \quad \text{and} \quad \frac{hc}{\lambda_2} = \Phi + K_{m2}\ .$$

The first equation yields $\Phi = (hc/\lambda_1) - K_{m1}$. When this is used to substitute for Φ in the second equation, the result is $(hc/\lambda_2) = (hc/\lambda_1) - K_{m1} + K_{m2}$. The solution for λ_2 is

$$\begin{aligned}\lambda_2 &= \frac{hc\lambda_1}{hc + \lambda_1(K_{m2} - K_{m1})} \\ &= \frac{(1240\,\text{eV}\cdot\text{nm})(491\,\text{nm})}{1240\,\text{eV}\cdot\text{nm} + (491\,\text{nm})(1.43\,\text{eV} - 0.710\,\text{eV})} \\ &= 382\ \text{nm}\ .\end{aligned}$$

Here $hc = 1240\,\text{eV}\cdot\text{nm}$, calculated in Exercise 3, is used.

(b) The first equation displayed above yields

$$\Phi = \frac{hc}{\lambda_1} - K_{m1} = \frac{1240\,\text{eV}\cdot\text{nm}}{491\,\text{nm}} - 0.710\,\text{eV} = 1.82\ \text{eV}\ .$$

28. (a) We calculate frequencies from the wavelengths (expressed in SI units) using Eq. 39-1. Our plot of the points and the line which gives the least squares fit to the data is shown below. The vertical axis is in volts and the horizontal axis, when multiplied by 10^{14}, gives the frequencies in Hertz.

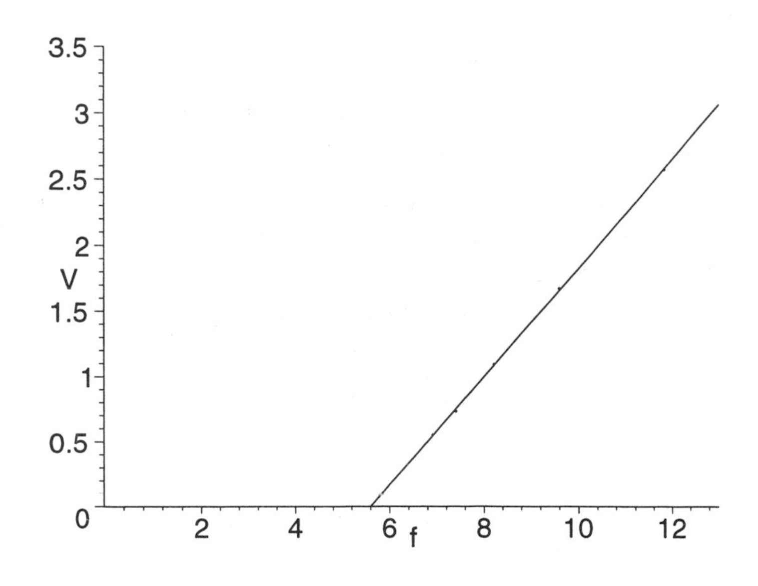

From our least squares fit procedure, we determine the slope to be 4.14×10^{-15} V·s, which is in very good agreement with the value given in Eq. 39-3 (once it has been multiplied by e).

(b) Our least squares fit procedure can also determine the y-intercept for that line. The y-intercept is the negative of the photoelectric work function. In this way, we find $\Phi = 2.31\,\text{eV}$.

29. Using the result of problem 3, the number of photons emitted from the laser per unit time is

$$R = \frac{P}{E_{\text{ph}}} = \frac{2.00 \times 10^{-3}\,\text{W}}{(1240\,\text{eV}\cdot\text{nm}/600\,\text{nm})(1.60 \times 10^{-19}\,\text{J/eV})} = 6.05 \times 10^{15}/\text{s} \ ,$$

of which $(1.0 \times 10^{-16})(6.05 \times 10^{15}/\text{s}) = 0.605/\text{s}$ actually cause photoelectric emissions. Thus the current is $i = (0.605/\text{s})(1.60 \times 10^{-19}\,\text{C}) = 9.68 \times 10^{-20}\,\text{A}$.

30. (a) Find the speed v of the electron from $r = m_e v/eB$: $v = rBe/m_e$. Thus

$$\begin{aligned} K_{\max} &= \frac{1}{2}m_e v^2 = \frac{1}{2}m_e\left(\frac{rBe}{m_e}\right)^2 = \frac{(rB)^2e^2}{2m_e} \\ &= \frac{(1.88 \times 10^{-4}\,\text{T}\cdot\text{m})^2(1.60 \times 10^{-19}\,\text{C})^2}{2(9.11 \times 10^{-31}\,\text{kg})(1.60 \times 10^{-19}\,\text{J/eV})} \\ &= 3.10\ \text{keV} \ . \end{aligned}$$

(b) Using the result of problem 3, the work done is

$$W = E_{\text{photon}} - K_{\max} = \frac{1240\,\text{eV}\cdot\text{nm}}{71 \times 10^{-3}\,\text{nm}} - 3.10\,\text{keV} = 14\ \text{keV} \ .$$

31. (a) When a photon scatters from an electron initially at rest, the change in wavelength is given by $\Delta\lambda = (h/mc)(1 - \cos\phi)$, where m is the mass of an electron and ϕ is the scattering angle. Now, $h/mc = 2.43 \times 10^{-12}\,\text{m} = 2.43\,\text{pm}$, so $\Delta\lambda = (2.43\,\text{pm})(1 - \cos 30°) = 0.326\,\text{pm}$. The final wavelength is $\lambda' = \lambda + \Delta\lambda = 2.4\,\text{pm} + 0.326\,\text{pm} = 2.73\,\text{pm}$.

(b) Now, $\Delta\lambda = (2.43\,\text{pm})(1 - \cos 120°) = 3.645\,\text{pm}$ and $\lambda' = 2.4\,\text{pm} + 3.645\,\text{pm} = 6.05\,\text{pm}$.

32. (a) The rest energy of an electron is given by $E = m_e c^2$. Thus the momentum of the photon in question is given by

$$\begin{aligned} p &= \frac{E}{c} = \frac{m_e c^2}{c} = m_e c \\ &= (9.11 \times 10^{-31}\,\text{kg})(2.998 \times 10^8\,\text{m/s}) \\ &= 2.73 \times 10^{-22}\ \text{kg}\cdot\text{m/s} \ . \end{aligned}$$

We may also express the momentum in terms of MeV/c: $p = m_e c^2/c = 0.511\,\text{MeV}/c$.

(b) From Eq. 39-7,

$$\lambda = \frac{h}{p} = \frac{6.63 \times 10^{-34}\,\text{J}\cdot\text{s}}{2.73 \times 10^{-22}\,\text{kg}\cdot\text{m/s}} = 2.43 \times 10^{-12}\,\text{m} = 2.43\ \text{pm} \ .$$

(c) Using Eq. 39-1,

$$f = \frac{c}{\lambda} = \frac{2.998 \times 10^8\,\text{m/s}}{2.43 \times 10^{-12}\,\text{m}} = 1.24 \times 10^{20}\ \text{Hz} \ .$$

33. (a) The x-ray frequency is

$$f = \frac{c}{\lambda} = \frac{2.998 \times 10^8\,\text{m/s}}{35.0 \times 10^{-12}\,\text{m}} = 8.57 \times 10^{18}\ \text{Hz} \ .$$

(b) The x-ray photon energy is

$$E = hf = (4.14 \times 10^{-15}\,\text{eV}\cdot\text{s})(8.57 \times 10^{18}\,\text{Hz}) = 3.55 \times 10^4 \text{ eV} .$$

(c) From Eq. 39-7,

$$p = \frac{h}{\lambda} = \frac{6.63 \times 10^{-34}\,\text{J}\cdot\text{s}}{35.0 \times 10^{-12}\,\text{m}} = 1.89 \times 10^{-23} \text{ kg}\cdot\text{m/s} .$$

34. (a) Eq. 39-11 yields

$$\Delta\lambda = \frac{h}{m_e c}(1 - \cos\phi) = (2.43\,\text{pm})(1 - \cos 180°) = +4.86 \text{ pm} .$$

(b) Using the result of problem 3, the change in photon energy is

$$\Delta E = \frac{hc}{\lambda'} - \frac{hc}{\lambda} = (1240\,\text{eV}\cdot\text{nm})\left(\frac{1}{0.01\,\text{nm} + 4.86\,\text{pm}} - \frac{1}{0.01\,\text{nm}}\right) = -41 \text{ keV} .$$

(c) From conservation of energy, $\Delta K = -\Delta E = 41\,\text{keV}$.

(d) The electron will move straight ahead after the collision, since it has acquired some of the forward linear momentum from the photon.

35. With no loss of generality, we assume the electron is initially at rest (which simply means we are analyzing the collision from its initial rest frame). If the photon gave all its momentum and energy to the (free) electron, then the momentum and the kinetic energy of the electron would become

$$p = \frac{hf}{c} \quad \text{and} \quad K = hf ,$$

respectively. Plugging these expressions into Eq. 38-51 (with m referring to the mass of the electron) leads to

$$\begin{aligned} (pc)^2 &= K^2 + 2Kmc^2 \\ (hf)^2 &= (hf)^2 + 2hfmc^2 \end{aligned}$$

which is clearly impossible, since the last term ($2hfmc^2$) is not zero. We have shown that considering total momentum and energy absorption of a photon by a free electron leads to an inconsistency in the mathematics, and thus cannot be expected to happen in nature.

36. (a) Using the result of problem 3, we find

$$\lambda = \frac{hc}{E} = \frac{1240\,\text{nm}\cdot\text{eV}}{0.511\,\text{MeV}} = 2.43 \times 10^{-3}\,\text{nm} = 2.43 \text{ pm} .$$

(b) Now, Eq. 39-11 leads to

$$\begin{aligned} \lambda' &= \lambda + \Delta\lambda = \lambda + \frac{h}{m_e c}(1 - \cos\phi) \\ &= 2.43\,\text{pm} + (2.43\,\text{pm})(1 - \cos 90.0°) = 4.86 \text{ pm} . \end{aligned}$$

(c) The scattered photons have energy equal to

$$E' = E\left(\frac{\lambda}{\lambda'}\right) = (0.511\,\text{MeV})\left(\frac{2.43\,\text{pm}}{4.86\,\text{pm}}\right) = 0.255 \text{ MeV} .$$

37. (a) Since the mass of an electron is $m = 9.109 \times 10^{-31}$ kg, its Compton wavelength is

$$\lambda_C = \frac{h}{mc} = \frac{6.626 \times 10^{-34}\,\text{J}\cdot\text{s}}{(9.109 \times 10^{-31}\,\text{kg})(2.998 \times 10^8\,\text{m/s})} = 2.426 \times 10^{-12}\,\text{m} = 2.43\,\text{pm} .$$

(b) Since the mass of a proton is $m = 1.673 \times 10^{-27}$ kg, its Compton wavelength is

$$\lambda_C = \frac{6.626 \times 10^{-34}\,\text{J}\cdot\text{s}}{(1.673 \times 10^{-27}\,\text{kg})(2.998 \times 10^8\,\text{m/s})} = 1.321 \times 10^{-15}\,\text{m} = 1.32\,\text{fm} .$$

(c) We use the formula developed in Exercise 3: $E = (1240\,\text{eV}\cdot\text{nm})/\lambda$, where E is the energy and λ is the wavelength. Thus for the electron, $E = (1240\,\text{eV}\cdot\text{nm})/(2.426 \times 10^{-3}\,\text{nm}) = 5.11 \times 10^5\,\text{eV} = 0.511\,\text{MeV}$.

(d) For the proton, $E = (1240\,\text{eV}\cdot\text{nm})/(1.321 \times 10^{-6}\,\text{nm}) = 9.39 \times 10^8\,\text{eV} = 939\,\text{MeV}$.

38. The $(1 - \cos\phi)$ factor in Eq. 39-11 is largest when $\phi = 180°$. Thus, using Table 38-3, we obtain

$$\Delta\lambda_{\text{max}} = \frac{hc}{m_p c^2}(1 - \cos 180°) = \frac{1240\,\text{MeV}\cdot\text{fm}}{938\,\text{MeV}}(1 - (-1)) = 2.6\,\text{fm}$$

where we have extended the result of problem 3 somewhat by noting that $hc = 1240\,\text{eV}\cdot\text{nm}$ can equivalently be written as $1240\,\text{MeV}\cdot\text{fm}$.

39. If E is the original energy of the photon and E' is the energy after scattering, then the fractional energy loss is

$$frac = \frac{E - E'}{E} .$$

Sample Problem 39-4 shows that this is

$$frac = \frac{\Delta\lambda}{\lambda + \Delta\lambda} .$$

Thus

$$\frac{\Delta\lambda}{\lambda} = \frac{frac}{1 - frac} = \frac{0.75}{1 - 0.75} = 3 .$$

A 300% increase in the wavelength leads to a 75% decrease in the energy of the photon.

40. (a) The fractional change is

$$\begin{aligned}\frac{\Delta E}{E} &= \frac{\Delta(hc/\lambda)}{hc/\lambda} = \lambda\Delta\left(\frac{1}{\lambda}\right) = \lambda\left(\frac{1}{\lambda'} - \frac{1}{\lambda}\right) = \frac{\lambda}{\lambda'} - 1 = \frac{\lambda}{\lambda + \Delta\lambda} - 1 \\ &= -\frac{1}{\lambda/\Delta\lambda + 1} = -\frac{1}{(\lambda/\lambda_C)(1 - \cos\phi)^{-1} + 1} .\end{aligned}$$

If $\lambda = 3.0\,\text{cm} = 3.0 \times 10^{10}\,\text{pm}$ and $\phi = 90°$, the result is

$$\frac{\Delta E}{E} = -\frac{1}{(3.0 \times 10^{10}\,\text{pm}/2.43\,\text{pm})(1 - \cos 90°)^{-1} + 1} = -8.1 \times 10^{-11} .$$

(b) Now $\lambda = 500\,\text{nm} = 5.00 \times 10^5\,\text{pm}$ and $\phi = 90°$, so

$$\frac{\Delta E}{E} = -\frac{1}{(5.00 \times 10^5\,\text{pm}/2.43\,\text{pm})(1 - \cos 90°)^{-1} + 1} = -4.9 \times 10^{-6} .$$

(c) With$\lambda = 25\,\text{pm}$ and $\phi = 90°$, we find

$$\frac{\Delta E}{E} = -\frac{1}{(25\,\text{pm}/2.43\,\text{pm})(1 - \cos 90°)^{-1} + 1} = -8.9 \times 10^{-2} .$$

(d) In this case, $\lambda = hc/E = 1240\,\text{nm}\cdot\text{eV}/1.0\,\text{MeV} = 1.24 \times 10^{-3}\,\text{nm} = 1.24\,\text{pm}$, so

$$\frac{\Delta E}{E} = -\frac{1}{(1.24\,\text{pm}/2.43\,\text{pm})(1-\cos 90°)^{-1}+1} = -0.66 \ .$$

(e) From the calculation above, we see that the shorter the wavelength the greater the fractional energy change for the photon as a result of the Compton scattering. Since $\Delta E/E$ is virtually zero for microwave and visible light, the Compton effect is significant only in the x-ray to gamma ray range of the electromagnetic spectrum.

41. The difference between the electron-photon scattering process in this problem and the one studied in the text (the Compton shift, see Eq. 39-11) is that the electron is in motion relative with speed v to the laboratory frame. To utilize the result in Eq. 39-11, shift to a new reference frame in which the electron is at rest before the scattering. Denote the quantities measured in this new frame with a prime ($'$), and apply Eq. 39-11 to yield

$$\Delta\lambda' = \lambda' - \lambda_0' = \frac{h}{m_e c}(1-\cos\pi) = \frac{2h}{m_e c} \ ,$$

where we note that $\phi = \pi$ (since the photon is scattered back in the direction of incidence). Now, from the Doppler shift formula (Eq. 38-25) the frequency f_0' of the photon prior to the scattering in the new reference frame satisfies

$$f_0' = \frac{c}{\lambda_0'} = f_0\sqrt{\frac{1+\beta}{1-\beta}} \ ,$$

where $\beta = v/c$. Also, as we switch back from the new reference frame to the original one after the scattering

$$f = f'\sqrt{\frac{1-\beta}{1+\beta}} = \frac{c}{\lambda'}\sqrt{\frac{1-\beta}{1+\beta}} \ .$$

We solve the two Doppler-shift equations above for λ' and λ_0' and substitute the results into the Compton shift formula for $\Delta\lambda'$:

$$\Delta\lambda' = \frac{1}{f}\sqrt{\frac{1-\beta}{1+\beta}} - \frac{1}{f_0}\sqrt{\frac{1-\beta}{1+\beta}} = \frac{2h}{m_e c^2} \ .$$

Some simple algebra then leads to

$$E = hf = hf_0\left(1 + \frac{2h}{m_e c^2}\sqrt{\frac{1+\beta}{1-\beta}}\right)^{-1} \ .$$

42. From Sample Problem 39-4, we have

$$\begin{aligned}\frac{\Delta E}{E} &= \frac{\Delta\lambda}{\lambda + \Delta\lambda} \\ &= \frac{(h/mc)(1-\cos\phi)}{\lambda'} \\ &= \frac{hf'}{mc^2}(1-\cos\phi)\end{aligned}$$

where we use the fact that $\lambda + \Delta\lambda = \lambda' = c/f'$.

43. (a) From Eq. 39-11, $\Delta\lambda = (h/m_e c)(1-\cos\phi)$. In this case $\phi = 180°$ (so $\cos\phi = -1$), and the change in wavelength for the photon is given by $\Delta\lambda = 2h/m_e c$. The energy E' of the scattered photon (whose initial energy is $E = hc/\lambda$) is then

$$\begin{aligned}E' &= \frac{hc}{\lambda + \Delta\lambda} = \frac{E}{1+\Delta\lambda/\lambda} = \frac{E}{1+(2h/m_e c)(E/hc)} = \frac{E}{1+2E/m_e c^2} \\ &= \frac{50.0\,\text{keV}}{1+2(50.0\,\text{keV})/0.511\,\text{MeV}} = 41.8\,\text{keV} \ .\end{aligned}$$

(b) From conservation of energy the kinetic energy K of the electron is given by $K = E - E' = 50.0\,\text{keV} - 41.8\,\text{keV} = 8.2\,\text{keV}$.

44. (a) From Eq. 39-11

$$\Delta\lambda = \frac{h}{m_e c}(1 - \cos\phi) = (2.43\,\text{pm})(1 - \cos 90^\circ) = 2.43\,\text{pm} .$$

(b) The fractional shift should be interpreted as $\Delta\lambda$ divided by the original wavelength:

$$\frac{\Delta\lambda}{\lambda} = \frac{2.425\,\text{pm}}{590\,\text{nm}} = 4.11 \times 10^{-6} .$$

(c) The change in energy for a photon with $\lambda = 590\,\text{nm}$ is given by

$$\begin{aligned}\Delta E_{\text{ph}} &= \Delta\left(\frac{hc}{\lambda}\right) \approx -\frac{hc\Delta\lambda}{\lambda^2} \\ &= -\frac{(4.14 \times 10^{-15}\,\text{eV}\cdot\text{s})(2.998 \times 10^8\,\text{m/s})(2.43\,\text{pm})}{(590\,\text{nm})^2} \\ &= -8.67 \times 10^{-6}\,\text{eV} .\end{aligned}$$

For an x ray photon of energy $E_{\text{ph}} = 50\,\text{keV}$, $\Delta\lambda$ remains the same (2.43 pm), since it is independent of E_{ph}. The fractional change in wavelength is now

$$\frac{\Delta\lambda}{\lambda} = \frac{\Delta\lambda}{hc/E_{\text{ph}}} = \frac{(50 \times 10^3\,\text{eV})(2.43\,\text{pm})}{(4.14 \times 10^{-15}\,\text{eV}\cdot\text{s})(2.998 \times 10^8\,\text{m/s})} = 9.78 \times 10^{-2} ,$$

and the change in photon energy is now

$$\Delta E_{\text{ph}} = hc\left(\frac{1}{\lambda + \Delta\lambda} - \frac{1}{\lambda}\right) = -\left(\frac{hc}{\lambda}\right)\frac{\Delta\lambda}{\lambda + \Delta\lambda} = -E_{\text{ph}}\left(\frac{\alpha}{1+\alpha}\right)$$

where $\alpha = \Delta\lambda/\lambda$. We substitute $E_{\text{ph}} = 50\,\text{keV}$ and $\alpha = 9.78 \times 10^{-2}$ to obtain $\Delta E_{\text{ph}} = -4.45\,\text{keV}$. (Note that in this case $\alpha \approx 0.1$ is not close enough to zero so the approximation $\Delta E_{\text{ph}} \approx hc\Delta\lambda/\lambda^2$ is not as accurate as in the first case, in which $\alpha = 4.12 \times 10^{-6}$. In fact if one were to use this approximation here, one would get $\Delta E_{\text{ph}} \approx -4.89\,\text{keV}$, which does not amount to a satisfactory approximation.)

45. The initial wavelength of the photon is (using the result of problem 3)

$$\lambda = \frac{hc}{E} = \frac{1240\,\text{eV}\cdot\text{nm}}{17500\,\text{eV}} = 0.07086\ \text{nm}$$

or 70.86 pm. The maximum Compton shift occurs for $\phi = 180^\circ$, in which case Eq. 39-11 (applied to an electron) yields

$$\Delta\lambda = \left(\frac{hc}{m_e c^2}\right)(1 - \cos 180^\circ) = \left(\frac{1240\,\text{eV}\cdot\text{nm}}{511 \times 10^3\,\text{eV}}\right)(1 - (-1)) = 0.00485\ \text{nm}$$

where Table 38-3 is used. Therefore, the new photon wavelength is $\lambda' = 0.07086\,\text{nm} + 0.00485\,\text{nm} = 0.0757\,\text{nm}$. Consequently, the new photon energy is

$$E' = \frac{hc}{\lambda'} = \frac{1240\,\text{eV}\cdot\text{nm}}{0.0757\,\text{nm}} = 1.64 \times 10^4\,\text{eV} = 16.4\ \text{keV} .$$

By energy conservation, then, the kinetic energy of the electron must equal $E' - E = 17.5\,\text{keV} - 16.4\,\text{keV} = 1.1\,\text{keV}$.

46. We rewrite Eq. 39-9 as

$$\frac{h}{m\lambda} - \frac{h}{m\lambda'}\cos\phi = \frac{v}{\sqrt{1-(v/c)^2}}\cos\theta\ ,$$

and Eq. 39-10 as

$$\frac{h}{m\lambda'}\sin\phi = \frac{v}{\sqrt{1-(v/c)^2}}\sin\theta\ .$$

We square both equations and add up the two sides:

$$\left(\frac{h}{m}\right)^2\left[\left(\frac{1}{\lambda} - \frac{1}{\lambda'}\cos\phi\right)^2 + \left(\frac{1}{\lambda'}\sin\phi\right)^2\right] = \frac{v^2}{1-(v/c)^2}\ ,$$

where we use $\sin^2\theta + \cos^2\theta = 1$ to eliminate θ. Now the right-hand side can be written as

$$\frac{v^2}{1-(v/c)^2} = -c^2\left[1 - \frac{1}{1-(v/c)^2}\right]\ ,$$

so

$$\frac{1}{1-(v/c)^2} = \left(\frac{h}{mc}\right)^2\left[\left(\frac{1}{\lambda} - \frac{1}{\lambda'}\cos\phi\right)^2 + \left(\frac{1}{\lambda'}\sin\phi\right)^2\right] + 1\ .$$

Now we rewrite Eq. 39-8 as

$$\frac{h}{mc}\left(\frac{1}{\lambda} - \frac{1}{\lambda'}\right) + 1 = \frac{1}{\sqrt{1-(v/c)^2}}\ .$$

If we square this, then it can be directly compared with the previous equation we obtained for $[1-(v/c)^2]^{-1}$. This yields

$$\left[\frac{h}{mc}\left(\frac{1}{\lambda} - \frac{1}{\lambda'}\right) + 1\right]^2 = \left(\frac{h}{mc}\right)^2\left[\left(\frac{1}{\lambda} - \frac{1}{\lambda'}\cos\phi\right)^2 + \left(\frac{1}{\lambda'}\sin\phi\right)^2\right] + 1\ .$$

We have so far eliminated θ and v. Working out the squares on both sides and noting that $\sin^2\phi + \cos^2\phi = 1$, we get

$$\lambda' - \lambda = \Delta\lambda = \frac{h}{mc}(1-\cos\phi)\ .$$

47. The magnitude of the fractional energy change for the photon is given by

$$\left|\frac{\Delta E_{\text{ph}}}{E_{\text{ph}}}\right| = \left|\frac{\Delta(hc/\lambda)}{hc/\lambda}\right| = \left|\lambda\,\Delta\left(\frac{1}{\lambda}\right)\right| = \lambda\left(\frac{1}{\lambda} - \frac{1}{\lambda+\Delta\lambda}\right) = \frac{\Delta\lambda}{\lambda+\Delta\lambda} = \beta$$

where $\beta = 0.10$. Thus $\Delta\lambda = \lambda\beta/(1-\beta)$. We substitute this expression for $\Delta\lambda$ in Eq. 39-11 and solve for $\cos\phi$:

$$\begin{aligned}\cos\phi &= 1 - \frac{mc}{h}\Delta\lambda = 1 - \frac{mc\lambda\beta}{h(1-\beta)} = 1 - \frac{\beta(mc^2)}{(1-\beta)E_{\text{ph}}}\\ &= 1 - \frac{(0.10)(511\,\text{keV})}{(1-0.10)(200\,\text{keV})} = 0.716\ .\end{aligned}$$

This leads to an angle of $\phi = 44°$.

48. Referring to Sample Problem 39-4, we see that the fractional change in photon energy is

$$\frac{E-E'}{E} = \frac{\Delta\lambda}{\lambda+\Delta\lambda} = \frac{h/mc(1-\cos\phi)}{(hc/E) + (h/mc(1-\cos\phi))}\ .$$

Energy conservation demands that $E - E' = K$, the kinetic energy of the electron. In the maximal case, $\phi = 180°$, and we find

$$\frac{K}{E} = \frac{h/mc(1 - \cos 180°)}{(hc/E) + (h/mc(1 - \cos 180°))} = \frac{h/mc(2)}{(hc/E) + (h/mc(2))} .$$

Multiplying both sides by E and simplifying the fraction on the right-hand side leads to

$$K = E\left(\frac{2/mc}{c/E + 2/mc}\right) = \frac{E^2}{mc^2/2 + E} .$$

49. We substitute the classical relationship between momentum p and velocity v, $v = p/m$ into the classical definition of kinetic energy, $K = \frac{1}{2}mv^2$, to obtain $K = p^2/2m$. Here m is the mass of an electron. Thus $p = \sqrt{2mK}$. The relationship between the momentum and the de Broglie wavelength λ is $\lambda = h/p$, where h is the Planck constant. Thus,

$$\lambda = \frac{h}{\sqrt{2mK}} .$$

If K is given in electron volts, then

$$\begin{aligned} \lambda &= \frac{6.626 \times 10^{-34}\,\text{J}\cdot\text{s}}{\sqrt{2(9.109 \times 10^{-31}\,\text{kg})(1.602 \times 10^{-19}\,\text{J/eV})K}} = \frac{1.226 \times 10^{-9}\,\text{m}\cdot\text{eV}^{1/2}}{\sqrt{K}} \\ &= \frac{1.226\,\text{nm}\cdot\text{eV}^{1/2}}{\sqrt{K}} . \end{aligned}$$

50. The de Broglie wavelength for the bullet is

$$\lambda = \frac{h}{p} = \frac{h}{mv} = \frac{6.63 \times 10^{-34}\,\text{J}\cdot\text{s}}{(40 \times 10^{-3}\,\text{kg})(1000\,\text{m/s})} = 1.7 \times 10^{-35}\ \text{m} .$$

51. We start with the result of Exercise 49: $\lambda = h/\sqrt{2mK}$. Replacing K with eV, where V is the accelerating potential and e is the fundamental charge, we obtain

$$\begin{aligned} \lambda &= \frac{h}{\sqrt{2meV}} = \frac{6.626 \times 10^{-34}\,\text{J}\cdot\text{s}}{\sqrt{2(9.109 \times 10^{-31}\,\text{kg})(1.602 \times 10^{-19}\,\text{C})(25.0 \times 10^{3}\,\text{V})}} \\ &= 7.75 \times 10^{-12}\,\text{m} = 7.75\ \text{pm} . \end{aligned}$$

52. (a) Using Table 38-3 and the result of problem 3, we obtain

$$\lambda = \frac{h}{p} = \frac{h}{\sqrt{2m_e K}} = \frac{hc}{\sqrt{2m_e c^2 K}} = \frac{1240\,\text{eV}\cdot\text{nm}}{\sqrt{2(511000\,\text{eV})(1000\,\text{eV})}} = 0.039\ \text{nm} .$$

(b) A photon's de Broglie wavelength is equal to its familiar wave-relationship value. Using the result of problem 3,

$$\lambda = \frac{hc}{E} = \frac{1240\,\text{eV}\cdot\text{nm}}{1.00\,\text{keV}} = 1.24\,\text{nm} .$$

(c) The neutron mass may be found in Appendix B. Using the conversion from electronvolts to Joules, we obtain

$$\lambda = \frac{h}{\sqrt{2m_n K}} = \frac{6.63 \times 10^{-34}\,\text{J}\cdot\text{s}}{\sqrt{2(1.675 \times 10^{-27}\,\text{kg})(1.6 \times 10^{-16}\,\text{J})}} = 9.1 \times 10^{-13}\ \text{m} .$$

53. We use the result of Exercise 49: $\lambda = (1.226\,\text{nm}\cdot\text{eV}^{1/2})/\sqrt{K}$, where K is the kinetic energy. Thus

$$K = \left(\frac{1.226\,\text{nm}\cdot\text{eV}^{1/2}}{\lambda}\right)^2 = \left(\frac{1.226\,\text{nm}\cdot\text{eV}^{1/2}}{590\,\text{nm}}\right)^2 = 4.32 \times 10^{-6}\ \text{eV}\ .$$

54. (a) We solve v from $\lambda = h/p = h/(m_p v)$:

$$v = \frac{h}{m_p\lambda} = \frac{6.63 \times 10^{-34}\,\text{J}\cdot\text{s}}{(1.675 \times 10^{-27}\,\text{kg})(0.100 \times 10^{-12}\,\text{m})} = 3.96 \times 10^6\ \text{m/s}\ .$$

(b) We set $eV = K = \frac{1}{2}m_p v^2$ and solve for the voltage:

$$V = \frac{m_p v^2}{2e} = \frac{(1.67 \times 10^{-27}\,\text{kg})(3.96 \times 10^6\,\text{m/s})^2}{2(1.60 \times 10^{-19}\,\text{C})} = 8.18 \times 10^3\ \text{V}\ .$$

55. (a) The average kinetic energy is

$$K = \frac{3}{2}kT = \frac{3}{2}(1.38 \times 10^{-23}\,\text{J/K})(300\,\text{K}) = 6.21 \times 10^{-21}\,\text{J} = 3.88 \times 10^{-2}\ \text{eV}\ .$$

(b) The de Broglie wavelength is

$$\begin{aligned}\lambda &= \frac{h}{\sqrt{2m_n K}} \\ &= \frac{6.63 \times 10^{-34}\,\text{J}\cdot\text{s}}{\sqrt{2(1.675 \times 10^{-27}\,\text{kg})(6.21 \times 10^{-21}\,\text{J})}} \\ &= 1.5 \times 10^{-10}\ \text{m}\ .\end{aligned}$$

56. (a) and (b) The momenta of the electron and the photon are the same:

$$p = \frac{h}{\lambda} = \frac{6.63 \times 10^{-34}\,\text{J}\cdot\text{s}}{0.20 \times 10^{-9}\,\text{m}} = 3.3 \times 10^{-24}\ \text{kg}\cdot\text{m/s}\ .$$

The kinetic energy of the electron is

$$K_e = \frac{p^2}{2m_e} = \frac{(3.3 \times 10^{-24}\,\text{kg}\cdot\text{m/s})^2}{2(9.11 \times 10^{-31}\,\text{kg})} = 6.0 \times 10^{-18}\,\text{J} = 38\ \text{eV}\ ,$$

while that for the photon is

$$K_{\text{ph}} = pc = (3.3 \times 10^{-24}\,\text{kg}\cdot\text{m/s})(2.998 \times 10^8\,\text{m/s}) = 9.9 \times 10^{-16}\,\text{J} = 6.2\ \text{keV}\ .$$

57. (a) The momentum of the photon is given by $p = E/c$, where E is its energy. Its wavelength is

$$\lambda = \frac{h}{p} = \frac{hc}{E} = \frac{1240\,\text{eV}\cdot\text{nm}}{1.00\,\text{eV}} = 1240\ \text{nm}\ .$$

See Exercise 3. The momentum of the electron is given by $p = \sqrt{2mK}$, where K is its kinetic energy and m is its mass. Its wavelength is

$$\lambda = \frac{h}{p} = \frac{h}{\sqrt{2mK}}\ .$$

According to Exercise 49, if K is in electron volts, this is

$$\lambda = \frac{1.226\,\text{nm}\cdot\text{eV}^{1/2}}{\sqrt{K}} = \frac{1.226\,\text{nm}\cdot\text{eV}^{1/2}}{\sqrt{1.00\,\text{eV}}} = 1.23\ \text{nm}\ .$$

(b) For the photon,

$$\lambda = \frac{hc}{E} = \frac{1240\,\text{eV}\cdot\text{nm}}{1.00\times10^{9}\,\text{eV}} = 1.24\times10^{-6}\,\text{nm}\,.$$

Relativity theory must be used to calculate the wavelength for the electron. According to Eq. 38-51, the momentum p and kinetic energy K are related by $(pc)^2 = K^2 + 2Kmc^2$. Thus,

$$\begin{aligned} pc &= \sqrt{K^2 + 2Kmc^2} \\ &= \sqrt{(1.00\times10^{9}\,\text{eV})^2 + 2(1.00\times10^{9}\,\text{eV})(0.511\times10^{6}\,\text{eV})} \\ &= 1.00\times10^{9}\,\text{eV}\,. \end{aligned}$$

The wavelength is

$$\lambda = \frac{h}{p} = \frac{hc}{pc} = \frac{1240\,\text{eV}\cdot\text{nm}}{1.00\times10^{9}\,\text{eV}} = 1.24\times10^{-6}\,\text{nm}\,.$$

58. (a) The average de Broglie wavelength is

$$\begin{aligned} \lambda_{\text{avg}} &= \frac{h}{p_{\text{avg}}} = \frac{h}{\sqrt{2mK_{\text{avg}}}} = \frac{h}{\sqrt{2m(3kT/2)}} = \frac{hc}{\sqrt{2(mc^2)kT}} \\ &= \frac{1240\,\text{eV}\cdot\text{nm}}{\sqrt{3(4)(938\,\text{MeV})(8.62\times10^{-5}\,\text{eV/K})(300\,\text{K})}} \\ &= 7.3\times10^{-11}\,\text{m} = 73\,\text{pm}\,. \end{aligned}$$

(b) The average separation is

$$\begin{aligned} d_{\text{avg}} &= \frac{1}{\sqrt[3]{n}} = \frac{1}{\sqrt[3]{p/kT}} \\ &= \sqrt[3]{\frac{(1.38\times10^{-23}\,\text{J/K})(300\,\text{K})}{1.01\times10^{5}\,\text{Pa}}} = 3.4\,\text{nm}\,. \end{aligned}$$

(c) Yes, since $\lambda_{\text{avg}} \ll d_{\text{avg}}$.

59. (a) The kinetic energy acquired is $K = qV$, where q is the charge on an ion and V is the accelerating potential. Thus $K = (1.60\times10^{-19}\,\text{C})(300\,\text{V}) = 4.80\times10^{-17}\,\text{J}$. The mass of a single sodium atom is, from Appendix F, $m = (22.9898\,\text{g/mol})/(6.02\times10^{23}\,\text{atom/mol}) = 3.819\times10^{-23}\,\text{g} = 3.819\times10^{-26}\,\text{kg}$. Thus, the momentum of an ion is

$$p = \sqrt{2mK} = \sqrt{2(3.819\times10^{-26}\,\text{kg})(4.80\times10^{-17}\,\text{J})} = 1.91\times10^{-21}\,\text{kg}\cdot\text{m/s}\,.$$

(b) The de Broglie wavelength is

$$\lambda = \frac{h}{p} = \frac{6.63\times10^{-34}\,\text{J}\cdot\text{s}}{1.91\times10^{-21}\,\text{kg}\cdot\text{m/s}} = 3.47\times10^{-13}\,\text{m}\,.$$

60. (a) We use the result of problem 3:

$$E_{\text{photon}} = \frac{hc}{\lambda} = \frac{1240\,\text{nm}\cdot\text{eV}}{1.00\,\text{nm}} = 1.24\,\text{keV}$$

and for the electron

$$K = \frac{p^2}{2m_e} = \frac{(h/\lambda)^2}{2m_e} = \frac{(hc/\lambda)^2}{2m_ec^2} = \frac{1}{2(0.511\,\text{MeV})}\left(\frac{1240\,\text{eV}\cdot\text{nm}}{1.00\,\text{nm}}\right)^2 = 1.50\,\text{eV}\,.$$

(b) In this case, we find

$$E_{\text{photon}} = \frac{1240\,\text{nm}\cdot\text{eV}}{1.00 \times 10^{-6}\,\text{nm}} = 1.24 \times 10^{9}\,\text{eV} = 1.24\ \text{GeV}\ ,$$

and for the electron (recognizing that $1240\,\text{eV}\cdot\text{nm} = 1240\,\text{MeV}\cdot\text{fm}$)

$$\begin{aligned} K &= \sqrt{p^2c^2 + (m_ec^2)^2} - m_ec^2 = \sqrt{(hc/\lambda)^2 + (m_ec^2)^2} - m_ec^2 \\ &= \sqrt{\left(\frac{1240\,\text{MeV}\cdot\text{fm}}{1.00\,\text{fm}}\right)^2 + (0.511\,\text{MeV})^2} - 0.511\,\text{MeV} \\ &= 1.24 \times 10^{3}\,\text{MeV} = 1.24\ \text{GeV}\ . \end{aligned}$$

We note that at short λ (large K) the kinetic energy of the electron, calculated with the relativistic formula, is about the same as that of the photon. This is expected since now $K \approx E \approx pc$ for the electron, which is the same as $E = pc$ for the photon.

61. We need to use the relativistic formula $p = \sqrt{(E/c)^2 - m_e^2c^2} \approx E/c \approx K/c$ (since $E \gg m_ec^2$). So

$$\lambda = \frac{h}{p} \approx \frac{hc}{K} = \frac{1240\,\text{eV}\cdot\text{nm}}{50 \times 10^{9}\,\text{eV}} = 2.5 \times 10^{-8}\ \text{nm}\ ,$$

which is about 200 times smaller than the radius of an average nucleus.

62. (a) Since $K = 7.5\,\text{MeV} \ll m_\alpha c^2 = 4(932\,\text{MeV})$, we may use the non-relativistic formula $p = \sqrt{2m_\alpha K}$. Using Eq. 38-43 (and recognizing that $1240\,\text{eV}\cdot\text{nm} = 1240\,\text{MeV}\cdot\text{fm}$), we obtain

$$\lambda = \frac{h}{p} = \frac{hc}{\sqrt{2m_\alpha c^2 K}} = \frac{1240\,\text{MeV}\cdot\text{fm}}{\sqrt{2(4\,\text{u})(931.5\,\text{MeV/u})(7.5\,\text{MeV})}} = 5.2\ \text{fm}\ .$$

(b) Since $\lambda = 5.2\,\text{fm} \ll 30\,\text{fm}$, to a fairly good approximation, the wave nature of the α particle does not need to be taken into consideration.

63. The wavelength associated with the unknown particle is $\lambda_p = h/p_p = h/(m_pv_p)$, where p_p is its momentum, m_p is its mass, and v_p is its speed. The classical relationship $p_p = m_pv_p$ was used. Similarly, the wavelength associated with the electron is $\lambda_e = h/(m_ev_e)$, where m_e is its mass and v_e is its speed. The ratio of the wavelengths is $\lambda_p/\lambda_e = (m_ev_e)/(m_pv_p)$, so

$$m_p = \frac{v_e\lambda_e}{v_p\lambda_p}m_e = \frac{9.109 \times 10^{-31}\,\text{kg}}{3(1.813 \times 10^{-4})} = 1.675 \times 10^{-27}\ \text{kg}\ .$$

According to Appendix B, this is the mass of a neutron.

64. (a) Setting $\lambda = h/p = h/\sqrt{(E/c)^2 - m_e^2c^2}$, we solve for $K = E - m_ec^2$:

$$\begin{aligned} K &= \sqrt{\left(\frac{hc}{\lambda}\right)^2 + m_e^2c^4} - m_ec^2 \\ &= \sqrt{\left(\frac{1240\,\text{eV}\cdot\text{nm}}{10 \times 10^{-3}\,\text{nm}}\right)^2 + (0.511\,\text{MeV})^2} - 0.511\,\text{MeV} \\ &= 0.015\,\text{MeV} = 15\ \text{keV}\ . \end{aligned}$$

(b) Using the result of problem 3,

$$E = \frac{hc}{\lambda} = \frac{1240\,\text{eV}\cdot\text{nm}}{10 \times 10^{-3}\,\text{nm}} = 1.2 \times 10^{5}\,\text{eV} = 120\ \text{keV}\ .$$

(c) The electron microscope is more suitable, as the required energy of the electrons is much less than that of the photons.

65. The same resolution requires the same wavelength, and since the wavelength and particle momentum are related by $p = h/\lambda$, we see that the same particle momentum is required. The momentum of a 100 keV photon is $p = E/c = (100 \times 10^3\,\text{eV})(1.60 \times 10^{-19}\,\text{J/eV})/(3.00 \times 10^8\,\text{m/s}) = 5.33 \times 10^{-23}\,\text{kg}\cdot\text{m/s}$. This is also the magnitude of the momentum of the electron. The kinetic energy of the electron is

$$K = \frac{p^2}{2m} = \frac{(5.33 \times 10^{-23}\,\text{kg}\cdot\text{m/s})^2}{2(9.11 \times 10^{-31}\,\text{kg})} = 1.56 \times 10^{-15}\ \text{J}\ .$$

The accelerating potential is

$$V = \frac{K}{e} = \frac{1.56 \times 10^{-15}\,\text{J}}{1.60 \times 10^{-19}\,\text{C}} = 9.76 \times 10^3\ \text{V}\ .$$

66. (a)

$$\begin{aligned} nn^* &= (a+ib)(a+ib)^* = (a+ib)(a^* + i^*b^*) = (a+ib)(a-ib) \\ &= a^2 + iba - iab + (ib)(-ib) = a^2 + b^2\ , \end{aligned}$$

which is always real since both a and b are real.

(b)

$$\begin{aligned} |nm| &= |(a+ib)(c+id)| \\ &= |ac + iad + ibc + (-i)^2 bd| \\ &= |(ac - bd) + i(ad + bc)| \\ &= \sqrt{(ac-bd)^2 + (ad+bc)^2} \\ &= \sqrt{a^2c^2 + b^2d^2 + a^2d^2 + b^2c^2}\ . \end{aligned}$$

But

$$\begin{aligned} |n|\,|m| &= |a+ib|\,|c+id| = \sqrt{a^2+b^2}\,\sqrt{c^2+d^2} \\ &= \sqrt{a^2c^2 + b^2d^2 + a^2d^2 + b^2c^2}\ , \end{aligned}$$

so $|nm| = |n|\,|m|$.

67. We plug Eq. 39-17 into Eq. 39-16, and note that

$$\frac{d\psi}{dx} = \frac{d}{dx}\left(Ae^{ikx} + Be^{-ikx}\right) = ikAe^{ikx} - ikBe^{-ikx}.$$

Also,

$$\frac{d^2\psi}{dx^2} = \frac{d}{dx}\left(ikAe^{ikx} - ikBe^{-ikx}\right) = -k^2Ae^{ikx} - k^2Be^{ikx}\ .$$

Thus,

$$\frac{d^2\psi}{dx^2} + k^2\psi = -k^2Ae^{ikx} - k^2Be^{ikx} + k^2\left(Ae^{ikx} + Be^{-ikx}\right) = 0\ .$$

68. (a) We use Euler's formula $e^{i\phi} = \cos\phi + i\sin\phi$ to re-write $\psi(x)$ as

$$\begin{aligned} \psi(x) &= \psi_0 e^{ikx} = \psi_0(\cos kx + i\sin kx) \\ &= (\psi_0\cos kx) + i(\psi_0\sin kx) = a + ib\ , \end{aligned}$$

where $a = \psi_0\cos kx$ and $b = \psi_0\sin kx$ are both real quantities.

(b)

$$\begin{aligned}\psi(x,t) &= \psi(x)e^{-i\omega t} = \psi_0 e^{ikx}\, e^{-i\omega t} = \psi_0 e^{i(kx-\omega t)} \\ &= [\psi_0\cos(kx-\omega t)] + i\,[\psi_0\sin(kx-\omega t)]\ .\end{aligned}$$

69. The angular wave number k is related to the wavelength λ by $k = 2\pi/\lambda$ and the wavelength is related to the particle momentum p by $\lambda = h/p$, so $k = 2\pi p/h$. Now, the kinetic energy K and the momentum are related by $K = p^2/2m$, where m is the mass of the particle. Thus $p = \sqrt{2mK}$ and

$$k = \frac{2\pi\sqrt{2mK}}{h}\ .$$

70. We note that $|e^{ikx}|^2 = (e^{ikx})^*(e^{ikx}) = e^{-ikx}e^{ikx} = 1$. Referring to Eq. 39-14, we see therefore that $|\psi|^2 = |\Psi|^2$.

71. For $U = U_0$, Schrödinger's equation becomes

$$\frac{d^2\psi}{dx^2} + \frac{8\pi^2 m}{h^2}\,[E-U_0]\,\psi = 0\ .$$

We substitute $\psi = \psi_0 e^{ikx}$. The second derivative is $d^2\psi/dx^2 = -k^2\psi_0 e^{ikx} = -k^2\psi$. The result is

$$-k^2\psi + \frac{8\pi^2 m}{h^2}\,[E-U_0]\,\psi = 0\ .$$

Solving for k, we obtain

$$k = \sqrt{\frac{8\pi^2 m}{h^2}\,[E-U_0]} = \frac{2\pi}{h}\sqrt{2m\,[E-U_0]}\ .$$

72. The wave function is now given by

$$\Psi(x,t) = \psi_0\, e^{-i(kx+\omega t)}\ .$$

This function describes a plane matter wave traveling in the negative x direction. An example of the actual particles that fit this description is a free electron with linear momentum $\vec{p} = -(hk/2\pi)\hat{\mathrm{i}}$ and kinetic energy $K = p^2/2m_e = h^2k^2/8\pi^2 m_e$.

73. (a) The wave function is now given by

$$\Psi(x,t) = \psi_0\left[e^{i(kx-\omega t)} + e^{-i(kx+\omega t)}\right] = \psi_0\, e^{-i\omega t}\left(e^{ikx} + e^{-ikx}\right)\ .$$

Thus

$$\begin{aligned}|\Psi(x,t)|^2 &= \left|\psi_0\, e^{-i\omega t}\left(e^{ikx}+e^{-ikx}\right)\right|^2 \\ &= \left|\psi_0\, e^{-i\omega t}\right|^2\left|e^{ikx}+e^{-ikx}\right|^2 \\ &= \psi_0^2\left|e^{ikx}+e^{-ikx}\right|^2 \\ &= \psi_0^2\left|(\cos kx + i\sin kx) + (\cos kx - i\sin kx)\right|^2 \\ &= 4\psi_0^2(\cos kx)^2 \\ &= 2\psi_0^2(1+\cos 2kx)\ .\end{aligned}$$

(b) Consider two plane matter waves, each with the same amplitude $\psi_0/\sqrt{2}$ and traveling in opposite directions along the x axis. The combined wave Ψ is a standing wave:

$$\begin{aligned}\Psi(x,t) &= \psi_0\, e^{i(kx-\omega t)} + \psi_0\, e^{-i(kx+\omega t)} = \psi_0\left(e^{ikx}+e^{-ikx}\right)e^{-i\omega t} \\ &= (2\psi_0\cos kx)\, e^{-i\omega t}\ .\end{aligned}$$

Thus, the squared amplitude of the matter wave is

$$|\Psi(x,t)|^2 = (2\psi_0 \cos kx)^2 \left|e^{-i\omega t}\right|^2 = 2\psi_0^2(1+\cos 2kx) \ ,$$

which is shown below.

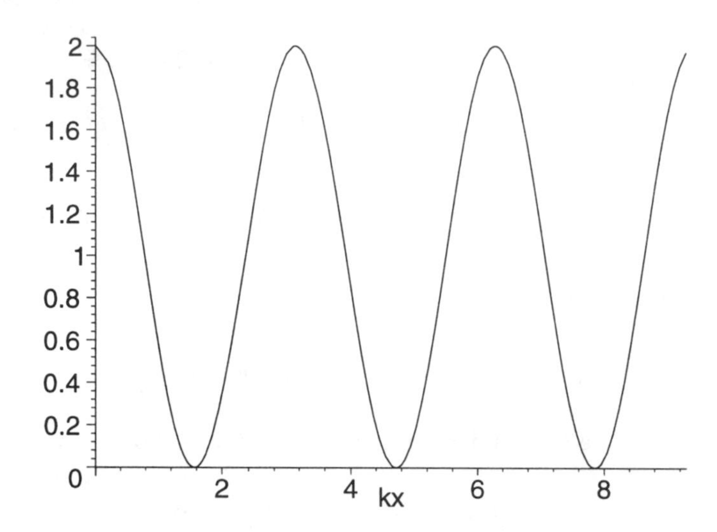

(c) We set $|\Psi(x,t)|^2 = 2\psi_0^2(1+\cos 2kx) = 0$ to obtain $\cos(2kx) = -1$. This gives

$$2kx = 2\left(\frac{2\pi}{\lambda}\right) = (2n+1)\pi \ , \qquad (n = 0, 1, 2, 3, \ldots)$$

We solve for x:

$$x = \frac{1}{4}(2n+1)\lambda \ .$$

(d) The most probable positions for finding the particle are where $|\Psi(x,t)| \propto (1+\cos 2kx)$ reaches its maximum. Thus $\cos 2kx = 1$, or

$$2kx = 2\left(\frac{2\pi}{\lambda}\right) = 2n\pi \ , \qquad (n = 0, 1, 2, 3, \ldots)$$

We solve for x:

$$x = \frac{1}{2}n\lambda \ .$$

74. (a) Since $p_x = p_y = 0$, $\Delta p_x = \Delta p_y = 0$. Thus from Eq. 39-20 both Δx and Δy are infinite. It is therefore impossible to assign a y or z coordinate to the position of an electron.

(b) Since it is independent of y and z the wave function $\Psi(x)$ should describe a plane wave that extends infinitely in both the y and z directions. Also from Fig. 39-11 we see that $|\Psi(x)|^2$ extends infinitely along the x axis. Thus the matter wave described by $\Psi(x)$ extends throughout the entire three-dimensional space.

75. The uncertainty in the momentum is $\Delta p = m\,\Delta v = (0.50\,\text{kg})(1.0\,\text{m/s}) = 0.50\,\text{kg}\cdot\text{m/s}$, where Δv is the uncertainty in the velocity. Solving the uncertainty relationship $\Delta x\,\Delta p \geq \hbar$ for the minimum uncertainty in the coordinate x, we obtain

$$\Delta x = \frac{\hbar}{\Delta p} = \frac{0.60\,\text{J}\cdot\text{s}}{2\pi(0.50\,\text{kg}\cdot\text{m/s})} = 0.19 \text{ m} \ .$$

76. If the momentum is measured at the same time as the position, then

$$\Delta p \approx \frac{\hbar}{\Delta x} = \frac{6.63 \times 10^{-34}\,\text{J}\cdot\text{s}}{2\pi(50\,\text{pm})} = 2.1 \times 10^{-24}\ \text{kg}\cdot\text{m/s}\ .$$

77. We use the uncertainty relationship $\Delta x\,\Delta p \ge \hbar$. Letting $\Delta x = \lambda$, the de Broglie wavelength, we solve for the minimum uncertainty in p:

$$\Delta p = \frac{\hbar}{\Delta x} = \frac{h}{2\pi\lambda} = \frac{p}{2\pi}$$

where the de Broglie relationship $p = h/\lambda$ is used. We use $1/2\pi = 0.080$ to obtain $\Delta p = 0.080p$. We would expect the measured value of the momentum to lie between $0.92p$ and $1.08p$. Measured values of zero, $0.5p$, and $2p$ would all be surprising.

78. (a) Using the result of problem 3,

$$E = \frac{hc}{\lambda} = \frac{1240\,\text{nm}\cdot\text{eV}}{10.0 \times 10^{-3}\,\text{nm}} = 124\,\text{keV}\ .$$

(b) The kinetic energy gained by the electron is equal to the energy decrease of the photon:

$$\begin{aligned}\Delta E &= \Delta\left(\frac{hc}{\lambda}\right) = hc\left(\frac{1}{\lambda} - \frac{1}{\lambda + \Delta\lambda}\right) = \left(\frac{hc}{\lambda}\right)\left(\frac{\Delta\lambda}{\lambda + \Delta\lambda}\right) = \frac{E}{1 + \frac{\lambda}{\Delta\lambda}} \\ &= \frac{E}{1 + \frac{\lambda}{\lambda_C(1-\cos\phi)}} = \frac{124\,\text{keV}}{1 + \frac{10.0\,\text{pm}}{(2.43\,\text{pm})(1-\cos 180^\circ)}} \\ &= 40.5\ \text{keV}\ .\end{aligned}$$

(c) It is impossible to "view" an atomic electron with such a high-energy photon, because with the energy imparted to the electron the photon would have knocked the electron out of its orbit.

79. (a) The transmission coefficient T for a particle of mass m and energy E that is incident on a barrier of height U and width L is given by

$$T = e^{-2kL}\ ,$$

where

$$k = \sqrt{\frac{8\pi^2 m(U - E)}{h^2}}\ .$$

For the proton,

$$\begin{aligned}k &= \sqrt{\frac{8\pi^2(1.6726 \times 10^{-27}\,\text{kg})(10\,\text{MeV} - 3.0\,\text{MeV})(1.6022 \times 10^{-13}\,\text{J/MeV})}{(6.6261 \times 10^{-34}\,\text{J}\cdot\text{s})^2}} \\ &= 5.8082 \times 10^{14}\ \text{m}^{-1}\ ,\end{aligned}$$

$kL = (5.8082 \times 10^{14}\,\text{m}^{-1})(10 \times 10^{-15}\,\text{m}) = 5.8082$, and

$$T = e^{-2\times 5.8082} = 9.02 \times 10^{-6}\ .$$

The value of k was computed to a greater number of significant digits than usual because an exponential is quite sensitive to the value of the exponent. The mass of a deuteron is $2.0141\,\text{u} = 3.3454 \times 10^{-27}\,\text{kg}$, so

$$\begin{aligned}k &= \sqrt{\frac{8\pi^2(3.3454 \times 10^{-27}\,\text{kg})(10\,\text{MeV} - 3.0\,\text{MeV})(1.6022 \times 10^{-13}\,\text{J/MeV})}{(6.6261 \times 10^{-34}\,\text{J}\cdot\text{s})^2}} \\ &= 8.2143 \times 10^{14}\ \text{m}^{-1}\ ,\end{aligned}$$

$kL = (8.2143 \times 10^{14}\,\text{m}^{-1})(10 \times 10^{-15}\,\text{m}) = 8.2143$, and

$$T = e^{-2\times 8.2143} = 7.33 \times 10^{-8}\ .$$

(b) Mechanical energy is conserved. Before the particles reach the barrier, each of them has a kinetic energy of 3.0 MeV and a potential energy of zero. After passing through the barrier, each again has a potential energy of zero, so each has a kinetic energy of 3.0 MeV.

(c) Energy is also conserved for the reflection process. After reflection, each particle has a potential energy of zero, so each has a kinetic energy of 3.0 MeV.

80. Letting

$$T \approx e^{-2kL} = \exp\left(-2L\sqrt{\frac{8\pi^2 m(U-E)}{h^2}}\right) ,$$

and using the result of Exercise 3 in Chapter 39, we solve for E:

$$\begin{aligned} E &= U - \frac{1}{2m}\left(\frac{h \ln T}{4\pi L}\right)^2 \\ &= 6.0\,\text{eV} - \frac{1}{2(0.511\,\text{MeV})}\left[\frac{(1240\,\text{eV}\cdot\text{nm})(\ln 0.001)}{4\pi(0.70\,\text{nm})}\right]^2 \\ &= 5.1\ \text{eV} . \end{aligned}$$

81. (a) If m is the mass of the particle and E is its energy, then the transmission coefficient for a barrier of height U and width L is given by

$$T = e^{-2kL} ,$$

where

$$k = \sqrt{\frac{8\pi^2 m(U-E)}{h^2}} .$$

If the change ΔU in U is small (as it is), the change in the transmission coefficient is given by

$$\Delta T = \frac{dT}{dU}\,\Delta U = -2LT\,\frac{dk}{dU}\,\Delta U .$$

Now,

$$\frac{dk}{dU} = \frac{1}{2\sqrt{U-E}}\sqrt{\frac{8\pi^2 m}{h^2}} = \frac{1}{2(U-E)}\sqrt{\frac{8\pi^2 m(U-E)}{h^2}} = \frac{k}{2(U-E)} .$$

Thus,

$$\Delta T = -LTk\,\frac{\Delta U}{U-E} .$$

For the data of Sample Problem 39-7, $2kL = 10.0$, so $kL = 5.0$ and

$$\frac{\Delta T}{T} = -kL\frac{\Delta U}{U-E} = -(5.0)\frac{(0.010)(6.8\,\text{eV})}{6.8\,\text{eV} - 5.1\,\text{eV}} = -0.20 .$$

There is a 20% decrease in the transmission coefficient.

(b) The change in the transmission coefficient is given by

$$\Delta T = \frac{dT}{dL}\,\Delta L = -2ke^{-2kL}\,\Delta L = -2kT\,\Delta L$$

and

$$\frac{\Delta T}{T} = -2k\,\Delta L = -2(6.67\times 10^9\,\text{m}^{-1})(0.010)(750\times 10^{-12}\,\text{m}) = -0.10 .$$

There is a 10% decrease in the transmission coefficient.

(c) The change in the transmission coefficient is given by

$$\Delta T = \frac{dT}{dE}\Delta E = -2Le^{-2kL}\frac{dk}{dE}\Delta E = -2LT\frac{dk}{dE}\Delta E \ .$$

Now, $dk/dE = -dk/dU = -k/2(U-E)$, so

$$\frac{\Delta T}{T} = kL\frac{\Delta E}{U-E} = (5.0)\frac{(0.010)(5.1\,\text{eV})}{6.8\,\text{eV} - 5.1\,\text{eV}} = 0.15 \ .$$

There is a 15% increase in the transmission coefficient.

82. (a) The rate at which incident protons arrive at the barrier is $n = 1.0\,\text{kA}/1.60\times10^{-19}\,\text{C} = 6.25\times10^{23}/\text{s}$. Letting $nTt = 1$, we find the waiting time t:

$$\begin{aligned} t &= (nT)^{-1} = \frac{1}{n}\exp\left(2L\sqrt{\frac{8\pi^2 m_p(U-E)}{h^2}}\right) \\ &= \left(\frac{1}{6.25\times10^{23}/\text{s}}\right)\exp\left(\frac{2\pi(0.70\,\text{nm})}{1240\,\text{eV}\cdot\text{nm}}\sqrt{8(938\,\text{MeV})(6.0\,\text{eV}-5.0\,\text{eV})}\right) \\ &= 3.37\times10^{111}\,\text{s} \approx 10^{104}\ \text{y}\ , \end{aligned}$$

which is much longer than the age of the universe.

(b) Replacing the mass of the proton with that of the electron, we obtain the corresponding waiting time for an electron:

$$\begin{aligned} t &= (nT)^{-1} = \frac{1}{n}\exp\left[2L\sqrt{\frac{8\pi^2 m_e(U-E)}{h^2}}\right] \\ &= \left(\frac{1}{6.25\times10^{23}/\text{s}}\right)\exp\left[\frac{2\pi(0.70\,\text{nm})}{1240\,\text{eV}\cdot\text{nm}}\sqrt{8(0.511\,\text{MeV})(6.0\,\text{eV}-5.0\,\text{eV})}\right] \\ &= 2.1\times10^{-19}\ \text{s}\ . \end{aligned}$$

The enormous difference between the two waiting times is the result of the difference between the masses of the two kinds of particles.

83. The kinetic energy of the car of mass m moving at speed v is given by $E = \frac{1}{2}mv^2$, while the potential barrier it has to tunnel through is $U = mgh$, where $h = 24\,\text{m}$. According to Eq. 39-21 and 39-22 the tunneling probability is given by $T \approx e^{-2kL}$, where

$$\begin{aligned} k &= \sqrt{\frac{8\pi^2 m(U-E)}{h^2}} = \sqrt{\frac{8\pi^2 m(mgh - \frac{1}{2}mv^2)}{h^2}} \\ &= \frac{2\pi(1500\,\text{kg})}{6.63\times10^{-34}\,\text{J}\cdot\text{s}}\sqrt{2\left[(9.8\,\text{m/s}^2)(24\,\text{m}) - \frac{1}{2}(20\,\text{m/s})^2\right]} \\ &= 1.2\times10^{38}\ \text{m}^{-1}\ . \end{aligned}$$

Thus, $2kL = 2(1.2\times10^{38}\,\text{m}^{-1})(30\,\text{m}) = 7.2\times10^{39}$. One can see that $T \approx e^{-2kL}$ is essentially zero.

Chapter 40

1. (a) This is computed in part (a) of Sample Problem 40-1.

   (b) With $m_p = 1.67 \times 10^{-27}$ kg, we obtain

$$E_1 = \left(\frac{h^2}{8mL^2}\right)n^2 = \left(\frac{(6.63 \times 10^{-34}\,\text{J}\cdot\text{s})^2}{8m_p(100 \times 10^{12}\,\text{m})^2}\right)(1)^2 = 3.29 \times 10^{-21}\,\text{J} = 0.0206\,\text{eV} .$$

2. According to Eq. 40-4 $E_n \propto L^{-2}$. As a consequence, the new energy level E'_n satisfies

$$\frac{E'_n}{E_n} = \left(\frac{L'}{L}\right)^{-2} = \left(\frac{L}{L'}\right)^2 = \frac{1}{2} ,$$

   which gives $L' = \sqrt{2}L$. Thus, the width of the potential well must be multiplied by a factor of $\sqrt{2}$.

3. To estimate the energy, we use Eq. 40-4, with $n = 1$, L equal to the atomic diameter, and m equal to the mass of an electron:

$$E = n^2\frac{h^2}{8mL^2} = \frac{(1)^2(6.63 \times 10^{-34}\,\text{J}\cdot\text{s})^2}{8(9.11 \times 10^{-31}\,\text{kg})(1.4 \times 10^{-14}\,\text{m})^2} = 3.07 \times 10^{-10}\,\text{J} = 1920\,\text{MeV} .$$

4. We can use the mc^2 value for an electron from Table 38-3 (511×10^3 eV) and the hc value developed in problem 3 of Chapter 39 by writing Eq. 40-4 as

$$E_n = \frac{n^2h^2}{8mL^2} = \frac{n^2(hc)^2}{8(mc^2)L^2} .$$

   For $n = 3$, we set this expression equal to 4.7 eV and solve for L:

$$L = \frac{n(hc)}{\sqrt{8(mc^2)E_n}} = \frac{3(1240\,\text{eV}\cdot\text{nm})}{\sqrt{8(511 \times 10^3\,\text{eV})(4.7\,\text{eV})}} = 0.85\,\text{nm} .$$

5. With $m_p = 1.67 \times 10^{-27}$ kg, we obtain

$$E_1 = \left(\frac{h^2}{8mL^2}\right)n^2 = \left(\frac{(6.63 \times 10^{-34}\,\text{J}\cdot\text{s})^2}{8m_p(100 \times 10^{12}\,\text{m})^2}\right)(1)^2 = 3.29 \times 10^{-21}\,\text{J} = 0.0206\,\text{eV} .$$

   Alternatively, we can use the mc^2 value for a proton from Table 38-3 (938×10^6 eV) and the $hc =$ 1240 eV·nm value developed in problem 3 of Chapter 39 by writing Eq. 40-4 as

$$E_n = \frac{n^2h^2}{8mL^2} = \frac{n^2(hc)^2}{8(m_pc^2)L^2} .$$

   This alternative approach is perhaps easier to plug into, but it is recommended that both approaches be tried to find which is most convenient.

6. Since $E_n \propto L^{-2}$ in Eq. 40-4, we see that if L is doubled, then E_1 becomes $(2.6\,\text{eV})(2)^{-2} = 0.65\,\text{eV}$.

7. We can use the mc^2 value for an electron from Table 38-3 (511×10^3 eV) and the hc value developed in problem 3 of Chapter 39 by writing Eq. 40-4 as

$$E_n = \frac{n^2h^2}{8mL^2} = \frac{n^2(hc)^2}{8(mc^2)L^2} \,.$$

The energy to be absorbed is therefore

$$\begin{aligned}\Delta E &= E_4 - E_1 = \frac{(4^2-1^2)h^2}{8m_eL^2} = \frac{15(hc)^2}{8(m_ec^2)L^2} \\ &= \frac{15(1240\,\text{eV}\cdot\text{nm})^2}{8(511\times10^3\,\text{eV})(0.250\,\text{nm})^2} = 90.3\ \text{eV}\,.\end{aligned}$$

8. (a) Let the quantum numbers of the pair in question be n and $n+1$, respectively. We note that

$$E_{n+1} - E_n = \frac{(n+1)^2h^2}{8mL^2} - \frac{n^2h^2}{8mL^2} = \frac{(2n+1)h^2}{8mL^2}$$

Therefore, $E_{n+1} - E_n = (2n+1)E_1$. Now

$$E_{n+1} - E_n = E_5 = 5^2E_1 = 25E_1 = (2n+1)E_1 \,,$$

which leads to $2n+1 = 25$, or $n = 12$.

(b) Now let

$$E_{n+1} - E_n = E_6 = 6^2E_1 = 36E_1 = (2n+1)E_1 \,,$$

which gives $2n+1 = 36$, or $n = 17.5$. This is not an integer, so it is impossible to find the pair that fits the requirement.

9. From Eq. 40-4

$$E_{n+2} - E_n = \left(\frac{h^2}{8mL^2}\right)(n+2)^2 - \left(\frac{h^2}{8mL^2}\right)n^2 = \left(\frac{h^2}{2mL^2}\right)(n+1)\,.$$

10. (a) Let the quantum numbers of the pair in question be n and $n+1$, respectively. Then $E_{n+1} - E_n = E_1(n+1)^2 - E_1n^2 = (2n+1)E_1$. Letting

$$E_{n+1} - E_n = (2n+1)E_1 = 3(E_4 - E_3) = 3(4^2E_1 - 3^2E_1) = 21E_1 \,,$$

we get $2n+1 = 21$, or $n = 10$.

(b) Now letting

$$E_{n+1} - E_n = (2n+1)E_1 = 2(E_4 - E_3) = 2(4^2E_1 - 3^2E_1) = 14E_1 \,,$$

we get $2n+1 = 14$, which does not have an integer-valued solution. So it is impossible to find the pair of energy levels that fits the requirement.

11. The energy levels are given by $E_n = n^2h^2/8mL^2$, where h is the Planck constant, m is the mass of an electron, and L is the width of the well. The frequency of the light that will excite the electron from the state with quantum number n_i to the state with quantum number n_f is $f = \Delta E/h = (h/8mL^2)(n_f^2 - n_i^2)$ and the wavelength of the light is

$$\lambda = \frac{c}{f} = \frac{8mL^2c}{h(n_f^2 - n_i^2)} \,.$$

We evaluate this expression for $n_i = 1$ and $n_f = 2$, 3, 4, and 5, in turn. We use $h = 6.626 \times 10^{-34}$ J·s, $m = 9.109 \times 10^{-31}$ kg, and $L = 250 \times 10^{-12}$ m, and obtain 6.87×10^{-8} m for $n_f = 2$, 2.58×10^{-8} m for $n_f = 3$, 1.37×10^{-8} m for $n_f = 4$, and 8.59×10^{-9} m for $n_f = 5$.

12. We can use the mc^2 value for an electron from Table 38-3 (511×10^3 eV) and the hc value developed in problem 3 of Chapter 39 by rewriting Eq. 40-4 as

$$E_n = \frac{n^2h^2}{8mL^2} = \frac{n^2(hc)^2}{8(mc^2)L^2} \ .$$

(a) The first excited state is characterized by $n = 2$, and the third by $n' = 4$. Thus,

$$\begin{aligned} \Delta E &= \frac{(hc)^2}{8(mc^2)L^2}\left(n'^2 - n^2\right) \\ &= \frac{(1240\,\text{eV}\cdot\text{nm})^2}{8(511 \times 10^3\,\text{eV})(0.250\,\text{nm})^2}\left(4^2 - 2^2\right) \\ &= (6.02\,\text{eV})(16 - 4) \end{aligned}$$

which yields $\Delta E = 72.2\,\text{eV}$.

(b) Now that the electron is in the $n' = 4$ level, it can "drop" to a lower level (n'') in a variety of ways. Each of these drops is presumed to cause a photon to be emitted of wavelength

$$\lambda = \frac{hc}{E_{n'} - E_{n''}} = \frac{8(mc^2)L^2}{hc\left(n'^2 - n''^2\right)} \ .$$

For example, for the transition $n' = 4$ to $n'' = 3$, the photon emitted would have wavelength

$$\lambda = \frac{8(511 \times 10^3\,\text{eV})(0.250\,\text{nm})^2}{(1240\,\text{eV}\cdot\text{nm})\left(4^2 - 3^2\right)} = 29.4 \text{ nm} \ ,$$

and once it is then in level $n'' = 3$ it might fall to level $n''' = 2$ emitting another photon. Calculating in this way all the possible photons emitted during the de-excitation of this system, we find $\lambda_{4\to1} = 13.7\,\text{nm}$, $\lambda_{4\to2} = 17.2\,\text{nm}$, $\lambda_{3\to1} = 25.8\,\text{nm}$, $\lambda_{4\to3} = 29.4\,\text{nm}$, $\lambda_{3\to2} = 41.2\,\text{nm}$, and $\lambda_{2\to1} = 68.7\,\text{nm}$.

(c) A system making the $4 \to 1$ transition will make no further transitions unless it is re-excited. If it makes the $4 \to 2$ transition, then that is likely to followed by the $2 \to 1$ transition. However, if it makes the $4 \to 3$ transition, then it could make either the $3 \to 1$ transition or the pair of transitions: $3 \to 2$ and $2 \to 1$.

(d) The possible transitions are shown below. The energy levels are not drawn to scale.

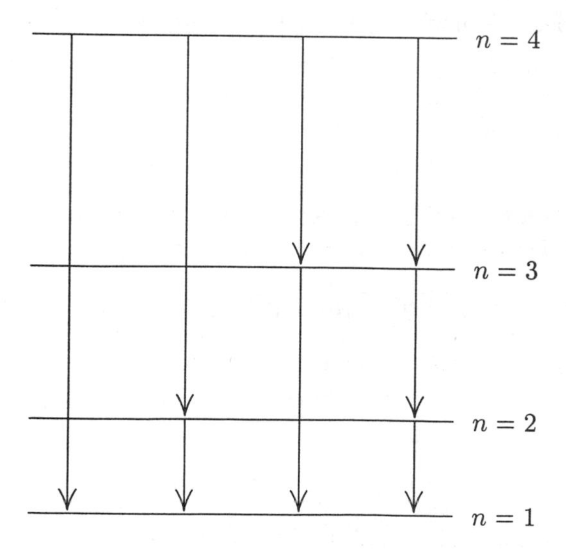

13. We can use the mc^2 value for an electron from Table 38-3 (511×10^3 eV) and the hc value developed in problem 3 of Chapter 39 by writing Eq. 40-4 as

$$E_n = \frac{n^2h^2}{8mL^2} = \frac{n^2(hc)^2}{8(mc^2)L^2} .$$

(a) With $L = 3.0 \times 10^9$ nm, the energy difference is

$$E_2 - E_1 = \frac{1240^2}{8(511 \times 10^3)(3.0 \times 10^9)^2}(2^2 - 1^2) = 1.3 \times 10^{-19} \text{ eV} .$$

(b) Since $(n+1)^2 - n^2 = 2n + 1$, we have

$$\Delta E = E_{n+1} - E_n = \frac{h^2}{8mL^2}(2n+1) = \frac{(hc)^2}{8(mc^2)L^2}(2n+1) .$$

Setting this equal to 1.0 eV, we solve for n:

$$\begin{aligned} n &= \frac{4(mc^2)L^2\Delta E}{(hc)^2} - \frac{1}{2} \\ &= \frac{4(511 \times 10^3 \text{ eV})(3.0 \times 10^9 \text{ nm})^2(1.0 \text{ eV})}{(1240 \text{ eV}\cdot\text{nm})^2} - \frac{1}{2} \\ &\approx 12 \times 10^{18} . \end{aligned}$$

(c) At this value of n, the energy is

$$E_n = \frac{1240^2}{8(511 \times 10^3)(3.0 \times 10^9)^2}\left(6 \times 10^{18}\right)^2 \approx 6 \times 10^{18} \text{ eV} .$$

(d) Since

$$\frac{E_n}{mc^2} = \frac{6 \times 10^{18} \text{ eV}}{511 \times 10^3 \text{ eV}} \gg 1 ,$$

the energy is indeed in the relativistic range.

14. (a) With Eq. 40-11, we compare the ψ_1^2 and ψ_2^2 graphs in Fig. 40-6. The former has a maximum at the center and the latter is zero there. Thus, the excitation of the system described in this problem implies the electron has become much less likely to be detected near the middle of the well.

(b) Examining the $0 \le x \le 25$ pm regions of those two graphs, we conclude that the excited state electron is somewhat more likely to be "near" (not "at") a well wall. Eq. 40-13 supports this conclusion in the sense that there is more "area" under the curve of ψ_2^2 in the $0 \le x \le 25$ pm region than under the ψ_1^2 curve for that region.

15. (a) The allowed energy values are given by $E_n = n^2h^2/8mL^2$. The difference in energy between the state n and the state $n+1$ is

$$\Delta E_{\text{adj}} = E_{n+1} - E_n = \left[(n+1)^2 - n^2\right]\frac{h^2}{8mL^2} = \frac{(2n+1)h^2}{8mL^2}$$

and

$$\frac{\Delta E_{\text{adj}}}{E} = \left[\frac{(2n+1)h^2}{8mL^2}\right]\left(\frac{8mL^2}{n^2h^2}\right) = \frac{2n+1}{n^2} .$$

As n becomes large, $2n + 1 \longrightarrow 2n$ and $(2n+1)/n^2 \longrightarrow 2n/n^2 = 2/n$.

(b) As $n \longrightarrow \infty$, ΔE_{adj} and E do not approach 0, but $\Delta E_{\text{adj}}/E$ does.

(c) See part (b).

(d) See part (b).

(e) $\Delta E_{\text{adj}}/E$ is a better measure than either ΔE_{adj} or E alone of the extent to which the quantum result is approximated by the classical result.

16. We follow Sample Problem 40-3 in the presentation of this solution. The integration result quoted below is discussed in a little more detail in that Sample Problem. We note that the arguments of the sine functions used below are in radians.

(a) The probability of detecting the particle in the region $0 \le x \le \frac{L}{4}$ is

$$\left(\frac{2}{L}\right)\left(\frac{L}{\pi}\right)\int_0^{\pi/4} \sin^2 y\, dy = \frac{2}{\pi}\left(\frac{y}{2} - \frac{\sin 2y}{4}\right)_0^{\pi/4} = 0.091 \;.$$

(b) As expected from symmetry,

$$\left(\frac{2}{L}\right)\left(\frac{L}{\pi}\right)\int_{\pi/4}^{\pi} \sin^2 y\, dy = \frac{2}{\pi}\left(\frac{y}{2} - \frac{\sin 2y}{4}\right)_{\pi/4}^{\pi} = 0.091 \;.$$

(c) For the region $\frac{L}{4} \le x \le \frac{3L}{4}$, we obtain

$$\left(\frac{2}{L}\right)\left(\frac{L}{\pi}\right)\int_{\pi/4}^{3\pi/4} \sin^2 y\, dy = \frac{2}{\pi}\left(\frac{y}{2} - \frac{\sin 2y}{4}\right)_{\pi/4}^{3\pi/4} = 0.82$$

which we could also have gotten by subtracting the results of part (a) and (b) from 1; that is, $1 - 2(0.091) = 0.82$.

17. The probability that the electron is found in any interval is given by $P = \int |\psi|^2\, dx$, where the integral is over the interval. If the interval width Δx is small, the probability can be approximated by $P = |\psi|^2\, \Delta x$, where the wave function is evaluated for the center of the interval, say. For an electron trapped in an infinite well of width L, the ground state probability density is

$$|\psi|^2 = \frac{2}{L}\sin^2\left(\frac{\pi x}{L}\right) \,,$$

so

$$P = \left(\frac{2\,\Delta x}{L}\right)\sin^2\left(\frac{\pi x}{L}\right) \,.$$

(a) We take $L = 100\,\text{pm}$, $x = 25\,\text{pm}$, and $\Delta x = 5.0\,\text{pm}$. Then,

$$P = \left[\frac{2(5.0\,\text{pm})}{100\,\text{pm}}\right]\sin^2\left[\frac{\pi(25\,\text{pm})}{100\,\text{pm}}\right] = 0.050 \;.$$

(b) We take $L = 100\,\text{pm}$, $x = 50\,\text{pm}$, and $\Delta x = 5.0\,\text{pm}$. Then,

$$P = \left[\frac{2(5.0\,\text{pm})}{100\,\text{pm}}\right]\sin^2\left[\frac{\pi(50\,\text{pm})}{100\,\text{pm}}\right] = 0.10 \;.$$

(c) We take $L = 100\,\text{pm}$, $x = 90\,\text{pm}$, and $\Delta x = 5.0\,\text{pm}$. Then,

$$P = \left[\frac{2(5.0\,\text{pm})}{100\,\text{pm}}\right]\sin^2\left[\frac{\pi(90\,\text{pm})}{100\,\text{pm}}\right] = 0.0095 \;.$$

18. (a) We recall that a derivative with respect to a dimensional quantity carries the (reciprocal) units of that quantity. Thus, the first term in Eq. 40-18 has dimensions of ψ multiplied by dimensions of x^{-2}. The second term contains no derivatives, does contain ψ, and involves several other factors that (as we show below) turn out to have dimensions of x^{-2}:

$$\frac{8\pi^2 m}{h^2}[E - U(x)] \implies \frac{\text{kg}}{(\text{J}\cdot\text{s})^2}[\text{J}]$$

assuming SI units. Recalling from Eq. 7-9 that J = $\text{kg}\cdot\text{m}^2/\text{s}^2$, then we see the above is indeed in units of m^{-2} (which means dimensions of x^{-2}).

(b) In one-dimensional Quantum Physics, the wavefunction has units of $\text{m}^{-1/2}$ as Sample Problem 40-2 shows. Thus, since each term in Eq. 40-18 has units of ψ multiplied by units of x^{-2}, then those units are $\text{m}^{-1/2}\cdot\text{m}^{-2} = \text{m}^{-2.5}$.

19. According to Fig. 40-9, the electron's initial energy is 109 eV. After the additional energy is absorbed, the total energy of the electron is 109 eV + 400 eV = 509 eV. Since it is in the region $x > L$, its potential energy is 450 eV (see Section 40-5), so its kinetic energy must be 509 eV − 450 eV = 59 eV.

20. From Fig. 40-9, we see that the sum of the kinetic and potential energies in that particular finite well is 280 eV. The potential energy is zero in the region $0 < x < L$. If the kinetic energy of the electron is detected while it is in that region (which is the only region where this is likely to happen), we should find $K = 280$ eV.

21. (a) and (b) Schrödinger's equation for the region $x > L$ is

$$\frac{d^2\psi}{dx^2} + \frac{8\pi^2 m}{h^2}[E - U_0]\,\psi = 0 ,$$

where $E - U_0 < 0$. If $\psi^2(x) = Ce^{-2kx}$, then $\psi(x) = C'e^{-kx}$, where C' is another constant satisfying $C'^2 = C$. Thus $d^2\psi/dx^2 = 4k^2C'e^{-kx} = 4k^2\psi$ and

$$\frac{d^2\psi}{dx^2} + \frac{8\pi^2 m}{h^2}[E - U_0]\,\psi = k^2\psi + \frac{8\pi^2 m}{h^2}[E - U_0]\,\psi .$$

This is zero provided that

$$k^2 = \frac{8\pi^2 m}{h^2}[U_0 - E] .$$

The quantity on the right-hand side is positive, so k is real and the proposed function satisfies Schrödinger's equation. If k is negative, however, the proposed function would be physically unrealistic. It would increase exponentially with x. Since the integral of the probability density over the entire x axis must be finite, ψ diverging as $x \to \infty$ would be unacceptable. Therefore, we choose

$$k = \frac{2\pi}{h}\sqrt{2m\,(U_0 - E)} > 0 .$$

22. (a) and (b) In the region $0 < x < L$, $U_0 = 0$, so Schrödinger's equation for the region is

$$\frac{d^2\psi}{dx^2} + \frac{8\pi^2 m}{h^2}E\psi = 0$$

where $E > 0$. If $\psi^2(x) = B\sin^2 kx$, then $\psi(x) = B'\sin kx$, where B' is another constant satisfying $B'^2 = B$. Thus $d^2\psi/dx^2 = -k^2B'\sin kx = -k^2\psi(x)$ and

$$\frac{d^2\psi}{dx^2} + \frac{8\pi^2 m}{h^2}E\psi = -k^2\psi + \frac{8\pi^2 m}{h^2}E\psi .$$

This is zero provided that

$$k^2 = \frac{8\pi^2 mE}{h^2} .$$

The quantity on the right-hand side is positive, so k is real and the proposed function satisfies Schrödinger's equation. In this case, there exists no physical restriction as to the sign of k. It can assume either positive or negative values. Thus

$$k = \pm\frac{2\pi}{h}\sqrt{2mE} .$$

23. Schrödinger's equation for the region $x > L$ is

$$\frac{d^2\psi}{dx^2} + \frac{8\pi^2 m}{h^2}\left[E - U_0\right]\psi = 0 .$$

If $\psi = De^{2kx}$, then $d^2\psi/dx^2 = 4k^2De^{2kx} = 4k^2\psi$ and

$$\frac{d^2\psi}{dx^2} + \frac{8\pi^2 m}{h^2}\left[E - U_0\right]\psi = 4k^2\psi + \frac{8\pi^2 m}{h^2}\left[E - U_0\right]\psi .$$

This is zero provided

$$k = \frac{\pi}{h}\sqrt{2m\left(U_0 - E\right)} .$$

The proposed function satisfies Schrödinger's equation provided k has this value. Since U_0 is greater than E in the region $x > L$, the quantity under the radical is positive. This means k is real. If k is positive, however, the proposed function is physically unrealistic. It increases exponentially with x and becomes large without bound. The integral of the probability density over the entire x axis must be unity. This is impossible if ψ is the proposed function.

24. We can use the mc^2 value for an electron from Table 38-3 (511×10^3 eV) and the hc value developed in problem 3 of Chapter 39 by writing Eq. 40-20 as

$$E_{nx,ny} = \frac{2h^2}{8m}\left(\frac{n_x^2}{L_x^2} + \frac{n_y^2}{L_y^2}\right) = \frac{(hc)^2}{8(mc^2)}\left(\frac{n_x^2}{L_x^2} + \frac{n_y^2}{L_y^2}\right) .$$

For $n_x = n_y = 1$, we obtain

$$E_{1,1} = \frac{(1240\,\text{eV}\cdot\text{nm})^2}{8(511\times 10^3\,\text{eV})}\left(\frac{1}{(0.800\,\text{nm})^2} + \frac{1}{(1.600\,\text{nm})^2}\right) = 0.73\ \text{eV} .$$

25. We can use the mc^2 value for an electron from Table 38-3 (511×10^3 eV) and the hc value developed in problem 3 of Chapter 39 by writing Eq. 40-21 as

$$E_{nx,ny,nz} = \frac{2h^2}{8m}\left(\frac{n_x^2}{L_x^2} + \frac{n_y^2}{L_y^2} + \frac{n_z^2}{L_z^2}\right) = \frac{(hc)^2}{8(mc^2)}\left(\frac{n_x^2}{L_x^2} + \frac{n_y^2}{L_y^2} + \frac{n_z^2}{L_z^2}\right) .$$

For $n_x = n_y = n_z = 1$, we obtain

$$E_{1,1} = \frac{(1240\,\text{eV}\cdot\text{nm})^2}{8(511\times 10^3\,\text{eV})}\left(\frac{1}{(0.800\,\text{nm})^2} + \frac{1}{(1.600\,\text{nm})^2} + \frac{1}{(0.400\,\text{nm})^2}\right) = 3.1\ \text{eV} .$$

26. We are looking for the values of the ratio

$$\frac{E_{nx,ny}}{h^2/8mL^2} = L^2\left(\frac{n_x^2}{L_x^2} + \frac{n_y^2}{L_y^2}\right) = \left(n_x^2 + \frac{1}{4}n_y^2\right)$$

and the corresponding differences.

(a) For $n_x = n_y = 1$, the ratio becomes $1 + \frac{1}{4} = 1.25$.

(b) For $n_x = 1$ and $n_y = 2$, the ratio becomes $1 + \frac{1}{4}(4) = 2.00$. One can check (by computing other (n_x, n_y) values) that this is the next to lowest energy in the system.

(c) The lowest set of states that are degenerate are $(n_x, n_y) = (1,4)$ and $(2,2)$. Both of these states have that ratio equal to $1 + \frac{1}{4}(16) = 5.00$.

(d) For $n_x = 1$ and $n_y = 3$, the ratio becomes $1 + \frac{1}{4}(9) = 3.25$. One can check (by computing other (n_x, n_y) values) that this is the lowest energy greater than that computed in part (b). The next higher energy comes from $(n_x, n_y) = (2,1)$ for which the ratio is $4 + \frac{1}{4}(1) = 4.25$. The difference between these two values is $4.25 - 3.25 = 1.00$.

27. The energy levels are given by

$$E_{n_x,n_y} = \frac{h^2}{8m}\left[\frac{n_x^2}{L_x^2} + \frac{n_y^2}{L_y^2}\right] = \frac{h^2}{8mL^2}\left[n_x^2 + \frac{n_y^2}{4}\right]$$

where the substitutions $L_x = L$ and $L_y = 2L$ were made. In units of $h^2/8mL^2$, the energy levels are given by $n_x^2 + n_y^2/4$. The lowest five levels are $E_{1,1} = 1.25$, $E_{1,2} = 2.00$, $E_{1,3} = 3.25$, $E_{2,1} = 4.25$, and $E_{2,2} = E_{1,4} = 5.00$. It is clear that there are no other possible values for the energy less than 5. The frequency of the light emitted or absorbed when the electron goes from an initial state i to a final state f is $f = (E_f - E_i)/h$, and in units of $h/8mL^2$ is simply the difference in the values of $n_x^2 + n_y^2/4$ for the two states. The possible frequencies are 0.75 (1,2 $\longrightarrow$ 1,1), 2.00 (1,3 $\longrightarrow$ 1,1), 3.00 (2,1 $\longrightarrow$ 1,1), 3.75 (2,2 $\longrightarrow$ 1,1), 1.25 (1,3 $\longrightarrow$ 1,2), 2.25 (2,1 $\longrightarrow$ 1,2), 3.00 (2,2 $\longrightarrow$ 1,2), 1.00 (2,1 $\longrightarrow$ 1,3), 1.75 (2,2 $\longrightarrow$ 1,3), 0.75 (2,2 $\longrightarrow$ 2,1), all in units of $h/8mL^2$.

28. We are looking for the values of the ratio

$$\frac{E_{n_x,n_y,n_z}}{h^2/8mL^2} = L^2\left(\frac{n_x^2}{L_x^2} + \frac{n_y^2}{L_y^2} + \frac{n_z^2}{L_z^2}\right) = (n_x^2 + n_y^2 + n_z^2)$$

and the corresponding differences.

(a) For $n_x = n_y = n_z = 1$, the ratio becomes $1 + 1 + 1 = 3.00$.

(b) For $n_x = n_y = 2$ and $n_z = 1$, the ratio becomes $4 + 4 + 1 = 9.00$. One can check (by computing other (n_x, n_y, n_z) values) that this is the third lowest energy in the system. One can also check that this same ratio is obtained for $(n_x, n_y, n_z) = (2,1,2)$ and $(1,2,2)$.

(c) For $n_x = n_y = 1$ and $n_z = 3$, the ratio becomes $1 + 1 + 9 = 11.00$. One can check (by computing other (n_x, n_y, n_z) values) that this is three "steps" up from the lowest energy in the system. One can also check that this same ratio is obtained for $(n_x, n_y, n_z) = (1,3,1)$ and $(3,1,1)$. If we take the difference between this and the result of part (b), we obtain $11.00 - 9.00 = 2.00$.

(d) For $n_x = n_y = 1$ and $n_z = 2$, the ratio becomes $1 + 1 + 4 = 6.00$. One can check (by computing other (n_x, n_y, n_z) values) that this is the next to the lowest energy in the system. One can also check that this same ratio is obtained for $(n_x, n_y, n_z) = (2,1,1)$ and $(1,2,1)$. Thus, three states (three arrangements of (n_x, n_y, n_z) values) have this energy.

(e) For $n_x = 1$, $n_y = 2$ and $n_z = 3$, the ratio becomes $1 + 4 + 9 = 14.00$. One can check (by computing other (n_x, n_y, n_z) values) that this is five "steps" up from the lowest energy in the system. One can also check that this same ratio is obtained for $(n_x, n_y, n_z) = (1,3,2), (2,3,1), (2,1,3), (3,1,2)$ and $(3,2,1)$. Thus, six states (six arrangements of (n_x, n_y, n_z) values) have this energy.

29. The ratios computed in problem 28 can be related to the frequencies emitted using $f = \Delta E/h$, where each level E is equal to one of those ratios multiplied by $h^2/8mL^2$. This effectively involves no more

than a cancellation of one of the factors of h. Thus, for a transition from the second excited state (see part (b) of problem 28) to the ground state (treated in part (a) of that problem), we find

$$f = (9.00 - 3.00)\left(\frac{h}{8mL^2}\right) = (6.00)\left(\frac{h}{8mL^2}\right) .$$

In the following, we omit the $h/8mL^2$ factors. For a transition between the fourth excited state and the ground state, we have $f = 12.00 - 3.00 = 9.00$. For a transition between the third excited state and the ground state, we have $f = 11.00 - 3.00 = 8.00$. For a transition between the third excited state and the first excited state, we have $f = 11.00 - 6.00 = 5.00$. For a transition between the fourth excited state and the third excited state, we have $f = 12.00 - 11.00 = 1.00$. For a transition between the third excited state and the second excited state, we have $f = 11.00 - 9.00 = 2.00$. For a transition between the second excited state and the first excited state, we have $f = 9.00 - 6.00 = 3.00$, which also results from some other transitions.

30. For $n = 1$

$$\begin{aligned} E_1 &= -\frac{m_e e^4}{8\varepsilon_0^2 h^2} \\ &= -\frac{(9.11\times10^{-31}\,\text{kg})(1.6\times10^{-19}\,\text{C})^4}{8(8.85\times10^{-12}\,\text{F/m})^2(6.63\times10^{-34}\,\text{J}\cdot\text{s})^2(1.60\times10^{-19}\,\text{J/eV})} \\ &= -13.6\ \text{eV} . \end{aligned}$$

31. From Eq. 40-6,

$$\Delta E = hf = (4.14\times10^{-15}\,\text{eV}\cdot\text{s})(6.2\times10^{14}\,\text{Hz}) = 2.6\ \text{eV} .$$

32. The difference between the energy absorbed and the energy emitted is

$$E_{\text{photon absorbed}} - E_{\text{photon emitted}} = \frac{hc}{\lambda_{\text{absorbed}}} - \frac{hc}{\lambda_{\text{emitted}}} .$$

Thus, using the result of problem 3 in Chapter 39, the net energy absorbed is

$$hc\Delta\left(\frac{1}{\lambda}\right) = (1240\,\text{eV}\cdot\text{nm})\left(\frac{1}{375\,\text{nm}} - \frac{1}{580\,\text{nm}}\right) = 1.17\ \text{eV} .$$

33. The energy E of the photon emitted when a hydrogen atom jumps from a state with principal quantum number u to a state with principal quantum number ℓ is given by

$$E = A\left(\frac{1}{\ell^2} - \frac{1}{u^2}\right)$$

where $A = 13.6\,\text{eV}$. The frequency f of the electromagnetic wave is given by $f = E/h$ and the wavelength is given by $\lambda = c/f$. Thus,

$$\frac{1}{\lambda} = \frac{f}{c} = \frac{E}{hc} = \frac{A}{hc}\left(\frac{1}{\ell^2} - \frac{1}{u^2}\right) .$$

The shortest wavelength occurs at the series limit, for which $u = \infty$. For the Balmer series, $\ell = 2$ and the shortest wavelength is $\lambda_B = 4hc/A$. For the Lyman series, $\ell = 1$ and the shortest wavelength is $\lambda_L = hc/A$. The ratio is $\lambda_B/\lambda_L = 4$.

34. (a) The energy level corresponding to the probability density distribution shown in Fig. 40-20 is the $n = 2$ level. Its energy is given by

$$E_2 = -\frac{13.6\,\text{eV}}{2^2} = -3.4\ \text{eV} .$$

(b) As the electron is removed from the hydrogen atom the final energy of the proton-electron system is zero. Therefore, one needs to supply at least 3.4 eV of energy to the system in order to bring its energy up from $E_2 = -3.4\,\text{eV}$ to zero. (If more energy is supplied, then the electron will retain some kinetic energy after it is removed from the atom.)

35. (a) Since energy is conserved, the energy E of the photon is given by $E = E_i - E_f$, where E_i is the initial energy of the hydrogen atom and E_f is the final energy. The electron energy is given by $(-13.6\,\text{eV})/n^2$, where n is the principal quantum number. Thus,

$$E = E_i - E_f = \frac{-13.6\,\text{eV}}{(3)^2} - \frac{-13.6\,\text{eV}}{(1)^2} = 12.1\ \text{eV} .$$

(b) The photon momentum is given by

$$p = \frac{E}{c} = \frac{(12.1\,\text{eV})(1.60 \times 10^{-19}\,\text{J/eV})}{3.00 \times 10^8\,\text{m/s}} = 6.45 \times 10^{-27}\ \text{kg}\cdot\text{m/s} .$$

(c) Using the result of problem 3 in Chapter 39, the wavelength is

$$\lambda = \frac{1240\,\text{eV}\cdot\text{nm}}{12.1\,\text{eV}} = 102\ \text{nm} .$$

36. (a) The "home-base" energy level for the Balmer series is $n = 2$. Thus the transition with the least energetic photon is the one from the $n = 3$ level to the $n = 2$ level. The energy difference for this transition is

$$\Delta E = E_3 - E_2 = -(13.6\,\text{eV})\left(\frac{1}{3^2} - \frac{1}{2^2}\right) = 1.889\,\text{eV} .$$

Using the result of problem 3 in Chapter 39, the corresponding wavelength is

$$\lambda = \frac{hc}{\Delta E} = \frac{1240\,\text{eV}\cdot\text{nm}}{1.889\,\text{eV}} = 658\ \text{nm} .$$

(b) For the series limit, the energy difference is

$$\Delta E = E_\infty - E_2 = -(13.6\,\text{eV})\left(\frac{1}{\infty^2} - \frac{1}{2^2}\right) = 3.40\ \text{eV} .$$

The corresponding wavelength is then

$$\lambda = \frac{hc}{\Delta E} = \frac{1240\,\text{eV}\cdot\text{nm}}{3.40\,\text{eV}} = 366\ \text{nm} .$$

37. If kinetic energy is not conserved, some of the neutron's initial kinetic energy is used to excite the hydrogen atom. The least energy that the hydrogen atom can accept is the difference between the first excited state ($n = 2$) and the ground state ($n = 1$). Since the energy of a state with principal quantum number n is $-(13.6\,\text{eV})/n^2$, the smallest excitation energy is $13.6\,\text{eV} - (13.6\,\text{eV})/(2)^2 = 10.2\,\text{eV}$. The neutron does not have sufficient kinetic energy to excite the hydrogen atom, so the hydrogen atom is left in its ground state and all the initial kinetic energy of the neutron ends up as the final kinetic energies of the neutron and atom. The collision must be elastic.

38. (a) We use Eq. 40-25. At $r = a$

$$\psi^2(r) = \left(\frac{1}{\sqrt{\pi}a^{3/2}}\,e^{-a/a}\right)^2 = \frac{1}{\pi a^3}\,e^{-2} = \frac{1}{\pi(5.29 \times 10^{-2}\,\text{nm})^3}\,e^{-2} = 291\ \text{nm}^{-3} .$$

(b) We use Eq. 40-31. At $r = a$

$$P(r) = \frac{4}{a^3}\,a^2 e^{-2a/a} = \frac{4e^{-2}}{a} = \frac{4e^{-2}}{5.29 \times 10^{-2}\,\text{nm}} = 10.2\ \text{nm}^{-1}\ .$$

39. (a) We use Eq. 40-31. At $r = 0$, $P(r) \propto r^2 = 0$.

(b) At $r = a$

$$P(r) = \frac{4}{a^3}\,a^2 e^{-2a/a} = \frac{4e^{-2}}{a} = \frac{4e^{-2}}{5.29 \times 10^{-2}\,\text{nm}} = 10.2\ \text{nm}^{-1}\ .$$

(c) At $r = 2a$

$$P(r) = \frac{4}{a^3}\,(2a)^2 e^{-4a/a} = \frac{16e^{-4}}{a} = \frac{16e^{-4}}{5.29 \times 10^{-2}\,\text{nm}} = 5.54\ \text{nm}^{-1}\ .$$

40. (a) $\Delta E = -(13.6\,\text{eV})(4^{-2} - 1^{-2}) = 12.8\,\text{eV}$.

(b) The values of the photon energies are:

$$\begin{aligned}
E_{4\to1} &= \Delta E_{\text{part (a)}} = 12.8\ \text{eV} \\
E_{3\to1} &= -(13.6\,\text{eV})(3^{-2} - 1^{-2}) = 12.1\ \text{eV} \\
E_{2\to1} &= -(13.6\,\text{eV})(2^{-2} - 1^{-2}) = 10.2\ \text{eV} \\
E_{4\to2} &= -(13.6\,\text{eV})(4^{-2} - 2^{-2}) = 2.55\ \text{eV} \\
E_{3\to2} &= -(13.6\,\text{eV})(3^{-2} - 2^{-2}) = 1.89\ \text{eV} \\
E_{4\to3} &= -(13.6\,\text{eV})(4^{-2} - 3^{-2}) = 0.66\ \text{eV}
\end{aligned}$$

The various photon energies correspond to the transitions between energy levels indicated below. The levels are not drawn to scale.

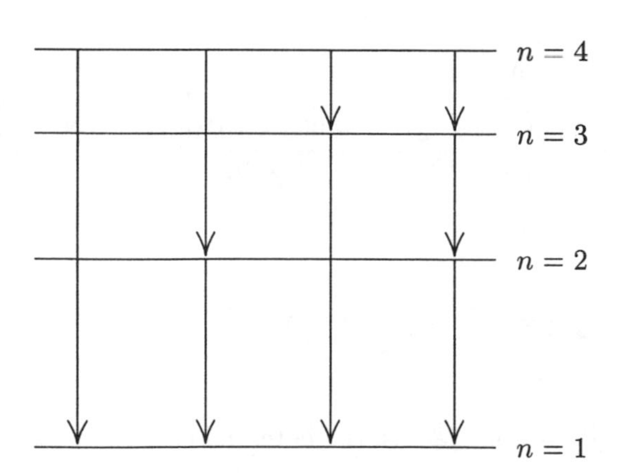

41. (a) We take the electrostatic potential energy to be zero when the electron and proton are far removed from each other. Then, the final energy of the atom is zero and the work done in pulling it apart is $W = -E_i$, where E_i is the energy of the initial state. The energy of the initial state is given by $E_i = (-13.6\,\text{eV})/n^2$, where n is the principal quantum number of the state. For the ground state, $n = 1$ and $W = 13.6\,\text{eV}$.

(b) For the state with $n = 2$, $W = (13.6\,\text{eV})/(2)^2 = 3.40\,\text{eV}$.

42. Conservation of linear momentum of the atom-photon system requires that

$$p_{\text{recoil}} = p_{\text{photon}} \implies m_p v_{\text{recoil}} = \frac{hf}{c}$$

where we use Eq. 39-7 for the photon and use the classical momentum formula for the atom (since we expect its speed to be much less than c). Thus, from Eq. 40-6 and Table 38-3,

$$\begin{aligned} v_{\text{recoil}} &= \frac{\Delta E}{m_p c} = \frac{E_4 - E_1}{(m_p c^2)/c} \\ &= \frac{(-13.6\,\text{eV})\left(4^{-2} - 1^{-2}\right)}{(938 \times 10^6\,\text{eV})/(2.998 \times 10^8\,\text{m/s})} \\ &= 4.1\ \text{m/s}\ . \end{aligned}$$

43. (a) and (b) Using Eq. 40-6 and the result of problem 3 in Chapter 39, we find

$$\Delta E = E_{\text{photon}} = \frac{hc}{\lambda} = \frac{1240\,\text{eV}\cdot\text{nm}}{486.1\,\text{nm}} = 2.55\ \text{eV}\ .$$

Referring to Fig. 40-16, we see that this must be one of the Balmer series transitions (this fact could also be found from Fig. 40-17). Therefore, $n_{\text{low}} = 2$, but what precisely is n_{high}?

$$\begin{aligned} E_{\text{high}} &= E_{\text{low}} + \Delta E \\ -\frac{13.6\,\text{eV}}{n^2} &= -\frac{13.6\,\text{eV}}{2^2} + 2.55\,\text{eV} \end{aligned}$$

which yields $n = 4$. Thus, the transition is from the $n = 4$ to the $n = 2$ state.

44. (a) The calculation is shown in Sample Problem 40-6. The difference in the values obtained in parts (a) and (b) of that Sample Problem is $122\,\text{nm} - 91.4\,\text{nm} \approx 31\,\text{nm}$.

(b) Fig. 40-17 shows that the width of the Balmer series is $656.3\,\text{nm} - 364.6\,\text{nm} \approx 292\,\text{nm}$. This can be confirmed with a calculation very much like the one shown in Sample Problem 40-6, but with the longest wavelength arising from the $3 \to 2$ transition, and the series limit obtained from the $\infty \to 2$ transition.

(c) We use Eq. 39-1. For the Lyman series,

$$\Delta f = \frac{2.998 \times 10^8\,\text{m/s}}{91.4 \times 10^{-9}\,\text{m}} - \frac{2.998 \times 10^8\,\text{m/s}}{122 \times 10^{-9}\,\text{m}} = 8.2 \times 10^{14}\ \text{Hz}$$

or 8.2×10^2 THz. For the Balmer series,

$$\Delta f = \frac{2.998 \times 10^8\,\text{m/s}}{364.6 \times 10^{-9}\,\text{m}} - \frac{2.998 \times 10^8\,\text{m/s}}{656.3 \times 10^{-9}\,\text{m}} = 3.65 \times 10^{14}\ \text{Hz}$$

which is equivalent to 365 THz.

45. Letting $a = 5.292 \times 10^{-11}$ m be the Bohr radius, the potential energy becomes

$$U = -\frac{e^2}{4\pi\varepsilon_0 a} = \frac{(8.99 \times 10^9\,\text{N}\cdot\text{m}^2/\text{C}^2)(1.602 \times 10^{-19}\,\text{C})^2}{5.292 \times 10^{-11}\,\text{m}} = -4.36 \times 10^{-18}\,\text{J} = -27.2\ \text{eV}\ .$$

The kinetic energy is $K = E - U = (-13.6\,\text{eV}) - (-27.2\,\text{eV}) = 13.6\,\text{eV}$.

46. (a) and (b) Using Eq. 40-6 and the result of problem 3 in Chapter 39, we find

$$\Delta E = E_{\text{photon}} = \frac{hc}{\lambda} = \frac{1240\,\text{eV}\cdot\text{nm}}{121.6\,\text{nm}} = 10.2\ \text{eV}\ .$$

Referring to Fig. 40-16, we see that this must be one of the Lyman series transitions. Therefore, $n_{\text{low}} = 1$, but what precisely is n_{high}?

$$\begin{aligned} E_{\text{high}} &= E_{\text{low}} + \Delta E \\ -\frac{13.6\,\text{eV}}{n^2} &= -\frac{13.6\,\text{eV}}{1^2} + 10.2\,\text{eV} \end{aligned}$$

which yields $n = 2$ (this is confirmed by the calculation found from Sample Problem 40-6). Thus, the transition is from the $n = 2$ to the $n = 1$ state.

47. (a) Since $E_2 = -0.85\,\text{eV}$ and $E_1 = -13.6\,\text{eV} + 10.2\,\text{eV} = -3.4\,\text{eV}$, the photon energy is $E_{\text{photon}} = E_2 - E_1 = -0.85\,\text{eV} - (-3.4\,\text{eV}) = 2.6\,\text{eV}$.

(b) From

$$E_2 - E_1 = (-13.6\,\text{eV})\left(\frac{1}{n_2^2} - \frac{1}{n_1^2}\right) = 2.6\ \text{eV}$$

we obtain

$$\frac{1}{n_2^2} - \frac{1}{n_1^2} = -\frac{2.6\,\text{eV}}{13.6\,\text{eV}} \approx -\frac{3}{16} = \frac{1}{4^2} - \frac{1}{2^2}\ .$$

Thus, $n_2 = 4$ and $n_1 = 2$. So the transition is from the $n = 4$ state to the $n = 2$ state. One can easily verify this by inspecting the energy level diagram of Fig. 40-16.

48. The wavelength λ of the photon emitted in a transition belonging to the Balmer series satisfies

$$E_{\text{ph}} = \frac{hc}{\lambda} = E_n - E_2 = -(13.6\,\text{eV})\left(\frac{1}{n^2} - \frac{1}{2^2}\right) \quad \text{where } n = 3,\ 4,\ 5,\ \ldots$$

Using the result of problem 3 in Chapter 39, we find

$$\lambda = \frac{4hcn^2}{(13.6\,\text{eV})(n^2 - 4)} = \frac{4(1240\,\text{eV}\cdot\text{nm})}{13.6\,\text{eV}}\left(\frac{n^2}{n^2 - 4}\right)\ .$$

Plugging in the various values of n, we obtain these values of the wavelength: $\lambda = 656\,\text{nm}$ (for $n = 3$), $\lambda = 486\,\text{nm}$ (for $n = 4$), $\lambda = 434\,\text{nm}$ (for $n = 5$), $\lambda = 410\,\text{nm}$ (for $n = 6$), $\lambda = 397\,\text{nm}$ (for $n = 7$), $\lambda = 389\,\text{nm}$ (for $n = 8$), etc. Finally for $n = \infty$, $\lambda = 365\,\text{nm}$. These values agree well with the data found in Fig. 40-17. [One can also find λ beyond three significant figures by using the more accurate values for m_e, e and h listed in Appendix B when calculating E_n in Eq. 40-24. Another factor that contributes to the error is the motion of the atomic nucleus. It can be shown that this effect can be accounted for by replacing the mass of the electron m_e by $m_e m_p/(m_p + m_e)$ in Eq. 40-24, where m_p is the mass of the proton. Since $m_p \gg m_e$, this is not a major effect.]

49. According to Sample Problem 40-8, the probability the electron in the ground state of a hydrogen atom can be found inside a sphere of radius r is given by

$$p(r) = 1 - e^{-2x}\left(1 + 2x + 2x^2\right)$$

where $x = r/a$ and a is the Bohr radius. We want $r = a$, so $x = 1$ and

$$p(a) = 1 - e^{-2}\,(1 + 2 + 2) = 1 - 5e^{-2} = 0.323\ .$$

The probability that the electron can be found outside this sphere is $1 - 0.323 = 0.677$. It can be found outside about 68% of the time.

50. Using Eq. 40-6 and the result of problem 3 in Chapter 39, we find

$$\Delta E = E_{\text{photon}} = \frac{hc}{\lambda} = \frac{1240\,\text{eV}\cdot\text{nm}}{102.6\,\text{nm}} = 12.09\ \text{eV}\ .$$

Referring to Fig. 40-16, we see that this must be one of the Lyman series transitions. Therefore, $n_{\text{low}} = 1$, but what precisely is n_{high}?

$$\begin{aligned} E_{\text{high}} &= E_{\text{low}} + \Delta E \\ -\frac{13.6\,\text{eV}}{n^2} &= -\frac{13.6\,\text{eV}}{1^2} + 12.09\,\text{eV} \end{aligned}$$

which yields $n = 3$. Thus, the transition is from the $n = 3$ to the $n = 1$ state.

51. The proposed wave function is

$$\psi = \frac{1}{\sqrt{\pi} a^{3/2}} e^{-r/a}$$

where a is the Bohr radius. Substituting this into the right side of Schrödinger's equation, our goal is to show that the result is zero. The derivative is

$$\frac{d\psi}{dr} = -\frac{1}{\sqrt{\pi} a^{5/2}} e^{-r/a}$$

so

$$r^2 \frac{d\psi}{dr} = -\frac{r^2}{\sqrt{\pi} a^{5/2}} e^{-r/a}$$

and

$$\frac{1}{r^2} \frac{d}{dr}\left(r^2 \frac{d\psi}{dr}\right) = \frac{1}{\sqrt{\pi} a^{5/2}} \left[-\frac{2}{r} + \frac{1}{a}\right] e^{-r/a} = \frac{1}{a}\left[-\frac{2}{r} + \frac{1}{a}\right]\psi .$$

The energy of the ground state is given by $E = -me^4/8\varepsilon_0^2 h^2$, and the Bohr radius is given by $a = h^2 \varepsilon_0/\pi m e^2$, so $E = -e^2/8\pi\varepsilon_0 a$. The potential energy is given by $U = -e^2/4\pi\varepsilon_0 r$, so

$$\begin{aligned} \frac{8\pi^2 m}{h^2}[E - U]\psi &= \frac{8\pi^2 m}{h^2}\left[-\frac{e^2}{8\pi\varepsilon_0 a} + \frac{e^2}{4\pi\varepsilon_0 r}\right]\psi = \frac{8\pi^2 m}{h^2}\frac{e^2}{8\pi\varepsilon_0}\left[-\frac{1}{a} + \frac{2}{r}\right]\psi \\ &= \frac{\pi m e^2}{h^2 \varepsilon_0}\left[-\frac{1}{a} + \frac{2}{r}\right]\psi = \frac{1}{a}\left[-\frac{1}{a} + \frac{2}{r}\right]\psi . \end{aligned}$$

The two terms in Schrödinger's equation cancel, and the proposed function ψ satisfies that equation.

52. From Sample Problem 40-8, we know that the probability of finding the electron in the ground state of the hydrogen atom inside a sphere of radius r is given by

$$p(r) = 1 - e^{-2x}\left(1 + 2x + 2x^2\right)$$

where $x = r/a$. Thus the probability of finding the electron between the two shells indicated in this problem is given by

$$\begin{aligned} p(a < r < 2a) &= p(2a) - p(a) \\ &= \left[1 - e^{-2x}\left(1 + 2x + 2x^2\right)\right]_{x=2} - \left[1 - e^{-2x}\left(1 + 2x + 2x^2\right)\right]_{x=1} \\ &= 0.44 . \end{aligned}$$

53. The radial probability function for the ground state of hydrogen is $P(r) = (4r^2/a^3)e^{-2r/a}$, where a is the Bohr radius. (See Eq. 40-31.) We want to evaluate the integral $\int_0^\infty P(r)\,dr$. Eq. 15 in the integral table of Appendix E is an integral of this form. We set $n = 2$ and replace a in the given formula with $2/a$ and x with r. Then

$$\int_0^\infty P(r)\,dr = \frac{4}{a^3}\int_0^\infty r^2 e^{-2r/a}\,dr = \frac{4}{a^3}\frac{2}{(2/a)^3} = 1 .$$

54. (a) The allowed values of l for a given n are $0, 1, 2, \ldots, n-1$. Thus there are n different values of l.

(b) The allowed values of m_l for a given l are $-l, -l+1, \cdots, l$. Thus there are $2l+1$ different values of m_l.

(c) According to part (a) above, for a given n there are n different values of l. Also, each of these l's can have $2l+1$ different values of m_l [see part (b) above]. Thus, the total number of m_l's is

$$\sum_{l=0}^{n-1}(2l+1) = n^2 \ .$$

55. Since Δr is small, we may calculate the probability using $p = P(r)\,\Delta r$, where $P(r)$ is the radial probability density. The radial probability density for the ground state of hydrogen is given by Eq. 40-31:

$$P(r) = \left(\frac{4r^2}{a^3}\right) e^{-2r/a}$$

where a is the Bohr radius.

(a) Here, $r = 0.500a$ and $\Delta r = 0.010a$. Then,

$$p = \left(\frac{4r^2\,\Delta r}{a^3}\right) e^{-2r/a} = 4(0.500)^2(0.010)\,e^{-1} = 3.68 \times 10^{-3} \ .$$

(b) We set $r = 1.00a$ and $\Delta r = 0.010a$. Then,

$$p = \left(\frac{4r^2\,\Delta r}{a^3}\right) e^{-2r/a} = 4(1.00)^2(0.010)\,e^{-2} = 5.41 \times 10^{-3} \ .$$

56. According to Fig. 40-23, the quantum number n in question satisfies $r = n^2 a$. Letting $r = 1.0\,\text{mm}$, we solve for n:

$$n = \sqrt{\frac{r}{a}} = \sqrt{\frac{1.0 \times 10^{-3}\,\text{m}}{5.29 \times 10^{-11}\,\text{m}}} \approx 4.3 \times 10^3 \ .$$

57. The radial probability function for the ground state of hydrogen is $P(r) = (4r^2/a^3)e^{-2r/a}$, where a is the Bohr radius. (See Eq. 40-31.) The integral table of Appendix E may be used to evaluate the integral $r_{\text{avg}} = \int_0^\infty rP(r)\,dr$. Setting $n = 3$ and replacing a in the given formula with $2/a$ (and x with r), we obtain

$$r_{\text{avg}} = \int_0^\infty rP(r)\,dr = \frac{4}{a^3}\int_0^\infty r^3 e^{-2r/a}\,dr = \frac{4}{a^3}\,\frac{6}{(2/a)^4} = 1.5a \ .$$

58. (a) The plot shown below for $|\psi_{200}(r)|^2$ is to be compared with the dot plot of Fig. 40-20. We note that the horizontal axis of our graph is labeled "r," but it is actually r/a (that is, it is in units of the parameter a). Now, in the plot below there is a high central peak between $r = 0$ and $r \sim 2a$, corresponding to the densely dotted region around the center of the dot plot of Fig. 40-20. Outside this peak is a region of near-zero values centered at $r = 2a$, where $\psi_{200} = 0$. This is represented in the dot plot by the empty ring surrounding the central peak. Further outside is a broader, flatter, low peak which reaches its maximum value at $r = 4a$. This corresponds to the outer ring with near-uniform dot density which is lower than that of the central peak.

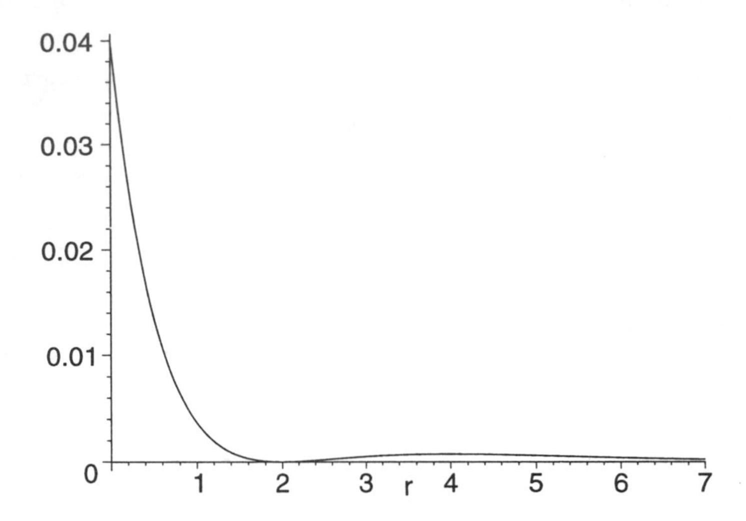

(b) The extrema of $\psi^2(r)$ for $0 < r < \infty$ may be found by squaring the given function, differentiating with respect to r, and setting the result equal to zero:

$$-\frac{1}{32}\frac{(r-2a)(r-4a)}{a^6\pi}\,e^{-r/a} = 0$$

which has roots at $r = 2a$ and $r = 4a$. We can verify directly from the plot above that $r = 4a$ is indeed a local maximum of $\psi_{200}^2(r)$. As discussed in part (a), the other root ($r = 2a$) is a local minimum.

(c) Using Eq. 40-30 and Eq. 40-28, the radial probability is

$$P_{200}(r) = 4\pi r^2\psi_{200}^2(r) = \frac{r^2}{8a^3}\left(2-\frac{r}{a}\right)^2 e^{-r/a} .$$

(d) Let $x = r/a$. Then

$$\begin{aligned}\int_0^\infty P_{200}(r)\,dr &= \int_0^\infty \frac{r^2}{8a^3}\left(2-\frac{r}{a}\right)^2 e^{-r/a}\,dr \\ &= \frac{1}{8}\int_0^\infty x^2(2-x)^2e^{-x}\,dx \\ &= \int_0^\infty (x^4-4x^3+4x^2)e^{-x}\,dx \\ &= \frac{1}{8}[4!-4(3!)+4(2!)] \\ &= 1\end{aligned}$$

where the integral formula

$$\int_0^\infty x^n\,e^{-x}\,dx = n!$$

is used.

59. (a) ψ_{210} is real. Squaring it, we obtain the probability density:

$$|\psi_{210}|^2 = \frac{r^2}{32\pi a^5}e^{-r/a}\cos^2\theta .$$

Each of the other functions is multiplied by its complex conjugate, obtained by replacing i with $-i$ in the function. Since $e^{i\phi}e^{-i\phi} = e^0 = 1$, the result is the square of the function without the exponential factor:

$$|\psi_{21+1}|^2 = \frac{r^2}{64\pi a^5} e^{-r/a} \sin^2\theta$$

and

$$|\psi_{21-1}|^2 = \frac{r^2}{64\pi a^5} e^{-r/a} \sin^2\theta \ .$$

The last two functions lead to the same probability density.

(b) The total probability density for the three states is the sum:

$$\begin{aligned} |\psi_{210}|^2 + |\psi_{21+1}|^2 + |\psi_{21-1}|^2 &= \frac{r^2}{32\pi a^5} e^{-r/a} \left[\cos^2\theta + \frac{1}{2}\sin^2\theta + \frac{1}{2}\sin^2\theta\right] \\ &= \frac{r^2}{32\pi a^5} e^{-r/a} \ . \end{aligned}$$

The trigonometric identity $\cos^2\theta + \sin^2\theta = 1$ is used. We note that the total probability density does not depend on θ or ϕ; it is spherically symmetric.

Chapter 41

1. One way to think of the units of h is that, because of the equation $E = hf$ and the fact that f is in cycles/second, then the "explicit" units for h should be J·s/cycle. Then, since 2π rad/cycle is a conversion factor for cycles → radians, $\hbar = h/2\pi$ can be thought of as the Planck constant expressed in terms of radians instead of cycles. Using the precise values stated in Appendix B,

$$\begin{aligned}\hbar &= \frac{h}{2\pi} = \frac{6.62606876 \times 10^{-34}\ \text{J}\cdot\text{s}}{2\pi} = 1.05457 \times 10^{-34}\ \text{J}\cdot\text{s} \\ &= \frac{1.05457 \times 10^{-34}\ \text{J}\cdot\text{s}}{1.6021765 \times 10^{-19}\ \text{J/eV}} = 6.582 \times 10^{-16}\ \text{eV}\cdot\text{s}\ .\end{aligned}$$

2. For a given quantum number l there are $(2l+1)$ different values of m_l. For each given m_l the electron can also have two different spin orientations. Thus, the total number of electron states for a given l is given by $N_l = 2(2l+1)$.

   (a) Now $l = 3$, so $N_l = 2(2 \times 3 + 1) = 14$.

   (b) In this case, $l = 1$, which means $N_l = 2(2 \times 1 + 1) = 6$.

   (c) Here $l = 1$, so $N_l = 2(2 \times 1 + 1) = 6$.

   (d) Now $l = 0$, so $N_l = 2(2 \times 0 + 1) = 2$.

3. (a) For a given value of the principal quantum number n, the orbital quantum number ℓ ranges from 0 to $n-1$. For $n = 3$, there are three possible values: 0, 1, and 2.

   (b) For a given value of ℓ, the magnetic quantum number m_ℓ ranges from $-\ell$ to $+\ell$. For $\ell = 1$, there are three possible values: -1, 0, and $+1$.

4. (a) We use Eq. 41-2:

$$L = \sqrt{l(l+1)}\,\hbar = \sqrt{3(3+1)}\,(1.055 \times 10^{-34}\ \text{J}\cdot\text{s}) = 3.653 \times 10^{-34}\ \text{J}\cdot\text{s}\ .$$

   (b) We use Eq. 41-7: $L_z = m_l\hbar$. For the maximum value of L_z set $m_l = l$. Thus

$$[L_z]_{\text{max}} = l\hbar = 3(1.055 \times 10^{-34}\ \text{J}\cdot\text{s}) = 3.165 \times 10^{-34}\ \text{J}\cdot\text{s}\ .$$

5. For a given quantum number n there are n possible values of l, ranging from 0 to $n-1$. For each l the number of possible electron states is $N_l = 2(2l+1)$ (see problem 2). Thus, the total number of possible electron states for a given n is

$$N_n = \sum_{l=0}^{n-1} N_l = 2\sum_{l=0}^{n-1}(2l+1) = 2n^2\ .$$

   (a) In this case $n = 4$, which implies $N_n = 2(4^2) = 32$.

(b) Now $n = 1$, so $N_n = 2(1^2) = 2$.

(c) Here $n = 3$, and we obtain $N_n = 2(3^2) = 18$.

(d) Finally, $n = 2 \rightarrow N_n = 2(2^2) = 8$.

6. Using Table 41-1, we find for $n = 4$ and $l = 3$: $m_l = +3, +2, +1, 0, -1, -2, -3$ and $m_s = \pm\frac{1}{2}$.

7. The principal quantum number n must be greater than 3. The magnetic quantum number m_ℓ can have any of the values $-3, -2, -1, 0, +1, +2$, or $+3$. The spin quantum number can have either of the values $-\frac{1}{2}$ or $+\frac{1}{2}$.

8. Using Table 41-1, we find $l = [m_l]_{\text{max}} = 4$ and $n = l_{\text{max}} + 1 \geq l + 1 = 5$. And, as usual, $m_s = \pm\frac{1}{2}$.

9. The principal quantum number n must be greater than 3. The magnetic quantum number m_l can have any of the values $-3, -2, -1, 0, +1, +2$, or $+3$. The spin quantum number can have either of the values $-\frac{1}{2}$ or $+\frac{1}{2}$.

10. For a given quantum number n there are n possible values of l, ranging from 0 to $n - 1$. For each l the number of possible electron states is $N_l = 2(2l + 1)$ (see problem 2). Thus the total number of possible electron states for a given n is

$$N_n = \sum_{l=0}^{n-1} N_l = 2\sum_{l=0}^{n-1}(2l + 1) = 2n^2 .$$

Thus, in this problem, the total number of electron states is $N_n = 2n^2 = 2(5)^2 = 50$.

11. (a) For $\ell = 3$, the magnitude of the orbital angular momentum is $L = \sqrt{\ell(\ell+1)}\hbar = \sqrt{3(3+1)}\hbar = \sqrt{12}\hbar$.

(b) The magnitude of the orbital dipole moment is $\mu_{\text{orb}} = \sqrt{\ell(\ell+1)}\mu_B = \sqrt{12}\mu_B$.

(c) We use $L_z = m_\ell\hbar$ to calculate the z component of the orbital angular momentum, $\mu_z = -m_\ell\mu_B$ to calculate the z component of the orbital magnetic dipole moment, and $\cos\theta = m_\ell/\sqrt{\ell(\ell+1)}$ to calculate the angle between the orbital angular momentum vector and the z axis. For $\ell = 3$, the magnetic quantum number m_ℓ can take on the values $-3, -2, -1, 0, +1, +2, +3$. Results are tabulated below.

m_ℓ	L_z	$\mu_{\text{orb},\ z}$	θ
-3	$-3\hbar$	$+3\mu_B$	150.0°
-2	$-2\hbar$	$+2\mu_B$	125°
-1	$-\hbar$	$+\mu_B$	107°
0	0	0	90.0°
1	$+\hbar$	$-\mu_B$	73.2°
2	$2\hbar$	$-2\mu_B$	54.7°
3	$3\hbar$	$-3\mu_B$	30.0°

12. (a) For $n = 3$ there are 3 possible values of l: 0, 1, and 2.

(b) We interpret this as asking for the number of distinct values for m_l (this ignores the multiplicity of any particular value). For each l there are $2l + 1$ possible values of m_l. Thus the number of possible m_l's for $l = 2$ is $(2l + 1) = 5$. Examining the $l = 1$ and $l = 0$ cases cannot lead to any new (distinct) values for m_l, so the answer is 5.

(c) Regardless of the values of n, l and m_l, for an electron there are always two possible values of m_s: $\pm\frac{1}{2}$.

(d) The population in the $n = 3$ shell is equal to the number of electron states in the shell, or $2n^2 = 2(3^2) = 18$.

(e) Each subshell has its own value of l. Since there are three different values of l for $n = 3$, there are three subshells in the $n = 3$ shell.

13. Since $L^2 = L_x^2 + L_y^2 + L_z^2$, $\sqrt{L_x^2 + L_y^2} = \sqrt{L^2 - L_z^2}$. Replacing L^2 with $\ell(\ell+1)\hbar^2$ and L_z with $m_\ell \hbar$, we obtain

$$\sqrt{L_x^2 + L_y^2} = \hbar\sqrt{\ell(\ell+1) - m_\ell^2} \,.$$

For a given value of ℓ, the greatest that m_ℓ can be is ℓ, so the smallest that $\sqrt{L_x^2 + L_y^2}$ can be is $\hbar\sqrt{\ell(\ell+1) - \ell^2} = \hbar\sqrt{\ell}$. The smallest possible magnitude of m_ℓ is zero, so the largest $\sqrt{L_x^2 + L_y^2}$ can be is $\hbar\sqrt{\ell(\ell+1)}$. Thus,

$$\hbar\sqrt{\ell} \le \sqrt{L_x^2 + L_y^2} \le \hbar\sqrt{\ell(\ell+1)} \,.$$

14. (a) The value of l satisfies $\sqrt{l(l+1)}\hbar \approx \sqrt{l^2}\hbar = l\hbar = L$, so $l \simeq L/\hbar \simeq 3 \times 10^{74}$.

(b) The number is $2l + 1 \approx 2(3 \times 10^{74}) = 6 \times 10^{74}$.

(c) Since

$$\cos\theta_{\text{min}} = \frac{m_{l\,\text{max}}\hbar}{\sqrt{l(l+1)}\hbar} = \frac{l}{\sqrt{l(l+1)}} \approx 1 - \frac{1}{2l} = 1 - \frac{1}{2(3 \times 10^{74})}$$

or $\cos\theta_{\text{min}} \simeq 1 - \theta_{\text{min}}^2/2 \approx 1 - 10^{-74}/6$, we have $\theta_{\text{min}} \simeq \sqrt{10^{-74}/3} = 6 \times 10^{-38}$ rad. The correspondence principle requires that all the quantum effects vanish as $\hbar \to 0$. In this case $\hbar/L$ is extremely small so the quantization effects are barely existent, with $\theta_{\text{min}} \simeq 10^{-38}$ rad $\simeq 0$.

15. The magnitude of the spin angular momentum is $S = \sqrt{s(s+1)}\,\hbar = (\sqrt{3}/2)\hbar$, where $s = \frac{1}{2}$ is used. The z component is either $S_z = \hbar/2$ or $-\hbar/2$. If $S_z = +\hbar/2$, the angle θ between the spin angular momentum vector and the positive z axis is

$$\theta = \cos^{-1}\left(\frac{S_z}{S}\right) = \cos^{-1}\left(\frac{1}{\sqrt{3}}\right) = 54.7^\circ \,.$$

If $S_z = -\hbar/2$, the angle is $\theta = 180^\circ - 54.7^\circ = 125.3^\circ$.

16. (a) From Fig. 41-10 and Eq. 41-18,

$$\Delta E = 2\mu_B B = \frac{2(9.27 \times 10^{-24}\,\text{J/T})(0.50\,\text{T})}{1.60 \times 10^{-19}\,\text{J/eV}} = 58\,\mu\text{eV} \,.$$

(b) From $\Delta E = hf$ we get

$$f = \frac{\Delta E}{h} = \frac{9.27 \times 10^{-24}\,\text{J}}{6.63 \times 10^{-34}\,\text{J}\cdot\text{s}} = 1.4 \times 10^{10}\,\text{Hz} = 14\text{ GHz} \,.$$

(c) The wavelength is

$$\lambda = \frac{c}{f} = \frac{2.998 \times 10^{8}\,\text{m/s}}{1.4 \times 10^{10}\,\text{Hz}} = 2.1\text{ cm} \,,$$

which is in the short radio wave region.

17. The acceleration is

$$a = \frac{F}{M} = \frac{(\mu \cos\theta)\,(dB/dz)}{M}\,,$$

where M is the mass of a silver atom, μ is its magnetic dipole moment, B is the magnetic field, and θ is the angle between the dipole moment and the magnetic field. We take the moment and the field to be parallel ($\cos\theta = 1$) and use the data given in Sample Problem 41-1 to obtain

$$a = \frac{(9.27 \times 10^{-24}\,\text{J/T})(1.4 \times 10^{3}\,\text{T/m})}{1.8 \times 10^{-25}\,\text{kg}} = 7.21 \times 10^{4}\ \text{m/s}^2\ .$$

18. (a) From Eq. 41-19,

$$F = \mu_B \left|\frac{dB}{dz}\right| = (9.27 \times 10^{-24}\,\text{J/T})(1.6 \times 10^{2}\,\text{T/m}) = 1.5 \times 10^{-21}\ \text{N}\ .$$

(b) The vertical displacement is

$$\begin{aligned} \Delta x &= \frac{1}{2}at^2 = \frac{1}{2}\left(\frac{F}{m}\right)\left(\frac{l}{v}\right)^2 \\ &= \frac{1}{2}\left(\frac{1.5 \times 10^{-21}\,\text{N}}{1.67 \times 10^{-27}\,\text{kg}}\right)\left(\frac{0.80\,\text{m}}{1.2 \times 10^{5}\,\text{m/s}}\right)^2 \\ &= 2.0 \times 10^{-5}\ \text{m}\ . \end{aligned}$$

19. The energy of a magnetic dipole in an external magnetic field $\vec{B}$ is $U = -\vec{\mu}\cdot\vec{B} = -\mu_z B$, where $\vec{\mu}$ is the magnetic dipole moment and μ_z is its component along the field. The energy required to change the moment direction from parallel to antiparallel is $\Delta E = \Delta U = 2\mu_z B$. Since the z component of the spin magnetic moment of an electron is the Bohr magneton μ_B, $\Delta E = 2\mu_B B = 2(9.274 \times 10^{-24}\,\text{J/T})(0.200\,\text{T}) = 3.71 \times 10^{-24}\,\text{J}$. The photon wavelength is

$$\lambda = \frac{c}{f} = \frac{hc}{\Delta E} = \frac{(6.63 \times 10^{-34}\,\text{J}\cdot\text{s})(3.00 \times 10^{8}\,\text{m/s})}{3.71 \times 10^{-24}\,\text{J}} = 5.36 \times 10^{-2}\ \text{m}\ .$$

20. We let $\Delta E = 2\mu_B B_{\text{eff}}$ (based on Fig. 41-10 and Eq. 41-18) and solve for B_{eff}:

$$B_{\text{eff}} = \frac{\Delta E}{2\mu_B} = \frac{hc}{2\lambda\mu_B} = \frac{1240\,\text{nm}\cdot\text{eV}}{2(21 \times 10^{-7}\,\text{nm})(5.788 \times 10^{-5}\,\text{eV/T})} = 51\ \text{mT}\ .$$

21. (a) Using the result of problem 3 in Chapter 39,

$$\Delta E = hc\left(\frac{1}{\lambda_1} - \frac{1}{\lambda_2}\right) = (1240\,\text{eV}\cdot\text{nm})\left(\frac{1}{588.995\,\text{nm}} - \frac{1}{589.592\,\text{nm}}\right) = 2.13\ \text{meV}\ .$$

(b) From $\Delta E = 2\mu_B B$ (see Fig. 41-10 and Eq. 41-18), we get

$$B = \frac{\Delta E}{2\mu_B} = \frac{2.13 \times 10^{-3}\,\text{eV}}{2(5.788 \times 10^{-5}\,\text{eV/T})} = 18\ \text{T}\ .$$

22. The total magnetic field, $B = B_{\text{local}} + B_{\text{ext}}$, satisfies $\Delta E = hf = 2\mu B$ (see Eq. 41-22). Thus,

$$B_{\text{local}} = \frac{hf}{2\mu} - B_{\text{ext}} = \frac{(6.63 \times 10^{-34}\text{J}\cdot\text{s})(34 \times 10^{6}\,\text{Hz})}{2(1.41 \times 10^{-26}\,\text{J/T})} - 0.78\,\text{T} = 19\ \text{mT}\ .$$

23. Because of the Pauli principle (and the requirement that we construct a state of lowest possible total energy), two electrons fill the $n = 1, 2, 3$ levels and one electron occupies the $n = 4$ level. Thus, using Eq. 40-4,

$$\begin{aligned} E_{\text{ground}} &= 2E_1 + 2E_2 + 2E_3 + E_4 \\ &= 2\left(\frac{h^2}{8mL^2}\right)(1)^2 + 2\left(\frac{h^2}{8mL^2}\right)(2)^2 + 2\left(\frac{h^2}{8mL^2}\right)(3)^2 + \left(\frac{h^2}{8mL^2}\right)(4)^2 \\ &= (2 + 8 + 18 + 16)\left(\frac{h^2}{8mL^2}\right) = 44\left(\frac{h^2}{8mL^2}\right) . \end{aligned}$$

24. Using Eq. 40-20 (see also problem 27 in Chapter 40) we find that the lowest four levels of the rectangular corral (with this specific "aspect ratio") are non-degenerate, with energies $E_{1,1} = 1.25$, $E_{1,2} = 2.00$, $E_{1,3} = 3.25$, and $E_{2,1} = 4.25$ (all of these understood to be in "units" of $h^2/8mL^2$). Therefore, obeying the Pauli principle, we have

$$E_{\text{ground}} = 2E_{1,1} + 2E_{1,2} + 2E_{1,3} + E_{2,1} = 2(1.25) + 2(2.00) + 2(3.25) + 4.25$$

which means (putting the "unit" factor back in) that the lowest possible energy of the system is $E_{\text{ground}} = 17.25(h^2/8mL^2)$.

25. (a) Promoting one of the electrons (described in problem 23) to a not-fully occupied higher level, we find that the configuration with the least total energy greater than that of the ground state has the $n = 1$ and 2 levels still filled, but now has only one electron in the $n = 3$ level; the remaining two electrons are in the $n = 4$ level. Thus,

$$\begin{aligned} E_{\text{first excited}} &= 2E_1 + 2E_2 + E_3 + 2E_4 \\ &= 2\left(\frac{h^2}{8mL^2}\right)(1)^2 + 2\left(\frac{h^2}{8mL^2}\right)(2)^2 + \left(\frac{h^2}{8mL^2}\right)(3)^2 + 2\left(\frac{h^2}{8mL^2}\right)(4)^2 \\ &= (2 + 8 + 9 + 32)\left(\frac{h^2}{8mL^2}\right) = 51\left(\frac{h^2}{8mL^2}\right) . \end{aligned}$$

(b) Now, the configuration which provides the next higher total energy, above that found in part (a), has the bottom three levels filled (just as in the ground state configuration) and has the seventh electron occupying the $n = 5$ level:

$$\begin{aligned} E_{\text{second excited}} &= 2E_1 + 2E_2 + 2E_3 + E_5 \\ &= 2\left(\frac{h^2}{8mL^2}\right)(1)^2 + 2\left(\frac{h^2}{8mL^2}\right)(2)^2 + 2\left(\frac{h^2}{8mL^2}\right)(3)^2 + \left(\frac{h^2}{8mL^2}\right)(5)^2 \\ &= (2 + 8 + 18 + 25)\left(\frac{h^2}{8mL^2}\right) = 53\left(\frac{h^2}{8mL^2}\right) . \end{aligned}$$

(c) The third excited state has the $n = 1, 3, 4$ levels filled, and the $n = 2$ level half-filled:

$$\begin{aligned} E_{\text{third excited}} &= 2E_1 + E_2 + 2E_3 + 2E_4 \\ &= 2\left(\frac{h^2}{8mL^2}\right)(1)^2 + \left(\frac{h^2}{8mL^2}\right)(2)^2 + 2\left(\frac{h^2}{8mL^2}\right)(3)^2 + 2\left(\frac{h^2}{8mL^2}\right)(4)^2 \\ &= (2 + 4 + 18 + 32)\left(\frac{h^2}{8mL^2}\right) = 56\left(\frac{h^2}{8mL^2}\right) . \end{aligned}$$

(d) The energy states of this problem and problem 23 are suggested in the sketch below:

third excited $56(h^2/8mL^2)$

second excited $53(h^2/8mL^2)$

first excited $51(h^2/8mL^2)$

ground state $44(h^2/8mL^2)$

26. (a) Using Eq. 40-20 (see also problem 27 in Chapter 40) we find that the lowest five levels of the rectangular corral (with this specific "aspect ratio") have energies $E_{1,1} = 1.25$, $E_{1,2} = 2.00$, $E_{1,3} = 3.25$, $E_{2,1} = 4.25$, and $E_{2,2} = 5.00$ (all of these understood to be in "units" of $h^2/8mL^2$). It should be noted that the energy level we denote $E_{2,2}$ actually corresponds to two energy levels ($E_{2,2}$ and $E_{1,4}$; they are degenerate), but that will not affect our calculations in this problem. The configuration which provides the lowest system energy higher than that of the ground state has the first three levels filled, the fourth one empty, and the fifth one half-filled:

$$E_{\text{first excited}} = 2E_{1,1} + 2E_{1,2} + 2E_{1,3} + E_{2,2} = 2(1.25) + 2(2.00) + 2(3.25) + 5.00$$

which means (putting the "unit" factor back in) the energy of the first excited state is $E_{\text{first excited}} = 18.00(h^2/8mL^2)$.

(b) The configuration which provides the next higher system energy has the first two levels filled, the third one half-filled, and the fourth one filled:

$$E_{\text{second excited}} = 2E_{1,1} + 2E_{1,2} + E_{1,3} + 2E_{2,1} = 2(1.25) + 2(2.00) + 3.25 + 2(4.25)$$

which means (putting the "unit" factor back in) the energy of the second excited state is $E_{\text{second excited}} = 18.25(h^2/8mL^2)$.

(c) Now, the configuration which provides the *next* higher system energy has the first two levels filled, with the next three levels half-filled:

$$E_{\text{third excited}} = 2E_{1,1} + 2E_{1,2} + E_{1,3} + E_{2,1} + E_{2,2} = 2(1.25) + 2(2.00) + 3.25 + 4.25 + 5.00$$

which means (putting the "unit" factor back in) the energy of the third excited state is $E_{\text{third excited}} = 19.00(h^2/8mL^2)$.

(d) The energy states of this problem and problem 24 are suggested in the sketch below:

third excited $19.00(h^2/8mL^2)$

second excited $18.25(h^2/8mL^2)$

first excited $18.00(h^2/8mL^2)$

ground state $17.25(h^2/8mL^2)$

27. In terms of the quantum numbers n_x, n_y, and n_z, the single-particle energy levels are given by

$$E_{n_x,n_y,n_z} = \frac{h^2}{8mL^2}\left(n_x^2 + n_y^2 + n_z^2\right) .$$

The lowest single-particle level corresponds to $n_x = 1$, $n_y = 1$, and $n_z = 1$ and is $E_{1,1,1} = 3(h^2/8mL^2)$. There are two electrons with this energy, one with spin up and one with spin down. The next lowest single-particle level is three-fold degenerate in the three integer quantum numbers. The energy is $E_{1,1,2} = E_{1,2,1} = E_{2,1,1} = 6(h^2/8mL^2)$. Each of these states can be occupied by a spin up and a spin down electron, so six electrons in all can occupy the states. This completes the assignment of the eight electrons to single-particle states. The ground state energy of the system is $E_{\text{gr}} = (2)(3)(h^2/8mL^2) + (6)(6)(h^2/8mL^2) = 42(h^2/8mL^2)$.

28. We use the results of problem 28 in Chapter 40. The Pauli principle requires that no more than two electrons be in the lowest energy level (at $E_{1,1,1} = 3(h^2/8mL^2)$), but – due to their degeneracies – as many as six electrons can be in the next three levels ($E' = E_{1,1,2} = E_{1,2,1} = E_{2,1,1} = 6(h^2/8mL^2)$, $E'' = E_{1,2,2} = E_{2,2,1} = E_{2,1,2} = 9(h^2/8mL^2)$, and $E''' = E_{1,1,3} = E_{1,3,1} = E_{3,1,1} = 11(h^2/8mL^2)$). Using Eq. 40-21, the level above those can only hold two electrons: $E_{2,2,2} = (2^2 + 2^2 + 2^2)(h^2/8mL^2) = 12(h^2/8mL^2)$. And the next higher level can hold as much as twelve electrons (see part (e) of problem 28 in Chapter 40) and has energy $E'''' = 14(h^2/8mL^2)$.

(a) The configuration which provides the lowest system energy higher than that of the ground state has the first level filled, the second one with one vacancy, and the third one with one occupant:

$$E_{\text{first excited}} = 2E_{1,1,1} + 5E' + E'' = 2(3) + 5(6) + 9$$

which means (putting the "unit" factor back in) the energy of the first excited state is $E_{\text{first excited}} = 45(h^2/8mL^2)$.

(b) The configuration which provides the next higher system energy has the first level filled, the second one with one vacancy, the third one empty, and the fourth one with one occupant:

$$E_{\text{second excited}} = 2E_{1,1,1} + 5E' + E'' = 2(3) + 5(6) + 11$$

which means (putting the "unit" factor back in) the energy of the second excited state is $E_{\text{second excited}} = 47(h^2/8mL^2)$.

(c) Now, there are a couple of configurations which provides the *next* higher system energy. One has the first level filled, the second one with one vacancy, the third and fourth ones empty, and the fifth one with one occupant:

$$E_{\text{third excited}} = 2E_{1,1,1} + 5E' + E''' = 2(3) + 5(6) + 12$$

which means (putting the "unit" factor back in) the energy of the third excited state is $E_{\text{third excited}} = 48(h^2/8mL^2)$. The other configuration with this same total energy has the first level filled, the second one with two vacancies, and the third one with one occupant.

(d) The energy states of this problem and problem 27 are suggested in the sketch below:

third excited $48(h^2/8mL^2)$

second excited $47(h^2/8mL^2)$

first excited $45(h^2/8mL^2)$

ground state $42(h^2/8mL^2)$

29. For a given shell with quantum number n the total number of available electron states is $2n^2$. Thus, for the first four shells ($n = 1$ through 4) the number of available states are 2, 8, 18, and 32 (see Appendix G). Since $2 + 8 + 18 + 32 = 60 < 63$, according to the "logical" sequence the first four shells would be completely filled in an europium atom, leaving $63 - 60 = 3$ electrons to partially occupy the $n = 5$ shell. Two of these three electrons would fill up the $5s$ subshell, leaving only one remaining electron in the only partially filled subshell (the $5p$ subshell). In chemical reactions this electron would have the tendency to be transferred to another element, leaving the remaining 62 electrons in chemically stable, completely filled subshells. This situation is very similar to the case of sodium, which also has only one electron in a partially filled shell (the $3s$ shell).

30. The first three shells ($n = 1$ through 3), which can accommodate a total of $2 + 8 + 18 = 28$ electrons, are completely filled. For selenium ($Z = 34$) there are still $34 - 28 = 6$ electrons left. Two of them go to the $4s$ subshell, leaving the remaining four in the highest occupied subshell, the $4p$ subshell. Similarly, for bromine ($Z = 35$) the highest occupied subshell is also the $4p$ subshell, which contains five electrons; and for krypton ($Z = 36$) the highest occupied subshell is also the $4p$ subshell, which now accommodates six electrons.

31. Without the spin degree of freedom the number of available electron states for each shell would be reduced by half. So the values of Z for the noble gas elements would become half of what they are now: $Z = 1, 5, 9, 18, 27$, and 43. Of this set of numbers, the only one which coincides with one of the familiar noble gas atomic numbers ($Z = 2, 10, 18, 36, 54$, and 86) is 18. Thus, argon would be the only one that would remain "noble."

32. When a helium atom is in its ground state, both of its electrons are in the $1s$ state. Thus, for each of the electrons, $n = 1$, $l = 0$, and $m_l = 0$. One of the electrons is spin up ($m_s = +\frac{1}{2}$), while the other is spin down ($m_s = -\frac{1}{2}$).

33. (a) All states with principal quantum number $n = 1$ are filled. The next lowest states have $n = 2$. The orbital quantum number can have the values $\ell = 0$ or 1 and of these, the $\ell = 0$ states have the lowest energy. The magnetic quantum number must be $m_\ell = 0$ since this is the only possibility if $\ell = 0$. The spin quantum number can have either of the values $m_s = -\frac{1}{2}$ or $+\frac{1}{2}$. Since there is

no external magnetic field, the energies of these two states are the same. Therefore, in the ground state, the quantum numbers of the third electron are either $n = 2$, $\ell = 0$, $m_\ell = 0$, $m_s = -\frac{1}{2}$ or $n = 2$, $\ell = 0$, $m_\ell = 0$, $m_s = +\frac{1}{2}$.

(b) The next lowest state in energy is an $n = 2$, $\ell = 1$ state. All $n = 3$ states are higher in energy. The magnetic quantum number can be $m_\ell = -1$, 0, or $+1$; the spin quantum number can be $m_s = -\frac{1}{2}$ or $+\frac{1}{2}$. If both external and internal magnetic fields can be neglected, all these states have the same energy.

34. (a) The number of different m_l's is $2l + 1 = 3$, and the number of different m_s's is 2. Thus, the number of combinations is $N = (3 \times 2)^2/2 = 18$.

(b) There are six states disallowed by the exclusion principle, in which both electrons share the quantum numbers

$$(n, l, m_l, m_s) = \left(2, 1, 1, \frac{1}{2}\right), \left(2, 1, 1, -\frac{1}{2}\right), \left(2, 1, 0, \frac{1}{2}\right), \left(2, 1, 0, -\frac{1}{2}\right), \left(2, 1, -1, \frac{1}{2}\right), \left(2, 1, -1, -\frac{1}{2}\right).$$

35. For a given value of the principal quantum number n, there are n possible values of the orbital quantum number ℓ, ranging from 0 to $n - 1$. For any value of ℓ, there are $2\ell + 1$ possible values of the magnetic quantum number m_ℓ, ranging from $-\ell$ to $+\ell$. Finally, for each set of values of ℓ and m_ℓ, there are two states, one corresponding to the spin quantum number $m_s = -\frac{1}{2}$ and the other corresponding to $m_s = +\frac{1}{2}$. Hence, the total number of states with principal quantum number n is

$$N = 2\sum_0^{n-1}(2\ell + 1) .$$

Now

$$\sum_0^{n-1} 2\ell = 2\sum_0^{n-1} \ell = 2\frac{n}{2}(n - 1) = n(n - 1) ,$$

since there are n terms in the sum and the average term is $(n - 1)/2$. Furthermore,

$$\sum_0^{n-1} 1 = n .$$

Thus $N = 2\,[n(n - 1) + n] = 2n^2$.

36. The kinetic energy gained by the electron is eV, where V is the accelerating potential difference. A photon with the minimum wavelength (which, because of $E = hc/\lambda$, corresponds to maximum photon energy) is produced when all of the electron's kinetic energy goes to a single photon in an event of the kind depicted in Fig. 41-15. Thus, using the result of problem 3 in Chapter 39,

$$eV = \frac{hc}{\lambda_{\text{min}}} = \frac{1240\,\text{eV}\cdot\text{nm}}{0.10\,\text{nm}} = 1.24 \times 10^4 \text{ eV} .$$

Therefore, the accelerating potential difference is $V = 1.24 \times 10^4\text{ V} = 12.4\,\text{kV}$.

37. We use $eV = hc/\lambda_{\text{min}}$ (see Eq. 41-23 and Eq. 39-4):

$$h = \frac{eV\lambda_{\text{min}}}{c} = \frac{(1.60 \times 10^{-19}\,\text{C})(40.0 \times 10^3\,\text{eV})(31.1 \times 10^{-12}\,\text{m})}{2.998 \times 10^8\,\text{m/s}} = 6.63 \times 10^{-34}\text{ J}\cdot\text{s} .$$

38. Letting $eV = hc/\lambda_{\text{min}}$ (see Eq. 41-23 and Eq. 39-4), we get

$$\lambda_{\text{min}} = \frac{hc}{eV} = \frac{1240\,\text{nm}\cdot\text{eV}}{eV} = \frac{1240\,\text{pm}\cdot\text{keV}}{eV} = \frac{1240\,\text{pm}}{V}$$

where V is measured in kV.

39. The initial kinetic energy of the electron is 50.0 keV. After the first collision, the kinetic energy is 25 keV; after the second, it is 12.5 keV; and after the third, it is zero. The energy of the photon produced in the first collision is 50.0 keV − 25.0 keV = 25.0 keV. The wavelength associated with this photon is

$$\lambda = \frac{1240\,\text{eV}\cdot\text{nm}}{25.0\times10^3\,\text{eV}} = 4.96\times10^{-2}\,\text{nm} = 49.6\,\text{pm}$$

where the result of Exercise 3 of Chapter 39 is used. The energies of the photons produced in the second and third collisions are each 12.5 keV and their wavelengths are

$$\lambda = \frac{1240\,\text{eV}\cdot\text{nm}}{12.5\times10^3\,\text{eV}} = 9.92\times10^{-2}\,\text{nm} = 99.2\,\text{pm}\ .$$

40. (a) and (b) Let the wavelength of the two photons be λ_1 and $\lambda_2 = \lambda_1 + \Delta\lambda$. Then,

$$eV = \frac{hc}{\lambda_1} + \frac{hc}{\lambda_1 + \Delta\lambda}\ ,$$

or

$$\lambda_1 = \frac{-(\Delta\lambda/\lambda_0 - 2) \pm \sqrt{(\Delta\lambda/\lambda_0)^2 + 4}}{2/\Delta\lambda}\ .$$

Here, $\Delta\lambda = 130\,\text{pm}$ and $\lambda_0 = hc/eV = 1240\,\text{keV}\cdot\text{pm}/20\,\text{keV} = 62\,\text{pm}$. The result of problem 3 in Chapter 39 is adapted to these units ($hc = 1240\,\text{eV}\cdot\text{nm} = 1240\,\text{keV}\cdot\text{pm}$). We choose the plus sign in the expression for λ_1 (since $\lambda_1 > 0$) and obtain

$$\lambda_1 = \frac{-(130\,\text{pm}/62\,\text{pm} - 2) + \sqrt{(130\,\text{pm}/62\,\text{pm})^2 + 4}}{2/62\,\text{pm}} = 87\ \text{pm}\ ,$$

and

$$\lambda_2 = \lambda_1 + \Delta\lambda = 87\,\text{pm} + 130\,\text{pm} = 2.2\times10^2\ \text{pm}\ .$$

The energy of the electron after its first deceleration is

$$K = K_i - \frac{hc}{\lambda_1} = 20\,\text{keV} - \frac{1240\,\text{keV}\cdot\text{pm}}{87\,\text{pm}} = 5.7\ \text{keV}.$$

The energies of the two photons are

$$E_1 = \frac{hc}{\lambda_1} = \frac{1240\,\text{keV}\cdot\text{pm}}{87\,\text{pm}} = 14\ \text{keV}$$

and

$$E_2 = \frac{hc}{\lambda_2} = \frac{1240\,\text{keV}\cdot\text{pm}}{130\,\text{pm}} = 5.7\ \text{keV}\ .$$

41. Suppose an electron with total energy E and momentum $\mathbf{p}$ spontaneously changes into a photon. If energy is conserved, the energy of the photon is E and its momentum has magnitude E/c. Now the energy and momentum of the electron are related by $E^2 = (pc)^2 + (mc^2)^2$, so $pc = \sqrt{E^2 - (mc^2)^2}$. Since the electron has non-zero mass, E/c and p cannot have the same value. Hence, momentum cannot be conserved. A third particle must participate in the interaction, primarily to conserve momentum. It does, however, carry off some energy.

42. (a) We use $eV = hc/\lambda_{\text{min}}$ (see Eq. 41-23 and Eq. 39-4). The result of problem 3 in Chapter 39 is adapted to these units ($hc = 1240\,\text{eV}\cdot\text{nm} = 1240\,\text{keV}\cdot\text{pm}$).

$$\lambda_{\text{min}} = \frac{hc}{eV} = \frac{1240\,\text{keV}\cdot\text{pm}}{50.0\,\text{keV}} = 24.8\ \text{pm}\ .$$

(b) and (c) The values of λ for the K_α and K_β lines do not depend on the external potential and are therefore unchanged.

43. (a) The cut-off wavelength λ_{min} is characteristic of the incident electrons, not of the target material. This wavelength is the wavelength of a photon with energy equal to the kinetic energy of an incident electron. According to the result of Exercise 3 of Chapter 39,

$$\lambda_{\text{min}} = \frac{1240\,\text{eV}\cdot\text{nm}}{35 \times 10^3\,\text{eV}} = 3.54 \times 10^{-2}\,\text{nm} = 35.4\ \text{pm}\ .$$

(b) A K_α photon results when an electron in a target atom jumps from the L-shell to the K-shell. The energy of this photon is $25.51\,\text{keV} - 3.56\,\text{keV} = 21.95\,\text{keV}$ and its wavelength is $\lambda_{\text{K}\alpha} = (1240\,\text{eV}\cdot\text{nm})/(21.95 \times 10^3\,\text{eV}) = 5.65 \times 10^{-2}\,\text{nm} = 56.5\,\text{pm}$.

(c) A K_β photon results when an electron in a target atom jumps from the M-shell to the K-shell. The energy of this photon is $25.51\,\text{keV} - 0.53\,\text{keV} = 24.98\,\text{keV}$ and its wavelength is $\lambda_{\text{K}\beta} = (1240\,\text{eV}\cdot\text{nm})/(24.98 \times 10^3\,\text{eV}) = 4.96 \times 10^{-2}\,\text{nm} = 49.6\,\text{pm}$.

44. The result of problem 3 in Chapter 39 is adapted to these units ($hc = 1240\,\text{eV}\cdot\text{nm} = 1240\,\text{keV}\cdot\text{pm}$). For the K_α line from iron

$$\Delta E = \frac{hc}{\lambda} = \frac{1240\,\text{keV}\cdot\text{pm}}{193\,\text{pm}} = 6.4\ \text{keV}\ .$$

We remark that for the hydrogen atom the corresponding energy difference is

$$\Delta E_{12} = -(13.6\,\text{eV})\left(\frac{1}{2^2} - \frac{1}{1^1}\right) = 10\ \text{eV}.$$

That this difference is much greater in iron is due to the fact that its atomic nucleus contains 26 protons, exerting a much greater force on the K- and L-shell electrons than that provided by the single proton in hydrogen.

45. Since the frequency of an x-ray emission is proportional to $(Z-1)^2$, where Z is the atomic number of the target atom, the ratio of the wavelength λ_{Nb} for the K_α line of niobium to the wavelength λ_{Ga} for the K_α line of gallium is given by $\lambda_{\text{Nb}}/\lambda_{\text{Ga}} = (Z_{\text{Ga}}-1)^2/(Z_{\text{Nb}}-1)^2$, where Z_{Nb} is the atomic number of niobium (41) and Z_{Ga} is the atomic number of gallium (31). Thus $\lambda_{\text{Nb}}/\lambda_{\text{Ga}} = (30)^2/(40)^2 = 9/16$.

46. The result of problem 3 in Chapter 39 is adapted to these units ($hc = 1240\,\text{eV}\cdot\text{nm} = 1240\,\text{keV}\cdot\text{pm}$). The energy difference $E_L - E_M$ for the x-ray atomic energy levels of molybdenum is

$$\Delta E = E_L - E_M = \frac{hc}{\lambda_L} - \frac{hc}{\lambda_M} = \frac{1240\,\text{keV}\cdot\text{pm}}{63.0\,\text{pm}} - \frac{1240\,\text{keV}\cdot\text{pm}}{71.0\,\text{pm}} = 2.2\ \text{keV}\ .$$

47. From the data given in the problem, we calculate frequencies (using Eq. 39-1), take their square roots, look up the atomic numbers (see Appendix F), and do a least-squares fit to find the slope: the result is 5.02×10^7 with the odd-sounding unit of a square root of a Hertz. We remark that the least squares procedure also returns a value for the y-intercept of this statistically determined "best-fit" line; that result is negative and would appear on a graph like Fig. 41-17 to be at about -0.06 on the vertical axis. Also, we can estimate the slope of the Moseley line shown in Fig. 41-17:

$$\frac{(1.95 - 0.50)10^9\,\text{Hz}^{1/2}}{40 - 11} \approx 5.0 \times 10^7\,\text{Hz}^{1/2}\ .$$

These are in agreement with the discussion in §41-10.

48. (a) From Fig. 41-14 we estimate the wavelengths corresponding to the K_α and K_β lines to be $\lambda_\alpha = 70.0\,\text{pm}$ and $\lambda_\beta = 63.0\,\text{pm}$, respectively. Using the result of problem 3 in Chapter 39, adapted to these units ($hc = 1240\,\text{eV}\cdot\text{nm} = 1240\,\text{keV}\cdot\text{pm}$),

$$E_\alpha = \frac{hc}{\lambda_\alpha} = \frac{1240\,\text{keV}\cdot\text{pm}}{70.0\,\text{pm}} = 17.7\ \text{keV}\ ,$$

and $E_\beta = (1240\,\text{keV}\cdot\text{nm})/(63.0\,\text{pm}) = 19.7\,\text{keV}$.

(b) Both Zr and Nb can be used, since $E_\alpha < 18.00\,\text{eV} < E_\beta$ and $E_\alpha < 18.99\,\text{eV} < E_\beta$. According to the hint given in the problem statement, Zr is the better choice.

49. (a) An electron must be removed from the K-shell, so that an electron from a higher energy shell can drop. This requires an energy of 69.5 keV. The accelerating potential must be at least 69.5 kV.

(b) After it is accelerated, the kinetic energy of the bombarding electron is 69.5 keV. The energy of a photon associated with the minimum wavelength is 69.5 keV, so its wavelength is

$$\lambda_{\text{min}} = \frac{1240\,\text{eV}\cdot\text{nm}}{69.5\times10^3\,\text{eV}} = 1.78\times10^{-2}\,\text{nm} = 17.8\,\text{pm}\ .$$

(c) The energy of a photon associated with the K_α line is $69.5\,\text{keV} - 11.3\,\text{keV} = 58.2\,\text{keV}$ and its wavelength is $\lambda_{K\alpha} = (1240\,\text{eV}\cdot\text{nm})/(58.2\times10^3\,\text{eV}) = 2.13\times10^{-2}\,\text{nm} = 21.3\,\text{pm}$. The energy of a photon associated with the K_β line is $69.5\,\text{keV} - 2.30\,\text{keV} = 67.2\,\text{keV}$ and its wavelength is $\lambda_{K\beta} = (1240\,\text{eV}\cdot\text{nm})/(67.2\times10^3\,\text{eV}) = 1.85\times10^{-2}\,\text{nm} = 18.5\,\text{pm}$. The result of Exercise 3 of Chapter 39 is used.

50. We use Eq. 37-31, Eq. 40-6, and the result of problem 3 in Chapter 39, adapted to these units ($hc = 1240\,\text{eV}\cdot\text{nm} = 1240\,\text{keV}\cdot\text{pm}$). Letting $2d\sin\theta = m\lambda = mhc/\Delta E$, where $\theta = 74.1°$, we solve for d:

$$d = \frac{mhc}{2\Delta E\sin\theta} = \frac{(1)(1240\,\text{keV}\cdot\text{nm})}{2(8.979\,\text{keV} - 0.951\,\text{keV})(\sin 74.1°)} = 80.3\ \text{pm}\ .$$

51. (a) According to Eq. 41-26, $f \propto (Z-1)^2$, so the ratio of energies is (using Eq. 39-2) $f/f' = [(Z-1)/(Z'-1)]^2$.

(b) We refer to Appendix F. Applying the formula from part (a) to $Z = 92$ and $Z' = 13$, we obtain

$$\frac{E}{E'} = \frac{f}{f'} = \left(\frac{Z-1}{Z'-1}\right)^2 = \left(\frac{92-1}{13-1}\right)^2 = 57.5\ .$$

(c) Applying this to $Z = 92$ and $Z' = 3$, we obtain

$$\frac{E}{E'} = \left(\frac{92-1}{3-1}\right)^2 = 2070\ .$$

52. (a) The transition is from $n = 2$ to $n = 1$, so Eq. 41-26 combined with Eq. 41-24 yields

$$f = \left(\frac{m_e e^4}{8\varepsilon_0^2 h^3}\right)\left(\frac{1}{1^2} - \frac{1}{2^2}\right)(Z-1)^2$$

so that the constant in Eq. 41-27 is

$$C = \sqrt{\frac{3m_e e^4}{32\varepsilon_0^2 h^3}} = 4.9673\times10^7\ \text{Hz}^{1/2}$$

using the values in the next-to-last column in the Table in Appendix B (but note that the power of ten is given in the middle column).

(b) We are asked to compare the results of Eq. 41-27 (squared, then multiplied by the accurate values of h/e found in Appendix B to convert to x ray energies) with those in the table of K_α energies (in eV) given at the end of the problem. We look up the corresponding atomic numbers in Appendix F. An example is shown below (for Nitrogen):

$$E_{\text{theory}} = \frac{h}{e}C^2(Z-1)^2 = \frac{6.6260688\times10^{-34}\,\text{J}\cdot\text{s}}{1.6021765\times10^{-19}\,\text{J/eV}}\left(4.9673\times10^7\ \text{Hz}^{1/2}\right)^2(7-1)^2 = 367.35\ \text{eV}$$

which is 6.4% lower than the experimental value of 392.4 eV. Progressing through the list, from Lithium to Magnesium , we find all the theoretical values are lower than the experimental ones by these percentages: 24.8%, 15.4%, 10.9%, 7.9%, 6.4%, 4.7%, 3.5%, 2.6%, 2.0%, and 1.5%.

(c) The trend is clear from the list given above: the agreement between theory and experiment becomes better as Z increases. One might argue that the most questionable step in §41-10 is the replacement $e^4 \to (Z-1)^2 e^4$ and ask why this could not equally well be $e^4 \to (Z-.9)^2 e^4$ or $e^4 \to (Z-.8)^2 e^4$? For large Z, these subtleties would not matter so much as they do for small Z, since $Z - \xi \approx Z$ for $Z \gg \xi$.

53. (a) The length of the pulse's wave train is given by $L = c\Delta t = (2.998 \times 10^8\,\text{m/s})(10 \times 10^{-15}\,\text{s}) = 3.0 \times 10^{-6}\,\text{m}$. Thus, the number of wavelengths contained in the pulse is

$$N = \frac{L}{\lambda} = \frac{3.0 \times 10^{-6}\,\text{m}}{500 \times 10^{-9}\,\text{m}} = 6.0\ .$$

(b) We solve for X from $10\,\text{fm}/1\,\text{m} = 1\,\text{s}/X$:

$$X = \frac{(1\,\text{s})(1\,\text{m})}{10 \times 10^{-15}\,\text{m}} = \frac{1\,\text{s}}{(10 \times 10^{-15})(3.15 \times 10^7\,\text{s/y})} = 3.2 \times 10^6\text{y}\ .$$

54. According to Sample Problem 41-6, $N_x/N_0 = 1.3 \times 10^{-38}$. Let the number of moles of the lasing material needed be n; then $N_0 = nN_A$, where N_A is the Avogadro constant. Also $N_x = 10$. We solve for n:

$$n = \frac{N_x}{(1.3 \times 10^{-38})\,N_A} = \frac{10}{(1.3 \times 10^{-38})(6.02 \times 10^{23})} = 1.3 \times 10^{15}\ \text{mol}\ .$$

55. The number of atoms in a state with energy E is proportional to $e^{-E/kT}$, where T is the temperature on the Kelvin scale and k is the Boltzmann constant. Thus the ratio of the number of atoms in the thirteenth excited state to the number in the eleventh excited state is

$$\frac{n_{13}}{n_{11}} = e^{-\Delta E/kT}\ ,$$

where ΔE is the difference in the energies: $\Delta E = E_{13} - E_{11} = 2(1.2\,\text{eV}) = 2.4\,\text{eV}$. For the given temperature, $kT = (8.62 \times 10^{-2}\,\text{eV/K})(2000\,\text{K}) = 0.1724\,\text{eV}$. Hence,

$$\frac{n_{13}}{n_{11}} = e^{-2.4/0.1724} = 9.0 \times 10^{-7}\ .$$

56. (a) The distance from the Earth to the Moon is $d_{em} = 3.82 \times 10^8\,\text{m}$ (see Appendix C). Thus, the time required is given by

$$t = \frac{2d_{em}}{c} = \frac{2(3.82 \times 10^8\,\text{m})}{2.998 \times 10^8\,\text{m/s}} = 2.55\ \text{s}\ .$$

(b) We denote the uncertainty in time measurement as δt and let $2\delta d_{es} = 15\,\text{cm}$. Then, since $d_{em} \propto t$, $\delta t/t = \delta d_{em}/d_{em}$. We solve for δt:

$$\delta t = \frac{t\delta d_{em}}{d_{em}} = \frac{(2.55\,\text{s})(0.15\,\text{m})}{2(3.82 \times 10^8\,\text{m})} = 5.0 \times 10^{-10}\ \text{s}\ .$$

57. From Eq. 41-29, $N_2/N_1 = e^{-(E_2-E_1)/kT}$. We solve for T:

$$T = \frac{E_2 - E_1}{k\ln(N_1/N_2)} = \frac{3.2\,\text{eV}}{(1.38 \times 10^{-23}\,\text{J/K})\ln(2.5 \times 10^{15}/6.1 \times 10^{13})} = 10000\ \text{K}\ .$$

58. Consider two levels, labeled 1 and 2, with $E_2 > E_1$. Since $T = -|T| < 0$,

$$\frac{N_2}{N_1} = e^{-(E_2-E_1)/kT} = e^{-|E_2-E_1|/(-k|T|)} = e^{|E_2-E_1|/k|T|} > 1\ .$$

Thus, $N_2 > N_1$; this is population inversion. We solve for T:

$$T = -|T| = -\frac{E_2 - E_1}{k\ln(N_2/N_1)} = -\frac{2.26\,\text{eV}}{(8.62 \times 10^{-5}\,\text{eV/K})\ln(1 + 0.100)} = -2.75 \times 10^5\ \text{K}\ .$$

59. (a) If t is the time interval over which the pulse is emitted, the length of the pulse is $L = ct = (3.00 \times 10^8\,\text{m/s})(1.20 \times 10^{-11}\,\text{s}) = 3.60 \times 10^{-3}\,\text{m}$.

(b) If E_p is the energy of the pulse, E is the energy of a single photon in the pulse, and N is the number of photons in the pulse, then $E_p = NE$. The energy of the pulse is $E_p = (0.150\,\text{J})/(1.602 \times 10^{-19}\,\text{J/eV}) = 9.36 \times 10^{17}\,\text{eV}$ and the energy of a single photon is $E = (1240\,\text{eV}\cdot\text{nm})/(694.4\,\text{nm}) = 1.786\,\text{eV}$. Hence,

$$N = \frac{E_p}{E} = \frac{9.36 \times 10^{17}\,\text{eV}}{1.786\,\text{eV}} = 5.24 \times 10^{17}\,\text{photons} .$$

60. Let the power of the laser beam be P and the energy of each photon emitted be E. Then, the rate of photon emission is

$$\begin{aligned} R &= \frac{P}{E} = \frac{P}{hc/\lambda} = \frac{P\lambda}{hc} \\ &= \frac{(2.3 \times 10^{-3}\,\text{W})(632.8 \times 10^{-9}\,\text{m})}{(6.63 \times 10^{-34}\,\text{J}\cdot\text{s})(2.998 \times 10^8\,\text{m/s})} \\ &= 7.3 \times 10^{15}\,\text{s}^{-1} . \end{aligned}$$

61. The Moon is a distance $R = 3.82 \times 10^8$ m from Earth (see Appendix C). We note that the "cone" of light has apex angle equal to 2θ. If we make the small angle approximation (equivalent to using Eq. 37-14), then the diameter D of the spot on the Moon is

$$\begin{aligned} D &= 2R\theta = 2R\left(\frac{1.22\lambda}{d}\right) \\ &= \frac{2(3.82 \times 10^8\,\text{m})(1.22)(600 \times 10^{-9}\,\text{m})}{0.12\,\text{m}} \\ &= 4.7 \times 10^3\,\text{m} = 4.7\,\text{km} . \end{aligned}$$

62. Let the range of frequency of the microwave be Δf. Then the number of channels that could be accommodated is

$$N = \frac{\Delta f}{10\,\text{MHz}} = \frac{(2.998 \times 10^8\,\text{m/s})[(450\,\text{nm})^{-1} - (650\,\text{nm})^{-1}]}{10\,\text{MHz}} = 2.1 \times 10^7 .$$

The higher frequencies of visible light would allow many more channels to be carried compared with using the microwave.

63. Let the power of the laser beam be P and the energy of each photon emitted be E. Then, the rate of photon emission is

$$\begin{aligned} R &= \frac{P}{E} = \frac{P}{hc/\lambda} = \frac{P\lambda}{hc} \\ &= \frac{(5.0 \times 10^{-3}\,\text{W})(0.80 \times 10^{-6}\,\text{m})}{(6.63 \times 10^{-34}\,\text{J}\cdot\text{s})(2.998 \times 10^8\,\text{m/s})} \\ &= 2.0 \times 10^{16}\,\text{s}^{-1} . \end{aligned}$$

64. For the nth harmonic of the standing wave of wavelength λ in the cavity of width L we have $n\lambda = 2L$, so $n\Delta\lambda + \lambda\Delta n = 0$. Let $\Delta n = \pm 1$ and use $\lambda = 2L/n$ to obtain

$$|\Delta\lambda| = \frac{\lambda|\Delta n|}{n} = \frac{\lambda}{n} = \lambda\left(\frac{\lambda}{2L}\right) = \frac{(533\,\text{nm})^2}{2(8.0 \times 10^7\,\text{nm})} = 1.8 \times 10^{-12}\,\text{m} = 1.8\,\text{pm} .$$

65. (a) If both mirrors are perfectly reflecting, there is a node at each end of the crystal. With one end partially silvered, there is a node very close to that end. We assume nodes at both ends, so there are an integer number of half-wavelengths in the length of the crystal. The wavelength in the crystal is $\lambda_c = \lambda/n$, where λ is the wavelength in a vacuum and n is the index of refraction of ruby. Thus $N(\lambda/2n) = L$, where N is the number of standing wave nodes, so

$$N = \frac{2nL}{\lambda} = \frac{2(1.75)(0.0600\,\text{m})}{694\times10^{-9}\,\text{m}} = 3.03\times10^{5}\ .$$

(b) Since $\lambda = c/f$, where f is the frequency, $N = 2nLf/c$ and $\Delta N = (2nL/c)\,\Delta f$. Hence,

$$\Delta f = \frac{c\,\Delta N}{2nL} = \frac{(2.998\times10^{8}\,\text{m/s})(1)}{2(1.75)(0.0600\,\text{m})} = 1.43\times10^{9}\ \text{Hz}\ .$$

(c) The speed of light in the crystal is c/n and the round-trip distance is $2L$, so the round-trip travel time is $2nL/c$. This is the same as the reciprocal of the change in frequency.

(d) The frequency is $f = c/\lambda = (2.998\times10^{8}\,\text{m/s})/(694\times10^{-9}\,\text{m}) = 4.32\times10^{14}\,\text{Hz}$ and the fractional change in the frequency is $\Delta f/f = (1.43\times10^{9}\,\text{Hz})/(4.32\times10^{14}\,\text{Hz}) = 3.31\times10^{-6}$.

66. (a) We denote the upper level as level 1 and the lower one as level 2. From $N_1/N_2 = e^{-(E_1-E_2)/kT}$ we get (using the result of problem 3 in Chapter 39)

$$\begin{aligned} N_1 &= N_2 e^{-(E_1-E_2)/kT} = N_2 e^{-hc/\lambda kT} \\ &= (4.0\times10^{20})e^{-(1240\,\text{eV}\cdot\text{nm})/[(580\,\text{nm})(8.62\times10^{-5}\,\text{eV/K})(300\,\text{K})]} \\ &= 5.0\times10^{-16} \ll 1\ , \end{aligned}$$

so practically no electron occupies the upper level.

(b) With $N_1 = 3.0\times10^{20}$ atoms emitting photons and $N_2 = 1.0\times10^{20}$ atoms absorbing photons, then the net energy output is

$$\begin{aligned} E &= (N_1 - N_2)\,E_{\text{photon}} = (N_1 - N_2)\,\frac{hc}{\lambda} \\ &= \left(2.0\times10^{20}\right)\frac{(6.63\times10^{-34}\,\text{J}\cdot\text{s})(2.998\times10^{8}\,\text{m/s})}{580\times10^{-9}\,\text{m}} \\ &= 68\ \text{J}\ . \end{aligned}$$

67. (a) The intensity at the target is given by $I = P/A$, where P is the power output of the source and A is the area of the beam at the target. We want to compute I and compare the result with $10^8\ \text{W/m}^2$. The beam spreads because diffraction occurs at the aperture of the laser. Consider the part of the beam that is within the central diffraction maximum. The angular position of the edge is given by $\sin\theta = 1.22\lambda/d$, where λ is the wavelength and d is the diameter of the aperture (see Exercise 61). At the target, a distance D away, the radius of the beam is $r = D\tan\theta$. Since θ is small, we may approximate both $\sin\theta$ and $\tan\theta$ by θ, in radians. Then, $r = D\theta = 1.22D\lambda/d$ and

$$\begin{aligned} I &= \frac{P}{\pi r^2} = \frac{Pd^2}{\pi(1.22D\lambda)^2} \\ &= \frac{(5.0\times10^{6}\,\text{W})(4.0\,\text{m})^2}{\pi\left[1.22(3000\times10^{3}\,\text{m})(3.0\times10^{-6}\,\text{m})\right]^2} \\ &= 2.1\times10^{5}\ \text{W/m}^2\ , \end{aligned}$$

not great enough to destroy the missile.

(b) We solve for the wavelength in terms of the intensity and substitute $I = 1.0 \times 10^8\,\mathrm{W/m^2}$:

$$\lambda = \frac{d}{1.22D}\sqrt{\frac{P}{\pi I}} = \frac{4.0\,\mathrm{m}}{1.22(3000 \times 10^3\,\mathrm{m})}\sqrt{\frac{5.0 \times 10^6\,\mathrm{W}}{\pi(1.0 \times 10^8\,\mathrm{W/m^2})}}$$
$$= 1.4 \times 10^{-7}\,\mathrm{m} = 140\ \mathrm{nm}\ .$$

68. (a) The radius of the central disk is

$$R = \frac{1.22 f\lambda}{d} = \frac{(1.22)(3.50\,\mathrm{cm})(515\,\mathrm{nm})}{3.00\,\mathrm{mm}} = 7.33\ \mu\mathrm{m}\ .$$

(b) The average power flux density in the incident beam is

$$\frac{P}{\pi d^2/4} = \frac{4(5.00\,\mathrm{W})}{\pi(3.00\,\mathrm{mm})^2} = 707\ \mathrm{kW/m^2}\ .$$

(c) The average power flux density in the central disk is

$$\frac{(0.84)P}{\pi R^2} = \frac{(0.84)(5.00\,\mathrm{W})}{\pi(7.33\,\mu\mathrm{m})^2} = 24.9\ \mathrm{GW/m^2}\ .$$

69. (a) In the lasing action the molecules are excited from energy level E_0 to energy level E_2. Thus the wavelength λ of the sunlight that causes this excitation satisfies

$$\Delta E = E_2 - E_0 = \frac{hc}{\lambda}\ ,$$

which gives (using the result of problem 3 in Chapter 39)

$$\lambda = \frac{hc}{E_2 - E_0} = \frac{1240\,\mathrm{eV\cdot nm}}{0.289\,\mathrm{eV} - 0} = 4.29 \times 10^3\,\mathrm{nm} = 4.29\,\mu\mathrm{m}\ .$$

(b) Lasing occurs as electrons jump down from the higher energy level E_2 to the lower level E_1. Thus the lasing wavelength λ' satisfies

$$\Delta E' = E_2 - E_1 = \frac{hc}{\lambda'}\ ,$$

which gives

$$\lambda' = \frac{hc}{E_2 - E_1} = \frac{1240\,\mathrm{eV\cdot nm}}{0.289\,\mathrm{eV} - 0.165\,\mathrm{eV}} = 1.00 \times 10^4\,\mathrm{nm} = 10.0\,\mu\mathrm{m}\ .$$

(c) Both λ and λ' belong to the infrared region of the electromagnetic spectrum.

70. (a) The energy difference between the two states 1 and 2 was equal to the energy of the photon emitted. Since the photon frequency was $f = 1666\,\mathrm{MHz}$, its energy was given by $hf = (4.14 \times 10^{-15}\,\mathrm{eV\cdot s})(1666\,\mathrm{MHz}) = 6.90 \times 10^{-6}\,\mathrm{eV}$. Thus,

$$E_2 - E_1 = hf = 6.9 \times 10^{-6}\,\mathrm{eV} = 6.9\,\mu\mathrm{eV}\ .$$

(b) The emission was in the *radio* region of the electromagnetic spectrum.

Chapter 42

1. The number of atoms per unit volume is given by $n = d/M$, where d is the mass density of copper and M is the mass of a single copper atom. Since each atom contributes one conduction electron, n is also the number of conduction electrons per unit volume. Since the molar mass of copper is $A = 63.54\,\text{g/mol}$, $M = A/N_A = (63.54\,\text{g/mol})/(6.022 \times 10^{23}\,\text{mol}^{-1}) = 1.055 \times 10^{-22}\,\text{g}$. Thus,

$$n = \frac{8.96\,\text{g/cm}^3}{1.055 \times 10^{-22}\,\text{g}} = 8.49 \times 10^{22}\,\text{cm}^{-3} = 8.49 \times 10^{28}\,\text{m}^{-3} .$$

2. We compute $\left(\frac{3}{16\sqrt{2}\pi}\right)^{2/3} \approx 0.121$.

3. We use the ideal gas law in the form of Eq. 20-9:

$$p = nkT = (8.43 \times 10^{28}\,\text{m}^{-3})(1.38 \times 10^{-23}\,\text{J/K})(300\,\text{K}) = 3.49 \times 10^{8}\,\text{Pa} = 3490\,\text{atm} .$$

4. We note that $n = 8.43 \times 10^{28}\,\text{m}^{-3} = 84.3\,\text{nm}^{-3}$. From Eq. 42-9,

$$E_F = \frac{0.121(hc)^2}{m_e c^2} n^{2/3} = \frac{0.121(1240\,\text{eV}\cdot\text{nm})^2}{511 \times 10^3\,\text{eV}} (84.3\,\text{nm}^{-3})^{2/3} = 7.0\,\text{eV}$$

where the result of problem 3 in Chapter 39 is used.

5. (a) For copper, Eq. 42-10 leads to

$$\frac{d\rho}{dT} = [\rho\alpha]_{\text{Cu}} = (2 \times 10^{-8}\,\Omega\cdot\text{m})(4 \times 10^{-3}\,\text{K}^{-1}) = 8 \times 10^{-11}\,\Omega\cdot\text{m/K} .$$

(b) For silicon,

$$\frac{d\rho}{dT} = [\rho\alpha]_{\text{Si}} = (3 \times 10^{3}\,\Omega\cdot\text{m})(-70 \times 10^{-3}\,\text{K}^{-1}) = -2.1 \times 10^{2}\,\Omega\cdot\text{m/K} .$$

6. We note that there is one conduction electron per atom and that the molar mass of gold is $197\,\text{g/mol}$. Therefore, combining Eqs. 42-2, 42-3 and 42-4 leads to

$$n = \frac{(19.3\,\text{g/cm}^3)(10^6\,\text{cm}^3/\text{m}^3)}{(197\,\text{g/mol})/(6.02 \times 10^{23}\,\text{mol}^{-1})} = 5.90 \times 10^{28}\,\text{m}^{-3} .$$

7. (a) Eq. 42-5 gives

$$N(E) = \frac{8\sqrt{2}\pi m^{3/2}}{h^3} E^{1/2}$$

for the density of states associated with the conduction electrons of a metal. This can be written

$$n(E) = CE^{1/2}$$

where

$$C = \frac{8\sqrt{2}\pi m^{3/2}}{h^3} = \frac{8\sqrt{2}\pi(9.109\times10^{-31}\,\text{kg})^{3/2}}{(6.626\times10^{-34}\,\text{J}\cdot\text{s})^3} = 1.062\times10^{56}\,\text{kg}^{3/2}/\text{J}^3\cdot\text{s}^3\ .$$

Now, $1\,\text{J} = 1\,\text{kg}\cdot\text{m}^2/\text{s}^2$ (think of the equation for kinetic energy $K = \frac{1}{2}mv^2$), so $1\,\text{kg} = 1\,\text{J}\cdot\text{s}^2\cdot\text{m}^{-2}$. Thus, the units of C can be written $(\text{J}\cdot\text{s}^2)^{3/2}\cdot(\text{m}^{-2})^{3/2}\cdot\text{J}^{-3}\cdot\text{s}^{-3} = \text{J}^{-3/2}\cdot\text{m}^{-3}$. This means

$$C = (1.062\times10^{56}\,\text{J}^{-3/2}\cdot\text{m}^{-3})(1.602\times10^{-19}\,\text{J/eV})^{3/2} = 6.81\times10^{27}\,\text{m}^{-3}\cdot\text{eV}^{-3/2}\ .$$

(b) If $E = 5.00\,\text{eV}$, then

$$n(E) = (6.81\times10^{27}\,\text{m}^{-3}\cdot\text{eV}^{-3/2})(5.00\,\text{eV})^{1/2} = 1.52\times10^{28}\,\text{eV}^{-1}\cdot\text{m}^{-3}\ .$$

8. We equate E_F with $\frac{1}{2}m_e v_\text{F}^2$ and write our expressions in such a way that we can make use of the electron mc^2 value found in Table 38-3:

$$v_\text{F} = \sqrt{\frac{2E_\text{F}}{m}} = c\sqrt{\frac{2E_\text{F}}{mc^2}} = (3.0\times10^5\,\text{km/s})\sqrt{\frac{2(7.0\,\text{eV})}{5.11\times10^5\,\text{eV}}} = 1.6\times10^3\,\text{km/s}\ .$$

9. (a) At absolute temperature $T = 0$, the probability is zero that any state with energy above the Fermi energy is occupied.

(b) The probability that a state with energy E is occupied at temperature T is given by

$$P(E) = \frac{1}{e^{(E-E_F)/kT}+1}$$

where k is the Boltzmann constant and E_F is the Fermi energy. Now, $E - E_F = 0.062\,\text{eV}$ and $(E - E_F)/kT = (0.062\,\text{eV})/(8.62\times10^{-5}\,\text{eV/K})(320\,\text{K}) = 2.248$, so

$$P(E) = \frac{1}{e^{2.248}+1} = 0.0956\ .$$

See Appendix B or Sample Problem 42-1 for the value of k.

10. We use the result of problem 7:

$$n(E) = CE^{1/2} = \left[6.81\times10^{27}\,\text{m}^{-3}\cdot(\text{eV})^{-2/3}\right](8.0\,\text{eV})^{1/2} = 1.9\times10^{28}\,\text{m}^{-3}\cdot\text{eV}^{-1}\ .$$

This is consistent with Fig.42-5.

11. According to Eq. 42-9, the Fermi energy is given by

$$E_F = \left(\frac{3}{16\sqrt{2}\pi}\right)^{2/3}\frac{h^2}{m}n^{2/3}$$

where n is the number of conduction electrons per unit volume, m is the mass of an electron, and h is the Planck constant. This can be written $E_F = An^{2/3}$, where

$$A = \left(\frac{3}{16\sqrt{2}\pi}\right)^{2/3}\frac{h^2}{m} = \left(\frac{3}{16\sqrt{2}\pi}\right)^{2/3}\frac{(6.626\times10^{-34}\,\text{J}\cdot\text{s})^2}{9.109\times10^{-31}\,\text{kg}} = 5.842\times10^{-38}\,\text{J}^2\cdot\text{s}^2/\text{kg}\ .$$

Since $1\,\text{J} = 1\,\text{kg}\cdot\text{m}^2/\text{s}^2$, the units of A can be taken to be $\text{m}^2\cdot\text{J}$. Dividing by $1.602\times10^{-19}\,\text{J/eV}$, we obtain $A = 3.65\times10^{-19}\,\text{m}^2\cdot\text{eV}$.

12. We reproduce the calculation of Exercise 6: Combining Eqs. 42-2, 42-3 and 42-4, the number density of conduction electrons in gold is

$$n = \frac{(19.3\,\text{g/cm}^3)(6.02\times10^{23}/\text{mol})}{(197\,\text{g/mol})} = 5.90\times10^{22}\,\text{cm}^{-3} = 59.0\ \text{nm}^{-3}\ .$$

Now, using the result of Exercise 3 in Chapter 39, Eq. 42-9 leads to

$$E_\text{F} = \frac{0.121(hc)^2}{(m_e c^2)} n^{2/3} = \frac{0.121(1240\,\text{eV}\cdot\text{nm})^2}{511\times10^3\,\text{eV}}(59.0\,\text{nm}^{-3})^{2/3} = 5.52\ \text{eV}\ .$$

13. Let $E_1 = 63\,\text{meV} + E_\text{F}$ and $E_2 = -63\,\text{meV} + E_\text{F}$. Then according to Eq. 42-6,

$$P_1 = \frac{1}{e^{(E_1-E_\text{F})/kT}+1} = \frac{1}{e^x+1}$$

where $x = (E_1 - E_\text{F})/kT$. We solve for e^x:

$$e^x = \frac{1}{P_1} - 1 = \frac{1}{0.090} - 1 = \frac{91}{9}\ .$$

Thus,

$$P_2 = \frac{1}{e^{(E_2-E_\text{F})/kT}+1} = \frac{1}{e^{-(E_1-E_\text{F})/kT}+1} = \frac{1}{e^{-x}+1} == \frac{1}{(91/9)^{-1}+1} = 0.91\ ,$$

where we use $E_2 - E_\text{F} = -63\,\text{meV} = E_F - E_1 = -(E_1 - E_\text{F})$.

14. (a) Eq. 42-6 leads to

$$\begin{aligned} E &= E_\text{F} + kT\,\ln(P^{-1}-1) \\ &= 7.0\,\text{eV} + (8.62\times10^{-5}\,\text{eV/K})(1000\,\text{K})\,\ln\left(\frac{1}{0.90}-1\right) \\ &= 6.8\ \text{eV}\ . \end{aligned}$$

(b) $n(E) = CE^{1/2} = (6.81\times10^{27}\,\text{m}^{-3}\cdot\text{eV}^{-3/2})(6.8\,\text{eV})^{1/2} = 1.77\times10^{28}\ \text{m}^{-3}\cdot\text{eV}^{-1}$.

(c) $n_\text{O}(E) = P(E)n(E) = (0.90)(1.77\times10^{28}\,\text{m}^{-3}\cdot\text{eV}^{-1}) = 1.6\times10^{28}\,\text{m}^{-3}\cdot\text{eV}^{-1}$.

15. The Fermi-Dirac occupation probability is given by $P_\text{FD} = 1/\left(e^{\Delta E/kT}+1\right)$, and the Boltzmann occupation probability is given by $P_\text{B} = e^{-\Delta E/kT}$. Let f be the fractional difference. Then

$$f = \frac{P_\text{B} - P_\text{FD}}{P_\text{B}} = \frac{e^{-\Delta E/kT} - \dfrac{1}{e^{\Delta E/kT}+1}}{e^{-\Delta E/kT}}\ .$$

Using a common denominator and a little algebra yields

$$f = \frac{e^{-\Delta E/kT}}{e^{-\Delta E/kT}+1}\ .$$

The solution for $e^{-\Delta E/kT}$ is

$$e^{-\Delta E/kT} = \frac{f}{1-f}\ .$$

We take the natural logarithm of both sides and solve for T. The result is

$$T = \frac{\Delta E}{k\ln\left(\frac{f}{1-f}\right)}\ .$$

(a) Letting f equal 0.01, we evaluate the expression for T:

$$T = \frac{(1.00\,\text{eV})(1.60 \times 10^{-19}\,\text{J/eV})}{(1.38 \times 10^{-23}\,\text{J/K})\ln\left(\frac{0.010}{1-0.010}\right)} = 2.5 \times 10^3\ \text{K}\ .$$

(b) We set f equal to 0.10 and evaluate the expression for T:

$$T = \frac{(1.00\,\text{eV})(1.60 \times 10^{-19}\,\text{J/eV})}{(1.38 \times 10^{-23}\,\text{J/K})\ln\left(\frac{0.10}{1-0.10}\right)} = 5.3 \times 10^3\ \text{K}\ .$$

16. According to Eq. 42-6,

$$P(E_\text{F} + \Delta E) = \frac{1}{e^{(E_\text{F}+\Delta E-E_\text{F})/kT}+1} = \frac{1}{e^{\Delta E/kT}+1} = \frac{1}{e^x+1}$$

where $x = \Delta E/kT$. Also,

$$P(E_\text{F} - \Delta E) = \frac{1}{e^{(E_\text{F}-\Delta E-E_\text{F})/kT}+1} = \frac{1}{e^{-\Delta E/kT}+1} = \frac{1}{e^{-x}+1}\ .$$

Thus,

$$P(E_\text{F} + \Delta E) + P(E_\text{F} - \Delta E) = \frac{1}{e^x+1} + \frac{1}{e^{-x}+1} = \frac{e^x+1+e^{-x}+1}{(e^{-x}+1)(e^x+1)} = 1\ .$$

A special case of this general result can be found in problem 13, where $\Delta E = 63\,\text{meV}$ and $P(E_\text{F} + 63\,\text{meV}) + P(E_\text{F} - 63\,\text{meV}) = 0.090 + 0.91 = 1.0$.

17. (a) The volume per cubic meter of sodium occupied by the sodium ions is

$$V_\text{Na} = \frac{(971\,\text{kg})(6.022 \times 10^{23}/\text{mol})(4\pi/3)(98 \times 10^{-12}\,\text{m})^3}{(23\,\text{g/mol})} = 0.100\ \text{m}^3\ ,$$

so the fraction available for conduction electrons is $1 - (V_\text{Na}/1.00\,\text{m}^3) = 1 - 0.100 = 0.900$.

(b) For copper,

$$V_\text{Cu} = \frac{(8960\,\text{kg})(6.022 \times 10^{23}/\text{mol})(4\pi/3)(135 \times 10^{-12}\,\text{m})^3}{63.5\,\text{g/mol}} = 0.876\ \text{m}^{-3}\ .$$

Thus, the fraction is $1 - (V_\text{Cu}/1.00\,\text{m}^3) = 1 - 0.876 = 0.124$.

(c) Sodium, because the electrons occupy a greater portion of the space available.

18. We use $N_\text{O} = N(E)P(E) = CE^{1/2}\left[e^{(E-E_\text{F})/kT}+1\right]^{-1}$, where C is given in problem 7(a). At $E = 4.00\,\text{eV}$,

$$\begin{aligned} n_\text{O} &= \frac{\left(6.8 \times 10^{27}\,\text{m}^{-3}\cdot(\text{eV})^{-3/2}\right)(4.00\,\text{eV})^{1/2}}{e^{(4.00\,\text{eV}-7.00\,\text{eV})/[(8.62\times10^{-5}\,\text{eV/K})(1000\,\text{K})]}+1} \\ &= 1.36 \times 10^{28}\ \text{m}^{-3}\cdot\text{eV}^{-1}\ , \end{aligned}$$

and at $E = 6.75\,\text{eV}$,

$$\begin{aligned} n_\text{O} &= \frac{\left(6.8 \times 10^{27}\,\text{m}^{-3}\cdot(\text{eV})^{-3/2}\right)(6.75\,\text{eV})^{1/2}}{e^{(6.75\,\text{eV}-7.00\,\text{eV})/[(8.62\times10^{-5}\,\text{eV/K})(1000\,\text{K})]}+1} \\ &= 1.67 \times 10^{28}\ \text{m}^{-3}\cdot\text{eV}^{-1}\ . \end{aligned}$$

Similarly at $E = 7.00$, 7.25 and 9.00 eV, the values of $n_\text{O}(E)$ are $9.0\times10^{27}\,\text{m}^{-3}\cdot\text{eV}^{-1}$, $9.5\times10^{26}\,\text{m}^{-3}\cdot\text{eV}^{-1}$ and $1.7 \times 10^{18}\,\text{m}^{-3}\cdot\text{eV}^{-1}$, respectively. We note that the latter value is effectively zero (relative to the other results).

19. (a) The ideal gas law in the form of Eq. 20-9 leads to $p = NkT/V = nkT$. Thus, we solve for the molecules per cubic meter:

$$n = \frac{p}{kT} = \frac{(1.0\,\text{atm})(1.0 \times 10^5\,\text{Pa/atm})}{(1.38 \times 10^{-23}\,\text{J/K})(273\,\text{K})} = 2.7 \times 10^{25}\ \text{m}^{-3}\ .$$

(b) Combining Eqs. 42-2, 42-3 and 42-4 leads to the conduction electrons per cubic meter in copper:

$$n = \frac{8.96 \times 10^3\,\text{kg/m}^3}{(63.54)(1.67 \times 10^{-27}\,\text{kg})} = 8.43 \times 10^{28}\ \text{m}^{-3}\ .$$

(c) The ratio is $(8.43 \times 10^{28}\,\text{m}^{-3})/(2.7 \times 10^{25}\,\text{m}^{-3}) = 3.1 \times 10^3$.

(d) We use $d_{\text{avg}} = n^{-1/3}$. For case (a), $d_{\text{avg}} = (2.7 \times 10^{25}\,\text{m}^{-3})^{-1/3}$ which equals 3.3 nm. For case (b), $d_{\text{avg}} = (8.43 \times 10^{28}\,\text{m}^{-3})^{-1/3} = 0.23\,\text{nm}$.

20. The molar mass of carbon is $m = 12.01115\,\text{g/mol}$ and the mass of the Earth is $M_e = 5.98 \times 10^{24}\,\text{kg}$. Thus, the number of carbon atoms in a diamond as massive as the Earth is $N = (M_e/m)N_{\text{A}}$, where N_{A} is the Avogadro constant. From the result of Sample Problem 42-1, the probability in question is given by

$$\begin{aligned} P &= Ne^{-E_g/kT} = \left(\frac{M_e}{m}\right) N_{\text{A}} e^{-E_g/kT} \\ &= \left(\frac{5.98 \times 10^{24}\,\text{kg}}{12.01115\,\text{g/mol}}\right)(6.02 \times 10^{23}/\text{mol})(3 \times 10^{-93}) = 9 \times 10^{-43}\ . \end{aligned}$$

21. (a) We evaluate $P(E) = 1/\left(e^{(E-E_F)/kT} + 1\right)$ for the given value of E, using

$$kT = \frac{(1.381 \times 10^{-23}\,\text{J/K})(273\,\text{K})}{1.602 \times 10^{-19}\,\text{J/eV}} = 0.02353\ \text{eV}\ .$$

For $E = 4.4\,\text{eV}$, $(E - E_F)/kT = (4.4\,\text{eV} - 5.5\,\text{eV})/(0.02353\,\text{eV}) = -46.25$ and

$$P(E) = \frac{1}{e^{-46.25} + 1} = 1.00\ .$$

Similarly, for $E = 5.4\,\text{eV}$, $P(E) = 0.986$, for $E = 5.5\,\text{eV}$, $P(E) = 0.500$, for $E = 5.6\,\text{eV}$, $P(E) = 0.0141$, and for $E = 6.4\,\text{eV}$, $P(E) = 2.57 \times 10^{-17}$.

(b) Solving $P = 1/\left(e^{\Delta E/kT} + 1\right)$ for $e^{\Delta E/kT}$, we get

$$e^{\Delta E/kT} = \frac{1}{P} - 1\ .$$

Now, we take the natural logarithm of both sides and solve for T. The result is

$$T = \frac{\Delta E}{k\ \ln\left(\frac{1}{P} - 1\right)} = \frac{(5.6\,\text{eV} - 5.5\,\text{eV})(1.602 \times 10^{-19}\,\text{J/eV})}{(1.381 \times 10^{-23}\,\text{J/K})\ln\left(\frac{1}{0.16} - 1\right)} = 699\ \text{K}\ .$$

22. The probability P_h that a state is occupied by a hole is the same as the probability the state is *unoccupied* by an electron. Since the total probability that a state is either occupied or unoccupied is 1, we have $P_h + P = 1$. Thus,

$$P_h = 1 - \frac{1}{e^{(E-E_F)/kT} + 1} = \frac{e^{(E-E_F)/kT}}{1 + e^{(E-E_F)/kT}} = \frac{1}{e^{-(E-E_F)/kT} + 1}\ .$$

23. Let N be the number of atoms per unit volume and n be the number of free electrons per unit volume. Then, the number of free electrons per atom is n/N. We use the result of Exercise 11 to find n: $E_F = An^{2/3}$, where $A = 3.65 \times 10^{-19}\,\text{m}^2 \cdot \text{eV}$. Thus,

$$n = \left(\frac{E_F}{A}\right)^{3/2} = \left(\frac{11.6\,\text{eV}}{3.65 \times 10^{-19}\,\text{m}^2 \cdot \text{eV}}\right)^{3/2} = 1.79 \times 10^{29}\,\text{m}^{-3} .$$

If M is the mass of a single aluminum atom and d is the mass density of aluminum, then $N = d/M$. Now, $M = (27.0\,\text{g/mol})/(6.022 \times 10^{23}\,\text{mol}^{-1}) = 4.48 \times 10^{-23}\,\text{g}$, so $N = (2.70\,\text{g/cm}^3)/(4.48 \times 10^{-23}\,\text{g}) = 6.03 \times 10^{22}\,\text{cm}^{-3} = 6.03 \times 10^{28}\,\text{m}^{-3}$. Thus, the number of free electrons per atom is

$$\frac{n}{N} = \frac{1.79 \times 10^{29}\,\text{m}^{-3}}{6.03 \times 10^{28}\,\text{m}^{-3}} = 2.97 .$$

24. Let the energy of the state in question be an amount ΔE above the Fermi energy E_F. Then, Eq. 42-6 gives the occupancy probability of the state as

$$P = \frac{1}{e^{(E_F+\Delta E-E_F)/kT}+1} = \frac{1}{e^{\Delta E/kT}+1} .$$

We solve for ΔE to obtain

$$\Delta E = kT\,\ln\left(\frac{1}{P}-1\right) = (1.38 \times 10^{23}\,\text{J/K})(300\,\text{K})\,\ln\left(\frac{1}{0.10}-1\right) = 9.1 \times 10^{-21}\,\text{J} ,$$

which is equivalent to $5.7 \times 10^{-2}\,\text{eV} = 57\,\text{meV}$.

25. (a) According to Appendix F the molar mass of silver is 107.870 g/mol and the density is $10.49\,\text{g/cm}^3$. The mass of a silver atom is

$$\frac{107.870 \times 10^{-3}\,\text{kg/mol}}{6.022 \times 10^{23}\,\text{mol}^{-1}} = 1.791 \times 10^{-25}\,\text{kg} .$$

We note that silver is monovalent, so there is one valence electron per atom (see Eq. 42-2). Thus, Eqs. 42-4 and 42-3 lead to

$$n = \frac{\rho}{M} = \frac{10.49 \times 10^3\,\text{kg/m}^3}{1.791 \times 10^{25}\,\text{kg}} = 5.86 \times 10^{28}\,\text{m}^{-3} .$$

(b) The Fermi energy is

$$\begin{aligned} E_F &= \frac{0.121h^2}{m}n^{2/3} = \frac{(0.121)(6.626 \times 10^{-34}\,\text{J}\cdot\text{s})^2}{9.109 \times 10^{-31}\,\text{kg}}(5.86 \times 10^{28}\,\text{m}^{-3})^{2/3} \\ &= 8.80 \times 10^{-19}\,\text{J} = 5.49\,\text{eV} . \end{aligned}$$

(c) Since $E_F = \frac{1}{2}mv_F^2$,

$$v_F = \sqrt{\frac{2E_F}{m}} = \sqrt{\frac{2(8.80 \times 10^{-19}\,\text{J})}{9.109 \times 10^{-31}\,\text{kg}}} = 1.39 \times 10^6\,\text{m/s} .$$

(d) The de Broglie wavelength is

$$\lambda = \frac{h}{mv_F} = \frac{6.626 \times 10^{-34}\,\text{J}\cdot\text{s}}{(9.109 \times 10^{-31}\,\text{kg})(1.39 \times 10^6\,\text{m/s})} = 5.23 \times 10^{-10}\,\text{m} .$$

26. (a) Combining Eqs. 42-2, 42-3 and 42-4 leads to the conduction electrons per cubic meter in zinc:

$$n = \frac{2(7.133\,\text{g/cm}^3)}{(65.37\,\text{g/mol})/(6.02 \times 10^{23}/\text{mol})} = 1.31 \times 10^{23}\,\text{cm}^{-3} = 1.31 \times 10^{29}\ \text{m}^{-3}\ .$$

(b) From Eq. 42-9,

$$E_\text{F} = \frac{0.121h^2}{m_e}n^{2/3} = \frac{0.121(6.63 \times 10^{-34}\,\text{J·s})^2(1.31 \times 10^{29}\,\text{m}^{-3})^{2/3}}{(9.11 \times 10^{-31}\,\text{kg})(1.60 \times 10^{-19}\,\text{J/eV})} = 9.43\ \text{eV}\ .$$

(c) Equating the Fermi energy to $\frac{1}{2}m_e v_F^2$, we find (using the $m_e c^2$ value in Table 38-3)

$$v_\text{F} = \sqrt{\frac{2E_\text{F}c^2}{m_e c^2}} = \sqrt{\frac{2(9.43\,\text{eV})(2.998 \times 10^8\,\text{m/s})^2}{511 \times 10^3\,\text{eV}}} = 1.82 \times 10^6\ \text{m/s}\ .$$

(d) The de Broglie wavelength is

$$\lambda = \frac{h}{m_e v_\text{F}} = \frac{6.63 \times 10^{-34}\,\text{J·s}}{(9.11 \times 10^{-31}\,\text{kg})(1.82 \times 10^6\,\text{m/s})} = 0.40\ \text{nm}\ .$$

27. (a) Setting $E = E_F$ (see Eq. 42-9), Eq. 42-5 becomes

$$N(E_F) = \frac{8\pi m\sqrt{2m}}{h^3}\left(\frac{3}{16\pi\sqrt{2}}\right)^{1/3}\frac{h}{\sqrt{m}}n^{1/3}\ .$$

Noting that $16\sqrt{2} = 2^4 2^{1/2} = 2^{9/2}$ so that the cube root of this is $2^{3/2} = 2\sqrt{2}$, we are able to simplify the above expression and obtain

$$N(E_F) = \frac{4m}{h^2}\sqrt[3]{3\pi^2 n}$$

which is equivalent to the result shown in the problem statement. Since the desired numerical answer uses eV units, we multiply numerator and denominator of our result by c^2 and make use of the mc^2 value for an electron in Table 38-3 as well as the hc value found in problem 3 of Chapter 39:

$$N(E_F) = \left(\frac{4mc^2}{(hc)^2}\sqrt[3]{3\pi^2}\right)n^{1/3} = \left(\frac{4(511 \times 10^3\,\text{eV})}{(1240\,\text{eV·nm})^2}\sqrt[3]{3\pi^2}\right)n^{1/3} = \left(4.11\ \text{nm}^{-2}\cdot\text{eV}^{-1}\right)n^{1/3}$$

which is equivalent to the value indicated in the problem statement.

(b) Since there are 10^{27} cubic nanometers in a cubic meter, then the result of problem 1 may be written

$$n = 8.49 \times 10^{28}\,\text{m}^{-3} = 84.9\ \text{nm}^{-3}\ .$$

The cube root of this is $n^{1/3} \approx 4.4/\text{nm}$. Hence, the expression in part (a) leads to

$$N(E_F) = \left(4.11\,\text{nm}^{-2}\cdot\text{eV}^{-1}\right)\left(4.4\,\text{nm}^{-1}\right) = 18\ \text{nm}^{-3}\cdot\text{eV}^{-1}\ .$$

If we multiply this by $10^{27}\,\text{m}^3/\text{nm}^3$, we see this compares very well with the curve in Fig. 42-5 evaluated at 7.0 eV.

28. (a) The derivative of $P(E)$ is

$$\left(\frac{-1}{\left(e^{(E-E_F)/kT}+1\right)^2}\right)\frac{d}{dE}e^{(E-E_F)/kT} = \left(\frac{-1}{\left(e^{(E-E_F)/kT}+1\right)^2}\right)\frac{1}{kT}e^{(E-E_F)/kT}\ .$$

Evaluating this at $E = E_F$ we readily obtain the desired result.

(b) The equation of a line may be written $y = m(x - x_o)$ where m is the slope (here: equal to $-1/kT$, from part (a)) and x_o is the x-intercept (which is what we are asked to solve for). It is clear that $P(E_F) = 2$, so our equation of the line, evaluated at $x = E_F$, becomes $2 = (-1/kT)(E_F - x_o)$, which leads to $x_o = E_F + 2kT$.

29. The average energy of the conduction electrons is given by

$$E_{\text{avg}} = \frac{1}{n}\int_0^\infty EN(E)P(E)\,dE$$

where n is the number of free electrons per unit volume, $N(E)$ is the density of states, and $P(E)$ is the occupation probability. The density of states is proportional to $E^{1/2}$, so we may write $N(E) = CE^{1/2}$, where C is a constant of proportionality. The occupation probability is one for energies below the Fermi energy and zero for energies above. Thus,

$$E_{\text{avg}} = \frac{C}{n}\int_0^{E_F} E^{3/2}\,dE = \frac{2C}{5n}E_F^{5/2} .$$

Now

$$n = \int_0^\infty N(E)P(E)\,dE = C\int_0^{E_F} E^{1/2}\,dE = \frac{2C}{3}E_F^{3/2} .$$

We substitute this expression into the formula for the average energy and obtain

$$E_{\text{avg}} = \left(\frac{2C}{5}\right)E_F^{5/2}\left(\frac{3}{2CE_F^{3/2}}\right) = \frac{3}{5}E_F .$$

30. Let the volume be $\mathcal{V} = 1.0 \times 10^{-6}\,\text{m}^3$. Then,

$$\begin{aligned} K_{\text{total}} &= NE_{\text{avg}} = n\mathcal{V}E_{\text{avg}} \\ &= (8.43 \times 10^{28}\,\text{m}^{-3})(1.0 \times 10^{-6}\,\text{m}^3)\left(\frac{3}{5}\right)(7.0\,\text{eV})(1.6 \times 10^{-19}\,\text{J/eV}) \\ &= 5.7 \times 10^4\,\text{J} = 57\text{ kJ} . \end{aligned}$$

31. (a) Using Eq. 42-4, the energy released would be

$$\begin{aligned} E &= NE_{\text{avg}} \\ &= \frac{(3.1\,\text{g})}{(63.54\,\text{g/mol})/(6.02 \times 10^{23}/\text{mol})}\left(\frac{3}{5}\right)(7.0\,\text{eV})(1.6 \times 10^{-19}\,\text{J/eV}) \\ &= 1.98 \times 10^4\,\text{J} \approx 20\text{ kJ} . \end{aligned}$$

(b) Keeping in mind that a Watt is a Joule per second, we have

$$\frac{1.98 \times 10^4\,\text{J}}{100\,\text{J/s}} = 198\text{ s} .$$

32. (a) At $T = 300\,\text{K}$

$$f = \frac{3kT}{2E_F} = \frac{3(8.62 \times 10^{-5}\,\text{eV/K})(300\,\text{K})}{2(7.0\,\text{eV})} = 5.5 \times 10^{-3} .$$

(b) At $T = 1000\,\text{K}$,

$$f = \frac{3kT}{2E_F} = \frac{3(8.62 \times 10^{-5}\,\text{eV/K})(1000\,\text{K})}{2(7.0\,\text{eV})} = 1.8 \times 10^{-2} .$$

(c) Many calculators and most math software packages (here we use MAPLE) have built-in numerical integration routines. Setting up ratios of integrals of Eq. 42-7 and canceling common factors, we obtain

$$frac = \frac{\int_{E_F}^{\infty} \sqrt{E}/(e^{(E-E_F)/kT}+1)\,dE}{\int_0^{\infty} \sqrt{E}/(e^{(E-E_F)/kT}+1)\,dE}$$

where $k = 8.62 \times 10^{-5}$ eV/K. We use the Fermi energy value for copper ($E_F = 7.0$ eV) and evaluate this for $T = 300$ K and $T = 1000$ K; we find $frac = 0.00385$ and $frac = 0.0129$, respectively.

33. The fraction f of electrons with energies greater than the Fermi energy is (approximately) given in Problem 42-32:

$$f = \frac{3kT/2}{E_F}$$

where T is the temperature on the Kelvin scale, k is the Boltzmann constant, and E_F is the Fermi energy. We solve for T:

$$T = \frac{2fE_F}{3k} = \frac{2(0.013)(4.7\,\text{eV})}{3(8.62 \times 10^{-5}\,\text{eV/K})} = 4.7 \times 10^2 \text{ K} .$$

It should be noted that the numerical approach, discussed briefly in part (c) of problem 32, would lead to a value closer to $T = 6.5 \times 10^2$ K.

34. If we use the approximate formula discussed in problem 32, we obtain

$$frac = \frac{3(8.62 \times 10^{-5}\,\text{eV/K})(961 + 273\,\text{K})}{2(5.5\,\text{eV})} \approx 0.03 .$$

The numerical approach is briefly discussed in part (c) of problem 32. Although the problem does not ask for it here, we remark that numerical integration leads to a fraction closer to 0.02.

35. (a) Since the electron jumps from the conduction band to the valence band, the energy of the photon equals the energy gap between those two bands. The photon energy is given by $hf = hc/\lambda$, where f is the frequency of the electromagnetic wave and λ is its wavelength. Thus, $E_g = hc/\lambda$ and

$$\lambda = \frac{hc}{E_g} = \frac{(6.63 \times 10^{-34}\,\text{J·s})(2.998 \times 10^8\,\text{m/s})}{(5.5\,\text{eV})(1.60 \times 10^{-19}\,\text{J/eV})} = 2.26 \times 10^{-7}\,\text{m} = 226\text{ nm} .$$

Photons from other transitions have a greater energy, so their waves have shorter wavelengths.

(b) These photons are in the ultraviolet portion of the electromagnetic spectrum.

36. Each Arsenic atom is connected (by covalent bonding) to four Gallium atoms, and each Gallium atom is similarly connected to four Arsenic atoms. The "depth" of their very non-trivial lattice structure is, of course, not evident in a flattened-out representation such as shown for Silicon in Fig. 42-9. Still we try to convey some sense of this (in the $[1,0,0]$ view shown below – for those who might be familiar with Miller indices) by using letters to indicate the depth: A for the closest atoms (to the observer), b for the next layer deep, C for further into the page, d for the last layer seen, and E (not shown) for the atoms that are at the deepest layer (and are behind the A's) needed for our description of the structure. The capital letters are used for the Gallium atoms, and the small letters for the Arsenic. Consider the Arsenic atom (with the letter b) near the upper left; it has covalent bonds with the two A's and the two C's near it. Now consider the Arsenic atom (with the letter d) near the upper right; it has covalent bonds with the two C's which are near it and with the two E's (which are behind the A's which are near

it).

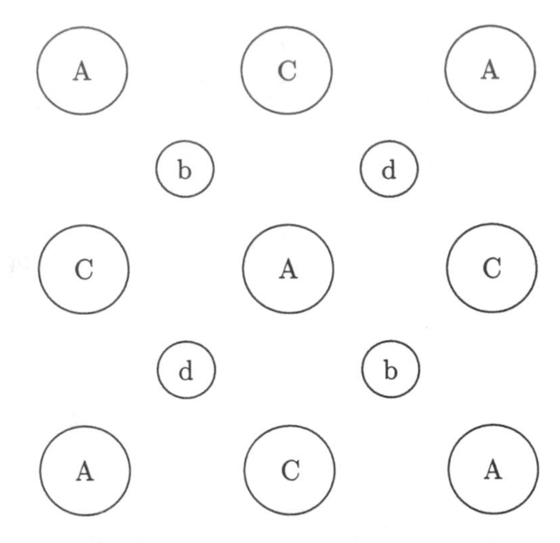

(a) The 3p, 3d and 4s subshells of both Arsenic and Gallium are filled. They both have partially filled 4p subshells. An isolated, neutral Arsenic atom has three electrons in the 4p subshell, and an isolated, neutral Gallium atom has one electron in the 4p subshell. To supply the total of eight shared electrons (for the four bonds connected to each ion in the lattice), not only the electrons from 4p must be shared but also the electrons from 4s. The core of the Arsenic ion has charge $q = +5e$ (due to the "loss" of the three 4p and two 4s electrons), and the charge of the Gallium ion has charge $q = +3e$ (due to the "loss" of its single 4p and two 4s electrons).

(b) As remarked in part (a), there are two electrons shared in each of the covalent bonds. This is the same situation that one finds for Silicon (see Fig. 42-9).

37. The description in the problem statement implies that an atom is at the centerpoint C of the regular tetrahedron, since its four *neighbors* are at the four vertices. The side length for the tetrahedron is given as $a = 388$ pm. Since each face is an equilateral triangle, the "altitude" of each of those triangles (which is not to be confused with the altitude of the tetrahedron itself) is $h' = \frac{1}{2}a\sqrt{3}$ (this is generally referred to as the "slant height" in the solid geometry literature). At a certain location along the line segment representing "slant height" of each face is the center C' of the face. Imagine this line segment starting at atom A and ending at the midpoint of one of the sides. Knowing that this line segment bisects the $60°$ angle of the equilateral face, then it is easy to see that C' is a distance $AC' = a/\sqrt{3}$. If we draw a line from C' all the way to farthest point on the tetrahedron (this will land on an atom we label B), then this new line is the altitude h of the tetrahedron. Using the Pythagorean theorem,

$$h = \sqrt{a^2 - (AC')^2} = \sqrt{a^2 - \left(\frac{a}{\sqrt{3}}\right)^2} = a\sqrt{\frac{2}{3}} .$$

Now we include coordinates: imagine atom B is on the $+y$ axis at $y_b = h = a\sqrt{2/3}$, and atom A is on the $+x$ axis at $x_a = AC' = a/\sqrt{3}$. Then point C' is the origin. The tetrahedron centerpoint C is on the y axis at some value y_c which we find as follows: C must be equidistant from A and B, so

$$\begin{aligned} y_b - y_c &= \sqrt{x_a^2 + y_c^2} \\ a\sqrt{\frac{2}{3}} - y_c &= \sqrt{\left(\frac{a}{\sqrt{3}}\right)^2 + y_c^2} \end{aligned}$$

which yields $y_c = a/2\sqrt{6}$.

(a) In unit vector notation, using the information found above, we express the vector starting at C and going to A as

$$\vec{r}_{ac} = x_a\,\hat{\mathrm{i}} + (-y_c)\,\hat{\mathrm{j}} = \frac{a}{\sqrt{3}}\,\hat{\mathrm{i}} - \frac{a}{2\sqrt{6}}\,\hat{\mathrm{j}}\ .$$

Similarly, the vector starting at C and going to B is $\vec{r}_{bc} = (y_b - y_c)\hat{\mathrm{j}} = \frac{a}{2}\sqrt{3/2}\,\hat{\mathrm{j}}$. Therefore, using Eq. 3-20,

$$\theta = \cos^{-1}\left(\frac{\vec{r}_{ac}\cdot\vec{r}_{bc}}{|\vec{r}_{ac}|\,|\vec{r}_{bc}|}\right) = \cos^{-1}\left(-\frac{1}{3}\right)$$

which yields $\theta = 109.5°$ for the angle between adjacent bonds.

(b) The length of vector $\vec{r}_{bc}$ (which is, of course, the same as the length of $\vec{r}_{ac}$) is

$$|\vec{r}_{bc}| = \frac{a}{2}\sqrt{\frac{3}{2}} = \frac{388\ \text{pm}}{2}\sqrt{\frac{3}{2}} = 237.6\ \text{pm}\ .$$

We note that in the solid geometry literature, the distance $\frac{a}{2}\sqrt{\frac{3}{2}}$ is known as the circumradius of the regular tetrahedron.

38. (a) At the bottom of the conduction band $E = 0.67\,\text{eV}$. Also $E_F = 0.67\,\text{eV}/2 = 0.335\,\text{eV}$. So the probability that the bottom of the conduction band is occupied is

$$\begin{aligned} P(E) &= \frac{1}{e^{(E-E_F)/kT}+1} = \frac{1}{e^{(0.67\,\text{eV}-0.335\,\text{eV})/[(8.62\times10^{-5}\,\text{eV/K})(290\,\text{K})]}+1} \\ &= 1.5\times10^{-6}\ . \end{aligned}$$

(b) At the top of the valence band $E = 0$, so the probability that the state is *unoccupied* is given by

$$\begin{aligned} 1-P(E) &= 1-\frac{1}{e^{(E-E_F)/kT}+1} = \frac{1}{e^{-(E-E_F)/kT}+1} \\ &= \frac{1}{e^{-(0-0.335\,\text{eV})/[(8.62\times10^{-5}\,\text{eV/K})(290\,\text{K})]}+1} \\ &= 1.5\times10^{-6}\ . \end{aligned}$$

39. (a) The number of electrons in the valence band is

$$N_{\text{ev}} = N_v P(E_v) = \frac{N_v}{e^{(E_v-E_F)/kT}+1}\ .$$

Since there are a total of N_v states in the valence band, the number of holes in the valence band is

$$\begin{aligned} N_{\text{hv}} &= N_v - N_{\text{ev}} = N_v\left[1-\frac{1}{e^{(E_v-E_F)/kT}+1}\right] \\ &= \frac{N_v}{e^{-(E_v-E_F)/kT}+1}\ . \end{aligned}$$

Now, the number of electrons in the conduction band is

$$N_{\text{ec}} = N_c P(E_c) = \frac{N_c}{e^{(E_c-E_F)/kT}+1}\ ,$$

Hence, from $N_{\text{ev}} = N_{\text{hc}}$, we get

$$\frac{N_v}{e^{-(E_v-E_F)/kT}+1} = \frac{N_c}{e^{(E_c-E_F)/kT}+1}\ .$$

(b) In this case, $e^{(E_c-E_F)/kT} \gg 1$ and $e^{-(E_v-E_F)/kT} \gg 1$. Thus, from the result of part (a),

$$\frac{N_c}{e^{(E_c-E_F)/kT}} \approx \frac{N_v}{e^{-(E_v-E_F)/kT}} ,$$

or $e^{(E_v-E_c+2E_F)/kT} \approx N_v/N_c$. We solve for E_F:

$$E_F \approx \frac{1}{2}(E_c + E_v) + \frac{1}{2}kT\ln\left(\frac{N_v}{N_c}\right) .$$

40. (a) n-type, since each phosphorous atom has one more valence electron than a silicon atom.
(b) The added charge carrier density is $n_P = 10^{-7}n_{Si} = 10^{-7}(5\times10^{28}\,\text{m}^{-3}) = 5\times10^{21}\ \text{m}^{-3}$.
(c) The ratio is $(5\times10^{21}\,\text{m}^{-3})/[2(5\times10^{15}\,\text{m}^{-3})] = 5\times10^5$. Here the factor of 2 in the denominator reflects the contribution to the charge carrier density from *both* the electrons in the conduction band *and* the holes in the valence band.

41. Sample Problem 42-6 gives the fraction of silicon atoms that must be replaced by phosphorus atoms. We find the number the silicon atoms in 1.0 g, then the number that must be replaced, and finally the mass of the replacement phosphorus atoms. The molar mass of silicon is 28.086 g/mol, so the mass of one silicon atom is $(28.086\,\text{g/mol})/(6.022\times10^{23}\,\text{mol}^{-1}) = 4.66\times10^{-23}\,\text{g}$ and the number of atoms in 1.0 g is $(1.0\,\text{g})/(4.66\times10^{-23}\,\text{g}) = 2.14\times10^{22}$. According to Sample Problem 42-6 one of every 5×10^6 silicon atoms is replaced with a phosphorus atom. This means there will be $(2.14\times10^{22})/(5\times10^6) = 4.29\times10^{15}$ phosphorus atoms in 1.0 g of silicon. The molar mass of phosphorus is 30.9758 g/mol so the mass of a phosphorus atom is $(30.9758\,\text{g/mol})/(6.022\times10^{-23}\,\text{mol}^{-1}) = 5.14\times10^{-23}\,\text{g}$. The mass of phosphorus that must be added to 1.0 g of silicon is $(4.29\times10^{15})(5.14\times10^{-23}\,\text{g}) = 2.2\times10^{-7}\,\text{g}$.

42. (a) Measured from the top of the valence band, the energy of the donor state is $E = 1.11\,\text{eV} - 0.11\,\text{eV} = 1.0\,\text{eV}$. We solve E_F from Eq. 42-6:

$$\begin{aligned} E_F &= E - kT\ln\left[P^{-1} - 1\right]) \\ &= 1.0\,\text{eV} - (8.62\times10^{-5}\,\text{eV/K})(300\,\text{K})\ln\left[(5.00\times10^{-5})^{-1} - 1\right] \\ &= 0.744\ \text{eV} . \end{aligned}$$

(b) Now $E = 1.11\,\text{eV}$, so

$$\begin{aligned} P(E) &= \frac{1}{e^{(E-E_F)/kT} + 1} = \frac{1}{e^{(1.11\,\text{eV}-0.744\,\text{eV})/[(8.62\times10^{-5}\,\text{eV/K})(300\,\text{K})]} + 1} \\ &= 7.13\times10^{-7} . \end{aligned}$$

43. (a) The probability that a state with energy E is occupied is given by

$$P(E) = \frac{1}{e^{(E-E_F)/kT} + 1}$$

where E_F is the Fermi energy, T is the temperature on the Kelvin scale, and k is the Boltzmann constant. If energies are measured from the top of the valence band, then the energy associated with a state at the bottom of the conduction band is $E = 1.11\,\text{eV}$. Furthermore, $kT = (8.62\times10^{-5}\,\text{eV/K})(300\,\text{K}) = 0.02586\,\text{eV}$. For pure silicon, $E_F = 0.555\,\text{eV}$ and $(E-E_F)/kT = (0.555\,\text{eV})/(0.02586\,\text{eV}) = 21.46$. Thus,

$$P(E) = \frac{1}{e^{21.46} + 1} = 4.79\times10^{-10} .$$

For the doped semiconductor, $(E - E_F)/kT = (0.11\,\text{eV})/(0.02586\,\text{eV}) = 4.254$ and

$$P(E) = \frac{1}{e^{4.254} + 1} = 1.40\times10^{-2} .$$

(b) The energy of the donor state, relative to the top of the valence band, is $1.11\,\text{eV} - 0.15\,\text{eV} = 0.96\,\text{eV}$. The Fermi energy is $1.11\,\text{eV} - 0.11\,\text{eV} = 1.00\,\text{eV}$. Hence, $(E - E_F)/kT = (0.96\,\text{eV} - 1.00\,\text{eV})/(0.02586\,\text{eV}) = -1.547$ and

$$P(E) = \frac{1}{e^{-1.547} + 1} = 0.824 \ .$$

44. (a) The vertical axis in the graph below is the current in nanoamperes:

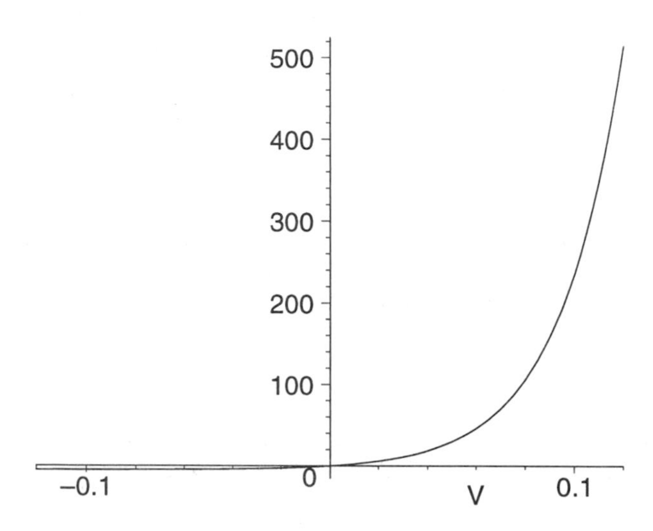

(b) The ratio is

$$\frac{i|_{v=+0.50\,\text{V}}}{i|_{v=-0.50\,\text{V}}} = \frac{i_0[e^{+0.50\,\text{eV}/[(8.62\times10^{-5}\,\text{eV/K})(300\,\text{K})]} - 1]}{i_0[e^{-0.50\,\text{eV}/[(8.62\times10^{-5}\,\text{eV/K})(300\,\text{K})]} - 1]} = 2.5 \times 10^8 \ .$$

45. The energy received by each electron is exactly the difference in energy between the bottom of the conduction band and the top of the valence band (1.1 eV). The number of electrons that can be excited across the gap by a single 662-keV photon is $N = (662 \times 10^3\,\text{eV})/(1.1\,\text{eV}) = 6.0 \times 10^5$. Since each electron that jumps the gap leaves a hole behind, this is also the number of electron-hole pairs that can be created.

46. Since (using the result of problem 3 in Chapter 39)

$$E_{\text{photon}} = \frac{hc}{\lambda} = \frac{1240\,\text{eV}\cdot\text{nm}}{140\,\text{nm}} = 8.86\,\text{eV} > 7.6\,\text{eV} \ ,$$

the light will be absorbed by the KCI crystal. Thus, the crystal is opaque to this light.

47. The valence band is essentially filled and the conduction band is essentially empty. If an electron in the valence band is to absorb a photon, the energy it receives must be sufficient to excite it across the band gap. Photons with energies less than the gap width are not absorbed and the semiconductor is transparent to this radiation. Photons with energies greater than the gap width are absorbed and the semiconductor is opaque to this radiation. Thus, the width of the band gap is the same as the energy of a photon associated with a wavelength of 295 nm. We use the result of Exercise 3 of Chapter 39 to obtain

$$E_{\text{gap}} = \frac{1240\,\text{eV}\cdot\text{nm}}{\lambda} = \frac{1240\,\text{eV}\cdot\text{nm}}{295\,\text{nm}} = 4.20\ \text{eV} \ .$$

48. We denote the maximum dimension (side length) of each transistor as ℓ_{max}, the size of the chip as A, and the number of transistors on the chip as N. Then $A = N\ell_{max}^2$. Therefore,

$$\ell_{max} = \sqrt{\frac{A}{N}} = \sqrt{\frac{(1.0\,\text{in.} \times 0.875\,\text{in.})(2.54 \times 10^{-2}\,\text{m/in.})^2}{3.5 \times 10^6}} = 1.3 \times 10^{-5}\,\text{m} = 13\,\mu\text{m} .$$

49. (a) According to Chapter 26, the capacitance is $C = \kappa\varepsilon_0 A/d$. In our case $\kappa = 4.5$, $A = (0.50\,\mu\text{m})^2$, and $d = 0.20\,\mu\text{m}$, so

$$C = \frac{\kappa\varepsilon_0 A}{d} = \frac{(4.5)(8.85 \times 10^{-12}\,\text{F/m})(0.50\,\mu\text{m})^2}{0.20\,\mu\text{m}} = 5.0 \times 10^{-17}\ \text{F} .$$

(b) Let the number of elementary charges in question be N. Then, the total amount of charges that appear in the gate is $q = Ne$. Thus, $q = Ne = CV$, which gives

$$N = \frac{CV}{e} = \frac{(5.0 \times 10^{-17}\,\text{F})(1.0\,\text{V})}{1.6 \times 10^{-19}\,\text{C}} = 3.1 \times 10^2 .$$

Chapter 43

1. In order for the α particle to penetrate the gold nucleus, the separation between the centers of mass of the two particles must be no greater than $r = r_{\text{Cu}} + r_\alpha = 6.23\,\text{fm} + 1.80\,\text{fm} = 8.03\,\text{fm}$. Thus, the minimum energy K_α is given by

$$\begin{aligned} K_\alpha &= U = \frac{1}{4\pi\varepsilon_0}\frac{q_\alpha q_{\text{Au}}}{r} = \frac{kq_\alpha q_{\text{Au}}}{r} \\ &= \frac{(8.99\times10^9\ \text{V}\cdot\text{m/C})(2e)(79)(1.60\times10^{-19}\ \text{C})}{8.03\times10^{-15}\ \text{m}} = 28.3\times10^6\ \text{eV}\ . \end{aligned}$$

We note that the factor of e in $q_\alpha = 2e$ was not set equal to 1.60×10^{-19} C, but was instead carried through to become part of the final units.

2. Our calculation is similar to that shown in Sample Problem 43-1. We set $K = 5.30\,\text{MeV} = U = (1/4\pi\varepsilon_0)(q_\alpha q_{\text{Cu}}/r_{\text{min}})$ and solve for the closest separation, r_{min} :

$$\begin{aligned} r_{\text{min}} &= \frac{q_\alpha q_{\text{Cu}}}{4\pi\varepsilon_0 K} = \frac{kq_\alpha q_{\text{Cu}}}{4\pi\varepsilon_0 K} \\ &= \frac{(2e)(29)(1.60\times10^{-19}\ \text{C})(8.99\times10^9\ \text{V}\cdot\text{m/C})}{5.30\times10^6\ \text{eV}} \\ &= 1.58\times10^{-14}\ \text{m} = 15.8\ \text{fm}\ . \end{aligned}$$

We note that the factor of e in $q_\alpha = 2e$ was not set equal to 1.60×10^{-19} C, but was instead allowed to cancel the "e" in the non-SI energy unit, electronvolt.

3. The conservation laws of (classical kinetic) energy and (linear) momentum determine the outcome of the collision. The results are given in Chapter 10, Eqs. 10-30 and 10-31. The final speed of the α particle is

$$v_{\alpha f} = \frac{m_\alpha - m_{\text{Au}}}{m_\alpha + m_{\text{Au}}}\, v_{\alpha i}\ ,$$

and that of the recoiling gold nucleus is

$$v_{\text{Au},f} = \frac{2m_\alpha}{m_\alpha + m_{\text{Au}}}\, v_{\alpha i}\ .$$

(a) Therefore, the kinetic energy of the recoiling nucleus is

$$\begin{aligned} K_{\text{Au},f} &= \frac{1}{2}m_{\text{Au}}v_{\text{Au},f}^2 \\ &= \frac{1}{2}m_{\text{Au}}\left(\frac{2m_\alpha}{m_\alpha + m_{\text{Au}}}\right)^2 v_{\alpha i}^2 = K_{\alpha i}\frac{4m_{\text{Au}}m_\alpha}{(m_\alpha + m_{\text{Au}})^2} \\ &= (5.00\ \text{MeV})\frac{4(197\ \text{u})(4.00\ \text{u})}{(4.00\ \text{u} + 197\ \text{u})^2} \\ &= 0.390\ \text{MeV}\ . \end{aligned}$$

(b) The final kinetic energy of the alpha particle is

$$\begin{aligned} K_{\alpha f} &= \frac{1}{2}m_\alpha v_{\alpha f}^2 \\ &= \frac{1}{2}m_\alpha\left(\frac{m_\alpha - m_{\rm Au}}{m_\alpha + m_{\rm Au}}\right)^2 v_{\alpha i}^2 = K_{\alpha i}\left(\frac{m_\alpha - m_{\rm Au}}{m_\alpha + m_{\rm Au}}\right)^2 \\ &= (5.00\,{\rm MeV})\left(\frac{4.00\,{\rm u} - 197\,{\rm u}}{4.00\,{\rm u} + 197\,{\rm u}}\right)^2 \\ &= 4.61\ {\rm MeV}\ . \end{aligned}$$

We note that $K_{\alpha f} + K_{{\rm Au},f} = K_{\alpha i}$ is indeed satisfied.

4. We solve for A from Eq. 43-3:

$$A = \left(\frac{r}{r_0}\right)^3 = \left(\frac{3.6\,{\rm fm}}{1.2\,{\rm fm}}\right)^3 = 27\ .$$

5. We locate a nuclide from Table 43-1 by finding the coordinate (N, Z) of the corresponding point in Fig. 43-4. It is clear that all the nuclides listed in Table 43-1 are stable except the last two, ^{227}Ac and ^{239}Pu.

6. We note that the mean density and mean radius for the Sun are given in Appendix C. Since $\rho = M/V$ where $V \propto r^3$, we get $r \propto \rho^{-1/3}$. Thus, the new radius would be

$$r = R_s\left(\frac{\rho_s}{\rho}\right)^{1/3} = (6.96\times 10^8\,{\rm m})\left(\frac{1410\,{\rm kg/m^3}}{2\times 10^{17}\,{\rm kg/m^3}}\right)^{1/3} = 1.3\times 10^4\ {\rm m}\ .$$

7. (a) 6 protons, since $Z = 6$ for carbon (see Appendix F).

(b) 8 neutrons, since $A - Z = 14 - 6 = 8$ (see Eq. 43-1).

8. The problem with Web-based services is that there are no guarantees of accuracy or that the webpage addresses will not change from the time this solution is written to the time someone reads this. Still, it is worth mentioning that a very accessible website for a wide variety of periodic table and isotope-related information is http://www.webelements.com. Two websites aimed more towards the nuclear professional are http://nucleardata.nuclear.lu.se/nucleardata and http://www.nndc.bnl.gov/nndc/ensdf, which are where some of the information mentioned below was obtained.

(a) According to Appendix F, the atomic number 60 corresponds to the element Neodymium (Nd). The first website mentioned above gives ^{142}Nd, ^{143}Nd, ^{144}Nd, ^{145}Nd, ^{146}Nd, ^{148}Nd, and ^{150}Nd in its list of naturally occurring isotopes. Two of these, ^{144}Nd and ^{150}Nd, are not perfectly stable, but their half-lives are much longer than the age of the universe (detailed information on their half-lives, modes of decay, etc are available at the last two websites referred to, above).

(b) In this list, we are asked to put the nuclides which contain 60 neutrons and which are recognized to exist but not stable nuclei (this is why, for example, ^{108}Cd is not included here). Although the problem does not ask for it, we include the half-lives of the nuclides in our list, though it must be admitted that not all reference sources agree on those values (we picked the ones we regarded as "most reliable"). Thus, we have ^{97}Rb (0.2 s), ^{98}Sr (0.7 s), ^{99}Y (2 s), ^{100}Zr (7 s), ^{101}Nb (7 s), ^{102}Mo (11 minutes), ^{103}Tc (54 s), ^{105}Rh (35 hours), ^{109}In (4 hours), ^{110}Sn (4 hours), ^{111}Sb (75 s), ^{112}Te (2 minutes), ^{113}I (7 s), ^{114}Xe (10 s), ^{115}Cs (1.4 s), and ^{116}Ba (1.4 s).

(c) We would include in this list: ^{60}Zn, ^{60}Cu, ^{60}Ni, ^{60}Co, ^{60}Fe, ^{60}Mn, ^{60}Cr, and ^{60}V.

9. Although we haven't drawn the requested lines in the following table, we can indicate their slopes: lines of constant A would have $-45°$ slopes, and those of constant $N-Z$ would have $45°$. As an example of the latter, the $N-Z=20$ line (which is one of "eighteen-neutron excess") would pass through Cd-114 at the lower left corner up through Te-122 at the upper right corner. The first column corresponds to $N=66$, and the bottom row to $Z=48$. The last column corresponds to $N=70$, and the top row to $Z=52$. Much of the information below (regarding values of $T_{1/2}$ particularly) was obtained from the websites http://nucleardata.nuclear.lu.se/nucleardata and http://www.nndc.bnl.gov/nndc/ensdf (we refer the reader to the remarks we made in the solution to problem 8).

^{118}Te 6.0 days	^{119}Te 16.0 h	^{120}Te 0.1%	^{121}Te 19.4 days	^{122}Te 2.6%
^{117}Sb 2.8 h	^{118}Sb 3.6 min	^{119}Sb 38.2 s	^{120}Sb 15.9 min	^{121}Sb 57.2%
^{116}Sn 14.5%	^{117}Sn 7.7%	^{118}Sn 24.2%	^{119}Sn 8.6%	^{120}Sn 32.6%
^{115}In 95.7%	^{116}In 14.1 s	^{117}In 43.2 min	^{118}In 5.0 s	^{119}In 2.4 min
^{114}Cd 28.7%	^{115}Cd 53.5 h	^{116}Cd 7.5%	^{117}Cd 2.5 h	^{118}Cd 50.3 min

10. (a) The atomic number $Z=39$ corresponds to the element Yttrium (see Appendix F and/or Appendix G), and $Z=53$ corresponds to Iodine.
    (b) A detailed listing of stable nuclides (such as the website http://nucleardata.nuclear.lu.se/nucleardata) shows that the stable isotope of Iodine has 74 neutrons, and that the stable isotope of Yttrium has 50 neutrons (this can also be inferred from the Molar Mass values listed in Appendix F).
    (c) The number of neutrons left over is $235-127-89=19$.

11. (a) For ^{239}Pu, $Q=94e$ and $R=6.64$ fm. Including a conversion factor for J $\to$ eV, we obtain

$$U = \frac{3Q^2}{20\pi\varepsilon_0 r} = \frac{3[94(1.60\times10^{-19}\,\mathrm{C})]^2(8.99\times10^9\,\mathrm{N\cdot m^2/C^2})}{5(6.64\times10^{-15}\,\mathrm{m})}\left(\frac{1\,\mathrm{eV}}{1.60\times10^{-19}\,\mathrm{J}}\right)$$
$$= 1.15\times10^9\,\mathrm{eV} = 1.15\ \mathrm{GeV}\ .$$

    (b) Since $Z=94$ and $A=239$, the electrostatic potential per nucleon is $1.15\,\mathrm{GeV}/239 = 4.81\,\mathrm{MeV/nucleon}$, and per proton is $1.15\,\mathrm{GeV}/94 = 12.2\,\mathrm{MeV/proton}$. These are of the same order of magnitude as the binding energy per nucleon.
    (c) The binding energy is significantly reduced by the electrostatic repulsion among the protons.

12. (a) For ^{55}Mn the mass density is

$$\rho_m = \frac{M}{V} = \frac{0.055\,\mathrm{kg/mol}}{(4\pi/3)[(1.2\times10^{-15}\,\mathrm{m})(55)^{1/3}]^3(6.02\times10^{23}/\mathrm{mol})} = 2.3\times10^{17}\ \mathrm{kg/m^3}\ ,$$

and for ^{209}Bi

$$\rho_m = \frac{M}{V} = \frac{0.209\,\mathrm{kg/mol}}{(4\pi/3)[(1.2\times10^{-15}\,\mathrm{m})(209)^{1/3}]^3(6.02\times10^{23}/\mathrm{mol})} = 2.3\times10^{17}\ \mathrm{kg/m^3}\ .$$

(b) For ^{55}Mn the charge density is

$$\rho_q = \frac{Ze}{V} = \frac{(25)(1.6 \times 10^{-19}\,\text{C})}{(4\pi/3)[(1.2 \times 10^{-15}\,\text{m})(55)^{1/3}]^3} = 1.0 \times 10^{25}\ \text{C/m}^3 \ ,$$

and for ^{209}Bi

$$\rho_q = \frac{Ze}{V} = \frac{(83)(1.6 \times 10^{-19}\,\text{C})}{(4\pi/3)[(1.2 \times 10^{-15}\,\text{m})(209)^{1/3}]^3} = 8.8 \times 10^{24}\ \text{C/m}^3 \ .$$

(c) Since $V \propto r^3 = (r_0 A^{1/3})^3 \propto A$, we expect $\rho_m \propto A/V \propto A/A \approx$ const. for all nuclides, while $\rho_q \propto Z/V \propto Z/A$ should gradually decrease since $A > 2Z$ for large nuclides.

13. The binding energy is given by $\Delta E_{\text{be}} = [Zm_H + (A - Z)m_n - M_{\text{Pu}}]\,c^2$, where Z is the atomic number (number of protons), A is the mass number (number of nucleons), m_H is the mass of a hydrogen atom, m_n is the mass of a neutron, and M_{Pu} is the mass of a $^{239}_{94}$Pu atom. In principle, nuclear masses should be used, but the mass of the Z electrons included in ZM_H is canceled by the mass of the Z electrons included in M_{Pu}, so the result is the same. First, we calculate the mass difference in atomic mass units: $\Delta m = (94)(1.00783\,\text{u}) + (239 - 94)(1.00867\,\text{u}) - (239.05216\,\text{u}) = 1.94101\,\text{u}$. Since 1 u is equivalent to 931.5 MeV, $\Delta E_{\text{be}} = (1.94101\,\text{u})(931.5\,\text{MeV/u}) = 1808\,\text{MeV}$. Since there are 239 nucleons, the binding energy per nucleon is $\Delta E_{\text{ben}} = E/A = (1808\,\text{MeV})/239 = 7.56\,\text{MeV}$.

14. (a) The mass number A is the number of nucleons in an atomic nucleus. Since $m_p \approx m_n$ the mass of the nucleus is approximately Am_p. Also, the mass of the electrons is negligible since it is much less than that of the nucleus. So $M \approx Am_p$.

(b) For ^{1}H, the approximate formula gives $M \approx Am_p = (1)(1.007276\,\text{u}) = 1.007276\,\text{u}$. The actual mass is (see Table 47-1) 1.007825 u. The percent error committed is then $\delta = (1.007825\,\text{u} - 1.007276\,\text{u})/1.007825\,\text{u} = 0.054\%$. Similarly, $\delta = 0.50\%$ for ^{7}Li, 0.81% for ^{31}P, 0.83% for ^{81}Br, 0.81% for ^{120}Sn, 0.78% for ^{157}Gd, 0.74% for ^{197}Au, 0.72% for ^{272}Ac, and 0.71% for ^{239}Pu.

(c) No. In a typical nucleus the binding energy per nucleon is several MeV, which is a bit less than 1% of the nucleon mass times c^2. This is comparable with the percent error calculated in part (b), so we need to use a more accurate method to calculate the nuclear mass.

15. (a) The de Broglie wavelength is given by $\lambda = h/p$, where p is the magnitude of the momentum. The kinetic energy K and momentum are related by Eq. 38-51, which yields

$$pc = \sqrt{K^2 + 2Kmc^2} = \sqrt{(200\,\text{MeV})^2 + 2(200\,\text{MeV})(0.511\,\text{MeV})} = 200.5\,\text{MeV} \ .$$

Thus,

$$\lambda = \frac{hc}{pc} = \frac{1240\,\text{eV}\cdot\text{nm}}{200.5 \times 10^6\,\text{eV}} = 6.18 \times 10^{-6}\,\text{nm} = 6.18\ \text{fm} \ .$$

(b) The diameter of a copper nucleus, for example, is about 8.6 fm, just a little larger than the de Broglie wavelength of a 200-MeV electron. To resolve detail, the wavelength should be smaller than the target, ideally a tenth of the diameter or less. 200-MeV electrons are perhaps at the lower limit in energy for useful probes.

16. We take the speed to be constant, and apply the classical kinetic energy formula:

$$\begin{aligned} t &= \frac{d}{v} = \frac{d}{\sqrt{2K/m}} = 2r\sqrt{\frac{m_n}{2K}} = \frac{r}{c}\sqrt{\frac{2mc^2}{K}} \\ &\approx \frac{(1.2 \times 10^{-15}\,\text{m})(100)^{1/3}}{3.0 \times 10^8\,\text{m/s}}\sqrt{\frac{2(938\,\text{MeV})}{5\,\text{MeV}}} \\ &\approx 10^{-22}\ \text{s} \ . \end{aligned}$$

17. We note that $hc = 1240\,\text{MeV}\cdot\text{fm}$ (see problem 3 of Chapter 39), and that the classical kinetic energy $\frac{1}{2}mv^2$ can be written directly in terms of the classical momentum $p = mv$ (see below). Letting $p \simeq \Delta p \simeq h/\Delta x \simeq h/r$, we get

$$E = \frac{p^2}{2m} \simeq \frac{(hc)^2}{2(mc^2)r^2} = \frac{(1240\,\text{MeV}\cdot\text{fm})^2}{2(938\,\text{MeV})[(1.2\,\text{fm})(100)^{1/3}]^2} \simeq 30\ \text{MeV} .$$

18. (a) In terms of the original value of u, the newly defined u is greater by a factor of 1.007825. So the mass of ^{1}H would be 1.000000 u, the mass of ^{12}C would be $(12.000000/1.007825)\text{u} = 11.90683\,\text{u}$, and the mass of ^{238}U would be $(238.050785/1.007825)\text{u} = 236.2025\,\text{u}$.

    (b) Defining the mass of ^{1}H to be exactly 1 does not result in any overall simplification.

19. (a) Since the nuclear force has a short range, any nucleon interacts only with its nearest neighbors, not with more distant nucleons in the nucleus. Let N be the number of neighbors that interact with any nucleon. It is independent of the number A of nucleons in the nucleus. The number of interactions in a nucleus is approximately NA, so the energy associated with the strong nuclear force is proportional to NA and, therefore, proportional to A itself.

    (b) Each proton in a nucleus interacts electrically with every other proton. The number of pairs of protons is $Z(Z-1)/2$, where Z is the number of protons. The Coulomb energy is, therefore, proportional to $Z(Z-1)$.

    (c) As A increases, Z increases at a slightly slower rate but Z^2 increases at a faster rate than A and the energy associated with Coulomb interactions increases faster than the energy associated with strong nuclear interactions.

20. (a) The first step is to add energy to produce $^4\text{He} \to p + {}^3\text{H}$, which – to make the electrons "balance" – may be rewritten as $^4\text{He} \to {}^1\text{H} + {}^3\text{H}$. The energy needed is $\Delta E_1 = (m_{^3\text{H}} + m_{^1\text{H}} - m_{^4\text{He}})c^2 = (3.01605\,\text{u} + 1.00783\,\text{u} - 4.00260\,\text{u})(931.5\,\text{MeV/u}) = 19.8\,\text{MeV}$. The second step is to add energy to produce $^3\text{H} \to n + {}^2\text{H}$. The energy needed is $\Delta E_2 = (m_{^2\text{H}} + m_n - m_{^3\text{H}})c^2 = (2.01410\,\text{u} + 1.00867\,\text{u} - 3.01605\,\text{u})(931.5\,\text{MeV/u}) = 6.26\,\text{MeV}$. The third step: $^2\text{H} \to p + n$, which – to make the electrons "balance" – may be rewritten as $^2\text{H} \to {}^1\text{H} + n$. The work required is $\Delta E_3 = (m_{^1\text{H}} + m_n - m_{^2\text{H}})c^2 = (1.00783\,\text{u} + 1.00867\,\text{u} - 2.01410\,\text{u})(931.5\,\text{MeV/u}) = 2.23\,\text{MeV}$.

    (b) The total binding energy is $\Delta E_{\text{be}} = \Delta E_1 + \Delta E_2 + \Delta E_3 = 19.8\,\text{MeV} + 6.26\,\text{MeV} + 2.23\,\text{MeV} = 28.3\,\text{MeV}$.

    (c) The binding energy per nucleon is $\Delta E_{\text{ben}} = \Delta E_{\text{be}}/A = 28.3\,\text{MeV}/4 = 7.07\,\text{MeV}$.

21. Let f_{24} be the abundance of ^{24}Mg, let f_{25} be the abundance of ^{25}Mg, and let f_{26} be the abundance of ^{26}Mg. Then, the entry in the periodic table for Mg is $24.312 = 23.98504 f_{24} + 24.98584 f_{25} + 25.98259 f_{26}$. Since there are only three isotopes, $f_{24} + f_{25} + f_{26} = 1$. We solve for f_{25} and f_{26}. The second equation gives $f_{26} = 1 - f_{24} - f_{25}$. We substitute this expression and $f_{24} = 0.7899$ into the first equation to obtain $24.312 = (23.98504)(0.7899) + 24.98584 f_{25} + 25.98259 - (25.98259)(0.7899) - 25.98259 f_{25}$. The solution is $f_{25} = 0.09303$. Then, $f_{26} = 1 - 0.7899 - 0.09303 = 0.1171$. 78.99% of naturally occurring magnesium is ^{24}Mg, 9.30% is ^{25}Mg, and 11.71% is ^{26}Mg.

22. (a) Table 43-1 gives the atomic mass of ^{1}H as $m = 1.007825\,\text{u}$. Therefore, the *mass excess* for ^{1}H is $\Delta = (1.007825\,\text{u} - 1.000000\,\text{u})(931.5\,\text{MeV/u}) = +7.29\,\text{MeV}$.

    (b) The mass of the neutron is given in Sample Problem 43-3. Thus, for the neutron, $\Delta = (1.008665\,\text{u} - 1.000000\,\text{u})(931.5\,\text{MeV/u}) = +8.07\,\text{MeV}$.

    (c) Appealing again to Table 43-1, we obtain, for ^{120}Sn, $\Delta = (119.902199\,\text{u} - 120.000000\,\text{u})(931.5\,\text{MeV/u}) = -91.10\,\text{MeV}$.

23. We first "separate" all the nucleons in one copper nucleus (which amounts to simply calculating the nuclear binding energy) and then figure the number of nuclei in the penny (so that we can multiply the

two numbers and obtain the result). To begin, we note that (using Eq. 43-1 with Appendix F and/or G) the copper-63 nucleus has 29 protons and 34 neutrons. We use the more accurate values given in Sample Problem 43-3:

$$\Delta E_{\text{be}} = (29(1.007825\,\text{u}) + 34(1.008665\,\text{u}) - 62.92960\,\text{u})\,(931.5\,\text{MeV/u}) = 551.4\text{ MeV} .$$

To figure the number of nuclei (or, equivalently, the number of atoms), we adapt Eq. 43-20:

$$N_{\text{Cu}} = \left(\frac{3.0\,\text{g}}{62.92960\,\text{g/mol}}\right)(6.02\times 10^{23}\,\text{atoms/mol}) \approx 2.9\times 10^{22}\text{ atoms} .$$

Therefore, the total energy needed is

$$N_{\text{Cu}}\Delta E_{\text{be}} = (551.4\,\text{MeV})\left(2.9\times 10^{22}\right) = 1.6\times 10^{25}\text{ MeV} .$$

24. It should be noted that when the problem statement says the "masses of the proton and the deuteron are ..." they are actually referring to the corresponding atomic masses (given to very high precision). That is, the given masses include the "orbital" electrons. As in many computations in this chapter, this circumstance (of implicitly including electron masses in what should be a purely nuclear calculation) does not cause extra difficulty in the calculation (see remarks in Sample Problems 43-4, 43-6, and 43-7). Setting the gamma ray energy equal to ΔE_{be}, we solve for the neutron mass (with each term understood to be in u units):

$$\begin{aligned} m_{\text{n}} &= M_{\text{d}} - m_{\text{H}} + \frac{E_\gamma}{c^2} \\ &= 2.0141019 - 1.007825035 + \frac{2.2233}{931.502} \\ &= 1.0062769 + 0.0023868 \end{aligned}$$

which yields $m_{\text{n}} = 1.0086637\,\text{u}$, where the last digit (7) is uncertain to within roughly ± 2 (but this depends on what precisely the uncertainties are in the given data).

25. If a nucleus contains Z protons and N neutrons, its binding energy is $\Delta E_{\text{be}} = (Zm_H + Nm_n - m)c^2$, where m_H is the mass of a hydrogen atom, m_n is the mass of a neutron, and m is the mass of the atom containing the nucleus of interest. If the masses are given in atomic mass units, then mass excesses are defined by $\Delta_H = (m_H - 1)c^2$, $\Delta_n = (m_n - 1)c^2$, and $\Delta = (m - A)c^2$. This means $m_H c^2 = \Delta_H + c^2$, $m_n c^2 = \Delta_n + c^2$, and $mc^2 = \Delta + Ac^2$. Thus $E = (Z\Delta_H + N\Delta_n - \Delta) + (Z + N - A)c^2 = Z\Delta_H + N\Delta_n - \Delta$, where $A = Z + N$ is used. For ${}^{197}_{79}\text{Au}$, $Z = 79$ and $N = 197 - 79 = 118$. Hence,

$$\Delta E_{\text{be}} = (79)(7.29\,\text{MeV}) + (118)(8.07\,\text{MeV}) - (-31.2\,\text{MeV}) = 1560\,\text{MeV} .$$

This means the binding energy per nucleon is $\Delta E_{\text{ben}} = (1560\,\text{MeV})/197 = 7.92\,\text{MeV}$.

26. (a) Since $60\,\text{y} = 2(30\,\text{y}) = 2T_{1/2}$, the fraction left is $2^{-2} = 1/4$.

    (b) Since $90\,\text{y} = 3(30\,\text{y}) = 3T_{1/2}$, the fraction that remains is $2^{-3} = 1/8$.

27. By the definition of half-life, the same has reduced to $\frac{1}{2}$ its initial amount after 140 d. Thus, reducing it to $\frac{1}{4} = (\frac{1}{2})^2$ of its initial number requires that two half-lives have passed: $t = 2T_{1/2} = 280\,\text{d}$.

28. We note that $t = 24\,\text{h}$ is four times $T_{1/2} = 6.5\,\text{h}$. Thus, it has reduced by half, four-fold:

$$\left(\frac{1}{2}\right)^4 (48\times 10^{19}) = 3\times 10^{19} .$$

29. (a) The decay rate is given by $R = \lambda N$, where λ is the disintegration constant and N is the number of undecayed nuclei. Initially, $R = R_0 = \lambda N_0$, where N_0 is the number of undecayed nuclei at that time. One must find values for both N_0 and λ. The disintegration constant is related to the half-life $T_{1/2}$ by $\lambda = (\ln 2)/T_{1/2} = (\ln 2)/(78\,\text{h}) = 8.89 \times 10^{-3}\,\text{h}^{-1}$. If M is the mass of the sample and m is the mass of a single atom of gallium, then $N_0 = M/m$. Now, $m = (67\,\text{u})(1.661 \times 10^{-24}\,\text{g/u}) = 1.113 \times 10^{-22}\,\text{g}$ and $N_0 = (3.4\,\text{g})/(1.113 \times 10^{-22}\,\text{g}) = 3.05 \times 10^{22}$. Thus $R_0 = (8.89 \times 10^{-3}\,\text{h}^{-1})(3.05 \times 10^{22}) = 2.71 \times 10^{20}\,\text{h}^{-1} = 7.53 \times 10^{16}\,\text{s}^{-1}$.

(b) The decay rate at any time t is given by

$$R = R_0\, e^{-\lambda t}$$

where R_0 is the decay rate at $t = 0$. At $t = 48\,\text{h}$, $\lambda t = (8.89 \times 10^{-3}\,\text{h}^{-1})(48\,\text{h}) = 0.427$ and

$$R = (7.53 \times 10^{16}\,\text{s}^{-1})\, e^{-0.427} = 4.91 \times 10^{16}\,\text{s}^{-1} \ .$$

30. (a) Replacing differentials with deltas in Eq. 43-11, we use the fact that $\Delta N = -12$ during $\Delta t = 1.0$ s to obtain

$$\frac{\Delta N}{N} = -\lambda \Delta t \implies \lambda = 4.8 \times 10^{-18}/\text{s}$$

where $N = 2.5 \times 10^{18}$, mentioned at the second paragraph of §43-3, is used.

(b) Eq. 43-17 yields $T_{1/2} = \ln 2/\lambda = 1.4 \times 10^{17}$ s, or about 4.6 billion years.

31. (a) The half-life $T_{1/2}$ and the disintegration constant are related by $T_{1/2} = (\ln 2)/\lambda$, so $T_{1/2} = (\ln 2)/(0.0108\,\text{h}^{-1}) = 64.2\,\text{h}$.

(b) At time t, the number of undecayed nuclei remaining is given by

$$N = N_0\, e^{-\lambda t} = N_0\, e^{-(\ln 2)t/T_{1/2}} \ .$$

We substitute $t = 3T_{1/2}$ to obtain

$$\frac{N}{N_0} = e^{-3 \ln 2} = 0.125 \ .$$

In each half-life, the number of undecayed nuclei is reduced by half. At the end of one half-life, $N = N_0/2$, at the end of two half-lives, $N = N_0/4$, and at the end of three half-lives, $N = N_0/8 = 0.125 N_0$.

(c) We use

$$N = N_0\, e^{-\lambda t} \ .$$

10.0 d is 240 h, so $\lambda t = (0.0108\,\text{h}^{-1})(240\,\text{h}) = 2.592$ and

$$\frac{N}{N_0} = e^{-2.592} = 0.0749 \ .$$

32. (a) We adapt Eq. 43-20:

$$N_{\text{Pu}} = \left(\frac{0.002\,\text{g}}{239\,\text{g/mol}}\right)(6.02 \times 10^{23}\,\text{nuclei/mol}) \approx 5 \times 10^{18}\ \text{nuclei} \ .$$

(b) Eq. 43-19 leads to

$$R = \frac{N \ln 2}{T_{1/2}} = \frac{5 \times 10^{18} \ln 2}{2.41 \times 10^4\,\text{y}} = 1.4 \times 10^{14}/\text{y}$$

which is equivalent to $4.6 \times 10^6/\text{s} = 4.6 \times 10^6$ Bq (the unit becquerel is defined in §43-3).

33. The rate of decay is given by $R = \lambda N$, where λ is the disintegration constant and N is the number of undecayed nuclei. In terms of the half-life $T_{1/2}$, the disintegration constant is $\lambda = (\ln 2)/T_{1/2}$, so

$$\begin{aligned} N &= \frac{R}{\lambda} = \frac{RT_{1/2}}{\ln 2} = \frac{(6000\,\text{Ci})(3.7\times10^{10}\,\text{s}^{-1}/\text{Ci})(5.27\,\text{y})(3.16\times10^{7}\,\text{s/y})}{\ln 2} \\ &= 5.33\times10^{22}\ \text{nuclei}\ . \end{aligned}$$

34. Using Eq. 43-14 and Eq. 43-17 (and the fact that mass is proportional to the number of atoms), the amount decayed is

$$\begin{aligned} |\Delta m| &= m|_{t_f=16.0\,\text{h}} - m|_{t_i=14.0\,\text{h}} \\ &= m_0(1-e^{-t_i \ln 2/T_{1/2}}) - m_0(1-e^{-t_f \ln 2/T_{1/2}}) \\ &= m_0(e^{-t_f \ln 2/T_{1/2}} - e^{-t_i \ln 2/T_{1/2}}) \\ &= (5.50\,\text{g})\left[e^{-(16.0\,\text{h}/12.7\,\text{h})\ln 2} - e^{-(14.0\,\text{h}/12.7\,\text{h})\ln 2}\right] \\ &= 0.256\ \text{g}\ . \end{aligned}$$

35. (a) We assume that the chlorine in the sample had the naturally occurring isotopic mixture, so the average mass number was 35.453, as given in Appendix F. Then, the mass of ^{226}Ra was

$$m = \frac{226}{226 + 2(35.453)}(0.10\,\text{g}) = 76.1\times10^{-3}\ \text{g}\ .$$

The mass of a ^{226}Ra nucleus is $(226\,\text{u})(1.661\times10^{-24}\,\text{g/u}) = 3.75\times10^{-22}\,\text{g}$, so the number of ^{226}Ra nuclei present was $N = (76.1\times10^{-3}\,\text{g})/(3.75\times10^{-22}\,\text{g}) = 2.03\times10^{20}$.

(b) The decay rate is given by $R = N\lambda = (N \ln 2)/T_{1/2}$, where λ is the disintegration constant, $T_{1/2}$ is the half-life, and N is the number of nuclei. The relationship $\lambda = (\ln 2)/T_{1/2}$ is used. Thus,

$$R = \frac{(2.03\times10^{20})\ln 2}{(1600\,\text{y})(3.156\times10^{7}\,\text{s/y})} = 2.79\times10^{9}\ \text{s}^{-1}\ .$$

36. (a) We use $R = R_0 e^{-\lambda t}$ to find t:

$$t = \frac{1}{\lambda}\ln\frac{R_0}{R} = \frac{T_{1/2}}{\ln 2}\ln\frac{R_0}{R} = \frac{14.28\,\text{d}}{\ln 2}\ln\frac{3050}{170} = 59.5\ \text{d}\ .$$

(b) The required factor is

$$\frac{R_0}{R} = e^{\lambda t} = e^{t\ln 2/T_{1/2}} = e^{(3.48\,\text{d}/14.28\,\text{d})\ln 2} = 1.18\ .$$

37. We label the two isotopes with subscripts 1 (for ^{32}P) and 2 (for ^{33}P). Initially, 10% of the decays come from ^{33}P, which implies that the initial rate $R_{02} = 9R_{01}$. Using Eq. 43-16, this means

$$R_{01} = \lambda_1 N_{01} = \frac{1}{9}R_{02} = \frac{1}{9}\lambda_2 N_{02}\ .$$

At time t, we have $R_1 = R_{01}e^{-\lambda_1 t}$ and $R_2 = R_{02}e^{-\lambda_2 t}$. We seek the value of t for which $R_1 = 9R_2$ (which means 90% of the decays arise from ^{33}P). We divide equations to obtain $(R_{01}/R_{02})e^{-(\lambda_1-\lambda_2)t} = 9$, and solve for t:

$$\begin{aligned} t &= \frac{1}{\lambda_1-\lambda_2}\ln\left(\frac{R_{01}}{9R_{02}}\right) = \frac{\ln(R_{01}/9R_{02})}{\ln 2/T_{1/2\,1} - \ln 2/T_{1/2\,2}} \\ &= \frac{\ln[(1/9)^2]}{\ln 2[(14.3\,\text{d})^{-1} - (25.3\,\text{d})^{-1}]} = 209\ \text{d}\ . \end{aligned}$$

38. We have one alpha particle (helium nucleus) produced for every plutonium nucleus that decays. To find the number that have decayed, we use Eq. 43-14, Eq. 43-17, and adapt Eq. 43-20:

$$N_0 - N = N_0\left(1 - e^{-t\ln 2/T_{1/2}}\right) = N_A \frac{12.0\,\text{g/mol}}{239\,\text{g/mol}}\left(1 - e^{-20000\ln 2/24100}\right)$$

where N_A is the Avogadro constant. This yields 1.32×10^{22} alpha particles produced. In terms of the amount of helium gas produced (assuming the α particles slow down and capture the appropriate number of electrons), this corresponds to

$$m_{\text{He}} = \left(\frac{1.32 \times 10^{22}}{6.02 \times 10^{23}/\text{mol}}\right)(4.0\,\text{g/mol}) = 87.9 \times 10^{-3}\ \text{g}\ .$$

39. The number N of undecayed nuclei present at any time and the rate of decay R at that time are related by $R = \lambda N$, where λ is the disintegration constant. The disintegration constant is related to the half-life $T_{1/2}$ by $\lambda = (\ln 2)/T_{1/2}$, so $R = (N\ln 2)/T_{1/2}$ and $T_{1/2} = (N\ln 2)/R$. Since 15.0% by mass of the sample is ^{147}Sm, the number of ^{147}Sm nuclei present in the sample is

$$N = \frac{(0.150)(1.00\,\text{g})}{(147\,\text{u})(1.661 \times 10^{-24}\,\text{g/u})} = 6.143 \times 10^{20}\ .$$

Thus

$$T_{1/2} = \frac{(6.143 \times 10^{20})\ln 2}{120\,\text{s}^{-1}} = 3.55 \times 10^{18}\,\text{s} = 1.12 \times 10^{11}\ \text{y}\ .$$

40. We note that $2.42\,\text{min} = 145.2\,\text{s}$. We are asked to plot (with SI units understood)

$$\ln R = \ln\left(R_0 e^{-\lambda t} + R_0' e^{-\lambda' t}\right)$$

where $R_0 = 3.1 \times 10^5$, $R_0' = 4.1 \times 10^6$, $\lambda = \ln 2/145.2$ and $\lambda' = \ln 2/24.6$. Our plot is shown below.

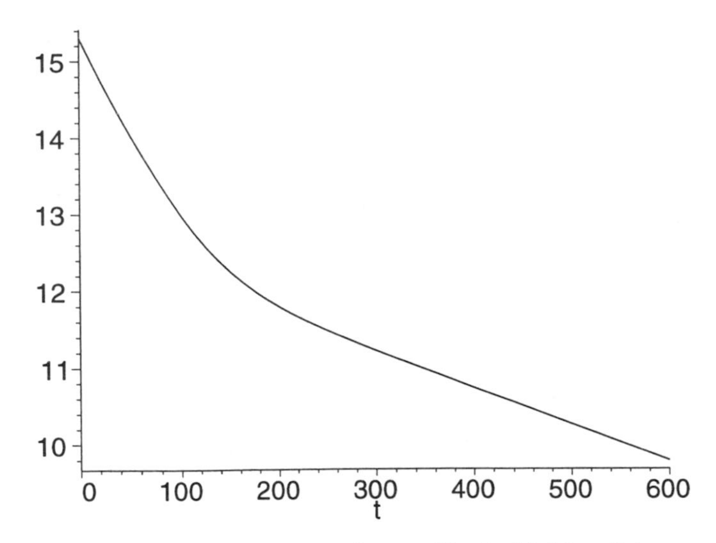

We note that the magnitude of the slope for small t is λ' (the disintegration constant for ^{110}Ag), and for large t is λ (the disintegration constant for ^{108}Ag).

41. If N is the number of undecayed nuclei present at time t, then

$$\frac{dN}{dt} = R - \lambda N$$

where R is the rate of production by the cyclotron and λ is the disintegration constant. The second term gives the rate of decay. Rearrange the equation slightly and integrate:

$$\int_{N_0}^{N} \frac{dN}{R - \lambda N} = \int_0^t dt$$

where N_0 is the number of undecayed nuclei present at time $t = 0$. This yields

$$-\frac{1}{\lambda} \ln \frac{R - \lambda N}{R - \lambda N_0} = t \, .$$

We solve for N:

$$N = \frac{R}{\lambda} + \left(N_0 - \frac{R}{\lambda}\right) e^{-\lambda t} \, .$$

After many half-lives, the exponential is small and the second term can be neglected. Then, $N = R/\lambda$, regardless of the initial value N_0. At times that are long compared to the half-life, the rate of production equals the rate of decay and N is a constant.

42. Combining Eqs. 43-19 and 43-20, we obtain

$$M_{\text{sam}} = N \frac{M_{\text{K}}}{N_{\text{A}}} = \left(\frac{RT_{1/2}}{\ln 2}\right) \left(\frac{40\,\text{g/mol}}{6.02 \times 10^{23}/\text{mol}}\right)$$

which gives 0.66 g for the mass of the sample once we plug in 1.7×10^5/s for the decay rate and $1.28 \times 10^9\,\text{y} = 4.04 \times 10^{16}\,\text{s}$ for the half-life.

43. (a) The sample is in secular equilibrium with the source and the decay rate equals the production rate. Let R be the rate of production of ^{56}Mn and let λ be the disintegration constant. According to the result of problem 41, $R = \lambda N$ after a long time has passed. Now, $\lambda N = 8.88 \times 10^{10}\,\text{s}^{-1}$, so $R = 8.88 \times 10^{10}\,\text{s}^{-1}$.

   (b) They decay at the same rate as they are produced, $8.88 \times 10^{10}\,\text{s}^{-1}$.

   (c) We use $N = R/\lambda$. If $T_{1/2}$ is the half-life, then the disintegration constant is $\lambda = (\ln 2)/T_{1/2} = (\ln 2)/(2.58\,\text{h}) = 0.269\,\text{h}^{-1} = 7.46 \times 10^{-5}\,\text{s}^{-1}$, so $N = (8.88 \times 10^{10}\,\text{s}^{-1})/(7.46 \times 10^{-5}\,\text{s}^{-1}) = 1.19 \times 10^{15}$.

   (d) The mass of a ^{56}Mn nucleus is $(56\,\text{u})(1.661 \times 10^{-24}\,\text{g/u}) = 9.30 \times 10^{-23}\,\text{g}$ and the total mass of ^{56}Mn in the sample at the end of the bombardment is $Nm = (1.19 \times 10^{15})(9.30 \times 10^{-23}\,\text{g}) = 1.11 \times 10^{-7}\,\text{g}$.

44. (a) The rate at which Radium-226 is decaying is

$$R = \lambda N = \left(\frac{\ln 2}{T_{1/2}}\right)\left(\frac{M}{m}\right) = \frac{(\ln 2)(1.00\,\text{mg})(6.02 \times 10^{23}/\text{mol})}{(1600\,\text{y})(3.15 \times 10^7\,\text{s/y})(226\,\text{g/mol})} = 3.66 \times 10^7\,\text{s}^{-1} \, .$$

   (b) Since $1600\,\text{y} \gg 3.82\,\text{d}$ the time required is $t \gg 3.82\,\text{d}$.

   (c) It is decaying at the same rate as it is produced, or $R = 3.66 \times 10^7\,\text{s}^{-1}$.

   (d) From $R_{\text{Ra}} = R_{\text{Rn}}$ and $R = \lambda N = (\ln 2/T_{1/2})(M/m)$, we get

$$\begin{aligned} M_{\text{Rn}} &= \left(\frac{T_{1/2\text{Rn}}}{T_{1/2\text{Ra}}}\right)\left(\frac{m_{\text{Rn}}}{m_{\text{Ra}}}\right) M_{\text{Ra}} \\ &= \frac{(3.82\,\text{d})(1.00 \times 10^{-3}\,\text{g})(222\,\text{u})}{(1600\,\text{y})(365\,\text{d/y})(226\,\text{u})} \\ &= 6.42 \times 10^{-9}\,\text{g} \, . \end{aligned}$$

45. Since the spreading is assumed uniform, the count rate $R = 74,000/\text{s}$ is given by $R = \lambda N = \lambda(M/m)(a/A)$, where $M = 400\,\text{g}$, m is the mass of the ^{90}Sr nucleus, $A = 2000\,\text{km}^2$, and a is the area in question. We solve for a:

$$\begin{aligned} a &= A\left(\frac{m}{M}\right)\left(\frac{R}{\lambda}\right) = \frac{AmRT_{1/2}}{M\ln 2} \\ &= \frac{(2000\times 10^6\,\text{m}^2)(90\,\text{g/mol})(29\,\text{y})(3.15\times 10^7\,\text{s/y})(74,000/\text{s})}{(400\,\text{g})(6.02\times 10^{23}/\text{mol})(\ln 2)} \\ &= 7.3\times 10^{-2}\,\text{m}^2 = 730\ \text{cm}^2\ . \end{aligned}$$

46. Eq. 25-43 gives the electrostatic potential energy between two uniformly charged spherical charges (in this case $q_1 = 2e$ and $q_2 = 90e$) with r being the distance between their centers. Assuming the "uniformly charged spheres" condition is met in this instance, we write the equation in such a way that we can make use of $k = 1/4\pi\varepsilon_0$ and the electronvolt unit:

$$U = k\frac{(2e)(90e)}{r} = \left(8.99\times 10^9\,\frac{\text{V}\cdot\text{m}}{\text{C}}\right)\frac{(3.2\times 10^{-19}\,\text{C})\,(90e)}{r} = \frac{2.59\times 10^{-7}}{r}\ \text{eV}$$

with r understood to be in meters. It is convenient to write this for r in femtometers, in which case $U = 259/r$ MeV. This is shown plotted below.

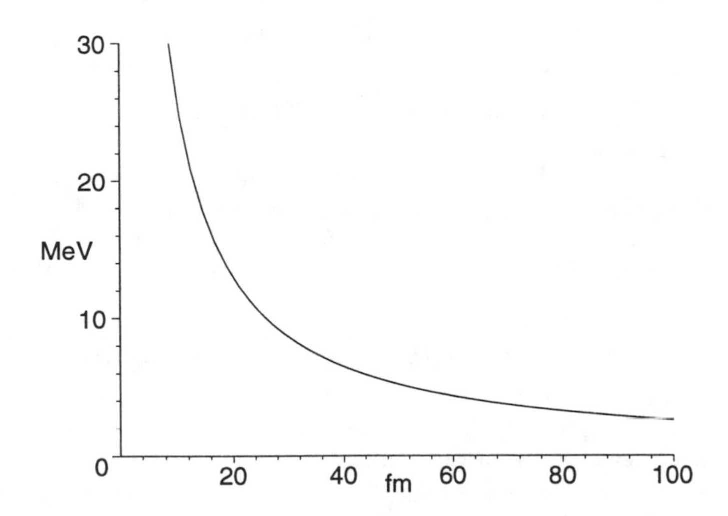

47. The fraction of undecayed nuclei remaining after time t is given by

$$\frac{N}{N_0} = e^{-\lambda t} = e^{-(\ln 2)t/T_{1/2}}$$

where λ is the disintegration constant and $T_{1/2}$ ($= (\ln 2)/\lambda$) is the half-life. The time for half the original ^{238}U nuclei to decay is 4.5×10^9 y. For ^{244}Pu at that time,

$$\frac{(\ln 2)t}{T_{1/2}} = \frac{(\ln 2)(4.5\times 10^9\,\text{y})}{8.2\times 10^7\,\text{y}} = 38.0$$

and

$$\frac{N}{N_0} = e^{-38.0} = 3.1\times 10^{-17}\ .$$

For ^{248}Cm at that time,

$$\frac{(\ln 2)t}{T_{1/2}} = \frac{(\ln 2)(4.5\times 10^9\,\text{y})}{3.4\times 10^5\,\text{y}} = 9170$$

and

$$\frac{N}{N_0} = e^{-9170} = 3.31 \times 10^{-3983} .$$

For any reasonably sized sample this is less than one nucleus and may be taken to be zero. A standard calculator probably cannot evaluate e^{-9170} directly. Our recommendation is to treat it as $(e^{-91.70})^{100}$.

48. (a) The nuclear reaction is written as $^{238}\text{U} \to {}^{234}\text{Th} + {}^{4}\text{He}$. The energy released is

$$\begin{aligned}\Delta E_1 &= (m_U - m_{He} - m_{Th})c^2 \\ &= (238.05079\,\text{u} - 4.00260\,\text{u} - 234.04363\,\text{u})(931.5\,\text{MeV/u}) \\ &= 4.25\ \text{MeV} .\end{aligned}$$

(b) The reaction series consists of $^{238}\text{U} \to {}^{237}\text{U} + n$, followed by

$$\begin{aligned}{}^{237}\text{U} &\to {}^{236}\text{Pa} + p \\ {}^{236}\text{Pa} &\to {}^{235}\text{Pa} + n \\ {}^{235}\text{Pa} &\to {}^{234}\text{Th} + p\end{aligned}$$

The net energy released is then

$$\begin{aligned}\Delta E_2 &= (m_{^{238}\text{U}} - m_{^{237}\text{U}} - m_n)c^2 + (m_{^{237}\text{U}} - m_{^{236}\text{Pa}} - m_p)c^2 \\ &\quad + (m_{^{236}\text{Pa}} - m_{^{235}\text{Pa}} - m_n)c^2 + (m_{^{235}\text{Pa}} - m_{^{234}\text{Th}} - m_p)c^2 \\ &= (m_{^{238}\text{U}} - 2m_n - 2m_p - m_{^{234}\text{Th}})c^2 \\ &= [238.05079\,\text{u} - 2(1.00867\,\text{u}) - 2(1.00783\,\text{u}) - 234.04363\,\text{u}](931.5\,\text{MeV/u}) \\ &= -24.1\ \text{MeV} .\end{aligned}$$

(c) This leads us to conclude that the binding energy of the α particle is

$$\left|(2m_n + 2m_p - m_{He})c^2\right| = |-24.1\,\text{MeV} - 4.25\,\text{MeV}| = 28.3\ \text{MeV} .$$

49. Energy and momentum are conserved. We assume the residual thorium nucleus is in its ground state. Let K_α be the kinetic energy of the alpha particle and K_{Th} be the kinetic energy of the thorium nucleus. Then, $Q = K_\alpha + K_{Th}$. We assume the uranium nucleus is initially at rest. Then, conservation of momentum yields $0 = p_\alpha + p_{Th}$, where p_α is the momentum of the alpha particle and p_{Th} is the momentum of the thorium nucleus. Both particles travel slowly enough that the classical relationship between momentum and energy can be used. Thus $K_{Th} = p_{Th}^2/2m_{Th}$, where m_{Th} is the mass of the thorium nucleus. We substitute $p_{Th} = -p_\alpha$ and use $K_\alpha = p_\alpha^2/2m_\alpha$ to obtain $K_{Th} = (m_\alpha/m_{Th})K_\alpha$. Consequently,

$$Q = K_\alpha + \frac{m_\alpha}{m_{Th}}K_\alpha = \left(1 + \frac{m_\alpha}{m_{Th}}\right)K_\alpha = \left(1 + \frac{4.00\,\text{u}}{234\,\text{u}}\right)(4.196\,\text{MeV}) = 4.27\ \text{MeV} .$$

50. (a) The disintegration energy for uranium-235 "decaying" into thorium-232 is

$$\begin{aligned}Q_3 &= \left(m_{^{235}\text{U}} - m_{^{232}\text{Th}} - m_{^{3}\text{He}}\right)c^2 \\ &= (235.0439\,\text{u} - 232.0381\,\text{u} - 3.0160\,\text{u})(931.5\,\text{MeV/u}) \\ &= -9.50\ \text{MeV} .\end{aligned}$$

(b) Similarly, the disintegration energy for uranium-235 decaying into thorium-231 is

$$\begin{aligned}Q_4 &= \left(m_{^{235}\text{U}} - m_{^{231}\text{Th}} - m_{^{4}\text{He}}\right)c^2 \\ &= (235.0439\,\text{u} - 231.0363\,\text{u} - 4.0026\,\text{u})(931.5\,\text{MeV/u}) \\ &= 4.66\ \text{MeV} .\end{aligned}$$

(c) Finally, the considered transmutation of uranium-235 into thorium-230 has a Q-value of

$$\begin{aligned} Q_5 &= \left(m_{^{235}\text{U}} - m_{^{230}\text{Th}} - m_{^{5}\text{He}}\right)c^2 \\ &= (235.0439\,\text{u} - 230.0331\,\text{u} - 5.0122\,\text{u})(931.5\,\text{MeV/u}) \\ &= -1.30\ \text{MeV}\ . \end{aligned}$$

Only the second decay process (the α decay) is spontaneous, as it releases energy.

51. (a) For the first reaction

$$\begin{aligned} Q_1 &= (m_{\text{Ra}} - m_{\text{Pb}} - m_{\text{C}})c^2 \\ &= (223.01850\,\text{u} - 208.98107\,\text{u} - 14.00324\,\text{u})(931.5\,\text{MeV/u}) \\ &= 31.8\ \text{MeV}\ , \end{aligned}$$

and for the second one

$$\begin{aligned} Q_2 &= (m_{\text{Ra}} - m_{\text{Rn}} - m_{\text{He}})c^2 \\ &= (223.01850\,\text{u} - 219.00948\,\text{u} - 4.00260\,\text{u})(931.5\,\text{MeV/u}) \\ &= 5.98\ \text{MeV}\ . \end{aligned}$$

(b) From $U \propto q_1 q_2/r$, we get

$$U_1 \approx U_2 \left(\frac{q_{\text{Pb}}\, q_C}{q_{\text{Rn}}\, q_{\text{He}}}\right) = (30.0\,\text{MeV})\frac{(82e)(6.0e)}{(86e)(2.0e)} = 86\ \text{MeV}\ .$$

52. (a) The mass number A of a radionuclide changes by 4 in an α decay and is unchanged in a β decay. If the mass numbers of two radionuclides are given by $4n + k$ and $4n' + k$ (where $k = 0,\ 1,\ 2,\ 3$), then the heavier one can decay into the lighter one by a series of α (and β) decays, as their mass numbers differ by only an integer times 4. If $A = 4n + k$, then after α-decaying for m times, its mass number becomes $A = 4n + k - 4m = 4(n - m) + k$, still in the same chain.

(b) $235 = 58 \times 4 + 3 = 4n_1 + 3$, $236 = 59 \times 4 = 4n_2$, $238 = 59 \times 4 + 2 = 4n_2 + 2$, $239 = 59 \times 4 + 3 = 4n_2 + 3$, $240 = 60 \times 4 = 4n_3$, $245 = 61 \times 4 + 1 = 4n_4 + 1$, $246 = 61 \times 4 + 2 = 4n_4 + 2$, $249 = 62 \times 4 + 1 = 4n_5 + 1$, $253 = 63 \times 4 + 1 = 4n_6 + 1$.

53. Let ${}^{A}_{Z}X$ represent the unknown nuclide. The reaction equation is

$${}^{A}_{Z}X + {}^{1}_{0}\text{n} \to {}^{0}_{-1}\text{e} + 2\,{}^{4}_{2}\text{He}\ .$$

Conservation of charge yields $Z + 0 = -1 + 4$ or $Z = 3$. Conservation of mass number yields $A + 1 = 0 + 8$ or $A = 7$. According to the periodic table in Appendix G (also see Appendix F), lithium has atomic number 3, so the nuclide must be ${}^{7}_{3}\text{Li}$.

54. (a) We recall that $mc^2 = 0.511$ MeV from Table 38-3, and note that the result of problem 3 in Chapter 39 can be written as $hc = 1240\,\text{MeV}\cdot\text{fm}$. Using Eq. 38-51 and Eq. 39-13, we obtain

$$\begin{aligned} \lambda &= \frac{h}{p} = \frac{hc}{\sqrt{K^2 + 2Kmc^2}} \\ &= \frac{1240\,\text{MeV}\cdot\text{fm}}{\sqrt{(1.0\,\text{MeV})^2 + 2(1.0\,\text{MeV})(0.511\,\text{MeV})}} = 9.0 \times 10^2\ \text{fm}\ . \end{aligned}$$

(b) $r = r_0 A^{1/3} = (1.2\,\text{fm})(150)^{1/3} = 6.4\,\text{fm}$.

(c) Since $\lambda \gg r$ the electron cannot be confined in the nuclide. We recall from Chapters 40 and 41, that at least $\lambda/2$ was needed in any particular direction, to support a standing wave in an "infinite well." A finite well is able to support *slightly* less than $\lambda/2$ (as one can infer from the ground state wavefunction in Fig. 40-8), but in the present case λ/r is far too big to be supported.

(d) A strong case can be made on the basis of the remarks in part (c), above.

55. Let $M_{\rm Cs}$ be the mass of one atom of ${}^{137}_{55}$Cs and $M_{\rm Ba}$ be the mass of one atom of ${}^{137}_{56}$Ba. To obtain the nuclear masses, we must subtract the mass of 55 electrons from $M_{\rm Cs}$ and the mass of 56 electrons from $M_{\rm Ba}$. The energy released is $Q = [(M_{\rm Cs} - 55m) - (M_{\rm Ba} - 56m) - m]\,c^2$, where m is the mass of an electron. Once cancellations have been made, $Q = (M_{\rm Cs} - M_{\rm Ba})c^2$ is obtained. Therefore,

$$Q = [136.9071\,\text{u} - 136.9058\,\text{u}]\,c^2 = (0.0013\,\text{u})c^2 = (0.0013\,\text{u})(931.5\,\text{MeV/u}) = 1.21\ \text{MeV}\ .$$

56. Assuming the neutrino has negligible mass, then

$$\Delta m\,c^2 = (\mathbf{m}_{\rm Ti} - \mathbf{m}_{\rm V} - m_e)\,c^2\ .$$

Now, since Vanadium has 23 electrons (see Appendix F and/or G) and Titanium has 22 electrons, we can add and subtract $22m_e$ to the above expression and obtain

$$\Delta m\,c^2 = (\mathbf{m}_{\rm Ti} + 22m_e - \mathbf{m}_{\rm V} - 23m_e)\,c^2 = (m_{\rm Ti} - m_{\rm V}\,)\,c^2\ .$$

We note that our final expression for Δmc^2 involves the *atomic* masses, and that this assumes (due to the way they are usually tabulated) the atoms are in the ground states (which is certainly not the case here, as we discuss below). The question now is: do we set $Q = -\Delta mc^2$ as in Sample Problem 43-7? The answer is "no." The atom is left in an excited (high energy) state due to the fact that an electron was captured from the lowest shell (where the absolute value of the energy, E_K, is quite large for large Z – see Eq. 41-25). To a very good approximation, the energy of the K-shell electron in Vanadium is equal to that in Titanium (where there is now a "vacancy" that must be filled by a readjustment of the whole electron cloud), and we write $Q = -\Delta mc^2 - E_K$ so that Eq. 43-27 still holds. Thus,

$$Q = (m_{\rm V} - m_{\rm Ti}\,)\,c^2 - E_K\ .$$

57. The decay scheme is $\text{n} \to \text{p} + \text{e}^- + \nu$. The electron kinetic energy is a maximum if no neutrino is emitted. Then, $K_{\rm max} = (m_n - m_p - m_e)c^2$, where m_n is the mass of a neutron, m_p is the mass of a proton, and m_e is the mass of an electron. Since $m_p + m_e = m_H$, where m_H is the mass of a hydrogen atom, this can be written $K_{\rm max} = (m_n - m_H)c^2$. Hence, $K_{\rm max} = (840 \times 10^{-6}\,\text{u})c^2 = (840 \times 10^{-6}\,\text{u})(931.5\,\text{MeV/u}) =$ 0.783 MeV.

58. We obtain

$$\begin{aligned} Q &= (m_{\rm V} - m_{\rm Ti}\,)\,c^2 - E_K \\ &= (48.94852\,\text{u} - 48.94787\,\text{u})\,(931.5\,\text{MeV/u}) - 0.00547\,\text{MeV} \\ &= 0.600\ \text{MeV}\ . \end{aligned}$$

59. (a) Since the positron has the same mass as an electron, and the neutrino has negligible mass, then

$$\Delta m\,c^2 = (\mathbf{m}_{\rm B} + m_e - \mathbf{m}_{\rm C})\,c^2\ .$$

Now, since Carbon has 6 electrons (see Appendix F and/or G) and Boron has 5 electrons, we can add and subtract $6m_e$ to the above expression and obtain

$$\Delta m\,c^2 = (\mathbf{m}_{\rm B} + 7m_e - \mathbf{m}_{\rm C} - 6m_e)\,c^2 = (m_{\rm B} + 2m_e - m_{\rm C}\,)\,c^2\ .$$

We note that our final expression for Δmc^2 involves the *atomic* masses, as well an "extra" term corresponding to two electron masses. From Eq. 38-47 and Table 38-3, we obtain

$$Q = (m_{\rm C} - m_{\rm B} - 2m_e)\,c^2 = (m_{\rm C} - m_{\rm B})\,c^2 - 2(0.511\,\text{MeV})\ .$$

(b) The disintegration energy for the positron decay of Carbon-11 is

$$Q = (11.011434\,\mathrm{u} - 11.009305\,\mathrm{u})\,(931.5\,\mathrm{MeV/u}) - 1.022\,\mathrm{MeV} = 0.961\ \mathrm{MeV}\ .$$

60. (a) The rate of heat production is

$$\begin{aligned}\frac{dE}{dt} &= \sum_{i=1}^{3} R_i Q_i = \sum_{i=1}^{3} \lambda_1 N_i Q_i = \sum_{i=1}^{3}\left(\frac{\ln 2}{T_{1/2_i}}\right)\frac{(1.00\,\mathrm{kg})f_i}{m_i}Q_i \\ &= \frac{(1.00\,\mathrm{kg})(\ln 2)(1.60\times10^{-13}\,\mathrm{J/MeV})}{(3.15\times10^{7}\,\mathrm{s/y})(1.661\times10^{-27}\,\mathrm{kg/u})}\left[\frac{(4\times10^{-6})(51.7\,\mathrm{MeV})}{(238\,\mathrm{u})(4.47\times10^{9}\,\mathrm{y})}\right. \\ &\qquad \left.+\frac{(13\times10^{-6})(42.7\,\mathrm{MeV})}{(232\,\mathrm{u})(1.41\times10^{10}\,\mathrm{y})}+\frac{(4\times10^{-6})(1.31\,\mathrm{MeV})}{(40\,\mathrm{u})(1.28\times10^{9}\,\mathrm{y})}\right] \\ &= 1.0\times10^{-9}\ \mathrm{W}\ .\end{aligned}$$

(b) The contribution to heating, due to radioactivity, is $P = (2.7\times10^{22}\,\mathrm{kg})(1.0\times10^{-9}\,\mathrm{W/kg}) = 2.7\times10^{13}\,\mathrm{W}$, which is very small compared to what is received from the Sun.

61. Since the electron has the maximum possible kinetic energy, no neutrino is emitted. Since momentum is conserved, the momentum of the electron and the momentum of the residual sulfur nucleus are equal in magnitude and opposite in direction. If p_e is the momentum of the electron and p_S is the momentum of the sulfur nucleus, then $p_S = -p_e$. The kinetic energy K_S of the sulfur nucleus is $K_S = p_S^2/2M_S = p_e^2/2M_S$, where M_S is the mass of the sulfur nucleus. Now, the electron's kinetic energy K_e is related to its momentum by the relativistic equation $(p_ec)^2 = K_e^2 + 2K_emc^2$, where m is the mass of an electron. See Eq. 38-51. Thus,

$$\begin{aligned}K_S &= \frac{(p_ec)^2}{2M_Sc^2} = \frac{K_e^2+2K_emc^2}{2M_Sc^2} = \frac{(1.71\,\mathrm{MeV})^2+2(1.71\,\mathrm{MeV})(0.511\,\mathrm{MeV})}{2(32\,\mathrm{u})(931.5\,\mathrm{MeV/u})} \\ &= 7.83\times10^{-5}\,\mathrm{MeV} = 78.3\ \mathrm{eV}\end{aligned}$$

where $mc^2 = 0.511\,\mathrm{MeV}$ is used (see Table 38-3).

62. We solve for t from $R = R_0e^{-\lambda t}$:

$$t = \frac{1}{\lambda}\ln\frac{R_0}{R} = \left(\frac{5730\,\mathrm{y}}{\ln 2}\right)\ln\left[\left(\frac{15.3}{63.0}\right)\left(\frac{5.00}{1.00}\right)\right] = 1.61\times10^{3}\ \mathrm{y}\ .$$

63. (a) The mass of a ^{238}U atom is $(238\,\mathrm{u})(1.661\times10^{-24}\,\mathrm{g/u}) = 3.95\times10^{-22}\,\mathrm{g}$, so the number of uranium atoms in the rock is $N_U = (4.20\times10^{-3}\,\mathrm{g})/(3.95\times10^{-22}\,\mathrm{g}) = 1.06\times10^{19}$. The mass of a ^{206}Pb atom is $(206\,\mathrm{u})(1.661\times10^{-24}\,\mathrm{g}) = 3.42\times10^{-22}\,\mathrm{g}$, so the number of lead atoms in the rock is $N_{\mathrm{Pb}} = (2.135\times10^{-3}\,\mathrm{g})/(3.42\times10^{-22}\,\mathrm{g}) = 6.24\times10^{18}$.

(b) If no lead was lost, there was originally one uranium atom for each lead atom formed by decay, in addition to the uranium atoms that did not yet decay. Thus, the original number of uranium atoms was $N_{U0} = N_U + N_{\mathrm{Pb}} = 1.06\times10^{19} + 6.24\times10^{18} = 1.68\times10^{19}$.

(c) We use

$$N_U = N_{U0}\,e^{-\lambda t}$$

where λ is the disintegration constant for the decay. It is related to the half-life $T_{1/2}$ by $\lambda = (\ln 2)/T_{1/2}$. Thus

$$t = -\frac{1}{\lambda}\ln\left(\frac{N_U}{N_{U0}}\right) = -\frac{T_{1/2}}{\ln 2}\ln\left(\frac{N_U}{N_{U0}}\right) = -\frac{4.47\times10^{9}\,\mathrm{y}}{\ln 2}\ln\left(\frac{1.06\times10^{19}}{1.68\times10^{19}}\right) = 2.97\times10^{9}\ \mathrm{y}\ .$$

64. The original amount of ^{238}U the rock contains is given by

$$m_0 = me^{\lambda t} = (3.70\,\text{mg})\, e^{(\ln 2)(260\times10^6\,\text{y})/(4.47\times10^9\,\text{y})} = 3.85 \text{ mg} .$$

Thus, the amount of lead produced is

$$m' = (m_0 - m)\left(\frac{m_{206}}{m_{238}}\right) = (3.85\,\text{mg} - 3.70\,\text{mg})\left(\frac{206}{238}\right) = 0.132 \text{ mg} .$$

65. We can find the age t of the rock from the masses of ^{238}U and ^{206}Pb. The initial mass of ^{238}U is

$$m_{\text{U}_0} = m_\text{U} + \frac{238}{206} m_\text{Pb} .$$

Therefore, $m_\text{U} = m_{\text{U}_0} e^{-\lambda_\text{U} t} = (m_\text{U} + m_{^{238}\text{Pb}}/206)e^{-(t\ln 2)/T_{1/2\text{U}}}$. We solve for t:

$$\begin{aligned} t &= \frac{T_{1/2\text{U}}}{\ln 2} \ln\left(\frac{m_\text{U} + (238/206)m_\text{Pb}}{m_\text{U}}\right) \\ &= \frac{4.47\times10^9\,\text{y}}{\ln 2} \ln\left[1 + \left(\frac{238}{206}\right)\left(\frac{0.15\,\text{mg}}{0.86\,\text{mg}}\right)\right] \\ &= 1.18\times10^9 \text{ y} . \end{aligned}$$

For the β decay of ^{40}K, the initial mass of ^{40}K is

$$m_{\text{K}_0} = m_\text{K} + (40/40)m_\text{Ar} = m_\text{K} + m_\text{Ar} ,$$

so

$$m_\text{K} = m_{\text{K}_0} e^{-\lambda_\text{K} t} = (m_\text{K} + m_\text{Ar})e^{-\lambda_\text{K} t} .$$

We solve for m_K:

$$\begin{aligned} m_\text{K} &= \frac{m_\text{Ar} e^{-\lambda_\text{K} t}}{1 - e^{-\lambda_\text{K} t}} = \frac{m_\text{Ar}}{e^{\lambda_\text{K} t} - 1} \\ &= \frac{1.6\,\text{mg}}{e^{(\ln 2)(1.18\times10^9\,\text{y})/(1.25\times10^9\,\text{y})} - 1} = 1.7 \text{ mg} . \end{aligned}$$

66. The becquerel (Bq) and curie (Ci) are defined in §43-3. Thus, $R = 8700/60 = 145$ Bq, and

$$R = \frac{145\,\text{Bq}}{3.7\times10^{10}\,\text{Bq/Ci}} = 3.92\times10^{-9} \text{ Ci} .$$

67. The decay rate R is related to the number of nuclei N by $R = \lambda N$, where λ is the disintegration constant. The disintegration constant is related to the half-life $T_{1/2}$ by $\lambda = (\ln 2)/T_{1/2}$, so $N = R/\lambda = RT_{1/2}/\ln 2$. Since $1\,\text{Ci} = 3.7\times10^{10}$ disintegrations/s,

$$N = \frac{(250\,\text{Ci})(3.7\times10^{10}\,\text{s}^{-1}/\text{Ci})(2.7\,\text{d})(8.64\times10^4\,\text{s/d})}{\ln 2} = 3.11\times10^{18} .$$

The mass of a ^{198}Au atom is $M = (198\,\text{u})(1.661\times10^{-24}\,\text{g/u}) = 3.29\times10^{-22}$ g, so the mass required is $NM = (3.11\times10^{18})(3.29\times10^{-22}\,\text{g}) = 1.02\times10^{-3}\,\text{g} = 1.02\,\text{mg}$.

68. The annual dose equivalent is $(20\,\text{h})(52\,\text{week/y})(7.0\,\mu\text{Sv/h}) = 7.3\,\text{mSv}$.

69. The dose equivalent is the product of the absorbed dose and the RBE factor, so the absorbed dose is (dose equivalent)/(RBE) $= (250\times10^{-6}\,\text{Sv})/(0.85) = 2.94\times10^{-4}\,\text{Gy}$. But $1\,\text{Gy} = 1\,\text{J/kg}$, so the absorbed dose is

$$(2.94\times10^{-4}\,\text{Gy})\left(1\,\frac{\text{J}}{\text{kg}\cdot\text{Gy}}\right) = 2.94\times10^{-4} \text{ J/kg} .$$

To obtain the total energy received, we multiply this by the mass receiving the energy: $E = (2.94\times10^{-4}\,\text{J/kg})(44\,\text{kg}) = 1.29\times10^{-2}\,\text{J}$.

70. (a) Using Eq. 43-31, the energy absorbed is

$$(2.4 \times 10^{-4}\,\text{Gy})(75\,\text{kg}) = 18\,\text{mJ} .$$

(b) The dose equivalent is

$$(2.4 \times 10^{-4}\,\text{Gy})(12) = 2.9 \times 10^{-3}\,\text{Sv} = 0.29\text{ rem}$$

where Eq. 43-32 is used in the last step.

71. (a) Adapting Eq. 43-20, we find

$$N_0 = \frac{(2.5 \times 10^{-3}\,\text{g})(6.02 \times 10^{23}/\text{mol})}{239\,\text{g/mol}} = 6.3 \times 10^{18} .$$

(b) From Eq. 43-14 and Eq. 43-17,

$$\begin{aligned} |\Delta N| &= N_0 \left[1 - e^{-t \ln 2/T_{1/2}}\right] \\ &= (6.3 \times 10^{18}) \left[1 - e^{-(12\,\text{h}) \ln 2/(24{,}100\,\text{y})(8760\,\text{h/y})}\right] \\ &= 2.5 \times 10^{11} . \end{aligned}$$

(c) The energy absorbed by the body is

$$(0.95)E_\alpha |\Delta N| = (0.95)\,(5.2\,\text{MeV})\,(2.5 \times 10^{11})\,(1.6 \times 10^{-13}\,\text{J/MeV}) = 0.20\text{ J} .$$

(d) On a per unit mass basis, the previous result becomes (according to Eq. 43-31)

$$\frac{0.20\,\text{mJ}}{85\,\text{kg}} = 2.3 \times 10^{-3}\,\text{J/kg} = 2.3\text{ mGy} .$$

(e) Using Eq. 43-32, $(2.3\,\text{mGy})(13) = 30\,\text{mSv}$.

72. From Eq. 20-24, we obtain

$$T = \frac{2}{3}\left(\frac{K_{\text{avg}}}{k}\right) = \frac{2}{3}\left(\frac{5.00 \times 10^{6}\,\text{eV}}{8.62 \times 10^{-5}\,\text{eV/K}}\right) = 3.9 \times 10^{10}\text{ K} .$$

73. (a) Following Sample Problem 43-10, we compute

$$\Delta E \approx \frac{\hbar}{t_{\text{avg}}} = \frac{(4.14 \times 10^{-15}\,\text{eV}\cdot\text{fs})/2\pi}{1.0 \times 10^{-22}\,\text{s}} = 6.6 \times 10^{6}\text{ eV} .$$

(b) In order to fully distribute the energy in a fairly large nucleus, and create a "compound nucleus" equilibrium configuration, about 10^{-15} s is typically required. A reaction state that exists no more than about 10^{-22} s does not qualify as a compound nucleus.

74. (a) We compare both the proton numbers (atomic numbers, which can be found in Appendix F and/or G) and the neutron numbers (see Eq. 43-1) with the magic nucleon numbers (special values of either Z or N) listed in §43-8. We find that ^{18}O, ^{60}Ni, ^{92}Mo, ^{144}Sm, and ^{207}Pb each have a filled shell for either the protons or the neutrons (two of these, ^{18}O and ^{92}Mo, are explicitly discussed in that section).

(b) Consider ^{40}K, which has $Z = 19$ protons (which is one less than the magic number 20). It has $N = 21$ neutrons, so it has one neutron outside a closed shell for neutrons, and thus qualifies for this list. Others in this list include ^{91}Zr, ^{121}Sb, and ^{143}Nd.

(c) Consider ^{13}C, which has $Z = 6$ and $N = 13 - 6 = 7$ neutrons. Since 8 is a magic number, then ^{13}C has a vacancy in an otherwise filled shell for neutrons. Similar arguments lead to inclusion of ^{40}K, ^{49}Ti, ^{205}Tl, and ^{207}Pb in this list.

75. A generalized formation reaction can be written X + x → Y, where X is the target nucleus, x is the incident light particle, and Y is the excited compound nucleus (^{20}Ne). We assume X is initially at rest. Then, conservation of energy yields

$$m_X c^2 + m_x c^2 + K_x = m_Y c^2 + K_Y + E_Y$$

where m_X, m_x, and m_Y are masses, K_x and K_Y are kinetic energies, and E_Y is the excitation energy of Y. Conservation of momentum yields

$$p_x = p_Y \ .$$

Now, $K_Y = p_Y^2/2m_Y = p_x^2/2m_Y = (m_x/m_Y)K_x$, so

$$m_X c^2 + m_x c^2 + K_x = m_Y c^2 + (m_x/m_Y)K_x + E_Y$$

and

$$K_x = \frac{m_Y}{m_Y - m_x}\left[(m_Y - m_X - m_x)c^2 + E_Y\right] \ .$$

(a) Let x represent the alpha particle and X represent the ^{16}O nucleus. Then, $(m_Y - m_X - m_x)c^2 = (19.99244\,\text{u} - 15.99491\,\text{u} - 4.00260\,\text{u})(931.5\,\text{MeV/u}) = -4.722\,\text{MeV}$ and

$$K_\alpha = \frac{19.99244\,\text{u}}{19.99244\,\text{u} - 4.00260\,\text{u}}(-4.722\,\text{MeV} + 25.0\,\text{MeV}) = 25.35\,\text{MeV} \ .$$

(b) Let x represent the proton and X represent the ^{19}F nucleus. Then, $(m_Y - m_X - m_x)c^2 = (19.99244\,\text{u} - 18.99841\,\text{u} - 1.00783\,\text{u})(931.5\,\text{MeV/u}) = -12.85\,\text{MeV}$ and

$$K_\alpha = \frac{19.99244\,\text{u}}{19.99244\,\text{u} - 1.00783\,\text{u}}(-12.85\,\text{MeV} + 25.0\,\text{MeV}) = 12.80\,\text{MeV} \ .$$

(c) Let x represent the photon and X represent the ^{20}Ne nucleus. Since the mass of the photon is zero, we must rewrite the conservation of energy equation: if E_γ is the energy of the photon, then $E_\gamma + m_X c^2 = m_Y c^2 + K_Y + E_Y$. Since $m_X = m_Y$, this equation becomes $E_\gamma = K_Y + E_Y$. Since the momentum and energy of a photon are related by $p_\gamma = E_\gamma/c$, the conservation of momentum equation becomes $E_\gamma/c = p_Y$. The kinetic energy of the compound nucleus is $K_Y = p_Y^2/2m_Y = E_\gamma^2/2m_Y c^2$. We substitute this result into the conservation of energy equation to obtain

$$E_\gamma = \frac{E_\gamma^2}{2m_Y c^2} + E_Y \ .$$

This quadratic equation has the solutions

$$E_\gamma = m_Y c^2 \pm \sqrt{(m_Y c^2)^2 - 2m_Y c^2 E_Y} \ .$$

If the problem is solved using the relativistic relationship between the energy and momentum of the compound nucleus, only one solution would be obtained, the one corresponding to the negative sign above. Since $m_Y c^2 = (19.99244\,\text{u})(931.5\,\text{MeV/u}) = 1.862 \times 10^4\,\text{MeV}$,

$$\begin{aligned} E_\gamma &= (1.862 \times 10^4\,\text{MeV}) - \sqrt{(1.862 \times 10^4\,\text{MeV})^2 - 2(1.862 \times 10^4\,\text{MeV})(25.0\,\text{MeV})} \\ &= 25.0\,\text{MeV} \ . \end{aligned}$$

The kinetic energy of the compound nucleus is very small; essentially all of the photon energy goes to excite the nucleus.

76. (a) From the decay series, we know that N_{210}, the amount of ^{210}Pb nuclei, changes because of two decays: the decay from ^{226}Ra into ^{210}Pb at the rate $R_{226} = \lambda_{226}N_{226}$, and the decay from ^{210}Pb into ^{206}Pb at the rate $R_{210} = \lambda_{210}N_{210}$. The first of these decays causes N_{210} to increase while the second one causes it to decrease. Thus,

$$\frac{dN_{210}}{dt} = R_{226} - R_{210} = \lambda_{226}N_{226} - \lambda_{210}N_{210} \ .$$

(b) We set $dN_{210}/dt = R_{226} - R_{210} = 0$ to obtain $R_{226}/R_{210} = 1$.

(c) From $R_{226} = \lambda_{226}N_{226} = R_{210} = \lambda_{210}N_{210}$, we obtain

$$\frac{N_{226}}{N_{210}} = \frac{\lambda_{210}}{\lambda_{226}} = \frac{T_{1/2_{226}}}{T_{1/2_{210}}} = \frac{1.60 \times 10^3\,\text{y}}{22.6\,\text{y}} = 70.8 \ .$$

(d) Since only 1.00% of the ^{226}Ra remains, the ratio R_{226}/R_{210} is 0.00100 of that of the equilibrium state computed in part (b). Thus the ratio is $(0.0100)(1) = 0.0100$.

(e) This is similar to part (d) above. Since only 1.00% of the ^{226}Ra remains, the ratio N_{226}/N_{210} is 1.00% of that of the equilibrium state computed in part (c), or $(0.0100)(70.8) = 0.708$.

(f) Since the actual value of N_{226}/N_{210} is 0.09, which much closer to 0.0100 than to 1, the sample of the lead pigment cannot be 300 years old. So *Emmaus* is not a *Vermeer*.

77. Using Eq. 43-14 with Eq. 43-17, we find the fraction remaining:

$$\frac{N}{N_0} = e^{-t\ln 2/T_{1/2}} = e^{-30\ln 2/29} = 0.49 \ .$$

78. Using Eq. 43-15 with Eq. 43-17, we find the initial activity:

$$R_0 = Re^{t\ln 2/T_{1/2}} = (7.4 \times 10^8\,\text{Bq})\,e^{24\ln 2/83.61} = 9.0 \times 10^8\,\text{Bq} \ .$$

79. (a) Molybdenum beta decays into Technetium:

$$^{99}_{42}\text{Mo} \rightarrow {}^{99}_{43}\text{Tc} + e^- + \nu$$

(b) Each decay corresponds to a photon produced when the Technetium nucleus de-excites [note that the de-excitation half-life is much less than the beta decay half-life]. Thus, the gamma rate is the same as the decay rate: 8.2×10^7/s.

(c) Eq. 43-19 leads to

$$N = \frac{RT_{1/2}}{\ln 2} = \frac{(38/\text{s})(6.0\,\text{h})(3600\,\text{s/h})}{\ln 2} = 1.2 \times 10^6 \ .$$

80. (a) Assuming a "target" area of one square meter, we establish a ratio:

$$\frac{\text{rate through you}}{\text{total rate upward}} = \frac{1\,\text{m}^2}{(2.6 \times 10^5\,\text{km}^2)(1000\,\text{m/km})^2} = 3.8 \times 10^{-12} \ .$$

The SI unit becquerel is equivalent to a disintegration per second. With half the beta-decay electrons moving upward, we find

$$\text{rate through you} = \frac{1}{2}\left(1 \times 10^{16}/\text{s}\right)\left(3.8 \times 10^{-12}\right) = 1.9 \times 10^4/\text{s}$$

which implies (converting s $\rightarrow$ h) the rate of electrons you would intercept is $R_0 = 7 \times 10^7/\text{h}$.

(b) Let D indicate the current year (2000, 2001, etc) Combining Eq. 43-15 and Eq. 43-17, we find

$$R = R_0 e^{-t\ln 2/T_{1/2}} = \left(7 \times 10^7/\text{h}\right) e^{-(D-1996)\ln 2/(30.2\,\text{y})} \ .$$

81. Eq. 43-19 leads to

$$\begin{aligned} R &= \frac{\ln 2}{T_{1/2}} N \\ &= \frac{\ln 2}{30.2\,\text{y}} \left(\frac{M_{\text{sam}}}{m_{\text{atom}}} \right) \\ &= \frac{\ln 2}{9.53 \times 10^8\,\text{s}} \left(\frac{0.0010\,\text{kg}}{137 \times 1.661 \times 10^{-27}\,\text{kg}} \right) \\ &= 3.2 \times 10^{12}\ \text{Bq} \ = \ 86\ \text{Ci}\ . \end{aligned}$$

82. The lines that lead toward the lower left are alpha decays, involving an atomic number change of $\Delta Z_\alpha = -2$ and a mass number change of $\Delta A_\alpha = -4$. The short horizontal lines toward the right are beta decays (involving electrons, not positrons) in which case A stays the same but the change in atomic number is $\Delta Z_\beta = +1$. Fig. 43-16 shows three alpha decays and two beta decays; thus,

$$Z_f = Z_i + 3\Delta Z_\alpha + 2\Delta Z_\beta \quad \text{and} \quad A_f = A_i + 3\Delta A_\alpha\ .$$

Referring to Appendix F or G, we find $Z_i = 93$ for Neptunium, so $Z_f = 93 + 3(-2) + 2(1) = 89$, which indicates the element Actinium. We are given $A_i = 237$, so $A_f = 237 + 3(-4) = 225$. Therefore, the final isotope is ^{225}Ac.

83. We note that every Calcium-40 atom and Krypton-40 atom found now in the sample was once one of the original number of Potassium atoms. Thus, using Eq. 43-13 and Eq. 43-17, we find

$$\begin{aligned} \ln\left(\frac{N_\text{K}}{N_\text{K} + N_\text{Ar} + N_\text{Ca}} \right) &= -\lambda t \\ \ln\left(\frac{1}{1 + 1 + 8.54} \right) &= -\frac{\ln 2}{T_{1/2}} t \end{aligned}$$

which (with $T_{1/2} = 1.26 \times 10^9$ y) yields $t = 4.3 \times 10^9$ y.

84. We note that 3.82 days is 330048 s, and that a becquerel is a disintegration per second (see §43-3). From Eq. 34-19, we have

$$\frac{N}{\mathcal{V}} = \frac{R}{\mathcal{V}} \frac{T_{1/2}}{\ln 2} = \left(1.55 \times 10^5\ \frac{\text{Bq}}{\text{m}^3} \right) \frac{330048\,\text{s}}{\ln 2} = 7.4 \times 10^{10}\ \frac{\text{atoms}}{\text{m}^3}$$

where we have divided by volume $\mathcal{V}$. We estimate $\mathcal{V}$ (the volume breathed in 48 h = 2880 min) as follows:

$$\left(2\,\frac{\text{Liters}}{\text{breath}} \right) \left(\frac{1\,\text{m}^3}{1000\,\text{L}} \right) \left(40\,\frac{\text{breaths}}{\text{min}} \right) (2880\,\text{min})$$

which yields $\mathcal{V} \approx 200\,\text{m}^3$. Thus, the order of magnitude of N is

$$\left(\frac{N}{\mathcal{V}} \right) (\mathcal{V}) \approx \left(7 \times 10^{10}\,\frac{\text{atoms}}{\text{m}^3} \right) (200\,\text{m}^3) \ \approx\ 10^{13}\ \text{atoms}\ .$$

85. Kinetic energy (we use the classical formula since v is much less than c) is converted into potential energy (see Eq. 25-43). From Appendix F or G, we find $Z = 3$ for Lithium and $Z = 90$ for Thorium; the charges on those nuclei are therefore $3e$ and $90e$, respectively. We manipulate the terms so that one of the factors of e cancels the "e" in the kinetic energy unit MeV, and the other factor of e is set equal to its SI value 1.6×10^{-19} C. We note that $k = 1/4\pi\varepsilon_0$ can be written as 8.99×10^9 V·m/C. Thus, from energy conservation, we have

$$K = U \implies r = \frac{kq_1q_2}{K} = \frac{(8.99 \times 10^9\,\frac{\text{V}\cdot\text{m}}{\text{C}})\,(3 \times 1.6 \times 10^{-19}\,\text{C})\,(90e)}{3.00 \times 10^6\,\text{eV}}$$

which yields $r = 1.3 \times 10^{-13}$ m (or about 130 fm).

86. Form Appendix F and/or G, we find $Z = 107$ for Bohrium, so this isotope has $N = A - Z = 262 - 107 = 155$ neutrons. Thus,

$$\Delta E_{\text{ben}} = \frac{(Zm_{\text{H}} + Nm_n - m_{\text{Bh}})\,c^2}{A} = \frac{((107)(1.007825\,\text{u}) + (155)(1.008665\,\text{u}) - 262.1231\,\text{u})\,(931.5\,\text{MeV/u})}{262}$$

which yields 7.3 MeV per nucleon.

87. Since R is proportional to N (see Eq. 43-16) then $N/N_0 = R/R_0$. Combining Eq. 43-13 and Eq.43-17 leads to

$$t = -\frac{T_{1/2}}{\ln 2}\ln\left(\frac{R}{R_0}\right) = -\frac{5730\,\text{y}}{\ln 2}\ln(0.020) = 3.2 \times 10^4\ \text{y}\ .$$

88. Adapting Eq. 43-20, we have

$$N_{\text{Kr}} = \frac{M_{\text{sam}}}{M_{\text{Kr}}}N_A = \left(\frac{20 \times 10^{-9}\,\text{g}}{92\,\text{g/mol}}\right)(6.02 \times 10^{23}\,\text{atoms/mol}) = 1.3 \times 10^{14}\ \text{atoms}\ .$$

Consequently, Eq. 43-19 leads to

$$R = \frac{N\ln 2}{T_{1/2}} = \frac{(1.3 \times 10^{14})\ln 2}{1.84\,\text{s}} = 4.9 \times 10^{13}\ \text{Bq}\ .$$

Chapter 44

1. (a) The mass of a single atom of ^{235}U is $(235\,\text{u})(1.661 \times 10^{-27}\,\text{kg/u}) = 3.90 \times 10^{-25}\,\text{kg}$, so the number of atoms in 1.0 kg is $(1.0\,\text{kg})/(3.90 \times 10^{-25}\,\text{kg}) = 2.56 \times 10^{24}$. An alternate approach (but essentially the same once the connection between the "u" unit and N_A is made) would be to adapt Eq. 43-20.

   (b) The energy released by N fission events is given by $E = NQ$, where Q is the energy released in each event. For 1.0 kg of ^{235}U, $E = (2.56 \times 10^{24})(200 \times 10^6\,\text{eV})(1.60 \times 10^{-19}\,\text{J/eV}) = 8.19 \times 10^{13}\,\text{J}$.

   (c) If P is the power requirement of the lamp, then $t = E/P = (8.19 \times 10^{13}\,\text{J})/(100\,\text{W}) = 8.19 \times 10^{11}\,\text{s} = 2.6 \times 10^4\,\text{y}$. The conversion factor $3.156 \times 10^7\,\text{s/y}$ is used to obtain the last result.

2. We note that the sum of superscripts (mass numbers A) must balance, as well as the sum of Z values (where reference to Appendix F or G is helpful). A neutron has $Z = 0$ and $A = 1$. Uranium has $Z = 92$.

   - Since xenon has $Z = 54$, then "Y" must have $Z = 92 - 54 = 38$, which indicates the element Strontium. The mass number of "Y" is $235 + 1 - 140 - 1 = 95$, so "Y" is ^{95}Sr.
   - Iodine has $Z = 53$, so "Y" has $Z = 92 - 53 = 39$, corresponding to the element Yttrium (the symbol for which, coincidentally, is Y). Since $235 + 1 - 139 - 2 = 95$, then the unknown isotope is ^{95}Y.
   - The atomic number of Zirconium is $Z = 40$. Thus, $92 - 40 - 2 = 52$, which means that "X" has $Z = 52$ (Tellurium). The mass number of "X" is $235 + 1 - 100 - 2 = 134$, so we obtain ^{134}Te.
   - Examining the mass numbers, we find $b = 235 + 1 - 141 - 92 = 3$.

3. If R is the fission rate, then the power output is $P = RQ$, where Q is the energy released in each fission event. Hence, $R = P/Q = (1.0\,\text{W})/(200 \times 10^6\,\text{eV})(1.60 \times 10^{-19}\,\text{J/eV}) = 3.12 \times 10^{10}\,\text{fissions/s}$.

4. Adapting Eq. 43-20, there are

$$N_{\text{Pu}} = \frac{M_{\text{sam}}}{M_{\text{Pu}}} N_{\text{A}} = \left(\frac{1000\,\text{g}}{239\,\text{g/mol}}\right)(6.02 \times 10^{23}/\text{mol}) = 2.5 \times 10^{24}$$

   plutonium nuclei in the sample. If they all fission (each releasing 180 MeV), then the total energy release is 4.5×10^{26} MeV.

5. At $T = 300\,\text{K}$, the average kinetic energy of the neutrons is (using Eq. 20-24)

$$K_{\text{avg}} = \frac{3}{2}kT = \frac{3}{2}(8.62 \times 10^{-5}\,\text{eV/K})(300\,\text{K}) \approx 0.04\,\text{eV} .$$

6. We consider the process $^{98}\text{Mo} \to {}^{49}\text{Sc} + {}^{49}\text{Sc}$. The disintegration energy is $Q = (m_{\text{Mo}} - 2m_{\text{Sc}})c^2 = [97.90541\,\text{u} - 2(48.95002\,\text{u})](931.5\,\text{MeV/u}) = +5.00\,\text{MeV}$. The fact that it is positive does not necessarily mean we should expect to find a great deal of Molybdenum nuclei spontaneously fissioning; the energy barrier (see Fig. 44-3) is presumably higher and/or broader for Molybdenum than for Uranium.

7. If M_{Cr} is the mass of a ^{52}Cr nucleus and M_{Mg} is the mass of a ^{26}Mg nucleus, then the disintegration energy is $Q = (M_{\text{Cr}} - 2M_{\text{Mg}})\,c^2 = [51.94051\,\text{u} - 2(25.98259\,\text{u})]\,(931.5\,\text{MeV/u}) = -23.0\,\text{MeV}$.

8. (a) Using Eq. 43-19 and adapting Eq. 43-20 to this sample, the number of fission-events per second is

$$\begin{aligned} R_{\text{fission}} &= \frac{N \ln 2}{T_{1/2\,\text{fission}}} = \frac{M_{\text{sam}} N_A \ln 2}{M_{\text{U}} T_{1/2\,\text{fission}}} \\ &= \frac{(1.0\,\text{g})(6.02 \times 10^{23}/\text{mol}) \ln 2}{(235\,\text{g/mol})(3.0 \times 10^{17}\,\text{y})(365\,\text{d/y})} = 16 \text{ fissions/day} . \end{aligned}$$

(b) Since $R \propto \frac{1}{T_{1/2}}$ (see Eq. 43-19), the ratio of rates is

$$\frac{R_\alpha}{R_{\text{fission}}} = \frac{T_{1/2\,\text{fission}}}{T_{1/2\,\alpha}} = \frac{3.0 \times 10^{17}\,\text{y}}{7.0 \times 10^{8}\,\text{y}} = 4.3 \times 10^{8} .$$

9. The energy released is

$$\begin{aligned} Q &= (m_{\text{U}} + m_n - m_{\text{Cs}} - m_{\text{Rb}} - 2m_n)c^2 \\ &= (235.04392\,\text{u} - 1.00867\,\text{u} - 140.91963\,\text{u} - 92.92157\,\text{u})(931.5\,\text{MeV/u}) \\ &= 181 \text{ MeV} . \end{aligned}$$

10. First, we figure out the mass of U-235 in the sample (assuming "3.0%" refers to the proportion by weight as opposed to proportion by number of atoms):

$$\begin{aligned} M_{\text{U-235}} &= (3.0\%) M_{\text{sam}} \left(\frac{(97\%) m_{238} + (3.0\%) m_{235}}{(97\%) m_{238} + (3.0\%) m_{235} + 2m_{16}} \right) \\ &= (0.030)(1000\,\text{g}) \left(\frac{0.97(238) + 0.030(235)}{0.97(238) + 0.030(235) + 2(16.0)} \right) = 26.4 \text{ g} . \end{aligned}$$

Next, this uses some of the ideas illustrated in Sample Problem 43-5; our notation is similar to that used in that example. The number of ^{235}U nuclei is

$$N_{235} = \frac{(26.4\,\text{g})(6.02 \times 10^{23}/\text{mol})}{235\,\text{g/mol}} = 6.77 \times 10^{22} .$$

If all the U-235 nuclei fission, the energy release (using the result of Eq. 44-6) is

$$N_{235} Q_{\text{fission}} = (6.77 \times 10^{22})\,(200\,\text{MeV}) = 1.35 \times 10^{25}\,\text{MeV} = 2.17 \times 10^{12} \text{ J} .$$

Keeping in mind that a Watt is a Joule per second, the time that this much energy can keep a 100-W lamp burning is found to be

$$t = \frac{2.17 \times 10^{12}\,\text{J}}{100\,\text{W}} = 2.17 \times 10^{10}\,\text{s} \approx 690 \text{ y} .$$

If we had instead used the $Q = 208\,\text{MeV}$ value from Sample Problem 44-1, then our result would have been 715 y, which perhaps suggests that our result is meaningful to just one significant figure ("roughly 700 years").

11. (a) If X represents the unknown fragment, then the reaction can be written

$$^{235}_{92}\text{U} + ^{1}_{0}\text{n} \rightarrow ^{83}_{32}\text{Ge} + ^{A}_{Z}X$$

where A is the mass number and Z is the atomic number of the fragment. Conservation of charge yields $92+0 = 32+Z$, so $Z = 60$. Conservation of mass number yields $235+1 = 83+A$, so $A = 153$. Looking in Appendix F or G for nuclides with $Z = 60$, we find that the unknown fragment is $^{153}_{60}$Nd.

(b) We neglect the small kinetic energy and momentum carried by the neutron that triggers the fission event. Then, $Q = K_{\rm Ge} + K_{\rm Nd}$, where $K_{\rm Ge}$ is the kinetic energy of the germanium nucleus and $K_{\rm Nd}$ is the kinetic energy of the neodymium nucleus. Conservation of momentum yields $\vec{p}_{\rm Ge} + \vec{p}_{\rm Nd} = 0$. Now, we can write the classical formula for kinetic energy in terms of the magnitude of the momentum vector:

$$K = \frac{1}{2}mv^2 = \frac{p^2}{2m}$$

which implies that $K_{\rm Nd} = (m_{\rm Ge}/m_{\rm Nd})K_{\rm Ge}$. Thus, the energy equation becomes

$$Q = K_{\rm Ge} + \frac{M_{\rm Ge}}{M_{\rm Nd}}K_{\rm Ge} = \frac{M_{\rm Nd} + M_{\rm Ge}}{M_{\rm Nd}}K_{\rm Ge}$$

and

$$K_{\rm Ge} = \frac{M_{\rm Nd}}{M_{\rm Nd} + M_{\rm Ge}}Q = \frac{153\,\rm u}{153\,\rm u + 83\,\rm u}(170\,\rm MeV) = 110\ \rm MeV\ .$$

Similarly,

$$K_{\rm Nd} = \frac{M_{\rm Ge}}{M_{\rm Nd} + M_{\rm Ge}}Q = \frac{83\,\rm u}{153\,\rm u + 83\,\rm u}(170\,\rm MeV) = 60\ \rm MeV\ .$$

(c) The initial speed of the germanium nucleus is

$$v_{\rm Ge} = \sqrt{\frac{2K_{\rm Ge}}{M_{\rm Ge}}} = \sqrt{\frac{2(110\times10^6\,\rm eV)(1.60\times10^{-19}\,\rm J/eV)}{(83\,\rm u)(1.661\times10^{-27}\,\rm kg/u)}} = 1.60\times10^7\ \rm m/s\ .$$

The initial speed of the neodymium nucleus is

$$v_{\rm Nd} = \sqrt{\frac{2K_{\rm Nd}}{M_{\rm Nd}}} = \sqrt{\frac{2(60\times10^6\,\rm eV)(1.60\times10^{-19}\,\rm J/eV)}{(153\,\rm u)(1.661\times10^{-27}\,\rm kg/u)}} = 8.69\times10^6\ \rm m/s\ .$$

12. (a) Consider the process $^{239}\rm U + n \rightarrow {}^{140}Ce + {}^{99}Ru + Ne$. We have $Z_f - Z_i = Z_{\rm Ce} + Z_{\rm Ru} - Z_{\rm U} = 58 + 44 - 92 = 10$. Thus the number of beta-decay events is 10.

(b) Using Table 38-3, the energy released in this fission process is

$$\begin{aligned} Q &= (m_{\rm U} + m_n - m_{\rm Ce} - m_{\rm Ru} - 10m_e)c^2 \\ &= (238.05079\,\rm u + 1.00867\,\rm u - 139.90543\,\rm u - 98.90594\,\rm u)(931.5\,\rm MeV/u) - 10(0.511\,\rm MeV) \\ &= 226\ \rm MeV\ . \end{aligned}$$

13. (a) The electrostatic potential energy is given by

$$U = \frac{1}{4\pi\varepsilon_0}\frac{Z_{\rm Xe}Z_{\rm Sr}e^2}{r_{\rm Xe} + r_{\rm Sr}}$$

where $Z_{\rm Xe}$ is the atomic number of xenon, $Z_{\rm Sr}$ is the atomic number of strontium, $r_{\rm Xe}$ is the radius of a xenon nucleus, and $r_{\rm Sr}$ is the radius of a strontium nucleus. Atomic numbers can be found either in Appendix F or Appendix G. The radii are given by $r = (1.2\,\rm fm)A^{1/3}$, where A is the mass number, also found in Appendix F. Thus, $r_{\rm Xe} = (1.2\,\rm fm)(140)^{1/3} = 6.23\,\rm fm = 6.23\times10^{-15}\,\rm m$ and $r_{\rm Sr} = (1.2\,\rm fm)(96)^{1/3} = 5.49\,\rm fm = 5.49\times10^{-15}\,\rm m$. Hence, the potential energy is

$$U = (8.99\times10^9\,\rm V\cdot m/C)\frac{(54)(38)(1.60\times10^{-19}\,\rm C)^2}{6.23\times10^{-15}\,\rm m + 5.49\times10^{-15}\,\rm m} = 4.08\times10^{-11}\,\rm J = 251\ \rm MeV\ .$$

(b) The energy released in a typical fission event is about 200 MeV, roughly the same as the electrostatic potential energy when the fragments are touching. The energy appears as kinetic energy of the fragments and neutrons produced by fission.

14. (a) The surface area a of a nucleus is given by $a \simeq 4\pi R^2 \simeq 4\pi[R_0 A^{1/3}]^2 \propto A^{2/3}$. Thus, the fractional change in surface area is

$$\frac{\Delta a}{a_i} = \frac{a_f - a_i}{a_i} = \frac{(140)^{2/3} + (96)^{2/3}}{(236)^{2/3}} - 1 = +0.25 \ .$$

(b) Since $V \propto R^3 \propto (A^{1/3})^3 = A$, we have

$$\frac{\Delta V}{V} = \frac{V_f}{V_i} - 1 = \frac{140 + 96}{236} - 1 = 0 \ .$$

(c) The fractional change in potential energy is

$$\begin{aligned} \frac{\Delta U}{U} &= \frac{U_f}{U_i} - 1 \\ &= \frac{Q_{\text{Xe}}^2/R_{\text{Xe}} + Q_{\text{Sr}}^2/R_{\text{Sr}}}{Q_{\text{U}}^2/R_{\text{U}}} - 1 \\ &= \frac{(54)^2(140)^{-1/3} + (38)^2(96)^{-1/3}}{(92)^2(236)^{-1/3}} - 1 \ = \ -0.36 \ . \end{aligned}$$

15. If P is the power output, then the energy E produced in the time interval Δt $(= 3\,\text{y})$ is $E = P\,\Delta t = (200 \times 10^6\,\text{W})(3\,\text{y})(3.156 \times 10^7\,\text{s/y}) = 1.89 \times 10^{16}\,\text{J}$, or $(1.89 \times 10^{16}\,\text{J})/(1.60 \times 10^{-19}\,\text{J/eV}) = 1.18 \times 10^{35}\,\text{eV} = 1.18 \times 10^{29}\,\text{MeV}$. At 200 MeV per event, this means $(1.18 \times 10^{29})/200 = 5.90 \times 10^{26}$ fission events occurred. This must be half the number of fissionable nuclei originally available. Thus, there were $2(5.90 \times 10^{26}) = 1.18 \times 10^{27}$ nuclei. The mass of a ^{235}U nucleus is $(235\,\text{u})(1.661 \times 10^{-27}\,\text{kg/u}) = 3.90 \times 10^{-25}\,\text{kg}$, so the total mass of ^{235}U originally present was $(1.18 \times 10^{27})(3.90 \times 10^{-25}\,\text{kg}) = 462\,\text{kg}$.

16. In Sample Problem 44-2, it is noted that the rate of consumption of U-235 by (nonfission) neutron capture is one-fourth as big as the rate of the rate of neutron-induced fission events. Consequently, the mass of ^{235}U should be larger than that computed in problem 15 by 25%: $(1.25)(462\,\text{kg}) = 5.8 \times 10^2\,\text{kg}$. If appeal is to made to other sources (other than Sample Problem 44-2), then it might be possible to argue for a factor other than 1.25 (we found others in our brief search) and thus to a somewhat different result.

17. When a neutron is captured by ^{237}Np it gains 5.0 MeV, more than enough to offset the 4.2 MeV required for ^{238}Np to fission. Consequently, ^{237}Np is fissionable by thermal neutrons.

18. (a) Using the result of problem 4, the TNT equivalent is

$$\frac{(2.50\,\text{kg})(4.54 \times 10^{26}\,\text{MeV/kg})}{2.6 \times 10^{28}\,\text{MeV}/10^6\,\text{ton}} = 4.4 \times 10^4\,\text{ton} = 44\,\text{kton} \ .$$

(b) Assuming that this is a fairly inefficiently designed bomb, then much of the remaining 92.5 kg is probably "wasted" and was included perhaps to make sure the bomb did not "fizzle." There is also an argument for having more than just the critical mass based on the short assembly-time of the material during the implosion, but this so-called "super-critical mass," as generally quoted, is much less than 92.5 kg, and does not necessarily have to be purely Plutonium.

19. If R is the decay rate then the power output is $P = RQ$, where Q is the energy produced by each alpha decay. Now $R = \lambda N = N \ln 2/T_{1/2}$, where λ is the disintegration constant and $T_{1/2}$ is the half-life. The relationship $\lambda = (\ln 2)/T_{1/2}$ is used. If M is the total mass of material and m is the mass of a single ^{238}Pu nucleus, then

$$N = \frac{M}{m} = \frac{1.00\,\text{kg}}{(238\,\text{u})(1.661 \times 10^{-27}\,\text{kg/u})} = 2.53 \times 10^{24} \ .$$

Thus,

$$P = \frac{NQ \ln 2}{T_{1/2}} = \frac{(2.53 \times 10^{24})(5.50 \times 10^6\,\text{eV})(1.60 \times 10^{-19}\,\text{J/eV})(\ln 2)}{(87.7\,\text{y})(3.156 \times 10^7\,\text{s/y})} = 558\,\text{W} \ .$$

20. (a) We solve Q_{eff} from $P = RQ_{\text{eff}}$:

$$\begin{aligned} Q_{\text{eff}} &= \frac{P}{R} = \frac{P}{N\lambda} = \frac{mPT_{1/2}}{M\ln 2} \\ &= \frac{(90.0\,\text{u})(1.66\times 10^{-27}\,\text{kg/u})(0.93\,\text{W})(29\,\text{y})(3.15\times 10^{7}\,\text{s/y})}{(1.00\times 10^{-3}\,\text{kg})(\ln 2)(1.60\times 10^{-13}\,\text{J/MeV})} \\ &= 1.2\ \text{MeV}\ . \end{aligned}$$

(b) The amount of ^{90}Sr needed is

$$M = \frac{150\,\text{W}}{(0.050)(0.93\,\text{W/g})} = 3.2\ \text{kg}\ .$$

21. Since Plutonium has $Z = 94$ and Uranium has $Z = 92$, we see that (to conserve charge) two electrons must be emitted so that the nucleus can gain a $+2e$ charge. In the beta decay processes described in Chapter 43, electrons and neutrinos are emitted. The reaction series is as follows:

$$\begin{aligned} {}^{238}\text{U} + \text{n} \rightarrow\ & {}^{239}\text{Np} \;\; + {}^{239}\text{U} + e + \nu \\ & {}^{239}\text{Np} \;\; \rightarrow {}^{239}\text{Pu} + e + \nu \end{aligned}$$

22. After each time interval t_{gen} the number of nuclides in the chain reaction gets multiplied by k. The number of such time intervals that has gone by at time t is t/t_{gen}. For example, if the multiplication factor is 5 and there were 12 nuclei involved in the reaction to start with, then after one interval 60 nuclei are involved. And after another interval 300 nuclei are involved. Thus, the number of nuclides engaged in the chain reaction at time t is $N(t) = N_0 k^{t/t_{\text{gen}}}$. Since $P \propto N$ we have

$$P(t) = P_0 k^{t/t_{\text{gen}}}\ .$$

23. (a) The energy yield of the bomb is $E = (66\times 10^{-3}\,\text{megaton})(2.6\times 10^{28}\,\text{MeV/megaton}) = 1.72\times 10^{27}\,\text{MeV}$. At 200 MeV per fission event, $(1.72\times 10^{27}\,\text{MeV})/(200\,\text{MeV}) = 8.58\times 10^{24}$ fission events take place. Since only 4.0% of the ^{235}U nuclei originally present undergo fission, there must have been $(8.58\times 10^{24})/(0.040) = 2.14\times 10^{26}$ nuclei originally present. The mass of ^{235}U originally present was $(2.14\times 10^{26})(235\,\text{u})(1.661\times 10^{-27}\,\text{kg/u}) = 83.7\,\text{kg}$.

(b) Two fragments are produced in each fission event, so the total number of fragments is $2(8.58\times 10^{24}) = 1.72\times 10^{25}$.

(c) One neutron produced in a fission event is used to trigger the next fission event, so the average number of neutrons released to the environment in each event is 1.5. The total number released is $(8.58\times 10^{24})(1.5) = 1.29\times 10^{25}$.

24. We recall Eq. 44-6: $Q \approx 200\,\text{MeV} = 3.2\times 10^{-11}$ J. It is important to bear in mind that Watts multiplied by seconds give Joules. From $E = Pt_{\text{gen}} = NQ$ we get the number of free neutrons:

$$N = \frac{Pt_{\text{gen}}}{Q} = \frac{(500\times 10^{6}\,\text{W})(1.0\times 10^{-3}\,\text{s})}{3.2\times 10^{-11}\,\text{J}} = 1.6\times 10^{16}\ .$$

25. Let P_0 be the initial power output, P be the final power output, k be the multiplication factor, t be the time for the power reduction, and t_{gen} be the neutron generation time. Then, according to the result of Problem 22,

$$P = P_0\, k^{t/t_{\text{gen}}}\ .$$

We divide by P_0, take the natural logarithm of both sides of the equation and solve for $\ln k$:

$$\ln k = \frac{t_{\text{gen}}}{t}\ln\frac{P}{P_0} = \frac{1.3\times 10^{-3}\,\text{s}}{2.6\,\text{s}}\ln\frac{350\,\text{MW}}{1200\,\text{MW}} = -0.0006161\ .$$

Hence, $k = e^{-0.0006161} = 0.99938$.

26. We use the formula from problem 22:

$$\begin{aligned} P(t) &= P_0 k^{t/t_{\text{gen}}} \\ &= (400\,\text{MW})(1.0003)^{(5.00\,\text{min})(60\,\text{s/min})/(0.00300\,\text{s})} \\ &= 8.03 \times 10^3\ \text{MW} . \end{aligned}$$

27. (a) Let v_{ni} be the initial velocity of the neutron, v_{nf} be its final velocity, and v_f be the final velocity of the target nucleus. Then, since the target nucleus is initially at rest, conservation of momentum yields $m_n v_{ni} = m_n v_{nf} + m v_f$ and conservation of energy yields $\frac{1}{2} m_n v_{ni}^2 = \frac{1}{2} m_n v_{nf}^2 + \frac{1}{2} m v_f^2$. We solve these two equations simultaneously for v_f. This can be done, for example, by using the conservation of momentum equation to obtain an expression for v_{nf} in terms of v_f and substituting the expression into the conservation of energy equation. We solve the resulting equation for v_f. We obtain $v_f = 2m_n v_{ni}/(m + m_n)$. The energy lost by the neutron is the same as the energy gained by the target nucleus, so

$$\Delta K = \frac{1}{2} m v_f^2 = \frac{1}{2} \frac{4m_n^2 m}{(m + m_n)^2} v_{ni}^2 .$$

The initial kinetic energy of the neutron is $K = \frac{1}{2} m_n v_{ni}^2$, so

$$\frac{\Delta K}{K} = \frac{4m_n m}{(m + m_n)^2} .$$

(b) The mass of a neutron is 1.0 u and the mass of a hydrogen atom is also 1.0 u. (Atomic masses can be found in Appendix G.) Thus,

$$\frac{\Delta K}{K} = \frac{4(1.0\,\text{u})(1.0\,\text{u})}{(1.0\,\text{u} + 1.0\,\text{u})^2} = 1.0 .$$

Similarly, the mass of a deuterium atom is 2.0 u, so $(\Delta K)/K = 4(1.0\,\text{u})(2.0\,\text{u})/(2.0\,\text{u}+1.0\,\text{u})^2 = 0.89$. The mass of a carbon atom is 12 u, so $(\Delta K)/K = 4(1.0\,\text{u})(12\,\text{u})/(12\,\text{u} + 1.0\,\text{u})^2 = 0.28$. The mass of a lead atom is 207 u, so $(\Delta K)/K = 4(1.0\,\text{u})(207\,\text{u})/(207\,\text{u} + 1.0\,\text{u})^2 = 0.019$.

(c) During each collision, the energy of the neutron is reduced by the factor $1 - 0.89 = 0.11$. If E_i is the initial energy, then the energy after n collisions is given by $E = (0.11)^n E_i$. We take the natural logarithm of both sides and solve for n. The result is

$$n = \frac{\ln(E/E_i)}{\ln 0.11} = \frac{\ln(0.025\,\text{eV}/1.00\,\text{eV})}{\ln 0.11} = 7.9 .$$

The energy first falls below 0.025 eV on the eighth collision.

28. Our approach is the same as that shown in Sample Problem 44-3. We have

$$\frac{N_5(t)}{N_8(t)} = \frac{N_5(0)}{N_8(0)} e^{-(\lambda_5 - \lambda_8)t} ,$$

or

$$\begin{aligned} t &= \frac{1}{\lambda_8 - \lambda_5} \ln\left[\left(\frac{N_5(t)}{N_8(t)}\right)\left(\frac{N_8(0)}{N_5(0)}\right)\right] \\ &= \frac{1}{(1.55 - 9.85)10^{-10}\,\text{y}^{-1}} \ln\left[(0.0072)(0.15)^{-1}\right] = 3.6 \times 10^9\ \text{y} . \end{aligned}$$

29. (a) $P_{\text{avg}} = (15 \times 10^9\,\text{W}\cdot\text{y})/(200{,}000\,\text{y}) = 7.5 \times 10^4\,\text{W} = 75\,\text{kW}$.

(b) Using the result of Eq. 44-6, we obtain

$$\begin{aligned} M &= \frac{m_{\rm U} E_{\rm total}}{Q} \\ &= \frac{(235\,{\rm u})(1.66\times10^{-27}\,{\rm kg/u})(15\times10^{9}\,{\rm W\cdot y})(3.15\times10^{7}\,{\rm s/y})}{(200\,{\rm MeV})(1.6\times10^{-13}\,{\rm J/MeV})} \\ &= 5.8\times10^{3}\ {\rm kg}\ . \end{aligned}$$

30. The nuclei of ^{238}U can capture neutrons and beta-decay. With large amount of neutrons available due to the fission of ^{235}U, the probability for this process is substantially increased, resulting in a much higher decay rate for ^{238}U and causing the depletion of ^{238}U (and relative enrichment of ^{235}U).

31. Let t be the present time and $t = 0$ be the time when the ratio of ^{235}U to ^{238}U was 3.0%. Let N_{235} be the number of ^{235}U nuclei present in a sample now and $N_{235,\,0}$ be the number present at $t = 0$. Let N_{238} be the number of ^{238}U nuclei present in the sample now and $N_{238,\,0}$ be the number present at $t = 0$. The law of radioactive decay holds for each specie, so

$$N_{235} = N_{235,\,0}\, e^{-\lambda_{235} t}$$

and

$$N_{238} = N_{238,\,0}\, e^{-\lambda_{238} t}\ .$$

Dividing the first equation by the second, we obtain

$$r = r_0\, e^{-(\lambda_{235}-\lambda_{238})t}$$

where $r = N_{235}/N_{238}$ $(= 0.0072)$ and $r_0 = N_{235,\,0}/N_{238,\,0}$ $(= 0.030)$. We solve for t:

$$t = -\frac{1}{\lambda_{235}-\lambda_{238}} \ln\frac{r}{r_0}\ .$$

Now we use $\lambda_{235} = (\ln 2)/T_{1/2_{235}}$ and $\lambda_{238} = (\ln 2)/T_{1/2_{238}}$ to obtain

$$t = -\frac{T_{1/2_{235}} T_{1/2_{238}}}{(T_{1/2_{238}} - T_{1/2_{235}})\ln 2} \ln\frac{r}{r_0} = -\frac{(7.0\times10^{8}\,{\rm y})(4.5\times10^{9}\,{\rm y})}{(4.5\times10^{9}\,{\rm y} - 7.0\times10^{8}\,{\rm y})\ln 2} \ln\frac{0.0072}{0.030} = 1.71\times10^{9}\ {\rm y}\ .$$

32. (a) Fig. 43-9 shows the barrier height to be about 30 MeV.

(b) The potential barrier height listed in Table 44-2 is roughly 5 MeV. There is some model-dependence involved in arriving at this estimate, and other values can be found in the literature (6 MeV is frequently cited).

33. The height of the Coulomb barrier is taken to be the value of the kinetic energy K each deuteron must initially have if they are to come to rest when their surfaces touch (see Sample Problem 44-4). If r is the radius of a deuteron, conservation of energy yields

$$2K = \frac{1}{4\pi\varepsilon_0}\frac{e^2}{2r}\ ,$$

so

$$K = \frac{1}{4\pi\varepsilon_0}\frac{e^2}{4r} = (8.99\times10^{9}\,{\rm V\cdot m/C})\frac{(1.60\times10^{-19}\,{\rm C})^2}{4(2.1\times10^{-15}\,{\rm m})} = 2.74\times10^{-14}\,{\rm J} = 170\ {\rm keV}\ .$$

34. We are given the energy release per fusion ($Q = 3.27\,{\rm MeV} = 5.24\times10^{-13}\,{\rm J}$) and that a pair of deuterium atoms are consumed in each fusion event. To find how many pairs of deuterium atoms are in the sample, we adapt Eq. 43-20:

$$N_{d\,\rm pairs} = \frac{M_{\rm sam}}{2M_d} N_{\rm A} = \left(\frac{1000\,{\rm g}}{2(2.0\,{\rm g/mol})}\right)(6.02\times10^{23}/{\rm mol}) = 1.5\times10^{26}\ .$$

Multiplying this by Q gives the total energy released: 7.9×10^{13} J. Keeping in mind that a Watt is a Joule per second, we have

$$t = \frac{7.9 \times 10^{13}\,\text{J}}{100\,\text{W}} = 7.9 \times 10^{11}\,\text{s} = 2.5 \times 10^4\ \text{y}\ .$$

35. (a) Our calculation is identical to that in Sample Problem 44-4 except that we are now using R appropriate to two deuterons coming into "contact," as opposed to the $R = 1.0\,\text{fm}$ value used in the Sample Problem. If we use $R = 2.1\,\text{fm}$ for the deuterons (this is the value given in problem 33), then our K is simply the K calculated in Sample Problem 44-4, divided by 2.1:

$$K_{d+d} = \frac{K_{p+p}}{2.1} = \frac{360\,\text{keV}}{2.1} \approx 170\ \text{keV}\ .$$

Consequently, the voltage needed to accelerate each deuteron from rest to that value of K is 170 kV.

(b) Not all deuterons that are accelerated towards each other will come into "contact" and not all of those that do so will undergo nuclear fusion. Thus, a great many deuterons must be repeatedly encountering other deuterons in order to produce a macroscopic energy release. An accelerator needs a fairly good vacuum in its beam pipe, and a very large number flux is either impractical and/or very expensive. Regarding expense, there are other factors that have dissuaded researchers from using accelerators to build a controlled fusion "reactor," but those factors may become less important in the future – making the feasibility of accelerator "add-on's" to magnetic and inertial confinement schemes more cost-effective.

36. Our calculation is very similar to that in Sample Problem 44-4 except that we are now using R appropriate to two Lithium-7 nuclei coming into "contact," as opposed to the $R = 1.0\,\text{fm}$ value used in the Sample Problem. If we use

$$R = r = r_0 A^{1/3} = (1.2\,\text{fm})\sqrt[3]{7} = 2.3\ \text{fm}$$

and $q = Ze = 3e$, then our K is given by (see Sample Problem 44-4)

$$K = \frac{Z^2e^2}{16\pi\varepsilon_0 r} = \frac{3^2(1.6 \times 10^{-19}\,\text{C})^2}{16\pi(8.85 \times 10^{-12}\,\text{F/m})(2.3 \times 10^{15}\,\text{m})}$$

which yields $2.3 \times 10^{-13}\,\text{J} = 1.4\,\text{MeV}$. We interpret this as the answer to the problem, though the term "Coulomb barrier height" as used here may be open to other interpretations.

37. From the expression for $n(K)$ given we may write $n(K) \propto K^{1/2}e^{-K/kT}$. Thus, with $k = 8.62 \times 10^{-5}\,\text{eV/K} = 8.62 \times 10^{-8}\,\text{keV/K}$, we have

$$\begin{aligned}\frac{n(K)}{n(K_{\text{avg}})} &= \left(\frac{K}{K_{\text{avg}}}\right)^{1/2} e^{-(K-K_{\text{avg}})/kT} \\ &= \left(\frac{5.00\,\text{keV}}{1.94\,\text{keV}}\right)^{1/2} e^{-(5.00\,\text{keV}-1.94\,\text{keV})/[(8.62\times10^{-8}\,\text{keV/K})(1.50\times10^{7}\,\text{K})]} \\ &= 0.151\ .\end{aligned}$$

38. (a) Rather than use $P(v)$ as it is written in Eq. 20-27, we use the more convenient nK expression given in problem 37 of this chapter [44]. The $n(K)$ expression can be derived from Eq. 20-27, but we do not show that derivation here. To find the most probable energy, we take the derivative of $n(K)$ and set the result equal to zero:

$$\left.\frac{dn(K)}{dK}\right|_{K=K_p} = \frac{1.13n}{(kT)^{3/2}}\left(\frac{1}{2K^{1/2}} - \frac{K^{3/2}}{kT}\right)e^{-K/kT}\Bigg|_{K=K_p} = 0\,,$$

which gives $K_p = \frac{1}{2}kT$. Specifically, for $T = 1.5 \times 10^7$ K we find

$$K_p = \frac{1}{2}kT = \frac{1}{2}(8.62 \times 10^{-5}\,\text{eV/K})(1.5 \times 10^7\,\text{K}) = 6.5 \times 10^2\ \text{eV}$$

or 0.65 keV, in good agreement with Fig. 44-10.

(b) Eq. 20-35 gives the most probable speed in terms of the molar mass M, and indicates its derivation (see also Sample Problem 20-6). Since the mass m of the particle is related to M by the Avogadro constant, then

$$v_p = \sqrt{\frac{2RT}{M}} = \sqrt{\frac{2RT}{m\,N_A}} = \sqrt{\frac{2kT}{m}}$$

using Eq. 20-7. With $T = 1.5 \times 10^7$ K and $m = 1.67 \times 10^{-27}$ kg, this yields $v_p = 5.0 \times 10^5$ m/s.

(c) The corresponding kinetic energy is

$$K_{v,p} = \frac{1}{2}mv_p^2 = \frac{1}{2}m\left(\sqrt{\frac{2kT}{m}}\right)^2 = kT$$

which is twice as large as that found in part (a). Thus, at $T = 1.5 \times 10^7$ K we have $K_{v,p} = 1.3$ keV, which is indicated in Fig. 44-10 by a single vertical line.

39. If M_{He} is the mass of an atom of helium and M_{C} is the mass of an atom of carbon, then the energy released in a single fusion event is

$$Q = [3M_{\text{He}} - M_{\text{C}}]\,c^2 = [3(4.0026\,\text{u}) - (12.0000\,\text{u})]\,(931.5\,\text{MeV/u}) = 7.27\,\text{MeV}\ .$$

Note that $3M_{\text{He}}$ contains the mass of six electrons and so does M_{C}. The electron masses cancel and the mass difference calculated is the same as the mass difference of the nuclei.

40. In Fig. 44-11, let $Q_1 = 0.42$ MeV, $Q_2 = 1.02$ MeV, $Q_3 = 5.49$ MeV and $Q_4 = 12.86$ MeV. For the overall proton-proton cycle

$$\begin{aligned} Q &= 2Q_1 + 2Q_2 + 2Q_3 + Q_4 \\ &= 2(0.42\,\text{MeV} + 1.02\,\text{MeV} + 5.49\,\text{MeV}) + 12.86\,\text{MeV} = 26.7\ \text{MeV}\ . \end{aligned}$$

41. (a) From $\rho_{\text{H}} = 0.35\rho = n_p m_p$, we get the proton number density n_p:

$$n_p = \frac{0.35\rho}{m_p} = \frac{(0.35)(1.5 \times 10^5\,\text{kg/m}^3)}{1.67 \times 10^{-27}\,\text{kg}} = 3.14 \times 10^{31}\ \text{m}^{-3}\ .$$

(b) From Chapter 20 (see Eq. 20-9), we have

$$\frac{N}{V} = \frac{p}{kT} = \frac{1.01 \times 10^5\,\text{Pa}}{(1.38 \times 10^{-23}\,\text{J/K})(273\,\text{K})} = 2.68 \times 10^{25}\ \text{m}^{-3}$$

for an ideal gas under "standard conditions." Thus,

$$\frac{n_p}{(N/V)} = \frac{3.14 \times 10^{31}\,\text{m}^{-3}}{2.44 \times 10^{25}\,\text{m}^{-3}} = 1.2 \times 10^6\ .$$

42. We assume the neutrino has negligible mass. The photons, of course, are also taken to have zero mass.

$$\begin{aligned} Q_1 &= (2m_p - m_2 - m_e)c^2 = [2(m_1 - m_e) - (m_2 - m_e) - m_e]c^2 \\ &= [2(1.007825\,\text{u}) - 2.014102\,\text{u} - 2(0.0005486\,\text{u})](931.5\,\text{MeV/u}) \\ &= 0.42\ \text{MeV} \end{aligned}$$

$$\begin{aligned} Q_2 &= (m_2 + m_p - m_3)c^2 = (m_2 + m_p - m_3)c^2 \\ &= (2.014102\,\text{u}) + 1.007825\,\text{u} - 3.016029\,\text{u})(931.5\,\text{MeV/u}) \\ &= 5.49\ \text{MeV} \end{aligned}$$

$$\begin{aligned} Q_3 &= (2m_3 - m_4 - 2m_p)c^2 = (2m_3 - m_4 - 2m_p)c^2 \\ &= [2(3.016029\,\text{u}) - 4.002603\,\text{u} - 2(1.007825\,\text{u})](931.5\,\text{MeV/u}) \\ &= 12.86\ \text{MeV}\ . \end{aligned}$$

43. (a) Let M be the mass of the Sun at time t and E be the energy radiated to that time. Then, the power output is $P = dE/dt = (dM/dt)c^2$, where $E = Mc^2$ is used. At the present time,

$$\frac{dM}{dt} = \frac{P}{c^2} = \frac{3.9 \times 10^{26}\,\text{W}}{(2.998 \times 10^8\,\text{m/s})^2} = 4.33 \times 10^9\ \text{kg/s}\ .$$

(b) We assume the rate of mass loss remained constant. Then, the total mass loss is $\Delta M = (dM/dt)\,\Delta t = (4.33 \times 10^9\,\text{kg/s})(4.5 \times 10^9\,\text{y})(3.156 \times 10^7\,\text{s/y}) = 6.15 \times 10^{26}\,\text{kg}$. The fraction lost is

$$\frac{\Delta M}{M + \Delta M} = \frac{6.15 \times 10^{26}\,\text{kg}}{2.0 \times 10^{30}\,\text{kg} + 6.15 \times 10^{26}\,\text{kg}} = 3.07 \times 10^{-4}\ .$$

44. (a) We are given the energy release per fusion (calculated in §44-7: $Q = 26.7\,\text{MeV} = 4.28 \times 10^{-12}\,\text{J}$) and that four protons are consumed in each fusion event. To find how many sets of four protons are in the sample, we adapt Eq. 43-20:

$$N_{4p} = \frac{M_{\text{sam}}}{4M_H} N_{\text{A}} = \left(\frac{1000\,\text{g}}{4(1.0\,\text{g/mol})}\right)(6.02 \times 10^{23}/\text{mol}) = 1.5 \times 10^{26}\ .$$

Multiplying this by Q gives the total energy released: 6.4×10^{14} J. It is not required that the answer be in SI units; we could have used MeV throughout (in which case the answer is 4.0×10^{27} MeV).

(b) The number of ^{235}U nuclei is

$$N_{235} = \left(\frac{1000\,\text{g}}{235\,\text{g/mol}}\right)(6.02 \times 10^{23}/\text{mol}) = 2.56 \times 10^{24}\ .$$

If all the U-235 nuclei fission, the energy release (using the result of Eq. 44-6) is

$$N_{235}Q_{\text{fission}} = \left(2.56 \times 10^{22}\right)(200\,\text{MeV}) = 5.1 \times 10^{26}\,\text{MeV} = 8.2 \times 10^{13}\ \text{J}\ .$$

We see that the fusion process (with regard to a unit mass of fuel) produces a larger amount of energy (despite the fact that the Q value per event is smaller).

45. (a) Since two neutrinos are produced per proton-proton cycle (see Eq. 44-10 or Fig. 44-11), the rate of neutrino production R_ν satisfies

$$R_\nu = \frac{2P}{Q} = \frac{2(3.9 \times 10^{26}\,\text{W})}{(26.7\,\text{MeV})(1.6 \times 10^{-13}\,\text{J/MeV})} = 1.8 \times 10^{38}\ \text{s}^{-1}\ .$$

(b) Let d_{es} be the Earth to Sun distance, and R be the radius of Earth (see Appendix C). Earth represents a small cross section in the "sky" as viewed by a fictitious observer on the Sun. The rate of neutrinos intercepted by that area (very small, relative to the area of the full "sky") is

$$R_{\nu,\text{Earth}} = R_\nu \left(\frac{\pi R_e^2}{4\pi d_{es}^2}\right) = \frac{(1.8 \times 10^{38}\,\text{s}^{-1})}{4}\left(\frac{6.4 \times 10^6\,\text{m}}{1.5 \times 10^{11}\,\text{m}}\right)^2 = 8.2 \times 10^{28}\ \text{s}^{-1}\ .$$

46. (a) The products of the carbon cycle are $2e^+ + 2\nu + {}^4\text{He}$, the same as that of the proton-proton cycle (see Eq. 44-10). The difference in the number of photons is not significant.

    (b) $Q_{\text{carbon}} = Q_1 + Q_2 + \cdots + Q_6 = (1.95 + 1.19 + 7.55 + 7.30 + 1.73 + 4.97)\,\text{MeV} = 24.7\,\text{MeV}$, which is the same as that for the proton-proton cycle (once we subtract out the electron-positron annihilations; see Fig. 44-11): $Q_{p-p} = 26.7\,\text{MeV} - 2(1.02\,\text{MeV}) = 24.7\,\text{MeV}$.

47. (a) The mass of a carbon atom is $(12.0\,\text{u})(1.661 \times 10^{-27}\,\text{kg/u}) = 1.99 \times 10^{-26}\,\text{kg}$, so the number of carbon atoms in 1.00 kg of carbon is $(1.00\,\text{kg})/(1.99 \times 10^{-26}\,\text{kg}) = 5.02 \times 10^{25}$. The heat of combustion per atom is $(3.3 \times 10^7\,\text{J/kg})/(5.02 \times 10^{25}\,\text{atom/kg}) = 6.58 \times 10^{-19}\,\text{J/atom}$. This is 4.11 eV/atom.

    (b) In each combustion event, two oxygen atoms combine with one carbon atom, so the total mass involved is $2(16.0\,\text{u}) + (12.0\,\text{u}) = 44\,\text{u}$. This is $(44\,\text{u})(1.661 \times 10^{-27}\,\text{kg/u}) = 7.31 \times 10^{-26}\,\text{kg}$. Each combustion event produces $6.58 \times 10^{-19}\,\text{J}$ so the energy produced per unit mass of reactants is $(6.58 \times 10^{-19}\,\text{J})/(7.31 \times 10^{-26}\,\text{kg}) = 9.00 \times 10^6\,\text{J/kg}$.

    (c) If the Sun were composed of the appropriate mixture of carbon and oxygen, the number of combustion events that could occur before the Sun burns out would be $(2.0 \times 10^{30}\,\text{kg})/(7.31 \times 10^{-26}\,\text{kg}) = 2.74 \times 10^{55}$. The total energy released would be $E = (2.74 \times 10^{55})(6.58 \times 10^{-19}\,\text{J}) = 1.80 \times 10^{37}\,\text{J}$. If P is the power output of the Sun, the burn time would be

$$t = \frac{E}{P} = \frac{1.80 \times 10^{37}\,\text{J}}{3.9 \times 10^{26}\,\text{W}} = 4.62 \times 10^{10}\,\text{s} = 1460\,\text{y} .$$

48. The mass of the hydrogen in the Sun's core is $m_{\text{H}} = 0.35(\frac{1}{8}M_{\text{Sun}})$. The time it takes for the hydrogen to be entirely consumed is

$$t = \frac{M_{\text{H}}}{dm/dt} = \frac{(0.35)(\frac{1}{8})(2.0 \times 10^{30}\,\text{kg})}{(6.2 \times 10^{11}\,\text{kg/s})(3.15 \times 10^7\,\text{s/y})} = 5 \times 10^9\ \text{y} .$$

49. Since the mass of a helium atom is $(4.00\,\text{u})(1.661 \times 10^{-27}\,\text{kg/u}) = 6.64 \times 10^{-27}\,\text{kg}$, the number of helium nuclei originally in the star is $(4.6 \times 10^{32}\,\text{kg})/(6.64 \times 10^{-27}\,\text{kg}) = 6.92 \times 10^{58}$. Since each fusion event requires three helium nuclei, the number of fusion events that can take place is $N = 6.92 \times 10^{58}/3 = 2.31 \times 10^{58}$. If Q is the energy released in each event and t is the conversion time, then the power output is $P = NQ/t$ and

$$t = \frac{NQ}{P} = \frac{(2.31 \times 10^{58})(7.27 \times 10^6\,\text{eV})(1.60 \times 10^{-19}\,\text{J/eV})}{5.3 \times 10^{30}\,\text{W}} = 5.07 \times 10^{15}\,\text{s} = 1.6 \times 10^8\ \text{y} .$$

50. (a) From $E = NQ = (M_{\text{sam}}/4m_p)Q$ we get the energy per kilogram of hydrogen consumed:

$$\frac{E}{M_{\text{sam}}} = \frac{Q}{4m_p} = \frac{(26.2\,\text{MeV})(1.60 \times 10^{-13}\,\text{J/MeV})}{4(1.67 \times 10^{-27}\,\text{kg})} = 6.3 \times 10^{14}\ \text{J/kg} .$$

    (b) Keeping in mind that a Watt is a Joule per second, the rate is

$$\frac{dm}{dt} = \frac{3.9 \times 10^{26}\,\text{W}}{6.3 \times 10^{14}\,\text{J/kg}} = 6.2 \times 10^{11}\ \text{kg/s} .$$

    This agrees with the computation shown in Sample Problem 44-5.

    (c) From the Einstein relation $E = Mc^2$ we get $P = dE/dt = c^2 dM/dt$, or

$$\frac{dM}{dt} = \frac{P}{c^2} = \frac{3.9 \times 10^{26}\,\text{W}}{(3.0 \times 10^8\,\text{m/s})^2} = 4.3 \times 10^9\ \text{kg/s} .$$

    This finding, that $\frac{dm}{dt} > \frac{dM}{dt}$, is in large part due to the fact that, as the protons are consumed, their mass is mostly turned into alpha particles (helium), which remain in the Sun.

(d) The time to lose 0.10% of its total mass is

$$t = \frac{0.0010M}{dM/dt} = \frac{(0.0010)(2.0 \times 10^{30}\,\text{kg})}{(4.3 \times 10^9\,\text{kg/s})(3.15 \times 10^7\,\text{s/y})} = 1.5 \times 10^{10}\ \text{y}\ .$$

51. (a) $Q = (5m_{^2\text{H}} - m_{^3\text{He}} - m_{^4\text{He}} - m_{^1\text{H}} - 2m_n)c^2 = [5(2.014102\,\text{u}) - 3.016029\,\text{u} - 4.002603\,\text{u} - 1.007825\,\text{u} - 2(1.008665\,\text{u})](931.5\,\text{MeV/u}) = 24.9\,\text{MeV}$.

(b) Assuming 30.0% of the deuterium undergoes fusion, the total energy released is

$$E = NQ = \left(\frac{0.300M}{5m_{^2\text{H}}}\right) Q\ .$$

Thus, the rating is

$$\begin{aligned} R &= \frac{E}{2.6 \times 10^{28}\,\text{MeV/megaton TNT}} \\ &= \frac{(0.300)(500\,\text{kg})(24.9\,\text{MeV})}{5(2.0\,\text{u})(1.66 \times 10^{-27}\,\text{kg/u})(2.6 \times 10^{28}\,\text{MeV/megaton TNT})} \\ &= 8.65\,\text{megaton TNT}\ . \end{aligned}$$

52. In Eq. 44-13,

$$\begin{aligned} Q &= (2m_{^2\text{H}} - m_{^3\text{He}} - m_n)c^2 \\ &= [2(2.014102\,\text{u}) - 3.016049\,\text{u} - 1.008665\,\text{u}](931.5\,\text{MeV/u}) \\ &= 3.27\,\text{MeV}\ . \end{aligned}$$

In Eq. 44-14,

$$\begin{aligned} Q &= (2m_{^2\text{H}} - m_{^3\text{H}} - m_{^1\text{H}})c^2 \\ &= [2(2.014102\,\text{u}) - 3.016049\,\text{u} - 1.007825\,\text{u}](931.5\,\text{MeV/u}) \\ &= 4.03\,\text{MeV}\ . \end{aligned}$$

Finally, in Eq. 44-15,

$$\begin{aligned} Q &= (m_{^2\text{H}} + m_{^3\text{H}} - m_{^4\text{He}} - m_n)c^2 \\ &= [2.014102\,\text{u} + 3.016049\,\text{u} - 4.002603\,\text{u} - 1.008665\,\text{u}](931.5\,\text{MeV/u}) \\ &= 17.59\,\text{MeV}\ . \end{aligned}$$

53. Since 1.00 L of water has a mass of 1.00 kg, the mass of the heavy water in 1.00 L is $0.0150 \times 10^{-2}\,\text{kg} = 1.50 \times 10^{-4}\,\text{kg}$. Since a heavy water molecule contains one oxygen atom, one hydrogen atom and one deuterium atom, its mass is $(16.0\,\text{u} + 1.00\,\text{u} + 2.00\,\text{u}) = 19.0\,\text{u}$ or $(19.0\,\text{u})(1.661 \times 10^{-27}\,\text{kg/u}) = 3.16 \times 10^{-26}\,\text{kg}$. The number of heavy water molecules in a liter of water is $(1.50 \times 10^{-4}\,\text{kg})/(3.16 \times 10^{-26}\,\text{kg}) = 4.75 \times 10^{21}$. Since each fusion event requires two deuterium nuclei, the number of fusion events that can occur is $N = 4.75 \times 10^{21}/2 = 2.38 \times 10^{21}$. Each event releases energy $Q = (3.27 \times 10^6\,\text{eV})(1.60 \times 10^{-19}\,\text{J/eV}) = 5.23 \times 10^{-13}\,\text{J}$. Since all events take place in a day, which is $8.64 \times 10^4\,\text{s}$, the power output is

$$P = \frac{NQ}{t} = \frac{(2.38 \times 10^{21})(5.23 \times 10^{-13}\,\text{J})}{8.64 \times 10^4\,\text{s}} = 1.44 \times 10^4\,\text{W} = 14.4\,\text{kW}\ .$$

54. Conservation of energy gives $Q = K_\alpha + K_\text{n}$, and conservation of linear momentum (due to the assumption of negligible initial velocities) gives $|p_\alpha| = |p_\text{n}|$. We can write the classical formula for kinetic energy in terms of momentum:

$$K = \frac{1}{2}mv^2 = \frac{p^2}{2m}$$

which implies that $K_n = (m_\alpha/m_n)K_\alpha$. Consequently, conservation of energy and momentum allows us to solve for kinetic energy of the alpha particle which results from the fusion:

$$K_\alpha = \frac{Q}{1+\frac{m_\alpha}{m_n}} = \frac{17.59\,\text{MeV}}{1+\frac{4.0015\,\text{u}}{1.008665\text{u}}} = 3.541\ \text{MeV}$$

where we have found the mass of the alpha particle by subtracting two electron masses from the ^{4}He mass (quoted several times in this and the previous chapter). Then, $K_n = Q - K_\alpha$ yields 14.05 MeV for the neutron kinetic energy.

Chapter 45

1. Using Table 45-1, the difference in mass between the muon and the pion is

$$\Delta m = \left(139.6\,\frac{\text{MeV}}{c^2} - 105.7\,\frac{\text{MeV}}{c^2}\right) = \frac{(33.9\,\text{MeV})(1.60\times10^{-13}\,\text{J/MeV})}{(2.998\times10^8\,\text{m/s})^2} = 6.03\times10^{-29}\ \text{kg}\ .$$

2. We establish a ratio, using Eq. 22-4 and Eq. 14-1:

$$\begin{aligned}\frac{F_{\text{gravity}}}{F_{\text{electric}}} &= \frac{Gm_e^2/r^2}{ke^2/r^2} = \frac{4\pi\varepsilon_0 Gm_e^2}{e^2} \\ &= \frac{(6.67\times10^{-11}\,\text{N}\cdot\text{m}^2/\text{C}^2)(9.11\times10^{-31}\,\text{kg})^2}{(9.0\times10^9\,\text{N}\cdot\text{m}^2/\text{C}^2)(1.60\times10^{-19}\,\text{C})^2} \\ &= 2.4\times10^{-43}\ .\end{aligned}$$

Since $F_{\text{gravity}} \ll F_{\text{electric}}$, we can neglect the gravitational force acting between particles in a bubble chamber.

3. Conservation of momentum requires that the gamma ray particles move in opposite directions with momenta of the same magnitude. Since the magnitude p of the momentum of a gamma ray particle is related to its energy by $p = E/c$, the particles have the same energy E. Conservation of energy yields $m_\pi c^2 = 2E$, where m_π is the mass of a neutral pion. According to Table 45-4, the rest energy of a neutral pion is $m_\pi c^2 = 135.0\,\text{MeV}$. Hence, $E = (135.0\,\text{MeV})/2 = 67.5\,\text{MeV}$. We use the result of Exercise 3 of Chapter 39 to obtain the wavelength of the gamma rays:

$$\lambda = \frac{1240\,\text{eV}\cdot\text{nm}}{67.5\times10^6\,\text{eV}} = 1.84\times10^{-5}\,\text{nm} = 18.4\ \text{fm}\ .$$

4. By charge conservation, it is clear that reversing the sign of the pion means we must reverse the sign of the muon. In effect, we are replacing the charged particles by their antiparticles. Less obvious is the fact that we should now put a "bar" over the neutrino (something we should also have done for some of the reactions and decays discussed in the previous two chapters, except that we had not yet learned about antiparticles). To understand the "bar" we refer the reader to the discussion in §45-4. The decay of the negative pion is $\pi^- \to \mu^- + \bar{\nu}$. A subscript can be added to the antineutrino to clarify what "type" it is, as discussed in §45-4.

5. The energy released would be twice the rest energy of Earth, or $E = 2mc^2 = 2(5.98\times10^{24}\,\text{kg})(2.998\times10^8\,\text{m/s})^2 = 1.08\times10^{42}\,\text{J}$. The mass of Earth can be found in Appendix C.

6. Since the density of water is $\rho = 1000\,\text{kg/m}^3 = 1\,\text{kg/L}$, then the total mass of the pool is $\rho\mathcal{V} = 4.32\times10^5\,\text{kg}$, where $\mathcal{V}$ is the given volume. Now, the fraction of that mass made up by the protons is 10/18 (by counting the protons versus total nucleons in a water molecule). Consequently, if we ignore

the effects of neutron decay (neutrons can beta decay into protons) in the interest of making an order-of-magnitude calculation, then the number of particles susceptible to decay via this $T_{1/2} = 10^{32}$ y half-life is

$$N = \frac{\frac{10}{18} M_{\text{pool}}}{m_p} = \frac{\frac{10}{18}(4.32 \times 10^5 \text{ kg})}{1.67 \times 10^{-27} \text{ kg}} = 1.44 \times 10^{32} .$$

Using Eq. 43-19, we obtain

$$R = \frac{N \ln 2}{T_{1/2}} = \frac{(1.44 \times 10^{32}) \ln 2}{10^{32} \text{ y}} \approx 1 \text{ decay/y} .$$

7. From Eq. 38-45, the Lorentz factor would be

$$\gamma = \frac{E}{mc^2} = \frac{1.5 \times 10^6 \text{ eV}}{20 \text{ eV}} = 75000 .$$

Solving Eq. 38-8 for the speed, we find

$$\gamma = \frac{1}{\sqrt{1-(v/c)^2}} \implies v = c\sqrt{1 - \frac{1}{\gamma^2}}$$

which implies that the difference between v and c is

$$c - v = c\left(1 - \sqrt{1 - \frac{1}{\gamma^2}}\right) \approx c\left(1 - \left(1 - \frac{1}{2\gamma^2} + \cdots\right)\right)$$

where we use the binomial expansion (see Appendix E) in the last step. Therefore,

$$c - v \approx c\left(\frac{1}{2\gamma^2}\right) = (299792458 \text{ m/s})\left(\frac{1}{2(75000)^2}\right) = 0.0266 \text{ m/s} .$$

8. From Eq. 38-49, the Lorentz factor is

$$\gamma = 1 + \frac{K}{mc^2} = 1 + \frac{80 \text{ MeV}}{135 \text{ MeV}} = 1.59 .$$

Solving Eq. 38-8 for the speed, we find

$$\gamma = \frac{1}{\sqrt{1-(v/c)^2}} \implies v = c\sqrt{1 - \frac{1}{\gamma^2}}$$

which yields $v = 0.778c$ or $v = 2.33 \times 10^8$ m/s. Now, in the reference frame of the laboratory, the lifetime of the pion is not the given τ value but is "dilated." Using Eq. 38-9, the time in the lab is

$$t = \gamma\tau = (1.59)\left(8.3 \times 10^{-17} \text{ s}\right) = 1.3 \times 10^{-16} \text{ s} .$$

Finally, using Eq. 38-10, we find the distance in the lab to be

$$x = vt = \left(2.33 \times 10^8 \text{ m/s}\right)\left(1.3 \times 10^{-16} \text{ s}\right) = 3.1 \times 10^{-8} \text{ m} .$$

9. Table 45-4 gives the rest energy of each pion as 139.6 MeV. The magnitude of the momentum of each pion is $p_\pi = (358.3 \text{ MeV})/c$. We use the relativistic relationship between energy and momentum (Eq. 38-52) to find the total energy of each pion:

$$E_\pi = \sqrt{(p_\pi c)^2 + (m_\pi c^2)^2} = \sqrt{(358.3 \text{ MeV})^2 + (139.6 \text{ MeV})^2} = 384.5 \text{ MeV} .$$

Conservation of energy yields $m_\rho c^2 = 2E_\pi = 2(384.5 \text{ MeV}) = 769 \text{ MeV}$.

10. (a) In SI units, $K = (2200\,\text{MeV})(1.6 \times 10^{-13}\,\text{J/MeV}) = 3.52 \times 10^{-10}\,\text{J}$. Similarly, $mc^2 = 2.85 \times 10^{-10}\,\text{J}$ for the positive tau. Eq. 38-51 leads to the relativistic momentum:

$$p = \frac{1}{c}\sqrt{K^2 + 2Kmc^2} = \frac{1}{2.998 \times 10^8}\sqrt{(3.52 \times 10^{-10})^2 + 2\,(3.52 \times 10^{-10})\,(2.85 \times 10^{-10})}$$

which yields $p = 1.90 \times 10^{-18}\,\text{kg}\cdot\text{m/s}$.

(b) According to problem 46 in Chapter 38, the radius should be calculated with the relativistic momentum:

$$r = \frac{\gamma m v}{|q|B} = \frac{p}{eB}$$

where we use the fact that the positive tau has charge $e = 1.6 \times 10^{-19}\,\text{C}$. With $B = 1.20\,\text{T}$, this yields $r = 9.9\,\text{m}$.

11. (a) Conservation of energy gives $Q = K_2 + K_3 = E_1 - E_2 - E_3$ where E refers here to the *rest* energies (mc^2) instead of the total energies of the particles. Writing this as $K_2 + E_2 - E_1 = -(K_3 + E_3)$ and squaring both sides yields

$$K_2^2 + 2K_2E_2 - 2K_2E_1 + (E_1 - E_2)^2 = K_3^2 + 2K_3E_3 + E_3^2 \ .$$

Next, conservation of linear momentum (in a reference frame where particle 1 was at rest) gives $|p_2| = |p_3|$ (which implies $(p_2c)^2 = (p_3c)^2$). Therefore, Eq. 38-51 leads to

$$K_2^2 + 2K_2E_2 = K_3^2 + 2K_3E_3$$

which we subtract from the above expression to obtain

$$-2K_2E_1 + (E_1 - E_2)^2 = E_3^2 \ .$$

This is now straightforward to solve for K_2 and yields the result stated in the problem.

(b) Setting $E_3 = 0$ in

$$K_2 = \frac{1}{2E_1}\left[(E_1 - E_2)^2 - E_3^2\right]$$

and using the rest energy values given in Table 45-1 readily gives the same result for K_μ as computed in Sample Problem 45-1.

12. (a) Eq. 45-14 conserves charge since both the proton and the positron have $q = +e$ (and the neutrino is uncharged).

(b) Energy conservation is not violated since $m_pc^2 > m_ec^2 + m_\nu c^2$.

(c) We are free to view the decay from the rest frame of the proton. Both the positron and the neutrino are able to carry momentum, and so long as they travel in opposite directions with appropriate values of p (so that $\sum \vec{p} = 0$) then linear momentum is conserved.

(d) If we examine the spin angular momenta, there does seem to be a violation of angular momentum conservation (Eq. 45-14 shows a spin-one-half particle decaying into two spin-one-half particles).

13. (a) The conservation laws considered so far are associated with energy, momentum, angular momentum, charge, baryon number, and the three lepton numbers. The rest energy of the muon is 105.7 MeV, the rest energy of the electron is 0.511 MeV, and the rest energy of the neutrino is zero. Thus, the total rest energy before the decay is greater than the total rest energy after. The excess energy can be carried away as the kinetic energies of the decay products and energy can be conserved. Momentum is conserved if the electron and neutrino move away from the decay in opposite directions with equal magnitudes of momenta. Since the orbital angular momentum is zero, we consider only spin angular momentum. All the particles have spin $\hbar/2$. The total angular momentum after the decay must be either $\hbar$ (if the spins are aligned) or zero (if the spins are antialigned). Since the spin before the

decay is $\hbar/2$, angular momentum cannot be conserved. The muon has charge $-e$, the electron has charge $-e$, and the neutrino has charge zero, so the total charge before the decay is $-e$ and the total charge after is $-e$. Charge is conserved. All particles have baryon number zero, so baryon number is conserved. The muon lepton number of the muon is $+1$, the muon lepton number of the muon neutrino is $+1$, and the muon lepton number of the electron is 0. Muon lepton number is conserved. The electron lepton numbers of the muon and muon neutrino are 0 and the electron lepton number of the electron is $+1$. Electron lepton number is not conserved. The laws of conservation of angular momentum and electron lepton number are not obeyed and this decay does not occur..

(b) We analyze the decay in the same way. We find that only charge is not conserved.

(c) Here we find that energy and muon lepton number cannot be conserved.

14. (a) Noting that there are two positive pions created (so, in effect, its decay products are doubled), then we count up the electrons, positrons and neutrinos: $2e^+ + e^- + 5\nu + 4\bar{\nu}$.

(b) The final products are all leptons, so the baryon number of A_2^+ is zero. Both the pion and rho meson have integer-valued spins, so A_2^+ is a meson (and a boson).

15. For purposes of deducing the properties of the antineutron, one may cancel a proton from each side of the reaction and write the equivalent reaction as

$$\pi^+ \to p + \bar{n} \, .$$

Particle properties can be found in Tables 45-3 and 45-4. The pion and proton each have charge $+e$, so the antineutron must be neutral. The pion has baryon number zero (it is a meson) and the proton has baryon number $+1$, so the baryon number of the antineutron must be -1. The pion and the proton each have strangeness zero, so the strangeness of the antineutron must also be zero. In summary, $q = 0$, $B = -1$, and $S = 0$ for the antineutron.

16. (a) Referring to Tables 45-3 and 45-4, we find the strangeness of K^0 is $+1$, while it is zero for both π^+ and π^-. Consequently, strangeness is not conserved in this decay; $K^0 \to \pi^+ + \pi^-$ does not proceed via the strong interaction.

(b) The strangeness of each side is -1, which implies that the decay is governed by the strong interaction.

(c) The strangeness or Λ^0 is -1 while that of $p+\pi^-$ is zero, so the decay is not via the strong interaction.

(d) The strangeness of each side is -1; it proceeds via the strong interaction.

17. (a) See the solution to Exercise 13 for the quantities to be considered, adding strangeness to the list. The lambda has a rest energy of 1115.6 MeV, the proton has a rest energy of 938.3 MeV, and the kaon has a rest energy of 493.7 MeV. The rest energy before the decay is less than the total rest energy after, so energy cannot be conserved. Momentum can be conserved. The lambda and proton each have spin $\hbar/2$ and the kaon has spin zero, so angular momentum can be conserved. The lambda has charge zero, the proton has charge $+e$, and the kaon has charge $-e$, so charge is conserved. The lambda and proton each have baryon number $+1$, and the kaon has baryon number zero, so baryon number is conserved. The lambda and kaon each have strangeness -1 and the proton has strangeness zero, so strangeness is conserved. Only energy cannot be conserved.

(b) The omega has a rest energy of 1680 MeV, the sigma has a rest energy of 1197.3 MeV, and the pion has a rest energy of 135 MeV. The rest energy before the decay is greater than the total rest energy after, so energy can be conserved. Momentum can be conserved. The omega and sigma each have spin $\hbar/2$ and the pion has spin zero, so angular momentum can be conserved. The omega has charge $-e$, the sigma has charge $-e$, and the pion has charge zero, so charge is conserved. The omega and sigma have baryon number $+1$ and the pion has baryon number 0, so baryon number is conserved. The omega has strangeness -3, the sigma has strangeness -1, and the pion has strangeness zero, so strangeness is not conserved.

(c) The kaon and proton can bring kinetic energy to the reaction, so energy can be conserved even though the total rest energy after the collision is greater than the total rest energy before. Momentum can be conserved. The proton and lambda each have spin $\hbar/2$ and the kaon and pion each have spin zero, so angular momentum can be conserved. The kaon has charge $-e$, the proton has charge $+e$, the lambda has charge zero, and the pion has charge $+e$, so charge is not conserved. The proton and lambda each have baryon number $+1$, and the kaon and pion each have baryon number zero; baryon number is conserved. The kaon has strangeness -1, the proton and pion each have strangeness zero, and the lambda has strangeness -1, so strangeness is conserved. Only charge is not conserved.

18. (a) From Eq. 38-47,

$$\begin{aligned} Q &= -\Delta m\, c^2 = (m_{\Sigma^+} + m_{\mathrm{K}^+} - m_{\pi^+} - m_p)c^2 \\ &= 1189.4\,\mathrm{MeV} + 493.7\,\mathrm{MeV} - 139.6\,\mathrm{MeV} - 938.3\,\mathrm{MeV} \\ &= 605\ \mathrm{MeV}\ . \end{aligned}$$

(b) Similarly,

$$\begin{aligned} Q &= -\Delta m\, c^2 = (m_{\Lambda^0} + m_{\pi^0} - m_{\mathrm{K}^-} - m_p)c^2 \\ &= 1115.6\,\mathrm{MeV} + 135.0\,\mathrm{MeV} - 493.7\,\mathrm{MeV} - 938.3\,\mathrm{MeV} \\ &= -181\ \mathrm{MeV}\ . \end{aligned}$$

19. Conservation of energy (see Eq. 38-44) leads to

$$\begin{aligned} K_f &= -\Delta m\, c^2 + K_i = (m_{\Sigma^-} - m_{\pi^-} - m_n)c^2 + K_i \\ &= 1197.3\,\mathrm{MeV} - 139.6\,\mathrm{MeV} - 939.6\,\mathrm{MeV} + 220\,\mathrm{MeV} \\ &= 338\ \mathrm{MeV}\ . \end{aligned}$$

20. The formula for T_z as it is usually written to include strange baryons is $T_z = q - (S + B)/2$. Also, we interpret the symbol q in the T_z formula in terms of elementary charge units; this is how q is listed in Table 45-3. In terms of charge q as we have used it in previous chapters, the formula is $T_z = \frac{q}{e} - \frac{1}{2}(B+S)$. For instance, $T_z = +\frac{1}{2}$ for the proton (and the neutral Xi) and $T_z = -\frac{1}{2}$ for the neutron (and the negative Xi). The baryon number B is $+1$ for all the particles in Fig. 45-4(a). Rather than use a sloping axis as in Fig. 45-4 (there it is done for the q values), one reproduces (if one uses the "corrected" formula for T_z mentioned above) exactly the same pattern using regular rectangular axes (T_z values along the horizontal axis and Y values along the vertical) with the neutral lambda and sigma particles situated at the origin.

21. (a) As far as the conservation laws are concerned, we may cancel a proton from each side of the reaction equation and write the reaction as $\mathrm{p} \to \Lambda^0 + x$. Since the proton and the lambda each have a spin angular momentum of $\hbar/2$, the spin angular momentum of x must be either zero or $\hbar$. Since the proton has charge $+e$ and the lambda is neutral, x must have charge $+e$. Since the proton and the lambda each have a baryon number of $+1$, the baryon number of x is zero. Since the strangeness of the proton is zero and the strangeness of the lambda is -1, the strangeness of x is $+1$. We take the unknown particle to be a spin zero meson with a charge of $+e$ and a strangeness of $+1$. Look at Table 45-4 to identify it as a K^+ particle.

(b) Similar analysis tells us that x is a spin-$\frac{1}{2}$ antibaryon ($B = -1$) with charge and strangeness both zero. Inspection of Table 45-3 reveals it is an antineutron.

(c) Here x is a spin-0 (or spin-1) meson with charge zero and strangeness -1. According to Table 45-4, it could be a $\overline{K}^0$ particle.

22. (a) From Eq. 38-47,

$$\begin{aligned} Q &= -\Delta m\, c^2 = (m_{\Lambda^0} - m_p - m_{\pi^-})c^2 \\ &= 1115.6\,\text{MeV} - 938.3\,\text{MeV} - 139.6\,\text{MeV} = 37.7\text{ MeV} . \end{aligned}$$

(b) We use the formula obtained in problem 11 (where it should be emphasized that E is used to mean the rest energy, not the total energy):

$$\begin{aligned} K_p &= \frac{1}{2E_\Lambda}\left[(E_\Lambda - E_p)^2 - E_\pi^2\right] \\ &= \frac{(1115.6\,\text{MeV} - 938.3\,\text{MeV})^2 - (139.6\,\text{MeV})^2}{2(1115.6\,\text{MeV})} = 5.35\text{ MeV} . \end{aligned}$$

(c) By conservation of energy,

$$K_{\pi^-} = Q - K_p = 37.7\,\text{MeV} - 5.35\,\text{MeV} = 32.4\text{ MeV} .$$

23. (a) We indicate the antiparticle nature of each quark with a "bar" over it. Thus, $\bar{\text{u}}\bar{\text{u}}\bar{\text{d}}$ represents an antiproton.

(b) Similarly, $\bar{\text{u}}\bar{\text{d}}\bar{\text{d}}$ represents an antineutron.

24. (a) The combination ddu has a total charge of $(-\frac{1}{3} - \frac{1}{3} + \frac{2}{3}) = 0$, and a total strangeness of zero. From Table 45-3, we find it to be a neutron (n).

(b) For the combination uus, we have $Q = +\frac{2}{3} + \frac{2}{3} - \frac{1}{3} = 1$ and $S = 0 + 0 - 1 = 1$. This is the Σ^+ particle.

(c) For the quark composition ssd, we have $Q = -\frac{1}{3} - \frac{1}{3} - \frac{1}{3} = -1$ and $S = -1 - 1 + 0 = -2$. This is a Ξ^-.

25. (a) Looking at the first three lines of Table 45-5, since the particle is a baryon, we determine that it must consist of three quarks. To obtain a strangeness of -2, two of them must be s quarks. Each of these has a charge of $-e/3$, so the sum of their charges is $-2e/3$. To obtain a total charge of e, the charge on the third quark must be $5e/3$. There is no quark with this charge, so the particle cannot be constructed. In fact, such a particle has never been observed.

(b) Again the particle consists of three quarks (and no antiquarks). To obtain a strangeness of zero, none of them may be s quarks. We must find a combination of three u and d quarks with a total charge of $2e$. The only such combination consists of three u quarks.

26. (a) Using Table 45-3, we find $q = 0$ and $S = -1$ for this particle (also, $B = 1$, since that is true for all particles in that table). From Table 45-5, we see it must therefore contain a strange quark (which has charge $-1/3$), so the other two quarks must have charges to add to zero. Assuming the others are among the lighter quarks (none of them being an antiquark, since $B = 1$), then the quark composition is $\bar{\text{u}}\bar{\text{s}}\bar{\text{d}}$.

(b) The reasoning is very similar to that of part (a). The main difference is that this particle must have two strange quarks. Its quark combination turns out to be $\bar{\text{u}}\bar{\text{s}}\bar{\text{s}}$.

27. If we were to use regular rectangular axes, then this would appear as a right triangle. Using the sloping q axis as the problem suggests, it is similar to an "upside down" equilateral triangle as we show below.

The leftmost slanted line is for the -1 charge, and the rightmost slanted line is for the $+2$ charge.

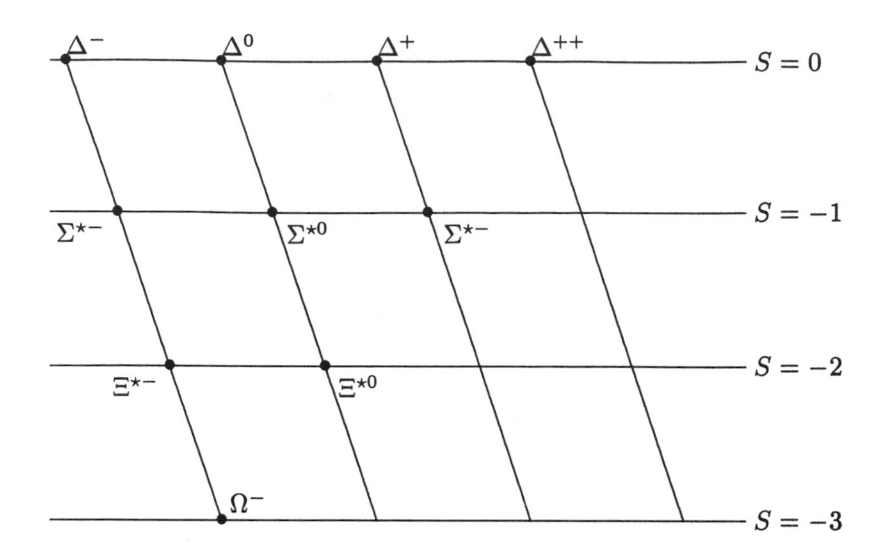

28. Since only the strange quark (s) has non-zero strangeness, in order to obtain $S = -1$ we need to combine s with some non-strange antiquark (which would have the negative of the quantum numbers listed in Table 45-5). The difficulty is that the charge of the strange quark is $-1/3$, which means that (to obtain a total charge of $+1$) the antiquark would have to have a charge of $+\frac{4}{3}$. Clearly, there are no such antiquarks in our list. Thus, a meson with $S = -1$ and $q = +1$ cannot be formed with the quarks/antiquarks of Table 45-5. Similarly, one can show that, since no quark has $q = -\frac{4}{3}$, there cannot be a meson with $S = +1$ and $q = -1$.

29. From $\gamma = 1 + K/mc^2$ (see Eq. 38-49) and $v = \beta c = c\sqrt{1-\gamma^{-2}}$ (see Eq. 38-8), we get

$$v = c\sqrt{1 - \left(1 + \frac{K}{mc^2}\right)^{-2}} \ .$$

Therefore, for the Σ^{*0} particle,

$$v = (2.9979 \times 10^8\,\text{m/s})\sqrt{1 - \left(1 + \frac{1000\,\text{MeV}}{1385\,\text{MeV}}\right)^{-2}} = 2.4406 \times 10^8 \text{ m/s} \ ,$$

and for Σ^0,

$$v' = (2.9979 \times 10^8\,\text{m/s})\sqrt{1 - \left(1 + \frac{1000\,\text{MeV}}{1192.5\,\text{MeV}}\right)^{-2}} = 2.5157 \times 10^8 \text{ m/s} \ .$$

Thus Σ^0 moves faster than Σ^{*0} by

$$\Delta v = v' - v = (2.5157 - 2.4406)(10^8\,\text{m/s}) = 7.51 \times 10^6 \text{ m/s} \ .$$

30. Letting $v = Hr = c$, we obtain

$$r = \frac{c}{H} = \frac{3.0 \times 10^8\,\text{m/s}}{0.0193\,\text{m/s}\cdot\text{ly}} = 1.6 \times 10^{10} \text{ ly} \ .$$

31. We apply Eq. 38-33 for the Doppler shift in wavelength:

$$\frac{\Delta\lambda}{\lambda} = \frac{v}{c}$$

where v is the recessional speed of the galaxy. We use Hubble's law to find the recessional speed: $v = Hr$, where r is the distance to the galaxy and H is the Hubble constant $(19.3 \times 10^{-3} \frac{\text{m}}{\text{s}\cdot\text{ly}})$. Thus, $v = [19.3 \times 10^{-3} \frac{\text{m}}{\text{s}\cdot\text{ly}}](2.40 \times 10^8\,\text{ly}) = 4.63 \times 10^6\,\text{m/s}$ and

$$\Delta\lambda = \frac{v}{c}\lambda = \left(\frac{4.63 \times 10^6\,\text{m/s}}{3.00 \times 10^8\,\text{m/s}}\right)(656.3\,\text{nm}) = 10.1\,\text{nm} .$$

Since the galaxy is receding, the observed wavelength is longer than the wavelength in the rest frame of the galaxy. Its value is $656.3\,\text{nm} + 10.1\,\text{nm} = 666.4\,\text{nm}$.

32. First, we find the speed of the receding galaxy from Eq. 38-30:

$$\begin{aligned}\beta &= \frac{1-(f/f_0)^2}{1+(f/f_0)^2} = \frac{1-(\lambda_0/\lambda)^2}{1+(\lambda_0/\lambda)^2} \\ &= \frac{1-(590.0\,\text{nm}/602.0\,\text{nm})^2}{1+(590.0\,\text{nm}/602.0\,\text{nm})^2} = 0.02013\end{aligned}$$

where we use $f = c/\lambda$ and $f_0 = c/\lambda_0$. Then from Eq. 45-19,

$$r = \frac{v}{H} = \frac{\beta c}{H} = \frac{(0.02013)(2.998 \times 10^8\,\text{m/s})}{19.3\,\text{mm/s}\cdot\text{ly}} = 3.13 \times 10^8\ \text{ly} .$$

(Note: if one uses the classical Doppler shift formula instead of the relativistic version in Eq. 38-30, one obtains $r = 31.7 \times 10^8\,\text{ly}$, which is reasonably close to the value we obtained above. This is to be expected since $\beta \approx 0.02 \ll 1$.)

33. (a) Letting $v(r) = Hr \le v_e = \sqrt{2GM/r}$, we get $M/r^3 \ge H^2/2G$. Thus,

$$\rho = \frac{M}{4\pi r^3/3} = \frac{3}{4\pi}\frac{M}{r^3} \ge \frac{3H^2}{8\pi G} .$$

(b) The density being expressed in H-atoms/m^3 is equivalent to expressing it in terms of $\rho_0 = m_{\text{H}}/\text{m}^3 = 1.67 \times 10^{-27}\,\text{kg/m}^3$. Thus,

$$\begin{aligned}\rho &= \frac{3H^2}{8\pi G\rho_0}\left(\text{H atoms/m}^3\right) = \frac{3(0.0193\,\text{m/s}\cdot\text{ly})^2(1.00\,\text{ly}/9.460 \times 10^{15}\,\text{m})^2(\text{H atoms/m}^3)}{8\pi(6.67 \times 10^{-11}\,\text{m}^3/\text{kg}\cdot\text{s}^2)(1.67 \times 10^{-27}\,\text{kg/m}^3)} \\ &= 4.5\ \text{H atoms/m}^3 .\end{aligned}$$

34. (a) From $f = c/\lambda$ and Eq. 38-30, we get

$$\lambda_0 = \lambda\sqrt{\frac{1-\beta}{1+\beta}} = (\lambda_0 + \Delta\lambda)\sqrt{\frac{1-\beta}{1+\beta}} .$$

Dividing both sides by λ_0 leads to

$$1 = (1+z)\sqrt{\frac{1-\beta}{1+\beta}} .$$

We solve for β:

$$\beta = \frac{(1+z)^2 - 1}{(1+z)^2 + 1} = \frac{z^2 + 2z}{z^2 + 2z + 2} .$$

(b) Now $z = 4.43$, so

$$\beta = \frac{(4.43)^2 + 2(4.43)}{(4.43)^2 + 2(4.43) + 2} = 0.934 \; .$$

(c) From Eq. 45-19,

$$r = \frac{v}{H} = \frac{\beta c}{H} = \frac{(0.943)(3.0 \times 10^8\,\text{m/s})}{0.0193\,\text{m/s}\cdot\text{ly}} = 1.5 \times 10^{10}\ \text{ly} \; .$$

35. (a) From Eq. 41-29, we know that $N_2/N_1 = e^{-\Delta E/kT}$. We solve for ΔE:

$$\begin{aligned}\Delta E &= kT \ln \frac{N_1}{N_2} = (8.62 \times 10^{-5}\,\text{eV/K})(2.7\,\text{K}) \ln\left(\frac{1-0.25}{0.25}\right) \\ &= 2.56 \times 10^{-4}\,\text{eV} = 256\,\mu\text{eV} \; .\end{aligned}$$

(b) Using the result of problem 3 in Chapter 39,

$$\lambda = \frac{hc}{\Delta E} = \frac{1240\,\text{eV}\cdot\text{nm}}{2.56 \times 10^{-4}\,\text{eV}} = 4.84 \times 10^6\ \text{nm} = 4.84\ \text{mm} \; .$$

36. From $F_{\text{grav}} = GMm/r^2 = mv^2/r$ we find $M \propto v^2$. Thus, the mass of the Sun would be

$$M_s' = \left(\frac{v_{\text{Mercury}}}{v_{\text{Pluto}}}\right)^2 M_s = \left(\frac{47.9\,\text{km/s}}{4.74\,\text{km/s}}\right)^2 M_s = 102 M_s \; .$$

37. (a) The mass M within Earth's orbit is used to calculate the gravitational force on Earth. If r is the radius of the orbit, R is the radius of the new Sun, and M_S is the mass of the Sun, then

$$M = \left(\frac{r}{R}\right)^3 M_S = \left(\frac{1.50 \times 10^{11}\,\text{m}}{5.90 \times 10^{12}\,\text{m}}\right)^3 (1.99 \times 10^{30}\,\text{kg}) = 3.27 \times 10^{25}\,\text{kg} \; .$$

The gravitational force on Earth is given by GMm/r^2, where m is the mass of Earth and G is the universal gravitational constant. Since the centripetal acceleration is given by v^2/r, where v is the speed of Earth, $GMm/r^2 = mv^2/r$ and

$$v = \sqrt{\frac{GM}{r}} = \sqrt{\frac{(6.67 \times 10^{-11}\,\text{m}^3/\text{s}^2\cdot\text{kg})(3.27 \times 10^{25}\,\text{kg})}{1.50 \times 10^{11}\,\text{m}}} = 1.21 \times 10^2\ \text{m/s} \; .$$

(b) The period of revolution is

$$T = \frac{2\pi r}{v} = \frac{2\pi(1.50 \times 10^{11}\,\text{m})}{1.21 \times 10^2\,\text{m/s}} = 7.82 \times 10^9\,\text{s} = 248\ \text{y} \; .$$

38. (a) The mass of the portion of the galaxy within the radius r from its center is given by $M' = (r/R)^3 M$. Thus, from $GM'm/r^2 = mv^2/r$ (where m is the mass of the star) we get

$$v = \sqrt{\frac{GM'}{r}} = \sqrt{\frac{GM}{r}\left(\frac{r}{R}\right)^3} = r\sqrt{\frac{GM}{R^3}} \; .$$

(b) In the case where $M' = M$, we have

$$T = \frac{2\pi r}{v} = 2\pi r\sqrt{\frac{r}{GM}} = \frac{2\pi r^{3/2}}{\sqrt{GM}} \; .$$

39. (a) We substitute $\lambda = (2898\,\mu\text{m}\cdot\text{K})/T$ into the result of Exercise 3 of Chapter 39: $E = (1240\,\text{eV}\cdot\text{nm})/\lambda$. First, we convert units: $2898\,\mu\text{m}\cdot\text{K} = 2.898 \times 10^6\,\text{nm}\cdot\text{K}$ and $1240\,\text{eV}\cdot\text{nm} = 1.240 \times 10^{-3}\,\text{MeV}\cdot\text{nm}$. Hence,

$$E = \frac{(1.240 \times 10^{-3}\,\text{MeV}\cdot\text{nm})T}{2.898 \times 10^6\,\text{nm}\cdot\text{K}} = (4.28 \times 10^{-10}\ \text{MeV/K})T\ .$$

(b) The minimum energy required to create an electron-positron pair is twice the rest energy of an electron, or $2(0.511\,\text{MeV}) = 1.022\,\text{MeV}$. Hence,

$$T = \frac{E}{4.28 \times 10^{-10}\,\text{MeV/K}} = \frac{1.022\,\text{MeV}}{4.28 \times 10^{-10}\,\text{MeV/K}} = 2.39 \times 10^9\ \text{K}\ .$$

40. (a) For the universal microwave background, Wien's law leads to

$$T = \frac{2898\,\mu\text{m}\cdot\text{K}}{\lambda_{\text{max}}} = \frac{2.898\,\text{mm}\cdot\text{K}}{1.1\,\text{mm}} = 2.6\ \text{K}\ .$$

(b) At "decoupling" (when the universe became approximately "transparent"),

$$\lambda_{\text{max}} = \frac{2898\,\mu\text{m}\cdot\text{K}}{T} = \frac{2898\,\mu\text{m}\cdot\text{K}}{10^5\,\text{K}} = 29\ \text{nm}\ .$$

41. (a) We use the relativistic relationship between speed and momentum:

$$p = \gamma m v = \frac{mv}{\sqrt{1-(v/c)^2}}\ ,$$

which we solve for the speed v:

$$\frac{v}{c} = \sqrt{1 - \frac{1}{\left(\frac{pc}{mc^2}\right)^2 + 1}}\ .$$

For an antiproton $mc^2 = 938.3\,\text{MeV}$ and $pc = 1.19\,\text{GeV} = 1190\,\text{MeV}$, so

$$v = c\sqrt{1 - \frac{1}{(1190\,\text{MeV}/938.3\,\text{MeV})^2 + 1}} = 0.785c\ .$$

For the negative pion $mc^2 = 193.6\,\text{MeV}$, and pc is the same. Therefore,

$$v = c\sqrt{1 - \frac{1}{(1190\,\text{MeV}/193.6\,\text{MeV})^2 + 1}} = 0.993c\ .$$

(b) See part (a).

(c) Since the speed of the antiprotons is about $0.78c$ but not over $0.79c$, an antiproton will trigger C1.

(d) Since the speed of the negative pions exceeds $0.79c$, a negative pion will trigger C2.

(e) and (f) We use $\Delta t = d/v$, where $d = 12\,\text{m}$. For an antiproton

$$\Delta t = \frac{12\,\text{m}}{0.785(2.998 \times 10^8\,\text{m/s})} = 5.1 \times 10^{-8}\,\text{s} = 51\ \text{ns}\ ,$$

and for a negative pion

$$\Delta t = \frac{12\,\text{m}}{0.993(2.998 \times 10^8\,\text{m/s})} = 4.0 \times 10^{-8}\,\text{s} = 40\ \text{ns}\ .$$

42. We note from track 1, and the quantum numbers of the original particle (A), that positively charged particles move in counterclockwise curved paths, and – by inference – negatively charged ones move along clockwise arcs. This immediately shows that tracks 1, 2, 4, 6, and 7 belong to positively charged particles, and tracks 5, 8 and 9 belong to negatively charged ones. Looking at the fictitious particles in the table (and noting that each appears in the cloud chamber once [or not at all]), we see that this observation (about charged particle motion) greatly narrows the possibilities:

$$\text{tracks } 2,4,6,7 \leftrightarrow \text{particles } C,F,H,J$$
$$\text{tracks } 5,8,9 \leftrightarrow \text{particles } D,E,G$$

This tells us, too, that the particle that does not appear at all is either B or I (since only one neutral particle "appears"). By charge conservation, tracks 2, 4 and 6 are made by particles with a single unit of positive charge (note that track 5 is made by one with a single unit of negative charge), which implies (by elimination) that track 7 is made by particle H. This is confirmed by examining charge conservation at the end-point of track 6. Having exhausted the charge-related information, we turn now to the fictitious quantum numbers. Consider the vertex where tracks 2, 3 and 4 meet (the Whimsy number is listed here as a subscript):

$$\text{tracks } 2,4 \leftrightarrow \text{particles } C_2, F_0, J_{-6}$$
$$\text{tracks } 3 \leftrightarrow \text{particle } B_4 \text{ or } I_6$$

The requirement that the Whimsy quantum number of the particle making track 4 must equal the sum of the Whimsy values for the particles making tracks 2 and 3 places a powerful constraint (see the subscripts above). A fairly quick trial and error procedure leads to the assignments: particle F makes track 4, and particles J and I make tracks 2 and 3, respectively. Particle B, then, is irrelevant to this set of events. By elimination, the particle making track 6 (the only positively charged particle not yet assigned) must be C. At the vertex defined by

$$A \rightarrow F + C + (\text{track } 5)_- ,$$

where the charge of that particle is indicated by the subscript, we see that Cuteness number conservation requires that the particle making track 5 has Cuteness $= -1$, so this must be particle G. We have only one decision remaining:

$$\text{tracks } 8,9 \leftrightarrow \text{particles } D,E$$

Re-reading the problem, one finds that the particle making track 8 must be particle D since it is the one with seriousness $= 0$. Consequently, the particle making track 9 must be E.

43. (a) During the time interval Δt, the light emitted from galaxy A has traveled a distance $c\Delta t$. Meanwhile, the distance between Earth and the galaxy has expanded from r to $r' = r + r\alpha\Delta t$. Let $c\Delta t = r' = r + r\alpha\Delta t$, which leads to

$$\Delta t = \frac{r}{c - r\alpha} .$$

(b) The detected wavelength λ' is longer then λ by $\lambda\alpha\Delta t$ due to the expansion of the universe: $\lambda' = \lambda + \lambda\alpha\Delta t$. Thus,

$$\frac{\Delta\lambda}{\lambda} = \frac{\lambda' - \lambda}{\lambda} = \alpha\Delta t = \frac{\alpha r}{c - \alpha r} .$$

(c) We use the binomial expansion formula (see Appendix E):

$$(1 \pm x)^n = 1 \pm \frac{nx}{1!} + \frac{n(n-1)x^2}{2!} + \cdots \qquad (x^2 < 1)$$

to obtain

$$\frac{\Delta\lambda}{\lambda} = \frac{\alpha r}{c - \alpha r} = \frac{\alpha r}{c}\left(1 - \frac{\alpha r}{c}\right)^{-1}$$

$$= \frac{\alpha r}{c}\left[1+\frac{-1}{1!}\left(-\frac{\alpha r}{c}\right)+\frac{(-1)(-2)}{2!}\left(-\frac{\alpha r}{c}\right)^2+\cdots\right]$$

$$\approx \frac{\alpha r}{c}+\left(\frac{\alpha r}{c}\right)^2+\left(\frac{\alpha r}{c}\right)^3 .$$

(d) When only the first term in the expansion for $\Delta\lambda/\lambda$ is retained we have

$$\frac{\Delta\lambda}{\lambda}\approx\frac{\alpha r}{c} .$$

(e) We set

$$\frac{\Delta\lambda}{\lambda}=\frac{v}{c}=\frac{Hr}{c}$$

and compare with the result of part (d) to obtain $\alpha = H$.

(f) We use the formula $\Delta\lambda/\lambda = \alpha r/(c-\alpha r)$ to solve for r:

$$r=\frac{c(\Delta\lambda/\lambda)}{\alpha(1+\Delta\lambda/\lambda)}=\frac{(2.998\times10^8\,\mathrm{m/s})(0.050)}{(0.0193\,\mathrm{m/s\cdot ly})(1+0.050)}=7.4\times10^8\ \mathrm{ly} .$$

(g) From the result of part (a),

$$\Delta t=\frac{r}{c-\alpha r}=\frac{(7.4\times10^8\,\mathrm{ly})(9.46\times10^{15}\,\mathrm{m/ly})}{2.998\times10^8\,\mathrm{m/s}-(0.0193\,\mathrm{m/s\cdot ly})(7.4\times10^8\,\mathrm{ly})}=2.5\times10^{16}\ \mathrm{s} ,$$

which is equivalent to 7.8×10^8 y.

(h) Letting $r = c\Delta t$, we solve for Δt:

$$\Delta t=\frac{r}{c}=\frac{7.4\times10^8\,\mathrm{ly}}{c}=7.4\times10^8\ \mathrm{y} .$$

(i) The distance is given by

$$r=c\Delta t=c(7.8\times10^8\,\mathrm{y})=7.8\times10^8\ \mathrm{ly} .$$

(j) From the result of part (f),

$$r_B=\frac{c(\Delta\lambda/\lambda)}{\alpha(1+\Delta\lambda/\lambda)}=\frac{(2.998\times10^8\,\mathrm{m/s})(0.080)}{(0.0193\,\mathrm{mm/s\cdot ly})(1+0.080)}=1.15\times10^9\ \mathrm{ly} .$$

(k) From the formula obtained in part (a),

$$\Delta t_B=\frac{r_B}{c-r_B\alpha}=\frac{(1.15\times10^9\,\mathrm{ly})(9.46\times10^{15}\,\mathrm{m/ly})}{2.998\times10^8\,\mathrm{m/s}-(1.15\times10^9\,\mathrm{ly})(0.0193\,\mathrm{m/s\cdot ly})}=3.9\times10^{16}\ \mathrm{s} ,$$

which is equivalent to 1.2×10^9 y.

(l) At the present time, the separation between the two galaxies A and B is given by $r_{now} = c\Delta t_B - c\Delta t_A$. Since $r_{now} = r_{then} + r_{then}\alpha\Delta t$, we get

$$r_{then}=\frac{r_{now}}{1+\alpha\Delta t}=4.4\times10^8\ \mathrm{ly} .$$

44. Assuming the line passes through the origin, its slope is $0.40c/(5.3\times10^9\,\mathrm{ly})$. Then,

$$T=\frac{1}{H}=\frac{1}{\mathrm{slope}}=\frac{5.3\times10^9\,\mathrm{ly}}{0.40c}=\frac{5.3\times10^9\,\mathrm{y}}{0.40}\approx 13\times10^9\ \mathrm{y} .$$

NOTES

NOTES

NOTES

NOTES

NOTES

NOTES

NOTES

NOTES